$E(X)$ — Expected value of the random variable X, i.e., the expected value of the probability distribution of X (3.9)

e — Specified sampling error in a determination of sample size (6.5)

F — F ratio: the ratio of the between-column variance to the between-row variance (8.3)

F — F ratio; the ratio of explained variance to unexplained variance (9.9)

$F(\nu_1, \nu_2)$ — F ratio with ν_1 and ν_2 degrees of freedom for the numerator and denominator, respectively (8.3)

f — Number of observations (frequency) in a class interval of a frequency distribution (1.8)

$f = \dfrac{n}{N}$ — Sampling fraction (5.5)

f_o — Observed frequency in a χ^2 test (8.1)

f_t — Theoretical, or expected, frequency in χ^2 test (8.1)

F_t — Forecast for time period t in exponential smoothing (10.6)

$F(x) = P(X \leq x)$ — Cumulative probability that random variable X is less than or equal to x (3.1)

f_{Md} — Frequency of the class containing the median (1.10)

f_p — Frequencies in classes preceding the one containing the median (1.10)

$f(x) = P(X = x)$ — Probability that random variable X is equal to the value x (3.1)

$f(Y|X)$ — Conditional probability distribution of Y given X in the linear regression model (9.1)

H — Value of the highest observation (1.2)

H_0 — Null hypothesis; basic hypothesis being tested (7.1)

H_1 — Alternative hypothesis; rejection of H_0 implies tentative acceptance of H_1 (7.1)

I — Effect of the irregular factor in time series analysis (10.4)

i — Size of the class interval (1.2)

K — Kruskal–Wallis test statistic in a one-factor analysis of variance by ranks (12.6)

k — Number of classes in a frequency distribution (1.2)

L — Value of the lowest observation (1.2)

$L(a_1|\theta_1)$ — Opportunity loss of act a_1 given state of nature θ_1 (17.1)

L_{Md} — Lower limit of class containing the median (1.10)

$L(\hat{\theta}; \theta)$ — Loss involved in estimating $\hat{\theta}$ when the parameter value is θ (17.3)

MA — Moving average figures in seasonal variation analysis (10.5)

Md — Median (1.10)

MS_c — Between-column mean square (8.3)

MS_r — Between-row mean square (8.3)

m — Midpoint of a class interval of a frequency distribution (1.8)

μ — Arithmetic mean of a population (1.8)

μ — Arithmetic mean of a probability distribution (3.7)

$\mu_{n\bar{p}}$ — Mean of sampling distribution of number of occurrences (5.2)

$\mu_{\bar{p}}$ — Mean of sampling distribution of a proportion (5.2)

$\mu_{\bar{p}_1 - \bar{p}_2}$ — Mean of sampling distribution of the difference between two proportions (6.3)

μ_r — Mean of sampling distribution of the number of runs, r (12.5)

μ_T — Mean of sampling distribution of T in the Wilcoxon matched-pairs signed rank test (12.3)

μ_U — Mean of sampling distribution of U in the Mann–Whitney U test (rank sum test) (12.4)

$\mu_{\bar{x}_1 - \bar{x}_2}$ — Mean of sampling distribution of the difference between two sample means (6.3)

$\mu_{Y.X}$ — Mean of conditional probability distribution of Y given X in the linear regression model; the population value that corresponds to the Y_c value computed from sample observations (9.1)

N — Number of observations in a population (1.8)

n — Number of observations in a sample (1.8)

$\dbinom{n}{x}$ — Number of combinations of n objects taken x at a time (2.4)

$_nP_x$ — Number of permutations of n objects taken x at a time (2.4)

Statistical Analysis
for Decision Making
Second Edition

Statistical Analysis for Decision Making
Second Edition

Morris Hamburg

The Wharton School
University of Pennsylvania

Harcourt Brace Jovanovich, Inc.
New York Chicago San Francisco Atlanta

Preface

The basic objectives of the second edition of *Statistical Analysis for Decision Making* are the same as those of the first edition. We have designed this book primarily for a first course in statistics for students of business and public administration, but the development of topics is also suitable for students in the social sciences and the liberal arts. With appropriate selection of chapters, this book may be used for a one-semester or two-semester course or for corresponding numbers of quarters. Our emphasis is on clear presentation of the fundamental concepts and methods of statistics. Many of the applications, examples, and exercises pertain to the analysis and solution of problems in managerial decision making, but we have also included problems in such fields as promotion, market research, quality control, accounting, polling, pollution control, consumer research, education, and psychology. The only mathematical background required is high school algebra. We do not include mathematical derivations in the body of the text, but some are given in footnotes and in Appendix C. Points for which an explanation in the language of calculus is particularly illuminating also appear in footnotes. Hence, the major emphasis is on showing the power of modern statistical reasoning and the scope and versatility of its methods, without inundating the student with the details of the underlying mathematical theory.

As in the first edition, the main topics covered here are probability and random variables, descriptive statistics, statistical inference, and statistical decision analysis. However, we have changed the order in which these subjects are presented and have considerably modified their details. Comments by users of the first edition have been particularly helpful in identifying the need for these changes.

Chapter 1 now deals with descriptive statistics, frequency distributions, and summary descriptive measures. This permits an easy and natural topical development, primarily because later the student can relate the concepts of probability measures and probability distributions to their less abstract and more easily understood analogs in descriptive statistics. We have also shortened and simplified the material on probability and random variables. These changes enable the instructor to spend less time on preliminaries while still providing the student with the background needed for the later work in statistical inference and statistical decision analysis. In discussing statistical inference, we cover estimation before hypothesis testing; this reverses the sequence used in the first edition and allows us to give these topics a more modern development. The addition of Chapter 12, "Nonparametric Statistics," also reflects our modern emphasis in statistical methodology. Chapter 9, "Regression and Correlation Analysis," has been substantially revised, and we have expanded the discussion of the use of high-speed electronic computers in multiple regression analysis. In Chapter 10, we have devoted less space to classical time series analysis and have added a discussion of exponential smoothing techniques. In our treatment of statistical decision analysis, we have deleted a few topics and added a few new ones, placing more emphasis on methods for capturing information on decision makers' betting odds and for using different amounts of empirical data in the construction of probability distributions. All the exercises, many of them new to this edition, now appear at the ends of the appropriate sections rather than at the ends of chapters. This provides the instructor with greater flexibility in selecting topics and in assigning exercises. Answers to odd-numbered exercises are given in the back of the book, with intermediate steps in their solutions often included. (Answers to all other exercises appear in the accompanying Solutions Manual.) The mathematical notation is simpler and more consistent throughout this edition. The endpapers contain a Glossary of Symbols, keyed to the section in which each symbol is first introduced. By placing the derivations of the shortcut formulas needed for descriptive statistics in an appendix, we have achieved an uncluttered presentation of descriptive statistics with more concentration on interpretation and less on computation. As in the first edition, we have tried to convey the idea that statistics is an exciting, fluid field that deals with a method of inquiry, a scientific method for acquiring, processing, and using knowledge in decision making.

Although only the author's name appears on this book, many individuals and organizations have made valuable contributions. My deep appreciation is again expressed to those who aided in the development of the first edition. Special thanks go to the reviewers of the second edition, C. Randall Byers, University of Idaho; Paul R. Merry, University of Denver; and James G. Morris, Kent State University; for their insightful comments and recommendations and their constructive criticisms. I have also benefited greatly from the comments of both teachers and students who have used the first edition. The following persons have been helpful: David Heinze, Rochester Institute of Technology; Donald R. Edwards, Air Force Institute of Technology; Byung T. Cho, Notre Dame University; John Howard, Jr., Western New England College, Massachusetts; Tom Alexandrin, Villanova University; Chloe I. Elmgren, Mount Mercy College, Iowa; Joe Hudson, Robert W. Brown, and Robert W. Bund, General Motors Institute; Raj Aggarwal, Indiana University, Gary; M. David Oh, Indiana University, Fort Wayne; Carl Nelson, Boston University; LaVerne Stanton, California State University at Fullerton; Y. K. Song, University of Detroit; William C. Struning, Seton Hall University, New Jersey. I wish to thank David Benson for his tireless and effective assistance in the preparation of exercises and solutions. An especially warm acknowledgment of gratitude goes to Mrs. Sylvia Balis and Mrs. Nita Innes for their cheerful, loyal, and exceptionally competent performance of the secretarial and typing work. Again, my thanks go to my colleagues in the Statistics Department of The Wharton School of the University of Pennsylvania, who were responsible for many of the original ideas involved in the exercises and who provided a thought-provoking, stimulating intellectual environment in which to work. I am very grateful to the staff of Harcourt Brace Jovanovich, Inc. Gary Burke provided wise counsel and sound overall management; and Judy Burke did a first-class job in the skillful improvements she brought to every chapter.

I am indebted to the Literary Executor of the late Sir Ronald A. Fisher, F.R.S., to Dr. Frank Yates, F.R.S., and to Oliver & Boyd Ltd., Edinburgh, for permission to reprint Tables III and IV from their book *Statistical Tables for Biological, Agricultural and Medical Research.* My gratitude also goes to the other authors and publishers whose generous permission to reprint tables or excerpts from tables has been acknowledged at appropriate places.

As in the first edition, I would like to dedicate this book to my wife, June, and my children, Neil and Barbara. I thank them for their continuing encouragement and assistance despite the competitive allure of other, less time-consuming family activities.

Morris Hamburg

Contents

Introduction 1

The Nature of Statistics *1*   The Role of Statistics *3*

1 Frequency Distributions and Summary Measures 5

1.1 Frequency Distributions *7*   1.2 Construction of a Frequency Distribution *9*   1.3 Class Limits *11*   1.4 Other Considerations in Constructing Frequency Distributions *13*   1.5 Graphic Presentation of Frequency Distributions *13*   1.6 Cumulative Frequency Distributions *15*   1.7 Descriptive Measures for Frequency Distributions *19*   1.8 The Arithmetic Mean *20* 1.9 The Weighted Arithmetic Mean *22*   1.10 The Median *26*   1.11 Characteristics and Uses of the Arithmetic Mean and Median *29*   1.12 The

Mode *30*   1.13 Dispersion: Distance Measures *37*   1.14 Dispersion: Average Deviation Methods *38*   1.15 Relative Dispersion: Coefficient of Variation *43*   1.16 Errors in Prediction *44*   1.17 Problems of Interpretation *45*

2 Introduction to Probability 49

2.1 The Meaning of Probability *49*   2.2 Elementary Probability Rules *60* 2.3 Bayes' Theorem *78*   2.4 Counting Principles and Techniques *83*

3 Discrete Random Variables and Probability Distributions 93

3.1 Random Variables *93*   3.2 Probability Distributions of Discrete Random Variables *104*   3.3 The Uniform Distribution *104*   3.4 The Binomial Distribution *106*   3.5 The Multinomial Distribution *118*   3.6 The Hypergeometric Distribution *121*   3.7 The Poisson Distribution *127*   3.8 Summary Measures for Probability Distributions *140*   3.9 Expected Value of a Random Variable *141*   3.10 Variance of a Random Variable *144*   3.11 Expected Value and Variance of Sums of Random Variables *146*   3.12 Joint Probability Distributions *150*

4 Statistical Investigations and Sampling 163

4.1 Formulation of the Problem *163*   4.2 Design of the Investigation *164* 4.3 Construction of Methodology *167*   4.4 Some Fundamental Concepts *168* 4.5 Fundamentals of Sampling *181*

5 Sampling Distributions 188

5.1 Sampling Distribution of Number of Occurrences *188*   5.2 Sampling Distribution of a Proportion *190*   5.3 Continuous Distributions *195*   5.4 The Normal Distribution *199*   5.5 Sampling Distribution of the Mean *212* 5.6 Other Probability Sample Designs *222*

6 Estimation 224

6.1 Point and Interval Estimation *225*   6.2 Criteria of Goodness of Estimation *226*   6.3 Confidence Interval Estimation (Large Samples) *231*

6.4 Confidence Interval Estimation (Small Samples) *244*   6.5 Determination of Sample Size *249*

7 Hypothesis Testing 257

7.1 The Rationale of Hypothesis Testing *258*   7.2 One-Sample Tests (Large Samples) *263*   7.3 Two-Sample Tests (Large Samples) *286*   7.4 The *t* Distribution: Small Samples with Unknown Population Standard Deviation(s) *299*   7.5 The *t* Test for Paired Observations *305*   7.6 Summary and a Look Ahead *308*

8 Chi-Square Tests and Analysis of Variance 312

8.1 Tests of Goodness of Fit *313*   8.2 Tests of Independence *328*   8.3 Analysis of Variance: Tests for Equality of Several Means *337*

9 Regression and Correlation Analysis 356

9.1 Introduction *356*   9.2 Scatter Diagrams *361*   9.3 Purposes of Regression and Correlation Analysis *364*   9.4 Estimation Using the Regression Line *365*   9.5 Confidence Intervals and Prediction Intervals in Regression Analysis *374*   9.6 Correlation Analysis: Measures of Association *388*   9.7 Inference About Population Parameters in Regression and Correlation *396*   9.8 Caveats and Limitations *401*   9.9 Multiple Regression and Correlation Analysis *410*   9.10 A Case Study *435*

10 Time Series 443

10.1 Introduction *443*   10.2 The Classical Time Series Model *444*   10.3 Description of Trend *447*   10.4 Fitting Trend Lines by the Method of Least Squares *450*   10.5 Measurement of Seasonal Variations *468*   10.6 Methods of Forecasting *482*

11 Index Numbers 491

11.1 The Need for and Use of Index Numbers *491*   11.2 Aggregative Price Indices *493*   11.3 Average of Relatives Indices *500*   11.4 General Problems of Index Number Construction *505*   11.5 Quantity Indices *507*

11.6 Deflation of Value Series by Price Indices *510*   11.7 Some Considerations in the Use of Index Numbers *514*

12 Nonparametric Statistics 518

12.1 Introduction *518*   12.2 The Sign Test *519*   12.3 Comparison of Two Populations Using Paired Observations (The Wilcoxon Matched-Pairs Signed Rank Test) *523*   12.4 Mann–Whitney *U* Test (Rank Sum Test) *528* 12.5 One-Sample Tests of Runs *532*   12.6 One-Factor Analysis of Variance by Ranks (Kruskal–Wallis Test) *535*   12.7 Rank Correlation *539*

13 Decision Making Using Prior Information 544

13.1 Introduction *544*   13.2 Structure of the Decision-Making Problem *545*   13.3 An Illustrative Example *546*   13.4 Criteria of Choice *548* 13.5 Expected Value of Perfect Information *556*   13.6 Representation by a Decision Diagram *559*   13.7 The Assessment of Probability Distributions *564*   13.8 Decision Making Based on Expected Utility *574*

14 Decision Making Using Both Prior and Sample Information 588

14.1 Introduction *588*   14.2 Posterior Analysis *589*

15 Devising Optimal Strategies Prior to Sampling 609

15.1 Introduction *609*   15.2 Preposterior Analysis *610*   15.3 Extensive-Form and Normal-Form Analyses *624*   15.4 Comparison of Extensive-Form and Normal-Form Analyses *630*   15.5 Sensitivity Analysis *633*   15.6 An Acceptance Sampling Example *635*   15.7 Optimal Sample Size *645*   15.8 General Comments *647*

16 Sequential Decision-Making Procedures 649

16.1 Introduction *649*   16.2 A New Product Development Problem *651*

17 Comparison of Classical and Bayesian Statistics 665

17.1 Introduction *665*   17.2 A Comparative Problem *668*   17.3 Classical and Bayesian Estimation *676*   17.4 Some Remarks on Classical and Bayesian Statistics *681*

Bibliography 684

Appendix A: Statistical Tables 689

Appendix B: Symbols, Subscripts, and Summations 719

Appendix C: Properties of Expected Values and Variances 724

Appendix D: Shortcut Formulas 729

Index 795

Introduction

The Nature of Statistics

This is a book about statistics. But what is statistics? One use of the term refers simply to numerical data. We are all familiar with collections of such statistics, pertaining to sports, population, the economy, the stock market, etc. However, in another use of the term, with which we are more concerned in this book, statistics is a body of theory and methods of analysis. The subject matter of statistics is very broad, extending from the planning of experiments and other studies that generate data to the collection, analysis, presentation, and interpretation of the data. Hence, numerical data constitute the raw material of this subject.

The most widely known statistical methods are those that summarize such data in terms of averages and other descriptive measures. For example, if we are interested in the incomes of a group of 1000 families chosen at random in a particular city, important characteristics of these incomes may be described by calculating an average income and a measure of the spread, or dispersion, of these incomes around the average. However, the essence of modern statistics is the theory and methodology of drawing inferences that extend beyond the particular set of data examined and of making decisions based on appropriate analyses of such data. Thus, in the preceding illustration, we are probably not so much interested in the incomes of the particular 1000 families included in the sample as in an *inference* about the incomes of *all* families in the city from which the sample was drawn. Such an inference might be in the form of a *test of a hypothesis* that the average income of all families in the city is $14,000 or less. The inference could also be in the form of a single figure, an *estimate* of the average income of *all* families in the city based on the average income observed in the sample of 1000 families. Or the marketing department of a company may want the information in order to *decide* among different types of advertising programs depending on whether it concludes that the city is a low-, medium-, or high-income area. The mathematical *theory of probability* provides the logical framework for the mental leap from the sample of data studied to the inference about all families in the city and for decisions such as the type of advertising program to be used.

A few points should be noted concerning this example. We may have wanted an inference about the incomes of *all* families in the city. However, since it would have been too expensive and too time-consuming to obtain the income data for every family in the city, only the sample of 1000 families was observed. The totality of families in the city (or, more generally, the totality of the elements about which the inference is desired) is referred to in statistics as the *universe* or *population*. The 1000 families, which represent a collection of only some elements of the universe, are referred to as a *sample*. In statistics, *sample data* are collected in order to make *inferences* or *decisions* concerning the *populations* from which samples are drawn.

Another point to note is that the sample was referred to as having been drawn "at random" from the population. A *random sample* is one drawn in such a way that the probability, or likelihood, of inclusion of every element in the population is known. However, even though these probabilities of inclusion may be known, the average income that would be observed for a random sample of 1000 families would vary from sample to sample. These sample-to-sample variations are known as *chance sampling fluctuations*. Although we cannot predict with certainty what the average income will be for any particular sample, the theory of probability, a branch of mathematics, enables us to compute how often these different sample results occur in the long run. It is an intriguing and remarkable fact that even though there is *uncertainty* concerning which particular sample may have been drawn, probability theory provides a rational basis for inference and decision making about the population or larger group from which the sample was taken.

Much of this text deals with the theory and methods by which such inferences and decisions are made.

The Role of Statistics

Statistical concepts and methods are widely applied in many areas of human activity. They are extensively used in the physical, natural, and social sciences, in business and public administration, and in many other fields.

In the sciences, the applications are far-ranging, extending from the design and analysis of experiments to the testing of new and competing hypotheses. In industry, statistics makes its contributions in short- and long-range planning and decision making. Hence, many firms use statistical methods to analyze patterns of change and to forecast economic trends for the firm, the industry, and the economy as a whole. Such forecasts often provide the foundation for corporate planning and control; areas such as purchasing, production, and inventory control depend on short-range forecasts, and capital investment and long-term development decisions depend on longer-range forecasts. Statistical methods are also employed in areas such as production control, inventory control, and quality control. In order to control the quality of manufactured products, for example, statistical methods are used to differentiate between variation attributable to chance causes and variation too great to be considered a result of chance. The latter type of variation can be analyzed and remedied. A large number of cases have been recorded in which applications of these statistical quality-control methods have resulted in substantial improvements in the quality of product and in lower costs because of reduction in rework and spoilage. Such statistical quality-control methods have been considered a major factor in the vast improvement in the quality of Japanese-manufactured products in the post-World War II period.

Over the past three decades in the fields of business and government, a body of quantitative techniques and procedures has been developed to aid and improve managerial decision making. The field of statistics has provided many of the fruitful ideas and techniques in this development. Currently, applications of statistics pervade virtually every activity of the business firm, including production, financial analysis, distribution analysis, market research, research and development, and accounting. Statistical methods have been used more often and with increasing sophistication in virtually every field in which they have been introduced. These methods constitute an integral part of the general development of more rational and more quantitative approaches to the solution of business problems. One outstanding characteristic of this development has been the increased adoption of scientific decision-making approaches using mathematical models. These models are mathematical formulas or equations that state the relationship among the important factors or variables in a problem or system. (For example, an equation may be developed to represent the relationship between a company's sales and the economic and other variables that influence sales.) Chapter 9 of this text discusses the methods for deriving one such type of mathematical model. The statistical methods discussed in that and other chapters bring a logical, objective, and systematic approach to decision making in business and other fields.

Extensive statistical activities are conducted by federal, state, and local governments. There are many applications of statistical ideas and methods in governmental administration, and governments collect and disseminate a great deal of statistical data. The most highly organized and extensive statistical information systems are those of the federal government. Such information systems, which include national income and product accounts, input–output accounts, flow of funds accounts, balance of payments accounts, and national balance sheets, depend on massive statistical collection and distribution systems. Statistical methods are applied to the resulting data to assess past trends and current status and to project future economic activity. These methods provide measures of human and physical resources and of economic growth, well-being, and potential. They are essential tools for appraising the performance and for analyzing the structure and behavior of an economy.

Massive data collection and dissemination activities are also carried out by governmental and private agencies in fields such as population, vital statistics, education, labor force, employment and earnings, business and trade, prices, housing, medical care, public health, agriculture, natural resources, welfare services, law enforcement, area and industrial development, construction, manufacturing, transportation, and communications.

Statistical analysis constitutes a body of theory and methods that plays an important role in this wide variety of human activities. It is extremely useful for communicating information, for drawing conclusions from data, and for the guidance of planning and decision making.

1

Frequency Distributions and Summary Measures

The methods to be discussed in this chapter are useful for describing patterns of variation in data. Variation is a basic fact of life. As individuals, we differ in age, sex, height, weight, and intelligence, in the quantities of the world's goods we possess, in the amount of our good or bad luck, and in myriad other characteristics. In the business world, variations are observed in the articles produced by manufacturing processes, in the yields of the economic factors of production, in production costs, financial costs, marketing costs, and so forth. Such variations occur both in data observed at a particular time and in data occurring over a period of time.

We begin our discussion by asking how you might go about summarizing the variation in a large body of numerical data. Suppose that data had been collected on the ages of all individuals in the United States and that you wanted to describe these approximately 200 million figures in some generally useful manner. How might you go about it, assuming that adequate resources for processing the data were available? Since it would be very difficult to see important characteristics of the data by merely listing them, you probably would try to group the age figures into classes. For example, you might set up classes of under five years, at least five but under ten years, and so forth. You could list the number of persons in each class, and if you divided these numbers by the total population, you would have the proportions of the population in each class. You would then find it relatively easy to summarize the general characteristics of the *age distribution* of the population. If you compared similar distributions, say, for the years 1915 and 1975, a number of important features would be observable without any further statistical analysis. For example, the range of ages in both distributions would be clear at a glance. The higher proportions of persons under the age of 20 and of persons over the age of 65 in 1975 than in 1915 would stand out. Also, smaller percentages of persons in the age categories 35 to 45 years would be observed for 1975, reflecting the decline in births during the 1930–1940 decade. Thus, the simple device of grouping the age figures into classes and recording the frequencies of occurrence in these classes would show us some of the underlying characteristics of the nation's age composition for each year. Generalizations about age patterns would thus become easier to make.

You might also want to continue your description of the age distributions by calculating one or more types of average. For example, you might be interested in computing an average age for the population in 1915 and in 1975 to determine whether this average had increased or decreased. Furthermore, if you wished to give a more exact description of the fact that in the later period there were heavier concentrations of persons in the younger and older age groups, you might attempt to construct a measure of how the ages were spread around an average age. Because of these heavier concentrations, the measure of spread or dispersion around the average would tend to be larger in 1975 than in the earlier period.

The types of statistical technique that might be used to summarize and describe the characteristics of the age data constitute the subject matter of this chapter. The table into which the data are grouped is referred to as a *frequency distribution*. The average or averages that can be computed are measures of *central tendency* or *central location* of the data, and the measure of spread around the average is a measure of *dispersion*. These and other techniques that group, summarize, and describe data are referred to as *descriptive statistics*. If the data treated by descriptive statistics represent a sample from a larger group or population, as noted on page 2, inferences may be desired about this larger group. Ways of making such statistical inferences are discussed in subsequent chapters.

The term *cross-sectional data* refers to data observed at a point in time, whereas *time series data* pertains to sets of figures that vary over a period of time. Frequency distribution analysis is concerned with cross-sectional data. In particular, such analysis deals with data where the order in which the observations were recorded is of no importance (for example, the ages of the present members of the labor force in the United States, the present wage distribution of employees in the automobile industry, or the distribution of U.S. corporations by net worth on a given date). On the other hand, if we recorded quality control data for a manufactured product, we would ordinarily be very much concerned with the order in which the articles were produced. For example, if a sudden run of defective articles was produced, we would be interested in knowing when this occurred and what was the general time pattern of production of defective and good articles. Similarly, in the study of economic growth, we might be interested in the variation over time of such data as real income per person or real gross national product per person. General methods of time series analysis are treated in Chapter 10.

1.1 FREQUENCY DISTRIBUTIONS

As indicated earlier, when we are confronted with large masses of ungrouped data (that is, listings of individual figures), it is very difficult to generalize about the information they contain. However, if a frequency distribution of the figures is formed, many of these features become readily discernible.

A *frequency distribution or frequency table records the number of cases that fall in each class of the data.* The numbers in each class are referred to as *frequencies;* hence the term "frequency distribution." When the numbers of items are expressed by their proportion in each class, the table is usually referred to as a *relative frequency distribution* or a *percentage distribution.*

How the classes of a frequency distribution are described depends on the nature of the data. In all cases, data are obtained either by counting or by measuring. For example, individuals have characteristics such as race, nationality, sex, and religion, and counts can be made of the number of persons who fall in each of the relevant categories. If a classification by nativity is used, the frequency distribution for residents of Centertown on January 1, 1977, may be shown as in Table 1-1.

Characteristics such as nativity, color, sex, and religion that can be expressed in qualitative classifications or categories are often referred to as *attributes* or *discrete variables.* It is always possible to encode the attribute classifications to make them numerical. Thus, in the preced-

TABLE 1-1
Distribution of Residents of Centertown on January 1,
1977, Classified by Nativity

Nativity	Number of Persons
Native born	91,436
Foreign born	10,322
Total	101,758

ing illustration, "native born" could have been denoted 0 and "foreign born" 1. In certain cases the data seem to fall naturally into simple numerical classifications. For example, families may be grouped according to number of children; the classes could be labeled 0, 1, 2, and so forth.

Data for qualitative characteristics or discrete variables can be presented graphically in terms of simple bar charts. Figure 1-1 gives a bar chart representation of the data given in Table 1-1.

To obtain *continuous variables,* or *continuous data* (that is, data that can assume any value in a given range), numerical measurements rather than counts are performed. When large numbers of measurements are made, it is convenient to use intervals or groupings of values and to list the number of cases in each class. With this procedure, a few problems have to be resolved concerning the number of class intervals to be, the size of these intervals, and the manner in which class limits should be stated.

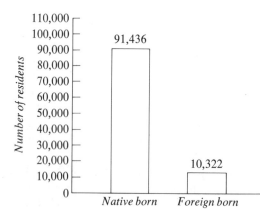

FIGURE 1-1
Number of residents of Centertown on January 1, 1977, classified by nativity.

1.2 CONSTRUCTION OF A FREQUENCY DISTRIBUTION

The decisions about the number and size of the classes in a frequency distribution are essentially arbitrary. However, these two choices are clearly interrelated; the smaller the intervals chosen, the more of them will be needed to cover the range of the scores. Frequency distributions generally are constructed with from five to twenty classes. When class intervals are of equal size, comparisons of classes are made easier and subsequent calculations from the distribution are simplified. However, this is not always a practical procedure. For example, with data on the annual incomes of families, in order to show the detail for the portion of the frequency distribution where the majority of incomes lie, class intervals of $3000 may be used up to about $15,000; then intervals of $5000 may be used up to $25,000 and a final class of $25,000 and over. It is clear that maintaining equal-sized classes of, say, $1000 throughout the entire range of income would result in too many classes. On the other hand, if much larger class intervals were used, too many families would be lumped together in the first one or two classes, and we would lose the information concerning how these incomes were distributed. The use of unequal class intervals and an open-ended interval for the highest class provides a simple way out of the dilemma.

An open-ended class interval is one that contains only one specific limit and an "open" or unspecified value at the other end, as for example *$25,000 and over* or *110 pounds and under*. The use of unequal class sizes and open-ended intervals generally becomes necessary when most of the data are concentrated within a certain range, when gaps appear in which relatively few items are observed, and finally when there are a very few extremely large or extremely small values. Open-ended intervals are sometimes also used to retain confidentiality of information. For example, the identity of the small number of individuals or companies in the highest class may be general knowledge, and stating an upper limit for the class might be considered excessively revealing.

We illustrate the construction of a frequency distribution by considering the figures in Table 1-2, which represent the earnings of 100 semiskilled employees of the Rowe Company for a particular week in 1977. Although the data have been arrayed from lowest earnings to highest, it is difficult to discern patterns in the ungrouped figures. However, when a frequency distribution is constructed, the nature of the data clearly emerges.

Let us assume that we would like to set up a frequency distribution with seven classes for the list of earnings figures shown in Table 1-2 and that we want the classes to be of equal size. A simple formula to obtain an estimate of the appropriate interval size is

$$i = \frac{H - L}{k}$$

TABLE 1-2
**Earnings of 100 Semiskilled Employees of the Rowe Company for One
Week in 1977**

$160.03	$181.92	$192.03	$197.03	$209.12
163.24	181.98	192.69	197.94	209.12
167.39	182.17	192.83	198.28	209.62
168.50	182.55	192.99	199.60	209.74
171.10	182.68	193.89	200.03	210.01
171.50	183.91	193.94	200.65	210.92
172.27	184.03	194.07	201.74	211.12
173.87	184.88	194.23	202.58	211.97
174.20	185.22	194.23	203.95	214.08
175.10	185.73	194.35	204.33	214.92
175.16	185.96	194.88	204.91	214.96
176.82	187.22	195.01	204.97	215.89
177.18	188.31	195.26	205.04	215.97
178.22	188.63	195.26	205.09	216.22
178.43	189.21	195.26	205.82	218.68
178.92	189.36	195.87	206.94	219.87
179.04	190.12	195.92	207.86	222.37
179.17	191.35	196.08	208.27	224.84
180.05	191.37	196.92	208.35	226.98
181.79	191.49	196.99	208.39	229.65

where i = the size of the class intervals
       H = the value of the highest item
       L = the value of the lowest item
       k = the number of classes

This formula for class interval size simply divides the total range of the
data (that is, the difference between the values of the highest and lowest
observations) by the number of classes. The resulting figure indicates
how large the class intervals would have to be in order to cover the entire range of the data in the desired number of classes. Other considerations involved in determining an appropriate number of classes are discussed in Section 1.4.

In the case of the weekly earnings data,

$$i = \frac{(\$229.65 - \$160.03)}{7} = \$9.95$$

Since it is desirable to have convenient sizes for class intervals, the $9.95
figure may be rounded to $10.00 and the distribution may be tentatively
set up on that basis. The frequency distribution shown in Table 1-3 results from a tally of the number of items that fall in each $10.00 class in-
terval.

Some important features of these data are immediately seen from
the frequency distribution. The approximate value of the range, or the

TABLE 1-3
Frequency Distribution of Earnings of 100 Semiskilled
Employees of the Rowe Company for One Week in 1977

Weekly Earnings	Number of Employees
$160.00 and under 170.00	4
170.00 and under 180.00	14
180.00 and under 190.00	18
190.00 and under 200.00	28
200.00 and under 210.00	20
210.00 and under 220.00	12
220.00 and under 230.00	4
Total	100

difference between the values of the highest and lowest items, is revealed. (Of course, since the identity of the individual items is lost in
the grouping process, we cannot tell from the frequency table alone what
the exact values of the highest and lowest items are.) Also, the frequency distribution gives at a glance some notion of how the elements
are clustered. For example, more of the weekly earnings figures fall in
the interval from $190.00 to $200.00 than in any other single class.
When the frequencies in the classes immediately preceding and following the $190.00 to $200.00 grouping are added to the 28 in that interval,
a total of 66, or $\frac{2}{3}$ of the 100 employees, are accounted for. Furthermore,
the distribution shows how the data are spread or dispersed throughout
the range from the lowest to the highest value. We can quickly determine whether the items are bunched near the center of the distribution
or spread rather evenly throughout. Also, we can see whether the frequencies fall away rather symmetrically on both sides of the center of the
distribution or whether they tend to fall mostly to one side of the center.
We now consider various statistical measures for describing these characteristics more precisely, but much information can be gained by simply studying the distribution itself. There is no single perfect frequency
distribution for a given set of data. Several alternative distributions
with different class interval sizes and different highest and lowest values
may be equally appropriate.

1.3 CLASS LIMITS

The way in which class limits of a frequency distribution are described
depends on the nature of the data. Figures on ages provide a good illustration of this point. Suppose that ages were recorded as of the *last*

birthday. Then a clear and unambiguous way of stating the class limits is as follows: 15 and under 20, 20 and under 25, etc. (Of course, there are other ways of wording the limits, such as "at least 15 but under 20," or "from 15 up to but not including 20."

Consider the first class interval, "15 and under 20." Since ages have been recorded as of the last birthday, this class encompasses individuals who have reached at least their fifteenth birthday but not their twentieth birthday. If you are 19.999 years of age, that is, a fraction of a day away from your twentieth birthday, you fall into the first class. However, upon attaining your twentieth birthday, you fall in the second class, "20 and under 25." Thus, these class intervals are five years in size. The midpoints of the classes, that is, the values located halfway between the class limits, are 17.5, 22.5, 27.5, 32.5, 37.5, and so on. These values are used in computations of statistical measures for the distribution. Note that with class limits established and stated this way, the *stated limits* are in fact the true boundaries, or *real limits*, of the classes.

Suppose, on the other hand, that age data were rounded to the *nearest* birthday. We could follow a widely used convention and state the class limits as follows: 15–19, 20–24, etc. Even though the stated limits in each class are only four years apart, it is important to realize that the size of these class intervals is still five years. For example, since the ages are given as of the nearest birthday, everyone between 14.5 and 19.5 years of age falls in the class 15–19. Thus, when data recorded to the nearest unit are grouped into frequency distribution classes, the lower real limit or lower boundary of any given class lies one half unit below the lower stated limit and the upper real limit or upper boundary lies one half unit above the upper stated limit. The midpoints of the class intervals may be obtained by averaging the lower and upper real limits or the lower and upper stated limits. For example, the midpoint of the class 15–19 is 17, which is the same figure obtained by averaging 14.5 and 19.5.

In summary, when raw data are rounded to the *last* unit, the stated class limits and real class limits are identical. When raw data are rounded to the *nearest* unit, the real limits are one half unit removed from the stated limits. With both types of data, the midpoints of classes are halfway between the stated limits or, equivalently, halfway between the real limits.

Of course, class intervals should always be mutually exclusive, and the class each item falls into should always be clear. If class limits are stated as 30–40, 40–50, etc., for example, it is not clear whether 40 belongs to the first class or the second.

Unfortunately, conventions are not universally observed. Often, one must use a frequency distribution constructed by others, and the nature of the raw data may not be clearly indicated. The producer of a frequency distribution should always indicate the nature of the underlying data.

1.4 OTHER CONSIDERATIONS IN CONSTRUCTING FREQUENCY DISTRIBUTIONS

A number of other points should be taken into account in the construction of a frequency distribution. If the data are such that there are concentrations of particular values, it is desirable that these values be at the midpoints of class intervals. For example, assume that data are collected on the amounts of the lunch checks in a student cafeteria. Suppose these checks predominantly occur in multiples of five cents, although not exclusively so. If class intervals are set up as $.70–.74, $.75–.79, etc., a preponderance of items would be concentrated at the lower limits. In calculating certain statistical measures from the frequency distribution, the assumption is made that the midpoints of classes are average (arithmetic mean) values of the items in these classes. If, in fact, most of the items lie at the lower limits of the respective classes, a systematic error will be introduced by this assumption, because the actual averages within classes will typically fall below the midpoints.

Another factor to be considered in constructing a frequency distribution is the desirability of having a relatively smooth progression of frequencies. In many frequency distributions of business and economic data, one class contains more items than any other single class and the frequencies drop off more or less gradually on either side of this class. Table 1-3 is an example of such a distribution. (As indicated in Section 1.2, the distribution may not be at all symmetrical.) However, erratic increases and decreases of frequencies from class to class tend to obscure the overall pattern, and such erratic variations often arise from the use of class intervals that are too small. Increasing the size of class intervals usually results in a smoother progression of frequencies, but wider classes reveal less detail than narrower classes. Thus, a compromise must be made in the construction of every frequency distribution. At one extreme, if we use class interval sizes of one unit each, every item of raw data is assigned to a separate class; at the other extreme, if we use only one class interval as wide as the range of the data, all items fall in the single class. Within the limits of these considerations, some freedom exists for the choice of an appropriate class interval size.

1.5 GRAPHIC PRESENTATION OF FREQUENCY DISTRIBUTIONS

The use of graphs for displaying frequency distributions will be illustrated for the data on weekly earnings shown in Table 1-3. One method is to represent the frequency of each class by a rectangle or bar. Such a

FIGURE 1-2
Histogram of frequency distribution of earnings of 100 semiskilled employees of the
Rowe Company for one week in 1977.

chart is generally referred to as a *histogram*. A histogram for the frequency table given in Table 1-3 is shown in Figure 1-2. In agreement with the usual convention, values of the variable are depicted on the horizontal axis and frequencies of occurrence are shown on the vertical axis.

An alternative method for the graphic presentation of a frequency distribution is the *frequency polygon*. In this type of graph, the frequency of each class is represented by a dot above the midpoint of each class at a height corresponding to the frequency of the class. The dots are joined by line segments to form a many-sided figure, or polygon. A frequency polygon can also be thought of as the line graph obtained by joining the midpoints of the tops of the bars in a histogram. By convention, the polygon is connected to the horizontal axis by line segments drawn from the dot representing the frequency in the lowest class to a point on the horizontal axis one half a class interval below the lower limit of the first class, and from the dot representing the frequency in the highest class to a point one half a class interval above the upper limit of

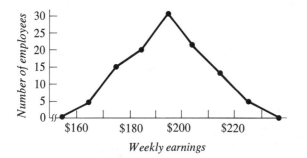

FIGURE 1-3
Frequency polygon for the distribution of earnings of 100 semiskilled employees of
the Rowe Company for one week in 1977.

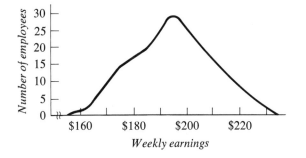

FIGURE 1-4
Frequency curve for the distribution of earnings of 100 semiskilled employees of the Rowe Company for one week in 1977.

the last class. A frequency polygon for the distribution given in Table 1-3 is shown in Figure 1-3. It is important to realize that the line segments are drawn only for convenience in reading the graph and that the only significant points are the plotted frequencies for the given midpoints. Interpolation for intermediate values between such points would be meaningless. Often, midpoints of classes are shown on the horizontal axis rather than class limits as were shown in Figure 1-3.

If the class sizes in a frequency distribution were gradually reduced and the number of items increased, the frequency polygon would approach a smooth curve more and more closely. Thus, as a limiting case, the variable of interest may be viewed as continuous rather than discrete, and the polygon would then assume the shape of a smooth curve. The frequency curve approached by the polygon for the weekly earnings would appear as shown in Figure 1-4.

1.6 CUMULATIVE FREQUENCY DISTRIBUTIONS

Sometimes interest centers on the number of cases that lie below or above specified values rather than within intervals. In such situations, it is convenient to use a *cumulative* frequency distribution rather than the usual frequency distribution. Table 1-4 shows a so-called "less than" cumulative distribution for the weekly earnings data shown in Table 1-2. The cumulative numbers of employees with earnings less than the lower class limits of $160.00, $170.00, and so on are given. Thus, there were no employees with weekly earnings of less than $160.00, four employees with earnings of less than $170.00, 4 + 14 = 18 employees with less than $180.00, and so on.

The graph of a cumulative frequency distribution is referred to as an

TABLE 1-4
Cumulative Frequency Distribution of Earnings of
100 Semiskilled Employees of the Rowe Company
for One Week in 1977

Weekly Earnings	Number of Employees
Less than $160.00	0
Less than 170.00	4
Less than 180.00	18
Less than 190.00	36
Less than 200.00	64
Less than 210.00	84
Less than 220.00	96
Less than 230.00	100

ogive (pronounced "ōjive"). The ogive for the cumulative distribution shown in Table 1-4 is given in Figure 1-5. The plotted points represent the number of employees having earnings less than the figure shown on the horizontal axis. The vertical coordinate of the last point represents the sum of the frequencies (100 in this case). The S-shaped configuration depicted in Figure 1-5 is quite typical of the appearance of a "less than" ogive. A "more than" ogive for the daily sales distribution would have class limits reading "more than $160.00" and so on. In this case, a reversed S-shaped figure would have been obtained, sloping downward from the upper left to the lower right.

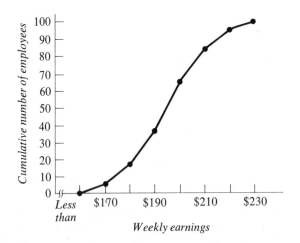

FIGURE 1-5
Ogive for the distribution of earnings of 100 semiskilled employees of the Rowe
Company for one week in 1977.

Exercises

1. The commissions paid to 21 Computron Corporation computer software sales representatives for a one-week period were

$153.17	$194.33	$146.89	$207.79	$214.80
183.04	222.22	188.88	276.18	
257.93	200.00	273.66	249.65	
244.56	282.49	199.23	265.01	
113.42	236.00	172.71	178.00	

Assuming that the data were recorded to the last cent, construct
a. a frequency table with about eight classes
b. a histogram
c. a frequency polygon

2. Daily sales of a small retail establishment are given below:

$ 97.60	$132.67
102.65	145.68
141.02	175.92
174.68	106.34
92.06	125.27
172.21	127.72
83.77	137.66
104.01	83.66
102.44	116.70
156.52	136.79
149.59	129.99
136.97	125.41
124.31	124.17
123.23	91.70
118.94	128.29

a. Construct a frequency distribution for sales during the period, assuming the data were recorded to the last cent. Use about five classes.
b. Sketch a frequency polygon showing the distribution in part (a).

3. The vice-president in charge of corporate credit for a large bank is reviewing the outstanding credit positions of the bank's largest corporate clients. The following list of the 25 largest clients and their credit positions has been compiled:

Company	Credit (in $Millions)	Company	Credit (in $Millions)
A	25.0	N	23.7
B	15.8	O	3.5
C	33.3	P	26.2
D	6.3	Q	13.9
E	13.5	R	5.1
F	9.2	S	11.0

(Continued)

Company	Credit (in $Millions)	Company	Credit (in $Millions)
G	18.6	T	14.4
H	2.0	U	29.3
I	10.4	V	22.0
J	15.7	W	19.9
K	8.0	X	12.5
L	23.0	Y	17.1
M	26.8		

    a. Set up a frequency table having seven classes and specify the midpoint of each class.

    b. Compute the cumulative frequency distribution.

4. Compute the cumulative frequency distribution for the distribution given below.

	Analysis of Ordinary Life Insurance
Policy Size	Number of Ordinary Life Insurance Policies in Force per 1000 Policies in a Certain Month
Under $1000	31
$1000–2499	181
2500–4999	108
5000–9999	213
10,000 or more	467

5. Comment critically on the following systems of designating class intervals:
    a. 83–102
       102–121
       etc.
    b. 33 and under 102
       103 and under 121
       122 and under 141
       etc.

6. The Tolkien Associates, a Philadelphia consulting firm, analyzed the market demand for Frosted Freakies, a breakfast cereal sold at many Philadelphia–Camden area retail food markets. This analysis was useful to the firm producing the cereal in its advertising and distribution planning decisions. Twenty-two Philadelphia food stores and fourteen Camden food stores were surveyed for a one-week period, and the number of boxes sold in each store during that period was recorded as follows:

        Philadelphia:
        2511   3180   1600   2752   2849   2466   753

```
3375  2754  3200  2115  2961  3162  2632
1447  3414  2648  2970  3334  2387  1789
2893
```
Camden:
```
1874  1280  1659  2062  1866  1731   832
2633  1509  1167  2374   514  1901  1326
```

a. Using six classes and "and under" upper class limits, construct and graph the frequency distribution for the total number of boxes sold in the two cities combined during the one-week period.
b. Using the same classes, construct and graph the frequency distribution for each city separately.
c. Which answer, that for part (a) or that for part (b), gives a better picture of the distribution of demand for Frosted Freakies? Explain your answer.

1.7 DESCRIPTIVE MEASURES FOR FREQUENCY DISTRIBUTIONS

As indicated in Section 1.2, once a frequency distribution is constructed from a set of figures, certain features of the data become readily apparent. For most purposes, however, it is necessary to have a more exact description of these characteristics than can be ascertained by a casual glance at the distribution. Thus, analytical measures are usually computed to describe such characteristics as the *central tendency, dispersion,* and *skewness* of the data. These measures are called *summary descriptions* of the frequency distribution, which is itself a summarization of the set of original data.

Averages are the measures used to describe the characteristic of *central tendency* or *central location* of data. One such measure is the arithmetic mean. This is doubtless the most familiar average; in fact, it is often referred to in common usage as "the average." For example, in ordinary conversation or print, we encounter such terms as "average income," "average growth rate," "average profit rate," and "average person." (Actually, several different types of averages, or measures of central tendency, are implied in these terms.) In this section, we consider only the most commonly employed and most generally useful averages. Averages attempt to convey with a single number the notion of "central location" or the "middle property" of a set of data. As we shall see, the type of average to be employed depends on the purpose of the application and the nature of the data being summarized.

Dispersion refers to the spread, or variability, in a set of data. One method of measuring this variability is in terms of the difference between the values of selected items in a distribution, such as the difference between the values of the highest and lowest items. Another more comprehensive method is in terms of some average of the deviations of all the items from an average. Dispersion is, of course, a very important characteristic of data, because we are frequently interested as much in the variability in a set of data as in its central tendency.

Skewness refers to the symmetry or lack of symmetry in the shape of a frequency distribution. This characteristic is of particular importance in judging the typicality of certain measures of central tendency.

We begin the discussion of averages or measures of location by considering the most familiar one, the arithmetic mean.

1.8 THE ARITHMETIC MEAN

Probably the most widely used and most generally understood way of describing the central tendency, or central location, of a set of data is the average known as the *arithmetic mean*. The arithmetic mean, or simply the *mean*, is the total of the values of a set of observations divided by the number of observations. For example, if $X_1, X_2, \ldots, X_n$ represent the values of n items or observations, the arithmetic mean of these items, denoted $\overline{X}$, is defined as

$$\overline{X} = \frac{X_1 + X_2 + \cdots + X_n}{n} = \frac{\sum_{i=1}^{n} X_i}{n}$$

For simplicity, subscript notation such as that given above will usually not be used in this book. (However, before continuing, you should turn to Appendix B and work out the examples given there.) Thus, when the subscripts are dropped, the formula becomes

(1.1)
$$\overline{X} = \frac{\sum X}{n}$$

where the symbol Σ (Greek capital *sigma*) means "the sum of."

For example, suppose that an accounting department established accounts receivable in the following amounts during a one-hour period: $600, $350, $275, $430, and $520. The arithmetic mean of the amounts of the accounts receivable is

$$\overline{X} = \frac{\$600 + \$350 + \$275 + \$430 + \$520}{5} = \frac{\$2175}{5} = \$435$$

The mean, $435, may be thought of as the size of each account receivable that would have been set up if the total of the five accounts was the same ($2175) but all the accounts were the same size. That is, the mean is the value each item would have if they were all identical and the total value and number of items remained unchanged.

Symbolism

In keeping with standard statistical practice, we will use the symbol $\bar{X}$ to denote the mean of a *sample* of observations. We will denote the number of observations in the sample by the lower case letter n. A value such as $\bar{X}$ computed from sample data is referred to as a *statistic*. A statistic may be used as an estimate of an analogous population measure, known as a *parameter*. Thus, the statistic $\bar{X}$ (the sample mean) may be thought of as an estimate of a parameter, the mean of the population from which the sample was drawn. It is conventional to denote population parameters by Greek letters and sample statistics by Roman letters. In keeping with this, we will denote a population mean by the Greek letter μ (*mu*).

If the population mean were calculated directly from the data collected from the entire population of, say, N members, then

(1.2)
$$\mu = \frac{X_1 + X_2 + \cdots + X_N}{N} = \frac{\Sigma\, X}{N}$$

In practice, the population mean μ is usually not calculated, because it is rarely feasible or advisable to accomplish a complete enumeration of the population.

Grouped Data

When data have been grouped into a frequency distribution, the arithmetic mean can be computed by a generalization of the definition for the mean of ungrouped data. As given in Equation 1.1, the formula for the mean of a set of ungrouped data is $\bar{X} = (\Sigma\, X)/n$. However, with grouped data, since the identity of the individual items has been lost, an estimate must be made of the total of the values of the observations, $\Sigma\, X$. This estimate is obtained by multiplying the midpoint of each class in the distribution by the frequency of that class and summing over all classes. In symbols, if m denotes the midpoint of a class and f the frequency, the arithmetic mean of a frequency distribution may be estimated from the following formula, known as the *direct method*:

(1.3)
$$\bar{X} = \frac{\Sigma\, fm}{n}$$

The computation of $\bar{X}$ for the frequency distribution of weekly earnings data shown in Table 1-3 is given in Table 1-5. The mean earnings figure of \$194.80 calculated from the frequency distribution is very close to the corresponding mean of \$194.91 for the ungrouped data given in Table 1-2. The small difference in these two figures illustrates the slight loss of accuracy involved in calculating statistical measures from frequency distributions rather than from ungrouped data. When

TABLE 1-5
Calculation of the Arithmetic Mean for Grouped Data by the Direct Method:
Weekly Earnings Data

Weekly Earnings	Number of Employees f	Midpoints m	fm
$160.00 and under $170.00	4	$165.00	$660
170.00 and under 180.00	14	175.00	2450
180.00 and under 190.00	18	185.00	3330
190.00 and under 200.00	28	195.00	5460
200.00 and under 210.00	20	205.00	4100
210.00 and under 220.00	12	215.00	2580
220.00 and under 230.00	4	225.00	900
	$n = \Sigma f = 100$		$19,480

$$\overline{X} = \frac{\Sigma fm}{n} = \frac{\$19,480}{100} = \$194.80$$

there is a large number of observations, this loss of accuracy is offset by
the fact that the calculations are far less tedious than when done from the
original data.

Shortcut formulas are often useful for calculating the arithmetic
mean and other measures for frequency distributions. One such shortcut formula, known as the step-deviation method, is explained in Appendix D.

1.9 THE WEIGHTED ARITHMETIC MEAN

In averaging a set of observations, it is often necessary to compute a
weighted average in order to arrive at the desired measure of central location. For example, suppose a company consisted of three divisions,
all selling different lines of products. The ratios of net profit to sales (expressed as percentages) for these divisions for the year 1976 were 5% for
Division A, 6% for Division B, and 7% for Division C. (These percentages are called *net profit to sales percentages*.) Assume that we want to
find the net profit to sales percentage *for the three divisions combined*,
or equivalently, *for the company as a whole*. This ratio is the figure that
results from dividing total net profits by total sales for the three divisions
combined. Clearly, if we have only the profit *percentages* for the three
divisions, we do not have enough information to compute the required
figure. However, if we are given the dollar sales for each of the three di-

TABLE 1-6
Calculation of a Weighted Arithmetic Mean: Net Profit to Sales Percentage for the
Three Divisions of a Company Combined, 1976

Division	Net Profit to Sales Percentage X	Sales w	Net Profit wX
A	5%	$10,000,000	$ 500,000
B	6%	10,000,000	600,000
C	7%	30,000,000	2,100,000
		$50,000,000	$3,200,000

$$\text{Weighted Mean} = \overline{X}_w = \frac{\Sigma\ wX}{\Sigma\ w} = \frac{\$3,200,000}{\$50,000,000} = 6.4\%$$

visions (that is, the denominators of the three ratios of net profit to sales
from which the percentages were computed), then these figures can be
used as "weights" in calculating the desired figure for the entire company. Specifically, the *weighted arithmetic mean* would be calculated
as shown in Table 1-6 by carrying out the following steps:

1. Multiply (that is, weight) the net profit to sales percentage for each di-
    vision by the sales of that division. As indicated in Table 1-6, the re-
    sulting figures are the net profits of the divisions.
2. Sum the net profits obtained in step 1 to obtain total net profit for the
    three divisions combined.
3. Sum the sales figures (the weights) to obtain total sales for the three
    divisions combined.
4. Calculate the desired average by dividing the total net profits figure
    obtained in step 2 by the total sales figure found in step 3.

Symbolically, the weighted arithmetic mean is given by the formula

(1.4)
$$\overline{X}_w = \frac{\Sigma\ wX}{\Sigma\ w}$$

where $\overline{X}_w$ = the weighted arithmetic mean
      X = the values of the observations to be averaged (net profits to
          sales percentages in this case)
      w = the weights applied to the X values (sales in this case)

Again, we note that the weighted average of 6.4% computed in
Table 1-6 is interpreted as the net profit to sales percentage for the three
divisions combined (that is, for the company as a whole). Since dollar
sales are in the denominator of the $\Sigma\ wX/\Sigma\ w$ ratio, the answer of 6.4%

may be interpreted in terms of dollars of profit per dollar of sales, that is, an average of $.064 profit per dollar of sales.

On the other hand, suppose we had asked instead, "What is the arithmetic mean net profit to sales ratio *per division*, without regard to the sales size of these divisions?" The answer is given by

$$\bar{X} = \frac{\Sigma X}{n} = \frac{5\% + 6\% + 7\%}{3} = \frac{18\%}{3} = 6\% \text{ per division}$$

where X = the percentages for the three divisions
$\quad\quad n$ = the number of the divisions

The result, $\bar{X}$, may be referred to as an *unweighted arithmetic mean* of the three ratios. In this computation, the net profit to sales percentages were totaled and the result divided by the number of divisions. Note that the result is therefore stated as 6% *per division*, because of the appearance of the number of divisions (3) in the denominator. Clearly, this calculation disregards differences that may exist in the amount of sales of the three divisions; however, this does not make the average meaningless. If we are interested in obtaining a "representative" or "typical" profit ratio and there are no extreme values to distort the representativeness of the unweighted mean, then this type of computation is a valid one. Of course, if one were seeking a typical or representative figure, it would be desirable to have more than the three observations present in this illustration, and other averages, such as the median or mode, may be preferable to the arithmetic mean in the determination of a typical or representative value. However, the unweighted arithmetic mean is not a meaningless figure; indeed, it is the correct answer to the question just posed.

Returning to weighted mean calculation, the weights that were applied to the three profit to sales percentages in this problem were the actual dollar amounts of sales for the three divisions. That is, the weights used were the values of the denominators of the original ratios. An alternative procedure would be to weight the ratios by a percentage breakdown of the denominators (in this case, a percentage breakdown of total sales). For example, in the computation shown in Table 1-6, weights of 20%, 20%, and 60% could have been applied instead of weights of $10,000,000, $10,000,000, and $30,000,000, and the same answer of 6.4% would have resulted. Indeed, any figures in the same proportions as $10,000,000, $10,000,000, and $30,000,000 would have led to the same numerical answer. Note that the reason the weighted arithmetic mean of 6.4% exceeded the unweighted mean of 6.0% was that in the weighted mean calculation, greater weight was applied to the 7% profit figure for Division C than to the corresponding 5% figure for Division A. This had the effect of pulling the weighted average up toward the 7% figure.

Exercises

1. The following refers to Exercise 1 on page 17.
   a. Calculate the arithmetic mean from the original data.
   b. Calculate the arithmetic mean using the frequency table.
   c. Which is the true arithmetic mean value? Explain your answer.

2. The Center City Bank reports bad debt ratios (dollar losses to total dollar credit extended) of 0.04 for personal loans and 0.02 for industrial loans in 1976. For the same year, the Neighborhood Bank reports bad debt ratios of 0.05 for personal loans and 0.03 for industrial loans. Can one conclude from this that Center City's overall bad debt ratio is less than Neighborhood's? Justify your response.

3. Mortimer Hutton, a wealthy investor who spends a large amount of time watching the quote machine in his broker's office, is weighing the relative merits of two periodic investment strategies. The first method is to buy the same number of shares of stock each investment period. The second requires an investment of a constant dollar amount each period, regardless of the stock price. His broker demonstrates the result of using each strategy on two stocks, United Aerodynamics and Mitton Industries, for a five-year period. The purchases in each case are made at midyear at the prevailing stock price. The constant dollar amount for the second strategy is $1000 (that is, the amount invested in the stock is as close to $1000 as possible).

First Strategy

	United Aerodynamics				Mitton Industries		
Year	Price per Share	Shares	Total Cost	Year	Price per Share	Shares	Total Cost
1	54	25	1350	1	25	25	625
2	50	25	1250	2	30	25	750
3	42	25	1050	3	43	25	1075
4	35	25	875	4	40	25	1000
5	30	25	750	5	49	25	1225

Second Strategy

	United Aerodynamics				Mitton Industries		
Year	Price per Share	Shares	Total Cost	Year	Price per Share	Shares	Total Cost
1	54	19	1026	1	25	40	1000
2	50	20	1000	2	30	33	990
3	42	24	1008	3	43	23	989
4	35	29	1015	4	40	25	1000
5	30	33	990	5	49	20	980

a. Calculate the average cost per share for each stock in each strategy.
b. Which strategy achieved the lower average cost for United Aerodynamics? For Mitton Industries?
c. Explain these differences in terms of the weights used in calculating the average cost per share for each stock and strategy.

4. The frequency distribution for the percentage return on sales of 150 U.S. companies for one year is as follows:

Percentage Return on Sales	Number of Companies
0% and under 5%	34
5% and under 10%	46
10% and under 15%	38
15% and under 20%	22
20% and under 25%	10

a. Compute the arithmetic mean of this distribution.
b. If one totaled the profit figures for the 150 firms and divided by the total sales of all the companies, would the resulting average equal the answer to part (a)?

5. Consider the following data for the percentage of the civilian labor force unemployed in a certain area:

County	Percentage Unemployed	Civilian Labor Force
A	3.6	114,395
B	3.8	214,758
C	2.5	206,324
D	6.5	843,160

a. What is the unweighted average of the percentage of the labor force unemployed per county?
b. What is the weighted average of the percentage of the labor force unemployed for the four counties combined?
c. Explain the reason for the difference in the figures obtained in parts (a) and (b).

1.10 THE MEDIAN

The *median* is another well known and widely used average. It has the connotation of the "middlemost" or "most central" value of a set of numbers. For ungrouped data, it is defined simply as the value of the central item when the data are arrayed by size. If there is an odd number of observations, the median is directly ascertainable. If there is an even number of items, there are two central values, and by convention, the value halfway between these two central observations is designated as the median.

For example, suppose that a test of a brand of gasoline in five new small economy cars yielded the following numbers of miles per gallon: 27, 29, 30, 32, and 33. Then the median number of miles per gallon would be 30. If another car were tested and the number of miles per gallon obtained was 34, the array would now read 27, 29, 30, 32, 33, 34. The median would be designated as 31, the number halfway between 30 and 32.

Another way of viewing the median is as a value below and above which lie an equal number of items. Thus, in the preceding illustration involving five observations, two lie above the median and two below. In the example involving six observations, three fall above and three fall below the median. Of course, in the case of an array with an even number of items, any value lying between the two central items may, strictly speaking, be referred to as a median. However, as indicated earlier, the convention is to use the midpoint between the two central items. In the case of tied values at the center of a set of observations, there may be no value such that equal numbers of items lie above and below it. Nevertheless, the central value, as defined in the preceding paragraph, is still designated as the median. For example, in the array 52, 60, 60, 60, 60, 61, 62, the number 60 is the median, although unequal numbers of items lie above and below this value.

In the case of a frequency distribution, since the identity of the original observations is not retained, the median is necessarily an estimated value. Because in a frequency distribution the data are arranged in order of magnitude, frequencies can be cumulated to determine the class in which the median observation falls. It is then necessary to make some assumption about how observations are distributed in that class. Conventionally, the assumption is made that observations are equally spaced, or evenly distributed, throughout the class containing the median. The value of the median is then established by a linear interpolation. The procedure is illustrated for the distribution of personal loans given in Table 1-7. First, the calculation of the median is explained without the use of symbols. Then the procedure is generalized by stating it as a formula.

There are 100 loans represented in the distribution shown in Table 1-7, so the median lies between the fiftieth and fifty-first loans. Since 49 loans occur prior to the class "$600 and under $800," the median must be in that class. Assuming that the 20 loans are evenly distributed between $600 and $800, we can determine the median observation by interpolating $\frac{1}{20}$ of the distance through this $200 class. The median is calculated by adding $\frac{1}{20}$ of $200 to the $600 lower limit of the class containing the median. That is,

TABLE 1-7
Calculation of the Median for a Frequency Distribution: Personal Loan Data

	Number of Loans f
$ 0 and under $ 200	6
200 and under 400	18
400 and under 600	25
	$\Sigma f_p = 49$
600 and under 800	20
800 and under 1000	17
1000 and under 1200	14
	100

$$Md = \$600 + \left(\frac{50 - 49}{20}\right)\$200 = \$600 + \left(\frac{1}{20}\right)\$200 = \$610.00$$

Thus, the formula for calculating the median of a frequency distribution is

(1.5) $$Md = L_{Md} + \left(\frac{n/2 - \Sigma f_p}{f_{Md}}\right)i$$

where Md = the median
 L_{Md} = the (real) lower limit of the class containing the median
 n = the total number of observations in the distribution
 Σf_p = the sum of the frequencies of the classes preceding the one containing the median
 f_{Md} = the frequency of the class containing the median
 i = the size of the class interval

It may seem that we have located the value of the fiftieth observation rather than one falling midway between the fiftieth and fifty-first. However, the value determined is indeed one lying halfway between the fiftieth and fifty-first observations. This can be ascertained by examining the assumption of an even distribution of items within the class in which the median falls.

If there are 20 observations in the $200 class from $600 to $800, we may think of the class as being divided into 20 equal subintervals of $10 each, as depicted in Figure 1-6. Since the items are assumed to be evenly distributed in the class from $600 to $800, they must be located at the midpoints of these subintervals. An interpolation of $\frac{1}{20}$ through the class interval brings us to the end of the first subinterval, $610, which is seen to be a value halfway between the first and second items. Since 49 frequencies preceded this class, the median of $610 is a value lying midway between the fiftieth and fifty-first observations.

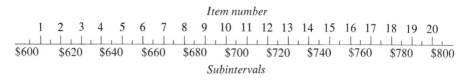

FIGURE 1-6
Diagram depicting the meaning of the assumption concerning an even distribution
of observations.

1.11   CHARACTERISTICS AND USES OF THE ARITHMETIC MEAN AND MEDIAN

The preceding sections have concentrated on the mechanics of calculating means and medians for ungrouped and grouped data. We now turn to a few of the characteristics and uses of these averages.

The arithmetic mean is doubtless the most widely used and most familiar measure of central tendency. Because it is defined as the total of the values of a set of observations divided by the number of these observations, it has the advantage of being a rigidly defined mathematical value that can be manipulated algebraically. For example, the means of two related distributions can be combined by suitable weighting. If the arithmetic mean income in 1976 of ten marketing executives of a company was $45,000 and the corresponding mean income of 19 marketing executives of another company was $55,000, then the arithmetic mean income for the 29 executives combined was, by Equation 1.4,

$$\bar{X}_w = \frac{10(\$45,000) + 19(\$55,000)}{10 + 19} = \$51,552$$

On the other hand, if we knew the median income of the ten executives from the first company and the median income of the 19 executives from the second company, there would be no way of averaging those two numbers to obtain the median income for the 29 executives combined. Because of such mathematical properties, the arithmetic mean is used more often than any other average in advanced statistical techniques.

A disadvantage of the mean is its tendency to be distorted by extreme values at either end of a distribution. In general, it is pulled in the direction of these extremes. Thus, the arithmetic mean of the five figures $110, $126, $132, $157, and $1000 is

$$\frac{\$110 + \$126 + \$132 + \$157 + \$1000}{5} = \$305$$

a value that is greater than four of the five items averaged. In such situations, the arithmetic mean may not be a very typical or representative figure.

The median is also a very useful measure of central tendency. Its relative freedom from distortion by skewness in a distribution makes it particularly useful for conveying the idea of a typical observation. It is primarily affected by the number of observations rather than their size. This can be seen by considering an array in which the median has been determined; if the largest item is multiplied by 100 (or any large number), the median remains unchanged. The arithmetic mean would, of course, be pulled toward the large extreme item.

The major disadvantage of the median is that it is an average of position and hence not a mathematical concept suitable for further algebraic treatment. Thus, as indicated earlier, if one knows the medians of each of two distributions, there is no algebraic way of averaging the two figures to obtain the median of the combined distribution.

1.12 THE MODE

Another average that is conceptually very useful but often not explicitly calculated is the *mode*. In French, to be "à la mode" is to be in fashion. The mode as a statistical average is the observation that occurs with the greatest frequency and thus is the most "fashionable" value. The mode is rarely determined for ungrouped data. The reason is that even when most of the data items are clustered toward the center of the array of observations, the item that occurs more often than any other may lie at the lower or upper end of the array and thus be a very unrepresentative figure. Therefore, determination of the mode is generally attempted only for grouped data.

When data are grouped into a frequency distribution, it is not possible to specify the single observation that occurs most frequently, since the identity of the individual items is lost. However, we can determine the *modal class*, the class that contains more observations than any other. Of course, class intervals should be of the same size when this determination is made. When the location of the modal class is considered along with the arithmetic mean and median, much useful information is generally conveyed not only about central tendency but also about the skewness of a frequency distribution.

Several formulas have been developed for determining the location of the mode within the modal class. These usually involve the use of frequencies in the classes preceding and following the modal class as weighting factors that tend to pull the mode up or down from the midpoint of the modal class. We shall not present any of these formulas here. For our purposes, the midpoint of the modal class may be taken as an estimate of the mode. To understand the meaning of the mode clearly, let us visualize the frequency polygon of a distribution and the

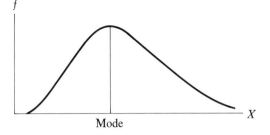

FIGURE 1-7
The location of the mode.[a]

[a] In this and subsequent graphs, the X on the horizontal axis denotes values of the observations and the *f* on the vertical axis denotes frequency of occurrence.

frequency curve approached as a limiting case when class size is gradually reduced. In the limiting situation, the variable under study may be thought of as continuous rather than as discrete. The mode may then be thought of as the value on the horizontal axis lying below the maximum point on the frequency curve (see Figure 1-7).

The mode of a frequency distribution has the connotation of a typical or representative value, a location in the distribution at which there is maximum clustering. In this sense, it serves as a standard against which to judge the representativeness or typicality of other averages. If a frequency distribution is symmetrical, the mode, median, and mean coincide. As noted earlier, extreme values in a distribution pull the arithmetic mean in the direction of these extremes. Stated somewhat differently, in a skewed distribution, the mean is pulled away from the mode toward the extreme values. The median also tends to be pulled away from the mode in the direction of skewness but is not affected as

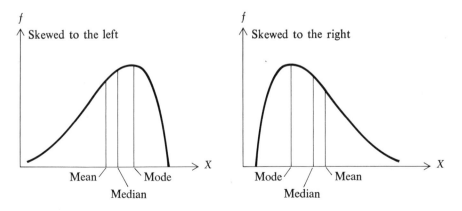

FIGURE 1-8
Skewed distributions depicting typical positions of averages.

much as the mean. If the mean exceeds the median,[1] a distribution is said to have *positive skewness* or to be *skewed to the right;* if the mean is less than the median, the terms *negative skewness* and *skewed to the left* are used. The order in which the averages fall in skewed distributions is shown in Figure 1-8.

Many distributions of economic data in the United States are skewed to the right. Examples include the distributions of incomes of individuals, savings of individuals, corporate assets, sizes of farms, and company sales within many industries. In many of these instances, the arithmetic mean is pulled so far from the median and mode as to be a very unrepresentative figure.

Multimodal Distributions

If more than one mode appears, the frequency distribution is referred to as *multimodal;* if there are two modes, it is referred to as *bimodal.* Extreme care must be exercised in analyzing such distributions. For example, consider a situation in which you want to compare the mean wages of workers in two different companies. Assume that the mean calculated for Company A exceeds that of Company B. If you conclude from this finding that workers in Company A earn higher wages, on the average, than those in Company B, without taking into account the fact that the wage distribution for each of these companies is bimodal, you may make serious errors of inference. To illustrate the principle involved, let us assume that the mean annual wage is $5000 for unskilled workers and $15,000 for skilled workers at each of these companies. Let us also assume that the individual distributions of wages of unskilled and skilled workers are symmetrical and that there are the same total number of workers in each company. Further, let us assume that these companies have workers only in the aforementioned two skill classifications. However, suppose 75% of the Company A workers are skilled, whereas only 50% of the Company B workers are skilled. Figure 1-9 shows the frequency curves of the distributions of annual wages at the two companies. Clearly, the mean annual wage of workers in Company A exceeds that in Company B, simply because there is a higher percentage of skilled workers at Company A. However, if you were ignorant of this fact, you might be tempted to infer that workers at Company A earn more than those at Company B. The fact of the matter is that unskilled workers at both companies earn the same wages, on the average, and the same holds true for the skilled workers. What is required here is to separate two wage distributions at each company, one for skilled workers and one for unskilled. A

---

[1] Sometimes the mode, rather than the median, is used for this comparison.

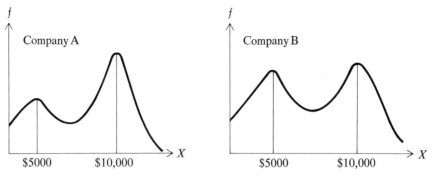

FIGURE 1-9
Bimodal frequency distributions: annual wages of workers in two companies.

comparison of the mean wages of unskilled workers (and of skilled workers) at the two companies would reveal their equality.

The principle involved here is one of *homogeneity* of the basic data. The fact that a wage distribution is bimodal (i.e., has two values around which frequencies are clustered) suggests that two different "cause systems" are present and that two distinct distributions should be recognized. The data on wages may be said to be *nonhomogeneous* with respect to skill level. Bimodal distributions would also result, for example, from the merging of height data for men and for women or weight data for men and for women.

If a basis for separating a bimodal distribution into two distributions cannot be found, then extreme care must be used in describing the data. In cases such as that shown for Company B in Figure 1-9, where the heights of the two modes are about equal, the arithmetic mean and median will probably fall between the modes and will not be representative of the large concentrations of values lying at the modes below and above these averages.

EXAMPLE 1-1

The following distribution gives the dollar cost per unit of output for 200 plants in the same industry:

Dollar Cost per Unit of Output	Number of Plants
$1.00 and under $1.02	6
1.02 and under 1.04	26
1.04 and under 1.06	52

(Continued)

Dollar Cost per Unit of Output	Number of Plants
1.06 and under 1.08	58
1.08 and under 1.10	39
1.10 and under 1.12	15
1.12 and under 1.14	3
1.14 and under 1.16	1
Total	200

a. Calculate the arithmetic mean of the distribution.
b. Calculate the median of the distribution.
c. Is the answer in (a) the same as the one you would have obtained by computing the following ratio?

$$\frac{\text{Total dollar cost for the 200 plants}}{\text{Total number of units of output of the 200 plants}}$$

d. Can you say that 50% of the units produced cost less than the median calculated in part (b)? Explain.
e. Suppose that the last class had been "$1.14 and over." What effect, if any, would this have had on your calculation of the arithmetic mean and of the median? Briefly justify your answers.

SOLUTION

a. $$\overline{X} = \frac{\$213.20}{200} = \$1.066$$

b. Median class = $1.06 and under $1.08

$$\text{Median} = \$1.06 + \left(\frac{100 - 84}{58}\right)(\$.02) = \$1.066$$

c. No. Conceptually, the ratio in (c) is a weighted mean of 200 cost per unit output ratios. The mean calculated in (a) is unweighted. It is an estimate of the figure that we would have obtained by adding up the original 200 cost per unit output ratios and dividing by 200 plants.
d. No. We have no way of telling how many units each plant produces. For example, the six plants whose costs are under $1.02 per unit might produce 60% of the total number of units in the industry. However, we can say that the grouped data procedure indicates that 50% of the *plants* had a cost per unit less than the median figure of $1.066.
e. The open-ended interval would have no effect on the median, which is concerned only with the order of the frequencies. However, the mean, which is concerned with the actual values, would have to be recalcu-

lated. A mean value for the items in the open interval would have to be assumed and multiplied by the frequency to obtain an estimate of the *fm* value for that class.

EXAMPLE 1-2

What is the proper average(s), if any, to use in the following situations? Briefly justify your answers.
  a. In determining which members of a class are in the upper half with respect to their overall grade average.
  b. In determining the average death rate for six cities combined, given the death rate of each.
  c. In a profit-sharing plan in which a firm wishes to find the average amount each worker is to receive to ensure equal distribution.
  d. In determining how high to make a bridge (not a drawbridge). The distribution of the heights of boats expected to pass under the bridge is known and is skewed to the left.
  e. In determining a typical wage figure for use in later arbitration for a company that employs 100 workers, several of whom are highly-paid specialists.
  f. In determining the average annual percentage rate of net profit to sales of a company over a ten-year period.

SOLUTION

  a. The median, since (generally) one-half of the grade averages will fall above and below this figure.
  b. The weighted arithmetic mean. The death rates of the six cities would be weighted by the population figures of these cities.
  c. The arithmetic mean. Divide the total profits to be shared (ΣX) by the number of workers (n).
  d. None of the averages would be appropriate, since they would result in a large number of boats being unable to pass under the bridge. The bridge should be high enough to allow all the expected traffic to pass under it.
  e. The median. Because of the tendency of the arithmetic mean to be distorted by a few extreme items, it would tend to be a less typical figure.
  f. The weighted arithmetic mean. The profit rates would be weighted by the sales figures for each year.

Exercises

1. The number of research projects initiated per month during 1975 by the marketing department of Sterling Pharmaceuticals was:

$$6, 4, 2, 2, 3, 0, 4, 3, 2, 5, 2, 4$$

Calculate the arithmetic mean, the mode, and the median.

2. Compute the mean and median of both the raw data and the frequency distribution in Exercise 2 on page 17.

3. Explain or criticize the following statement: The frequency distributions of family income, size of business, and wages of skilled employees all tend to be skewed to the right.

4. The following table presents a frequency distribution of the bank deposits in the First Metropolis Bank and Trust Company at the end of 1976:

Deposit Size	Number of Customers
$0–$499	7178
500– 999	8296
1000–1499	8342
1500–1999	9347
2000–2499	9546
2500–2999	6147
3000–3499	4239
3500–3999	2103
4000–4499	1445
4500 and over	676

a. Why do you think open-ended intervals are used in this distribution?
b. Which is the modal class?
c. Which is the median class?

5. The following data represent chainstore prices paid by farmers for two products during a certain month:

Composition Roofing (Price Range for a 90-Pound Roll)	Number of Purchases Reported	Douglas and Inland Firs, 2′ × 4′s Standard Grade or Better (Price Range per Thousand Pound-Foot)	Number of Purchases Reported
$2.35 and under $2.75	1	$ 76–85	2
2.75 and under 3.15	6	86–95	4
3.15 and under 3.55	33	96–105	8
3.55 and under 3.95	51	106–115	7
3.95 and under 4.35	121	116–125	28
4.35 and under 4.75	50	126–135	48
4.75 and under 5.15	44	136–145	57
5.15 and under 5.55	13	146–155	71
5.55 and under 5.95	5	156–165	45
		166–175	20
		176–185	1

a. Graph the frequency distribution for each product.
b. Indicate the median and modal class for each distribution.
c. Calculate and graph the cumulative distribution for each frequency distribution.
d. Compare the skewness of the two distributions. (No calculations are required.)

6. The following are the earnings for all employees of the Impecunious Corporation for a certain month:

Earnings for Month	Number of Employees with Given Earnings
$600 and under $650	13
650 and under 700	24
700 and under 750	29
750 and under 800	36
800 and under 850	34
850 and under 900	21
900 and under 950	11
	168

a. Compute the arithmetic mean of the distribution.
b. Is the answer to (a) the same as the one you would have obtained had you calculated the following ratio?

$$\frac{\text{Total earnings of all employees for the month}}{\text{Total number of employees for the month}}$$

Why or why not?
c. Compute the median of the distribution.
d. In which direction are these data skewed?
e. The comptroller of the company stated that the total payroll for the month was $129,832. Do you have any reason to doubt this statement? Support your position very briefly.
f. Do you think the arithmetic mean computed in (a) provides a satisfactory description of the typical earnings of these 168 employees for the given month? Why or why not?

1.13 DISPERSION: DISTANCE MEASURES

Central tendency, as measured by the various averages already discussed, is an important descriptive characteristic of statistical data. However, although two sets of data may have similar averages, they may differ considerably with respect to the spread, or dispersion, of the individual observations. Measures of dispersion describe this variation in numerical observations.

There are two types of measures of dispersion. Those of the first type, which may be referred to as *distance measures,* describe the spread of data in terms of the distance between the values of selected observations. The simplest of such measures is the *range,* or the difference between the values of the highest and lowest items. For example, if the loans extended by a bank to five corporate customers are $82,000, $125,000, $140,000, $212,000, and $245,000, the range of these loans is $245,000 − $82,000 = $163,000. Such a measure of dispersion may be useful for obtaining a rough notion of the spread in a set of data, but it is certainly inadequate for most analytical purposes. A disadvantage of the range is that it describes dispersion in terms of only two selected values in a set of observations. Thus, it ignores the nature of the variation among all other observations, which may be either tightly clustered in one small interval or spread out rather evenly between the extreme values. Furthermore, the two numbers used, the highest and the lowest, are extreme rather than typical values.

Other distance measures of dispersion employ more typical values. For example, the *interquartile range* is the difference between the *third quartile* and the *first quartile* values. The third quartile is a figure such that three quarters of the observations lie below it; the first quartile is a figure such that one quarter of the observations lie below it. Thus, the distance between these two numbers measures the spread between the values that bound the middle 50% of the values in a distribution. However, a main disadvantage of such a measure is again that it does not describe the variation *among* the middle 50% of the items (nor among the lower and upper one fourth of the values). The method of calculating quartile values will not be explicitly discussed here, but for frequency distribution data, its calculation proceeds in a manner completely analogous to that of the median, which is itself the *second quartile* value (that is, two quarters of the observations in a distribution lie below the median and two quarters lie above it). Quartiles are special cases of general measures known as *fractiles,* which refer to values that exceed specified fractions of the data. Thus, the ninth decile exceeds $\frac{9}{10}$ of the items, the ninety-ninth percentile exceeds $\frac{99}{100}$ of the items in a distribution, etc. Clearly, many arbitrary distance measures of dispersion could be developed, but they are infrequently used in practical applications.

1.14 DISPERSION: AVERAGE DEVIATION METHODS

The most comprehensive descriptions of dispersion are those in terms of the *average deviation* from some measure of central tendency. The most important such measures are the variance and the standard deviation.

The *variance* of the observations in a population, denoted σ^2 (Greek lower case *sigma*), is the arithmetic mean of the squared deviations from the population mean. In symbols, if $X_1, X_2, \ldots, X_N$ represent the values of the N observations in a population and μ is the arithmetic mean of these values, the population variance is defined by

$$(1.6) \qquad \sigma^2 = \frac{(X_1 - \mu)^2 + (X_2 - \mu)^2 + \cdots + (X_N - \mu)^2}{N}$$

$$= \frac{\Sigma\,(X - \mu)^2}{N}$$

As usual, subscripts have been dropped in the simplified form on the right-hand side of Equation 1.6.

Although the variance measures the extent of variation in the values of a set of observations, it is in units of squared deviations or squares of the original numbers. In order to obtain a measure of dispersion in terms of the units of the original data, the square root of the variance is taken. The resulting measure is known as the standard deviation. Thus, the *standard deviation* of a population is given by

$$(1.7) \qquad \sigma = \sqrt{\frac{\Sigma\,(X - \mu)^2}{N}}$$

By convention, the positive square root is used. The standard deviation is a measure of the spread in a set of observations. If all the values in a population were identical, each deviation from the mean would be 0, and the standard deviation would thus be equal to 0, its minimum value. On the other hand, as items are dispersed more and more widely from the mean, the standard deviation becomes larger and larger.

If we now consider the corresponding measures for a sample of n observations, it would seem logical to substitute the sample mean $\bar{X}$ for the population mean μ and the sample number of observations n for the population number N in Equations 1.6 and 1.7. However, it can be shown that when the sample variance and standard deviation are defined with $n - 1$ in the divisors, better estimates are obtained of the corresponding population parameters. Hence, in keeping with modern usage, we define the sample variance and sample standard deviation, respectively, as

Sample variance:

$$(1.8) \qquad s^2 = \frac{\Sigma\,(X - \bar{X})^2}{n - 1}$$

and

Sample standard deviation:

$$(1.9) \qquad s = \sqrt{\frac{\Sigma\,(X - \bar{X})^2}{n - 1}}$$

The term s^2 is usually referred to simply as the "sample variance" and s as the "sample standard deviation."

A brief justification will be given here for the division by $n - 1$. It can be shown mathematically that for an infinite population, when the sample variance is defined with divisor $n - 1$ as in formula (1.8), it is a so-called "unbiased estimator" of the population parameter, σ^2. This means that if all possible samples of size n were drawn from a given population and the variances of these samples were averaged (using the arithmetic mean), this average would be equal to the population variance, σ^2. Thus, when the sample variance is defined with divisor $n - 1$, on the average, it correctly estimates the population variance. Of course, for large samples, the difference in results obtained by using n rather than $n - 1$ as the divisor would tend to be very slight, but for small samples the difference can be rather substantial.

We will now concentrate on methods of computing the standard deviation, both for ungrouped and for grouped data. Then we discuss how this measure of dispersion is used. However, the major uses of the standard deviation are in connection with sampling theory and statistical inference, which are discussed in subsequent chapters.

The same accounts receivable data used to illustrate the arithmetic mean will be employed in the calculation of the standard deviation.

In Section 1.8, the following ungrouped data were given as the accounts receivable established by an accounting department during a one-hour period: $600, $350, $275, $430, and $520. The arithmetic mean was $435. The calculation of the standard deviation using Equation 1.9 is illustrated in Table 1-8.

The resulting standard deviation of $129.71 is an absolute measure

TABLE 1-8
Calculation of the Standard Deviation for Ungrouped Data by the Direct Method: Accounts Receivable Data

Amount of Accounts Receivable X	Deviation from Mean $X - \bar{X}$	Squared Deviations $(X - \bar{X})^2$
$600	$600 - $435 = $165	27,225
350	350 - 435 = -85	7225
275	275 - 435 = 160	25,600
430	430 - 435 = -5	25
520	520 - 435 = 85	7225
$\bar{X} = $435	$0	67,300

$$s = \sqrt{\frac{\Sigma(X - \bar{X})^2}{n-1}} = \sqrt{\frac{67,300}{4}} = \sqrt{16,825} = \$129.71$$

of dispersion, which means that it is stated in the units of the original data. Whether this is a great deal or only a small amount of dispersion cannot be immediately determined. This sort of judgment is based on the particular type of data analyzed (in this case, accounts receivable data). Furthermore, as we shall see in Section 1.15, relative measures of dispersion are preferable to absolute measures for comparative purposes.

In calculating the standard deviation for data grouped into frequency distributions, it is merely necessary to adjust the foregoing formulas to take account of this grouping. The defining formula (1.9) generalizes to

$$(1.10) \qquad\qquad s = \sqrt{\frac{\Sigma f(m - \bar{X})^2}{n - 1}}$$

where, as usual for grouped data, m represents the midpoint of a class, f the frequency in a class, $\bar{X}$ the arithmetic mean, and n the total number of observations. This calculation is illustrated in Table 1-9 for the frequency distribution of weekly earnings previously given in Tables 1-3 and 1-5. As shown in Table 1-9, the standard deviation is equal to $14.70.

Calculation of the sample standard deviation by the defining for-

TABLE 1-9
Calculation of the Standard Deviation for Grouped Data by the Direct Method:
Weekly Earnings Data

Weekly Earnings	Number of Employees f	Midpoints m	$m - \bar{X}$	Deviation $(m - \bar{X})^2$	$f(m - \bar{X})^2$
$160.00 and under 170.00	4	$165	−29.80	888.04	3552.16
170.00 and under 180.00	14	175	−19.80	392.04	5488.56
180.00 and under 190.00	18	185	−9.80	96.04	1728.72
190.00 and under 200.00	28	195	0.20	.04	1.12
200.00 and under 210.00	20	205	10.20	104.04	2080.80
210.00 and under 220.00	12	215	20.20	408.04	4896.48
220.00 and under 230.00	4	225	30.20	912.04	3648.16
	100				21,396.00

$$\bar{X} = 194.80$$

$$s = \sqrt{\frac{\Sigma f(m - \bar{X})^2}{n - 1}} = \sqrt{\frac{21,396}{99}} = \sqrt{216.1212}$$

$$s = \$14.70$$

mula can be tedious, particularly if the class midpoints and frequencies contain several digits and the arithmetic mean is not a round number. A shortcut method of calculation often useful in such cases is given in Appendix D.

Uses of the Standard Deviation

The standard deviation of a frequency distribution is very useful in describing the general characteristics of the data. For example, in the so-called normal distribution (a bell-shaped curve), which is discussed extensively in Chapters 5, 6, and 7, the standard deviation is used in conjunction with the mean to indicate the percentage of items that fall within specified ranges. Hence, if a population is in the form of a normal distribution, the following relationships apply:

$$\mu \pm \sigma \quad \text{includes 68.3\% of all of the items}$$
$$\mu \pm 2\sigma \quad \text{includes 95.5\% of all of the items}$$
$$\mu \pm 3\sigma \quad \text{includes 99.7\% of all of the items}$$

For example, if a production process is known to produce items that have a mean length of $\mu = 10$ inches and a standard deviation of 1 inch, then we can infer that 68.3% of the items have lengths between $10 - 1 = 9$ inches and $10 + 1 = 11$ inches. About 95.5% have lengths between $10 - 2 = 8$ and $10 + 2 = 12$ inches, and 99.7% have lengths between $10 - 3 = 7$ and $10 + 3 = 13$ inches. Thus, a range of $\mu \pm 3\sigma$ includes virtually all the items in a normal distribution. As discussed in Chapter 5, the normal distribution is perfectly symmetrical. If the departure from a symmetrical distribution is not too great, the rough generalization that virtually all the items are included within a range from 3σ below the mean to 3σ above the mean still holds.

The standard deviation is also useful in describing how far individual items in a distribution depart from the mean of the distribution. Suppose the population of students who took a certain aptitude test displayed a mean score of $\mu = 100$ with a standard deviation of $\sigma = 20$. Then a score of 80 on the examination can be described as lying one standard deviation below the mean. The terminology usually employed is that the *standard score* is -1; that is, if the examination score is denoted X, then

$$\frac{X - \mu}{\sigma} = \frac{80 - 100}{20} = -1$$

The standard score of an observation is simply the number of standard deviations the observation lies below or above the mean of the distribution. Hence, the score of 80 deviates from the mean by -20 units, which is equal to -1 in terms of units of standard deviations away from the mean. If standard scores are computed from sample rather than universe data, the formula $(X - \bar{X})/s$ would be used instead.

Comparisons can thus be made for items in distributions that differ in order of magnitude or in the units employed. For example, if a student scored 120 on an examination in which the mean was $\mu = 150$ and $\sigma = 30$, the standard score would be $(120 - 150)/30 = -1$. Thus, the score of 120 is the same number of standard deviations below the mean as the 80 in the preceding example. We could also compare standard scores in a distribution of wages with comparable figures in a distribution of length of employment service, and so on.

The standard deviation is doubtless the most widely used measure of dispersion, and considerable use is made of it in later chapters of this text.

1.15 RELATIVE DISPERSION: COEFFICIENT OF VARIATION

Although the standard score discussed earlier is useful for determining how far an *item* lies from the mean of a set of data, we often are interested in comparing the dispersion of *an entire set of data* with the dispersion of another set. As observed earlier, the standard deviation is an absolute measure of dispersion, whereas a relative measure is required for purposes of comparison. This is essential whenever the sets of data to be compared are expressed in different units, or even when the data are in the same units but are of different orders of magnitude. Such a relative measure is obtained by expressing the standard deviation as a percentage of the arithmetic mean. The resulting figure, referred to as the *coefficient of variation* (*CV*), is defined symbolically as

(**1.11**) $$CV = \frac{s}{\bar{X}}$$

Thus, for the frequency distribution of weekly earnings data of semi-skilled workers, the standard deviation is $14.70 with a mean of $194.80. The coefficient of variation is

$$CV = \frac{\$14.70}{\$194.80} = 7.5\%$$

Let us assume that the corresponding figures for the earnings of a group of highly skilled workers revealed a standard deviation of $18 with an arithmetic mean of $300. The coefficient of variation for this set of data is $CV = \$18/300 = 6\%$. Therefore, the earnings of the highly skilled group were relatively more uniform—that is, displayed relatively less variation than did the earnings of the semiskilled group.

Note that the earnings of the highly skilled group had the larger standard deviation, but because of the higher average weekly earnings, relative dispersion was less.

Both absolute and relative measures of dispersion are widely used in practical sampling problems. To give just one example, a question frequently arises about the sample size required to yield an estimate of a universe parameter with a specified degree of precision. For example, a finance company may want to know how large a random sample of its loans it must study in order to estimate the average dollar size of all its loans. If the company wants this estimate accurate within a specified number of *dollars*, an absolute measure of dispersion is appropriate. On the other hand, if the company wants the estimate to be within a specified *percentage* of the true average figure, a relative measure of dispersion would be used.

1.16 ERRORS IN PREDICTION

In this chapter, we have discussed descriptive measures for empirical frequency distributions, with emphasis on measures of central tendency and dispersion. Some interesting relationships between these two types of measure are observable when certain problems of prediction are considered. Suppose we want to guess or "predict" the value of an observation picked at random from a frequency distribution. Let us refer to the penalty of an incorrect prediction as the "cost of error." If there were a *fixed* cost of error on each prediction, no matter what the size of the error, we should guess the mode as the value of the random observation. This would give us the highest probability of guessing the *exact value* of the unknown observation. Assuming repeated trials of this prediction experiment, we would thus minimize the average (arithmetic mean) cost of error.

Suppose, on the other hand, that the cost of error varies directly with the size of error regardless of its sign, that is, regardless of whether the actual observation is above or below the predicted value. In this case, we would want a prediction that minimizes the average *absolute error*. The median would be the "best guess," since it minimizes average absolute deviations. The mean deviation about the median would be a measure of this minimum cost of error.

Finally, suppose the cost of error varies according to the square of the error (for example, an error of two units costs four times as much as an error of one unit). In this situation, the mean should be the predicted value, since it can be demonstrated mathematically that the average of the squared deviations about it is less than around any other figure. Here the variance, which may be interpreted as the average cost of error per observation, would represent a measure of this minimum error. Another point previously observed for the mean is that the average amount of error, taking account of sign, would be 0.

A practical business application of these ideas is in the determination of the optimum size of inventory to be maintained. Let us assume a situation in which the cost of overstocking a unit (cost of overage) is equal to the cost of being short one unit (cost of underage). Further, it may be assumed that the cost of error varies directly with the absolute amount of error. For example, having two units in excess of demand costs twice as much as one unit. In this situation, the optimum stocking level is the median of the frequency distribution of numbers of units demanded.

1.17 PROBLEMS OF INTERPRETATION

Many of the most common misinterpretations and misuses of statistics involve measures and concepts such as those discussed in this chapter: averages, dispersion, and skewness. Sometimes misleading interpretations are drawn from the use of averages that are not "typical" or "representative." Reference was made in Section 1.11 to the possible distortion of the typicality of the arithmetic mean because of the presence of extreme items at one end of a distribution. An interesting example of this distortion effect occurred in the case of a survey conducted by a popular periodical. One of the purposes of the survey was to determine the current status of persons who had graduated from college during the early Depression years. Among those included were the graduates of Princeton University for three successive years during the early 1930's. The results of the Princeton survey indicated that the arithmetic mean income of the respondents in the class that graduated in the second year was far higher than the corresponding mean income for the first- and third-year classes. The analysts attempted to rationalize this result in various ways. However, a reexamination of the data yielded a very simple explanation, which precluded potential misinterpretations. It turned out that one of the graduates of the second-year class was a member of one of the wealthiest families in the United States and was an heir to an immense fortune. His very large income exerted an obvious upward pull on the mean income of his class, making it an unrepresentative average.

Misinterpretations of averages often arise because dispersion is not taken into account. Prospective college students are sometimes discouraged when they observe the mean scholastic aptitude test scores of classes admitted to colleges or universities in which they are interested. Admissions officers have commented that students sometimes erroneously assume they will not be admitted to a school because their test scores are somewhat below the published mean scores for that school. Of course, such students fail to take into account dispersion around this average. Assuming a roughly symmetrical distribution, about one half of the admitted students on whom the published means were based had test scores that fell below that average.

Because of the shape of the underlying frequency distribution, sometimes no average will be typical. In Section 1.12, reference was made to bimodal distributions of wages of workers. Arithmetic means or medians for such distributions tend to fall somewhere between the two modes. Hence, they are not typical of the groups characterized by either of the modes. Of course, as indicated in Section 1.12, the solution when nonhomogeneous data are present is to separate the distinct distributions. However, sometimes U-shaped frequency distributions are encountered where the separation into different distributions is not warranted. In such distributions, frequencies are concentrated at both low and high values of the variable under consideration. For example, suppose the test scores of a mathematics class yield grades that are either very high, say in the 90s, or very low, say in the 60s. Means or medians, which might be about 75, would clearly be unrepresentative of the concentrations at either end of the distribution. When averages are presented for such distributions, without some indication of the nature of the underlying data, misinterpretations can occur quite easily.

Exercises

1. The closing prices of two common stocks traded on the New York Stock Exchange for a week in January 1976 were

	Highfly	*Stabil*
Monday	$28	$28
Tuesday	34	26
Wednesday	18	22
Thursday	20	24
Friday	25	25

   a. Compare the two stocks simply on the basis of measures of central tendency.
   b. Compute the standard deviation for each of the two stocks. What information do the standard deviations give concerning the price movements of the two stocks?

2. The standard deviation of sales for the past five years for Eastern Sports Shop

was $6782 and for House of Sports $23,221. Can we conclude that sales are more stable for Eastern than for House of Sports?

3. The following is a distribution of the scores achieved on a special graduate business aptitude examination by first-year students at a midwestern graduate school of business:

Score	Frequency
480 and under 510	3
510 and under 540	7
540 and under 570	12
570 and under 600	26
600 and under 630	35
630 and under 660	31
660 and under 690	18
690 and under 720	11
720 and under 750	7
	150

    a. Compute the standard deviation and the arithmetic mean of the distribution of scores.

    b. Can one compare the variability in the test scores of the students with the variability in their undergraduate grade point averages? If yes, how? If no, why not?

4. The following is a distribution of lifetimes, in hours, of 100 high-intensity light bulbs.

Lifetime (Hours)	Frequency
100 and under 200	12
200 and under 300	28
300 and under 400	20
400 and under 500	18
500 and under 600	14
600 and under 700	8
	100

    a. Compute the mean and coefficient of variation for the distribution.

    b. Compute and *interpret* the median.

    c. Suppose that the last class was "600 and over." What effect, if any, would this have on your calculations in (a) and (b)?

5. For the period 1967–1976, the annual earnings per share for the Tiny Tot Toy Company and the Gigantic Game Corporation are as follows:

	Tiny Tot	Gigantic Game
1967	$ .50	$6.40
1968	.80	7.00
1969	.90	6.80
1970	1.20	7.60
1971	1.00	8.00
1972	.80	8.30
1973	1.20	7.90
1974	1.40	8.50
1975	1.50	8.60
1976	1.70	8.90

a. Compute the arithmetic mean and standard deviation of the earnings per share for each firm. Which firm showed the greater absolute variation in earnings per share?

b. Compute the coefficient of variation for each firm. Which firm showed relatively greater variation in earnings per share?

Introduction to Probability

In Chapter 1, we discussed the frequency distribution as a device for summarizing the variation in sets of numerical observations in convenient tabular form. It is often very useful to describe and draw generalizations about patterns of variation using the concept of probability. The discussions of probability in this chapter and in Chapter 3 lay the foundation for our treatment of statistical analysis for decision making.

2.1 THE MEANING OF PROBABILITY

The development of a mathematical theory of probability began during the seventeenth century when the French nobleman Antoine Gombauld, known as the Chevalier de Méré, raised certain questions about games of chance. Specifically, he was puzzled about the probability of obtaining two sixes at least once in twenty-four rolls of a pair of dice. (This is a problem you should have little difficulty solving after reading this chapter.) De Méré posed the question to Blaise Pascal, a young French mathematician, who solved the problem. Subsequently, Pascal discussed this and other puzzlers raised by de Méré with the famous mathematician Pierre de Fermat. In the course of their correspondence, the mathematical theory of probability was born.

The several different methods of measuring probabilities represent different conceptual approaches and reveal some of the current controversy concerning the foundations of probability theory. In this chapter we discuss three conceptual approaches: *classical probability, relative frequency of occurrence*, and *subjective probability*. Regardless of the definition of probability used, the same mathematical rules apply in performing the calculations (that is, measures of probability are always added or multiplied under the same general circumstances).

Classical Probability

Since probability theory had its origin in games of chance, it is not surprising that the first method developed for measuring probabilities was particularly appropriate for gambling situations. According to the so-called *classical* concept of probability, the probability of an event A is defined as follows: If there are a possible outcomes favorable to the occurrence of the event A and b possible outcomes unfavorable to the occurrence of A, and if all outcomes are equally likely and mutually exclusive, then the probability that A will occur, denoted $P(A)$, is

$$P(A) = \frac{a}{a+b} = \frac{\text{number of outcomes favorable to occurrence of } A}{\text{total number of possible outcomes}}$$

Thus, if a fair coin with two faces, denoted head and tail, is tossed into the air, the probability that it will fall with the head uppermost is $P(\text{Head}) = 1/(1 + 1) = \frac{1}{2}$. In this case, there is one outcome favorable to the occurrence of the event "head" and one unfavorable outcome. (The extremely unlikely situation that the coin will stand on end is defined out of the problem; that is, it is not classified as an outcome for the purpose of the probability calculation.)

The equation above can also be used to determine the probability that a certain face will show when a true die is rolled. (A die is a small cube with 1, 2, 3, 4, 5, or 6 dots on each of its faces.) A *true* die is one that is equally likely to show any of the six numbers on its uppermost face when rolled. The probability of obtaining a 1 if such a die is rolled is $P(1) = 1/(1 + 5) = \frac{1}{6}$. Here, there is one outcome favorable to the event "1" and five unfavorable outcomes.

Some of the terms used in classical probability require further explanation. The *event* whose probability is sought consists of one or more possible outcomes of the given activity (tossing a coin, rolling a die, or drawing a card). These activities are known in modern terminology as *experiments*, a term referring to processes that result in different possible outcomes or observations. The term *equally likely* in referring to possible outcomes is considered intuitively clear. Two or more outcomes are said to be *mutually exclusive* if when one of the outcomes occurs, the others cannot. Thus, the appearances of a 1 and a 2 on a die are mutually exclusive events, since if a 1 results, a 2 cannot. All possible results of an experiment are conceived of as a complete, or *exhaustive*, set of mutually exclusive outcomes.

These classical probability measures have two very interesting characteristics. First, the objects referred to as *fair* coins, *true* dice, or *fair* decks of cards are abstractions in the sense that no real-world object possesses exactly the features postulated. For example, in order to be a *fair* coin (equally likely to fall "head" or "tail") the object would have to be a perfectly flat, homogeneous disk—an unlikely object. Second, in order to determine the probabilities in the above examples, no coins had to be tossed, no dice rolled, nor cards shuffled. That is, no experimental data had to be collected; the probability calculations were based entirely on logic.

In the context of this definition of probability, if it is *impossible* for an event A to occur, the probability of that event is said to be 0. For example, if the event A is the appearance of a 7 when a single die is rolled, then $P(A) = 0$. A probability of 1 is assigned to an event that is *certain* to occur. Thus, if the event A is the appearance of any one of the numbers 1, 2, 3, 4, 5, or 6 on a single roll of a die, then $P(A) = 1$. According to the classical definition, as well as all others, the probability of an event is a number between 0 and 1, and the sum of the probabilities that the event will occur and that it will not occur is 1.

Relative Frequency of Occurrence

Although the classical concept of probability is useful for solving problems involving games of chance, serious difficulties occur with a wide range of other types of problems. For example, it is inadequate for determining the probabilities that (a) a black male American, aged 30, will die within the next year, (b) a consumer in a certain metropolitan area will purchase a company's product during the next month, (c) a production process used by a particular firm will produce a defective item. In none of these situations is it feasible to establish a complete set of mutually exclusive outcomes, each equally likely to occur. For example, in (a), only two occurrences are possible: the individual will either live or die during the ensuing year. The likelihood that he will die is, of course, much smaller than the likelihood that he will live; but how much smaller? The probability that a 30-year-old black male American will live through the next year is greater than the corresponding probability for a 30-year-old black male inhabitant of India. However, how much greater is it and precisely what do these probabilities mean? Questions of this type require reference to data.

We know that the life insurance industry establishes mortailty rates by observing how many of a sample, of, say, 100,000 black American males, aged 30, die within a one-year period. In this instance, the number of deaths divided by 100,000 is the *relative frequency of occurrence* of death for the 100,000 individuals studied. It may also be viewed as an estimate of the *probability* of death for Americans in the given color-sex-age group. This relative frequency of occurrence concept can also be illustrated by a simple coin-tossing example.

Suppose you are asked to toss a coin known to be biased (that is, not a fair coin). You are not told whether the coin is more likely to produce a head or a tail, but you are asked to determine the probability of the appearance of a head by means of many tosses of the coin. Assume that 10,000 tosses of the coin result in 7000 heads and 3000 tails. Another way of stating this is that the relative frequency of occurrence of heads is 7000/10,000 or 0.70. It certainly seems reasonable to assign a probability of 0.70 to the appearance of a head with this particular coin. On the other hand, if the coin had been tossed only three times and one head resulted, you would have little confidence in assigning a probability of $\frac{1}{3}$ to the occurrence of a head.

As a working definition, the relative frequency concept of probability may be interpreted as the *proportion of times an event occurs in the long run under uniform or stable conditions.* In practice, past relative frequencies of occurrence are often used as probabilities. Hence, in the mortality illustration, if 800 of the 100,000 individuals of the given group died during the year, the relative frequency of death or probability of death is said to be 800/100,000 for individuals in the group.

Subjective Probability

The *subjective* or *personalistic* concept of probability is a relatively recent development.[1] Its application to statistical problems has occurred almost entirely in the post-World War II period. According to this concept, the probability of an event is the *degree of belief or degree of confi-*

---

[1] The concept was first introduced in 1926 by Frank Ramsey, who presented a formal theory of personal probability in F. P. Ramsey, *The Foundation of Mathematics and Other Logical Essays* (London: Kegan Paul; New York: Harcourt Brace Jovanovich, 1931). The theory was developed primarily by B. de Finetti, B. O. Koopman, I. J. Good, and L. J. Savage.

dence placed in the occurrence of an event by a particular individual based on the evidence available. This evidence may consist of data on relative frequency of occurrence and any other quantitative or nonquantitative information. The individual who considers it unlikely that an event will occur assigns a probability close to 0 to it; if one believes it is very likely an event will occur, one assigns it a probability close to 1. Thus, for example, in a consumer survey, an individual may assign a probability of $\frac{1}{2}$ to the event of purchasing an automobile during the next year. An industrial purchaser may assert a probability of $\frac{4}{5}$ that a future incoming shipment will have 2% or fewer defective items.

Subjective probabilities should be assigned on the basis of all objective and subjective evidence currently available and should reflect the decision maker's current degree of belief. Reasonable persons might arrive at different probability assessments because of differences in experience, attitudes, values, etc. Furthermore, these probability assignments may be made for events that will occur only once, in situations where neither classical probabilities nor relative frequencies appear to be appropriate.

This approach is thus a very broad and flexible one, permitting probability assignments to events for which there are no objective data or for which there is a combination of objective and subjective data. However, the assignments of the probabilities must be consistent. For example, if the purchaser assigns a probability of $\frac{4}{5}$ to the event that a shipment will have no more than 2% defective items, then a probability of $\frac{1}{5}$ must be assigned to the event that a shipment will have more than 2% defective items. In this book we accept the concept of subjective, or personal, probability as a reasonable and useful one, particularly in the context of business decision making.

Sample Spaces and Experiments

The concept of an *experiment* is a central one in probability and statistics. In this connection, an experiment is simply any process of measurement or observation of different outcomes. The experiment may be real or conceptual. The collection or totality of the possible outcomes of an experiment is referred to as its *sample space*. Thus, the collection of outcomes of the experiment of tossing a coin once (or twice, or any number of times) is a sample space. The objects that comprise the sample space are referred to as its *elements*. The elements are usually enclosed within braces, and the symbol S is conventionally used to denote a sample space.

On a single toss of a coin, there are two possible outcomes, tail (T) and head (H). Thus,

$$S = \{T, H\}$$

If the coin is tossed twice, there are four possibilities:

$$S = \{(T, T), (T, H), (H, T), (H, H)\}$$

In the first case, the experiment consists of *one trial*, a single toss of the coin; in the second case, the experiment contains *two trials*, the two tosses of the coin. In these examples, a physical experiment may actually be performed, or we may easily conceive of the possible outcomes of such an experiment.

In other situations, although no sequence of repetitive trials is involved, we may conceive of a set of outcomes as an experiment. These outcomes may simply be the result of an observational process and need not bear any resemblance to a laboratory experiment, as long as they are well defined. Thus, we may think of each of the following two-way classifications as constituting sample spaces:

0	1
Customer was granted credit.	Customer was not granted credit.
Employee elected a stock purchase plan.	Employee did not elect a stock purchase plan.
The merger will take place.	The merger will not take place.
The company uses direct mail advertising.	The company does not use direct mail advertising.

The elements in these two-element sample spaces may be designated 0 and 1, as indicated by the column headings. Therefore, each of the four illustrative sample spaces may be conveniently symbolized as

$$S = \{0, 1\}$$

There are at least two methods of graphically depicting sample spaces: (1) graphs using the conventional rectangular coordinate system and (2) tree diagrams. These methods are most useful for sample spaces with relatively small numbers of sample points (or elements). Tree diagrams are more manageable because of the obvious graphic difficulties encountered by the coordinate system method beyond three dimensions. These methods are illustrated in Examples 2-1 and 2-2.

EXAMPLE 2-1

Depict graphically the sample space generated by the experiment of tossing a coin twice.

SOLUTION

The sample space was earlier designated as

$$S = \{(T, T), (T, H), (H, T), (H, H)\}$$

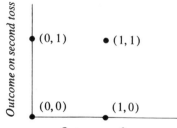

FIGURE 2-1
Graph for coin-tossing experiment.

Let $T = 0$ and $H = 1$. The sample space may now be written

$$S = \{(0, 0), (0, 1), (1, 0), (1, 1)\}$$

and graphed in two dimensions as shown in Figure 2-1. A tree diagram for this sample space is shown in Figure 2-2.

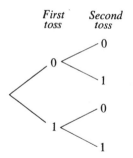

FIGURE 2-2
Tree diagram for coin-tossing experiment.

EXAMPLE 2-2

In a certain city, a market research firm studied the differences among consumers, classified by income groups and by whether or not they purchased a given product during a one-month period. The income groups used were "low," "middle," and "high." Consumers were classified as (a) did not purchase or (b) purchased the product at least once. Show this situation graphically.

SOLUTION

Let 0 stand for "low" income, 1 for "middle" income, and 2 for "high" income. For purchasing activity, let 0 stand for "did not purchase" and 1 for

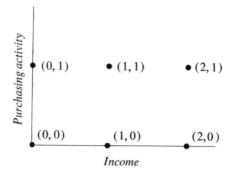

FIGURE 2-3
Graph for classification of consumers by income and purchasing activity.

"purchased at least once." The sample space may be expressed as the set of ordered pairs

$$S = \{(0, 0), (1, 0), (2, 0), (0, 1), (1, 1), (2, 1)\}$$

The graph is given in Figure 2-3, and the tree diagram in Figure 2-4. To illustrate how a tree diagram is interpreted, let us consider the element (0, 1). This element is depicted in the tree by starting at the left-hand side and following the uppermost branch to the "0," then continuing from this fork down the branch leading to a "1." The element (0, 1) denotes a consumer classified as "low" for income and "purchased at least once" for buying activity.

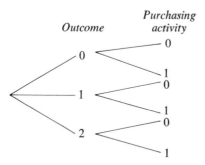

FIGURE 2-4
Tree diagram for classification of consumers by income and purchasing activity.

Events

The meaning of the term "event" as used in ordinary conversation is usually clear. However, since the concept of an event is fundamental to probability theory, it requires explicit definition. Once a sample space S has been specified, an event may be defined as a collection of elements each of which is also an element of S. An *elementary event* is a single possible outcome of an experiment. It is thus an event that cannot be further subdivided into other events.

For example, a single roll of a die constitutes an experiment. The sample space generated is

$$S = \{1, 2, 3, 4, 5, 6\}$$

We may define an event E, say, as the "appearance of a 2 or a 3," which may be expressed as

$$E = \{2, 3\}$$

If either a 2 or a 3 appears on the upper face of the die when it is rolled, the event E is said to have occurred. Note that E is not an elementary event, since it can be subdivided into the two elementary events, $\{2\}$ and $\{3\}$.

The complement of an event A in the sample space S is the collection of elements that are not in A. We will use the symbol $\overline{A}$, read "A bar" for the complement of A. For example, in the experiment of rolling a die once, the complement of the event that a 1 appears on the uppermost face is the event that a 2, 3, 4, 5, or 6 appears.

When a sample space S has been defined, S itself is referred to as the *certain event* or the *sure event*, since in any single trial of the experiment that generates S, one or another of its elements must occur. Hence, in the experiment of rolling a die once, the certain event is that either a 1, 2, 3, 4, 5, or 6 appears.

Two events A_1 and A_2 are said to be *mutually exclusive events* if when one of these events occurs, the other cannot. On a single roll of a die, the event that a 1 appears and the event that a 2 appears are mutually exclusive. On the other hand, the appearance of a 1 and the appearance of an odd number (1, 3, or 5) are not mutually exclusive events, since 1 is an odd number.

Probability and Sample Spaces

Probabilities may be thought of as numbers assigned to points in a sample space. These probabilities must have the following characteristics:

1. They are numbers greater than or equal to 0
2. They must add up to 1

EXAMPLE 2-3

In a study of the loans that were made to corporate borrowers and that matured during the past year, an officer of a commercial bank classified these loans into the following categories with respect to collection experience:

excellent, good, fair, poor, and bad.   The following table gives the number of loans in each category:

Collection Experience of Loans	Number of Loans
Excellent	580
Good	418
Fair	140
Poor	42
Bad	20
	1200

When the number of loans in each category was divided by the total number of loans, the following table of relative frequencies resulted:

Collection Experience of Loans	Relative Frequencies
Excellent	0.48
Good	0.35
Fair	0.12
Poor	0.03
Bad	0.02
	1.00

Then the loan officer could use these relative frequencies as probabilities for the collection experience for next year's loans to corporate borrowers. In the classification of next year's loans, there would thus be five elementary events (excellent, good, fair, poor, and bad) with respective probabilities of 0.48, 0.35, 0.12, 0.03, and 0.02.   The two characteristics of probabilities mentioned above are satisfied, because each probability is greater than or equal to 0 and because the probabilities add up to 1.   However, the events in a sample space need not be elementary events.   If a "profitable loan" is defined as an excellent, good, or fair loan and an "unprofitable loan" as a poor or bad one, then the events "profitable loan" and "unprofitable loan" make up a sample space with P(profitable loan) $= 0.48 + 0.35 + 0.12 = 0.95$ and P(unprofitable loan) $= 0.03 + 0.02 = 0.05$. This classification of loans might be of more interest to the loan officer than the original one.   Of course, the events defining the sample space must be a complete set of mutually exclusive events, and the sum of the probabilities of occurrence of an event A and its complement $\bar{A}$ must be equal to 1 (i.e., $P(A) + P(\bar{A}) = 1$).

EXAMPLE 2-4

A corporation economist was concerned about the surging of credit demand and the effect of inflationary pressures in the economy on the company's activities during the next six-month period. The economist established the following subjective probability distribution concerning the possible action of the Federal Reserve System in the area of monetary policy during the next half year:

Events E_i	Probability of Events $P(E_i)$
E_1: Tighten monetary policy	0.5
E_2: Leave current monetary situation unchanged	0.4
E_3: Relax monetary policy	<u>0.1</u>
	1.0

Thus, this economist assigned the value 0.9 to the probability that the Federal Reserve System will either tighten monetary policy or leave the current monetary situation unchanged.

Odds Ratios

Regardless of the definition used, we sometimes prefer to express probabilities in terms of odds. Thus, if the probability that a 6 will appear on the roll of a die is $\frac{1}{6}$, then the odds that it will appear are one to five, written $1:5$. If the economist in Example 2-4 assesses the probability that the Federal Reserve System will adopt a tightened monetary policy as $0.5/1 = \frac{1}{2}$, then this probability expressed in terms of odds is $1:1$.

If the probability that an event A will occur is

$$P(A) = \frac{a}{n}$$

then the odds in favor of the occurrence of A are

$$\text{odds in favor of } A = \frac{a}{n-a} = a:(n-a)$$

The odds against A are

$$\text{odds against } A = \frac{n-a}{a} = (n-a):a$$

Another way of viewing the odds ratio is as the ratio of the probability of the occurrence of an event A to the probability of its complement, $\overline{A}$. Thus,

$$\text{odds in favor of } A = \frac{P(A)}{P(\overline{A})} = \frac{a/n}{(n-a)/n} = \frac{a}{n-a}$$

$$\text{odds against } A = \frac{P(\overline{A})}{P(A)} = \frac{(n-a)/n}{a/n} = \frac{n-a}{a}$$

Exercises

1. If 0 represents a nonresponse to a mailed questionnaire and 1 represents a response, depict the set of outcomes representing responses to four out of five questionnaires.

2. A certain manufacturing process produces parachutes. A worker tests each parachute as it is produced and continues testing until a defective one is found. Specify the sample space of possible outcomes for the testing process.

3. An investment company is planning to add two new stocks to its portfolio. Its research group recommends four stocks: F. B. Richgood, General Thrills, Pacific Pie, and Multiplex. Specify the sample space representing the possible choices.

4. An econometric model predicts whether the gross national product (GNP) will increase, decrease, or remain the same the following year. Let X represent "the model's prediction, coded" and Y "the actual movement of GNP, coded." Graph the possible outcomes of X and Y.

5. A special electronic part is ordered by a firm in Omaha, Nebraska, from a firm in Düsseldorf, Germany. The Düsseldorf firm can fly the part to Kansas City, St. Louis, Chicago, or Minneapolis. Once it reaches one of these cities, it can be sent by train or truck to Omaha. Another alternative is to ship the part by train or truck to München from Düsseldorf; then it can be flown directly to Omaha. Use a tree diagram to find all possible shipping routes.

6. Draw a tree diagram depicting the possible outcomes resulting from flipping a coin three times.

7. Which of the following pairs of events are mutually exclusive?
    a. (1) General Rotors common stock closes higher on a given day, and (2) General Rotors common stock closes lower on the same day.
    b. In a shipment of two relays, (1) exactly one is defective, and (2) exactly two are defective.
    c. Six people (Joe, Harry, Pete, Frank, Steve, and John) apply for two job openings. (1) Frank is hired, and (2) Steve is hired.
    d. On two rolls of a die, (1) a 5 occurs, and (2) the sum of the two faces is 8.
    e. On two rolls of a die, (1) a 3 occurs, and (2) the sum of the two faces is 3.

2.2 ELEMENTARY PROBABILITY RULES

In most applications of probability theory, we are interested in combining probabilities of events that are related in some important way. In this section, we discuss two fundamental ways of combining probabilities: *addition* and *multiplication*.

Before considering the combining of probabilities by addition, we will define two new terms. The symbol $P(A_1$ or $A_2)$ refers to the probability that *either* event A_1 *or* event A_2 occurs. For example, if A_1 refers to the event that an individual is a male and A_2 refers to the event that the individual is a college graduate, then the symbol $P(A_1$ or $A_2)$ denotes the probability that the individual is *either* a male *or* a college graduate. Here the term "or" is used inclusively; that is, it includes the case of a person who is both a male and a college graduate.

The symbol $P(A_1$ and $A_2)$ is used to denote the probability that both events A_1 *and* A_2 will occur. $P(A_1$ and $A_2)$ is called the *joint probability* of the events A_1 and A_2. If A_1 and A_2 have the interpretations given in the preceding paragraph, then $P(A_1$ and $A_2)$ is the probability that the individual is both a male *and* a college graduate.

We now state the general addition rule for any two events A_1 and A_2 in a sample space S.

ADDITION RULE FOR ANY TWO EVENTS A_1 AND A_2

(2.1) $$P(A_1 \text{ or } A_2) = P(A_1) + P(A_2) - P(A_1 \text{ and } A_2)$$

If A_1 and A_2 are mutually exclusive events, that is, if they cannot both occur, then

$$P(A_1 \text{ and } A_2) = 0$$

This leads to a special case of the general addition rule.

ADDITION RULE FOR TWO MUTUALLY EXCLUSIVE EVENTS A_1 AND A_2

(2.2) $$P(A_1 \text{ or } A_2) = P(A_1) + P(A_2)$$

As an application of Equation 2.2 to the rolling of dice, note that the probability that either a 1 or a 2 will occur on a single roll is $P(A_1$ or $A_2) = P(A_1) + P(A_2) = \frac{1}{6} + \frac{1}{6} = \frac{1}{3}$. In Example 2-5 we consider an application of Equation 2.1 for two events that are not mutually exclusive.

EXAMPLE 2-5

What is the probability of obtaining a 6 on the first or second roll of a die or on both? Another way of wording this question is, "What is the probability of obtaining a 6 at least once in two rolls of a die?"

SOLUTION

Let A_1 denote the appearance of a 6 on the first roll and A_2 the appearance of a 6 on the second roll. We want to find the value of $P(A_1$ or $A_2)$. (As explained earlier, because of the inclusive meaning of "or," the symbol $P(A_1$ or $A_2)$ means the probability that a 6 appears either on the first or on the second roll or on both rolls.) Consider the sample space of 36 equally likely elements listed below. These are all possible outcomes of two rolls of the die; the numbers in each element represent the outcomes on the first and second rolls, respectively.

1,1	2,1	3,1	4,1	5,1	6,1
1,2	2,2	3,2	4,2	5,2	6,2
1,3	2,3	3,3	4,3	5,3	6,3
1,4	2,4	3,4	4,4	5,4	6,4
1,5	2,5	3,5	4,5	5,5	6,5
1,6	2,6	3,6	4,6	5,6	6,6

The probability that a 6 will appear on both the first and second rolls is $\frac{1}{36}$, i.e., $P(A_1 \text{ and } A_2) = \frac{1}{36}$. The probability that a 6 will appear on the first roll is $P(A_1) = \frac{1}{6}$; and on the second roll, $P(A_2) = \frac{1}{6}$. Hence, applying the addition rule, we have

$$P(A_1 \text{ or } A_2) = P(A_1) + P(A_2) - P(A_1 \text{ and } A_2)$$

$$= \frac{1}{6} + \frac{1}{6} - \frac{1}{36}$$

$$= \frac{11}{36}$$

The term $P(A_1 \text{ and } A_2)$ must be subtracted in this calculation in order to avoid double counting. That is, if we incorrectly solved the problem by using the addition theorem for mutually exclusive events, computing $P(A_1 \text{ or } A_2) = P(A_1) + P(A_2) = \frac{1}{6} + \frac{1}{6} = \frac{12}{36}$, we would have counted the event (6, 6) twice, because (6, 6) is an elementary event both of A_1 (6 on the first roll) and of A_2 (6 on the second roll). Note that eleven points in the sample space listed above represent the event "6 at least once in two trials." Therefore, the same result could have been obtained by using the classical concept of probability. The ratio of outcomes favorable to the event "6 at least once" to the total number of outcomes is $\frac{11}{36}$.

The ideas involved in the use of the addition rule are portrayed in Figures 2-5(a) and (b). The interiors of the rectangles represent the

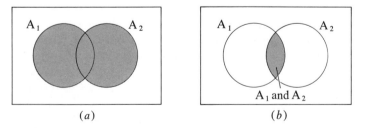

(a)   (b)

FIGURE 2-5
Symbolic portrayal of the addition rule where A_1 and A_2 are not mutually exclusive; $P(A_1 \text{ or } A_2) = P(A_1) + P(A_2) - P(A_1 \text{ and } A_2)$.

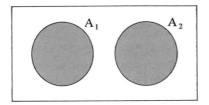

FIGURE 2-6
Symbolic portrayal of the addition rule where A_1 and A_2 are mutually exclusive; $P(A_1 \text{ or } A_2) = P(A_1) + P(A_2)$.

sample space, and the two events A_1 and A_2 are displayed as circles. The event "A_1 or A_2," whose probability is to be found, is shown as the tinted region in Figure 2-5(a). If we added all the outcomes in A_1 to those in A_2, which would be implied when we add $P(A_1)$ and $P(A_2)$, we would count the outcomes associated with the event "A_1 and A_2" twice; the latter event is shown in Figure 2-5(b) as the tinted region where the circles overlap. That is why we must subtract the term $P(A_1 \text{ and } A_2)$ from the sum of $P(A_1)$ and $P(A_2)$ in order to yield the desired probability, $P(A_1 \text{ or } A_2)$.

On the other hand, if A_1 and A_2 are mutually exclusive events, they cannot occur jointly. This is indicated in Figure 2-6 by depicting these events as circles that do not overlap. The event "A_1 or A_2" is represented by the tinted area of the sample space. Since the circles do not intersect, there is no double counting when $P(A_1)$ is added to $P(A_2)$ to obtain $P(A_1 \text{ or } A_2)$. (Note again that since A_1 and A_2 cannot occur together, $P(A_1 \text{ and } A_2) = 0$.)

The addition rule can, of course, be extended to more than two events. The generalization for n mutually nonexclusive events will not be given here because of its complexity. If the n events $A_1, A_2, \ldots, A_n$ are mutually exclusive, then we have the following general rule:

ADDITION RULE FOR n MUTUALLY EXCLUSIVE EVENTS

(2.3) $\qquad P(A_1 \text{ or } A_2 \text{ or } \ldots A_n) = P(A_1) + P(A_2) + \cdots + P(A_n)$

This generalized addition rule is applicable whenever we are interested in the probability that any one of several mutually exclusive events will occur. For example, returning to Example 2-3, the computation of the probability that the collection experience of a profitable loan will be excellent, good, or fair illustrates this generalized rule. In probability notation, we have

$$\begin{aligned} P(\text{excellent or good or fair}) &= P(\text{excellent}) + P(\text{good}) + P(\text{fair}) \\ &= 0.48 + 0.35 + 0.12 = 0.95 \end{aligned}$$

Joint Probability Tables

In many applications, we are interested in the probability of the *joint occurrence* of two or more events. To illustrate these joint probabilities, consider the data in Table 2-1. These figures represent the results of a market research survey in which 1000 persons were asked which of two competitive products they preferred, product ABC or product XYZ. To simplify the discussion, as shown in Table 2-1, A_1, A_2, B_1, and B_2 refer, respectively to "male," "female," "prefers product ABC," and "prefers product XYZ." Hence, the joint outcome that an individual is both male and prefers product ABC is denoted as "A_1 and B_1," and the joint probability that a randomly selected individual is male and prefers product ABC is $P(A_1 \text{ and } B_1)$. Analogous notation is used for the other possible joint outcomes and joint probabilities.

Joint probabilities may be illustrated in the following manner. If a person is selected at random from this group of 1000, the joint probability that the individual is a male and prefers product ABC is

$$P(A_1 \text{ and } B_1) = \frac{200}{1000} = 0.20$$

Similarly, the probability that a randomly selected person is a female and prefers product XYZ is

$$P(A_2 \text{ and } B_2) = \frac{400}{1000} = 0.40$$

It is useful to construct a so-called *joint probability table* by dividing all entries in Table 2-1 by the total number of individuals (1000). The resulting joint probability table is shown as Table 2-2. The figures in the table are examples of probabilities calculated as relative frequencies of occurrence, as discussed in Section 2.1.

TABLE 2-1
1000 Persons Classified by Sex and Product Preference

Sex	Prefers Product ABC B_1	Prefers Product XYZ B_2	Total
A_1 Male	200	300	500
A_2 Female	100	400	500
	300	700	1000

TABLE 2-2
Joint Probability Table for 1000 Persons Classified by Sex and Product Preference

Sex	Prefers Product ABC B_1	Prefers Product XYZ B_2	Marginal Probabilities
A_1 Male	0.20	0.30	0.50
A_2 Female	0.10	0.40	0.50
Marginal probabilities	0.30	0.70	1.00

Marginal Probabilities

In addition to the joint probabilities mentioned earlier, we can also separately obtain from Table 2-2 probabilities for each of the two classifications "sex" and "product preference." These probabilities, which are shown in the margins of the joint probability table, are referred to as *marginal probabilities* or *unconditional probabilities*. For example, the marginal probability that a randomly chosen individual is a male is $P(A_1) = 0.50$, and the marginal probability that a person prefers product ABC is $P(B_1) = 0.30$.

The marginal probabilities for each classification are obtained by summing the appropriate joint probabilities. For example, the marginal probability that an individual prefers product ABC is 0.30. The event B_1, "prefers product ABC," consists of two mutually exclusive parts, "A_1 and B_1" ("male and prefers product ABC") and "A_2 and B_1" ("female and prefers product ABC"). Hence, we have

$$P(B_1) = P[(A_1 \text{ and } B_1) \text{ or } (A_2 \text{ and } B_1)]$$
$$= P(A_1 \text{ and } B_1) + P(A_2 \text{ and } B_1)$$
$$= 0.20 + 0.10 = 0.30$$

Thus, to obtain marginal probabilities for product preference, we add the appropriate joint probabilities over the other classification (in this case, sex). This is an application of the addition rule for mutually exclusive events. All other marginal probabilities in Table 2-2 may be similarly calculated.

Conditional Probabilities

Often we are interested in how certain events are related to the occurrence of other events. In particular, we may be interested in the probability of the occurrence of an event given that another related event has occurred. Such probabilities are referred to as *conditional probabilities*. For instance, returning to the events discussed in Table 2-2, we may be interested in the probability that an individual prefers product ABC given that the individual is a male. This conditional probability is denoted $P(B_1|A_1)$ and is read "the probability of B_1 given A_1." The vertical line is read "given," and the event following the line, in this case A_1, is the one known to have occurred. We can now formally define a conditional probability.

Conditional Probability of B_1 Given A_1

Let A_1 and B_1 be two events in a sample space S. Then the conditional probability of B_1 given A_1, denoted $P(B_1|A_1)$, is

(2.4) $$P(B_1|A_1) = \frac{P(A_1 \text{ and } B_1)}{P(A_1)} \quad \text{where } P(A_1) > 0$$

The statement that $P(A_1) > 0$ is included in order to rule out the possibility of dividing by 0.

This definition may be illustrated by the example given in Table 2-2. The conditional probability that a person prefers product ABC given that the person is male is

$$P(B_1|A_1) = \frac{P(A_1 \text{ and } B_1)}{P(A_1)} = \frac{0.20}{0.50} = 0.40$$

We note then that the conditional probability of event B_1 given A_1 is found by dividing the joint probability of A_1 and B_1 by the marginal probability of A_1. The rationale of this procedure becomes clear by returning to Table 2-1, where we see that the proportion of males who prefer product ABC is $\frac{200}{500} = 0.40$. Thus, the conditional probability of B_1 given A_1 is simply the proportion of times that B_1 occurs out of the total number of times that A_1 occurs.

It may be noted that a chronological order is not necessarily implied in conditional probability. That is, in $P(B_1|A_1)$, event A_1 does not necessarily precede B_1 in time. In fact, using the same method of definition given in Equation 2.4, we would have

(2.5) $$P(A_1|B_1) = \frac{P(A_1 \text{ and } B_1)}{P(B_1)} \quad \text{where } P(B_1) > 0$$

Thus, in Table 2-2 the conditional probability that an individual is a male given that the person prefers product ABC is

$$P(A_1|B_1) = \frac{P(A_1 \text{ and } B_1)}{P(B_1)} = \frac{0.20}{0.30} = 0.67$$

Conditional probabilities are important concepts in our everyday affairs and in managerial decision making. For instance, we may be interested in the probability that a friend will arrive at an appointment punctually given that he has said that he will appear at a certain time. A university is concerned about the probability that an applicant for admission would have a satisfactory academic performance at the university given that the applicant has certain aptitude test scores. A marketing executive would be interested in the probability that the sales volume for one of her company's products will increase given that she has made a commitment for an expensive sales promotion campaign for that product. In the illustration given in Table 2-2, the users of the market research survey data would be interested in the relationship between sex and product preference. Indeed, the following conditional probabilities are very revealing:

$$P(B_1|A_1) = \frac{0.20}{0.50} = 0.40 \quad \text{and} \quad P(B_1|A_2) = \frac{0.10}{0.50} = 0.20$$

That is, the conditional probability of preference for product ABC given that the person is a male is 40%, whereas the corresponding conditional probability for a female is only 20%. Such sex differences in product preferences may be very important, for example, in choosing types of promotional effort to be used in selling the products.

Multiplication Rule

The multiplication rule for two events A_1 and B_1 follows immediately from the definition of conditional probability given in Equation 2.4. Multiplying both sides of Equation 2.4 by $P(A_1)$, we obtain the multiplication rule.

MULTIPLICATION RULE FOR ANY TWO EVENTS A_1 AND B_1

(2.6) $$P(A_1 \text{ and } B_1) = P(A_1)P(B_1|A_1)$$

Equivalently, from Equation 2.5, the multiplication rule may be stated as

(2.7) $$P(A_1 \text{ and } B_1) = P(B_1)P(A_1|B_1)$$

Hence, $P(A_1 \text{ and } B_1)$ can be computed by either Equation 2.6 or 2.7. As an example, let us apply these equations to the data of Table 2-2 to find the joint probability that a randomly selected person would be male and would prefer product ABC. By Equation 2.6, we have

$$P(A_1 \text{ and } B_1) = P(A_1)P(B_1|A_1)$$
$$= (0.50)(0.40) = 0.20$$

and by Equation 2.7, we obtain

$$P(A_1 \text{ and } B_1) = P(B_1)P(A_1|B_1)$$
$$= (0.30)(0.67) = 0.20$$

Of course, these calculations are given only to illustrate the principles involved in the multiplication rule. One would ordinarily not compute joint probabilities this way if the joint probabilities were already available as in Table 2-2.

The generalization of the multiplication rule to three or more events is straightforward.

MULTIPLICATION RULE FOR ANY THREE EVENTS A_1, A_2, AND A_3

(2.8) $P(A_1 \text{ and } A_2 \text{ and } A_3) = P(A_1)P(A_2|A_1)P(A_3|A_2 \text{ and } A_1)$

MULTIPLICATION RULE FOR ANY n EVENTS A_1, A_2, . . . , A_n

(2.9) $P(A_1 \text{ and } A_2 \text{ and } . . . \text{ and } A_n)$
$= P(A_1)P(A_2|A_1)P(A_3|A_2 \text{ and } A_1) . . . P(A_n|A_{n-1} \text{ and } . . . \text{ and } A_1)$

This notation means that the joint probability of the n events is given by the product of the probability that the first event A_1 has occurred, the conditional probability of the second event A_2 given that A_1 has occurred, the conditional probability of third event A_3 given that both A_2 and A_1 have occurred, and so on. Of course, the n events can be numbered arbitrarily; any one of them may be the first event, any of the remaining $n - 1$ may be second, and so forth.

Let us consider a couple of examples of these ideas. Suppose the probability that a sales representative following up on a lead will make the sale is 0.2. Past experience indicates that 40% of such sales are for amounts in excess of $100. What is the probability that the representative will make a sale in excess of $100?

To solve this problem, let us use the following symbols:

$P(A_1)$:   probability that a sale is made
$P(A_2|A_1)$: probability that the sale is in excess of $100 given that a sale is made

The required probability is given by

$$P(A_1 \text{ and } A_2) = P(A_1)P(A_2|A_1)$$
$$= (0.2)(0.4) = 0.08$$

It should be noted that $P(A_2|A_1)$ is a conditional probability because the 0.40 probability that a sale will exceed $100 depends on the sale actually being made.

Our second example is typical of many situations involving the sampling of human populations. Suppose we had a list of ten individuals, five of whom reside in New York, three in Pennsylvania, and two in New Jersey. Suppose we select three names at random from the list, one at a time, so that at each drawing all remaining names have an equal chance of being selected. (The actual techniques involved in such sampling procedures are discussed in Chapter 4.) What is the probability of the joint occurrence of the following events?

A_1: the first name was of a New York resident
A_2: the second name was of a Pennsylvania resident
A_3: the third name was of a New Jersey resident

The required joint probability is

$$P(A_1 \text{ and } A_2 \text{ and } A_3) = P(A_1)P(A_2|A_1)P(A_3|A_2 \text{ and } A_1)$$

$$= \left(\frac{5}{10}\right)\left(\frac{3}{9}\right)\left(\frac{2}{8}\right) = 0.042$$

Let us consider the factors on the right side of this equation. $P(A_1) = \frac{5}{10}$ because five of the ten names were those of New York residents. $P(A_2|A_1) = \frac{3}{9}$ because given that a New York resident's name was selected, there were nine remaining names, three of which were of Pennsylvania residents. Similarly, after the Pennsylvania name was chosen, there were eight remaining, two of which were of New Jersey residents.

An important point concerning the use of the multiplication rule may be noted from the preceding example. The probability calculation pertained to three names being drawn *successively*. On the other hand, what is the probability of obtaining the names of one New York, one Pennsylvania, and one New Jersey resident if the three names were drawn *simultaneously* from the list? The answer to this question is exactly the same as in the preceding computation. We can think of one of the three names as the "first," another as the "second," and another as the "third," although they have been drawn together. A_1 is again used to denote the event "the first name was of a resident of New York," etc.

It can now be seen that since the *joint* probability of the *simultaneous* occurrence of events A_1, A_2, and A_3 is the same as for their *successive* occurrence, the order in which the names appear is of no importance in the probability calculation. For example, if we had originally specified

A_1: the first name was of a Pennsylvania resident
A_2: the second name was of a New Jersey resident
A_3: the third name was of a New York resident

the required joint probability would be

$$P(A_1 \text{ and } A_2 \text{ and } A_3) = \left(\frac{3}{10}\right)\left(\frac{2}{9}\right)\left(\frac{5}{8}\right) = 0.042$$

Of course, this is the same numerical result as was obtained previously. We can state the following generalization about the use of the multiplication rule. *The multiplication rule may be used to obtain the joint probability of the successive or simultaneous occurrence of two or more events.*

The example of sampling names from a list has interesting implications. In the example, we drew a sample of three names at random *without replacement* of the sampled elements. In human populations, sampling is usually carried out without replacement; that is, after the necessary data are obtained, an individual drawn into the sample is usually not replaced prior to the choosing of another individual. We have seen how the partial exhaustion of the population because of sampling had to be taken into account in the calculation of conditional probabilities. If the sampling had been *with replacement,* then the basic probabilities of selection remain unchanged after each item is replaced in the population. For example, if the sampling from the list had been with replacement, then the respective probabilities for New York, Pennsylvania, and New Jersey residents would have been $P(A_1) = \frac{5}{10}$, $P(A_2) = \frac{3}{10}$, $P(A_3) = \frac{2}{10}$. Hence the joint probability of the occurrence of A_1, A_2, and A_3 would have been

$$P(A_1 \text{ and } A_2 \text{ and } A_3) = \left(\frac{5}{10}\right)\left(\frac{3}{10}\right)\left(\frac{2}{10}\right) = 0.03$$

In sampling *without replacement,* we are dealing with *dependent* events; in sampling *with replacement,* we are dealing with *independent* events. The meaning of these concepts and the multiplication rules for independent events are discussed next.

Statistical Independence

In our discussion of Table 2-2, we saw that there was a relationship between product preference and sex. For example, we found that the probability that a male preferred product ABC was 0.40, whereas the corresponding probability for a female was 0.20. In other words, product preference *depends* on sex. Hence, the corresponding events involved (for example, "male" and "prefers product ABC") are said to be *dependent.* Since it is important in analysis and decision making to detect such relationships, it is correspondingly important to know when we are dealing with *independent* events.

We now turn to the concept of independence, usually referred to as *statistical independence.* If two events, say A_1 and B_1, are statistically independent, then knowing that one of them has occurred does not affect the probability that the other will occur; in such a case, $P(B_1|A_1) = P(B_1)$. For example, in tossing a fair coin twice, suppose we use the following notation for events

A_1: head on first toss
B_1: head on second toss

The marginal, or unconditional, probability of obtaining a head on the second toss is $P(B_1) = \frac{1}{2}$. The conditional probability of obtaining a head on the second toss given that a head was obtained on the first toss is $P(B_1|A_1) = \frac{1}{2}$. Of course, the probability of obtaining a head on the second toss does not depend on whether a head or tail was obtained on the first toss. Thus, we note

$$P(B_1|A_1) = P(B_1) = \frac{1}{2}$$

We now define statistical independence of two events.

Statistical Independence

Two events A_1 and B_1 are *statistically independent* if

$$P(B_1|A_1) = P(B_1)$$

When this equality holds, it is also true[2] that

$$P(A_1|B_1) = P(A_1)$$

When two events A_1 and B_1 are statistically independent, the multiplication rule given in Equations 2.6 and 2.7 can be simplified.

MULTIPLICATION RULE FOR TWO STATISTICALLY INDEPENDENT EVENTS A_1 AND B_1

(2.10) $$P(A_1 \text{ and } B_1) = P(A_1)P(B_1)$$

It may be noted that if two events A_1 and B_1 are statistically independent, Equation 2.10 holds. The converse is also true. That is, if two events A_1 and B_1 are related according to Equation 2.10, the two events are statistically independent; furthermore, $P(B_1|A_1) = P(B_1)$ and $P(A_1|B_1) = P(A_1)$.

The generalization of the multiplication rule for n collectively independent events (any one event is independent of any combination of the others) is given next.

---

[2] If several events are *collectively independent* (or *mutually independent*), then every possible conditional probability for every combination of events must be equal to the corresponding unconditional probability. For example, for three events $A_1, A_2,$ and A_3 to be collectively independent, the necessary and sufficient conditions are the following:

$$P(A_1) = P(A_1|A_2) \qquad P(A_1) = P(A_1|A_2 \text{ and } A_3)$$
$$P(A_2) = P(A_2|A_3) \qquad P(A_2) = P(A_2|A_1 \text{ and } A_3)$$
$$P(A_3) = P(A_3|A_1) \qquad P(A_3) = P(A_3|A_1 \text{ and } A_2)$$

**MULTIPLICATION RULE FOR _n_ COLLECTIVELY
INDEPENDENT EVENTS $A_1, A_2, \ldots, A_n$**

(2.11) $P(A_1 \text{ and } A_2 \text{ and } \ldots \text{ and } A_n) = P(A_1)P(A_2)P(A_3) \ldots P(A_n)$

As Examples 2-6 through 2-8 show, it is possible to solve a variety of probability problems using only the addition and multiplication rules.

EXAMPLE 2-6

A fair coin is tossed twice. What is the probability of obtaining exactly one head?

SOLUTION

One way of arriving at the solution is through the combined use of the addition and multiplication rules. Denote the appearance of a head by H and a tail by T. The event "exactly one head" in two trials may occur by obtaining a head on the first trial followed by a tail on the second or a tail followed by a head. These two events are mutually exclusive. Thus, by the addition rule,

$P(\text{exactly one head}) = P((H \text{ and } T) \text{ or } (T \text{ and } H))$
$$= P(H \text{ and } T) + P(T \text{ and } H)$$

The appearance of a head on the first toss and a tail on the second are independent events, as are a tail on the first toss and a head on the second. Thus, by the multiplication rule,

$$P(H \text{ and } T) = P(H)P(T) = \frac{1}{2} \times \frac{1}{2} = \frac{1}{4}$$

and

$$P(T \text{ and } H) = P(T)P(H) = \frac{1}{2} \times \frac{1}{2} = \frac{1}{4}$$

Hence,

$$P(\text{exactly one head}) = \frac{1}{4} + \frac{1}{4} = \frac{1}{2}$$

Another way of obtaining the solution is to consider the sample space of equally probable elements

$$S = \{(T, T), (T, H), (H, T), (H, H)\}$$

Since two of the four sample elements represent the occurrence of exactly one head, we have

$$P(\text{exactly one head}) = \frac{2}{4} = \frac{1}{2}$$

EXAMPLE 2-7

A national franchising firm is interviewing prospective buyers in the Morganville area. The probability that a prospective buyer will offer to buy the franchise is 0.1. Assuming statistical independence, what is the probability that in interviewing five prospective buyers, the firm will receive at least one offer to buy the franchise?

SOLUTION

Let O and F represent, respectively, an offer and a failure to offer to buy the franchise. Then, for any prospective buyer,

$$P(O) = 0.1 \text{ and } P(F) = 0.9$$

Let E denote the event "at least one offer to buy the franchise in interviewing five prospective buyers." Then $\bar{E}$, the complement of E, denotes the event "no offers to buy the franchise in interviewing five prospective buyers." Since the event $\bar{E}$ represents the successive occurrence of five failures to offer to buy the franchise and since we have assumed independence of events, we obtain by the multiplication rule for independent events

$$P(\bar{E}) = P(F \text{ and } F \text{ and } F \text{ and } F \text{ and } F) = P(F)P(F)P(F)P(F)P(F)$$
$$= (0.9)(0.9)(0.9)(0.9)(0.9) = 0.59$$

Therefore,

$$P(E) = 1 - P(\bar{E}) = 1 - 0.59 = 0.41$$

EXAMPLE 2-8

The following table refers to the 2500 employees of the Johnson Company, classified by sex and by opinion on a proposal to emphasize fringe benefits rather than wage increases in an impending contract discussion.

Sex	In Favor	Opinion Neutral	Opposed	Total
Male	900	200	400	1500
Female	300	100	600	1000
Total	1200	300	1000	2500

a. Calculate the probability that an employee selected from this group will be

1. a female opposed to the proposal

2. neutral
3. opposed to the proposal, given that the employee selected is a female
4. either a male or opposed to the proposal

b. Are opinion and sex independent for these employees?

SOLUTION

We use the following representation of events

A_1: Male B_1: In favor
A_2: Female B_2: Neutral
                          B_3: Opposed

In (a), we have

1. $P(A_2 \text{ and } B_3) = 600/2500 = 0.24$
2. $P(B_2) = 300/2500 = 0.12$

3. $P(B_3|A_2) = \dfrac{P(B_3 \text{ and } A_2)}{P(A_2)} = \dfrac{600/2500}{1000/2500} = 0.60$

4. $P(A_1 \text{ or } B_3) = P(A_1) + P(B_3) - P(A_1 \text{ and } B_3)$

$$= \frac{1500}{2500} + \frac{1000}{2500} - \frac{400}{2500}$$

$$= \frac{2100}{2500} = 0.84$$

In (b), in order for opinion and sex to be statistically independent, the joint probability of each pair of A events and B events would have to be equal to the product of the respective unconditional probabilities. That is, the following equalities would have to hold:

$P(A_1 \text{ and } B_1) = P(A_1)P(B_1)$ $P(A_2 \text{ and } B_1) = P(A_2)P(B_1)$
$P(A_1 \text{ and } B_2) = P(A_1)P(B_2)$ $P(A_2 \text{ and } B_2) = P(A_2)P(B_2)$
$P(A_1 \text{ and } B_3) = P(A_1)P(B_3)$ $P(A_2 \text{ and } B_3) = P(A_2)P(B_3)$

Clearly, these equalities do not hold; for example,

$$P(A_1 \text{ and } B_1) \neq P(A_1)P(B_1)$$

$$\frac{900}{2500} \neq \frac{1500}{2500} \times \frac{1200}{2500}$$

Another way of viewing the problem is that each conditional probability would have to be equal to the corresponding unconditional probability. Thus, the following equalities would have to hold:

$P(A_1|B_1) = P(A_1)$ $P(A_1|B_2) = P(A_1)$ $P(A_1|B_3) = P(A_1)$
$P(A_2|B_1) = P(A_2)$ $P(A_2|B_2) = P(A_2)$ $P(A_2|B_3) = P(A_2)$

These equalities do not hold; for example,

$$P(A_1|B_1) \neq P(A_1)$$

$$\frac{900}{1200} \neq \frac{1500}{2500}$$

The nature of the dependence (lack of independence) can be summarized briefly as follows: The proportion of males declines as we move from favorable to opposed opinions. This type of relationship is sometimes described by saying that there is a *direct* relationship between the proportion of males and favorableness of opinion. Correspondingly, there is an *inverse* relationship between the proportion of females and favorableness of opinion. The discovery and interpretation of such dependence lays the groundwork for improved decision making.

Exercises

1. Let $P(A) = 0.5$, $P(B) = 0.4$, and $P(A \text{ and } B) = 0.2$.
   a. Are A and B mutually exclusive events? Why?
   b. Are A and B independent events? Why?

2. There are two major reasons for classifying a bottle of soda defective; either the filler (a machine) overfills the bottle or it underfills the bottle. The machine underfills 2% of the time and overfills 1% of the time. What is the probability that a bottle will be rejected because of the filler?

3. a. In a certain city, 30% of the households have electric dryers and 40% have electric stoves. If 25% of the households have both electric stoves and electric dryers, what is the probability that a household selected at random will have an electric stove or an electric dryer?
   b. Are electric stove ownership and electric dryer ownership independent?

4. A certain family has three children. Male and female children are equally probable, and sexes of successive children are independent. Let M stand for male and F for female.
   a. List all elements in the sample space.
   b. What are the probabilities of the ordered events MMM and MFM?
   c. Let A be the event that both sexes appear. Find $P(A)$.
   d. Let B be the event that there is at most one girl. Find $P(B)$.
   e. Prove that A and B are independent.

5. Events A and B have the following probability structure:

$$P(A \text{ and } B) = \frac{1}{8}$$

$$P(A \text{ and } \bar{B}) = \frac{7}{12}$$

$$P(\bar{A} \text{ and } B) = \frac{1}{4}$$

   a. What is the probability of $\bar{A}$ and $\bar{B}$?
   b. Are A and B independent events?

6. If the probability that Company A will buy Company B is 0.6, what are the odds that Company A will not buy Company B?

7. In roulette, there are 38 slots in which a ball may land. There are numbers 0, 00, and 1 through 36. The odd numbers are red, the even numbers are black, and the zeros are green. A ball is thrown randomly into a slot.

a. What is the probability that it is red?

b. What is the probability it is number 27?

c. What is the probability it is either red or the number 27?

d. What are the odds in favor of black?

e. What are the odds in favor of the number 27?

f. If you play black an infinite number of times, what fraction of times will you win? What fraction of times will you lose?

8. An investment firm purchases three stocks for one-week trading purposes. It assesses the probabilities that the stocks will increase in value over the week as 0.85, 0.8, and 0.5. What is the probability that all three stocks will increase, assuming that the movements of these stocks are independent? Is this a reasonable assumption?

9. The probability that a life insurance sales representative following up a magazine lead will make a sale is 0.4. Sales representative Maria Gianelli has four leads on a certain day. Assuming independence, what is the probability that

a. she will sell all four?

b. she will sell none?

c. she will sell exactly one?

10. In a group of 15 persons, there are 7 males and 10 Republicans. Furthermore, there are 4 male Republicans. If a random selection is made, what is the probability of selecting a female who is not a Republican?

11. A firm is recruiting recent graduates for seven engineering positions. In the past, 35% of the college students who were offered similar positions have turned them down. The firm offers positions to eight graduates. Is the firm justified in doing so? Explain, assuming independence between the decisions of individual students.

12. A national franchising company is interviewing prospective buyers in the Tulleytown area. The probability that an interviewee will buy the franchise is 0.1. Assuming independence, what is the probability the firm will have to interview more than five people before making a sale?

13. A census of a company's 1000 employees in regard to a certain proposal showed 240 of its 350 white-collar workers in favor of the proposal and a total of 300 workers opposed to the proposal. (All the workers can be classified as either white collar or blue collar.)

a. What is the probability that an employee selected at random will be a blue-collar worker opposed to the proposal?

b. What is the probability that an employee selected at random will be in favor of the proposal?

c. What is the probability that a particular blue-collar worker chosen will be in favor of the proposal?

d. Are job type (blue or white collar) and opinion on the proposal independent? Prove your answer and give a short statement about the implication of this finding.

14. The following information pertains to new-car dealers in the United States:

| Type of Dealership | Region of Dealer | | | | |
	North	South	Midwest	West	Total
Admiral Motors	155	50	135	110	450
Bord Motors	90	65	40	90	285
Shysler Motors	50	50	30	85	215
U.S. Motors	35	35	15	65	150
Total	330	200	220	350	1100

If a name is selected randomly from the American Automobile Dealers Association list of all U.S. dealers handling the four American manufacturers' brands, what is the probability that it is the name of

a. a Southern Admiral Motors dealer?
b. a Southern dealer?
c. an Admiral Motors dealer?
d. a Southern dealer if the name is known to be that of an Admiral Motors dealer?

Are type of dealership and region independent?

15. A company's management made a certain proposal to all its sales representatives in different sales regions. Questionnaries were sent to each representative, and the results were as follows:

| Opinion | | Region | | |
	East	Midwest	West	Total
Opposed	55	65	60	180
Not opposed	95	135	190	420
Total	150	200	250	600

a. What is the probability that a questionnarie selected at random is that of a Western sales representative in favor of the proposal?
b. What is the probability that a questionnaire selected at random is that of a Midwestern sales representative?
c. If a questionnaire is selected at random from the group that responded unfavorably to the proposal, what is the probability that the respondent comes from the Eastern region?
d. Are regional district and opinion on the proposal independent? If yes, prove your assertion. If no, specify what the numbers in the cells of the table would have been had the two factors been independent.

2.3 BAYES' THEOREM

The Reverend Thomas Bayes (1702–1761), an English Presbyterian minister and mathematician, considered the question of how one might make inferences from observed sample data about the larger groups from which the data were drawn. His motivation was his desire to prove the existence of God by examining the sample evidence of the world about him. Mathematicians had previously concentrated on the problem of deducing the consequences of specified hypotheses. Bayes was interested in the inverse problem of drawing conclusions about hypotheses from observations of consequences. He derived a theorem that calculated probabilities of "causes" based on the observed "effects." The theorem may also be thought of as a means of revising probabilities of events based on additional information. In the period since World War II, a body of knowledge known as *Bayesian decision theory* has been developed to solve problems involving decision making under uncertainty.

Bayes' theorem is really nothing more than a statement of conditional probabilities. The following problem illustrates the nature of the theorem. Assume that 1% of the inhabitants of a country suffer from a certain disease. Let A_1 represent the event "has the disease" and A_2 denote "does not have the disease." Now suppose a person is selected at random from this population. What is the probability that this individual has the disease? Since 1% of the population has the disease and it is equally likely that any individual would be selected, we assign a probability of 0.01 to the event "has the disease." This probability, $P(A_1) = P(\text{has the disease}) = 0.01$, is referred to as a *prior probability* in the sense that it is assigned prior to the observation of any empirical information. Of course, $P(A_2) = 0.99$ is the corresponding prior probability that the individual does not have the disease.

Now let us assume that a new but imperfect diagnostic test has been developed, and let B denote the event "test indicates the disease is present." Suppose that through past experience it has been determined that the conditional probability that the test indicates the disease is present, given that the person has the disease, is

$$P(B|A_1) = 0.97$$

and the corresponding probability, given that the person does *not* have the disease, is

$$P(B|A_2) = 0.05$$

Suppose a person is selected at random and given the test and the test indicates that the disease is present. What is the probability that this person actually has the disease? In symbols, we want the conditional probability $P(A_1|B)$. It is important to note the nature of this question. The probability P(has the disease|test indicates the disease is present) $= P(A_1|B)$ is referred to as a *posterior probability* or a *revised probability*, because it is assigned after the observation of empirical or additional information. The posterior probability is the type of probability computed with the help of Bayes' theorem. We will derive the value of $P(A_1|B)$ from probability principles developed in this chapter.

By the definition of conditional probability given in Equations 2.4 and 2.5, we have

$$\textbf{(2.12)} \qquad P(A_1|B) = \frac{P(A_1 \text{ and } B)}{P(B)}$$

We compute the joint probability $P(A_1 \text{ and } B)$ in the numerator of Equation 2.12 by the multiplication rule given in Equation 2.6.

$$\textbf{(2.13)} \qquad P(A_1 \text{ and } B) = P(A_1)P(B|A_1)$$

To compute the denominator of Equation 2.12, we observe that

$$P(B) = P[(A_1 \text{ and } B) \text{ or } (A_2 \text{ and } B)]$$

Hence,

$$\textbf{(2.14)} \qquad P(B) = P(A_1 \text{ and } B) + P(A_2 \text{ and } B)$$

since the two joint events $(A_1 \text{ and } B)$ and $(A_2 \text{ and } B)$ are mutually exclusive.

If we express the joint probabilities $P(A_1 \text{ and } B)$ and $P(A_2 \text{ and } B)$ according to the multiplication rule, Equation 2.14 becomes

$$\textbf{(2.15)} \qquad P(B) = P(A_1)P(B|A_1) + P(A_2)P(B|A_2)$$

Now, substituting into the numerator of the right side of Equation 2.12 the expression for the joint probability $P(A_1 \text{ and } B)$ given in Equation 2.13, and substituting into the denominator the marginal probability $P(B)$ given in Equation 2.15, we obtain the result known as Bayes' Theorem.

BAYES' THEOREM FOR TWO BASIC EVENTS, A_1 AND A_2

$$\textbf{(2.16)} \qquad P(A_1|B) = \frac{P(A_1)P(B|A_1)}{P(A_1)P(B|A_1) + P(A_2)P(B|A_2)}$$

In terms of the disease example, we find, by substituting the known values into the Bayes' theorem formula in Equation 2.16, that

$$P(A_1|B) = \frac{(0.01)(0.97)}{(0.01)(0.97) + (0.99)(0.05)} = \frac{0.0097}{0.0592} = 0.16$$

Hence, the posterior probability that the individual has the disease given that the test indicated the presence of the disease is 0.16. In summary, if a person were selected at random from the population of this country, the prior probability assignment that the individual has the disease is 0.01. On the other hand, after we have the empirical information that the test indicated the disease is present, we revise the probability that the individual has the disease upward to 0.16.

Although the posterior probability (0.16) is sixteen times as large as the prior probability (0.01), 0.16 is still a surprisingly low probability that the individual has the disease given that the test indicated the disease was present. This results from the fact that there are a large number of persons in this population who do not have the disease but for whom the test would (falsely) indicate that the disease is present.

As we have seen, Bayes' theorem weights prior information with empirical evidence. The manner in which it does this may be seen by laying out the calculations in a form such as Table 2-3. The first column of Table 2-3 gives the basic events of interest, "has the disease" and "does not have the disease." The second column shows the prior probability assignments to these basic events. The third column shows the conditional probabilities of the additional information given the basic events. As noted in the caption, such conditional probabilities are referred to as "likelihoods." In the illustrative problem these are, respectively, $P(B|A_1) = P$(test indicates disease is present|has the disease) and $P(B|A_2) = P$(test indicates disease is present|does not have the disease). The fourth column gives the joint probabilities of the basic events and the additional information. It may be noted that as indicated in Equation 2.15, the sum of these joint probabilities is the marginal probability $P(B)$. When the joint probabilities are divided by their total, $P(B)$ (in this case, 0.0592), the results are the revised probabilities shown in the last column. The first probability shown in the last column (0.0097/0.0592 = 0.16) is the Bayes' theorem calculation required in the illustrative problem.

TABLE 2-3
Bayes' Theorem Calculations for Illustrative Problem

Events A_i	Prior Probabilities $P(A_i)$	Likelihoods $P(B\|A_i)$	Joint Probabilities $P(A_i)P(B\|A_i)$	Revised Probabilities $P(A_i\|B)$
A_1: Has the disease	0.01	0.97	0.0097	0.0097/0.0592 = 0.16
A_2: Does not have the disease	0.99	0.05	0.0495	0.0495/0.0592 = 0.84
	1.00		$P(B) = 0.0592$	1.00

In the problem given above, there were two basic events (A_i). If there are more than two basic events, then correspondingly, additional terms appear in the denominator of Equation 2.16. Hence, we can make the following formal statement of Bayes' theorem for n basic events. Assume a set of complete and mutually exclusive events A_1, A_2,

. . . , A_n. The appearance of one of the A_i events is a necessary condition for the occurrence of another event B, which is observed. The probabilities $P(A_i)$ and $P(B|A_i)$ are known. The posterior probability of event A_1 given that B has occurred is given by Bayes' theorem.

BAYES' THEOREM FOR n BASIC EVENTS, A_1, A_2, . . . , A_n

$$(2.17) \quad P(A_1|B) = \frac{P(A_1)P(B|A_1)}{P(A_1)P(B|A_1) + P(A_2)P(B|A_2) + \cdots + P(A_n)P(B|A_n)}$$

Although the theorem was stated in Equations 2.16 and 2.17 for a particular one of the A_i, namely A_1, it is perfectly general, since any of the n events A_i can be designated A_1. In most applications of the theorem to decision problems, the A_i represent events that precede the occurrence of the observed event B. In this connection, we can think of the theorem as answering the question "Given that event B has occurred, what is the probability that it was preceded by event A_1?" Or, as indicated earlier, another way of looking at it is that $P(A_1|B)$ is the revised probability assigned to event A_1 after event B is observed.

In modern Bayesian decision theory, subjective prior probability assignments are made in many applications. It is argued that it is meaningful to assign prior probabilities concerning hypotheses based on degree of belief. Bayes' theorem is then viewed as a means of revising these probability assignments. In business applications, this has meant that executives' intuitions, subjective judgments, and present quantitative knowledge are captured in the form of prior probabilities; these figures undergo revision as relevant empirical data are collected. This procedure seems sensible and fruitful for a wide variety of applications. Examples 2-9 and 2-10 suggest some of the many possible types of applications of this very interesting theorem.

EXAMPLE 2-9

A corporation uses a "selling aptitude test" to aid it in the selection of its sales force. Past experience has shown that only 65% of all persons applying for a sales position achieved a classification of "satisfactory" in actual selling, whereas the remainder were classified "unsatisfactory." Of those classified as "satisfactory," 80% had scored a passing grade on the aptitude test. Only 30% of those classified "unsatisfactory" had passed the test. On the basis of this information, what is the probability that a candidate would be a "satisfactory" salesperson given a passing grade on the aptitude test?

SOLUTION

If A_1 stands for a "satisfactory" classification as a salesperson and B stands for "passes the test," then the probability that a candidate would be a "satisfactory" salesperson, given a passing grade on the aptitude test, is

$$P(A_1|B) = \frac{(0.65)(0.80)}{(0.65)(0.80) + (0.35)(0.30)} = 0.83$$

Thus, the tests are of some value in screening candidates. Assuming no change in the type of candidates applying for the selling positions, the probability that a random applicant would be satisfactory is 65%. On the other hand, if the company accepts only applicants who pass the test, this probability increases to 0.83.

EXAMPLE 2-10

A certain company is planning to market a new product. The company's marketing vice-president is particularly concerned about the product's superiority over the closest competitive product, which is sold by another company. The marketing vice-president assessed the probability of the new product's superiority to be 0.7. This executive then ordered a market survey, which indicated that the new product was superior to its competitor. Assume the market survey has the following reliability: If the product is really superior, the probability that the survey will indicate "superior" is 0.8. If the product is really worse than its competitor, the probability that the survey will indicate "superior" is 0.3. After completion of the market survey, what should be this executive's revised probability assignment to the event "new product is superior to its competitor"?

SOLUTION

Let A_1 represent the event "new product is superior to its competitor" and B the event "market survey indicates that the new product is superior to its competitor." Then

$$P(A_1|B) = \frac{(0.7)(0.8)}{(0.7)(0.8) + (0.3)(0.3)} = 0.86$$

Exercises

1. A firm is contemplating changing the packaging sizes of its product; it wishes to eliminate the 3-ounce size and offer a 4-ounce size at a slightly higher price. The marketing manager feels the probability that this change will increase profits is 70%. The change is tried in a limited test area and results in reduced profits. The probability of this result occurring even if the change would actually increase profits nationally is 0.4, whereas if it would not increase profits nationally, the probability of this occurrence is 0.8. What should be the manager's revised estimate of the probability of the profitability of the change?

2. An investor feels the probability that a certain stock will go up in value during the next month is 0.65. Value-Dow, an investment advisory firm, predicts that the stock will not go up over the period. Over a long period of time, Value-Dow has proven to be correct in 90% of its predictions. What probability should the investor assign to the stock going up in the light of Value-Dow's prediction?

3. Fifteen percent of the items produced by a certain machine are defective. The company hires an inspector to check each item before shipment. The probability that the inspector will ship a defective item is 0.2, and the probability that the inspector will not ship a good item is 0.1. If an item is shipped by this inspector, what is the probability that it is good?

4. TEC Exploration Company is involved in a mining exploration in northern Canada. The chief engineer originally feels that there is a 50:50 chance that a significant mineral find will occur. The results of a first test drilling are favorable. The probability that the test drilling would give misleading results is 0.3. What should be the engineer's revised estimate of the probability that a significant mineral find will occur?

5. A student of finance is closely studying the financial structure, history, and decision-making processes of a large corporation. The corporation is presently considering a rather large investment proposal, the adoption of which is heavily dependent on the firm's profits for the coming year. The student assigns the following subjective probabilities to the adoption of the proposal: 0.9 if the firm's profits increase during the coming year, 0.2 if profits decrease, and 0.6 if profits remain constant. The student assesses the probabilities of the different profit pictures as follows:

> Profits increase: 0.4
> Profits constant: 0.5
> Profits decrease: 0.1

At the end of the year, the corporation decides not to adopt the investment proposal. Based on the above information, what is the probability that profits increased during the year?

2.4 COUNTING PRINCIPLES AND TECHNIQUES

In the problems we have encountered so far, the pertinent sample spaces have been comparatively simple. However, in many situations the numbers of points in the appropriate sample spaces are so great that efficient methods are needed to count these points, in order to arrive at required probabilities or answer other questions of interest. In this connection, it is useful to return to the concept of sequences of experimental trials to specify a simple but important fundamental principle.

The Multiplication Principle

If an experiment can result in n_1 distinct outcomes on the first trial, n_2 distinct outcomes on the second trial, and so forth for k sequential trials, then the total number of different sequences of outcomes in the k trials is $(n_1)(n_2) \cdot \cdot \cdot (n_k)$.

It is sometimes helpful to think in terms of the sequential performance of tasks rather than trials of an experiment. Thus, using somewhat different language than was employed in the context of experimental trials, if the first of a sequence of tasks can be performed in n_1 ways, the second in n_2 ways, and so forth for k tasks, then the sequence of k tasks can be carried out in $(n_1)(n_2) \cdot \cdot \cdot (n_k)$ ways.

Thus, if a coin is tossed and then a card is drawn at random from a standard deck of cards, there are $2 \times 52 = 104$ possible different sequences. For example, one such sequence of outcomes might be head, king of spades.

If a die is rolled three times, there are $6 \times 6 \times 6 = 216$ different sequences.

If it is possible to go from Philadelphia to Baltimore in two different ways and from Baltimore to Washington in three different ways, then there are $2 \times 3 = 6$ ways of going from Philadelphia to Washington via Baltimore.

A tree diagram is often helpful in thinking about the total possible number of sequences. For example, in the case of the trip from Philadelphia to Washington, if A_1 and A_2 denote the two ways of going to Baltimore and $B_1, B_2,$ and B_3 the three ways of proceeding from Baltimore to Washington, then the total number of possible sequences is indicated by the total number of different paths through the tree from left to right (see Figure 2-7).

In the following sections, the multiplication principle is used in many different types of problems.

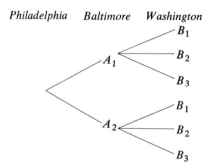

FIGURE 2-7
Tree diagram for trip from Philadelphia to Washington.

Permutations

In order to handle the problem of counting points in complicated sample spaces, the counting techniques of combination and permutation are used. In this connection, it is helpful to think in terms of objects that occur in groups. These groups may be characterized by type of object, the number belonging to each type, and the way in which the objects are arranged. For example, consider the letters *a, b, c, d, e*. There are five objects, one of each type. If we have the letters *a, b, b, c, c*, there are five objects, one of type *a*, two of *b*, and two of *c*. Returning to the first group of objects, *a b c d e, b a c d e*, and *c d e a b* differ in the *order* in which the five objects are arranged, but each of these groups contains the same number of objects of each type.

Suppose we have a group of *n* different objects. In how many ways can these *n* objects be arranged in order in a line? Applying the multiplication principle, we see that any one of the *n* objects can occupy the first position, any of the *n* − 1 remaining objects can occupy the second position, and so forth until we have only one possible object to occupy the *n*th position. Thus, the number of different possible arrangements of the *n* objects in a line consisting of *n* positions is

$$n! = (n)(n - 1) \cdot \cdot \cdot (2)(1)$$

The symbol *n*! is read "*n* factorial." We shall be concerned only with cases for which *n* is a nonnegative integer. By definition, 0! = 1.

Some examples of factorials are

$$1! = 1$$
$$2! = 2 \times 1 = 2$$
$$3! = 3 \times 2 \times 1 = 6$$
$$\cdot$$
$$\cdot$$
$$\cdot$$
$$10! = 10 \times 9 \times \cdot \cdot \cdot \times 2 \times 1 = 3,628,800$$

It is useful to note that $n! = n(n - 1)!$. Thus, $10! = 10 \times 9!$, etc. We see from this relation that it makes sense to define 0! = 1, since if we let *n* = 1, we have

$$1! = 1 \times 0!$$

and 0! = 1. This enables us to maintain a consistent definition of factorials for all nonnegative integers.

Factorials obviously increase in size very rapidly. For example, how many different arrangements can be made of a deck of 52 cards if the cards are placed in a line? The answer, 52!, is a number that contains 68 digits.[3]

Frequently we are interested in choosing and arranging in order some subgroup of n different objects. If x of the objects ($x \leq n$) are to be selected and arranged in order, as in a line, then each such arrangement is said to be a *permutation* of the n objects taken x at a time. The number of such permutations is denoted $_nP_x$. For example, suppose there are 50 persons competing in a contest for three rankings, first, second, and third. How many permutations of the 50 people taken three at a time are possible, i.e., how many different rankings are possible? The answer is

$$_{50}P_3 = 50 \times 49 \times 48 = 117,600$$

This is true because any one of the 50 persons can occupy the first position, any of the remaining 49 the second position, and any of 48 could fill the third place. By the multiplication principle, the number of different sequences of first, second, and third rankings is obtained by the indicated product.

We can now generalize this procedure to obtain a convenient formula for the number of permutations of n different objects taken x at a time.

$$_nP_x = (n)(n-1) \cdots [n - (x-1)]$$
$$= (n)(n-1) \cdots (n - x + 1)$$
$$= \frac{(n)(n-1) \cdots (n - x + 1)(n - x)!}{(n-x)!}$$

$$\textbf{(2.18)} \qquad _nP_x = \frac{n!}{(n-x)!}$$

It can be seen, in general, that if there are x positions to be filled, the first position can be filled in n ways; after one object has been placed in the first position, $x - 1$ positions remain. The second can be filled in $n - 1$ ways, the third in $n - 2$ ways, and so forth down to the xth or last position, which can be filled in $n - (x - 1)$ ways. In writing down the factors that must be multiplied together, 0 is subtracted from n in the first position, 1 is subtracted from n in the second position, and so forth down to $(x - 1)$ subtracted from n in the xth position. The formula $n!/(n - x)!$ follows from the definition of a factorial, since $(n - x)!$ cancels out all

---

[3] Warren Weaver points out in *Lady Luck* (Garden City, N.Y.: Doubleday, 1963), p. 88, about the number of possible arrangements in 52!, that, "If every human being on earth counted a million of these arrangements per second for twenty-four hours a day for lifetimes of eighty years each, they would have made only a negligible start in the job of counting all these arrangements—not a billionth of a billionth of one percent of them!"

factors after $(n - x + 1)$ in the numerator. Thus, in the contest problem, where $n = 50$ and $x = 3$, we have

$$_{50}P_3 = \frac{50!}{(50 - 3)!} = \frac{50!}{47!} = \frac{50 \times 49 \times 48 \times 47!}{47!} = 50 \times 49 \times 48$$

A special case of the formula for permutations occurs when all of the n objects are considered together. In this situation, we are concerned with the number of permutations of n different objects taken n at a time, which is

(2.19) $$_{n}P_n = \frac{n!}{(n - n)!} = \frac{n!}{0!} = n!$$

For example, if a consumer were given one cup of coffee of each of five brands and asked to rank these according to preference, the total number of possible rankings (excluding the possibility of ties) would be

$$_{5}P_5 = 5! = 120$$

Combinations

In the case of *permutations* of objects, the *order* in which the objects are arranged *is of importance. Where order is not important,* we are concerned with *combinations* of objects rather than permutations. A simple example will illustrate the difference. Suppose the president of a company is interested in setting up a very small finance committee of two people and plans to select them from a group of three executives named Brown, Jones, and Smith. How many possible committees could be formed? It is clear that order is of no importance in this situation. That is, a committee consisting of Brown and Jones is no different from a committee of Jones and Brown. Using first letters to symbolize the three names, there are three possible committees: BJ, BS, and JS.

This is an example of *combinations* of three objects taken two at a time. The terminology is similar to that used for permutations, so in general, we refer to the number of combinations that can be made of n different objects taken x at a time.

Using the same group of letters and treating them merely as symbols, if order of arrangement were important, it is clear that the number of permutations of the three objects taken two at a time would exceed the number of combinations of three objects taken two at a time. The following six permutations can be made in this case:

BJ	JB
BS	SB
JS	SJ

To develop a formula for combinations, we need merely to consider the relationship between numbers of combinations and numbers of permutations for the same group of n objects taken x at a time. Thus, fixing attention for the moment on any particular combination, there are x objects filling x positions. How many permutations can be made of these x objects in the x positions? Clearly, any one of the x objects may fill the first position, $x - 1$ the second, and so forth down to one object for the xth position. Thus, $x!$ distinct permutations can be formed of the x objects in x positions. Therefore, the number of permutations that can be formed of n different objects taken x at a time is $x!$ times the number of combinations of these n objects taken x at a time. The symbol for the number of combinations of n objects taken x at a time is $\binom{n}{x}$. Thus,

$$\binom{n}{x} x! = {}_nP_x$$

Solving for $\binom{n}{x}$ yields the following formula for the number of combinations that can be made of a group of n different objects taken x at a time:

(2.20) $$\binom{n}{x} = \frac{{}_nP_x}{x!} = \frac{n!}{x!(n - x)!}$$

Returning to the committee illustration, the number of combinations of the three people taken two at a time is

$$\binom{3}{2} = \frac{3!}{2!1!} = 3$$

which was the number previously listed. Similarly, the number of permutations of three objects taken two at a time is seen to be

$$ {}_3P_2 = \frac{3!}{1!} = 6$$

which was the number of ordered arrangements listed earlier.

EXAMPLE 2-11

A brief market research questionnaire requires the respondent to answer each of ten successive questions with either a "yes" or a "no." The sequence of ten yes–no responses is defined as the respondent's "profile." How many different possible profiles are there?

SOLUTION

There are two possible responses for each question. Therefore, by the multiplication principle, there are $2 \times 2 \times \cdots \times 2 = 2^{10} = 1024$ different profiles.

EXAMPLE 2-12

A manufacturing firm wants to locate five warehouses in the 48 continental states of the United States. It wants only one warehouse to be located in any state. If it wanted to examine the desirability of every possible combination of locations, how many locations would the firm have to consider?

SOLUTION

$$\binom{48}{5} = \frac{48!}{5!43!} = \frac{48 \times 47 \times 46 \times 45 \times 44}{5 \times 4 \times 3 \times 2 \times 1} = 1,712,304$$

An important practical principle emerges from consideration of this example. Obviously, it is not feasible to examine explicitly every possible combination of locations. Most geographical locations that would work out advantageously for the placement of warehouses are at or near concentrations of demand. Therefore, even assuming the company is correct in wanting to locate one warehouse in each of five states, the search for a solution can be reduced to a small fraction of the total number of possible locations.

EXAMPLE 2-13

There are six different operations in a manufacturing process. Let us refer to them as A, B, C, D, E, and F. A must be performed first, and F must be performed last. All other operations may be performed in any order. How many different sequences of operations are possible?

SOLUTION

Since A and F are fixed, we need only be concerned with B, C, D, and E. Any of these four may be performed first, any of the remaining three may come second, etc. Therefore, the number of possible different sequences is given by $_4P_4 = 24$.

EXAMPLE 2-14

An underwriting syndicate is to be formed from a group of investment banking firms, each of which is classified either as type A or as type B. There are six type A and eight type B firms. In how many ways can a syndicate of six firms be formed if
a. it must consist of three firms of type A and three of type B ?
b. it must consist of at least three firms of type A and at least one of type B ?

SOLUTION

In (a), we can think of the sequential selection of three firms from the six type A firms followed by a selection of three firms from the eight type B firms, where the order of selection is unimportant. Hence, the number of possible syndicates is

$$\binom{6}{3} \times \binom{8}{3} = 1120$$

The tabular arrangement below gives the solution to part (b):

Different Methods of Forming the Syndicate	Number of Ways of Forming the Syndicate
3A, 3B	$\binom{6}{3}\binom{8}{3} = 1120$
4A, 2B	$\binom{6}{4}\binom{8}{2} = 420$
5A, 1B	$\binom{6}{5}\binom{8}{1} = 48$
	Total 1588

Exercises

1. A motivational researcher shows a woman 12 projected colors for new spring clothes and asks her to pick out her four favorites.
   a. Give a specific possible outcome of the experiment.
   b. How many such outcomes are there?
   c. If one of the color choices is azure, how many possible outcomes will contain that color?
   d. What is the probability that the woman will choose azure as one of her four favorites?
   e. What is the probability that she will not choose either azure or carmine (another color available) in her selection of favorites?

2. In a determination of preference of package design, a panel of consumers was given four different packaging designs and asked to rank them. How many different possible rankings could the panel have given (excluding ties)?

3. Professor Tom Robbins anticipates teaching the same course for the next few years. In order not to become bored with his own jokes, he decides to tell a set of exactly three jokes each year. He may repeat one or two jokes from year to year, but he vows never to repeat the same set of three jokes. How many years can he last with a repertoire of seven jokes?

4. A company has three supervisors and seven regular employees. Each week,

a skeleton force of one supervisor and three regular employees is chosen at random to work on Saturday.

    a. If supervisor Alice Beaumont and regulars Karl Busch, Charles Lowengrub, and Emily Rossi always play bridge during their lunch hour and never at any other time, what is the probability they will play bridge on a given Saturday?

    b. How many possible sets of Saturday work forces will contain supervisor Alice Beaumont?

    c. How many possible sets of Saturday work forces will contain regular employee Karl Busch?

5. A committee consists of eight union and six nonunion workers. In how many ways can a subcommittee of six workers, three union and three nonunion, be formed? What would the answer be if all the subcommittee members to be chosen had to be

    a. union workers?

    b. nonunion workers?

6. A personnel director is evaluating 11 college graduates for four training positions in the company: one position in the marketing division, one in the research department, one in the finance area, and one in personnel. If a college degree is the only qualification necessary for any of these positions, in how many different ways can the openings be filled?

7. Lemon Motors orders seven different upholstering colors for its cars and twelve different colors of body paint.

    a. How many different color combinations of body and upholstering are available to the customer?

    b. If Lemon Motors allows the customer to order a roof color different from the basic body color, how many additional different color coordinations of body, roof, and upholstering are available to the customer?

8. Rittleman Furs, Inc. has just purchased an electronic computer to handle its accounts receivable. Each data card contains 80 columns in which a number from 0 to 9 or a letter may be punched to represent information about an account. It is decided that each account will be assigned four identification symbols, which will be punched in the first four columns of the data card to identify each card with a particular account.

    a. If only numbers are to be used, how many accounts can be handled by this method?

    b. If the first column is to be a letter and the next three are to be numbers, how many accounts can be handled?

    c. If either a letter or a number may be punched in each column, how many accounts can be handled?

9. Suppose you face a ten-question true–false examination fully aware that you are unprepared for it. If your strategy is to mark five questions true and five questions false, in how many ways can you mark your examination?

10. The board of directors of the Crosscheck Hockey Corporation, principal owners of the Moose Jaw Manglers hockey club, meets once a month in its lavish board room overlooking beautiful downtown Moose Jaw, Saskatchewan. Peter Luxon, the chairman of the board, sits at the head of the rec-

tangular conference table, and the other 15 directors sit around the three remaining sides in no special order.

   a. How many seating arrangements are possible?

   b. Mr. Luxon does not want the two executive vice-presidents of the company, who are also on the board, to sit together. How many possible arrangements are there if these two do not sit together?

   c. If the president of the company must sit at the foot of the table, how many possible arrangements are there (assuming that the remaining 14 directors can sit in any of the remaining 14 seats)?

11. A firm desires to build six new factories, two in the 13 Southern states, one in the six Middle Atlantic states, one in the four Far Western states, and two in the eight Midwest states. If the firm wants to study the desirability of each possible combination of locations, how many combinations would the firm have to consider?

12. Financial analyst George Serajian claims that he can tell from reading a company's annual report whether its gross sales will increase the following year. He is given the reports of ten companies from last year and is told that five had higher gross sales this year and five did not. Then he is asked to separate the ten reports into two groups according to whether he believes they increased gross sales or not. Three of the five that he places in the "increase" pile actually did increase gross sales. If one were to make the selections randomly, what is the probability that one would place at least three correctly in the increase group? Based on this result, do you believe Mr. Serajian can actually predict increases in gross sales?

13. Vivian Chan, the financial vice-president of the Petro Chemical Company, is considering five similar investment proposals for the upcoming fiscal period. After analyzing the financial condition of the firm and estimating the firm's cash flow for the period, she decides that all portfolios consisting of three proposals are equally feasible and desirable in the long run.

   a. Assuming that the three investments in the adopted portfolio are made simultaneously, how many different portfolios are there from which to choose?

   b. If the three adopted proposals are implemented sequentially and it is determined that, due to differing cash payback periods, the order of implementation is a distinguishing factor in the comparison of otherwise identical portfolios, how many portfolio arrangements are there?

   c. Under the assumptions of (b) and assuming random selection, what is the probability that the portfolio selected will include proposals A, B, and C?

14. A certain organization consists of eight men and eight women, from whom a committee of seven people is to be formed.

   a. What is the probability that there will be exactly one woman on the committee?

   b. Given that Mr. Greene is to be chairman of the committee, what is the probability that of the six remaining to be selected, exactly one will be a woman?

   c. Given that Mr. Greene is to be chairman of the committee, what is the probability that all of the six remaining to be selected will be women?

   d. Recalculate the probability for part (a) given that the organization consists of nine men and nine women.

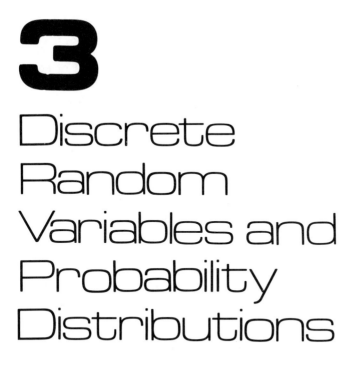

3

Discrete
Random
Variables and
Probability
Distributions

3.1 RANDOM VARIABLES

Managerial decisions are ordinarily made under conditions of uncertainty. A corporation treasurer makes investment decisions in the face of uncertainties concerning future movements of interest rates and stock market prices. A corporate executive committee may make a decision concerning expansion of manufacturing facilities despite uncertainty about future levels of demands for the company's products. An advertising manager makes decisions on advertising expenditures in various media without being certain of the sales that will be generated by these outlays. In each of these cases, the outcomes of concern, such as interest rates, stock market prices, levels of demand, and sales that result from advertising, may assume a number of values; hence, we refer to them as *variables*. In statistical analysis, such variables are usually called *random variables.* A random variable may be defined roughly as a variable that takes on different numerical values because of chance.[1]

In this chapter, we will be concerned primarily with *probability distributions* of random variables. This concept is central to all of statistics, and although we introduce it here in the context of the business decision problem, it is used in every field in which statistical methods are applied. Examples 3-1, 3-2, and 3-3 introduce the idea of a probability distribution.

EXAMPLE 3-1

Harnett Industries, a large corporation, was interested in diversifying its product line. In that connection, Harnett was negotiating the acquisition of Chase Products, a smaller company. Mary Prescott, manager of mergers and acquisitions at Harnett Industries, contemplated the purchase of Chase Products in the very near future, but she was uncertain about the price Harnett would have to pay per share for Chase Products common stock. Ms. Prescott set up the probability distribution shown in Table 3-1 for the stock price.

The two columns shown in Table 3-1 constitute the probability distribution for the random variable "price of Chase Products common stock." The following type of symbolism is conventionally used: If we let the symbol X stand for the random variable (in this example price of Chase

TABLE 3-1
Probability Distribution for the Price of Chase Products Common Stock

Price of Chase Products Common Stock (to the Nearest Dollar) x	Probability $f(x)$
$33	0.10
34	0.25
35	0.50
36	0.10
37	0.05

[1] From a mathematical viewpoint, a random variable is a function consisting of the elements of a sample space and the numbers assigned to these sample elements. Ordinarily, a shortcut method of referring to a random variable is used. For example, we may refer to the random variables "the price of XYZ stock at some specified future time," "annual volume of sales of product ABC," or, in a coin-tossing example, "number of heads obtained in two tosses of a coin."

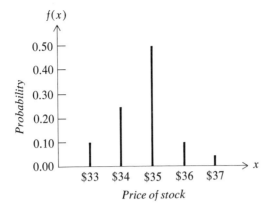

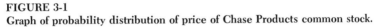

FIGURE 3-1
Graph of probability distribution of price of Chase Products common stock.

Products common stock), then we represent the values the random variable can assume by x. The probability that the random variable X will take on the value x is symbolized as $P(X = x)$ or simply $f(x)$. In Table 3-1, the values of the random variable are listed under the column headed x and the probabilities of these values are shown under $f(x)$. Thus,

$$P(X = \$33) = f(\$33) = 0.10$$
$$P(X = \$34) = f(\$34) = 0.25$$
$$P(X = \$35) = f(\$35) = 0.50$$
$$P(X = \$36) = f(\$36) = 0.10$$
$$P(X = \$37) = f(\$37) = 0.05$$

Note that these probabilities sum to 1.

   This is an example of a subjective probability distribution. We observe that by assigning a probability of 0.50 to a price of $35, Ms. Prescott indicated that she felt the odds were 50:50 that the stock would be priced at that figure as opposed to any other figure. She felt that a price of $35 was twice as likely to occur as a price of $34, to which she assigned a probability of 0.25. She felt that a price of $33, $36, or $37 was rather unlikely.

   When a probability distribution is graphed, it is conventional to display the values of the random variable on the horizontal axis and their probabilities on the vertical scale. The graph of the probability distribution for the common stock price example is shown in Figure 3-1.

EXAMPLE 3-2

A department store has classified its list of charge customers into two mutually exclusive categories: (1) high-volume purchasers, and (2) nonhigh-volume purchasers. Twenty percent of the customers are high-volume purchasers. Assume that a sample of four customers is drawn at random from the list. What is the probability distribution of the random variable

TABLE 3-2
Elements of the Sample Space for the
Experiment of Drawing a Random Sample of
Four Customers

$\bar{A}\,\bar{A}\,\bar{A}\,\bar{A}$	$A\,A\,\bar{A}\,\bar{A}$	$A\,A\,A\,\bar{A}$
	$A\,\bar{A}\,A\,\bar{A}$	$A\,A\,\bar{A}\,A$
$A\,\bar{A}\,\bar{A}\,\bar{A}$	$A\,\bar{A}\,\bar{A}\,A$	$A\,\bar{A}\,A\,A$
$\bar{A}\,A\,\bar{A}\,\bar{A}$	$\bar{A}\,A\,A\,\bar{A}$	$\bar{A}\,A\,A\,A$
$\bar{A}\,\bar{A}\,A\,\bar{A}$	$\bar{A}\,A\,\bar{A}\,A$	
$\bar{A}\,\bar{A}\,\bar{A}\,A$	$\bar{A}\,\bar{A}\,A\,A$	$A\,A\,A\,A$

TABLE 3-3
Probability Distribution of Number of
High-Volume Purchasers

x	$f(x)$
0	0.4096
1	0.4096
2	0.1536
3	0.0256
4	0.0016

"number of high-volume purchasers"? It may be assumed that the list of charge customers is so large that even though the sample is drawn without replacement, there is only a negligible loss of accuracy if the computations are performed as though the sampling were carried out with replacement. That is, the partial exhaustion of the list because drawing of the items in the sample is so small that for practical purposes, the probabilities of obtaining the two types of purchasers remain unchanged.

Let A represent the occurrence of a high-volume purchaser and $\bar{A}$ the occurrence of a nonhigh-volume purchaser. The elements of the sample space for the experiment of drawing the sample of four customers are listed in Table 3-2.

We denote by X the random variable "number of high-volume purchasers." X can take on the values 0, 1, 2, 3, 4. As can be seen from Table 3-2, one sample element corresponds to the occurrence of no high-volume purchasers; 4 elements to one high-volume purchaser; 6 elements to two high-volume purchasers; 4 elements to three high-volume purchasers; and 1 element to four high-volume purchasers. However, the sample elements are not equally likely. The probability of a high-volume purchaser is 0.2, and that of a nonhigh-volume purchaser is 0.8. Considering one sample element each for zero, one, two, three, and four high-volume purchasers, we have the following probabilities:

$$P(\bar{A}\,\bar{A}\,\bar{A}\,\bar{A}) = (0.8)(0.8)(0.8)(0.8) = (0.8)^4$$
$$P(A\,\bar{A}\,\bar{A}\,\bar{A}) = (0.2)(0.8)(0.8)(0.8) = (0.2)(0.8)^3$$
$$P(A\,A\,\bar{A}\,\bar{A}) = (0.2)(0.2)(0.8)(0.8) = (0.2)^2(0.8)^2$$
$$P(A\,A\,A\,\bar{A}) = (0.2)(0.2)(0.2)(0.8) = (0.2)^3(0.8)$$
$$P(A\,A\,A\,A) = (0.2)(0.2)(0.2)(0.2) = (0.2)^4$$

For any given value of the random variable "number of high-volume purchasers," each elementary event has the same probability. For example, in the case of one high-volume purchaser, each of the four elementary events (sample elements) has a probability of $(0.2)(0.8)^3$. If we multiply the specified probabilities for elementary events by the number of such elements in the composite events "no high-volume purchasers," "one high-volume purchaser," and so forth, we have

$$P(X = 0) = f(0) = 1(0.8)^4 = 0.4096$$
$$P(X = 1) = f(1) = 4(0.8)^3(0.2) = 0.4096$$

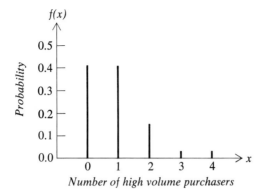

FIGURE 3-2
Graph of probability distribution of number of high-volume purchasers.

$$P(X = 2) = f(2) = 6(0.8)^2(0.2)^2 = 0.1536$$
$$P(X = 3) = f(3) = 4(0.8)(0.2)^3 = 0.0256$$
$$P(X = 4) = f(4) = 1(0.2)^4 = 0.0016$$

These values are summarized in Table 3-3 in the form of a probability distribution. Note that the probabilities add up to 1.

Actually, it would not have been necessary to list all the elements in the sample space in order to derive this probability distribution. A much briefer method for computing the required probabilities is explained in Section 3.4, where the binomial distribution is discussed.

A graph of this probability distribution is given in Figure 3-2.

EXAMPLE 3-3

Raymond Taylor, a corporation economist, develops a subjective probability distribution for the change that will take place in GNP during the following year. He establishes the following five categories of change and defines a random variable by letting the indicated numbers correspond to each category.

Change in GNP	Number Assigned
Down more than 5%	−2
Down 5% or less	−1
Unchanged	0
Up 5% or less	+1
Up more than 5%	+2

On the basis of all information available to him, Mr. Taylor assigns probabilities to each of these possible events as indicated in Table 3-4.

TABLE 3-4
Subjective Probability Distribution of Change in GNP

x	$f(x)$
-2	0.1
-1	0.1
0	0.2
$+1$	0.4
$+2$	0.2

Types of Random Variables

Random variables are classified as either *discrete* or *continuous*. A *discrete random variable* is one that can take on only a finite or countable number of distinct values. The three preceding examples illustrated probability distributions of discrete random variables.

A random variable is said to be *continuous* in a given range if it can assume any value in that range. The term "continuous random variable" implies that variation takes place along a *continuum*. Variables of this type can be *measured* to some degree of accuracy. Examples of continuous variables include weight, length, velocity, rate of production, dosage of a drug, and the length of life of a given product.

However, it may be argued that in the real world, all data are discrete. For example, if we want to measure weight and the measuring instrument only permits a determination to the nearest thousandth of a pound, then the resulting data will be discrete in units of thousandths of a pound. Despite this discreteness of data caused by limitations of measuring instruments, it is nevertheless useful in many instances to use mathematical models that treat certain variables as continuous. Furthermore, although measured data are essentially discrete in the real world, the variable under measurement is often continuous. Thus, if we use a continuous mathematical model of heights of individuals, where the underlying data are measured, or discrete, we may conceive of this model not as a convenient approximation, but rather as a model of reality that is more accurate than the discrete data from which the model was derived.

On the other hand, we often find it convenient to convert a variable that is conceptually continuous into a discrete one. Thus, in the case of heights of individuals, rather than using measurements along a continuous scale, we may set up classifications such as tall, medium, and short. In Example 3-3, GNP was treated as a discrete random variable with five distinct categories. Conceptually, it may be viewed as a continuous variable and is often so treated in econometric models.

It is sometimes said that one indication of progress in science is the extent to which discrete variables can be converted into continuous variables. Thus, the physicist treats color in terms of the continuous variable of wavelengths rather than the discrete classification of names of colors. Measurement of oral temperature by means of a thermometer treats human body temperature as varying along a continuous scale (although the resulting measurements are discrete), rather than as discrete (as when temperature is judged by placing a hand on the forehead of another person and classifying the temperature as "normal," "high," or in some other category). However, in applied problems, where a probability model is used to represent a real-world situation, we may work in terms of either discrete or continuous random variables, whichever appear to be most appropriate for the problem or decision-making situation in question. Only probability distributions of discrete random variables will be discussed in the remainder of this chapter.

Characteristics of Probability Distributions

In the three preceding examples, we saw that the sum of the probabilities in each probability distribution was equal to 1. It is possible to summarize the characteristics of probability distributions somewhat more formally.

A probability distribution of a discrete random variable X whose value at x is $f(x)$ possesses the following properties:[2]

1. $f(x) \geqslant 0$ for all real values of X

2. $\sum_{x} f(x) = 1$

Property (1) simply states that probabilities are greater than or equal to 0. The second property states that the sum of the probabilities in a probability distribution is equal to 1. The notation $\sum_{x} f(x)$ means "sum the values of $f(x)$ for all values that x takes on."

---

[2] A somewhat simplified notation is used here. A mathematically more elegant notation would represent the values that the random variable X could assume as $x_1, x_2, \ldots, x_n$ with associated probabilities $f(x_1), f(x_2), \ldots, f(x_n)$. Then the two properties would appear as

1. $f(x_i) \geqslant 0$ for all i

2. $\sum_{i=1}^{n} f(x_i) = 1$

If X takes on a countably infinite number of values, then the second property would appear as

$$\sum_{i=1}^{\infty} f(x_i) = 1$$

Other terms are also in use for probability distributions. "Probability function" and "probability distribution" are employed synonymously. Probability distributions of *discrete* random variables are often referred to as "probability mass functions," since the probabilities are "massed" at distinct points along the, say, x axis. The corresponding term for *continuous* random variables is "probability density function." The abbreviated terms "mass function" and "density function" are frequently used. We will ordinarily use the term "probability distribution" in referring to both discrete and continuous variables.

Cumulative Distribution Functions

Frequently, we are interested in the probability that a random variable is less than or equal to some specified value, or greater than a given value. The *cumulative distribution function* is particularly useful in this connection. We may define this function as follows.

Given a random variable X, the value of the cumulative distribution function at x, denoted $F(x)$, is *the probability that X takes on values less than or equal to x.* Hence,

(3.1) $$F(x) = P(X \leqslant x)$$

In the case of a discrete random variable, it is clear that

(3.2) $$F(c) = \sum_{x \leqslant c} f(x)$$

The symbol $\sum_{x \leqslant c} f(x)$ means "sum the values of $f(x)$ for all values of x less than or equal to c."

EXAMPLE 3-4

We return to Example 3-1, involving the probability distribution for the price of Chase Products common stock. The probability that the price would be \$33 or less is $P(X \leqslant \$33) = F(\$33) = 0.10$; \$34 or less, $P(X \leqslant \$34) = F(\$34) = 0.35$; and so on. In Table 3-5, the probability distribution and cumulative distribution function for the random variable "price of Chase Products common stock" are shown.

A graph of the cumulative distribution function is given in Figure 3-3. This graph is a *step function;* that is, the values change in discrete "steps" at the indicated integral values of the random variable, X. Thus, $F(x)$ takes the value 0 to the left of the point $x = \$33$, steps up to $F(x) = 0.10$ at $x = \$33$, and so on. The dot shown at the left of each horizontal line segment indicates the probability for the integral values of x. At these points, the values of the cumulative distribution function are read from the *upper* line segments.

TABLE 3-5
Probability Distribution and Cumulative Distribution Function for the
Price of Chase Products Common Stock

Price of Stock x	Probability f(x)	Cumulative Probability F(x)
$33	0.10	0.10
34	0.25	0.35
35	0.50	0.85
36	0.10	0.95
37	0.05	1.00

We note the following relations in this problem, which follow from the definition of a cumulative distribution function:

$$F(\$33) = f(\$33) = 0.10$$
$$F(\$34) = f(\$33) + f(\$34) = 0.10 + 0.25 = 0.35$$
$$F(\$35) = f(\$33) + f(\$34) + f(\$35) = 0.10 + 0.25 + 0.50 = 0.85$$
$$\text{and so on.}$$

The probabilities that the price would be more than $33, $34, and $35 are, respectively,

$$1 - F(\$33) = 0.90$$
$$1 - F(\$34) = 0.65$$
$$1 - F(\$35) = 0.15$$

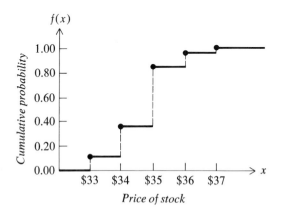

FIGURE 3-3
Graph of cumulative distribution function of the price of Chase Products common stock.

EXAMPLE 3-5

Let us return to Example 3-3, which discussed an economist's subjective probability distribution of change in GNP. A few questions will illustrate some uses of the cumulative distribution function.

What was the probability assigned by Mr. Taylor to

a. the event that the change in GNP will not exceed an increase of 5%?

$$F(1) = f(-2) + f(-1) + f(0) + f(1) = 0.8$$

or

$$F(1) = 1 - f(2) = 1 - 0.2 = 0.8$$

b. the event that GNP will not decline? The event "GNP will not decline" is the event "GNP will remain unchanged or will increase" and is thus the complement of the event "GNP will decrease." Thus, it is given by

$$1 - F(-1) = 1 - [f(-2) + f(-1)] = 1 - 0.2 = 0.8$$

or

$$1 - F(-1) = f(0) + f(1) + f(2) = 0.8$$

c. a change in GNP of 5% or less?

$$f(-1) + f(0) + f(1) = 0.7$$

or

$$F(1) - F(-2) = 0.8 - 0.1 = 0.7$$

Exercises

1. State whether the following random variables are discrete or continuous:
   a. Strength of a steel beam in pounds per inch
   b. The actual weight of a supposedly 16-ounce box of breakfast cereal
   c. X equals 0 if the weight of a supposedly 16-ounce box of cereal is less than 16 ounces and 1 if the weight is 16 ounces or more
   d. The number of defective batteries in a lot of 1000

2. The marketing department of a large firm is making a survey of consumers in a large metropolitan area to determine interest in a new product the firm is developing. The consumers to be surveyed are drawn from names in a telephone book serving the area, so we can assume that the population sampled is large enough for us to treat the sampling as having been done with replacement. It is known that 40% of the consumers listed in the telephone book in this area already use Miracle Worker, a similar product now sold by the firm. Assume that a sample of three names is drawn.
   a. Show the elements of the sample space for the drawing. Let M represent the occurrence of a Miracle Worker customer and $\bar{M}$ represent a noncustomer.

b. Determine the probability distribution of the random variable "number of Miracle Worker customers."

c. Graph this probability distribution.

3. Which of the following are valid probability functions?

a. $f(x) = \dfrac{x + 2}{10}$ $x = -1, 0, 1, 2$

b. $f(x) = \dfrac{x^2 - x}{2}$ $x = 0, 1, 2$

c. $f(x) = x/2$ $x = -1, 0, 1, 2$

d. $f(x) = \dfrac{x^2 + 1}{40}$ $x = 0, 1, \ldots, 4$

4. Find k such that the following are probability functions:

a. $\dfrac{k}{x^2}$ $x = 1, 2, 3$

b. kx $x = \frac{1}{4}, \frac{1}{2}, \frac{3}{4}$

c. $\dfrac{k}{x}$ $x = 1, 2, 3, 4$

5. A corporation economist, in building a model to predict a company's sales, developed the following categories of change for U.S. GNP:

> GNP down more than 4%
> GNP down 4% or less
> GNP unchanged
> GNP up 4% or less
> GNP up more than 4%

Let X be a random variable associated with change in GNP. Assign subjective probabilities to form a probability function based on your estimate of the change in this year's GNP as compared to last year's. Then graph the probability function and the cumulative probability function.

6. The probability distribution of sales of a new drug during the first month is as follows:

	x	$f(x)$
At least $15,000	2	0.2
At least $10,000 but less than $15,000	1	0.3
At least $5000 but less than $10,000	0	0.4
Less than $5000	−1	0.1

Find and graph the cumulative distribution.

7. Consider the following random variable for the lifetime of a particular machine:

> $X = 1$ if the machine lasts less than two years before wearing out
> $X = 2$ if the machine lasts at least two but less than three years

$X = 3$ if the machine lasts at least three but less than four years
$X = 4$ if the machine lasts at least four but less than five years
$X = 5$ if the machine lasts more than five years.

Let

$$F(x) = \frac{x^2}{25}$$

Find $f(x)$ in tabular form.

8. Let X be the number of minutes it takes to drain a soda filler of a particular flavor in order to change over to a new flavor. The probability distribution for X is

$$f(x) = \frac{x}{15} \qquad x = 1, 2, \ldots , 5$$

a. Prove that $f(x)$ is a probability function.
b. What is the probability that it will take exactly three minutes to drain the filler?
c. What is the probability that it will take at least two minutes but not more than four minutes?
d. Find the cumulative distribution (in tabular form).
e. What is the probability that the draining process will take at most three minutes?
f. What is the probability that it will take more than two minutes?

3.2 PROBABILITY DISTRIBUTIONS OF DISCRETE RANDOM VARIABLES

In many situations, it is useful to represent the probability distribution of a random variable by a general algebraic expression. Probability calculations can then be conveniently made by substituting appropriate values into the algebraic model. The mathematical expression is a compact summary of the process that has generated the probability distribution. Thus, the statement that a particular probability distribution is appropriate in a given situation contains a considerable amount of information concerning the nature of the underlying process. In the following sections, we discuss the uniform, binomial, multinomial, hypergeometric, and Poisson probability distributions of discrete random variables.

3.3 THE UNIFORM DISTRIBUTION

Sometimes, equal probabilities are assigned to all the possible values a random variable may assume. Such a probability distribution is referred to as a *uniform distribution*. For example, suppose a fair die is rolled

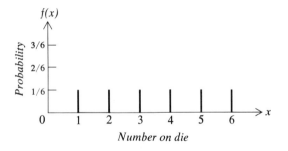

Number on die

FIGURE 3-4
Graph of probability mass function of numbers obtained in a roll of a fair die.

once. The probability is 1/6 that the die will show any given number on its uppermost face. The probability mass function in this case may be written as

$$f(x) = 1/6 \quad \text{for } x = 1, 2, \ldots, 6$$

A graph of this distribution is given in Figure 3-4.

As another illustration, let us consider the case of the Amgar Power Company, which produces electrical energy from geothermal steam fields. It takes about two years to build a production facility. In planning its production strategies, the company concludes that it is equally likely that demand two years hence will be 80,000 90,000, 100,000, 110,000, or 120,000 kilowatts. Therefore, the probability distribution established by Amgar Power Company for this future demand is

$$f(x) = 0.20 \quad \text{for } x = 80,000, 90,000, \ldots, 120,000$$

Later, we will examine how such information is used in decision-making procedures.

Exercises

1. Construct a situation in which the random variable can be said to have a uniform probability distribution.

2. In a study of the records of Southeast Airlines, it has been found that the actual arrival time of the scheduled 2 P.M. flight from Chicago, due at Kennedy Airport, New York at 6:30 P.M., is uniformly distributed by minutes in the range 6:05 P.M. to 7:40 P.M. Let $X = 1$ represent arrival at 6:05 P.M., $X = 2$ arrival at 6:06 P.M., and so on.
   a. Write out the mathematical expression for $f(x)$.
   b. What is the probability that the plane will be late?
   c. What is the probability that the plane will arrive after 7 P.M.?
   d. What is the probability the plane will arrive at or after 7 P.M.?
   e. What is the probability that the plane will arrive before 6:30 P.M.?

3. Let X denote the number of bricks a mason will lay in one hour, and assume that X is uniformly distributed in the range 150 to 200. If a certain project is 170 bricks short of completion and a second project is waiting to be started as soon as this one is finished,
   a. What is the probability that the mason will start the second project in the next hour?
   b. What is the probability that more than 25 bricks will have been laid on the second project at the end of the next hour?
   c. What is the probability that the first project will be more than ten bricks short of completion at the end of the hour?
   d. What is the probability that the mason will lay exactly 175 bricks during the next hour?

3.4   THE BINOMIAL DISTRIBUTION

The binomial distribution is undoubtedly the most widely applied probability distribution of a discrete random variable. It has been used to describe a large variety of processes in business and the social sciences as well as other areas. The type of process that gives rise to this distribution is usually referred to as a *Bernoulli trial* or a *Bernoulli process.*[3] The mathematical model for a Bernoulli process is developed from a very specific set of assumptions involving the concept of a series of experimental trials.

Let us envision a process or experiment characterized by repeated trials taking place under the following conditions:

1. On each trial, there are two mutually exclusive possible outcomes, which are referred to as "success" and "failure." In somewhat different language, the sample space of possible outcomes on each experimental trial is $S = \{$failure, success$\}$.

2. The probability of a success, denoted p, remains constant from trial to trial. The probability of a failure, denoted q, is equal to $1 - p$.

3. The trials are independent. That is, the outcomes on any given trial or sequence of trials do not affect the outcomes on subsequent trials.

The outcome on any specific trial is determined by chance. Processes having this characteristic are referred to as *random processes* or *stochastic processes*, and Bernoulli trials are one example of such processes.

Our aim is to develop a formula for the probability of x successes in n trials of a Bernoulli process. We start with a simple, specific case of a series of Bernoulli trials, five tosses of a coin. We calculate the probabil-

---

[3] Named after James Bernoulli (1654–1705), a member of a family of Swiss mathematicians and scientists, who did some of the early significant work on the binomial distribution.

ity of obtaining exactly two heads in five tosses; the resulting expression can then be generalized.

If we are tossing a fair coin five times, we may treat each toss as one Bernoulli trial. The possible outcomes on any particular trial are a head and a tail. Assume that the appearance of a head is a success. (Of course, the classification of one of the two possible outcomes as a "success" is completely arbitrary, and there is no necessary implication of desirability or goodness involved. For example, we may choose to refer to the appearance of a defective item in a production process as a success; or, if a series of births is treated as a Bernoulli process, the appearance of a female (male) may be classified as a success.) Suppose that the sequence of outcomes is

$$H\ T\ H\ T\ T$$

where H and T denote head and tail, as usual. We now introduce a convenient coding device for outcomes on Bernoulli trials. Let

$x_i = 0$ if the outcome on the ith trial is a failure, and
$x_i = 1$ if the outcome on the ith trial is a success.

Then the outcomes of the sequence of tosses given above may be written as

$$1\ 0\ 1\ 0\ 0$$

Since the probability of a success and a failure on a given trial are, respectively, p and q, the probability of this particular sequence of outcomes is, by the multiplication rule,

$$P(1, 0, 1, 0, 0) = pqpqq = q^3p^2$$

In this notation, for simplicity, commas have been used to separate the outcomes of the successive trials. Actually, though, this is the joint probability of the events that occurred on the five trials, that is, the probability of obtaining the specific sequence of successes and failures in the order in which they occurred. However, we are interested not in any specific order of results, but rather in the probability of obtaining a given number of successes in n trials. What then is the probability of obtaining exactly two successes in five Bernoulli trials? Nine other sequences satisfy the condition of exactly two successes in five trials:

$$1\ 1\ 0\ 0\ 0$$
$$1\ 0\ 0\ 1\ 0$$
$$1\ 0\ 0\ 0\ 1$$
$$0\ 1\ 1\ 0\ 0$$
$$0\ 1\ 0\ 1\ 0$$
$$0\ 1\ 0\ 0\ 1$$

$$0\ 0\ 1\ 1\ 0$$
$$0\ 0\ 1\ 0\ 1$$
$$0\ 0\ 0\ 1\ 1$$

By the same reasoning used earlier, each of these sequences has the same probability, q^3p^2. We can obtain the number of such sequences from the formula for the number of combinations of n objects taken x at a time given in Equation 2.20. Thus, the number of possible sequences in which two 1s can occur is $\binom{5}{2}$. We indicated in Equation 2.20 that

$$\binom{n}{x} = \frac{n!}{x!(n-x)!}$$

Thus,

$$\binom{5}{2} = \frac{5!}{2!3!} = 10$$

and we may write

$$P(\text{exactly 2 successes}) = \binom{5}{2} q^3p^2$$

In the case of the fair coin example, we assign a probability of $\frac{1}{2}$ to p and $\frac{1}{2}$ to q. Hence,

$$P(\text{exactly 2 heads}) = \binom{5}{2}\left(\frac{1}{2}\right)^3\left(\frac{1}{2}\right)^2 = \frac{10}{32} = \frac{5}{16}$$

This result may be generalized to obtain the probability of (exactly) x successes in n trials of a Bernoulli process. Let us assume $n - x$ failures occurred followed by x successes, in that order. We may then represent this sequence as

$$\underbrace{0\ 0\ 0\ \cdots\ 0}_{\substack{n-x \\ \text{failures}}}\ \underbrace{1\ 1\ 1\ \cdots\ 1}_{\substack{x \\ \text{successes}}}$$

The probability of this particular sequence is $q^{n-x}p^x$. The number of possible sequences of n trials resulting in exactly x successes is $\binom{n}{x}$.[4]

---

[4] Because the combination notation is universally used in connection with the binomial probability distribution, that convention is followed here. However, conceptually, we have here the number of distinct permutations that can be formed of n objects, $n - x$ of which are of one type and x of the other. Since the number of such permutations turns out to be equal to $\binom{n}{x}$, the combination notation may be used instead of that for permutations.

Therefore, the probability of obtaining x successes in n trials of a Bernoulli process is given by[5]

(3.3) $$f(x) = \binom{n}{x} q^{n-x} p^x \quad \text{for } x = 0, 1, 2, \ldots, n$$

If we denote by X the random variable "number of successes in these n trials," then

$$f(x) = P(X = x)$$

The fact that this is a probability distribution is verified by noting that (1) $f(x) \geqslant 0$ for all real values of x and (2) $\sum_x f(x) = 1$. The first condition is verified by noting that since p and n are nonnegative numbers, $f(x)$ cannot be negative. The second condition is true because (as can be shown mathematically)

$$\sum_x \binom{n}{x} q^{n-x} p^x = (q + p)^n = 1^n = 1$$

Therefore, the term "binomial probability distribution," or simply "binomial distribution," is usually used to refer to the probability distribution resulting from a Bernoulli process. In summary, in problems where the assumptions of a Bernoulli process are met, we can obtain the probabilities of zero, one, and so forth successes in n trials from the respective terms of the binomial expansion of $(q + p)^n$, where q and p denote the probabilities of failure and success on a single trial and n is the number of trials.

The binomial distribution has two parameters,[6] n and p. Each pair of values for these parameters establishes a different distribution. Thus, the binomial distribution is actually a family of probability distributions. Since computations become laborious for large values of n, it is advisable to make use of special tables. Selected values of the binomial cumulative distribution function are given in Table A-1 of Appendix A.

---

[5] The following method of writing the mathematical expression for such a probability distribution is often used:

$$f(x) = \binom{n}{x} q^{n-x} p^x \text{ for } x = 0, 1, 2, \ldots, n$$

$$= 0, \text{ elsewhere}$$

In this and other places where it is clear that $f(x)$ is equal to 0 for values of the random variable other than the specified ones, the notation on the last line will not be included.

[6] In this context, the term "parameters" refers to numerical quantities that are sufficient to specify a probability distribution. When particular values are assigned to the parameters of a probability function, a specific distribution in the family of possible distributions is defined. For example, $n = 10$, $p = \frac{1}{2}$ specifies a particular binomial distribution; $n = 20$, $p = \frac{1}{2}$ specifies another.

The values of

$$F(c) = P(X \leq c) = \sum_{x \leq c} f(x) \quad \text{for } x = 0, 1, 2, \ldots, n$$

are shown in that table for $n = 2$ to $n = 20$ and $p = 0.05$ to $p = 0.50$ in multiples of 0.05. Values of $f(c)$, cumulative probabilities for p values greater than 0.50, and probabilities that x is greater than a given value or lies between two values can be obtained by appropriate manipulation of these tabulated values. Some of the examples that follow illustrate the use of the table. More extensive tables have been published by the National Bureau of Standards and Harvard University, but even such tables usually do not go beyond $n = 50$ or $n = 100$. For large values of n, approximations are available for the binomial distribution, and the exact values generally need not be determined.

It is important to note that in the case of the binomial distribution, as with any other mathematical model, the correspondence between the real-world situation and the model must be carefully established. In many cases, the underlying assumptions of a Bernoulli process are not met. For example, suppose that in a production process, items produced by a certain machine tool are tested as to whether they meet specifications. If the items are tested in the order in which they are produced, then the assumption of independence would doubtless be violated. That is, whether an item meets specifications would not be independent of whether the preceding item(s) did. If the machine tool had become subject to wear, it is quite likely that if it produced an item that did not meet specifications, the next item would fail to conform to specifications in a similar way. Thus, whether or not an item is defective would *depend* on the characteristics of preceding items. In the coin-tossing illustration, on the other hand, we imagined an experiment in which a head or tail on a particular toss did not affect the outcome on the next toss.

It can be seen from the assumptions underlying a Bernoulli process that the binomial distribution is applicable to the situations of *sampling from a finite population with replacement* or *sampling from an infinite population* with or without replacement. In either of these cases, the probability of success may be viewed as remaining constant from trial to trial. If the population size is large relative to sample size, that is, if the sample constitutes only a small fraction of the population, and if p is not very close to 0 or to 1 in value, the binomial distribution is often sufficiently accurate, even though sampling may be carried out from a finite population without replacement. It is difficult to give universal rules of thumb on appropriate ratios of population size to sample size for this purpose. Some practitioners suggest a population size at least ten times the sample size. However, the purpose of the calculations must determine the required degree of accuracy.

EXAMPLE 3-5

The tossing of a fair coin five times was used earlier as an example of a Bernoulli process; the probability of obtaining two heads (successes) was calculated. Compute the probabilities of all possible numbers of heads and thus establish the particular binomial distribution that is appropriate in this case.

SOLUTION

This problem is an application of the binomial distribution for $p = \frac{1}{2}$ and $n = 5$. Letting X represent the random variable "number of heads," the probability distribution is as follows:

x	$f(x)$
0	$\binom{5}{0}\left(\frac{1}{2}\right)^5\left(\frac{1}{2}\right)^0 = \frac{1}{32}$
1	$\binom{5}{1}\left(\frac{1}{2}\right)^4\left(\frac{1}{2}\right)^1 = \frac{5}{32}$
2	$\binom{5}{2}\left(\frac{1}{2}\right)^3\left(\frac{1}{2}\right)^2 = \frac{10}{32}$
3	$\binom{5}{3}\left(\frac{1}{2}\right)^2\left(\frac{1}{2}\right)^3 = \frac{10}{32}$
4	$\binom{5}{4}\left(\frac{1}{2}\right)^1\left(\frac{1}{2}\right)^4 = \frac{5}{32}$
5	$\binom{5}{5}\left(\frac{1}{2}\right)^0\left(\frac{1}{2}\right)^5 = \frac{1}{32}$
	$\overline{1}$

EXAMPLE 3-6

Calculate the probability of obtaining at least one 6 in two rolls of a die (or in one roll of two dice) using the binomial distribution.

SOLUTION

We view the two rolls of the die as Bernoulli trials. If we define the appearance of a 6 as a success, $p = \frac{1}{6}$, $q = \frac{5}{6}$, and $n = 2$. It is instructive to examine the entire probability distribution.

x	$f(x)$
0	$\binom{2}{0}\left(\frac{5}{6}\right)^2\left(\frac{1}{6}\right)^0 = \left(\frac{5}{6}\right)^2$

(Continued)

x	$f(x)$
1	$\binom{2}{1}\left(\frac{5}{6}\right)^1\left(\frac{1}{6}\right)^1 = 2\left(\frac{5}{6}\right)\left(\frac{1}{6}\right)$
2	$\binom{2}{2}\left(\frac{5}{6}\right)^0\left(\frac{1}{6}\right)^2 = \left(\frac{1}{6}\right)^2$
	$\overline{1}$

The expressions at the right side of the $f(x)$ column are in the form with which the student is probably most familiar for the terms in the expansion of $(\frac{5}{6} + \frac{1}{6})^2$.

The required probability is

$$P\left(\begin{array}{c}\text{at least}\\ \text{one 6}\end{array}\right) = f(1) + f(2) = 2(\tfrac{5}{6})(\tfrac{1}{6}) + (\tfrac{1}{6})^2 = \tfrac{11}{36}$$

EXAMPLE 3-7

An interesting correspondence took place in 1693 between Samuel Pepys, author of the famous *Diary*, and Isaac Newton, in which Pepys posed a probability problem to the eminent mathematician. The question as originally stated by Pepys was:[7]

A has six dice in a box, with which he is to fling a six
B has in another box 12 dice, with which he is to fling two sixes
C has in another box 18 dice, with which he is to fling three sixes
(Question)—Whether B and C have not as easy a task as A at even luck?

In rather flowery seventeenth century English, Newton replied and said, essentially, "Sam, I do not understand your question." Newton asked whether individuals A, B, and C were to throw independently and whether the question pertained to the obtaining of *exactly* one, two, or three sixes or *at least* one, two, or three sixes.

After an exchange of letters, in which Pepys supplied little help in answering these queries, Newton decided to frame the question himself. In modern language, Newton's wording would appear somewhat as follows:

If A, B, and C toss dice independently, what are the probabilities that:
A will obtain at least one 6 in a roll of six dice?
B will obtain at least two 6s in a roll of twelve dice?
C will obtain at least three 6s in a roll of eighteen dice?

Newton's reply to these questions involved some rather tortuous arith-

---

[7] Schell, Emil D., "Samuel Pepys, Isaac Newton and Probability," *The American Statistician*, October 1960, pp. 27–30.

metic. His work doubtless represented a very respectable intellectual feat, considering the infantile state of probability theory at that time. Today, almost any beginning student of probability theory, standing on the shoulders of the giants who came before, would immediately see the application of the binomial distribution to the problem. Let us denote by $P(A)$, $P(B)$, and $P(C)$ the probabilities that A, B, and C would obtain the specified events. Then

$$P(A) = 1 - \binom{6}{0} (5/6)^6 (1/6)^0 \approx 0.67$$

$$P(B) = 1 - \binom{12}{0} (5/6)^{12}(1/6)^0 - \binom{12}{1} (5/6)^{11}(1/6)^1 \approx 0.62$$

$$P(C) = 1 - \binom{18}{0} (5/6)^{18}(1/6)^0 - \binom{18}{1} (5/6)^{17}(1/6)^1$$

$$- \binom{18}{2} (5/6)^{16}(1/6)^2 \approx 0.60$$

Thus, $P(A) > P(B) > P(C)$.

Pepys admitted frankly that he did not understand Newton's calculations and furthermore that he didn't believe the answer. He argued that since B throws twice as many dice as A, why can't he simply be considered two A's? Thus, he would have at least as great a probability of success as A. Of course, Pepys' question indicated that he was rather confused. There is no reason why the probability of at least two 6s in a roll of twelve dice should be twice the probability of at least one 6 in a roll of six dice, and, as seen by the above calculations, indeed it is not.

EXAMPLE 3-8

A project manager has determined that a certain subcontractor fails to deliver certain standard orders on schedule about 25% of the time. If this situation is viewed as a Bernoulli process, determine from Table A-1 of Appendix A the probabilities that in ten orders the subcontractor

a. will fail to deliver three or fewer orders on time
b. will fail to deliver between three and five (inclusive) orders on time
c. will deliver three or more orders on time
d. will deliver at most eight orders on time
e. will fail to deliver exactly two orders on time
f. will fail to deliver seven or more orders on time

SOLUTION

Let $p = 0.25$ stand for the probability that an order will *not* be delivered on time. (We define the probability of a success this way because Table A-1 gives p values only up to $p = .50$.) Then $q = 0.75$ and $n = 10$. X represents the number of orders not delivered on time. Note that a failure to deliver an order on time is considered a "success" in this problem despite the undesirability of this outcome.

a. $P(X \leq 3) = F(3) = \sum_{x=0}^{3} \binom{10}{x} (.75)^{10-x} (.25)^x$. From Table A-1 of Ap-

pendix A, $F(3) = 0.7759$

b. The probability of obtaining three, four, or five successes is given by the difference between "five or fewer successes" and "two or fewer successes." Thus,

$$P(3 \leq X \leq 5) = F(5) - F(2) = 0.9803 - 0.5256 = 0.4547$$

c. The event "three or more failures" is the same as the event "seven or fewer successes." Hence,

$$P(X \leq 7) = F(7) = 0.9996$$

d. The event "at most eight failures" is the same as "eight or fewer failures" or "two or more successes."

$$P(2 \leq X \leq 10) = F(10) - F(1) = 1.0000 - 0.2440 = 0.7560$$

e. The probability of "exactly two successes" is given by the difference between the probabilities of "two or fewer successes" and "one or fewer successes."

$$P(X = 2) = F(2) - F(1) = 0.5256 - 0.2440 = 0.2816$$

f. "Seven or more successes" is the complement of the event "six or fewer successes." Therefore,

$$P(X \geq 7) = 1 - P(X \leq 6) = 1 - F(6) = 1 - 0.9965 = 0.0035$$

An important property of the binomial distribution is that when $p = 0.50$, the distribution is symmetrical. For example, see Figure 3-5, where $p = 0.50$ and $n = 10$. When $p \neq 0.50$, the distribution is asym-

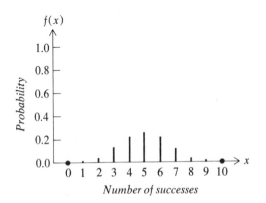

Number of successes

FIGURE 3-5
Graph of binomial distribution for $p = 0.5$, $n = 10$.

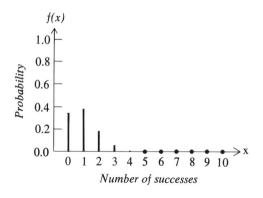

FIGURE 3-6
Graph of binomial distribution for $p = 0.1$, $n = 10$.

metrical (skewed). This property is illustrated in Figures 3-6 and 3-7, where the binomial distributions for $p = 0.1$, $n = 10$ and for $p = 0.9$, $n = 10$ are plotted.   •

Exercises

1. A quality control inspector is concerned about the number of defective items being produced by three (identical) machines. It is assumed that "in the long run" all three produce the same proportion, p, of defective items. Ten items are made on each machine, representing independent trials.
   a. What is the probability distribution of the number of defective items produced on the first machine?
   b. What is the probability distribution of the total number of defective items produced?
2. A certain delicate manufacturing process produces 25% defective items. In

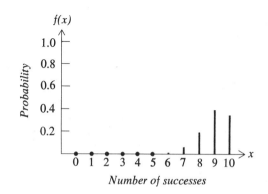

FIGURE 3-7
Graph of binomial distribution for $p = 0.9$, $n = 10$.

testing a new process, a sample of 50 items is produced. The new process will be installed if the sample yields ten or fewer defective items. Assume that the production of 50 items corresponds to 50 independent trials of the process. Write a symbolic expression, specifying all numerical values involved, for the exact probability of ten or fewer defective items, if the new process also produces 25% defective items on the average. Do not carry out the arithmetic.

3. A firm bills its accounts at a 1% discount for payment within ten days and for the full amount due after ten days. In the past, 30% of all invoices have been paid within ten days. Assuming independence of payments, if the firm sends out eight invoices during the first week of July, what is the probability that
   a. no one receives the discount?
   b. everyone receives the discount?
   c. three receive the discount?
   d. at least three receive the discount?

4. A certain portfolio consists of ten stocks. The investor feels that the probability of any one stock going down in price is 0.45 and that the price movements of the stocks are independent. What is the probability that exactly five will decline? That five or more will decline? Does the assumption of independence here seem logical? If not, is the binomial distribution the appropriate probability distribution for this problem?

5. In the past, a certain type of plastic bag has burst under 15 pounds of pressure 20% of the time. If a prospective buyer tests five bags chosen at random, what is the probability that exactly one will burst? Assume independence.

6. A manufacturer of 60-second development film advertises that 95 out of 100 prints will develop. Suppose you buy a roll of twelve prints and find that two do not develop. If the manufacturer's claim is true, what is the probability that two or more prints will not develop? Assume independence.

7. Two employees of a company were arguing whether a large barrel of widgets had been produced by a particular machine. The first employee, L. B. Jones, said he was sure it could not have come from that machine, for he had taken a random sample of 20 widgets from the barrel and found them all perfect. Since it is known that the machine in question produces 10% defective widgets, Jones stated that it was almost certain that a random sample of 20 items would contain at least one defective one, in view of the fact that a binomial distribution was applicable. Do you agree with Jones? Use figures to support your position.

8. An oil exploration firm is formed with enough capital to finance nine ventures. The probability of any exploration being successful is 0.15. Assuming independence, what are the firm's chances of
   a. exactly one successful exploration?
   b. at least one successful exploration?
   c. going bankrupt in the nine ventures?

9. A fifteen-question true–false examination, on which each correct answer is

worth $6\frac{2}{3}$ points, is given. If a student guesses randomly on each question, what is the probability of
a. scoring 60 or above if nothing is deducted for wrong answers?
b. scoring 60 or above if $6\frac{2}{3}$ points are deducted for each wrong answer? (Possible scores thus range from -100 to $+100$.)
c. scoring less than 0 if $6\frac{2}{3}$ points are deducted for each wrong answer?

10. The sales manager of a small electronics firm has just hired a new sales representative for the firm's line of calculators. From experience, the manager knows that an average sales representative will make one sale for every five customers approached, an above average one will make one sale in four attempts, and a below average one will have a sales ratio of $1:10$. In 16 attempts, the new employee makes one sale. Assuming independence, what is the probability of this event if
a. the employee is below average?
b. the employee is average?
c. the employee is above average?

11. Used-car dealer Laura Nielson claims that the odds are $3:1$ against her selling a car to any particular customer. If she attempts to sell automobiles to nine customers on a given day, what is the probability that she will make at least two sales? Assume independence.

12. An electronics engineer estimates that 25% of the components in a certain batch are defective and will fail when tested. Assuming independence, what is the probability that a device requiring three of these components will work properly when tested if
a. all three are required for operation of the device?
b. two of the three are backup units (that is, only one component need work properly for the device to operate)?

13. Solomon Horowitz, a small manufacturer of electrical wire, submits a bid on each of four different government contracts. In the past, his bid has been the low bid (i.e., he was awarded the contract) 15% of the time. Assuming that this relative frequency is a correct probability assignment and that the decisions on the four contracts are independent, what is the probability that the firm will not obtain any of the four contracts?

14. An accountant makes mistakes on 35% of the tax returns he prepares. What is the probability that he makes no mistakes on the first six returns he prepares for 1977? Assume independence.

15. Graph the binomial distributions for $n = 11$ and $p = 0.2, 0.5$, and 0.8. What can be said about skewness of the binomial distribution as the value of p departs from 0.5?

16. Sales representative Marcia Gomez plans to visit ten plant managers to try to sell an innovative product her company has developed. She estimates that her chance of landing a sale with any one plant manager is 0.6. Assuming independence, what is the probability
a. that she will meet her quota of seven sales?
b. that she will make no sales?
c. that she will make ten sales?

3.5 THE MULTINOMIAL DISTRIBUTION

In the case of the binomial distribution, there were two possible outcomes on each experimental trial. The multinomial distribution represents a straightforward generalization of the binomial distribution for the situation where there are more than two possible outcomes on each trial.

The assumptions underlying the multinomial distribution are completely analogous to those of the binomial distribution.

1. On each trial, there are k mutually exclusive possible outcomes, which may be referred to as $E_1, E_2, \ldots, E_k$. Therefore, the sample space of possible outcomes on each trial is $S = \{E_1, E_2, \ldots, E_k\}$.
2. The probabilities of outcomes $E_1, E_2, \ldots, E_k$, denoted $p_1, p_2, \ldots, p_k$, remain constant from trial to trial.
3. The trials are independent.

Under these assumptions, the probability that there will be x_1 occurrences of E_1, x_2 occurrences of E_2, . . . , x_k occurrences of E_k in n trials is given by

(3.4) $$f(x_1, x_2, \ldots, x_k) = \frac{n!}{x_1! \, x_2! \cdots x_k!} \, p_1^{x_1} p_2^{x_2} \cdots p_k^{x_k}$$

where $x_1 + x_2 + \cdots + x_k = n$ and $p_1 + p_2 + \cdots + p_k = 1$.

The expression $f(x_1 \, x_2, \ldots, x_n)$ is the general term of the multinomial distribution

$$(p_1 + p_2 + \cdots + p_k)^n$$

Analogously, in the binomial distribution, the probability of x successes in n trials is given by

$$f(x) = \frac{n!}{x! \, (n-x)!} \, q^{n-x} p^x$$

which is the general term of $(q + p)^n$. In terminology similar to that of the multinomial distribution, $(q + p)^n$ may be written

$$(p_1 + p_2)^n$$

where $p_1 + p_2 = 1$.

EXAMPLE 3-9

If eight dice are rolled, what is the probability that there will be three 1s, two 2s, no 3s, two 4s, no 5s, and one 6?

SOLUTION

Applying the multinomial distribution, we find that this probability is

$$f(3, 2, 0, 2, 0, 1)$$

$$= \frac{8!}{3! \ 2! \ 0! \ 2! \ 0! \ 1!} \left(\frac{1}{6}\right)^3 \left(\frac{1}{6}\right)^2 \left(\frac{1}{6}\right)^0 \left(\frac{1}{6}\right)^2 \left(\frac{1}{6}\right)^0 \left(\frac{1}{6}\right)^1$$

EXAMPLE 3-10

The Taylor Distributing Company has classified its accounts receivable into three categories, A, B, and C. There were 40% type A, 40% type B, and 20% type C accounts. A sample of three accounts was drawn at random from the entire list of accounts receivable. Give the probability distribution of the numbers of each type of account in the sample, assuming that the multinomial distribution is applicable.

SOLUTION

If x_1, x_2, and x_3 stand for the numbers of type A, B, and C accounts, then the appropriate multinomial distribution is

$$f(x_1, x_2, x_3) = \frac{3!}{x_1! \ x_2! \ x_3!} (0.40)^{x_1} (0.40)^{x_2} (0.20)^{x_3}$$

$$x_1 + x_2 + x_3 = 3 \quad \text{and} \quad x_i = 0, 1, 2, 3$$

This probability distribution is given in Table 3-6.

TABLE 3-6
Multinomial Distribution for Numbers of Type
A, B, and C Accounts Receivable; $p_1 = 0.40$,
$p_2 = 0.40$, $p_3 = 0.20$, and $n = 3$

(x_1, x_2, x_3)	$f(x_1, x_2, x_3)$
3, 0, 0	0.064
0, 3, 0	0.064
0, 0, 3	0.008
2, 1, 0	0.192
2, 0, 1	0.096
1, 2, 0	0.192
1, 1, 1	0.192
1, 0, 2	0.048
0, 2, 1	0.096
0, 1, 2	0.048
	1.000

Exercises

1. A quality control system installed by Electronic Components, Inc. attempts to reduce loss due to production of defective items by categorizing components as good or defective, with subcategories describing the degree of defectiveness. The following probabilities apply to this scheme:

X	P(X)
1 = good	0.75
2 = defective; recheck	0.10
3 = defective but repairable	0.03
4 = defective but with salvage value	0.07
5 = defective; no value	0.05

   a. Assuming independence, if eight parts are selected and tested on a given morning, what is the probability of obtaining four good components and one of each type of defective?
   b. On a given morning, what is the probability that four of the eight items tested are good?
   c. What is the name of the distribution used in (a) and (b)?

2. Wilbur Lobianco, a highly successful investment counselor, is reviewing the portfolios of his five most important clients. He has decided to add a stock to each portfolio and will make his choices with the use of his most successful technique. Hanging on the wall of his office is a circular target divided into five zones corresponding to his five favorite stocks. The area of each zone as a percentage of the total area of the target is as follows:

Stock	Percentage of Area
A	10
B	15
C	20
D	25
E	30

   By throwing five appropriately labeled darts at the target randomly, Mr. Lobianco will choose a stock for each portfolio. Assuming that the throws are independent and that all the darts hit the target,
   a. What is the probability that each stock will be chosen for one portfolio? (Each dart lands in a different zone.)
   b. What is the probability that stock A will be chosen for two portfolios, stocks B, C, and D for one portfolio each, and stock E for none?
   c. What is the probability that stock C will be chosen for all five portfolios?
   d. What is the probability that exactly three portfolios will acquire stock E?

3. A stationery store orders 6 reams of paper from a large paper manufacturer. The order is filled randomly from a large warehouse stock consisting of 1,000,000 reams of which 50% are high-grade, 30% are medium-grade, and 20% are low-grade paper. Assuming that the selections can be treated as having been carried out with replacement, what is the probability that the stationery store will receive 2 reams of each grade of paper?

3.6   THE HYPERGEOMETRIC DISTRIBUTION

In Section 3.4, the binomial distribution was discussed as the appropriate probability distribution for situations in which the underlying assumptions of a Bernoulli process were met. A major application of the binomial distribution was seen to be the computation of probabilities for the *sampling of finite populations with replacement.* In most practical situations, sampling is carried out *without replacement.* For example, if a sample of families is selected in a city in order to estimate the average income of all families in the city, sampling units are ordinarily not replaced prior to the selection of subsequent ones. That is, families are not replaced in the original population and thus given an opportunity to appear more than once in the sample. In fact, such samples are usually drawn in a single operation, without any possibility of drawing the same family twice. Moreover, in a sample drawn from a production process, articles are generally not replaced and given an opportunity to reappear in the sample. Thus, for both human populations and universes of physical objects, sampling is ordinarily carried out without replacement. In this section, we discuss the hypergeometric distribution as the appropriate model for sampling without replacement.

Suppose we have a list of 1000 persons, 950 of whom are adults and 50 of whom are children. Numbers from 1 to 1000 are assigned to these individuals. These numbers are printed on 1000 identical discs, which are placed in a large bowl. A sample of five chips is drawn at random from the bowl *without replacement.* These five chips may be drawn simultaneously or successively. For simplicity, we shall refer to this situation as the drawing of a random sample of five persons from the group of 1000, although, of course, the chips rather than persons are sampled. The process of numbering chips to correspond to persons and sampling the chips is simply a device to insure randomness in sampling the population of 1000 persons.

What is the probability that none of the five persons in the sample is a child? An alternative way of wording the question is, "What is the probability that all five persons in the sample are adults?" The sample space of possible outcomes in this experiment is the total number of samples of five persons that can be drawn from the population of 1000 persons. This is the number of combinations that can be formed of 1000 objects taken five at a time, $\binom{1000}{5}$. We can compute the required probability by obtaining the ratio of the number of sample points favorable to the event "none of the five persons is a child" to the total number of points in the sample space. The number of ways five adults can be drawn from the 950 adults is $\binom{950}{5}$. The number of ways no children can be selected from the 50 children is $\binom{50}{0}$. Therefore, the total number of ways of selecting five adults and no children from the population of 950 adults and 50 children is $\binom{950}{5}\binom{50}{0}$, by the multiplication principle.

Hence, the probability of obtaining no children (and five adults) in this sample of five persons is

$$\frac{\binom{950}{5}\binom{50}{0}}{\binom{1000}{5}}$$

If we carry out the arithmetic, we see a very interesting fact.

$$\frac{\binom{950}{5}\binom{50}{0}}{\binom{1000}{5}} = \frac{\dfrac{950!}{5!\,945!} \times \dfrac{50!}{0!\,50!}}{\dfrac{1000!}{5!\,995!}}$$

$$= \frac{950 \times 949 \times 948 \times 947 \times 946}{1000 \times 999 \times 998 \times 997 \times 996} = 0.7734$$

Grouping the product obtained as a multiplication of five factors, we have

$$P(\text{no children}) = \left(\frac{950}{1000}\right)\left(\frac{949}{999}\right)\left(\frac{948}{998}\right)\left(\frac{947}{997}\right)\left(\frac{946}{996}\right)$$

which is the result we would have arrived at if we had simply solved the original problem in terms of conditional probabilities. That is, the probability of obtaining an adult on the first draw is 950/1000; the probability of obtaining an adult on the second draw given that an adult was obtained on the first draw is 949/999, and so on. Therefore, the joint probability of obtaining no children (five adults) if the sampling is carried out without replacement is given by the multiplication of the five factors shown.

We can now state the general nature of this type of problem and the hypergeometric distribution as a solution to it. Suppose there is a population containing N elements, X of which are of one type, termed "successes," $N - X$ of another type, denoted "failures." The corresponding terminology for a random sample of n elements drawn without replacement is that we require x successes and therefore $n - x$ failures. The data of this general problem are tabulated below:

Population	*Required Sample*
X = number of successes[8]	x = number of successes
$N - X$ = number of failures	$n - x$ = number of failures
N = total number in population	n = total number in sample

Then the *hypergeometric distribution*, which gives the probability of x successes in a random sample of n elements drawn *without replacement*, is

$$(3.5) \qquad f(x) = \frac{\binom{N - X}{n - x}\binom{X}{x}}{\binom{N}{n}} \quad \text{for } x = 0, 1, 2, \ldots, [n, X]$$

where the symbol $[n, X]$ means the smaller of n and X. For example, in the illustration given above, if there had been only ten children in the population (X) and the sample size had been 50 (n), the largest value that the number of children in the sample (x) could take on would be ten (X). On the other hand, if X exceeded n, clearly x could be as large as n.

The hypergeometric distribution bears a very interesting relationship to the binomial distribution. Suppose, in the case of the population containing 950 adults and 50 children, we had been interested in the same probability (of obtaining no children in a random sample of five persons) but the sample was randomly drawn *with replacement*. Then, letting $q = 0.95$, $p = 0.05$, and $n = 5$ in the binomial distribution, we have

$$f(0) = \binom{5}{0}(0.95)^5(0.05)^0 = (0.95)^5 = 0.7738$$

Just as in the case of the hypergeometric distribution, where the required probability could have been computed by using the multiplication rule for *dependent* events, here in the case of the binomial distribu-

---

[8] Note that in order to maintain parallel notation for the population and sample in this case, the symbol X does *not* denote the *random variable* for number of successes in the sample, but is instead the total number of successes in the population being sampled.

tion, the probability could have been computed by simply using the multiplication rule for *independent* events

$$P(\text{no children}) = \left(\frac{950}{1000}\right)\left(\frac{950}{1000}\right)\left(\frac{950}{1000}\right)\left(\frac{950}{1000}\right)\left(\frac{950}{1000}\right)$$

It may be noted that the hypergeometric and binomial probability values are extremely close in this illustration, agreeing exactly in the first three decimal places. It can be shown that when N increases without limit, the hypergeometric distribution approaches the binomial distribution. Therefore, the binomial probabilities may be used as approximations to hypergeometric probabilities when n/N is small. A frequently used rule of thumb is that the population size should be at least ten times the sample size ($N > 10n$) for the approximations to be used. However, the governing considerations, as usual, include such matters as the purpose of the calculations, whether a sum of terms rather than a single term is being approximated, and whether terms near the center or the extremes of the distribution are involved.

Just as the multinomial distribution represents the generalization of the binomial distribution when there are more than two possible classifications of outcomes, the hypergeometric distribution can be similarly extended. No special name is given to the more general distribution; it also is referred to as the hypergeometric distribution. Assume a population that contains N elements, X_1 of type one, X_2 of type two, . . . , X_k of type k, and suppose we require, in a sample of n elements drawn without replacement, that there be x_1 elements of type one, x_2 of type two, . . . , x_k of type k. Tabulating the data in an analogous fashion to the two-outcome case, we have

Population (Number of Elements)	*Required Sample* (Number of Elements)
X_1 of type one	x_1 of type one
X_2 of type two	x_2 of type two
.	.
.	.
.	.
X_k of type k	x_k of type k

The *hypergeometric distribution*, which gives the probability of obtaining x_1 occurrences of type one, x_2 occurrences of type two, . . . , x_k occurrences of type k in a random sample of n elements drawn without replacement, is

(3.6)

$$f(x_1, x_2, \ldots, x_k) = \frac{\binom{X_1}{x_1}\binom{X_2}{x_2} \cdots \binom{X_k}{x_k}}{\binom{N}{n}}$$

$$\text{for } x_i = 0, 1, 2, \ldots, [n, X_i]$$

where

$$\sum_{i=1}^{k} X_i = N \quad \text{and} \quad \sum_{i=1}^{k} x_i = n$$

EXAMPLE 3-11

The Humblest Oil Corporation has 100 service stations in a certain community; it has classified them according to merit of geographic location as follows:

Merit of Location	Number of Stations
Excellent	22
Good	38
Fair	27
Poor	10
Disastrous	3
	100

The corporation has a computer program for drawing random samples (without replacement) of its service stations. In a random sample of 20 of these stations, what is the joint probability of obtaining six excellent, six good, four fair, three poor, and one disastrous station?

SOLUTION

$$f(6, 6, 4, 3, 1) = \frac{\binom{22}{6}\binom{38}{6}\binom{27}{4}\binom{10}{3}\binom{3}{1}}{\binom{100}{20}}$$

EXAMPLE 3-12

What is the probability that in a hand of cards dealt in a bridge game from a well shuffled deck you will have either all spades or all clubs?

SOLUTION

There are $\binom{52}{13}$ possible bridge hands. Using the hypergeometric distribution and the addition rule, the required probability is

$$\frac{\binom{13}{13}\binom{13}{0}\binom{13}{0}\binom{13}{0}}{\binom{52}{13}} + \frac{\binom{13}{0}\binom{13}{13}\binom{13}{0}\binom{13}{0}}{\binom{52}{13}}$$

$$= \frac{1}{\binom{52}{13}} + \frac{1}{\binom{52}{13}} \approx \frac{2}{635 \text{ billion}}$$

If you are dealt such a hand, it is fair to say you have observed a rare event.

Exercises

1. Certain machine parts are shipped in lots of 15 items. Four parts are selected randomly from each lot and tested, and the lot is considered acceptable if no defectives are among the four tested. What is the probability that a lot containing five defectives will be accepted?

2. In a small Southern town, there are 800 white and 400 nonwhite citizens registered to vote. Trial juries of 12 people are supposedly selected at random from the list of registered voters, with the possibility that a person can serve more than once. In the past month, there have been five trials, and the composition of each jury was 11 white citizens and one nonwhite citizen. What would be the probability of this occurrence if juries were actually selected at random? Write a mathematical expression that will yield the required probability when evaluated. You need not carry out the arithmetic.

3. A hardware store has ten accounts receivable with open balances. Of these, five have balances in excess of $50. An auditor selects at random four accounts receivable for audit. What is the probability that exactly half of these audited accounts will have balances in excess of $50?

4. The bylaws of the Spendthrift Bank and Trust Company stipulate that membership on committees of the board of directors shall be determined by lottery and that committees may make recommendations only when there exists unanimous agreement. A three-person committee on managerial appointments is to be selected from the full board of 20 directors to recommend a new vice-president to head the trust department. Twelve members of the full board favor selection of Mrs. Worth, while eight members favor Mr. Hedge. Assuming that the directors never change their minds, what is the probability that the committee will recommend the appointment of Mr. Hedge? What would be the probability of a committee recommendation of Mr. Hedge if only a majority vote of the committee were required?

5. Six oil companies each send four executives to a conference on new developments in the energy field. Six representatives are chosen at random to lead

discussion groups on various aspects of oil demand and supply in future years. What is the probability that

a. exactly one representative from each company is a discussion leader?

b. Company A has four discussion leaders?

c. Company B has no discussion leaders?

6. Sales representative Nina Waterton receives 100 leads a day, five of which are excellent sales prospects. She draws five at random to visit on a particular day from the 100 leads received that day. What is the exact probability that she does not draw any of the five excellent prospects? Calculate an approximate probability.

7. An investment banking firm plans to hire five summer corporate research trainees from a pool of 18 first-year MBA students. Nine of the 18 students attend the Cambridge Graduate School of Business, five attend the Philadelphia School of Finance, and the other four are students at the Palo Alto Business School. If the firm decides that all 18 applicants are equally qualified and therefore will select the potential trainees randomly from the pool, what is the probability that

a. all five trainees selected are Cambridge students?

b. no Palo Alto business students are selected?

c. two trainees are Cambridge students, two are Philadelphia students, and one is a Palo Alto student?

d. three trainees are Philadelphia students and the other two attend Palo Alto Business School?

3.7 THE POISSON DISTRIBUTION

Another very useful probability function is the Poisson distribution, named for a Frenchman who developed it during the first half of the nineteenth century.[9] The distribution can be used in its own right and also for approximation of binomial probabilities. The former use is by far the more important one and in this context has had many fruitful applications in a wide variety of fields. We will discuss the Poisson distribution first as a distribution in its own right and then as an approximation to the binomial distribution.

The Poisson Distribution Considered in Its Own Right

The Poisson distribution has been usefully employed to describe the probability functions of phenomena such as product demand; demands for service; numbers of telephone calls that come through a switchboard;

[9] Siméon Denis Poisson (1781–1840), was particularly noted for his applications of mathematics to the fields of electrostatics and magnetism. He wrote treatises in probability, calculus of variations, Fourier's series, and other areas.

numbers of accidents; numbers of traffic arrivals (such as trucks at terminals, airplanes at airports, ships at docks, and passenger cars at toll stations); and numbers of defects observed in various types of lengths, surfaces, or objects.

All these illustrations have certain elements in common. The given occurrences can be described in terms of a discrete random variable, which takes on values 0, 1, 2, and so forth. For example, product demand can be characterized by 0, 1, 2, etc., units purchased in a specified time period; number of defects can be counted as 0, 1, 2, etc., in a specified length of electrical cable. Product demand may be viewed as a process that produces random occurrences in continuous time, and the defects example pertains to random occurrences in a continuum of space. In cases such as the latter, the continuum may be one of area or volume as well as length. Thus, there may be a count of the number of blemishes in areas of sheetmetal used for aircraft or the number of a certain type of microscopic particles in a unit of volume such as a cubic centimeter of a solution. In all these cases, there is some rate (number of occurrences per interval of time or space) that characterizes the process producing the outcome.

Using as an example the occurrences of defects in a length of electrical cable, we can indicate the general nature of the process that produces a Poisson probability distribution. The length of cable has some rate of defects per interval, say, two defects per meter. Now suppose that if the entire length of cable is subdivided into very small subintervals, say, of one millimeter each, then (1) the probability that exactly one defect occurs in this subinterval is a very small number and is constant for each such subinterval, (2) the probability of two or more defects in a millimeter is so small that it may be considered to be 0, (3) the number of defects that occur in a millimeter does not depend on where that subinterval is located, and (4) the number of defects that occur in a subinterval does not depend on the number of events in any other nonoverlapping subinterval.

In the foregoing example, the subinterval was a unit of length. Analogous sets of conditions would characterize examples in which the subinterval is a unit of area, volume, or time.

The Nature of the Poisson Distribution

As indicated previously, the Poisson distribution results from occurrences that can be described by a discrete random variable. This random variable, denoted X, can take on values $x = 0, 1, 2, \ldots$ (where the three dots mean "*ad infinitum*"). That is, X can take on the values of all nonnegative integers. The probability of exactly x occurrences in the Poisson distribution is

(3.7) $$f(x) = \frac{\mu^x e^{-\mu}}{x!} \quad \text{for } x = 0, 1, 2, \ldots$$

where μ is the mean number of occurrences per interval and $e = 2.71828$ . . . (the base of the Naperian or natural logarithm system).

As can be seen from (3.7), the Poisson distribution has a single parameter symbolized by the Greek letter μ (*mu*). If we know the value of μ, we can write out the entire probability distribution. The parameter μ can be interpreted as the average number of occurrences per interval of time or space that characterizes the process producing the Poisson distribution. (The average referred to here is the arithmetic mean.) Hence, in the case of the Poisson distribution, μ can be interpreted as the mean number of occurrences per interval, or in other words, as an average rate of occurrences per interval. Thus, μ may represent an average of three units of demand per day, 5.3 demands for service per hour, 1.2 aircraft arrivals per five minutes, 1.5 defects per ten feet of electrical cable, and so on.

In order to illustrate how probabilities are calculated in the Poisson distribution, we consider the following example. A study revealed that the number of telephone calls per minute coming through a certain switchboard between 10:00 A.M. and 11:00 A.M. on business days is distributed according to the Poisson probability function with an average μ of 0.4 calls per minute. What is the probability distribution of the number of telephone calls per minute during the specified time period?

Let X represent the random variable "number of telephone calls per minute" during the given time period. Then $\mu = 0.4$ calls per minute is the parameter of the Poisson probability distribution of this random variable. The probability that no calls will occur (come through the switchboard) in a given minute is obtained by substituting $x = 0$ in the Poisson probability function, Equation 3.7. Hence,

(3.8) $$P(X = 0) = f(0) = \frac{(0.4)^0 e^{-0.4}}{0!}$$

Since $(0.4)^0 = 1$ and $0! = 1$, Equation 3.8 becomes simply

(3.9) $$f(0) = e^{-0.4} = 0.670$$

The value 0.670 for $f(0)$ can be found in Table A-11 of Appendix A, where exponential functions of the form e^x and e^{-x} are tabulated for values of x from 0.00 to 6.00 at intervals of 0.10.

Continuing with the calculation of the Poisson probability distribution, we find the probability of exactly one call in a given minute by substituting $x = 1$ in (3.7). Hence, $f(1)$ is given by

(3.10) $$P(X = 1) = f(1) = \frac{(0.4)^1 e^{-0.4}}{1!} = (.4)(.670) = 0.268$$

To find the other values of $f(x)$, we can use Table A-3 of Appendix A, which lists values of the cumulative distribution function for the Poisson distribution. That is, values of $F(c) = P(X \leq c) = \sum_{x=0}^{c} f(x)$, or the probabilities of c or fewer occurrences, are provided for selected values of the parameter μ. As in Table A-1 for the binomial cumulative distribution, probabilities such as $1 - F(c)$ or $a \leq f(x) \leq b$ can be obtained by appropriate manipulation of the tabulated values.

The use of Table A-3 will be illustrated in terms of the present example. To obtain the probability of no calls in a given minute, using $c = 0$ and $\mu = 0.4$ in Table A-3, we find the value of $F(0) = 0.670$. Of course, this is also the value of $f(0)$, since the probability of zero or fewer occurrences equals the probability of zero occurrences. Therefore, as before, $f(0) = 0.670$.

We find the probability of exactly one telephone call per minute by subtracting the probability of no calls from the probability of one or fewer calls, that is,

$$f(1) = F(1) - F(0) = 0.938 - 0.670 = 0.268$$

Similarly, using Table A-3, we find the values of $f(2), f(3)$, and $f(4)$:

$$f(2) = F(2) - F(1) = 0.992 - 0.938 = 0.054$$
$$f(3) = F(3) - F(2) = 0.999 - 0.992 = 0.007$$
$$f(4) = F(4) - F(3) = 1.000 - 0.999 = 0.001$$

Although, as indicated earlier, the random variable X in the Poisson distribution takes on the values $0, 1, 2, \ldots, F(4) = 1.00$ in this problem. This means that the probabilities of $5, 6, \ldots,$ occurrences are so small that they would appear as 0 when rounded to three decimal places.

TABLE 3-7
Poisson Probability Distribution of the Number of Telephone Calls Coming Through a Certain Switchboard Between 10:00 A.M. and 11:00 A.M. on Business Days

Number of Calls x	Probability $f(x)$
0	0.670
1	0.268
2	0.054
3	0.007
4	0.001
	1.000

The required probability distribution for this problem is given in Table 3-7. Several other illustrations of the use of the Poisson distribution are given in Examples 3-13, 3-14, and 3-15.

EXAMPLE 3-13

It has been observed that on weekdays at a certain small airport, airplanes arrive at an average rate of three for the one-hour period 1:00 P.M. to 2:00 P.M. If these arrivals are distributed according to the Poisson probability distribution, what are the probabilities that
a. exactly zero airplanes will arrive between 1:00 P.M. and 2:00 P.M. next Monday?
b. either one or two airplanes will arrive between 1:00 P.M. and 2:00 P.M. next Monday?
c. a total of exactly two airplanes will arrive between 1:00 and 2:00 P.M. during the next three weekdays?

SOLUTION

In this problem, we may use the parameter $\mu = 3$ arrivals per day for the time period 1:00 P.M.–2:00 P.M. Let X represent the random variable "number of arrivals during the specified time period." The mathematical solutions are given for (a), (b), and (c) to illustrate the theory involved. However, the answers may also be determined by looking up values in Table A-3 of Appendix A as indicated.
a. The random variable X follows the Poisson distribution with the parameter $\mu = 3$. Thus,

$$P(X = 0) = f(0) = \frac{3^0 e^{-3}}{0!} = 0.050$$

This value may be obtained from Table A-3 of Appendix A, for $\mu = 3$, $c = 0$. We note that $f(0) = F(0)$.
b. Since exactly one arrival and exactly two arrivals are mutually exclusive events, we have, by the addition rule,

$$P(X = 0 \text{ or } X = 1) = f(1) + f(2) = \frac{3^1 e^{-3}}{1!} + \frac{3^2 e^{-3}}{2!} = 0.373$$

This value can be obtained from Table A-3 of Appendix A for $\mu = 3$. The required probability is $F(2) - F(0) = 0.423 - 0.050 = 0.373$.
c. A total of exactly two arrivals in three weekdays during the time period 1:00 P.M.–2:00 P.M. can be obtained, for example, by having two arrivals on the first day, none on the second day and none on the third day during the specified one-hour period. The total number of ways in which the event in question can occur is shown in Table 3-8.
Let P_2 represent the required probability. Using the multiplication and addition rules, and again using the parameter $\mu = 3$ arrivals per day

TABLE 3-8
Possible Ways of Obtaining a Total of Exactly Two
Arrivals in Three Weekdays

Day 1	Number of Arrivals	
	Day 2	*Day 3*
2	0	0
0	2	0
0	0	2
1	1	0
1	0	1
0	1	1

during the period 1:00 P.M.–2:00 P.M., we have

$$P_2 = 3[f(2)][f(0)]^2 + 3[f(1)]^2[f(0)]$$

$$= 3\left(\frac{3^2 e^{-3}}{2!}\right)\left(\frac{3^0 e^{-3}}{0!}\right)^2 + 3\left(\frac{3^1 e^{-3}}{1!}\right)^2\left(\frac{3^0 e^{-3}}{0!}\right)$$

$$= \frac{81}{2} e^{-9} = 0.005$$

The solution is greatly simplified if we change the time interval for which the parameter μ is stated. This has the effect of changing the random variable in the problem. Thus, if μ = three arrivals *per day* during the time period 1:00 P.M.–2:00 P.M., then μ = nine arrivals *per three days* during the same time period. The probability of exactly two arrivals in three weekdays during the given one-hour period can then be obtained by computing $P(X = 2)$, where X is a Poisson-distributed random variable denoting the number of arrivals *per three days*. The required probability is, therefore, obtained by simply computing $f(2)$ in a Poisson distribution with the parameter $\mu = 9$.

$$P_2 = f(2) = \frac{9^2 e^{-9}}{2!} = \frac{81}{2} e^{-9} = 0.005$$

This value can be obtained from Table A-3 of Appendix A for $\mu = 9$. The probability is given by $F(2) - F(1) = 0.006 - 0.001 = 0.005$.

This problem illustrates the point that considerable simplification of computations for Poisson processes can often be accomplished by convenient choice of parameters.

We should note a few points concerning the appropriateness of the Poisson distribution in the preceding example. It was stated at the beginning of the problem that the airplane arrivals were distributed according to the Poisson distribution. Whether it is appropriate to consider the past arrival distribution as a Poisson distribution during the specified time periods depends on the nature of the past data. Actual relative arrival frequencies can be tabulated and compared with the theoretical probabilities given by a Poisson distribution. Tests of "goodness of fit" for judging the closeness of actual and theoretical frequencies are discussed in Chapter 8.

In practical work, the question often arises whether a given mathematical model is likely to be applicable in a certain situation. This requires careful examination of whether the underlying assumptions of the model are likely to be fulfilled by the real-world phenomena. For example, in this problem, suppose certain cargo deliveries are made either on Mondays or Tuesdays between 1:00 P.M. and 2:00 P.M. Assuming that if a delivery is made on Monday, it will not be made on Tuesday, the independence assumption (4) of a Poisson process is clearly violated. That is, the number of arrivals during the one-hour time period on Tuesday *depends* on the number of arrivals during the corresponding time period on Monday, and vice versa. Furthermore, if the nature of the aircraft arrivals is such that Monday and Tuesday always have more arrivals between 1:00 P.M. and 2:00 P.M. than do other weekdays, then assumption (3) is violated. That is, if we were to count arrivals for the one-hour period for a given day (or two days and so forth), then the number of occurrences obtained clearly would depend upon the day on which the count was begun. It is indeed a rare event when the assumptions of a probability distribution are perfectly met by a real-world process. Experience in a given field aids considerably in judging whether so great a departure from assumptions has occurred that a model may no longer be applicable.[10] In the final analysis, actual comparison of the data generated by a process with the probabilities of the theoretical distribution is the best way of determining the appropriateness of the distribution. Of course, even if a given mathematical model (or other type of model) has provided a good description of past data, there is no guarantee that this state of affairs will continue. The analyst must be alert to changes in the environment that would make the model inapplicable.

EXAMPLE 3-14

A department store has determined in connection with its inventory control system that the demand for a certain brand of portable radio was Poisson-distributed with the parameter $\mu = 4$ per day.

a. Determine the probability distribution of the daily demand for this item.

---

[10] An appropriate thought here is perhaps contained in the anonymous bit of advice, "Good judgment comes from experience, and experience comes from poor judgment."

b. If the store stocks five of these items on a particular day, what is the probability that the demand will be greater than the supply?

SOLUTION

Let X represent the random variable "number of portable radios of this brand sold per day."

a. The probability distribution of X is

x	$f(x)$
0	0.018
1	0.074
2	0.146
3	0.195
4	0.196
5	0.156
6	0.104
7	0.060
8	0.030
9	0.013
10	0.005
11	0.002
.	.
.	.
.	.

The sum of the probabilities for demand from zero through eleven units is 0.999. Therefore, the sum of the probabilities for twelve or more units is only 0.001.

The probabilities can be obtained from Table A-3 of Appendix A using the relationship $f(x) = F(x) - F(x - 1)$.

b. The probability that demand will be greater than five units is the complement of the probability that it will be five units or less. Thus, from Table A-3 of Appendix A we have

$$P(x > 5) = 1 - F(5) = 1 - 0.785 = 0.215$$

EXAMPLE 3-15

In connection with an auditing investigation, it was discovered that the number of entries made in each of six accounts receivable was distributed according to the Poisson probability distribution with parameter $\mu = 1$ per day. Entries in the accounts may be assumed to be independent. What is the probability that on a specified day

a. none of the six accounts will receive any entries?
b. each of the six accounts will receive at least one entry?
c. exactly three accounts will receive no entries?

SOLUTION

This problem illustrates a situation in which two different probability distributions must be used to provide a solution. In this case, the Poisson and binomial distributions are applicable.

a. Since the number of entries in a given account is Poisson-distributed with an average of one entry per day ($\mu = 1$), the probability that a given account will receive no entries on a specified day is

$$P(\text{no entry in a given account}) = f(0) = \frac{1^0 e^{-1}}{0!} = e^{-1}$$

Entries in different accounts are independent events. Therefore, by the multiplication rule, we have

$$P(\text{no entry in all six accounts}) = (e^{-1})^6 = e^{-6} = 0.002$$

b. The event that a given account will receive at least one entry is the complement of the event that the account receives no entries. Therefore,

$$P(\text{a given account receives at least one entry}) = 1 - e^{-1}$$

By the multiplication rule for independent events,

$$P(\text{at least one entry in each of the six accounts}) = (1 - e^{-1})^6 = 0.064$$

c. Let p be the probability that a given account receives no entries on a specified day. From part (a), $p = e^{-1}$. We may think of p as the probability of success in a Bernoulli trial. Thus, $q = 1 - e^{-1}$ and $n = 3$ in this problem.

$$P(\text{no entries in exactly three accounts}) = \binom{6}{3}(1 - e^{-1})^3 (e^{-1})^3 = 0.252$$

The Poisson Distribution as an Approximation to the Binomial Distribution

The foregoing discussion concerned the use of the Poisson probability function as a distribution in its own right. We turn now to a consideration of the Poisson distribution as an approximation to the binomial distribution.

We saw that the Bernoulli process gives rise to a two-parameter probability function, the binomial distribution. Since computations involving the binomial distribution become quite tedious when n is large, it is useful to have a simple method of approximation. The Poisson distribution is particularly suitable as an approximation when n is large and p is small.

Assume in the expression for $f(x)$ of the binomial distribution that as n is permitted to increase without bound, p approaches 0 in such a way that np remains constant. Let us denote this constant value for np as μ (which denotes the mean number of successes in n trials). Under these assumptions, it can be shown that the binomial expression for $f(x)$ approaches the value

$$f(x) = \frac{\mu^x e^{-\mu}}{x!}$$

where $\mu = np$ and e is the base of the natural logarithm system. Thus, as can be seen from expression (3.7), the value approached by the binomial distribution under the given conditions is the value of the Poisson distribution. Hence, the Poisson distribution can be used as an approximation to the binomial probability function. In this context, the Poisson distribution is similar to the binomial distribution, because it gives the probability of observing x successes in n trials of an experiment, where p is the probability of success on a single trial. That is, x, n, and p are interpreted in the same way as in the binomial distribution.

Because of the assumptions underlying the derivation of the Poisson distribution from the binomial distribution, the approximations to binomial probabilities are best when n is large and p is small. A frequently used rule of thumb is that the approximation is appropriate when $p \leqslant 0.05$ and $n \geqslant 20$. However, the Poisson distribution sometimes provides surprisingly close approximations even in cases where n is not large nor p very small. As an illustration of how these approximations may be carried out, we return to the problem of sampling the population consisting of 950 adults and 50 children. The probability of observing no children in a random sample of five persons drawn with replacement was previously computed from the binomial distribution. We now compute the same probability using the Poisson distribution. Since n is only 5 in this problem, this is not an ideal situation for the use of the Poisson distribution for approximating binomial probabilities. Rather, it is an example of the surprisingly small errors observed in certain cases, even though n is small, and it is used here simply to carry out the arithmetic for a familiar illustration.

The binomial parameters in this problem were $p = 0.05$ and $n = 5$. Therefore,

$$\mu = np = 5 \times 0.05 = 0.25$$

Thus, in the Poisson distribution, the probability of no successes (children) is

$$f(0) = \frac{(0.25)^0 e^{-0.25}}{0!} = e^{-0.25}$$

From Table A-3 of Appendix A with $c = 0$ and $\mu = 0.25$, we find the value of $F(0)$, which in this case is equal to $f(0)$ (since the probability of zero or fewer successes equals the probability of zero successes). Therefore,

$$f(0) = 0.779$$

This figure is the same in the first two decimal places as the corresponding number (0.7738) obtained from the binomial probability (see Section 3.6).

    However, the percentage errors would be much larger for the other terms of the binomial distribution, representing probabilities of one, two, and so forth, successes. It is recommended, therefore, that the Poisson approximations not be used unless the conditions for n and p in the rule of thumb mentioned above are met.

    The parameter $\mu = np$ can be interpreted as the average number of successes per sample of size n. Since p is the probability of success per trial and n is the number of trials, multiplication of n by p gives the average number of successes per n trials. In terms of the foregoing problem, we can interpret μ as a long-run relative frequency. The proportion of children in the population is $p = 0.05$, and a random sample of $n = $ five persons was drawn with replacement from this population. Suppose samples of size $n = 5$ were repeatedly drawn with replacement from the same population and the number of children was recorded for each sample. It can be proven mathematically, and it seems intuitively reasonable, that the *average proportion* of children per sample of five persons is equal to $p = 0.05$. Furthermore, it follows that the *average number* of children per sample of five persons is equal to $np = 5(.05) = 0.25$ children. The average referred to here is the arithmetic mean, obtained by totaling the proportions or numbers of children for all samples and dividing by the number of samples.

    The following example represents a more justifiable use of the Poisson approximation to binomial probabilities than the above illustration, which involved a small sample size.

EXAMPLE 3-16

An oil exploration firm is formed with enough capital to finance 20 ventures. The probability of any exploration being successful is 0.10. What are the firm's chances of
a. exactly one successful exploration?
b. at least one successful exploration?
c. two or fewer successful explorations?
d. three or more successful explorations?
Assume that the population of possible explorations is sufficiently large to warrant binomial probability calculations.

SOLUTION

a. Let $p = 0.10$ stand for the probability that an exploration will be successful and X for the number of successful explorations in ten trials. Using the binomial distribution with parameters $p = 0.10$ and $n = 20$, we find that the probability of exactly one successful exploration is

$$P(X = 1) = f(1) = \binom{20}{1} (0.9)^{19}(0.1)^1$$

This probability may be determined from Table A-1 of Appendix A as

$$P(X = 1) = F(1) - F(0) = 0.3917 - 0.1216 = 0.2701$$

An approximation to this probability is given by the Poisson distribution with parameter

$$\mu = np = 20(0.10) = 2$$

The Poisson probability of exactly one success is

$$f(1) = \frac{2^1 e^{-2}}{1!}$$

which may be determined from Table A-3 of Appendix A as

$$P(X = 1) = F(1) - F(0) = 0.406 - 0.135 = 0.271$$

Thus, the percentage error is about 1 in 270, or about 0.4%.

b. Using Tables A-1 and A-3 of Appendix A for parts (b), (c), and (d), we have

Binomial

$$P(X \geqslant 1) = 1 - F(0) = 1 - 0.1216 = 0.8784$$

Poisson

$$P(X \geqslant 1) = 1 - F(0) = 1 - 0.135 = 0.865$$

c. *Binomial*

$$P(X \leqslant 2) = F(2) = 0.6769$$

Poisson

$$P(X \leqslant 2) = F(2) = 0.677$$

d. *Binomial*

$$P(X \geqslant 3) = 1 - F(2) = 1 - 0.6769 = 0.3231$$

Poisson

$$P(X \geqslant 3) = 1 - F(2) = 1 - 0.677 = 0.323$$

Exercises

1. The number of car accidents between 8:00 A.M. and 9:00 A.M. on Monday on the Surekill Expressway is distributed according to the Poisson distribution with a mean of 4. What is the probability that on a given Monday there will be no accidents during the indicated time interval?

2. It is estimated that the number of taxicabs waiting to pick up customers in front of the Reading Terminal is Poisson-distributed with a mean of 3.2 cabs per unit of observed time. What is the probability that on a random observation more than five cabs will be waiting?

3. The average number of thread defects in a standard bolt of cloth produced by Burmont Mills is six. Assuming that a Poisson distribution is an appropriate model, what is the probability that
   a. a bolt will have eight or more thread defects?
   b. a bolt will have no thread defects?
   c. five bolts selected at random will all have eight or more thread defects? Assume independence.

4. The average number of misprints on a page of the Pleasant Valley *Gazette* is three. Assuming that the Poisson distribution is appropriate for describing the incidence of misprints on a page, what is the probability that each of the eight pages in the Monday morning edition has at most four misprints? Assume independence.

5. The number of lathes breaking down between the hours 8:00 A.M. and 4:00 P.M. per weekday in the machine shop of Eastinghouse Corporation has a Poisson distribution with a mean of 6. What is the probability that more than ten machines will break down on any given weekday during the indicated time period?

6. In a very large industrial complex, the average number of fatal accidents per month is 0.35. The number of accidents per month is adequately described by a Poisson distribution. What is the probability that five months will pass without a fatal accident? Assume independence.

7. A certain department store's telephone sales department receives 2000 calls on a given day, and the probability that any call will result in a sale is 0.01. Use the Poisson approximation to find the probability that the 2000 calls result in
   a. exactly 15 sales
   b. more than 15 sales

8. A check of 1000 50-pound crates of oranges from Great Groves revealed the following data on the number of spoiled oranges in each crate:

Spoiled Oranges in Crate	Number of Crates
0	390
1	365
2	175
3	57
4	11
5	2
	1000

It is hypothesized that the number of spoiled oranges per crate has a Poisson distribution with a mean of 0.95. Assuming that this hypothesis is correct, calculate the probabilities of getting zero, one, two, three, four, and five spoiled oranges in a randomly selected crate and compare these results with the above data. Would you be apt to agree or disagree with the hypothesis?

9. The complaint department of the customer service division of the Top Quality Appliance Store receives an average of 1.9 phone calls in its first 15 minutes (9:00–9:15 A.M.) of operation each weekday morning. Assume a Poisson probability distribution for the incoming calls during this 15-minute period each day.
   a. What is the probability of having at least one call during this period on any given day?
   b. Assuming independence, what is the probability of the department receiving at least one call during the given period on exactly one day of a five-day work week?

10. A certain production process produces 1% defective items. Use the Poisson approximation to calculate the probability that in a random sample of 1000 articles,
    a. exactly eight are defective
    b. at most eight are defective

11. A firm's office contains 20 typewriters. The probability that any one typewriter will not work on a given day is 0.05.
    a. Assume that the binomial distribution is an appropriate model.
       1. What is the probability that exactly one will not work on a given day?
       2. What is the probability that at least two will not work on a given day?
    b. Use the Poisson approximation to solve part (a) and compare your results.

3.8 SUMMARY MEASURES FOR PROBABILITY DISTRIBUTIONS

In Chapter 1 we discussed summary, or descriptive, measures for empirical frequency distributions. We now turn to the corresponding measures for theoretical frequency distributions, that is, for probability distributions. These descriptive measures for probability distributions are essential components of modern quantitative techniques employed as aids for decision making under conditions of uncertainty.

Earlier in this chapter, we considered certain probability distributions for discrete random variables as the appropriate mathematical models for real-world situations under specific sets of assumptions. Sometimes, from the nature of a problem, it is relatively easy to specify a suitable probability model. In other situations, the appropriate model is suggested only after substantial numbers of observations have been taken and empirical frequency distributions have been constructed. However, whatever the method by which we arrive at probability distributions, it is essential to be able to capture their salient properties in a few summary measures. These measures are the subject of the next two sections.

3.9 EXPECTED VALUE OF A RANDOM VARIABLE

Suppose the following game of chance were proposed to you: A fair coin is tossed. If it lands "heads," you win $10.00; if it lands "tails," you lose $5.00. What is the average amount that you would win per toss?

On any particular toss, you will either win $10.00 or lose $5.00. However, let us think in terms of a repeated experiment in which we toss the coin and play the game many times. Since the probability assigned to the event "head" is $\frac{1}{2}$ and to the event "tail" $\frac{1}{2}$, in the long run, you would win $10.00 on half of the tosses and lose $5.00 on half. Therefore, the average (arithmetic mean) winnings per toss would be obtained by weighting the outcome $10.00 by $\frac{1}{2}$ and $-$\$5.00 by $\frac{1}{2}$ to yield a weighted mean of $2.50 per toss. In terms of Equation 1.4, the weighted mean is

$$\bar{X}_w = \frac{\Sigma\, wX}{\Sigma\, w} = \frac{\$10(\frac{1}{2}) + (-\$5)(\frac{1}{2})}{\frac{1}{2} + \frac{1}{2}} = \$2.50 \text{ per toss}$$

Of course, when the weights are probabilities in a probability distribution, as in this case, the formula can be written without showing the division by the sum of the weights.

The average of $2.50 per toss is referred to as the *expected value* of the winnings. Note that on a single toss, only two outcomes are possible, namely, win $10.00 or lose $5.00. If these two possible winnings are viewed as the possible values of a random variable, which occur with probabilities of $\frac{1}{2}$ each, then the expected value of the random variable can be seen to be the mean of its probability distribution. More formally, if X is a discrete random variable that takes on the value x with probability $f(x)$, then the expected value of X, denoted $E(X)$, is

$$(3.11) \qquad\qquad E(X) = \sum_x xf(x)$$

That is, to obtain the expected value of a discrete random variable, each value the random variable can assume is multiplied by the probability of occurrence of that value, and then all these products are totaled. Since the expected value is used so frequently as a measure of central tendency for probability distributions, it is given a simpler symbol, μ.

TABLE 3-9
Calculation of the Expected Value for the Coin Tossing Problem

Value of Winnings x	Probability f(x)	Weighted Winnings xf(x)
$10	$\frac{1}{2}$	$5.00
−5	$\frac{1}{2}$	− 2.50
	1	$2.50

$$\mu = E(X) = \sum_x xf(x) = \$2.50$$

Hence, Equation 3.11 can be rewritten as

$$\mu = E(X) = \sum_x xf(x).$$

The calculation of the expected value for the problem of tossing a coin is shown in Table 3-9. As can be seen, this calculation is very similar to that for the mean of an empirical frequency distribution.

The expected value has a wide variety of applications in situations involving uncertain outcomes. Example 3-17 gives a simple illustration, in which the uncertainty is summarized in terms of a probability distribution of death for a particular type of insurance policy holder. Such probability distributions are ordinarily based on a large sample of observed experience, that is, on past relative frequencies of mortality. On the other hand, business problems often involve "one-time" decisions; with such problems, uncertainties can be summarized in terms of relevant subjective probability distributions. Example 3-18 illustrates such a situation.

EXAMPLE 3-17

Suppose an insurance company offers a 45-year-old man a $1000 one-year term insurance policy for an annual premium of $12. Assume that the number of deaths per 1000 is five for persons in this age group. What is the expected gain for the insurance company on a policy of this type?

We may think of this problem as representing a chance situation in which there are two possible outcomes, (1) the policy purchaser lives or (2) he dies during the year. Let X be a random variable denoting the dollar gain to the insurance company for these two outcomes. The probability that the man will live through the year is 0.995. In this case, the insurance company collects the premium of $12. The probability that the policy pur-

TABLE 3-10
Calculation of Expected Gain for an Insurance Company on a One-Year Term Policy

Outcome	x	$f(x)$	$xf(x)$
Policy holder lives	$12	0.995	$11.94
Policy holder dies	−$988	0.005	−$4.94
		1.000	$7.00

$$E(X) = \sum_x xf(x) = \$7.00$$

chaser will die during the year is 0.005. In this case, the company has collected a premium of $12 but must pay the claim of $1000, for a net gain of − $988. Thus, X takes on the values $12 and − $988 with respective probabilities 0.995 and 0.005. The calculation of expected gain for the insurance company is displayed in Table 3-10.

A couple of points may be noted. First, in insuring only one person, the company would not realize the expected gain but would have either the gain of $12 or the loss of $988. Hence, in order to realize expected gains, insurance companies "play the averages" by insuring large numbers of individuals. Second, in setting a premium for this policy, the insurance company would have to take into account usual expenses of doing business as well as the expected gain calculation.

EXAMPLE 3-18

Lawrence Carlton, a financial manager for a corporation, is considering two competing investment proposals. For each of these proposals, he has carried out an analysis in which he has determined various net profit figures and has assigned subjective probabilities to the realization of these returns. For proposal A, his analysis shows net profits of $20,000, $30,000, or $50,000 with respective probabilities 0.2, 0.4, and 0.4. For proposal B, he concludes that there is a 50% chance of a successful investment, estimated as producing net profits of $100,000, and of an unsuccessful investment, estimated as a break-even situation involving $0 of net profit. Assuming that each proposal requires the same dollar investment, which is preferable from the standpoint of expected monetary return?

Denoting the expected net profit on these proposals as $E(A)$ and $E(B)$, we have

$$E(A) = (\$20{,}000)(0.2) + (\$30{,}000)(0.4) + (\$50{,}000)(0.4) = \$36{,}000$$
$$E(B) = (\$0)(0.5) + (\$100{,}000)(0.5) = \$50{,}000$$

Mr. Carlton would maximize expected net profit by accepting proposal B.

3.10 VARIANCE OF A RANDOM VARIABLE

As we have seen, the expected value of a random variable is analogous to the arithmetic mean of a frequency distribution of data. Similarly, the variability of a random variable may be measured in the same general way as the variability of a frequency distribution. In Section 1.14 the variability of the observations in a population of data was referred to as the *variance*; the square root of the variance was called the *standard deviation*. The same terms are used for random variables. The variance of a random variable is the average (expected value) of the squared deviations from the expected value. The standard deviation of a random variable is the square root of the variance. The variance of a random variable may be written as $E(X - \mu)^2$ and may be calculated for discrete random variables by the following expression:

$$(3.12) \qquad \sigma^2 = \sum_x (x - \mu)^2 f(x)$$

A mathematically equivalent formula that is easier for purposes of calculation is

$$(3.13) \qquad \sigma^2 = E(X^2) - [E(X)]^2$$

where

$$E(X^2) = \sum_x x^2 f(x)$$

and

$$E(X) = \sum_x x f(x)$$

We illustrate the calculation of the variance and standard deviation for a few random variables. The calculation of the expected value as an intermediate step is also shown.

EXAMPLE 3-19

On the basis of past experience, a store manager assesses the probability distribution for the number of units of a particular product sold daily. The distribution is as shown in the first two columns of Table 3-11. What are the expected value and standard deviation of the number of units sold daily?

SOLUTION

The calculation of the expected value and standard deviation from the definitional formula, Equation 3.12, is given in Table 3-11. Also shown is the alternative calculation for the standard deviation, following from (3.13).

TABLE 3-11
Expected Value and Standard Deviation of Number of Units of a Product Sold

Number of Units Sold x	Probability $f(x)$	$xf(x)$	$x - \mu$	$(x - \mu)^2$	$(x - \mu)^2 f(x)$
0	0.15	0.00	−1.85	3.4225	0.513375
1	0.25	0.25	−0.85	0.7225	0.180625
2	0.30	0.60	0.15	0.0225	0.006750
3	0.20	0.60	1.15	1.3225	0.264500
4	0.10	0.40	2.15	4.6225	0.462250
	1.00	1.85			1.427500

$$\mu = E(X) = \sum_{x=0}^{4} xf(x) = 1.85 \text{ units sold}$$

$$\sigma^2 = E(X - \mu)^2 = \sum_{x=0}^{4} (x - \mu)^2 f(x) = 1.427500$$

$$\sigma = \sqrt{\sigma^2} = \sqrt{1.427500} = 1.19 \text{ units sold}$$

Alternative Calculation of the Standard Deviation

$$E(X^2) = \sum_{x=0}^{4} x^2 f(x) = 0(0.15) + 1(0.25) + 4(0.30) + 9(0.20) + 16(0.10) = 4.85$$

$$\sigma^2 = E(X^2) - [E(X)]^2 = 4.85 - (1.85)^2 = 1.4275$$

$$\sigma = \sqrt{1.4275} = 1.19 \text{ units sold}$$

EXAMPLE 3-20

Compute the mean and variance for the total obtained on the uppermost faces in a roll of two unbiased dice.

Let X denote the specified total on the two dice. Then

$$\mu = \sum_{x=2}^{12} (x)f(x) = 2(\tfrac{1}{36}) + 3(\tfrac{2}{36}) + 4(\tfrac{3}{36}) + 5(\tfrac{4}{36}) + 6(\tfrac{5}{36})$$
$$+ 7(\tfrac{6}{36}) + 8(\tfrac{5}{36}) + 9(\tfrac{4}{36}) + 10(\tfrac{3}{36}) + 11(\tfrac{2}{36}) + 12(\tfrac{1}{36}) = 7$$

$$\sigma^2 = \sum_{x=2}^{12} x^2 f(x) - \mu^2 = 4(\tfrac{1}{36}) + 9(\tfrac{2}{36}) + 16(\tfrac{3}{36}) + 25(\tfrac{4}{36}) + 36(\tfrac{5}{36})$$
$$+ 49(\tfrac{6}{36}) + 64(\tfrac{5}{36}) + 81(\tfrac{4}{36}) + 100(\tfrac{3}{36}) + 121(\tfrac{2}{36}) + 144(\tfrac{1}{36})$$
$$- (7)^2 = 54\tfrac{5}{6} - 49 = 5\tfrac{5}{6}$$

EXAMPLE 3-21

The Amiable Conglomerate Company estimates the net profit on a new product it is launching to be $3,000,000 during the first year if it is "successful," $1,000,000 if it is "moderately successful," and a loss of

$1,000,000 if it is "unsuccessful." The firm assigns the following probabilities to first-year prospects for this product: successful, 0.15; moderately successful, 0.60; and unsuccessful, 0.25. What are the expected value and standard deviation of first-year net profit for this product?

Let X denote first-year net profit in millions of dollars; then

$$E(X) = (3)(0.15) + (1)(0.60) + (-1)(0.25)$$
$$= \$0.80 \text{ million or } \$800,000$$

$$\sigma^2 = E(X^2) - [E(X)]^2 = 9(0.15) + 1(0.60) + 1(0.25) - (0.80)^2$$
$$= 2.20 - 0.64 = 1.56$$

$$\sigma = \sqrt{1.56} \approx 1.25 \text{ million or } \$1,250,000$$

3.11 EXPECTED VALUE AND VARIANCE OF SUMS OF RANDOM VARIABLES

We now discuss some of the important properties of expected values and variances of random variables. It can be shown that

(3.14) $$E(X_1 + X_2 + \cdots + X_N) = E(X_1) + E(X_2) + \cdots + E(X_N)$$

That is, the expected value of a sum of N random variables is equal to the sum of the expected values of these random variables.

A somewhat analogous relationship holds for variances of *independent* random variables. If $X_1, X_2, \ldots, X_N$ are N *independent* random variables, then

(3.15) $$\sigma^2(X_1 + X_2 + \cdots + X_N) = \sigma^2(X_1) + \sigma^2(X_2) + \cdots + \sigma^2(X_N)$$

That is, the variance of the sum of N *independent* random variables is equal to the sum of their variances. Note that there is no restriction of independence in the case of (3.14). That is, Equation 3.14 holds whether or not the variables are independent. Other properties of expected values and variances are summarized in Appendix C.

The following example illustrates the properties of expected values and variances discussed in this section.

EXAMPLE 3-22

Nancy Schultz is in charge of a project to develop a scale model of an assembly to be placed in production at a later time. She has scheduled the following activities to complete this project: (1) design of equipment, (2) training of personnel, (3) assembly of equipment. The times required to complete these activities are independent of one another. Based on her experience with similar projects in the past, she estimates the means (expected times) and standard deviations of the completion times in weeks for each activity as follows:

Activity	Expected Time (in Weeks)	Standard Deviation (in Weeks)
1. Design of equipment	10	4
2. Training of personnel	5	1
3. Assembly of equipment	6	3

Estimate the expected value and standard deviation of completion time for the entire project.

Let μ_1, μ_2, and μ_3 denote the expected values and σ_1, σ_2, and σ_3 the standard deviations of the completion times of the three activities. Then the expected value and standard deviation of completion time for the entire project are:

$$\mu = \mu_1 + \mu_2 + \mu_3 = 10 + 5 + 6 = 21 \text{ weeks}$$
$$\sigma = \sqrt{\sigma_1^2 + \sigma_2^2 + \sigma_3^2} = \sqrt{4^2 + 1^2 + 3^2} = 5.1 \text{ weeks}$$

Exercises

1. Let $X = -\$.15$ if a customer does not purchase from a telephone sales promotion and $X = +\$4.26$ if a customer does purchase from a telephone sales promotion. If the probability that a customer will purchase is 4%, what is the expected value of X? Interpret the result.

2. Let $X = 0$ if an investment proposal made by a research committee is rejected by the company's financial vice-president and $X = 1$ if the proposal is approved. If 20% of all proposals are rejected, what is the expected value of X? Interpret the result.

3. The demand for a certain product is a discrete random variable and is uniformly distributed in the range of 10 to 20 items per day. State the probability distribution of this random variable. What is the expected demand per day?

4. The probability distribution for the number of clients telephoning the senior partner in charge of management services for a Big Eight public accounting firm each day is as follows:

x	$f(x)$	x	$f(x)$
7	0.08	11	0.13
8	0.13	12	0.10
9	0.21	13	0.06
10	0.24	14	0.05

What is the expected number of clients telephoning the partner on a given day?

5. Suppose an insurance company offers a particular homeowner a $15,000 one-year term fire insurance policy on his home. Assume that the number of fires per 1000 dwellings for this particular home safety classification is two. For simplicity, assume that any fire will cause damages of at least $15,000. If the insurance company charges a premium of $36 a year, what is its expected gain from this policy? Does this mean that the company will earn that many dollars on *this particular policy?*

6. The Discount King department store chain has 12 stores in the Detroit metropolitan area. The net profit per month from each has an expected value of $6500 and a variance of 30,000. What is the expected total net profit per month and the standard deviation of total net profit for all 12 stores combined? Assume independence.

7. To relieve the strain of decision-making pressures on its executives, the Barker Brothers Game and Novelty Company has set up a roulette wheel in the executive dining room. A roulette wheel has 38 slots, numbered 1 through 36, 0, and 00. On a spin of the wheel, it is equally probable that a ball will rest on any of the 38 numbers. During lunch hour, any executive can place a $5 bet on any number on the wheel. If the ball rests on the chosen number, the payoff is $125. Twelve executives from the financial and marketing divisions have formed a group that bets on numbers 1 to 12 inclusive each day. Each person contributes $5 for every spin of the wheel and shares equally in the winnings, if any. What is the expected profit (or loss) for the group on each spin? What is the expected profit (or loss) for 500 spins of the wheel if the group places its usual bet combination each time?

8. An operations research analyst developed the following probability distribution for a discrete variable in a managerial application:

$$f(x) = \tfrac{1}{18}(x^3 + 3), \quad x = 0, 1, 2$$

Calculate the variance by two different methods.

9. After completing a study on the possible side effects of a new pain reliever for headaches, the research and marketing divisions of the Aspiring Medical Drug Company are trying to decide whether to test the drug further or place it on the market now. They have determined the following probabilities for the percentage of users who will experience an unpleasant side effect of the new drug:

Percentage of Users Affected	*Probability*	*Percentage of Users Affected*	*Probability*
0	0.03	6	0.14
1	0.06	7	0.10
2	0.07	8	0.09
3	0.11	9	0.07
4	0.17	10	0.02
5	0.14		

Find the expected value and standard deviation of percentage of users affected.

10. You are considering buying one of two stocks, Volatile or Stable, both now priced at $46, for a one-month trading venture. The probability distribution for the closing prices of the two stocks (rounded to nearest $1) one month hence is estimated as follows:

	Volatile		Stable	
Price	f (Price)	Price	f (Price)	
44	0.1	44	0.005	
45	0.1	45	0.015	
46	0.1	46	0.030	
47	0.1	47	0.100	
48	0.1	48	0.350	
49	0.1	49	0.350	
50	0.1	50	0.100	
51	0.1	51	0.030	
52	0.1	52	0.015	
53	0.1	53	0.005	

Find the expected values and risk of each stock. (Financial analysts often refer to variance as "risk.") Which stock would you purchase and why?

11. As assistant to the marketing manager of the Dull Oil Company, you must submit a report on which of two new petroleum products to introduce this year. Your research group has given you the following probability distributions for the net profits of each product:

	Product A		Product B	
Net Profit	f (Net Profit)	Net Profit	f (Net Profit)	
−$5000	0.1	$ 0	0.2	
0	0.2	+ 1000	0.3	
+ 3000	0.4	+ 3000	0.3	
+ 6000	0.3	+ 5000	0.2	

Find the expected net profit and standard deviation of net profit for each product.

12. A certain machinist produces five to eight finished pieces during an eight-hour shift. An efficiency expert wants to assess the value of this machinist, where value is defined as value added minus the machinist's labor cost. The value added for the work the machinist does is estimated at $9 per item, and the machinist earns $5 per hour. From past records, the machinist's output per eight-hour day is known to have the following probability distribution:

Number of Pieces Produced	Probability
5	0.2
6	0.4
7	0.3
8	0.1

What is the expected monetary value of the machinist to the company per eight-hour day?

13. The Amstar Motor Company is in the process of planning its new model line for 1977. The company has defined four stages in the development of a new automobile model. First, the research and marketing people decide what market they are trying to enter or expand their sales in and, hence, what general type of automobile should be built. Then the designers and engineers, with the goals of the marketing division in mind, develop the new product from drawing board to full-scale model. Next, the production division retools the plant where the new car will actually be built. Finally, the quality control people test the first automobiles off the line and recommend improvements and adjustments. After completion of this last step, the company is ready to begin production of the new model. Each stage must be completed before the next step is started. In an attempt to get an estimate of when the new automobile will be ready for introduction, senior management has obtained the following information from the various division heads in charge of the four stages:

Stage	Expected Completion Time of Stage (in Weeks)	Standard Deviation (in Weeks)
I	5	2
II	14	5
III	8	3
IV	3	1

What are the expected value and standard deviation of the total time for the new model's complete development process? Assume that the times for completion of the stages are independent.

3.12 JOINT PROBABILITY DISTRIBUTIONS

The discussion to this point has been concerned with probability distributions of discrete random variables considered one at a time. These are often referred to as *univariate probability distributions* or *uni-*

TABLE 3-12
Joint Probability Distribution for 1000 Persons Classified by Sex
and Product Preference

| | Product Preference | | |
Sex	Prefers Product ABC	Prefers Product XYZ	Marginal Probabilities
Male	0.20	0.30	0.50
Female	0.10	0.40	0.50
Marginal Probabilities	0.30	0.70	1.00

variate probability mass functions. In such distributions, probabilities
are assigned to events pertaining to a single random variable. In most
realistic decision-making situations, however, more than one factor must
be taken into account at a time. Frequently, the joint effects of several
variables, some or all of which are interdependent, must be analyzed,
often in terms of what their impact is on some objective the decision
maker wishes to achieve.

In this section, we consider the joint probability distributions of discrete random variables, or, stated differently, probability distributions of
two or more discrete random variables. Such functions are frequently
referred to as *multivariate probability distributions,* the term *bivariate
probability distribution* being used for the two-variable case.

We return to Table 2-2, the joint probability table for 1000 persons
classified by sex and product preference, for a simple example of a bivariate probability distribution. Table 2-2 is reproduced here for convenience as Table 3-12.

We now introduce some symbolism for the discussion of joint bivariate probability distributions. Instead of using the terminology of
events, as we did earlier in discussing Table 2-2, we now use the language of random variables. Let X represent the random variable "product preference" and Y the random variable "sex." We let X and Y take
on the following values:

$$X = 1 \text{ for "prefers product ABC"}$$
$$X = 2 \text{ for "prefers product XYZ"}$$

$$Y = 1 \text{ for "male"}$$
$$Y = 2 \text{ for "female"}$$

Then $P(X = x \text{ and } Y = y)$ denotes the joint probability that X takes on
the value x and Y takes on the value y.[11] For example, $P(X = 1$ and
$Y = 1)$ is the joint probability $P(\text{prefers product ABC and male}) = 0.20$,

[11] The notation $P(X = x \text{ and } Y = y)$ should really read $P\{(X = x) \text{ and } (Y = y)\}$. The simplified symbolism is in common use and will be employed in this book.

TABLE 3-13
Bivariate Probability Distribution Corresponding to Table 3-12

		x		
y	1	2	P(Y = y)
1	0.20	0.30	0.50
2	0.10	0.40	0.50
P(X = x)	0.30	0.70	1.00

and so on. The bivariate probability distribution of X and Y is given in Table 3-13.

Marginal Probability Distributions

The values in the cells of Table 3-13 are the joint probabilities of the respective outcomes denoted by the column and row headings for X and Y. Also displayed in the table are the separate univariate probability distributions of X and Y. Earlier, we referred to the probabilities in the margins of the table as marginal probabilities. These probabilities form *marginal probability distributions*. The marginal distribution of X consists of the values of X shown in the column headings and the column totals at the bottom; the marginal distribution of Y consists of the values of Y shown in the row headings and the row totals at the right side of the table. These distributions are shown in Table 3-14. As indicated, the symbols $P(X = x)$ and $P(Y = y)$ are used to denote the probabilities for the marginal probability distributions of X and Y.

Graph of Bivariate Probability Distribution

The probability distribution of a single discrete random variable is graphed by displaying the values of the random variable along the horizontal axis and the corresponding probabilities along the vertical axis.

TABLE 3-14
Marginal Probability Distributions of X and Y

Product Preference		Sex	
x	P(X = x)	y	P(Y = y)
1	0.30	1	0.50
2	0.70	2	0.50
	1.00		1.00

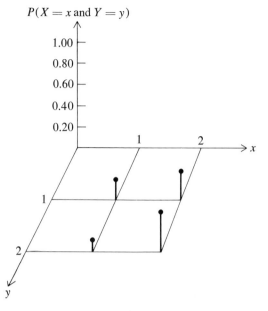

FIGURE 3-8
Graph of bivariate probability distribution shown in Table 3-13.

In the case of a bivariate distribution, two axes are required for the values of the random variables and a third for the measurement of probabilities. Usually, the joint values of the two variables are depicted on a plane (the x–y plane) and the associated probabilities are read along an axis perpendicular to the plane. A graph of the joint probability distribution of Table 3-13 is shown in Figure 3-8.

Conditional Probability Distributions

Another important type of probability distribution obtainable from a joint probability distribution is the *conditional probability distribution.* Using Equations 2.4 and 2.5, we may calculate the conditional probability that X takes on a particular value (say, $X = 2$) given that Y takes on a particular value (say, $Y = 1$) by dividing the joint probability that $X = 2$ and $Y = 1$ by the marginal probability that $Y = 1$. For example, the conditional probability $P(\text{prefers product XYZ}|\text{male}) = P(X = 2|Y = 1)$ is computed as follows:

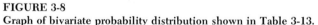

$$P(X = 2|Y = 1) = \frac{P(X = 2 \text{ and } Y = 1)}{P(Y = 1)} = \frac{0.30}{0.50} = 0.60$$

We can also find the conditional probability $P(X = 1|Y = 1)$, thus forming a conditional probability distribution for X given $Y = 1$. This distribution is shown in Table 3-15.

TABLE 3-15
Conditional Probability Distribution for
Product Preference (X) Given That the
Individual is Male (Y = 1)

x	$P(X = x \mid Y = 1)$
1	0.20/0.50 = 0.40
2	0.30/0.50 = <u>0.60</u>
	1.00

Similarly, we can obtain the conditional probability distribution for X given $Y = 2$ by dividing the joint probabilities in the row $Y = 2$ by the row total, $P(Y = 2) = 0.50$.

Corresponding calculations give us the conditional probability distributions for Y given particular values of X. For example, the conditional probability distribution of sex (Y) given that the individual prefers product XYZ $(X = 2)$ is shown in Table 3-16.

Independence

Returning to the terminology of events, we saw in Section 2-2 that if two events A_1 and B_1 are statistically independent, $P(B_1 \mid A_1) = P(B_1)$ and $P(A_1 \mid B_1) = A_1$. Using random variable notation, the analogous statement is that if X and Y are two *independent* random variables, then

$$P(X = x \mid Y = y) = P(X = x)$$

and

$$P(Y = y \mid X = x) = P(Y = y)$$

for all pairs of outcomes (x, y). For example, the first of these statements means that the conditional probability that X takes on a particular value

TABLE 3-16
Conditional Probability Distribution for Sex
(Y) Given That the Individual Prefers Product
XYZ (X = 2)

y	$P(Y = y \mid X = 2)$
1	0.30/0.70 = 0.43
2	0.40/0.70 = <u>0.57</u>
	1.00

x, given that Y takes on a particular value y, is equal to the marginal, or unconditional, probability that X takes on the particular value x.

Again returning to the terminology of events, we observed in Section 2.2 that by the multiplication rule, if two events A_1 and B_1 are statistically independent, $P(A_1 \text{ and } B_1) = P(A_1)P(B_1)$. The analogous statement for random variables is that if X and Y are two independent random variables, then

$$P(X = x \text{ and } Y = y) = P(X = x)P(Y = y)$$

for all pairs of outcomes (x, y).

We illustrate this definition of independence by returning to Table 3-13. Suppose we consider the outcome pair $(1, 1)$, that is, $X = 1$ and $Y = 1$. In this case,

$$P(X = 1 | Y = 1) = \frac{0.20}{0.50} = 0.40$$

and

$$P(X = 1) = 0.50$$

Since $P(X = 1 | Y = 1)$ is not equal to $P(X = 1)$, this suffices to show that X and Y are not independent random variables. Alternatively, we observe

$$P(X = 1 \text{ and } Y = 1) = 0.20$$
$$P(X = 1) = 0.30$$
$$P(Y = 1) = 0.50$$

We note that

$$P(X = 1 \text{ and } Y = 1) = 0.20 \neq P(X = 1)P(Y = 1) = (0.30)(0.50) = 0.15$$

Again, we conclude that X and Y are not independent random variables. Our reasoning is based on the fact that for X and Y to be independent random variables, the independence conditions stated earlier must hold for *all* pairs of outcomes (x, y). Since the conditions did not hold for the pair $(1, 1)$, that is, for $(X = 1, Y = 1)$, then X and Y are not independent.

If we retain the same marginal probability distributions shown in Table 3-13, what would the values of the joint probabilities have been if product preference (X) and sex (Y) had been *independent* random variables? To answer this question, we consider the marginal probability distributions in Table 3-13.

y	x 1	2	$P(Y = y)$
1			0.50
2			0.50
$P(X = x)$	0.30	0.70	1.00

TABLE 3-17
Joint Probability Distribution for Two Independent
Random Variables X and Y

	x		
y	1	2	$P(Y = y)$
1	.15	.35	.50
2	.15	.35	.50
$P(X = x)$	.30	.70	1.00

We compute the required joint probabilities as follows:

$$P(X = 1 \text{ and } Y = 1) = P(X = 1)P(Y = 1) = (0.30)(0.50) = 0.15$$
$$P(X = 2 \text{ and } Y = 1) = P(X = 2)P(Y = 1) = (0.70)(0.50) = 0.35$$

and so on.

In Table 3-17, we show the resulting joint probability distribution if product preference and sex had been independent random variables.

We now observe that $P(X = x|Y = y) = P(X)$ and $P(Y = y|X = x) = P(Y)$ for all pairs of outcomes (x, y). For example, for the outcome $(1, 1)$, $P(X = 1|Y = 1) = 0.15/0.50 = 0.30 = P(X = 1)$. Also, we observe that $P(X = x \text{ and } Y = y) = P(X = x)P(Y = y)$ for all pairs of outcomes (x, y). For example, for the outcome $(1, 1)$, $P(X = 1 \text{ and } Y = 1) = 0.15 = P(X = 1)P(Y = 1) = (0.30)(0.50) = 0.15$.

It may be noted from Table 3-17 that 30% $(0.15/0.50 = 0.30)$ of the males prefer product ABC and 30% of the females prefer product ABC. Correspondingly, 70% of the males prefer product XYZ, and the same percentage applies for females as well. Hence, we can say that product preference is *independent* of sex.

EXAMPLE 3-23

Table 3-18 presents the results of a sample survey of the employment status of the labor force by age in a certain community:
a. If a person is drawn at random from this labor force, what is
   1. the joint probability that the individual is under 25 years of age and unemployed?
   2. the marginal probability that the individual is under 25 years of age?
   3. the conditional probability of being unemployed, given that the individual is under 25 years of age?
b. What are the marginal probability distributions of age and employment status?
c. What is the conditional probability distribution of age given that an individual in this labor force is unemployed?
d. Is there evidence of dependence between employment status and age in this labor force?

TABLE 3-18
Sample of the Labor Force in a Certain Community Classified by Age Group and Employment Status

| Employment Status | Age Group (in Years) | | | |
	Under 25	25 and Under 45	45 and Over	All Ages
Unemployed	120	250	130	500
Employed	1880	4750	2870	9500
Labor force	2000	5000	3000	10,000

e. If the same marginal totals are retained, what would the number of persons in each cell of the table have to be for independence to exist between age and employment status?

SOLUTION

In order to answer these questions, we convert the table of numbers of occurrences (absolute frequencies) to probabilities (relative frequencies), by dividing each number in Table 3-18 by 10,000, the total number of persons in the sample. These relative frequencies are given in Table 3-19.

Let X and Y be random variables representing age group and employment status, respectively, and taking on values as follows:

Age Group	x	Employment Status	y
Under 25	0	Unemployed	0
25 and under 45	1	Employed	1
45 and over	2		

The joint probability distribution now appears as given in Table 3-20. The answers to the questions asked may now be given.

TABLE 3-19
Relative Frequencies of Occurrence for a Sample of the Labor Force of a Certain Community Classified by Age Group and Employment Status

| Employment Status | Age Group (in Years) | | | |
	Under 25	25 and Under 45	45 and Over	All Ages
Unemployed	0.012	0.025	0.013	0.050
Employed	0.188	0.475	0.287	0.950
Labor force	0.200	0.500	0.300	1.000

TABLE 3-20
Joint Probability Distribution Derived From Tables 3-18 and 3-19

y \ x	0	1	2	$P(Y=y)$
0	0.012	0.025	0.013	0.050
1	0.188	0.475	0.287	0.950
$P(X=x)$	0.200	0.500	0.300	1.000

a. 1. P (under 25 and unemployed) $= P(X = 0 \text{ and } Y = 0) = 0.012$
   2. P (under 25) $= P(X = 0) = 0.200$

   3. P (unemployed|under 25) $= P(Y = 0|X = 0) = \dfrac{0.012}{0.200} = 0.06$

b. The marginal probability distributions given in the margins of the probability table are

x	$P(X=x)$	y	$P(Y=y)$
0	0.200	0	0.050
1	0.500	1	0.950
2	0.300		1.000
	1.000		

c. This conditional probability distribution is given by

$$P(X = x|Y = 0) = \frac{P(X = x \text{ and } Y = 0)}{P(Y = 0)} \text{, where } x = 0, 1, 2$$

   Thus, each of the joint probabilities in the first row of Table 3-19 or 3-20 is divided by the marginal probability, or total, of that row. That is, the joint probabilities of being unemployed and in the specified age groups are divided by the marginal probability of being unemployed. Therefore, the required conditional probability distribution is

x	$P(X=x|Y=0)$
0	0.012/0.050 = 0.24
1	0.025/0.050 = 0.50
2	0.013/0.050 = 0.26
	1.00

d. Employment status and age are *not independent* random variables.

That is, it is not true that $P(X = x|Y = y) = P(Y = y)P(X = x)$ for *all* values of x and y. For example,

$$P(X = 0 \text{ and } Y = 0) \neq P(X = 0)P(Y = 0)$$

Numerically,

$$(0.012) \neq (0.200)(0.050)$$

It may be noted that for age group "25 and under 45" $(X = 1)$, the joint probabilities do factor into the product of the respective marginal probabilities. For example,

$$P(X = 1 \text{ and } Y = 0) = P(X = 1)P(Y = 0)$$

or

$$0.025 = (0.500)(0.050)$$

However, this is not sufficient. The equality must hold for *all* values of X and Y for these variables to be considered independent.

Another way of indicating that age and employment status are not independent random variables is to note that the marginal distributions are not equal to the corresponding conditional distributions. Thus, for age, the conditional and marginal probability distributions are

x	$P(X = x|Y = 0)$	$P(X = x|Y = 1)$	$P(X = x)$
0	0.240	0.198	0.200
1	0.500	0.500	0.500
2	0.260	0.302	0.300
	1.000	1.000	1.000

The specific nature of the dependence between age and employment status is that the percentage of unemployed individuals decreases with age. The unemployment rates are: under 25, 6.0%; 25 and under 45, 5.0%; 45 and over, 4.3%.

It is important to observe that the relationship between age and employment status depends to a certain extent on the arbitrary classifications used for the two variables. For example, if narrower age classifications had been used, it might have been found that the heaviest unemployment rates in the "under 25" group were among the teenagers and that although the unemployment rate was low for the "45 and over" group, the rate was quite high for persons 60 years and older. Similarly, if a narrower breakdown for employment had been used (including, for instance, part-time employment and multiple-job-holding classifications), greater insights might have been gained into the underlying relationships between age and employment status.

e. As indicated in (d), if age and employment status were independent, the

TABLE 3-21
Classification of Persons Under the Assumption of Independence Between Age and Employment Status

Employment Status	Under 25	Age Group (in Years) 25 and Under 45	45 and Over	All Ages
Unemployed	100	250	150	500
Employed	1900	4750	2850	9500
Labor force	2000	5000	3000	10,000

marginal probability distributions would be equal to the corresponding conditional probability distributions. One way of interpreting this in terms of the problem is to observe that since 0.05 of persons in all age groups were unemployed ($P(Y = 0) = 0.050$), then under the assumption of independence, the same proportion of individuals in *each* age group would be unemployed. Thus, 5% of 2000, 5000, and 3000 persons would be unemployed in the three respective age groups. In terms of marginal totals, the arithmetic for each age group would be:

Under 25: $\dfrac{500}{10,000} \times 2000 = 100$ unemployed

25 and under 45: $\dfrac{500}{10,000} \times 5000 = 250$ unemployed

45 and over: $\dfrac{500}{10,000} \times 3000 = 150$ unemployed

Therefore, the numbers of persons in each cell of the table under the assumption of independence are as given in Table 3-21.

EXERCISES

1. The following probability distributions pertain to the number of freight and passenger ships arriving on a given day at a certain port, where X is the number of freight ships and Y is the number of passenger ships.

x	$f(x)$	y	$g(y)$
0	0.1	0	0.3
1	0.3	1	0.3
2	0.4	2	0.2
3	0.2	3	0.2

If the number of freight ships has no effect on the number of passenger ships arriving, what is the joint distribution of X and Y?

2. An operations research analyst developed the following mathematical expression for the joint distribution of two discrete economic variables:

$$f(x, y) = \frac{1}{56}(2x + y^2) \quad x = 0, 1, 2, 3, \quad y = 0, 1, 2$$

   a. Display the joint bivariate distribution of X and Y.
   b. State the marginal probability distributions of X and Y.
   c. State the conditional probability distributions of the Y variable.
   d. Are X and Y independent?

3. The following table gives the results of a sample survey of sales of different types of packaging in 200 test communities:

		Type of Packaging			
Sales	Ordinary Bottle	Ordinary Cans	Flip-Top Cans	Screw-Top Bottle	Total
Under 2000 cases	45	40	30	25	140
At least 2000 cases	15	20	10	15	60
	60	60	40	40	200

   a. If a test community is selected at random for further study, what is
      1. the probability that it had sales of at least 2000 cases and was tested with flip-top cans?
      2. the marginal probability that it was tested with flip-top cans?
      3. the conditional probability that it had sales of at least 2000 cases given that it was tested with flip-top cans?
   b. What are the marginal distributions of sales and type of packaging?
   c. What is the conditional distribution of sales given that an ordinary bottle is used?
   d. Is there evidence of dependence between packaging and sales?
   e. If we retain the same marginal totals, what would the numbers in the cells of the table have to be for exact independence to exist between type of packaging and sales?

4. Random variables are defined for sales and advertising expenditures as follows:

$$X = \begin{cases} 0 & \text{if sales are less than \$10,000 per month} \\ 1 & \text{if sales are at least \$10,000 but less than 20,000 per month} \\ 2 & \text{if sales are at least \$20,000 per month} \end{cases}$$

$$Y = \begin{cases} 0 & \text{if advertising expenditures are less than \$10,000 per year} \\ 1 & \text{if advertising expenditures are at least \$10,000 per year but less than \$20,000 per year} \\ 2 & \text{if advertising expenditures are at least \$20,000 per year} \end{cases}$$

An operations research team has found the following joint probability relationship for a certain industry:

y \ x	0	1	2
0	0.08	0.07	0.05
1	0.11	0.18	0.12
2	0.01	0.07	0.31

    a. What is $f(1, 2)$?
    b. Find the marginal probability distributions.
    c. Find $f(x|Y = 0), f(x|Y = 1), f(x|Y = 2)$
    d. Are X and Y independent? What does the answer tell you?

5. A poll of the employees of Bajo, Inc. on a particular labor proposal gave the following results:

Opinion	Type of Position		Total
	Skilled	Unskilled	
For	275	225	500
Against	200	500	700
No opinion	125	75	200
Total	600	800	1400

Does a person's type of position affect his or her opinion on the labor proposal? Interpret your answer.

4

Statistical Investigations and Sampling

4.1 FORMULATION OF THE PROBLEM

A statistical investigation arises out of the need to solve some sort of a problem. Problems may be classified in a variety of ways. One such classification is (1) the problem of choosing among alternative courses of action and (2) the problem of reporting information. The problems of managerial decision making typically fall under category (1). For example, an industrial corporation wishes to choose a particular plant site from several alternatives; a financial vice-president wishes to decide among alternative methods of financing planned increases in productive capacity; or an advertising manager must select advertising media from many possible choices. Under heading (2), many illustrations can be given of statistical data collected for reporting purposes. For example, a trade association may report to its member firms on the characteristics of these companies. A research organization may publish data on the relationship between achievement of school children and the socioeconomic characteristics of their parents. An economist may report data on the frequency distribution of family incomes in a particular city. Even in the case of informational reporting, the data collected should have an ultimate decision-making purpose for someone.

Throughout our lives, we are involved in answering questions and solving problems. For many of these questions and problems, careful, detailed investigations are simply inappropriate. For example, to answer the question, "What clothes should I wear today?" or the question, "What type of transportation should I take to get to a friend's house?" does not require painstaking, objective, scientific investigation. This book, on the other hand, is concerned with the investigation of problems that do require careful planning and an objective, scientific approach to arrive at meaningful solutions. Many of these problems originally arise as rather vague questions. These questions must be translated into a series of other questions, which then form the basis of the investigation. In most carefully planned investigations, the problem will be defined and redefined many times. The purposes and importance of an investigation will determine the type of study to be conducted. There are instances where the objectives are particularly hard to define because of the many uses that will be made of the data and the very large number of research consumers who will utilize the results of the study. For example, the U.S. Bureau of the Census publishes a wide variety of data on population, housing, manufactures, and retail trade. It cannot specify in advance the many uses that will be made of these data. However, in all studies it is critical to spell out as meticulously as possible the purposes and objectives of the investigation. All subsequent analysis and interpretation depend upon these objectives, and it is only by stating very carefully what these objectives are that we can know what questions have been answered by the inquiry.

4.2 DESIGN OF THE INVESTIGATION

The types of investigations in which we are interested are what may be referred to as "controlled inquiries." We are all familiar with the scientist who controls variables in the laboratory. The scientist exercises control by manipulating the things and the events being investigated. For example, a chemist may hold the temperature of a gas constant while varying pressure and observing changes in volume.

Observational Studies

In most statistical investigations in business and economics, it is not possible to manipulate people and events as directly as a physical scientist manipulates experimental materials. For example, if we want to investigate the effect of income on a person's expenditure pattern, it would not be feasible for us to vary this individual's income. On the other hand, we can observe the different expenditure patterns of people who fall in different income groups, and therefore we can make statistical generalizations about how expenditures vary with differences in income. This would be an example of a so-called *observational* study. In this type of study, the analyst essentially examines historical relationships among variables of interest. By observing the important and relevant properties of the group under investigation, the study can be carried out in a controlled manner. For example, if we are interested in how family expenditures vary with family income and race, we can record data on family expenditures, family incomes, and race and then tabulate data on expenditures by income and racial classifications, such as white or black. Moreover, if we observe the differences in family expenditures for white and black families within the same income group, we have, in effect, "controlled" for the factor of income. That is, since the families observed are in the same income group, income cannot account for any differences in the expenditures observed.

If observational data represent historical relationships, it may be particularly difficult to ferret out causes and effects. For example, suppose we observe past data on the advertising expenses and sales of a particular company. Let us also assume that both of these series have been increasing over time. It may be quite incorrect to assume that it is the changes in advertising expenditures that have caused sales to increase. If a company's practice in the past had been to budget 3% of last year's sales for advertising expense, one may state that advertising expenses depend upon sales with a one-year lag. However, in this situation sales might be increasing quite independently of changes in advertising expenses. Therefore, one certainly would not be justified in concluding that changes in advertising expenses cause changes in sales. The point should also be made that many factors other than advertising may have influenced changes in sales. If data were not available on these other factors, it would not be possible to infer cause-and-effect relationships from these past observational data. The specific difficulty in attempting to derive cause-and-effect relationships in mathematical terms from historical data is that the various pertinent environmental factors will not ordinarily have been controlled or have remained stable.

Direct Experimentation Studies

Direct experimentation studies are being increasingly used in fields other than the physical sciences, where they have traditionally been employed. In such studies, the investigator directly controls or manipulates factors that affect a variable of interest. For example, a marketing experimenter may vary the amounts of direct mail exposure to a particular consumer audience. It is also possible to use different types of periodical advertising and observe the effects upon some experimental group. Various combinations of these direct mail exposures and periodical advertising, as well as such other types of promotional expenditure as a sales force, may be used. Thus, the investigator may be able to observe from the experiment that high levels of periodical advertising produce high sales effects only if there is a high concentration of sales force activity. Such scientifically controlled experiments for generating statistical data, to which only brief reference is being made here, can be very efficiently utilized to reduce the effect of uncontrolled variations. The real importance of this type of planning or design is that it gives greater assurance that the statistical investigation will yield valid and useful results.

Ideal Research Design

An important concept of a statistical inquiry is that of the ideal research design. The investigators should think through at the design stage what the ideal research experiment would be without reference to the limitations of data available, or to what data can feasibly be collected. Then if compromises must be made because of the practicalities of the real-world situation, the investigator will at least be completely aware of the specific compromises and expedients that have been employed. As an example, suppose we wanted to answer the question whether women or men are better automobile drivers. Clearly, it would be incorrect simply to obtain past data on the accident rates of men versus women. First of all, men may drive under quite different conditions than do women. For example, driving may constitute a large proportion of the work that many men do, whereas a larger proportion of women than men may drive primarily in connection with activities associated with the care of a home and family. The conditions of such driving differ considerably with respect to exposure to accident hazards. Many other reasons may be indicated for differences in accident rates between men and women apart from the essential driving ability of these two groups. Thus, as a first approximation to the ideal research design, perhaps we would like to have data for quite homogeneous groups of men and women, for example, women and men of essentially the same age, driving under essentially the same conditions, using the same types of automobiles. It may not be within the resources of a particular statistical investigation to gather data of this sort. However, once the ideal data required for a meaningful answer to the question have been thought through, the limitations of other, somewhat more practical sets of data become apparent.

Exercises

1. What is the difference between an observational study and a "controlled study"?
2. Why is it ordinarily very difficult to determine cause-and-effect relationships from historical data?
3. Why is it desirable to think through the ideal research design at the beginning stage of a statistical investigation?

4.3 CONSTRUCTION OF METHODOLOGY

An important phase of a statistical investigation is the construction of the conceptual or mathematical model to be used. A model is simply a representation of some aspect of the real world. Mechanical models are very profitably used in industry as well as other fields of endeavor. For example, airplane models may be tested in a wind tunnel, or ship models may be tested in experimental water basins. Various types of experiments may be carried out by varying certain factors and observing the effect of these variations on the mechanical models employed. Thus, we can manipulate and experiment with the models and draw corresponding inferences about their real-world counterparts. The advantages of this procedure are obvious as compared to attempting to manipulate an experiment using the real-world counterparts, such as actual airplanes or ships, after they are constructed. In statistical investigations, mathematical models are often used to state in mathematical terms the relationships among the relevant variables. These models are conceptual abstractions that attempt to describe, to predict, and often to control real-world phenomena. For example, the law of gravity describes and predicts the relationship between the distance an object falls and the time elapsed. Such models can be tested by physical experimentation.

In well designed statistical investigations, the nature of the model or models to be employed should be carefully thought through in the planning phases of the study. In fact, the nature of these models provides the conceptual framework that dictates the type of statistical data to be collected. Let us consider a few simple examples. Suppose a market research group wants to investigate the relationship between expenditures for a particular product and income and several other socioeconomic variables. The investigators may want to use a mathematical model such as a regression equation, to be discussed later in this book, which states in mathematical form the relationship among the variables. When the investigators determine the variables that are most logically related to the expenditures for the product, they also determine the types of data that will have to be collected in order to construct their model.

Even in the case of relatively simple informational reporting, there is a conceptual model involved. For example, suppose an agency wishes to determine the unemployment rate in a given community. Also, assume that the agency must gather the data by means of a sample survey of the labor force in this community. The ratio "proportion unemployed" is itself a model. It states a mathematical relationship between the numerator (number of persons unemployed) and the denominator (total number of persons in the labor force). The agency may wish to go further and state the range within which it is highly confident that the true unemployment rate falls. In such a situation, as we shall see later when we study estimation of population values, there is an implicit model, namely the probability distribution of a sample proportion.

Suppose a company wishes to establish a systematic procedure for accepting or rejecting shipments received from a particular supplier. Various types of models have been used to solve this sort of problem. The company may decide to accept or reject shipments on the basis of testing some hypothesis concerning the percentage of defective items observed. On the other hand, it may decide to base its acceptance procedure on the arithmetic mean value of some characteristic that is considered important. Other procedures are possible; for example, a formal decision model may be constructed. For these types of models, the probability distribution of the percentage of defective items produced by this company in the past may be required as well as data on the percentage of defective articles observed in a sample drawn from the particular incoming shipment in question. Obviously, the nature of the data to be observed and the nature of the analysis to be carried out will flow from the type of conceptual model used in the investigation.

4.4 SOME FUNDAMENTAL CONCEPTS

Statistical Universe

In the problem formulation stage, it is necessary to define very carefully the relevant *statistical universe* of observations. The universe, or *population*, consists of the total collection of items or elements that fall within the scope of a statistical investigation. The purpose of defining a statistical population is to provide very explicit limits for the data collection process and for the inferences and conclusions that may be drawn from the study. The items or elements that comprise the population may be individuals, families, employees, schools, corporations, and so on. Time and space limitations must be specified, and it should be clear whether or not any particular element falls within or outside the universe.

In survey work, a listing of all the elements in the population is referred to as the *frame,* or *sampling frame.* A *census* is a survey that attempts to include every element in the universe. The word "attempts" is used here because often in surveys of very large populations, despite every effort to do so, complete coverage may not be effected. Thus, for example, the Bureau of the Census readily admits that its national "censuses" of population invariably result in underenumerations. Strictly speaking, any partial enumeration of a population constitutes a *sample,* but the term "census" is used as indicated here. In most practical applications, it is not even feasible to attempt complete enumerations of populations, and therefore, typically, only samples of items are drawn. If the population is well defined in space and time, the problem of selecting a sample of elements from it is considerably simplified.

Let us illustrate some of the above ideas by means of a simple example. Suppose we draw a sample of 1000 families in a large city to estimate the arithmetic mean family income of all families in the city. The aggregate of all families in the city constitutes the universe, and each family is an element of the universe. The income of the family is a characteristic of the unit. A listing of all families in the city would comprise a frame. If, instead of drawing the sample of 1000 families, an attempt had been made to include all families in the city, a census would have been conducted. The definition of the universe would have to be specific as to the geographic boundaries that constitute the city and also the time period for which income would be observed. The terms "family" and "income" would also have to be rigidly specified. Of course, the precise definitions of all of these concepts would depend upon the underlying purposes of the investigation.

The terms *universe* and *sample* are relative. An aggregate of elements that constitutes a population for one purpose may merely be a sample for another. Thus, if one wants to determine the average weight of students in a particular classroom, the students in that room would represent the population. However, if one were to use the average weight of these students as an estimate of the corresponding average for all students in the school, then the students in the one room would be a sample of the larger population. The sample might not be a good one from a variety of viewpoints (such as representativeness), but nevertheless, it is a sample.

If the number of elements in the population is fixed, that is, if it is possible to count them and come to an end, the population is said to be *finite.* Such universes may range from a very small to a very large number of elements. For example, a small population might consist of the three vice-presidents of a corporation; a large population might be the retail transactions occurring in a large city during a one-year period. A point of interest concerning these two examples is that the vice-presidents represent a fixed and unchanging population, whereas the retail transactions illustrate a dynamic population that might differ considerably over time and space.

Infinite Populations

An *infinite population* is composed of an infinitely large number of items. Usually, such populations are conceptual constructs in which data are generated by processes that may be thought of as repeating indefinitely, such as the rolling of dice and the repeated measurement of the weight of an object. Sometimes, the population sampled is finite but so large that it makes little practical difference if it is considered to be infinite. For example, suppose that a population consisting of 1,000,000 manufactured articles contains 10,000 defectives. Thus, 1% of the articles are defective. If two articles are randomly drawn from the lot in succession, without replacing the first article after it is drawn, the probability of obtaining two defectives is

$$\left(\frac{10,000}{1,000,000}\right)\left(\frac{9999}{999,999}\right)$$

For practical purposes, this product is equal to $(0.01)(0.01) = 0.0001$. If the population is considered infinite with 1% defective articles, the probability of obtaining two defectives is exactly 0.0001. Frequently, in situations where a finite population is very large relative to sample size, it is simpler to treat this population as infinite. Since a finite population is depleted by sampling without replacement whereas an infinite population is inexhaustible, then if the depletion causes the population to change only slightly, it may be simpler for computational purposes to consider the population infinite.

Sometimes, an infinite population may be a *process* that produces finite populations. For example, a company may draw a sample from a lot from a particular supplier in order to decide whether to purchase from this supplier in the future. Thus, the purchaser makes a decision concerning the *manufacturing process* that produces future lots. The particular lot sampled for test purposes is a finite population. The process that produces the particular lot may be viewed as an infinite population. Care must be exercised in such situations to ensure that the manufacturing process is indeed a stable one and may validly be viewed as a single universe. Future testing may in fact reveal differences of such a magnitude that the conceptual universe should be viewed as having changed.

Target Populations

Another useful concept is the *target population,* or the universe about which inferences are desired. Sometimes in statistical work, it is impractical or perhaps impossible to draw a sample directly from this *target population,* but it is possible to obtain a sample from a very closely related one. That is, the list of elements that constitutes the sampled *frame* may be related to but is definitely different from the list of elements comprising the target population. For example, suppose it is desired to predict the winner in a forthcoming municipal election by means of a polling technique. The target population is the collection of individuals who will cast votes on election day. However, it is not possible to draw a sample directly from this population, since the specific individuals who will show up at the polls on election day are unknown. It may be possible to draw a sample from a closely related population, such as the eligible voting population. In this case, the list of eligible voters constitutes the sampling frame. The percentage of eligible voters who would vote for a given candidate may differ from the corresponding figure for the election day population. Furthermore, the percentage of the eligible voter population who would vote for a given candidate will probably change as the election date approaches. Thus, we have a situation in which the population that can be sampled changes over time and is different from that about which inferences are to be made. In the case of election polling, the situation is further complicated by the fact that at the time the sample is taken, many individuals may not have made up their minds concerning the candidate for whom they will vote. Therefore, some assumption must be made about how these "undecideds" will break down as to voting preferences. It is common to use "in-depth" interviews, in which questions are asked of the undecided voters concerning the issues and individuals in the campaign, to help determine for whom the respondents will probably vote. In carefully run election polls, numerous sample surveys are taken, spaced through time, including some investigations near the election date. Then, trends can be determined in voting composition, and inferences can be made from populations that are defined very close in time to election day. Some instances of incorrect predictions in national elections have resulted from failures to deal properly with the problem of undecided voters and from cessation of sampling too long before election day. There is no easy answer to the question of how to adjust for the fact that the populations sampled are different from the election day population. For example, one approach to the problem of nonvoters is to conduct post-election surveys to determine the composition of the nonvoting group and to estimate their probable voting pattern had they shown up at the polls. Historical information of this sort could conceivably be used to adjust future polls of eligible voters. However, this is an expensive procedure, and the appropriate method of adjustment is fraught with problems.

In many statistical investigations, the target population coincides with a population that can be sampled. However, in any situation where one must sample a past statistical universe and yet make estimates for a future universe, the problem of inference about the target universe is present.

Control Groups

Probably the most familiar setting involving the concept of a control group is the situation in which an experimental group is given some type of treatment. In order to determine the effect of the treatment, another group is included in the experiment but is not given the treatment. These "no-treatment" cases are known as the *control group*. The effect of the treatment can then be determined by comparing the relevant measures for the "treatment" and "control" groups. For example, in testing the effectiveness of an inoculation against a particular disease, the inoculation may be administered to a group of school children (the treatment group) and not given to another group of school children (the control group).[1] The effectiveness of the inoculation can then be determined by comparing the incidences of the disease in the two groups. The experiment should be designed so that there is no systematic difference between the two groups at the outset that would make one group more susceptible to the disease than the other. Therefore, such experiments are sometimes designed with so-called "matched pairs," where pairs of persons having similar characteristics, one from the treatment group and one from the control group, are drawn into the experiment. For example, if age and health are thought to have some effect on incidence of the disease, pairs of school children may be drawn who are similar with respect to these characteristics, and the treatment is given to one child of each pair and not to the other. Since the children are of similar ages and have the same health backgrounds, these factors cannot explain the fact that one child contracts the disease whereas the other does not. In the language of experimental design, age and general health conditions are said to have been *designed out* of the experiment. Numerous other techniques are employed in experimental design to ensure that treatment effects can be properly measured.

The concept of a control group is important in many statistical investigations in business and economics. In fact, in many instances, the results of an investigation may be uninterpretable unless one or more suit-

---

[1] Difficult ethical questions arise in cases of this sort involving human experimentation. If the inoculation is indeed effective, its use clearly should not be withheld from anyone who wants it. In cases where the effectiveness of a new treatment is highly questionable, yet human experimentation appears necessary, the treatment group often is composed entirely of volunteers.

able control groups have been included in the study. It is a sad fact that often after statistical investigations are completed at considerable expense it is found that because of faulty design and inadequate planning, the results cannot be meaningfully interpreted or the data collected are inappropriate for testing the hypotheses in question. It is of paramount importance that during the planning stage, the investigators think ahead to the completion of the study. They should ask, "If the collected data show thus-and-so, what conclusions can we reach?" This simple yet critical procedure will often highlight difficulties connected with the study design.

A few examples follow to illustrate the use of control groups in statistical studies. Suppose a mail-order firm decided to conduct a study to determine the characteristics of its high-volume purchasers. Its purpose is to determine the distinguishing characteristics of these customers in order to direct future campaigns to noncustomers who have similar attributes. Assume that the firm decides to do this by studying all its high-volume customers. At the conclusion of the investigation, it will be able to make statements such as, "The mean income of high-volume customers is so-many dollars." Or it may calculate that $X\%$ of these purchasers have a certain characteristic. Such population figures will be of virtually no use unless the company has an appropriate control group against which to compare them. What the company wants is to be able to isolate the distinguishing characteristics of high-volume purchasers. Thus, in studying its customers, the company should have separated them into two groups, "high-volume" and "nonhigh-volume." If it studied both groups, it would be in a position to determine those properties that are different between the two groups. Thus, if the company found that the high-volume and nonhigh-volume customers had the *same* mean incomes and that in *both* groups $X\%$ possessed a certain characteristic, it could not use these properties to distinguish between the two groups. The properties that differed most between the two groups would obviously be the most useful ones for spelling out the distinguishing characteristics of high-volume purchasers. In summary, the firm could have used the nonhigh-volume customers as a control group against which to compare the properties of the high-volume group, which in the terminology used earlier would represent the "treatment group."

Care must be used in the selection of the properties of the two groups to be observed. These properties should bear some logical relationship to the characteristic of high-volume versus nonhigh-volume purchases. Otherwise, the properties may be spurious indicators of the distinguishing characteristics of the two groups. For example, income level would be logically related to purchasing volume; if the high-volume purchaser group had a substantially higher income than the nonhigh-volume group, then income would evidently be a reasonable distinguishing characteristic. On the other hand, suppose the high-volume purchaser group happened to have a higher percentage of persons who wore black shoes at the time of the survey than did the nonhigh-volume group. This characteristic of shoe color would *not* seem to be logically related to volume of purchases. Hence, we would not be surprised if the relationship between shoe color and volume of purchases disappeared in subsequent investigations or even reversed itself.

A couple of comments may be made on the construction of control groups. If the treatment group is symbolized as A and the control group as B, then an alternative control group to the one used would have been the treatment and control groups combined, or $A + B$. Thus, in the above example, if the relevant data had been available for the high-volume and nonhigh-volume customers combined (that is, for all customers), this group could have constituted the control. For example, let us assume for simplicity that there were equal numbers of high-volume and nonhigh-volume customers. Suppose that 90% of high-volume customers possessed characteristic X, whereas only 50% of nonhigh-volume customers had this characteristic. The same information would be given by the statements that 90% of high-volume customers possessed characteristic X, whereas 70% of *all* customers possessed this characteristic. (The 70% figure, of course, is the weighted mean of 90% and 50%.) With the knowledge of equal numbers of persons in the high-volume and nonhigh-volume groups, it can be inferred that 50% of nonhigh-volume customers had the property in question. This point is of importance, because sometimes historical data may be available for an entire group $A + B$, whereas available resources may permit a study only of the treatment group A or (the more usual case) only a sample of this group. However, drawing conclusions about the present from a historical control group is dangerous, because systematic changes may have taken place in the treatment and control groups over time or in the surrounding conditions of the experiment. Therefore, the more scientifically desirable procedure is to design the treatment and control groups for the specific investigation in question.

The general objectives of an investigation determine the control groups to be used. Thus, individuals who are not customers of a firm could constitute an appropriate control group for an experiment to determine the distinguishing characteristics of the firm's customers; or customers who have not purchased a specific product could constitute the control group in an experiment to find the particular traits of the customers who purchase the product. It may also be noted that time considerations and available resources usually permit only drawing of samples from the treatment and control populations rather than complete enumerations of these populations.

Some other brief examples of the use of control groups will be given here. A national commission wanted to investigate insurance conditions in cities in which civil disturbances in the form of riots had occurred. Specifically, the commission wished to study cancellation rates for burglary, fire, and theft policies in sections of these cities ("riot areas") primarily affected by the riots. The main purpose was to determine whether individuals and businesses in these areas were having difficulty retaining such policies because of cancellations by insurance companies. It became clear in the planning stages of the study that it would not be sufficient merely to measure cancellation rates in the cities in which riots had occurred, because it would not be possible in the absence of other information to judge whether these rates were low, average, or high. Therefore, a sample of individuals and businesses in cities that had not experienced riots was used as a control group. Another control group was established consisting of individuals and businesses in the "nonriot areas" of the cities that had experienced riots. Thus, the data on cancellation rates could be meaningfully interpreted. Comparisons were made between cities that had experienced riots and those that had not. Further comparisons were made between cancellation rates in riot areas and nonriot areas in cities where these disturbances had been present. The data disclosed that burglary, fire, and theft insurance cancellation rates were higher in cities in which riots had occurred. Furthermore, within cities in which these civil disturbances had been present, cancellation rates were higher in sections where riots had been experienced than in nonriot sections. It may be noted, parenthetically, that if the company policies on cancellations were known and data were available in suitable form, the same information could have been obtained from the company records. However, such information was not available, so the sample survey was required to obtain the indicated data.

Another illustration is the case of a company that wished to determine whether its labor costs were very different from such costs throughout the company's industry group. It obtained the ratio of labor costs to total operating costs for the company and compared these to a published distribution of such ratios for all firms in the industry for the same time period. In this situation, all firms in the industry constituted the control group. This is an illustration of a rather obvious need for and choice of a control group. In many situations, the need and choice are somewhat more subtle.

Types of Errors

The concept of error is central throughout all statistical work. Wherever we have measurement, inference, or decision making, the possibility of error is present. In this section, we deal with errors of measurement. The concepts of errors of inference and decision making are treated in subsequent chapters.

It is useful to distinguish two different types of errors that may be present in statistical measurements, namely, *systematic errors* and *random errors*. Systematic errors, as the term implies, cause a measurement to be incorrect in some systematic way. They are errors involved in the procedures of a statistical investigation and may occur in the planning stages or during or after the collection process. Examples of causes of systematic error (or *bias*) are faulty design of a questionnaire (such as misleading or ambiguous questions), systematic mistakes in planning and carrying out the collection and processing of the data, nonresponse and refusals by respondents to provide information, and too great a discrepancy between the sampling frame and the target universe. If observations have arisen from a sample drawn from a statistical universe, systematic errors are those that persist even when the sample size is increased. As a generalization, these errors may be viewed as arising primarily from inaccuracies or deficiencies in the measuring instrument.

On the other hand, *random errors,* or *sampling errors,* may be viewed as arising from the operation of a large number of uncontrolled factors, conveniently described by the term "chance." As an example of this type of error, if repeated random samples of the same size are drawn from a statistical universe (with replacement of each sample after it is drawn), then a particular statistic, such as an arithmetic mean, will differ from sample to sample. These sample means tend to distribute themselves below and above the "true" population parameter (arithmetic mean), with small deviations between the statistic and the parameter occurring relatively frequently and large deviations occurring relatively infrequently. The word "true" has quotation marks around it because it refers to the figure that would have been obtained through equal complete coverage of the universe, that is, a complete census using the same definitions and procedures as had been used in the samples. The difference between the mean of a particular sample and the population mean is said to be a *random error* (or a *sampling error,* as it is termed in later chapters). The complete collection of factors that could explain why the sample mean differed from the population mean is unknown, but we can conveniently lump them all together and refer to the difference as a random or chance error. Random errors are those that arise from differences found between the outcomes of trials (or samples) and the corresponding universe value using the same measurement procedures and instruments. The sizes of the differences are indications of reliability or precision. Random errors decrease on the average as sample size is increased. It is precisely for this reason that we prefer a larger sample of observations to a smaller one, all other things being equal; that is, since sampling errors are on the average smaller for larger samples, the results are more reliable or more precise.

Systematic and random errors may occur in experiments where the variables are manipulated by the investigator or in survey work where observations are made on the elements of a population without any explicit attempt to manipulate directly the variables involved. A few examples will be given here of how bias, or systematic error, may be present in a statistical investigation. The problems of how random errors are measured and what constitute suitable models for the description of such errors represent central topics of statistical methods and are discussed extensively later in this chapter and Chapters 5 through 8.

Systematic Error: Biased Measurements

The possible presence of biased measurements in an experimental situation may be illustrated by a simple example. Suppose that a group of individuals measured the length of a 36-inch table top using the same yardstick. Let us further assume that the yardstick, although calibrated as though it were 36 inches long, was in fact 35 inches long and this fact was unknown to the individuals making the measurements. There would then be a systematic error of one inch present in each of the measurements, and a statistic such as an arithmetic mean of the readings would reflect this bias. In this type of situation, the systematic error could be detected if another, correctly calibrated, yardstick were used as a standard against which to test the incorrect one. This is an important methodological point. Often, systematic error can be discovered through the use of an independent measuring instrument. Even if the independent instrument is inaccurate, a comparison of the two measuring instruments may give clues about where the search for sources of bias should be made. The variation among the individual measurements made with the incorrect yardstick would be a measure of random error (also referred to as *experimental error*), that is, differences among individual observations that are not attributable to specific causes of variation. Note that the observations may have been very precise (although inaccurate), in the sense that each person's measurement was very close to that of every other person. Thus, the random errors would be small, and there would be good *repeatability*, because in repeating the experiment, each measurement would be close to preceding measurements. These random or chance errors may be assumed to be compensating, in the sense that some observations would tend to be too large and some too small. Since the table top is 36 inches long, the measurements would tend to cluster around a value about one inch greater than the true length of the table top. In summary, we have a model in which each individual measurement may be viewed as the sum of three components: (1) the true value, (2) systematic error, and (3) random error. This relationship is stated in equation form below.

(**4.1**) $$\text{Individual measurement} = \text{True value} + \text{Systematic error} + \text{Random error}$$

Systematic Error: *Literary Digest* Poll

A classic case of systematic error in survey sampling procedure is that of the *Literary Digest* prediction of the presidential election of 1936. During the election campaign between Franklin D. Roosevelt and Alf Landon, the *Literary Digest* magazine sent questionnaire ballots to a very large list of persons whose names appeared in telephone directories and automobile registration lists. Over two million ballots, amounting to about one fifth of the total number sent out, were returned by the respondents. On the basis of these replies, the *Literary Digest* erroneously predicted that Landon would be the next president of the United States. The reasons why the results of this survey were so severely biased are rather clear. In 1936, during the Great Depression, the presidential vote was cast largely along economic lines. The group of the electorate that did not own telephones or automobiles did not have an opportunity to be included in the sample. This group, which represented a lower economic level than owners of telephones and automobiles, voted predominantly for Roosevelt, the Democratic candidate. A second reason stemmed from the nonresponse group, which represented about four fifths of those polled. Typically, individuals of higher educational and higher economic status are more apt to respond to voluntary questionnaires than those with lower economic and educational status. Therefore, the nonresponse group doubtless contained a higher percentage of this lower status group than did the group that responded to the questionnaire. Again, this factor added a bias due to underrepresentation of Democratic votes. In summary, the sample used for prediction purposes contained a greater proportion of persons of higher socioeconomic status than were present in the target population, namely, those who cast votes on election day. Since this factor of socioeconomic status was related to the way people voted, a systematic overstatement of the Republican vote was present in the sample data.

Two methodological lessons can be derived from this example. First, as earlier noted, it is a dangerous procedure to sample a frame that differs considerably from the target population. Second, procedures must be established to deal with the problem of nonresponse in statistical surveys. Clearly, even if the proper target universe had been sampled in this case, the problem of nonresponse would still have to be properly handled.

Systematic Error: Method of Data Collection

Another example of bias will now be given to illustrate that the direction of systematic error may be associated with the nature of the agency that collects the data as well as the method by which the data are collected.

The alumni society of a large Eastern university decided to gather information from the graduates of that institution to determine a number of characteristics, including their current economic status. One of the questions of interest was the amount of last year's gross income, suitably defined. A mail questionnaire was sent to a random sample of graduates, and the results were tabulated from the returns. When frequency distributions were made and averages were calculated by year of graduation, it became clear that the income figures were unusually high as compared to virtually any existing external data that could be examined; in other words, the income figures were clearly biased in an upward direction. It is fairly easy in this case to speculate on the causes of this upward systematic error. In this type of mail questionnaire, a higher nonresponse rate could be expected from those graduates whose incomes were relatively low than from those with higher incomes. That is, it appears reasonable that those with higher incomes would have a greater propensity to respond than others. Furthermore, if there were instances of misreporting of incomes, it is probable that these tended to be overstatements rather than understatements, because of the desire to appear relatively economically successful.

On the other hand, let us consider the same sort of data as reported to the Internal Revenue Service on annual income tax returns. Doubtless, it is safe to say that there is relatively little overstatement of gross incomes. Indeed, it seems reasonable to suppose that there is a downward bias in these data in the aggregate. It may be noted that since responses to the Internal Revenue Service are mandatory, the effect of nonresponse may be considered negligible. Thus, the interesting situation is presented here of the same type of data being gathered by two different agencies, one set being biased in an upward direction, the other in the opposite direction. Therefore, in using secondary statistical information, it is clear that informed critical judgment must be exercised to extract meaningful inferences. This judgment must include practical considerations such as methods of data collection and auspices under which studies are conducted. Of course, false reporting is not easily overcome, particularly in situations in which no independent objective data are available against which the reported information may be checked.

Exercises

1. Explain the difference between a sample and a census.
2. Give an example of a situation where the population of interest would be
   a. an infinite population
   b. considered an infinite population yet in reality is a finite population

c. considered an infinite population, since it is generated by a finite population repeated indefinitely into the future

d. a target population

3. If one is interested in the percentage of consumers in the New York City area who would buy a certain type of men's suit at varying prices, what is the population of interest? Is this a fixed and unchanging population or a dynamic population? Explain your answer.

4. Discuss the need for and advantage of having a control group in a study, and give an example of a study in which use of a control group would be of value.

5. In each of the following situations, state whether a control group would be of use and, if so, what the control group would be.

   a. You are interested in investigating the accident rates in low-income urban areas, and you have data on numbers of accidents occurring in low-income urban areas, number of cars registered in low-income areas, actual area of low-income areas, and other similar information.

   b. You are interested in evaluating the effectiveness of a new safety lighting program to be installed in your plant.

6. The personnel director of a large manufacturing concern wished to determine employee opinion on the annual Christmas party. The party had been for employees only in the past, but the director thought husbands and wives of employees might be included for the next party. A random sample of 100 persons who had attended the party the previous year was selected and asked their opinions. Eighty wanted the party to be for employees only, but 20 requested that husbands and wives be included. Based on the results, the director decided to continue the present practice. Briefly evaluate the procedure employed by the director, indicating any possible sources of bias and nonhomogeneity.

7. Distinguish clearly between systematic errors and random errors. Explain which error will decrease with a larger sample size, which will not, and why. Which error can and should be eliminated?

8. State whether each of the following errors should be considered random, systematic, or both, and why:

   a. In a study that attempted to estimate the percentage of students who smoke, the first 100 students who entered the student lounge, the only area in the building where students are permitted to smoke, were asked if they smoked. The study resulted in an overestimate of the true percentage.

   b. In a study to estimate the average life of a certain type of vacuum tube, five tubes were purchased from five different stores in five different wholesale sales regions. The average life of the five tubes tested was shorter than the "true" average life.

   c. In a study to determine the true weight of a process that fills one-pound cans, 50 cans were selected randomly and weighed on a scale that measured 0.1 ounce too heavy. The process actually filled the cans on the average with 1 pound of material, but the 50 cans averaged 1.06 pounds.

   d. One thousand questionnaires were sent out asking the respondents to rate the community services (police, fire, garbage, and so on) as bad, adequate, or good. Of the 87 responses, a majority rated the services either bad or good. Yet the majority of the people in the community considered the services adequate.

4.5 FUNDAMENTALS OF SAMPLING

Purposes

Sampling is important in most applications of quantitative methods to managerial and other business problems. There are a wide variety of reasons why this is so. In certain instances, sampling may represent the only possible or practicable method of obtaining the desired information. For example, in the case of processes, such as manufacturing, where the universe is conceptually infinite (including all future as well as current production), it is not possible to accomplish a complete enumeration of the population. On the other hand, if sampling is a destructive process, it may be possible to effect a complete enumeration of the universe, but it would not be practical to do so. For example, if a military procurement agency wanted to test a shipment of bombs, it could detonate all of the bombs in a destructive testing procedure and obtain complete information concerning the quality of the shipment. However, since there would be no usable product remaining, a sampling procedure is clearly the only practical way to assess the quality of the shipment.

Sampling procedures are often employed for overall effectiveness, cost, timeliness, and other reasons. A complete census, although it does not have sampling error introduced by a partial enumeration of the universe, nevertheless often contains greater total error than does a sample survey, because greater care can usually be exercised in a sample survey than in carrying out censuses. Errors in collection, classification, and processing of information may be considerably smaller in the case of sample surveys, which can be carried out under far more carefully controlled conditions than substantially larger-scale complete censuses. For example, it may be possible to reduce response errors arising from lack of information, misunderstood questions, faulty recall, and other reasons only by intensive and expensive interviewing and measurement methods, which may be feasible in the case of a sample but prohibitively costly for a complete enumeration.

The employment of sampling rather than censuses for purposes of timeliness occurs in a wide variety of areas. A notable example is the wide array of government data on economic matters such as income, employment, and prices, which are collected on a sample basis at periodic intervals. Here, timeliness of publication of the results is of considerable importance. The more rapid collection and processing of data afforded by sampling procedures represents an important advantage over corresponding census methods.

Random and Nonrandom Selection

Items can be selected from statistical universes in a variety of ways. It is useful to distinguish random from nonrandom methods of selection. In this book, attention is focused on random, or probability, sampling, that is, sampling in which the probability of inclusion of every element in the universe is *known*. Nonrandom sampling methods are referred to as "judgment sampling," that is, selection methods in which judgment is exercised in deciding which elements of a universe to include in the sample. Such judgment samples may be drawn by choosing "typical" elements or groups of elements to represent the population. They may even involve random selection at one stage but allow the exercise of judgment in another. For example, areas may be selected at random in a given city, and interviewers may be instructed to obtain specified numbers of persons of given types within these areas but may be permitted to make their own decisions as to which individuals are brought into the sample.

This book deals only with random, or probability, sampling methods rather than judgment sampling because of the clear-cut superiority of probability selection techniques. The basic reason random sampling is preferable to judgment sampling is that in judgment selection there is no objective method of measuring the precision or reliability of estimates made from the sample. On the other hand, in random sampling, the precision with which estimates of population values can be made is obtainable from the sample itself. This is a very important advantage, since random sampling techniques thus provide an objective basis for measuring errors due to the sampling process and for stating the degree of confidence to be placed upon estimates of population values.

Judgment samples can sometimes be usefully employed in the planning and design of probability samples. For example, where expert judgment is available, a pilot sample may be selected on a judgment basis in order to obtain information that will aid in the development of an appropriate sampling frame for a probability sample.

Simple Random Sampling

We have seen that a *random sample* or *probability sample* is a sample drawn in such a way that the probability of inclusion of every element in the population *is known*. There are a wide variety of types of such probability samples, particularly in the area of sample surveys. Ex-

perts in survey sampling have developed a large body of theory and practice aimed toward the optimal design of probability samples. This is a
highly specialized area to which an entire course or two in a graduate
program in statistics is often allocated. We will concentrate upon the
simplest and most fundamental probability sampling method, namely,
simple random sampling. The major body of statistical theory is based
on this method of sampling.

We first define a simple random sample for the case of a finite population of N elements. A simple random sample of n elements is a sample
drawn in such a way that *every combination of n elements has an equal
chance of being the sample selected.* Since most practical sampling situations involve sampling *without replacement,* it is useful to think of
this type of sample as one in which each of the N population elements
has an equal probability, $1/N$, of being the one selected on the first draw,
each of the remaining $N - 1$ has an equal probability, $1/(N - 1)$, of being
selected on the second draw, and so on until the nth sample item has
been drawn. Since there are $\binom{N}{n}$ possible samples of n items, the probability that any sample of size n will be the one drawn is $1/\binom{N}{n}$.

This concept of a simple random sample may be illustrated by the
following example. Let the population consist of three elements, A, B,
and C. Thus, $N = 3$. Suppose we wish to draw a simple random
sample of two elements; then $n = 2$. Using Equation 2–20 for the
number of combinations that can be formed of N objects taken n at a
time, we find that the number of possible samples is

$$\binom{3}{2} = \frac{3!}{2!1!} = 3$$

These three possible samples contain the following pairs of elements:
(A, B), (A, C), and (B, C). The probability that any one of these three
samples will be the one selected is $\frac{1}{3}$.

It is a property of simple random sampling that every element in the
population has an equal probability of being included in the sample.
However, many other sample designs possess this property as well,
as for example, certain stratified sample and cluster sample procedures, discussed in Section 5.6.

Simple random samples were defined above for the case of sampling
a finite population without replacement. If a finite population is sampled *with replacement,* the same element could appear more than once
in the sample. Since, for practical purposes, this type of sampling is virtually never employed, it will not be discussed any further here.

On the other hand, simple random sampling of *infinite populations*
is of importance, particularly in the context of sampling of processes.
The following definition is often given corresponding to the one for

finite populations. For an infinite population, a simple random sample is one in which *on every selection, each element of the population has an equal probability of being the one drawn.* This is difficult to visualize, in terms of actual sampling from a physical population. Therefore it is helpful to take a more formal approach and to use the language of random variables. Thus, we view the drawing of the sample as an experiment in which observations of values of a random variable are generated, and the successive sample observations or elements are the outcomes of trials of the experiment. Then a simple random sample of *n* observations is defined by the presence of two conditions: (1) the *n* successive trials of the experiment are independent, and (2) the probability of a particular outcome of a trial defined as a success remains constant from trial to trial. In terms of sampling a physical population, we may interpret these as meaning that (1) the *n* successive sample observations are independent, and (2) the population composition remains constant from trial to trial.

To aid in the interpretation of the above definition, let us consider the case of drawing a simple random sample of *n* observations from a Bernoulli process. To make the illustration concrete, assume a situation in which a fair coin is tossed. A simple random sample of *n* observations would be the sample consisting of the outcomes on *n independent* tosses of the coin. Thus, if the number 0 denotes the appearance of a tail and 1 the appearance of a head, the following notation might designate a particular simple random sample of five observations, in which two tails and three heads were obtained in the indicated order: (0, 1, 0, 1, 1). The random variable in this illustration may be designated as "number of heads (or tails) obtained in five tosses of a fair coin." In summary, we note that (1) the tosses of the coin were statistically independent, and (2) the probability of obtaining a head (or tail) remained constant from trial to trial. It is conventional to use an abbreviated method of referring to such a sample as "a sample of five independent observations from a Bernoulli process," or "a sample of five independent observations from a binomial distribution."

The term "random sample," although it properly refers to a sample drawn with known probabilities, is often used to mean "simple random sample." The student should be aware of this alternative usage.

Methods of Simple Random Sampling

Although it is easy to state the definition of a simple random sample, it is not always obvious how such a sample is to be drawn from an actual population. The following are useful methods:

Drawing Chips from a Bowl. We first restrict our attention to the most straightforward situation, in which the population is finite and the elements are easily identified and can be numbered. For example, suppose there are 100 students in a college freshman class and we wish to draw a simple random sample of ten of these students without replacement. We could assign numbers from 1 to 100 to each of the students and place these numbers on physically similar disks (or balls, slips of paper, etc.), which could then be placed in a bowl. We shake the bowl to accomplish a thorough mixing of the disks and then proceed to draw the sample. The first disk is drawn, and we record the number written on it. We then shake the bowl again, draw the second disk, and record the result. The process is repeated until we have drawn ten numbers. The students corresponding to these ten numbers constitute the required simple random sample.

Tables of Random Numbers. If the population size is very large, the above procedure can become quite unwieldy and time-consuming. Furthermore, it may introduce biases if the disks are not thoroughly mixed. Therefore, in recent years, there has been a marked tendency to use tables of random digits for the purpose of drawing such samples. These tables are useful for the selection of other types of probability samples, as well.

A table of random digits is simply a table of digits that have been generated by a random process. For ease of use, the digits are usually combined into groups, for example, of five digits each. Thus, a table of random digits could be generated by the process of drawing chips from a bowl similar to the one just described. The digits 0, 1, 2, . . . , 9 could be written on disks and the disks placed in a bowl and then drawn, one at a time, *replacing the selected disk after each drawing.* Thus, on each selection, the population would consist of the ten digits. The recorded digits would constitute a particular sequence of random digits. These tables are now usually produced by a computer that has been programmed to generate random sequences of digits.

We now illustrate the use of random digits using Table 4-1. Suppose there were 9241 undergraduates at a large university and we wished to draw a simple random sample of 300 of these students. Each of the 9241 students could be assigned a four-digit number, say, from 0001 to 9241. This list of names and numbers would constitute the sampling frame. We now turn to a table of random digits in order to select a simple random sample of 300 such four-digit numbers. We may begin on any page in the table and proceed in any systematic manner to draw the sample. Assume we decided to use the first four columns of each group of five digits, beginning at the upper left and reading downward. Starting with the first group of digits, we find the sequence 98389. Since we are using the first four digits, we have the number 9838. This exceeds the largest number in our population, 9241, so we ignore this number and read down to pick up the next four-digit number, 1724. This is the number of the first student in the sample. Reading

TABLE 4-1
Random Digits[a]

98389	95130	36323	33381	98930	60278	33338	45778	86643	78214
17245	58145	89635	19473	61690	33549	70476	35153	41736	96170
01289	68740	70432	43824	98577	50959	36855	79112	01047	33005
98182	43535	79938	72575	13602	44115	11316	55879	78224	96740
59266	39490	21582	09389	93679	26320	51754	42930	93809	06815
42162	43375	78976	89654	71446	77779	95460	41250	01551	42552
50357	15046	27813	34984	32297	57063	65418	79579	23870	00982
11326	67204	56708	28022	80243	51848	06119	59285	86325	02877
55636	06783	60962	12436	75218	38374	43797	65961	52366	83357
31149	06588	27838	17511	02935	69747	88322	70380	77368	04222
25055	23402	60275	81173	21950	63463	09389	83095	90744	44178
35150	34706	08126	35809	57489	51799	01665	13834	97714	55167
61486	33467	28352	58951	70174	21360	99318	69504	65556	02724
44444	86623	28371	23287	36548	30503	76550	24593	27517	63304
14825	81523	62729	36417	67047	16506	76410	42372	55040	27431
59079	46755	72348	69595	53408	92708	67110	68260	79820	91123
48391	76486	60421	69414	37271	89276	07577	43880	08133	09898
67072	33693	81976	68018	89363	39340	93294	82290	95922	96329
86050	07331	89994	36265	62934	47361	25352	61467	51683	43833
84426	40439	57595	37715	16639	06343	00144	98294	64512	19201
41048	26126	02664	23909	50517	65201	07369	79308	79981	40286
30335	84930	99485	68202	79272	91220	76515	23902	29430	42049
33524	27659	20526	52412	86213	60767	70235	36975	28660	90993
26764	20591	20308	75604	49285	46100	13120	18694	63017	85112
85741	22843	16202	48470	97412	65416	36996	52391	81122	95157

[a] SOURCE: RAND Corporation, *A Million Random Digits with One Hundred Thousand Normal Deviates* (Glencoe, Ill.: Free Press, 1955), excerpt from page 387. Used by permission.

down consecutively, we find 0128, 9818 (which we ignore), 5926, and so on until 300 four-digit numbers between 0001 and 9241 have been specified. If any previously selected number is repeated, we simply ignore the repeated appearance and continue. In this illustration, we read downward on the page, but we could have read laterally, diagonally, or in any other systematic fashion. The important point is that each four-digit number has an equal probability of selection, regardless of what systematic method of drawing is used, and regardless of what numbers have already preceded.

Methods are available for drawing other types of samples than simple random samples and even for situations where the elements have not been listed beforehand. Many of the tables include instructions on their use, and we will not pursue the subject further here.

Exercises

1. The school board of Lower Fenwick wished to determine voter opinion concerning a special assessment to permit the expansion of school services. Lower Fenwick is an industrial community on the fringe of a metropolitan area and has a population of 25,000. There are 5000 students enrolled in the public schools of the community. The board selected a random sample of these students and sent questionnaires to their parents.
   a. Identify the statistical universe from which the above sample was drawn.
   b. Is the sample chosen a simple random sample of parents in Lower Fenwick? Of parents of public school children in Lower Fenwick? Why or why not?
   c. If you had been asked to assist the board, would you have approved the universe it studied? Defend your position.

2. Give at least three possible reasons why sampling procedures may be preferable to a census in certain situations.

3. State whether each of the following statements is true or false and explain your answer.
   a. Judgment sampling is good, since we can get an objective measure of the random error.
   b. When costs permit a census to be taken, a census is usually preferable to sampling.
   c. Systematic errors can often be reduced by better procedures, while random errors can only be reduced by larger sample sizes.

5

Sampling Distributions

In Chapter 1, we examined how to compute the arithmetic mean and standard deviation from the data contained in a sample. We now consider how such statistics differ from sample to sample if repeated simple random samples of the same size are drawn from a statistical population. The probability distribution of such a statistic is referred to as its *sampling distribution*. Thus, we may have a sampling distribution of a proportion, a sampling distribution of a mean, etc. These sampling distributions are the foundations of statistical inference and are of considerable importance in modern statistical decision theory as well. We begin our discussion by considering the sampling distributions of numbers of occurrences and proportions of occurrences.

5.1 SAMPLING DISTRIBUTION OF NUMBER OF OCCURRENCES

We can illustrate the meaning and properties of the sampling distribution of a number of occurrences by means of a typical example. Let us assume that the probability that any given customer who enters Johnson's Supermarket will purchase ice cream is 10%. We conceive of the purchasing process as an infinite population. Thus, we may view the successive purchases and failures to purchase as outcomes of a series of Bernoulli trials. That is, the three requirements of a Bernoulli process may be interpreted in terms of this problem as follows: (1) there are two possible outcomes on each trial: customer purchases ice cream or customer fails to purchase ice cream, (2) the probability of a purchase of ice cream remains constant from trial to trial, and (3) the trials are independent.

Suppose that on a particular day, we were to draw a simple random sample of five customers who entered the supermarket that day and note the number of purchasers of ice cream in the sample. This number is a random variable that can take on the values 0, 1, 2, 3, 4, or 5. Since we are dealing with a Bernoulli process, the probabilities of obtaining these numbers of purchasers may be computed by means of a binomial distribution with $p = 0.10$, $q = 0.90$, and $n = 5$. Therefore, the respective probabilities are given by the expansion of the binomial $(0.9 + 0.1)^5$. This probability distribution is shown in Table 5-1, using the same notation as in Chapter 3.

TABLE 5-1
Probability Distribution of the Number of Ice Cream Purchasers in a Simple Random Sample of Five Customers: $p = P$ (Ice Cream Purchaser) $= 0.10$

Number of Ice Cream Purchasers x	Probability $f(x)$
0	$\binom{5}{0}(0.9)^5(0.1)^0 = 0.59$
1	$\binom{5}{1}(0.9)^4(0.1)^1 = 0.33$
2	$\binom{5}{2}(0.9)^3(0.1)^2 = 0.07$
3	$\binom{5}{3}(0.9)^2(0.1)^3 = 0.01$
4	$\binom{5}{4}(0.9)^1(0.1)^4 \approx 0.00$
5	$\binom{5}{5}(0.9)^0(0.1)^5 \approx 0.00$
	$\overline{1.00}$

We can also interpret the probability distribution given in Table 5-1 as a sampling distribution. Since the number of ice cream purchasers observed in a sample of five customers is a sample statistic, Table 5-1 displays the probability distribution of this sample statistic. If we took repeated simple random samples of five customers each and the probability that any customer would purchase ice cream was 10%, then in the long run, we would observe no ice cream purchasers in 59% of these samples, one ice cream purchaser in 33% of the samples, and so forth. In summary, the probability distribution may now be called a *sampling distribution of number of occurrences.*

5.2 SAMPLING DISTRIBUTION OF A PROPORTION

Frequently, it is convenient to consider proportion of occurrences rather than number of occurrences. We can convert the numbers of occurrences to proportions by dividing by the sample size. The sample proportion, denoted $\bar{p}$ (pronounced p-bar), may be calculated from

(5.1) $$\bar{p} = \frac{x}{n}$$

where x is the number of occurrences of interest and n is the sample size. In the above example, $\bar{p}$ takes on the possible values $0/5 = 0.00$, $1/5 = 0.20, \ldots, 5/5 = 1.00$ with the same probabilities as the corresponding numbers of ice cream purchasers. (Note that the number of occurrences

TABLE 5-2
Sampling Distribution of the Proportion of Ice Cream Purchasers in a Simple Random Sample of Five Customers: $p = P(\text{Ice Cream Purchaser}) = 0.10$

Proportion of Ice Cream Purchasers $\bar{p}$	Probability $f(\bar{p})$
0.00	0.59
0.20	0.33
0.40	0.07
0.60	0.01
0.80	0.00
1.00	0.00
	1.00

in a sample of size n is given by $x = n\bar{p}$. This may be seen by multiplying both sides of (5.1) by n.) The sampling distribution of $\bar{p}$ is given in Table 5-2. In keeping with the usual convention, the probabilities are denoted $f(\bar{p})$.

Properties of the Sampling Distributions of $\bar{p}$ and $n\bar{p}$

We turn now to the properties of the sampling distributions of number of occurrences, $n\bar{p}$, and proportion of occurrences, $\bar{p}$. The means and standard deviations of these distributions are of particular interest in statistical inference. The calculation of these two measures is given in Table 5-3 for number of ice cream purchasers and in Table 5-4 for proportion of ice cream purchasers. Subscripts are used to indicate the random variable for which these measures are computed. For example, $\mu_{n\bar{p}}$ denotes the mean of the random variable $n\bar{p}$. The definitional formulas (3.11) and (3.12) were used for these computations. In actual applications, calculations such as those in Tables 5-3 and 5-4 are never made to obtain the mean and standard deviation of a binomial distribution because convenient computational formulas are available. The calculations are given here only to aid in understanding the meaning of sampling distributions. General formulas for the mean and standard deviation can be derived by substituting $\binom{n}{x} q^{n-x} p^{x}$ for $f(x)$ and $f(\bar{p})$ in the definitional formulas and performing appropriate manipulations. The results of these derivations are summarized in Table 5-5.

TABLE 5-3
Calculation of the Mean and Standard Deviation of the Sampling Distribution of the Number of Ice Cream Purchasers in a Simple Random Sample of Five Customers:
$p = P(\text{Ice Cream Purchaser}) = 0.10$

$x = n\bar{p}$	$f(x)$	$xf(x)$	$x - \mu_x$	$(x - \mu_x)^2$	$(x - \mu_x)^2 f(x)$
0	0.59	0.00	−0.5	0.25	0.1475
1	0.33	0.33	+0.5	0.25	0.0825
2	0.07	0.14	+1.5	2.25	0.1575
3	0.01	0.03	+2.5	6.25	0.0625
4	0.00	0.00	+3.5	12.25	0.0000
5	0.00	0.00	+4.5	20.25	0.0000
	1.00	0.50			0.4500

$\mu_{n\bar{p}} = \mu_x = \Sigma x f(x) = 0.50$ ice cream purchasers
$\sigma_{n\bar{p}} = \sigma_x = \sqrt{\Sigma(x - \mu)^2 f(x)} = \sqrt{0.4500} = 0.67$ ice cream purchasers

TABLE 5-4
Calculation of the Mean and Standard Deviation of the Sampling Distribution of the Proportion of Ice Cream Purchasers in a Simple Random Sample of Five Customers:
$p = P$ **(Ice Cream Purchaser) = 0.10**

$\dfrac{x}{n} = \bar{p}$	$f(\bar{p})$	$(\bar{p})f(\bar{p})$	$\bar{p} - \mu_{\bar{p}}$	$(\bar{p} - \mu_{\bar{p}})^2$	$(\bar{p} - \mu_{\bar{p}})^2 f(\bar{p})$
0.00	0.59	0.000	−0.10	0.01	0.0059
0.20	0.33	0.066	+0.10	0.01	0.0033
0.40	0.07	0.028	+0.30	0.09	0.0063
0.60	0.01	0.006	+0.50	0.25	0.0025
0.80	0.00	0.000	+0.70	0.49	0.0000
1.00	0.00	0.000	+0.90	0.81	0.0000
		0.100			0.0180

$\mu_{\bar{p}} = \mu_{x/n} = \Sigma \bar{p}f(\bar{p}) = 0.10 = 10\%$ ice cream purchasers

$\sigma_{\bar{p}} = \sigma_{x/n} = \sqrt{\Sigma(\bar{p} - \mu_{\bar{p}})^2 f_{(\bar{p})}} = \sqrt{0.0180} = 0.134 = 13.4\%$ ice cream purchasers

Let us illustrate the use of the formulas given in Table 5-5 for the above distributions of number and proportion of ice cream purchasers. Substituting $p = 0.10$, $q = 0.90$, and $n = 5$, we obtain

Number of Ice Cream Purchasers

Mean: $\mu_{n\bar{p}} = np = 5 \times 0.1 = 0.50$ ice cream purchasers

Standard
Deviation: $\sigma_{n\bar{p}} = \sqrt{npq} = \sqrt{5 \times 0.1 \times 0.9} = \sqrt{0.45}$
$= 0.67$ ice cream purchasers

Proportion of Ice Cream Purchasers

Mean: $\mu_{\bar{p}} = p = 0.10 = 10\%$ ice cream purchasers

Standard
Deviation: $\sigma_{\bar{p}} = \sqrt{\dfrac{pq}{n}} = \sqrt{\dfrac{0.10 \times 0.90}{5}} = \sqrt{0.0180} = 0.134$
$= 13.4\%$ ice cream purchasers

TABLE 5-5
Formulas for the Mean and Standard Deviation of a Binomial Distribution

Random Variable	*Mean*	*Standard Deviations*
Number of occurrences $(n\bar{p})$	$\mu_{n\bar{p}} = np$	$\sigma_{n\bar{p}} = \sqrt{npq}$
Proportion of occurrences $(\bar{p})$	$\mu_{\bar{p}} = p$	$\sigma_{\bar{p}} = \sqrt{\dfrac{pq}{n}}$

Of course, these are the same results obtained in the longer calculations shown in Tables 5-3 and 5-4. Let us interpret these results in terms of the appropriate sampling distributions. The mean of the binomial distribution in this example, where the sample statistic is *number* of ice cream purchasers $n\bar{p}$, is $\mu_{n\bar{p}} = n \cdot p = (5)(0.10) = 0.50$ ice cream purchasers. This says that if we take repeated simple random samples of five customers each and the probability is 10% that any customer would purchase ice cream, then on the average, there will be one half an ice cream purchaser per sample. (If the sample size were $n = 200$ with $p = 0.10$, then on the average, we would expect to obtain $(200)(0.10) = 20$ ice cream purchasers per sample.) The standard deviation, $\sigma_{n\bar{p}} = \sqrt{npq} = 0.67$, is a measure of the variation in *number* of ice cream purchasers attributable to the chance effects of random sampling.

When the sample statistic is *proportion* of ice cream purchasers $\bar{p}$, the mean $\mu_{\bar{p}} = p = 0.10$, or 10% purchasers. This means that if we draw simple random samples of five customers each, on the average, we will observe 10% ice cream purchasers in the samples. The standard deviation, $\sigma_{\bar{p}} = \sqrt{pq/n} = 0.134$ or 13.4%, is again a measure of variation attributable to the chance effects of sampling. Here the variation is expressed in terms of *proportion* of ice cream purchasers.

The binomial distribution in this example is skewed, as shown in Figure 5-1. Two horizontal scales are shown in this graph to depict corresponding values of $n\bar{p}$ and $\bar{p}$.

As we saw in Section 3.4, if p is less than 0.5, the distribution tails off to the right as in Figure 5-1. On the other hand, if p exceeds 0.5, the skewness is to the left. If p is held fixed and the sample size n is made larger, the sampling distributions of $n\bar{p}$ and $\bar{p}$ become more and more symmetrical. This is an important property of the binomial distribution from the standpoint of sampling theory and practice, and we examine it further in Section 5.3. An illustration of this property is given in Figure 5-2 in which p is held fixed at 0.20. The sampling distribution of $\bar{p}$ is shown for sample sizes of $n = 5$, 10, and 100.

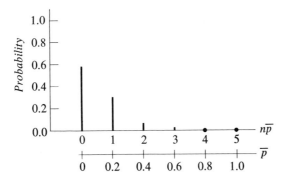

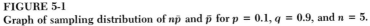

FIGURE 5-1
Graph of sampling distribution of $n\bar{p}$ and $\bar{p}$ for $p = 0.1$, $q = 0.9$, and $n = 5$.

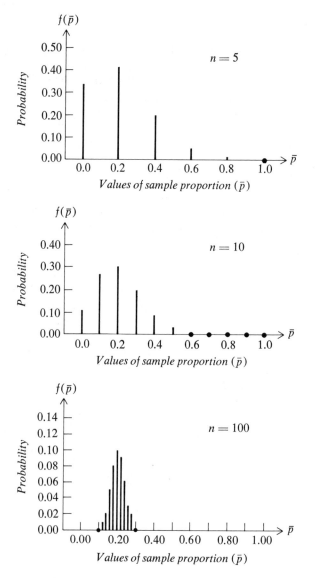

FIGURE 5-2
Binomial distributions for different values of n, with $p = 0.20$.

Exercises

1. An inspector of a bottling company's assembly line draws ten filled bottles at random for inspection. If a bottle contains within half an ounce of the proper amount, it is classified as good; otherwise, it is classified as defective. List the possible numbers of defective bottles that the inspector could obtain in a sample, and do the same for all possible proportions of defectives. Assuming

that the manufacturing process produces 10% defectives, assign probabilities to each of the possible sample outcomes. Calculate the mean and variance for the number of defectives and the proportion of defectives.

2. An econometric forecasting service makes the following claim to Nationwide Shoe Distributors: "We can predict with 85% accuracy whether consumer demand for shoes in the United States will increase or decrease next month as compared to the present month." The forecasting service makes demand predictions for shoes at the ends of six successive months. Assume independence regarding the accuracy of successive monthly predictions.
   a. List all possible proportions of correct predictions the service can have for its six predictions.
   b. Assign probabilities to each possible outcome under the assumption that the service's claim is correct.
   c. Assign probabilities to each possible outcome under the assumption that the service makes random guesses (and thus that the probability that any answer is correct is 0.5).

3. Assume that 40% of a large population favors a certain type of gun control legislation. If ten people selected at random are questioned in regard to the legislation, what are the mean and standard deviation of the sampling distribution of the number favoring the legislation? If 20 people are questioned? If 40 people are questioned? Assume independence.

4. The Podunk Poll interviews 1200 people drawn at random from a very large population and determines that X of the 1200 approve of a reduction in the corporate income tax rate. Assume independence.
   a. What is the form of the distribution of X?
   b. If 70% of the overall population from whom the 1200 were chosen favor the tax rate reduction, what is the standard deviation of X?

5. In 1976, 5% of the companies headquartered in a certain industrialized nation lost money from normal operations. If a random sample of 20 companies is drawn from this large population, what is the mean and standard deviation of the sampling distribution of the proportion of money losers in the random sample? Perform the same calculation for a random sample of 40 companies. Assume independence.

5.3 CONTINUOUS DISTRIBUTIONS

Thus far, we have dealt solely with probability distributions of *discrete* random variables. Probability distributions of continuous random variables are also of considerable importance in statistical theory. We turn therefore to an examination of such distributions, with particular emphasis on the meaning of their graphs. It is suggested that the reader review the definitions of discrete and continuous variables given in Section 3.1.

The binomial distribution that we have been discussing in this chapter is an example of a probability distribution of a *discrete* random variable. We have graphed such distributions by erecting *ordinates*

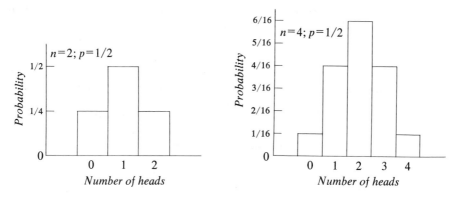

FIGURE 5-3
Histograms of the binomial distribution for $n = 2$, $p = \frac{1}{2}$, and $n = 4$, $p = \frac{1}{2}$.

(vertical lines) at distinct values along the horizontal axis. To gain better insight into the meaning of a graph of the probability distribution of a continuous random variable, let us begin by graphing a binomial distribution as a *histogram*. We assume a situation in which a fair coin is tossed twice and the random variable of interest is the number of heads obtained. The probabilities of zero, one, and two heads are respectively $\frac{1}{4}$, $\frac{1}{2}$, and $\frac{1}{4}$. This is an illustration of a binomial distribution in which $p = \frac{1}{2}$ and $n = 2$. If the coin is tossed four times, the probabilities of 0, 1, 2, 3, and 4 heads are, respectively, $\frac{1}{16}$, $\frac{4}{16}$, $\frac{6}{16}$, $\frac{4}{16}$, and $\frac{1}{16}$. This is a binomial distribution in which $p = \frac{1}{2}$ and $n = 4$. Graphs of these distributions in the form of histograms are given in Figure 5-3. In these histograms, let us now interpret 0, 1, 2, 3, and 4 heads not as discrete values, but rather as midpoints of classes whose respective limits are $-\frac{1}{2}$ to $\frac{1}{2}$, $\frac{1}{2}$ to $1\frac{1}{2}$, $1\frac{1}{2}$ to $2\frac{1}{2}$, and so on. The probabilities or relative frequencies associated with these classes are represented on the graphs by the areas of the rectangles or bars. Thus, in the graph for $n = 4$, since the rectangle for the class interval $2\frac{1}{2}$ to $3\frac{1}{2}$ has four times the area of that from $3\frac{1}{2}$ to $4\frac{1}{2}$, it represents four times the probability. If we were to represent the histogram for the case $n = 4$ by means of a smooth continuous curve, the curve would pass through the rectangle for three heads as shown in Figure 5-4(a). In Figure 5-4(b), the curve is simplified to a straight line, and it is clear that the shaded area under the curve for the class interval $2\frac{1}{2}$ to $3\frac{1}{2}$ is approximately equal to the area of the rectangle representing the probability of three heads, because the included area ABC is about equal to the excluded area CDE. In summary, in the approximation of a histogram by a smooth curve, the area under the curve bounded by the class limits for any given class represents the probability of occurrence of that class. In the foregoing illustration, if we had increased n greatly, say to 50 or 100, and decreased the width of the rectangles, we would see that

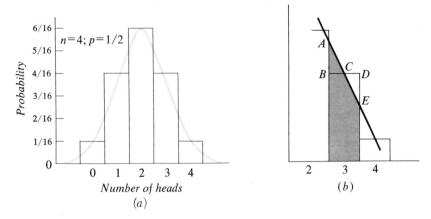

FIGURE 5-4
Approximation of a histogram by a continuous curve.

the corresponding shape of the histogram would appear to approach that of a continuous curve more and more closely. Since the total area of the rectangles in a histogram representing a probability distribution of a discrete random variable is equal to 1, the total area under a continuous curve representing the probability distribution of a continuous random variable is correspondingly equal to 1. Furthermore, the area under the curve lying between the two vertical lines erected at points a and b on the x axis represents the probability that the random variable X takes on values in the interval a to b.[1] This is depicted in Figure 5-5.

In the continuous case, since there are an infinite number of points between a and b, the probability that X lies between a and b may be viewed as the sum of an infinite number of ordinates erected from a to b.

---

[1] Let the value of the probability distribution of a random variable X at x be denoted $f(x)$. If X is discrete, the probability that X lies between a and b inclusive (in the closed interval $[a, b]$) is

$$P(a \leq X \leq b) = \sum_{x=a}^{b} f(x)$$

If X is continuous, the probability that X lies between a and b is

$$P(a \leq X \leq b) = \int_{a}^{b} f(x)\, dx$$

The reader acquainted with integral calculus can see that this definition in the continuous case is the counterpart of the summation in the discrete case. Also, it can be seen that the graphical interpretation of the probability in the continuous case is the area bounded by the curve whose value at x is $f(x)$, the X axis, and the ordinates at a and b. If the probability distribution is continuous at a and b, it makes no difference whether we consider $P(a \leq X \leq b)$ or $P(a < X < b)$, because the probability that X is exactly equal to a or exactly equal to b is 0.

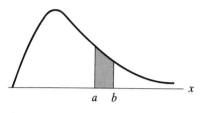

FIGURE 5-5
Graph of a continuous distribution: the shaded area represents the probability that the random variable X lies between a and b.

Intuitively, this sum can be seen to be identical with the area bounded by the curve, the horizontal axis, and the ordinates at a and b. In the discrete case, $f(x)$ denotes the probability that a random variable X takes on the value x. In the continuous case, $f(x)$ cannot be interpreted as the probability of an event x, since there is an infinite number of x values and the probability of any one of them must thus be considered 0.[2] For continuous random variables, probabilities can be interpreted graphically only in terms of *areas between* two values.

By mathematics beyond the scope of this text, it can be shown that if, in the binomial distribution, p is held fixed while n is increased without limit, the distribution approaches a particular continuous distribution, referred to as the *normal distribution, normal curve,* or *Gaussian distribution* (after the mathematician and astronomer Karl Friedrich Gauss). Although our illustration has been for the case $p = \frac{1}{2}$, this is not a necessary condition for the proof. Even if the binomial distribution is not symmetrical (that is, $p \neq \frac{1}{2}$), it still approaches the normal distribution as n increases. The shape of the normal curve is shown in Figure 5-6. Actually, in the early mathematical derivations, the binomial variable x was expressed in so-called *standard units,* that is, as $(x - \mu_{n\bar{p}})/\sigma_{n\bar{p}} = (x - np)/\sqrt{npq}$, and n was assumed to increase without limit. More modern proofs use other approaches to arrive at the same result.

FIGURE 5-6
Graph of the normal distribution.

[2] For example, $P(X = a) = \int_{a}^{a} f(x)\,dx = 0.$

A brief comment on standard units is useful at this point, because such units are so widely employed, particularly in sampling theory and statistical inference. Standard units are merely an example of the previously mentioned standard score (see Section 1.14). The standard score is the deviation of a value from the mean stated in units of the standard deviation. In general, it is of the form $(x - \mu)/\sigma$, where x denotes the value of the item and μ and σ are the mean and standard deviation of the distribution. In the case of a binomially distributed random variable, as indicated in Table 5-5, the mean and standard deviation of X, the number of successes in n trials, are respectively np and $\sqrt{npq}$. Hence, the standard score, or standard unit, is $(x - np)/\sqrt{npq}$. Other terms used to refer to standard scores or standard units include *standardized unit, standardized form,* and *standard form.*

5.4 THE NORMAL DISTRIBUTION

The normal distribution plays a central role in statistical theory and practice, particularly in the area of statistical inference. Because of the relationship we mentioned in Section 5.3, the normal distribution is very useful as an approximation to the binomial distribution in many instances where the latter is the theoretically correct one. As we will see, calculations involving the normal curve are generally much easier than those involving the binomial distribution because of the simple, compact form of tables of areas under the normal curve.

In addition to its use as an approximation to the binomial distribution, the normal distribution is important in its own right in sampling applications. Before we consider such applications, let us examine the basic properties of the distribution.

Properties of the Normal Curve

Probability distributions of continuous random variables can be described by the same types of measures (such as means, medians, and standard deviations) as are used for discrete random variables. One of the important characteristics of the normal curve is that we need know only the mean and standard deviation to be able to compute the entire distribution.

The normal probability distribution is defined by the equation

$$(5.2) \qquad f(x) = \frac{1}{\sqrt{2\pi}\,\sigma}\, e^{-(\frac{1}{2})[(x-\mu)/\sigma]^2}$$

In this equation, the mean and standard deviation, which determine the location and spread of the distribution, are respectively denoted by μ

and σ, which are said to be the two parameters of the normal distribution. This is analogous to the situation for the binomial distribution, in which the parameters are n and p. (The numbers π and e are simply constants that arise in the mathematical derivation; their approximate values are $\pi = 3.1416$ and $e = 2.7183$. π is the familiar quantity that appears in numerous mathematical formulas, such as the expression for the area of a circle, $A = \pi r^2$, where A denotes the area and r the radius. The constant e is the base of the natural logarithm system, as indicated in the discussion of the Poisson distribution in Section 3.7.) Thus, for given values of μ and σ, if we substitute a value of x into Equation 5.2, we can compute the corresponding value of $f(x)$. Following the usual convention, the values x of the random variable of interest are plotted along the horizontal axis and the corresponding ordinates $f(x)$ along the vertical axis. In Figure 5-7 are shown three normal probability distributions, differing in their locations and spread. Distribution (c) has the largest mean (50), and distribution (a) has the smallest mean (30). On the other hand, the standard deviation of (c), which is 10, is the smallest of the three, while that of (a), which is 20, is the largest. Thus, the normal distribution defined by Equation (5.2) represents a family of distributions, each specific member of that family being determined by particular values of the parameters μ and σ.

Graphically, a normal curve is bell-shaped and symmetrical around the ordinate erected at the mean, which lies at the center of the distribution. Recall that the total area under the graph of a continuous probability distribution is equal to 1; since one half of the area (representing probability) lies to the left (right) of the mean, the probability is 0.5 that a value of x will fall below (above) the mean. The values of x range from minus infinity to plus infinity. As we move further away from the mean, either to the right or to the left, the ordinates $f(x)$ get smaller and smaller. Thus, moving in either direction from the mean, the curve is *asymptotic* to the x axis; that is, the curve gets closer and closer to the horizontal axis

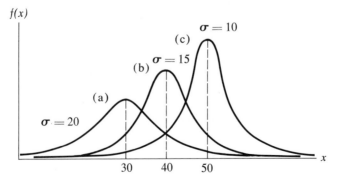

FIGURE 5-7
Three normal probability distributions.

but never reaches it. However, for practical purposes, we rarely need to consider x values lying beyond three or four standard deviations from the mean, since virtually the entire area is included within this range. Stated differently, there is virtually no area in the tails of a normal distribution beyond 3 or 4 standard deviations from the mean.

Areas under the Normal Curve

We now turn to the use of the areas under the normal curve. Although it was important to define the distribution as in Equation 5.2 in order to observe the relationship between x values and $f(x)$ values, in most applications in statistical inference we are not interested in the ordinates of the curve. Rather, since the normal curve is a continuous distribution and since it is a very useful probability distribution, we are interested in the areas under the curve.

It is convenient to use the term "normally distributed" for variables that have normal probability distributions, and we shall do so here. Of course, the term "normal" does not imply that other distributions are in some sense "abnormal"; it merely refers to probability distributions describable by Equation 5.2. Normally distributed variables occur in a variety of units, such as dollars, pounds, inches, and hours. It would be convenient to be able to transform any normally distributed variable into such a form that a single table of areas under the normal curve would be applicable, regardless of the units of the original data. The transformation used for this purpose is that of the standard unit, or, as it is often called in the case of a normal distribution, the *standard normal deviate*. As we noted earlier, to express an observation of a variable in standard units, we obtain the deviation of this observation from the mean of the distribution and then state this deviation in multiples of the standard deviation. For example, suppose a variable was normally distributed with mean 100 pounds and standard deviation 10 pounds. If one observed value of this variable is 120 pounds, what is this number in standardized units? The deviation of 120 pounds from 100 pounds is $+20$ pounds, in units of the original data. Dividing $+20$ pounds by 10 pounds, we obtain $+2$. Thus, a deviation of $+20$ pounds from the mean lies two standard deviations above the mean if one standard deviation equals 10 pounds.

Let us state this notion in general form. The number of standard units z for an observation x from a probability distribution is defined by

(5.3)
$$z = \frac{\text{value} - \text{mean}}{\text{standard deviation}} = \frac{x - \mu}{\sigma}$$

where x = the value of the observation
μ = the mean of the distribution
σ = the standard deviation of the distribution

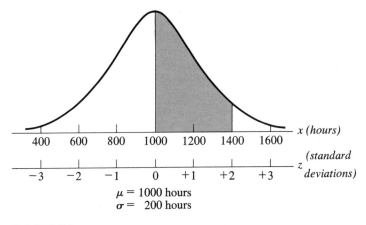

FIGURE 5-8
Relationship between *x* values and *z* values.

In the illustration, $z = (120 - 100)/10 = 20/10 = +2$. The $+2$ indicates a value lying two standard deviations *above* the mean. If the observation had been 80, we would have $z = (80 - 100)/10 = -20/10 = -2$. The -2 denotes a value lying two standard deviations *below* the mean.

We now turn to another example to illustrate the use of a table of areas under the normal curve. The Watts Happening Corporation has a manufacturing process that produces light bulbs whose lifetimes are normally distributed with an arithmetic mean of 1000 hours and a standard deviation of 200 hours. Figure 5-8 shows the relationship between values of the original variable (*x* values) and values in standard units (*z* values).

Suppose we wish to determine the proportion of light bulbs produced by this process with lifetimes between 1000 and 1400 hours, as indicated by the shaded area in Figure 5-8. We can obtain this value from Table A-5 of Appendix A, which gives areas under the normal curve lying between vertical lines erected at the mean and at specified points above the mean stated in multiples of standard deviations (*z* values). The left column of the table gives *z* values to one decimal place. The column headings give the second decimal place of the *z* value. The entries in the body of the table represent the area included between the vertical line at the mean and the line at the specified *z* value. Thus, returning to our example, the *z* value for 1400 hours is $z = (1400 - 1000)/200 = +2$. In Table A-5, we find the value 0.4772; hence, 47.72% of the area in a normal distribution lies between the mean and a value two standard deviations above the mean. We conclude that 0.4772 is the proportion of light bulbs produced by this process with lifetimes between 1000 and 1400 hours.

We now note a general point about the distribution of *z* values.

Comparing the x scales and z scales in Figure 5-8, we see that for a value at the mean in the distribution of x, $z = 0$. If an x value is at $\mu + \sigma$ (that is, one standard deviation above the mean), $z = +1$, and so on. Therefore, the probability distribution of z values, referred to as the *standard normal distribution,* is simply a normal distribution with mean 0 and standard deviation 1.[3]

EXAMPLE 5-1

What is the proportion of light bulbs produced by the above process with lifetimes between 600 and 1400 hours?

First, we transform these values to deviations from the mean in units of the standard deviation.

$$\text{If } x = 1400, z = \frac{1400 - 1000}{200} = +2$$

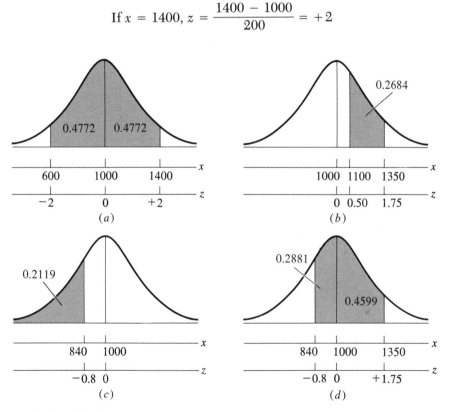

FIGURE 5-9
Areas corresponding to Examples 5-1 through 5-4.

---

[3] We note that Table A-5 gives values of the integral $\int_0^{z_0} f(z)\ dz$, where $z_0 = (x_0 - \mu)/\sigma$ and $f(z) = (1/\sqrt{2\pi})\ e^{-z^2/2}$.

$$\text{If } x = 600, z = \frac{600 - 1000}{200} = -2$$

Thus, we want to determine the area in a normal distribution that lies within two standard deviations of the mean. Table A-5 of Appendix A gives entries only for positive z values. However, since the normal distribution is symmetrical, the area between the mean and a value two standard deviations below the mean is the same as the area between the mean and a value two standard deviations above the mean. Hence, we double the area previously determined to obtain $2(0.4772) = 0.9544$ as the required area. We see that about 95.5% of the light bulbs produced by this process have lifetimes between 600 and 1400 hours. Note that in general, about 95.5% of the area in a normal distribution lies within two standard deviations of the mean. The required area is shown in Figure 5-9(a).

EXAMPLE 5-2

What is the proportion of light bulbs produced by this process with lifetimes between 1100 and 1350 hours?

Both 1100 and 1350 lie above the mean of 1000 hours. We can determine the required probability by obtaining (1) the area between the mean and 1350 and (2) the area between the mean and 1100, and then subtracting (2) from (1).

$$\text{If } x = 1350, z = \frac{1350 - 1000}{200} = \frac{350}{200} = 1.75$$

$$\text{If } x = 1100, z = \frac{1100 - 1000}{200} = \frac{100}{200} = 0.50$$

Table A-5 gives 0.4599 as the area corresponding to $z = 1.75$ and 0.1915 for $z = 0.50$. Subtracting 0.1915 from 0.4599 yields 0.2684 or 26.84% as the result. This area is shown in Figure 5-9(b).

EXAMPLE 5-3

What is the proportion of light bulbs produced by this process with lifetimes of less than 840 hours?

The observation 840 hours lies below the mean. We solve this problem by determining the area between the mean and 840 and subtracting this value from 0.5000, which is the entire area to the left of the mean.

$$\text{If } x = 840, z = \frac{840 - 1000}{200} = -0.80$$

Since only positive z values are shown in Table A-5, we look up $z = 0.80$ and find 0.2881. This is also the area between the mean and $z = -0.80$. Subtracting 0.2881 from 0.5000 gives the desired result, 0.2119. The area corresponding to this probability is shown in Figure 5-9(c).

EXAMPLE 5-4

What is the proportion of light bulbs produced by this process with lifetimes between 840 and 1350 hours?

Since 840 lies below the mean and 1350 lies above the mean, we determine (1) the area lying between 840 and the mean and (2) the area lying between 1350 and the mean, and add (1) to (2). The respective z values for 840 and 1350 were previously determined as -0.80 and $+1.75$, with corresponding areas of 0.2881 and 0.4599. Adding these two figures, we obtain 0.7480 as the proportion of light bulbs with lifetimes between 840 and 1350 hours. The corresponding area is shown in Figure 5-9(d).

It was stated earlier that in the normal distribution, the range of the x variable extends from minus infinity to plus infinity. Yet, in the problems just considered, negative lifetimes were impossible. This illustrates the point that a variable may be said to be normally distributed provided that the normal curve constitutes a good fit to its empirical frequency distribution within a range of about three standard deviations from the mean. Since virtually all the area is included in this range, the situation in the tails of the distribution is considered negligible.

It is useful to note the percentages of area that lie within integral numbers of standard deviations from the mean of a normal distribution. These values have been tabulated in Table 5-6. Hence, as was observed in Example 5-1, about 95.5% of the area in a normal distribution lies within plus or minus two standard deviations from the mean. The reader should verify the other figures from Table A-5. Let us restate these probability figures in terms of rough statements of odds. Since about two thirds of the area lies within one standard deviation, the odds are about two to one that in a normal distribution an observation will fall within that range. Correspondingly, the odds are about 95 : 5, or 19 : 1, for the two standard deviation range and 997 : 3, or about 332 : 1, for three standard deviations.

TABLE 5-6
Percentages of Area that Lie within Specified
Intervals Around the Mean in a Normal
Distribution

Interval	Percentage of Area
$\mu \pm \sigma$	68.3
$\mu \pm 2\sigma$	95.5
$\mu \pm 3\sigma$	99.7

The Normal Curve as an Approximation to the Binomial Distribution

In Section 5.3, we indicated that since the binomial distribution approaches the normal distribution when n becomes large, the normal curve can be used as an approximation to the binomial distribution for the calculation of probabilities for which the binomial distribution is the theoretically correct one. In general, the approximations are better when the value of p in the binomial distribution is close to $\frac{1}{2}$ than when p is close to 0 or 1, because for $p = \frac{1}{2}$ the binomial distribution is symmetrical, and as we have seen the normal curve is a symmetrical distribution. However, the normal distribution often provides surprisingly good approximations even when $p \neq \frac{1}{2}$ and even when n is not very large. A popular rule is that the normal distribution is an appropriate approximation to the binomial distribution when both $np \geq 5$ and $n(1 - p) \geq 5$. Under these conditions, the binomial distribution can be closely approximated by a normal curve with the same mean and standard deviation. We illustrate the use of the normal curve as an approximation to the binomial distribution by two examples. The first example illustrates the approximation of the probability of a single term in the binomial distribution by a normal curve calculation. The second example illustrates a corresponding calculation for a sum of terms in the binomial distribution.

EXAMPLE 5-5

Assume that a State Revenue Department has found from experience that 25% of the individual state income tax returns filed contain arithmetic errors. What is the probability that a randomly drawn sample of 20 returns will contain exactly four with arithmetic errors?

Using Equation 3.3 for the binomial distribution with $n = 20$, $p = 0.25$, and $q = 0.75$, we have $P(X = 4) = f(4) = \binom{20}{4} (0.75)^{16}(0.25)^{4}$.

This probability is evaluated from Table A-1 of Appendix A as

$$P(X = 4) = F(4) - F(3) = 0.4148 - 0.2252 = 0.1896$$

In order to obtain the normal curve approximation to this probability of exactly four returns with arithmetic errors, we set up a normal curve with the same mean and standard deviation as the given binomial distribution and find the area between 3.5 and 4.5, as shown in Figure 5-10. The reason for obtaining the area between 3.5 and 4.5 is that the random variable in the binomial distribution is discrete, whereas in the case of the normal curve it is continuous. Hence, as shown in Figure 5-11, if the binomial probabilities are depicted graphically in the form of a histogram, the true probability of exactly four occurrences is given by the area of the rectangle centered at 4. To approximate this area by a corresponding area under the normal curve, four erroneous tax returns can be treated as the midpoint of a

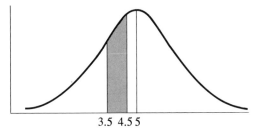

3.5 4.5 5

Number of tax returns with arithmetic errors

FIGURE 5-10
Area under the normal curve for the probability of obtaining exactly four tax returns with arithmetic errors in a randomly drawn sample of 20 returns: $p = 0.25$.

class whose limits are 3.5 and 4.5. The mean and standard deviation of the binomial distribution in this problem are

$$\mu = np = (20)(0.25) = 5$$
$$\sigma = \sqrt{npq} = \sqrt{(20)(0.25)(0.75)} = 1.94$$

Using these numbers as the mean and standard deviation of the approximating normal curve, we calculate the z values for 3.5 and 4.5 as follows:

$$z_1 = \frac{3.5 - 5}{1.94} = -0.77$$

$$z_2 = \frac{4.5 - 5}{1.94} = -0.26$$

The areas for these z values are, respectively, 0.2794 and 0.1026; their difference yields the desired approximation, $0.2794 - 0.1026 = 0.1768$. Hence, 0.1768 is the normal curve approximation to the true binomial probability, which is 0.1896.

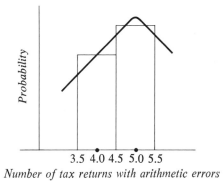

3.5 4.0 4.5 5.0 5.5

Number of tax returns with arithmetic errors

FIGURE 5-11
Representation of a binomial distribution as a histogram and the corresponding normal curve approximation.

EXAMPLE 5-6

In the same context as in Example 5-5, what is the probability that a randomly drawn sample of 20 tax returns will contain four or more with arithmetical errors?

Summing the appropriate terms in Equation (3.3), we find

$$P(X \geq 4) = \sum_{x=4}^{20} f(x) = \sum_{x=4}^{20} \binom{20}{x} (0.75)^{20-x}(0.25)^x$$

This probability is evaluated from Table A-1 of Appendix A as

$$P(X \geq 4) = 1 - F(3) = 1 - 0.2252 = 0.7748$$

The corresponding normal curve approximation is shown graphically in Figure 5-12. As indicated in Example 5-5, the z value for 3.5 is -0.77. Therefore, the desired area is $0.2794 + 0.5000 = 0.7794$. The closeness of this approximation to the true binomial probability of 0.7748 illustrates the fact that normal curve approximations involving sums of terms usually are closer to the corresponding true probabilities than are approximations for individual terms in the binomial distribution.

When, as in Example 5-5, the probability of a *single term* in the binomial distribution is desired, it is always necessary to use an interval one unit wide centered on that term, because the binomial distribution is discrete whereas the normal curve is continuous. On the other hand, this correction is often dispensed with when *sums of terms* in the binomial distribution are desired and the *sample size is large*. Thus, in Example 5-6, if the sample size had been, say, 100, we could have used the z value for 4 rather than for 3.5 in calculating the probability of four or more defectives. Problems at the end of this section requiring the use of this so-called "continuity correction" are identified there.

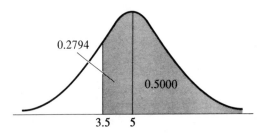

0.2794

0.5000

3.5   5

FIGURE 5-12
Area under the normal curve for the probability of obtaining four or more tax returns with errors in a randomly drawn sample of 20 returns: $p = 0.25$.

Exercises

1. What distinguishes a continuous variable from a discrete variable?

2. Two companies, A and B, produce a certain type of specialized steel. Suppose the thickness of this steel is normally distributed. Roughly sketch the probability distributions for the two companies on the same graph for each of the following cases:

   a. A has a mean of 3 inches and a standard deviation of $\frac{1}{2}$ inch; B has a mean of 3 inches and a standard deviation of 1 inch.

   b. A has a mean of 3 inches and a standard deviation of 1 inch; B has a mean of 4 inches and a standard deviation of 1 inch.

3. The actual weight of the 20-ounce box of Sugar Tops produced by the Carbohydrate Cereal Corporation is normally distributed with a mean of 20 ounces and a standard deviation of 0.5 ounces. For each of the following questions, using graphs with both X and Z axes, indicate the corresponding area under the normal curve. If a 20-ounce box of Sugar Tops is bought at random from a store, what is the probability that it will weigh

   a. under 20 ounces?
   b. more than 21 ounces?
   c. between 19.25 and 20.75 ounces?
   d. either more than 20.6 ounces or less than 19.4?
   e. at least 19.8 ounces?
   f. at most 20.4 ounces?

4. Let X represent the strength of a certain type of hemlock beam produced by Outland Lumber Company. Assume that X is normally distributed with a mean of 2000 pounds per square inch (psi) and a standard deviation of 100 psi. A beam is drawn at random, and its strength is tested. Change each of the following probability statements made in terms of X into statements about the standardized variable Z. What is the probability that the tested strength will be

   a. at least 2150 psi?
   b. less than 1825 psi?
   c. between 1875 and 2115 psi?
   d. between 1795 and 1905 psi?
   e. either more than 2250 or less than 1800 psi?
   f. at most 2100 psi?

5. Let X, the starting salary of an MBA from the Vance School of Finance, be normally distributed with a mean of $18,000 per year and a standard deviation of $1000. State whether each of the following statements about Vance MBAs is true or false and why.

   a. It is as probable that an MBA will receive a starting salary below $16,000 as that he or she will receive a starting salary above $20,000.

   b. It is more probable that an MBA's starting salary will be between $18,000 and $19,500 than that it will be between $16,500 and $18,000.

   c. It is as probable that an MBA's starting salary will be below $18,000 as it is that it will be above $18,000.

   d. The starting salary of an MBA will be below $20,000 95.5% of the time.

6. The amount of sewage dumped into the Deadfish River by the Could-Care-

Less Chemical Company each day is normally distributed with a mean of 200,000 gallons and a standard deviation of 15,000 gallons. If the local Environmental Standards Board has issued a restriction order limiting the amount of sewage the chemical firm can legally dump into the river to 239,000 gallons per day, what is the probability that there will be an illegal sewage discharge by the firm on any given day?

7. One form of the random walk hypothesis for stock market prices says that successive price changes in individual securities are independent and normally distributed. Thus, if the price of a stock at time t is P_t, this characteristic of price changes can be expressed as $P_t = P_{t-1} + \epsilon_t$, where ϵ_t has a normal distribution and is independent of ϵ_{t+k}, $k \neq 0$. Suppose this distribution has a mean of 0 and a standard deviation of $1 for a particular stock.
   a. What is the probability that the stock price will increase from time $t - 1$ to time t?
   b. What is the probability that the stock price will change by at least $1.50?
   c. What is the probability that the stock price P_t will be at most $P_{t-1} + $1.75?

8. Suppose that you are studying a group of children and that you have two measures of the aggression manifested by any child in a given aggression-inducing situation. You have a theory that the first measure of aggression X is a normally distributed random variable with mean 3 and variance 1. The second measure Y is also thought to be normally distributed, but with mean 2 and variance 4. Suppose that your theories are correct and that X and Y are independent random variables.
   a. find $P(1.5 < X < 3.5)$
   b. find $P(X < 2.5$ and $1.5 < Y < 3.5)$
   c. find $P(X < 2.5$ and $Y > 1.4)$
   d. find $P(X < 2.5$ or $Y > 1.4)$
   e. Between what two numbers do the middle 51% of the scores lie as measured by X, the first measure of aggression?

9. The weight of Greenbay Brand cookies packed in a box is normally distributed with $\mu = 12$ ounces and $\sigma = 0.5$ ounce. If an inspector from the FDA weighs one box chosen at random, what is the probability that the weight observed is
   a. more than 12 ounces?
   b. more than 13 ounces?
   c. between 11.5 and 13 ounces?

10. An efficiency expert has been hired by the Klondike Manufacturing Company to monitor the time required by the chassis assembly station of the transistor radio division to complete its assigned task. The results of the study indicate that the time required for the station to produce the chassis and send it along to the next station is normally distributed with a mean of 16 minutes and a standard deviation of 2 minutes.
   a. What is the probability that the station will complete its task in less than 18.5 minutes?
   b. What is the probability that the station does as well as the optimal efficiency time of 14.2 minutes?

 c. The probability is 0.119 that the required task time will be under a certain value; find this value.

11. A certain manufacturing process turns out 10% defective items. If a sample of 20 items is drawn with replacement and is inspected, what is the probability that
 a. exactly two items are defective?
 b. more than two are defective?
 c. fewer than two are defective?
 (Use the normal approximation to the binomial distribution with the continuity correction factor.)

12. According to a certain study, the probability that a reader will read any particular advertisement in *Scanners Digest* is 0.2. Multiroyal, Inc., in a large advertising campaign, places 25 advertisements in *Scanners Digest* in three months. Assuming independence, use the normal approximation to the binomial distribution with the continuity correction factor to find the approximate probability that a person reads
 a. exactly one Multiroyal advertisement
 b. exactly two Multiroyal advertisements
 c. exactly three Multiroyal advertisements
 d. exactly five Multiroyal advertisements

13. A statistical consultant has set up a mathematical model to classify an expensive solution as either "of workable quality" or "not of workable quality," on the basis of certain chemical tests. The model is such that it will classify a workable solution correctly 95% of the time. If 400 workable solutions are tested, what is the probability that more than 7% will be misclassified? (Use a normal distribution.)

14. The Crude Oil Company has two main oil fields, one in the Alaskan tundra region and one in the Texas Panhandle. The company has enjoyed 25% success in finding retrievable oil deposits with its drillings in the Alaskan field, whereas its successful drillings in Texas have been only 20% of the total drillings. In 1976, the company plans 400 new drillings in each field. If the success percentages stay the same, what is the probability of the company establishing 90 or more new successful oil wells in Alaska? In Texas? Assume independence.

15. The probability that any customer who enters a certain store will purchase a box of Alpine Milk is 0.2. If 2500 customers enter the store, what is the minimum number of boxes of Alpine Milk the store must have on hand if the probability that it will run out of this product is to be at most 1%? Assume independence.

16. According to Pulp Paper Company, 5% of all trees cut down cannot be used in the processing of paper. Assume independence, and suppose that 20 trees are cut down in a given period.
 a. Find the exact probability that two or more trees will not be usable for paper production.
 b. Use the normal approximation to the binomial distribution with the continuity correction factor to obtain an approximation to the exact probability.

5.5 SAMPLING DISTRIBUTION OF THE MEAN

In Sections 5.1 and 5.2, we discussed sampling distributions of numbers of occurrences and percentages of occurrences. We now turn to another important probability distribution, namely, the sampling distribution of the arithmetic mean. For brevity, we shall use the term "the sampling distribution of the mean," or simply "the sampling distribution of $\bar{x}$."[4] To illustrate the nature of this distribution, let us return to the Watts Happening Corporation manufacturing process that produces light bulbs whose lifetime is normally distributed with an arithmetic mean of 1000 hours and a standard deviation of 200 hours. We now interpret this distribution as an infinite population from which simple random samples can be drawn. It is possible for us to draw a large number of such samples of a given size, say $n = 5$, and compute the arithmetic mean lifetime of the five light bulbs in each sample. In accordance with our usual terminology, each such sample mean is referred to as a statistic. Since these statistics will usually differ from one another, we can consider them values of a random variable for which we can construct a frequency distribution. The universe mean of 1000 hours is the parameter around which these sample statistics will be distributed, with some sample means lying below 1000 and some lying above it. If we take any finite number of samples, the sampling distribution is referred to as an *empirical sampling distribution*. On the other hand, if we conceive of drawing all possible samples of the given size, the resulting sampling distribution is a *theoretical sampling distribution*. Statistical inference is based on these theoretical sampling distributions, which are nothing more than probability distributions of the relevant statistics. In most practical situations, only one sample is drawn from a statistical population in order to test a hypothesis or to estimate the value of a parameter. The work implied in generating a sampling distribution by drawing repeated samples of the same size is virtually never carried out, except perhaps as a learning experience. However, it is important for the reader to realize that the sampling distribution provides the underlying theoretical structure for decisions based on single samples.

Sampling from Normal Populations

What are the salient characteristics of the sampling distribution of the mean, if samples of the same size are drawn from a population in which values are normally distributed? To obtain an answer to this question,

---

[4] Although we used the symbol $\bar{X}$ in Chapter 3 to denote a sample mean, we will henceforth use instead the symbol $\bar{x}$ in sampling theory and statistical inference. The lower case notation is more convenient because of the use of $\bar{x}$ as a subscript.

let us begin by assuming that the sample size is five. Interpreting this in terms of our problem, let us assume that a random sample of five light bulbs is drawn from the above-mentioned population, and the mean life-time of these five bulbs, denoted $\bar{x}_1$, is determined. Then, another sample of five bulbs is drawn, and the mean $\bar{x}_2$ is determined. Let us assume that the first mean is 990 hours, which falls below the population mean, and that the second mean is 1022 hours, which lies above the population mean. The theoretical frequency distribution of $\bar{x}$ values of all such simple random samples of five bulbs would constitute the sampling distribution of the mean for samples of size five. Intuitively, we can see what some of the characteristics of such a distribution might be. A sample mean would be just as likely to lie above the population mean of 1000 hours as below it. Small deviations from 1000 hours would occur more frequently than large deviations. Furthermore, because of the effect of averaging, we would expect less dispersion or spread among these sample means than among the values of the individual items in the original population; that is, the standard deviation of the sampling distribution of the mean should be less than the standard deviation of the values of individual items in the population.

Other characteristics of sampling distributions of the mean might be noted. If samples of size 50 rather than five had been drawn, another sampling distribution of the mean would be generated. Again we would expect the means of these samples to cluster around the population mean of 1000 hours. However, we would expect to find even less dispersion among these sample means than in the case of samples of size five, because the larger the sample, the closer the sample mean is likely to be to the population mean. Thus, the standard deviation of the sampling distribution, which measures chance error inherent in the process of using samples to approximate population values, would decrease with increasing sample size. Another characteristic of these sampling distributions, which is not at all intuitively obvious but can be proved mathematically, is that if the original population distribution is normal, sampling distributions of the mean will also be normal. Figure 5-13 displays the relationships we have just discussed for the case of a normal population. For the population distribution, the horizontal axis represents values of individual items (x values). For the sampling distributions, the horizontal axis represents the means of samples of size five and 50. Since all three of the distributions are probability distributions of continuous random variables, the vertical axis pertains to probability densities.

The foregoing material introduces the following theorem:

THEOREM 5.1

If a random variable X is normally distributed with mean μ and standard deviation σ, then the random variable "the mean $\bar{x}$ of a simple random

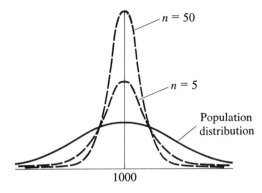

FIGURE 5-13
Relationship between a normal population distribution and normal sampling distributions of the mean for $n = 5$ and $n = 50$.

sample of size n"[5] is also normally distributed with mean $\mu_{\bar{x}} = \mu$ and standard deviation

$$\sigma_{\bar{x}} = \frac{\sigma}{\sqrt{n}}$$

In this statement of the theorem, we have used somewhat more formal language than in the preceding discussion. Instead of saying that the values of individual items in a population are normally distributed, we refer to normal distributions of the random variables X and $\bar{x}$.

A very interesting aspect of the theorem is that the expected value (mean) of the sampling distribution of the mean, symbolized μ_x, is equal to the original population mean μ. This relationship is proved in Rule 13 of Appendix C for the even more general case of simple random samples of size n from *any* infinite population. The standard deviation of the sampling distribution of the mean, usually referred to as the *standard error of the mean* and denoted $\sigma_{\bar{x}}$, is given by

$$(5.4) \qquad\qquad \sigma_{\bar{x}} = \frac{\sigma}{\sqrt{n}}$$

This relationship is proved in Rule 12 of Appendix C, again for the more general case of sampling from *any* infinite population.

Very important implications follow from Equation 5.4. We can think of any sample mean $\bar{x}$ as an estimate of the population mean $\mu_{\bar{x}}$. The difference between the statistic $\bar{x}$ and the parameter μ, $\bar{x} - \mu$, is referred to as a *sampling error*. (For example, if $\bar{x}$ were exactly equal to μ and were used as an estimate of μ, there would be no sampling error.) Therefore, $\sigma_{\bar{x}}$, which is a measure of the spread of the $\bar{x}$ values around μ, is a measure of *average sampling error;* that is, it measures the amount by which $\bar{x}$ can be expected to vary from sample to sample. Another interpretation is that $\sigma_{\bar{x}}$ is a measure of the *precision* with which μ can be

---

[5] See footnote 4.

estimated using $\bar{x}$. Referring to Equation 5.4, we see that $\sigma_{\bar{x}}$ varies directly with the dispersion in the original population, σ, and inversely with the square root of the sample size n. Thus, as might be expected, the greater the dispersion among the items in the original population, the greater the expected sampling error in using x as an estimate of μ; similarly, the smaller the population dispersion, the smaller the expected sampling error. In the limiting case in which every item in the population has the same value, the population standard deviation is 0; therefore, the standard error of the mean is also 0. This indicates that the mean of a sample from such a population would be a perfect estimate of the corresponding population mean, since there could be no sampling error. (For example, if every item in the population weighed 100 pounds, the population mean weight would be 100 pounds. All the items in any sample would weigh 100 pounds, and the sample mean would be 100 pounds; thus, the sample mean would estimate the population mean with perfect precision.) As the population dispersion increases, estimation precision decreases.

The fact that the standard error of the mean varies inversely with the square root of sample size means that there is a certain type of diminishing return in sampling effort. Quadrupling sample size only halves the standard error of the mean; multiplying sample size by 9 cuts the standard error only to one third its previous value.

Sampling from Nonnormal Populations

We mentioned in the foregoing discussion that if a population is normally distributed, the sampling distribution of $\bar{x}$ is also normal. However, many population distributions of business and economic data are not normally distributed. What then is the nature of the sampling distribution of $\bar{x}$? It is a remarkable fact that for almost all types of population distributions, the sampling distribution of $\bar{x}$ is approximately normal for sufficiently large samples.

This relationship between the shapes of the population distribution and of the sampling distribution of the mean has been summarized in what is often called the most important theorem of statistical inference, the *Central Limit Theorem*. The theorem is stated in terms of the z variable for the sampling distribution of the mean and the approach of the distribution of this variable to the standard normal distribution. The formal statement of the theorem is as follows.

Central Limit Theorem

THEOREM 5.2

If a random variable X, either discrete or continuous, has a mean μ and a finite standard deviation σ, then the probability distribution of

$z = (\bar{x} - \mu)/\sigma_{\bar{x}}$ approaches the standard normal distribution as n increases without limit.

This theorem is indeed a very general one, since it makes no restrictions on the shape of the original population distribution. The requirement of a finite standard deviation is not a practical restriction at all, since virtually all distributions involved in real-world problems satisfy this condition. The reader should also note that the z variable in this theorem is a transformation in terms of deviations of the sample mean $\bar{x}$ from the mean of the sampling distribution of means, μ, stated in multiples of the standard deviation of that distribution, $\sigma_{\bar{x}}$. Thus, it is exactly the same type of transformation that is,

$$z = \frac{\text{value} - \text{mean}}{\text{standard deviation}}$$

as was used in the illustrative examples given earlier in this chapter. Here, however, all the values pertain to the sampling distribution of the mean.

It is useful at this point to restate Theorems 5.1 and 5.2 in the less formal context of sampling applications and to summarize the importance of the concepts involved.

THEOREM 5.1 RESTATED

If a population distribution is normal, the sampling distribution of $\bar{x}$ is also normal for samples of all sizes.

THEOREM 5.2 RESTATED

If a population distribution is nonnormal, the sampling distribution of $\bar{x}$ may be considered approximately normal for a large sample.

The first of these theorems states that the sampling distribution of $\bar{x}$ will be *exactly* normal if the population is normal. The second, the Central Limit Theorem, assures us that no matter what the shape of the population distribution, the sampling distribution of $\bar{x}$ approaches normality as the sample size increases. The important point is that for a wide variety of population types, samples do not even have to be very large for the sampling distribution of $\bar{x}$ to be approximately normal. For example, only in the case of very highly skewed populations would the sampling distribution of $\bar{x}$ be appreciably skewed for samples larger than about 20. For most types of populations, the approach to normality is quite rapid as n increases.

Finite Population Multiplier

In our discussion of sampling distributions, we have dealt with infinite populations. However, many of the populations in practical problems are finite, as for example, the employees in a given industry, the households in a city, and the counties in the United States. It turns out that as a practical matter, formulas already obtained for infinite populations can be applied in most cases to finite populations as well. In those cases where the results for infinite populations are not directly applicable, a simple correction factor applied to the formula for the standard deviation of the relevant sampling distribution is all that is required.

In simple random sampling from an *infinite population*, we have seen that the sampling distribution of $\bar{x}$ has a mean $\mu_{\bar{x}}$ equal to the population mean, μ, and a standard deviation $\sigma_{\bar{x}}$ equal to $\sigma/\sqrt{n}$. In sampling from a *finite population*, the mean $\mu_{\bar{x}}$ of the sampling distribution of $\bar{x}$ again is equal to the population mean, μ, but the standard deviation (standard error of the mean) is given by the following formula.

STANDARD ERROR OF THE MEAN FOR FINITE POPULATIONS

(5.5)
$$\sigma_{\bar{x}} = \sqrt{\frac{N-n}{N-1}}\,\frac{\sigma}{\sqrt{n}}$$

Here N is the number of elements in the population and n is the number in the sample. The quantity $\sqrt{(N-n)/(N-1)}$ is usually referred to as the *finite population correction* or the *finite correction factor*. Thus, we see that in the case of a finite population, the standard error of the mean is equal to the finite population correction multiplied by $\sigma/\sqrt{n}$, which is the standard error of the mean in the infinite case. The finite population correction is approximately equal to 1 when the population size N is large relative to the sample size n; therefore, when we are choosing samples of size n from a much larger (but finite) population, the standard error of the mean $\sigma_{\bar{x}}$ is for practical purposes equal to $\sigma/\sqrt{n}$, as was the case therefore, when we sampled an infinite population.

To see why the finite correction factor is approximately 1 when population size is large relative to sample size, note that the factor $\sqrt{(N-n)/(N-1)}$ is approximately equal to $\sqrt{(N-n)/N}$ for large populations, since the subtraction of 1 in the denominator is negligible. We can now write

$$\sqrt{\frac{N-n}{N}} = \sqrt{1-\frac{n}{N}} = \sqrt{1-f}$$

where $f = n/N$ is referred to as the "sampling fraction," since it measures the fraction of the population contained in the sample. Thus, if the population size is $N = 1000$ and the sample size is $n = 10$, then $f = 10/1000 = 1/100$. In such a case, the finite population correction is very close to 1. Here, for example, $\sqrt{1 - 1/100} \approx 1$ (the symbol $\approx$ means "is approximately equal to"). In summary, in this case, $\sigma_{\bar{x}}$ is practically equal to $\sigma/\sqrt{n}$.

A generally employed rule of thumb is that the formula $\sigma_{\bar{x}} = \sigma/\sqrt{n}$ may be used whenever the size of the population is at least 20 times that of the sample, or in other words, whenever the sample represents 5% or less of the population.

A very striking implication of Equation 5.5 is that as long as the population is large relative to the sample, sampling precision becomes a function of sample size alone and does not depend on the proportion of the population sampled. Of course, it is assumed in this statement that the population standard deviation is constant. For example, let us assume a situation in which we draw a simple random sample of size $n = 100$ from each of two populations. Each population has a standard deviation equal to 200 units ($\sigma = 200$). In order to observe the effect of increasing the number of elements in the population, we further assume the populations are of different sizes, namely, $N = 10,000$ and $N = 1,000,000$. The standard error of the mean for the population of 10,000 elements is, by Equation 5.5,

$$\sigma_{\bar{x}} = \sqrt{\frac{10,000 - 100}{10,000 - 1}} \left(\frac{200}{\sqrt{100}}\right) \approx \sqrt{1}\left(\frac{200}{10}\right) \approx 20$$

For the population of 1,000,000 elements, we have

$$\sigma_{\bar{x}} = \sqrt{\frac{1,000,000 - 100}{1,000,000 - 1}} \left(\frac{200}{\sqrt{100}}\right) \approx \sqrt{1}\left(\frac{200}{10}\right) \approx 20$$

Thus, increasing the population size from 10,000 to 1,000,000 has virtually no effect on the standard error of the mean, since the finite population correction is approximately equal to 1 in both instances. Indeed, if the population size were increased to infinity, the same result would again be obtained for the standard error.

The finding that it is the absolute size of the sample and not the proportion of the population sampled that basically determines sampling precision is difficult for many people to accept intuitively. In fact, prior to the introduction of statistical quality control procedures in American industry, arbitrary methods such as sampling 10% of the items of incoming shipments, regardless of shipment size, were quite common. There tended to be a vague feeling in these cases that approximately the same sampling precision was obtained by maintaining a constant sampling fraction. However, it is clear that widely different standard errors resulted from large variations in the absolute sizes of the samples. The interesting principle that emerges from this discussion, for cases in which the populations are large relative to the samples, is that it is the absolute amount of work done (sample size), not the amount of work that might conceivably have been done (population size), that is important in determining sampling precision. We leave it to the reader's judgment whether this finding can be applied to other areas of human activity as well.

In our subsequent discussion of statistical inference, we will be concerned with measures of sampling error for proportions as well as for means. Therefore, we note the corresponding formula for the standard error of a proportion in sampling a finite population. In Section 5.2, it was indicated that the standard deviation of the sampling distribution of a proportion, which we <u>now</u> refer to as the *standard error of a proportion*, is given by $\sigma_{\bar{p}} = \sqrt{pq/n}$. Since our discussion referred to sampling as a *Bernoulli process*, it pertained to the sampling of an infinite population. The corresponding formula for the standard error of a proportion for a simple random sample of size n from a finite population is as follows.

STANDARD ERROR OF A PROPORTION FOR FINITE POPULATIONS

(5.6)
$$\sigma_{\bar{p}} = \sqrt{\frac{N - n}{N - 1}} \sqrt{\frac{pq}{n}}$$

The same sorts of approximation considerations discussed in the case of the mean are pertinent here as well. Hence, if the size of the population is at least 20 times that of the sample, the formula for infinite populations may be used.

Other Sampling Distributions

In this chapter, we have discussed sampling distributions of numbers and proportions of occurrences and sampling distributions of the mean. Just as we were able to use the binomial distribution as a sampling distribution of numbers of occurrences under the appropriate conditions, other distributions may similarly be used under other sets of conditions. However, it is frequently far simpler to use normal curve methods, based on the operation of the Central Limit Theorem. Two other continuous sampling distributions, which we have not yet examined, are important in elementary statistical methods: the student t distribution and the chi-square distribution. They will be discussed at the appropriate places in connection with statistical inference.

Exercises

1. You are employed by a firm that must buy ashburners from a foreign distributor. It is known that in the past, the mean life of the ashburners received from this distributor has been 300 hours with a standard deviation of 30 hours. The foreign distributor has assured your management that manufacturing processes and quality control procedures have been stabilized to the point where there will be an average life of at least 300 hours. Acting on this statement, your management places an order for 100,000 ashburners. A random sample of 900 is selected from the shipment, and the mean life is found to be 303 hours.

    a. If the true mean life was 300 hours, what is the probability that the mean life of ashburners in a random sample of this size would exceed 303 hours?

    b. Is the foreign distributor correct in asserting that the mean life of the ashburners is at least 300 hours?

2. The mean salary earned by civil servants in the five lowest salary levels in a large municipality is $10,500 with a standard deviation of $1138.

    a. Would it be correct to say that approximately 99.7% of all civil servants in these five salary levels earn between $7086 and $13,914? Why or why not?

    b. Would it be correct to say that if simple random samples of 100 civil servants each were repeatedly drawn from these classifications, approximately 99.7% of the time the average salary of the group would be between $10,158.60 and $10,841.40?

    c. Would it be correct to say that if simple random samples of 10,000 civil servants each were repeatedly drawn from these classifications, approximately 99.7% of the time the average salary of the group would be between $10,272.40 and $10,727.60? Why or why not?

    d. As an approximation, would (b) or (c) be more likely to be correct?

3. A specification calls for a drug to have a therapeutic effectiveness for a mean period of 50 hours. The standard deviation of the distribution of the period of effectiveness is known to be 16 hours. Shipments of the drug are to be accepted if the mean period exceeds 48 hours in a sample of 64 items drawn at random. Suppose the actual mean period of effectiveness of the drug in a given shipment is 44 hours. What is the probability that the shipment will be accepted when, in fact, it should not be?

4. The following is known about the clients of a successful stockbroker: (1) Fifteen percent of the clients are "heavy traders" and therefore generate rather large commissions. (2) The dollar sizes of the clients' accounts are normally distributed with an arithmetic mean of $20,000 and a standard deviation of $1500.

    a. If ten accounts are randomly selected, what is the probability of obtaining exactly one "heavy trader"?

    b. If one account is selected, what is the probability that its size will be between $18,500 and $23,000?

    c. If a random sample of 36 accounts is selected, what is the probability that the sample mean will be between $20,200 and $20,400?

    d. If one account is randomly selected, the probability is 0.209 that its size will exceed a certain dollar amount. What is this dollar amount?

5. a. According to a 1976 survey, there are 257 automotive dealers in State A and the average amount of sales per dealer is $502,680. Assume that the standard deviation is $78,000. Find the mean and standard deviation of the sampling distribution of the average amount of sales per dealer for random samples of 100 dealers.

    b. According to the same source, there are 8216 dealers in State B and the average amount of sales per dealer is $626,540. Assume the standard deviation is also $78,000. Find the mean and standard deviation of the sampling distribution of the average amount of sales per dealer for random samples of 100 dealers.

c. Compare the standard deviations of the sampling distributions in parts (a) and (b). Why are they different?

6. Do you agree or disagree with the following statements? Explain your answer.

a. The probability that there are ten defective items in a sample of 200 is the same as the probability that 95% of the items in a sample of 200 are not defective.

b. In tossing a fair coin, since the probability that the proportion of heads equals $\frac{1}{2}$ converges to 1 as the number of tosses increases to infinity, one could be reasonably certain that in flipping a coin 10,000 billion times, one would get exactly 5000 billion heads.

c. According to the Central Limit Theorem, if the mean and variance of a variable are known, we can use the normal distribution to approximate the probability that the variable will exceed some number.

d. If X is normally distributed, the only information we need know about X to answer probability statements about it is its mean and standard deviation.

e. The mean of a sample is always exactly normally distributed.

7. What is the probability of drawing a simple random sample of size 100 with a mean of 30 or more from a population with a mean of 28? The population variance is 81. Do we have to assume that the population is normal?

8. The life of an Ever-Steady spark plug is a normally distributed variable with a mean of 36,000 miles and a standard deviation of 2500 miles under normal driving conditions.

a. What is the probability that a spark plug will last at least 34,000 miles?

b. What is the probability that a spark plug will last more than 40,000 miles?

c. If you buy eight spark plugs for your car, what is the probability that the average life of the eight plugs exceeds 34,000 miles?

9. A clothing manufacturer has sales offices in Boston, New York, Washington, and Atlanta, and each office has 25 sales representatives. The weekly sales for the representatives are normally distributed with a mean of $1200 and a standard deviation of $200. Within what range about the mean is the probability 0.9973 that

a. a given representative's weekly sales will fall?

b. the average weekly sales per representative of the Atlanta sales office will fall?

c. the average weekly sales per representative of the company will fall?

10. The weekly demand for Snoozeburgers at a local Hamburger Queen franchise is normally distributed with a mean of 4500 and a standard deviation of 450. Calculate the range in which 95.5% of the weekly demand figures will fall. If demand figures are averaged every month (that is, the arithmetic mean is calculated for four weekly demands), calculate the range in which 95.5% of these average weekly demand figures would fall.

11. The mean salary of the presidents of 100 different small electronic controls companies is $38,900 with a standard deviation of $4210. A certain business magazine decides to make a study of the presidents of these 100 firms.

a. Suppose 25 presidents are selected at random. What are the mean and standard deviation of the distribution of average salaries for all possible samples of size 25?

    b. Suppose 50 presidents are selected at random. What are the mean and standard deviation of the distribution of average salaries for all possible samples of size 50?

    c. Suppose 100 presidents are selected at random. What are the mean and standard deviation of the distribution of average salaries for all possible samples of size 100?

12. The average amount of time spent per week in meetings by the 15 top executives of a certain company is 15 hours with a standard deviation of two hours. Five executives are selected at random and asked various questions, one of which is amount of time per week spent in meetings. The answers from the executives will be used for the orientation of newly hired management aspirants. The mean for the five executives questioned was 14.8 hours with a standard deviation of 1.5 hours.

    a. Are the 14.8 hours and 1.5 hours the mean and standard deviation of the sampling distribution of samples of five executives selected from the population of 15?

    b. Define the sampling distribution of the mean time spent per week in meetings for samples of five executives selected from the 15. Calculate the mean and standard deviation of the sampling distribution.

5.6 OTHER PROBABILITY SAMPLE DESIGNS

The discussion so far has been based on *simple random samples.* However, in many practical situations, other sample designs may be preferable to simple random sampling, because they may achieve greater precision of estimation at the same cost, or the same precision at lower cost. The subject of sample survey theory and methods is very specialized, and matters such as the selection of the optimal type of sample design and estimation method require a high level of expertise. It is usually wise to obtain the advice or active involvement of a knowledgeable sampling specialist in the planning and implementation of such projects.

    Two frequently used alternatives to simple random sampling are *stratified sampling* and *cluster sampling.* In the simplest form of stratified sampling, the population is classified into a complete set of mutually exclusive subgroups, or *strata,* and simple random samples are drawn independently from each of these strata. Sample statistics from these strata can be combined to yield an overall estimate of a population parameter. For example, assume that we had a list of the accounts receivable of a large company and we were interested in estimating the mean account size by sampling. We could group these accounts, say into five classes, ranging from the largest accounts in stratum 1 to the smallest accounts in stratum 5. Suppose we draw a simple random sample of accounts from each stratum and calculate the sample mean size of account for each of these strata. Then, we weight these $\bar{x}$ values (the weights are the population numbers of accounts in each stratum) to obtain an estimate of the population mean. It can be shown that this estimate tends, on the average, to be closer to the actual population mean than the one that would be obtained under simple random sampling of the entire population. In other words, stratification can be used to obtain greater precision in estimation. Furthermore, sample means may be compared with one another to reveal differences between strata that may be of interest.

In our example, the property used to stratify the population, namely size of account, was also the characteristic we were interested in estimating. In more realistic cases, for example, in studying consumer expenditures, we might stratify household units by characteristics such as disposable income, size of household unit, and geographic location. In estimating unemployment of the labor force, we might stratify by race, sex, education, and age. Stratification results in greater precision than simple random sampling, particularly when there are extreme items that can be grouped into separate strata.

Another widely used technique is *cluster sampling*, in which the population is subdivided into groups, or *clusters*, and a probability sample of these clusters is drawn and studied. For example, suppose we wished to conduct a survey of expenditures on durable goods for all households in the United States. We might draw a simple random sample of counties in the United States, and then draw a simple random sample of households within the sampled counties. The counties are referred to as clusters, since for purposes of analysis, households are conceived of as being clustered into county units. Several successive stages of sampling might include clusters such as counties, political subdivisions within counties, blocks, and households. Generally, expenditures on such things as the listing of population elements, travel, interviewing, and supervision are smaller when cluster sampling is used than when simple random sampling is used. In terms of cost required to obtain a fixed level of precision, a well-designed cluster sample is generally far superior to a simple random sample.

Stratified sampling and cluster sampling may be combined in a variety of ways. For example, suppose that in the survey of expenditures on durable goods for all households in the United States referred to in the preceding paragraph, we wanted estimates for each of the individual states as well as national figures. The states would then constitute 50 strata. A simple random sample of counties (clusters) could be drawn within each state, and then simple random samples of households could be drawn within the sampled counties.

In the chapters that follow, simple random sampling is assumed. In more complex types of sampling, although the estimation formulas and methods are more complicated, the basic principles and formulas of simple random sampling represent important components of the overall methodology.

6
Estimation

As we said at the beginning of Chapter 3, decisions must often be made when only incomplete information is available and there is uncertainty concerning the outcomes that must be considered by the decision maker. The remainder of this book deals with methods by which rational decisions can be made under such circumstances. In our brief introduction to probability theory, we have begun to see how probability concepts can be used to cope with problems of uncertainty. *Statistical inference* uses this theory as a basis for making reasonable decisions from incomplete data. Statistical inference treats two different classes of problems: (1) estimation, which is discussed in this chapter, and (2) hypothesis testing, which is examined in Chapters 7 and 8. In both cases, inferences are made about population characteristics from information contained in samples.

6.1 POINT AND INTERVAL ESTIMATION

The need to estimate population parameters from sample data stems from the fact that it is usually too expensive or not feasible to enumerate complete populations to obtain the required information. The cost of complete censuses of finite populations may be prohibitive; complete enumerations of infinite populations are impossible. Statistical estimation procedures provide us with the means of obtaining estimates of population parameters with desired degrees of precision. Numerous examples can be given of the need to obtain estimates of pertinent population parameters in business and economics. A marketing organization may be interested in estimates of average income and other socioeconomic characteristics of the consumers in a metropolitan area; a retail chain may want an estimate of the average number of pedestrians per day who pass a certain corner; or a bank may want an estimate of average interest rates on mortgages in a certain section of the country. Undoubtedly, in all of these cases, exact accuracy is not required, and estimates derived from sample data would probably provide appropriate information to meet the demands of the practical situation.

Two different types of estimates of population parameters are of interest: *point estimates* and *interval estimates*. A point estimate is a single number used as an estimate of the unknown population parameter. For example, the arithmetic mean income of a sample of families in a metropolitan area may be used as a point estimate of the corresponding population mean for all families in that metropolitan area; or the percentage of a sample of eligible voters in a political opinion poll who state that they would vote for a particular candidate may be used as an estimate of the corresponding unknown percentage in the relevant population.

A distinction can be made between an *estimate* and an *estimator*. Let us consider the illustration of estimating the population figure for arithmetic mean income of all families in a metropolitan area from the corresponding sample mean. The numerical value of the sample mean is said to be an *estimate* of the population mean figure. On the other hand, the statistical measure used (that is, the *method* of estimation) is referred to as an *estimator*. For example, the sample mean $\bar{x}$ is an estimator of the population mean. When a specific number is calculated for the sample mean, say \$8000, that number is an *estimate* of the population mean figure.

Whether one uses a point estimate rather than an interval estimate depends on the purpose of the investigation. For example, for planning purposes, a marketing department may estimate a single figure for annual sales of one of its company's products and may then break that figure down into monthly sales estimates. These figures may be passed on to the production department for the planning of production requirements. The production department may in turn convert its production requirements into materials purchasing plans for the purchasing department. If the marketing department estimated annual sales as a range, say from $10,000,000 to $12,000,000, rather than as a single figure, this could unduly complicate the subsequent steps of obtaining monthly breakdowns and planning production and purchasing requirements. However, for many practical purposes, it is not sufficient to have merely a single point estimate of a population parameter. Any single point estimate will be either right or wrong. It would certainly seem useful, and perhaps even necessary, to have in addition to a point estimate, some notion of the degree of error that might be involved in using this estimate. *Interval estimation* is useful in this connection. Roughly speaking, an interval estimate of a population parameter is a statement of two values between which we have some confidence that the parameter lies. Thus, an interval estimate in the example of the population arithmetic mean income of families in a metropolitan area might be $14,100 to $15,900. An interval estimate for the percentage of defectives in a shipment might be 3% to 5%. We may have a great deal of confidence or very little confidence that the population parameter is included in the range of the interval estimate, so it is necessary to attach some sort of probabilistic statement to the interval.

The procedure used to handle this problem is *confidence interval estimation*. The confidence interval is an interval estimate of the population parameter. A confidence coefficient such as 90% or 95% is attached to this interval to indicate the degree of confidence or credibility to be placed upon the estimated interval.

6.2 CRITERIA OF GOODNESS OF ESTIMATION

Numerous criteria have been developed by which to judge the goodness of point estimators of population parameters. A rigorous discussion of these criteria requires some complex mathematics that falls outside the scope of this text. However, it is possible to gain an appreciation of the nature of these criteria in an intuitive, nonrigorous way.

Let us return to our illustration of estimating the arithmetic mean income of families in a metropolitan area. This arithmetic mean, assuming suitable definitions of income, family, and metropolitan area, is an unknown population parameter that we designate as μ. Suppose we took a simple random sample of families from this population and calculated the arithmetic mean, $\bar{x}$, the median, Md, and the mid-range $(x_{max} + x_{min})/2$, where x_{max} and x_{min} are the largest and smallest sample observations. Which method would be the best estimator of the population mean? Probably, your answer would be that the sample mean $\bar{x}$ is the best estimator. In fact, if this question had not been raised, it might not even have occurred to you to use any statistic other than the sample mean as an estimator of the population mean. However, why do you think the sample mean represents the best estimator? It may not be very easy to articulate your answer to that question. As it turns out, the sample mean is preferable to the other estimators by the generally utilized criteria of goodness of estimation of classical statistical inference. Let us briefly examine the nature of a few of these criteria, namely, *unbiasedness, consistency,* and *efficiency.*

Unbiasedness. An estimator, such as a sample arithmetic mean, is a random variable, because it may take on different values, depending upon which population elements are drawn into the sample. As is proven in Rule 13 of Appendix C, the expected value of this random variable is the population mean. In symbols, we have

(6.1) $$E(\bar{x}) = \mu$$

where $\bar{x}$ = the sample mean
μ = the population mean

If the expected value of a sample statistic is equal to the population parameter for which the statistic is an estimator, the statistic (or estimator) is said to be *unbiased.* In symbols, if θ is a parameter to be estimated and $\hat{\theta}$ is a sample statistic used to estimate θ, then $\hat{\theta}$ is said to be an *unbiased estimator* of θ if

(6.2) $$E(\hat{\theta}) = \theta$$

Hence, if we say that a given estimator is unbiased, we are simply saying that this method of estimation is correct, *on the average.* That is, if the method is employed repeatedly, the average of all estimates obtained from this estimator is equal to the value of the population parameter. Clearly, if an estimator is unbiased, this does not guarantee useful individual estimates. The differences of these individual estimates from the value of the population parameter may represent large errors. The simple fact that the bias, or long-run average of these errors, is 0 may be of little practical importance. Furthermore, if we have two unbiased estimators, we require additional criteria in order to make a choice between them.

Just as the sample mean is an unbiased estimator of the population mean, the sample variance as defined in Equation 1.8 is an unbiased estimator of the population variance.[1] In symbols, as proven in Rule 14 of

---

[1] As can be seen from the nature of the proof in Rule 14 of Appendix C, $\Sigma (x - \bar{x})^2/(n - 1)$ is not an unbiased estimator of the variance of a *finite* population. Perhaps surprisingly, both $\sqrt{\Sigma (x - \bar{x})^2/n}$ and $\sqrt{\Sigma (x - \bar{x})^2/(n - 1)}$ are biased estimators of the population standard deviation.

Appendix C, we have

(6.3) $$E(s^2) = E\left[\frac{\Sigma(x - \bar{x})^2}{n - 1}\right] = \sigma^2$$

Equation 6.3 is a mathematical restatement of the point made in Section 1.14 that when the sample variance is defined with divisor $n - 1$, it is an unbiased estimator of the population variance.

Consistency. It is clear from the preceding discussion that knowing only that an estimator is unbiased gives us insufficient information as to the goodness of that method of estimation. It seems that closeness of the estimator to the parameter is of primary importance. Both the concepts of consistency and efficiency deal with this property of closeness. Consider the sample mean $\bar{x}$ as an estimator of the population parameter μ for an infinite population. What happens to the possible values of $\bar{x}$ as the sample size n increases? On an intuitive basis, we would certainly expect $\bar{x}$ to lie closer to μ, as n becomes larger and larger. Loosely speaking, if an estimator, say $\hat{\theta}$, approaches closer and closer to the parameter θ as the sample size n increases, $\hat{\theta}$ is said to be a *consistent estimator* of θ.

Interpreting this idea of consistency in terms of sampling, it means that the sampling distribution of the estimator becomes more and more "tightly packed" around the population parameter as the sample size increases. Figure 6-1 illustrates this concept for the sample mean as an estimator of μ, the mean of an infinite population. The graph represents the respective sampling distributions of the sample mean $\bar{x}$ for samples of size n_1, of size n_2, and of size n_3 drawn from the same population; n_3 is larger than n_2, which is larger than n_1. We know from the relationship $\sigma_{\bar{x}}^2 = \sigma^2/n$ that the sampling variance $\sigma_{\bar{x}}^2$ decreases as n increases. Note that all three sampling distributions center on the population parameter μ, since $\bar{x}$ is an unbiased estimator of μ.

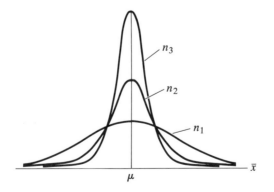

FIGURE 6-1
Sampling distributions of $\bar{x}$ as the sample size increases; $n_3 > n_2 > n_1$.

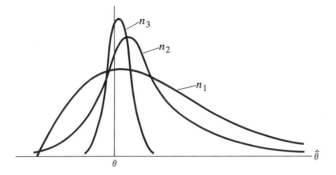

FIGURE 6-2
Sampling distributions of a biased but consistent estimator $\hat{\theta}$; $n_3 > n_2 > n_1$.

Figure 6-2 illustrates the concept of consistency for $\hat{\theta}$, a *biased esti-mator* of a parameter θ. As in Figure 6-1, the graph represents the respective sampling distributions of an estimator, in this case $\hat{\theta}$, for samples of size n_1, of size n_2, and of size n_3 drawn from the same popula-tion, with $n_3 > n_2 > n_1$. None of the distributions are centered on the population parameter θ. Since $E(\hat{\theta})$ is not equal to θ, $\hat{\theta}$ is a biased esti-mator of θ. However, as can be seen from the graph, as the sample size increases, the sampling distribution becomes increasingly "tightly packed" around θ.

Efficiency. The concept of efficiency refers to the sampling vari-ability of an estimator. If two competing estimators are both unbiased, the one with the smaller variance (for a given sample size) is said to be relatively more efficient. Somewhat more formally, if $\hat{\theta}_1$ and $\hat{\theta}_2$ are two unbiased estimators of θ, their relative efficiency is defined by the ratio

(6.4)
$$\frac{\sigma_{\hat{\theta}_2}^2}{\sigma_{\hat{\theta}_1}^2}$$

where $\sigma_{\hat{\theta}_1}^2$ is the smaller variance.

Let us consider as an example the situation of a simple random sample of size n drawn from a normal population with mean μ and variance σ^2. Suppose we want to consider the relative efficiency of the sample mean $\bar{x}$ and the sample median Md as estimators of the popula-tion mean μ. Both estimators are unbiased. We know that the variance of the sample mean $\bar{x}$ is $\sigma_{\bar{x}}^2 = \sigma^2/n$. The variance of the sample median Md is approximately $\sigma_{Md}^2 = 1.57\sigma^2/n$. Therefore, the relative efficiency of $\bar{x}$ with respect to Md is $\sigma_{Md}^2/\sigma_{\bar{x}}^2 = (1.57\sigma^2/n)/(\sigma^2/n) = 1.57$. We can in-terpret this result in terms of sample sizes. If the sample median rather than the sample mean were used as an estimator of the mean μ of a normal population, then in order to obtain the same precision provided by the sample mean, a sample size 57% larger would be required.

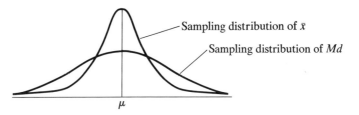

FIGURE 6-3
Sampling distributions of the mean $\bar{x}$ and median Md for the same sample size.

Stated differently, the required sample size for the sample median would be 157% of that for the sample mean. Figure 6-3 shows the sampling distributions for these two estimators.

Other Criteria. Other criteria for goodness of estimation may be found in standard texts on mathematical statistics. As we have seen, sampling error decreases with increasing sample size, and biased but consistent estimators approach the population parameter as sample size increases. However, greater cost is incurred with larger sample sizes. Therefore, most practical estimation situations involve tradeoffs between these considerations.

Two problems of interest in this chapter are the estimation of a population proportion and a population mean. In the case of estimating a population proportion, if we assume a Bernoulli process generating n sample observations, it can be shown that the observed sample proportion of successes, $\bar{p}$ = number of successes/number of sample observations, is an unbiased, efficient, and consistent estimator. Similarly, the mean $\bar{x}$ of a simple random sample of n observations is an unbiased, efficient, and consistent estimator of the population mean. Thus, it is not surprising that in many applications of statistical methods, sample proportions or sample means are used as the "best" point estimators of the corresponding population parameters. Perhaps the reader has had occasion to calculate a sample proportion or mean and has found it intuitively appealing to use such a figure as an estimator of the corresponding population parameter. If so, this intuitive approach was supported by sound statistical theory.

Exercises

1. Is each of the following statements about an estimator or an estimate? If the statement is about an estimate, is it a point or an interval estimate?
   a. The sales manager feels that February's sales will be between $40,000 and $50,000.
   b. In certain situations, the median of a class may be a better measure of central tendency than the mean. For example, in a research study on the

savings habits of American families, the large savings held by the wealthy may distort the mean.

   c. A bank auditor, after examining a sample of the bank's loan portfolio, says that 4% of all loans in the portfolio will have to be written off as bad debts.

   d. A security analyst predicts that the Dow Jones average will be between 900 and 1100 by the end of the year.

2. List three criteria used to judge goodness of a point estimate, and discuss the meaning of each.

3. State whether each of the following statements is true or false, and explain your answer.

   a. If a statistic is an unbiased estimator of a parameter, then it is the "best" estimator of the parameter.

   b. If an unbiased estimator of σ^2 is desired, it is best to calculate the sample variance as $\Sigma (x - \bar{x})^2/(n - 1)$.

   c. If $\hat{\theta}_1$ is a biased estimator of θ and $\hat{\theta}_2$ is an unbiased estimator of θ, and the ratio $\sigma^2_{\hat{\theta}_2}/\sigma^2_{\hat{\theta}_1}$ is greater than 1, then $\hat{\theta}_1$ is relatively more efficient than $\hat{\theta}_2$.

6.3 CONFIDENCE INTERVAL ESTIMATION (LARGE SAMPLES)

As indicated in Section 6.1, for many practical purposes, it is not sufficient merely to have a single point estimate of a population parameter. It is usually necessary to have an estimation procedure that gives some measure of the degree of precision involved. In classical statistical inference, the standard procedure for this purpose is confidence interval estimation. We will explain the rationale of confidence interval estimation in terms of an example in which a population mean is the parameter to be estimated.

Interval Estimation of a Mean: Rationale

Suppose a manufacturer has a very large production run of a certain brand of tires and is interested in obtaining an estimate of their arithmetic mean lifetime by drawing a simple random sample of 100 tires and subjecting them to a forced life test. Let us assume that from long experience in manufacturing this brand of tires, the manufacturer knows that the population standard deviation for a production run is $\sigma = 3000$ miles. (Of course, ordinarily the standard deviation of a population is not known exactly and must be estimated from a sample, just as are the mean and other parameters. However, let us assume in this case that the population standard deviation is indeed known.) When the sample of 100 tires was drawn, a mean lifetime of 22,500 miles was observed. Thus, we denote $\bar{x} = 22,500$ miles. This sample mean is our best *point*

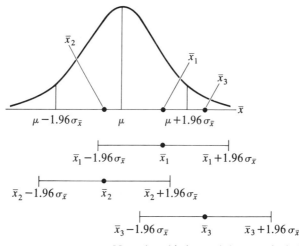

Note that this interval does not include
the parameter μ

FIGURE 6-4
Sampling distribution of the mean and confidence interval estimates for three illustrative samples.

estimate of the population mean lifetime, that is, of the mean lifetime of all tires in the production run. Additionally, we would like an *interval* estimate of the population mean lifetime. That is, we would like to be able to state that the population mean is between two limits, say $\bar{x} - 2\sigma_{\bar{x}}$ and $\bar{x} + 2\sigma_{\bar{x}}$ where $\bar{x} - 2\sigma_{\bar{x}}$ is the lower limit of the interval and $\bar{x} + 2\sigma_{\bar{x}}$ is the upper limit. Furthermore, we would like to have a high degree of confidence that the true population mean is included in this interval.

The procedure in confidence interval estimation is based on the concept of the sampling distribution. In this example, since we are dealing with the estimation of a mean, the appropriate distribution is the sampling distribution of the mean. We will review some fundamentals of this distribution to lay the foundation for confidence interval estimation. Figure 6-4 shows the sampling distribution of the mean for simple random samples of size $n = 100$ from a population with an unknown mean, denoted μ, and a standard deviation $\sigma = 3000$. We assume that the sample is large enough so that by the Central Limit Theorem, stated in Section 5.5, the sampling distribution may be assumed to be normal, even if the population is nonnormal. The standard error of the mean, which is the standard deviation of this sampling distribution, equals $\sigma_{\bar{x}} = \sigma/\sqrt{n}$.[2]

[2] Strictly speaking, the finite population correction should be shown in this formula, but we will assume the population is so large relative to sample size that for practical purposes the correction factor is equal to 1. The mean of the sampling distribution $\mu_{\bar{x}}$ is equal to the population mean μ.

In our work with the normal sampling distribution of the mean, we have learned how to make probability statements about sample means, given the value of the population mean. Thus, for example, in terms of the data of this problem, we can state that if repeated simple random samples of 100 tires each were drawn from the production run, 95% of the sample means $\bar{x}$ would lie within 1.96 standard error units of the mean of the sampling distribution (the population mean), or between $\mu - 1.96\sigma_{\bar{x}}$ and $\mu + 1.96\sigma_{\bar{x}}$. This range is indicated on the horizontal axis of the sampling distribution in Figure 6-4. For emphasis, vertical lines show the end points of this range. As usual, we determine the 1.96 figure from Table A-5 of Appendix A, where we find that 47.5% of the area in a normal distribution is included between the mean of a normal distribution and a value 1.96 standard deviations to the right of the mean; thus, by symmetry, 95% of the area is included in a range of ± 1.96 standard deviations from the mean. In terms of relative frequency, 95% of the $\bar{x}$ values of samples of size 100 would lie in this range if repeated samples were drawn from the given population.

How then might we construct the desired interval estimate for the population parameter? Let us consider again the repeated simple random samples of size 100 from the population represented by the production run of tires. Suppose our first sample yields a mean that exceeds μ but falls between μ and $\mu + 1.96\sigma_{\bar{x}}$. The position of this sample mean, denoted $\bar{x}_1$, is shown on the horizontal axis of Figure 6-4. Suppose now that we set up an interval from $\bar{x}_1 - 1.96\sigma_{\bar{x}}$ to $\bar{x}_1 + 1.96\sigma_{\bar{x}}$. This interval is shown immediately below the graph in Figure 6-4. As can be seen in this figure, this interval, which may be written as $\bar{x}_1 \pm 1.96\sigma_{\bar{x}}$, includes the population parameter, μ. This follows from the fact that $\bar{x}_1$ fell less than $1.96\sigma_{\bar{x}}$ from the mean of the sampling distribution, μ.

Now, let us assume our second sample from the same population yields the mean $\bar{x}_2$, which lies on the horizontal axis to the left of μ, but again at a distance less than $1.96\sigma_{\bar{x}}$ from μ. Again we set up an interval of the sample mean plus or minus $1.96\sigma_{\bar{x}}$, or from $\bar{x}_2 - 1.96\sigma_{\bar{x}}$ to $\bar{x}_2 + 1.96\sigma_{\bar{x}}$. This interval, shown below the graph in Figure 6-4, includes the population mean μ.

Finally, suppose a third sample is drawn from the same population, with the mean $\bar{x}_3$ shown on the horizontal axis of Figure 6-4. This sample mean lies to the right of μ, but at a distance *greater than* $1.96\sigma_{\bar{x}}$ above μ. Now, when we set up the range $\bar{x}_3 - 1.96\sigma_{\bar{x}}$ to $\bar{x}_3 + 1.96\sigma_{\bar{x}}$, this interval does not include μ.

We can imagine a continuation of this sampling procedure. Since 95% of the sample means fall within $1.96\sigma_{\bar{x}}$ of μ, we can assert that 95% of the intervals of the type $\bar{x} \pm 1.96\sigma_{\bar{x}}$ include the population parameter μ. Now, we can get to the crux of confidence interval estimation. In the problem originally posed, as in most practical applications, only one sample was drawn from the population, not repeated samples. On the basis of the single sample, we were required to estimate the population parameter. The procedure is simply to establish the interval $\bar{x} \pm 1.96\sigma_{\bar{x}}$ and attach a suitable statement to it. The interval itself is referred to as a *confidence interval*. Thus, for example, in our original problem the required confidence interval is

$$\bar{x} \pm 1.96\sigma_{\bar{x}} = \bar{x} \pm 1.96 \frac{\sigma}{\sqrt{n}} = 22{,}500 \pm 1.96 \frac{3000}{\sqrt{100}}$$

$$= 22{,}500 \pm 588 = 21{,}912 \text{ to } 23{,}088 \text{ miles}$$

We must be very careful how we interpret this confidence interval. It is incorrect to make a probability statement about this *specific* interval. For example, it is incorrect to state that the probability is 95% that the mean lifetime, μ, of all tires falls in this interval. *The population mean is not a random variable;* hence, probability statements cannot be made about it. The unknown population mean μ either lies in the interval or it does not. We must return to the line of argument used in explaining the method and indicate that the values of the random variable are the intervals of the form $\bar{x} \pm 1.96\sigma_{\bar{x}}$, not μ. Thus, the interpretation is that if repeated simple random samples of the same size were drawn from this population and the interval $\bar{x} \pm 1.96\sigma_{\bar{x}}$ were constructed from each of them, then 95% of the statements that the interval contains the population mean μ would be correct. Another way of putting it is that in 95 samples out of 100, the mean μ would lie within intervals constructed by this procedure. The 95% figure is referred to as a *confidence coefficient* to distinguish it from the type of probability calculated when deductive statements are made about sample values from known population parameters.[3]

Interval Estimation: Interpretation and Use

Despite the above interpretation of the meaning of a confidence interval, where the probability pertains to the estimation procedure rather than to the specific interval constructed from a single sample, the fact remains that we ordinarily must make an inference on the basis of the single sample drawn. We will not draw the repeated samples implied by the interpretational statement. For example, in the tire illustration, an inference was required about the production run based on the particular sample of 100 tires in hand, and an interval estimate of a mean lifetime of 21,912 to 23,088 miles was obtained. If the confidence coefficient at-

---

[3] This paragraph presents the standard interpretation of confidence intervals provided by classical statistics. Bayesian statistics (discussed in Chapters 13 through 17) disputes this interpretation, arguing that if μ is unknown, it may be treated as a random variable. Hence, in Bayesian statistics, "prior probability" statements may be made about μ based on subjective assessments, and "posterior probability" statements may be made that combine the prior probabilities and information obtained from sampling.

tached to the interval estimate is high, then the investigator will behave as though the interval estimate is correct. This interval may or may not encompass the actual value of the population parameter μ. However, since 95% of intervals so constructed would include the value of the mean lifetime μ of all tires in the production run, we will behave as though this particular interval does include the actual value. It is desirable to obtain a relatively narrow interval with a high confidence coefficient associated with it. One without the other is not particularly useful. Thus, for example, in estimation of a proportion (say, the proportion of persons in the labor force who are unemployed), we can assert even without sample data that the percentage lies somewhere between 0 and 100% with a confidence coefficient of 100%. Obviously, this statement is neither very profound nor very useful, because the interval is too wide. On the other hand, if the interval is very narrow but has a low associated confidence coefficient, say 10%, the statement would again have little practical utility.

Confidence coefficients such as 0.90, 0.95, and 0.99 and limits of two or three standard errors (such as 0.955 or 0.997) are conventionally used. Limits of two or three standard errors are those obtained by making an estimate of a population parameter and adding and subtracting two or three standard errors to establish confidence intervals. For a fixed confidence coefficient and population standard deviation, the only way to narrow a confidence interval and thus increase the precision of the statement is to increase the sample size. This is readily apparent from the way the confidence interval was constructed in the tire example. We computed $\bar{x} \pm 1.96\sigma_{\bar{x}}$, where $\sigma_{\bar{x}} = \sigma/\sqrt{n}$. If the 1.96 figure and σ are fixed, we can decrease the width of the interval only by increasing the sample size, n, since $\sigma_{\bar{x}}$ is inversely related to $\sqrt{n}$. Thus, the marginal benefit of increased precision must be measured against the increased cost of sampling. In Section 6.3, we discuss a method of determining the sample size required for a specified degree of precision.

One final point may be made before turning to confidence interval estimation of different types of population parameters. Ordinarily, as was indicated in the tire example, the standard deviation of the population σ is unknown. Therefore, it is not possible to calculate $\sigma_{\bar{x}}$, the standard error of the mean. However, we can estimate the standard deviation of the population from a sample and use this figure to calculate an estimated standard error of the mean. We use this estimation technique in the examples that follow.

Interval Estimation of a Mean (Large Samples)

We will use examples to discuss confidence interval estimation and will concentrate first on situations where the sample size is large. Our discussion will then focus, in turn, on interval estimation of a mean, a proportion, the difference between means, and the difference between proportions. Finally, we will briefly treat corresponding estimation procedures for small samples.

As an illustration of interval estimation of a mean from a large sample, let us look at Example 6-1.

EXAMPLE 6-1

A group of students working on a summer project with a social agency took a simple random sample of 120 families in a well defined "poverty area" of a large city in order to determine the mean annual family income of this area. The sample results were $\bar{x} = \$2810$, $s = \$780$, and $n = 120$. What would be the 99% confidence interval for the mean income of all families in this poverty area?

SOLUTION

The only way this problem differs from the illustration of the mean lifetime of tires is that here the population standard deviation is unknown. The usual procedure for large samples ($n > 30$) is simply to use the sample standard deviation as an estimate of the corresponding population standard deviation. Using s as an estimator of σ, we can compute an estimated standard error of the mean, $s_{\bar{x}}$. We have

$$s_{\bar{x}} = \frac{s}{\sqrt{n}} = \frac{\$780}{\sqrt{120}} = \$71.20$$

Hence, we may use $s_{\bar{x}}$ as an estimator of $\sigma_{\bar{x}}$, and because n is large we invoke the Central Limit Theorem to argue that the sampling distribution of $\bar{x}$ is approximately normal. Again, we have assumed the finite population correction to be equal to 1. The confidence interval, in general, is given by

(6.5)
$$\bar{x} \pm z s_{\bar{x}}$$

where z is the multiple of standard errors and $s_{\bar{x}}$ now replaces $\sigma_{\bar{x}}$, which was used when the population standard deviation was known. For a 99% confidence coefficient, $z = 2.58$. Therefore, the required interval is

$$\$2810 \pm 2.58(\$71.20) = \$2810 \pm \$183.70$$

so the population mean is roughly between $2626 and $2994 with a 99% confidence coefficient. The same sort of interpretation given earlier for confidence intervals again applies here.

Interval Estimation of a Proportion (Large Samples)

In many situations, it is important to estimate a proportion of occurrences in a population from sample observations. For example, it may be of interest to estimate the proportion of unemployed persons in a certain city, the proportion of eligible voters who intend to vote for a particular political candidate, or the proportion of students at a university who are in favor of changing the grading system. In all these cases, the corresponding proportions observed in simple random samples may be used to estimate the population proportions. Before turning to a description of how this estimation is accomplished, we will briefly establish the conceptual underpinnings of the procedure.

In Chapter 5, we saw that under certain conditions, the binomial distribution was the appropriate sampling distribution for the number of successes, x, in a simple random sample of size n. Furthermore, we noted in Section 5.5 that if p, the proportion to be estimated, is not too close to 0 or 1, the binomial distribution can be closely approximated by a normal curve with the same mean and standard deviation, that is, $\mu = np$ and $\sigma = \sqrt{npq}$.

It is a simple matter to convert from a sampling distribution of *number of successes* to the corresponding distribution of *proportion of successes*. If x is the number of successes in a sample of n observations, then the proportion of successes in the sample is $\bar{p} = x/n$. Hence, dividing the formulas in the preceding paragraph by n, we find that the mean and standard deviation of the sample proportion become

(6.6)
$$\mu_{\bar{p}} = p$$

and

(6.7)
$$\sigma_{\bar{p}} = \sqrt{\frac{pq}{n}}$$

Note that the subscript $\bar{p}$ has been used on the left sides of Equations 6.6 and 6.7 in keeping with the symbolism used for the corresponding values for the sampling distribution of the mean, namely, $\mu_{\bar{x}}$ and $\sigma_{\bar{x}}$. Therefore, in summary, if p is not too close to 0 or 1, the sampling distribution of $\bar{p}$ can be closely approximated by a normal curve with the mean and standard deviation given in (6.6) and (6.7). It is important to observe in this connection that the Central Limit Theorem holds for sample proportions as well as for sample means.

Note that the above discussion pertains to cases where the population size is large compared to the sample size. Otherwise, Equation 6.7 should be multiplied by the finite population correction, $\sqrt{(N - n)/(N - 1)}$. As we mentioned in Chapter 5, $\sigma_{\bar{p}}$ is referred to as the *standard error of a proportion*.

To illustrate confidence interval estimation for a proportion, we shall make assumptions similar to those in the preceding example. In Example 6-2, we assume a large simple random sample drawn from a population that is very large compared to the sample size.

EXAMPLE 6-2

In the town of Smallsville, a simple random sample of 800 automobile owners revealed that 480 would like to see the size of automobiles reduced. What are the 95.5% confidence limits for the proportion of all automobile owners in Smallsville who would like to see car size reduced?

SOLUTION

In this problem, we want a confidence interval estimate for p, a population proportion. We have obtained the sample statistic $\bar{p} = 480/800 = 0.60$, which is the sample proportion who wish to see car size reduced. As noted earlier, for large sample sizes and for p values not too close to 0 or 1, the sampling distribution of $\bar{p}$ may be approximated by a normal distribution with mean $\mu_{\bar{p}} = p$ and $\sigma_{\bar{p}} = \sqrt{pq/n}$. Here we encounter the same type of problem as in interval estimation of the mean. The formula for the exact standard error of a proportion, $\sigma_{\bar{p}} = \sqrt{pq/n}$, requires the values of the unknown population parameters p and q. Hence, we use an estimation procedure similar to that used in the case of the mean. Just as we used s to approximate σ, we can substitute the corresponding sample statistics $\bar{p}$ and $\bar{q}$ for the parameters p and q in the formula for $\sigma_{\bar{p}}$ in order to calculate an estimated standard error of a proportion, $s_{\bar{p}} = \sqrt{\bar{p}\bar{q}/n}$. Using the same type of reasoning as that for interval estimation of the mean, we can state a two-sided confidence interval estimate for a population proportion as

(6.8)
$$\bar{p} \pm z s_{\bar{p}}$$

In this problem, $z = 2$, since the confidence coefficient is 95.5%. Hence, substituting into Equation 6.8, we obtain our interval estimate of the proportion of all automobile owners in Smallsville who would like to see car size reduced:

$$0.60 \pm 2 \sqrt{\frac{0.60 \times 0.40}{800}} = 0.60 \pm 0.0346$$

Thus, the population proportion is estimated to be included in the interval 0.5654 to 0.6346, or roughly between 56.5% and 63.5% with a 95.5% confidence coefficient.

Interval Estimation
of the Difference between Two Means (Large Samples)

The foregoing examples of *estimation* of a population mean and proportion are based on single samples. We now examine interval estimation of the difference between means and the difference between proportions based on data obtained from two independent large samples. First, we examine an example of confidence interval estimation of the difference between two population means.

EXAMPLE 6-3

A large department store chain was interested in the difference between the average dollar amount of its delinquent charge accounts in the Northeastern and Western regions of the country for a certain year. The store took two independent simple random samples of these delinquent charge accounts, one from each region. The mean and standard deviation of the dollar amounts of these delinquent accounts were calculated to the nearest dollar with the following results.

Sample 1	Sample 2
$\bar{x}_1 = \$ \ 76$	$\bar{x}_2 = \$ \ 65$
$s_1 = \$ \ 25$	$s_2 = \$ \ 22$
$n_1 = \ 100$	$n_2 = \ 100$

The Northeastern region is denoted as 1 and the Western region as 2.

The analysts decided to establish 99.7% confidence limits for $\mu_1 - \mu_2$, where μ_1 and μ_2 denote the respective population mean sizes of delinquent accounts. Of course, a point estimate of $\mu_1 - \mu_2$ is given by $\bar{x}_1 - \bar{x}_2$. The required theory for the interval estimate is based on the fact that the sampling distribution of $\bar{x}_1 - \bar{x}_2$ for two large independent samples is exactly normal, if the population of differences is normal, with mean and standard deviation

(6.9) $$\mu_{\bar{x}_1 - \bar{x}_2} = \mu_1 - \mu_2$$

and

(6.10) $$\sigma_{\bar{x}_1 - \bar{x}_2} = \sqrt{\frac{\sigma_1^2}{n_1} + \frac{\sigma_2^2}{n_2}}$$

where σ_1 and σ_2 represent the respective population standard deviations of sizes of delinquent accounts.[4]

Since the population standard deviation σ_1 and σ_2 are unknown, and since the sample sizes are large, the sample standard deviations may be substituted into the formula for $\sigma_{\bar{x}_1 - \bar{x}_2}$ to give an estimated standard error of the difference between two means,

---

[4] Equation 6.9 follows from Rule 4 of Appendix C. By Rule 11 of Appendix C, the variance of the difference $\bar{x}_1 - \bar{x}_2$ is $\sigma_{\bar{x}_1 - \bar{x}_2}^2 = \sigma_{\bar{x}_1}^2 + \sigma_{\bar{x}_2}^2$. This is an application of the principle that the variance of the difference between two *independent* random variables is equal to the sum of the variances of these variables. Taking the square root of both sides of this equation, we obtain $\sigma_{\bar{x}_1 - \bar{x}_2} = \sqrt{\sigma_{\bar{x}_1}^2 + \sigma_{\bar{x}_2}^2}$, where $\sigma_{\bar{x}_1}^2$ and $\sigma_{\bar{x}_2}^2$ are simply the variances of the sampling distributions of $\bar{x}_1$ and $\bar{x}_2$. Substituting $\sigma_{\bar{x}_1}^2 = \sigma_1^2/n_1$ and $\sigma_{\bar{x}_2}^2 = \sigma_2^2/n_2$, we get Equation 6.10.

$$s_{\bar{x}_1 - \bar{x}_2} = \sqrt{\frac{s_1^2}{n_1} + \frac{s_2^2}{n_2}}$$

As usual with problems of this type, the population of differences may not be normal, and the population standard deviations are unknown. However, since the samples are large, we can use the Central Limit Theorem to assert that the sampling distribution of $\bar{x}_1 - \bar{x}_2$ is approximately normal. The required confidence limits are given by

(6.11) $$(\bar{x}_1 - \bar{x}_2) \pm z s_{\bar{x}_1 - \bar{x}_2}$$

The calculation for $s_{\bar{x}_1 - \bar{x}_2}$ in this problem is

$$s_{\bar{x}_1 - \bar{x}_2} = \sqrt{\frac{(25)^2}{100} + \frac{(22)^2}{100}} = \$3.33$$

Since a 99.7% confidence interval is desired, the value of z is 3. Therefore, substituting into (6.11) gives

$$(\$76 - \$65) \pm 3(\$3.33) = \$11 \pm \$9.99$$

Hence, to the nearest dollar, confidence limits for $\bar{x}_1 - \bar{x}_2$ are $1 and $21. It is a worthwhile exercise for the reader to attempt to express in words specifically what this confidence interval means.

Interval Estimation of the Difference between Two Proportions (Large Samples)

The procedure for constructing a confidence interval estimate for the difference between two proportions is analogous to the technique for means.

EXAMPLE 6-4

A credit reference service investigated two simple random samples of customers who applied for charge accounts in two different department stores. The service was interested in the proportion of applicants in each store who had annual incomes exceeding $10,000. It was decided to establish 90% confidence limits for the difference $p_1 - p_2$, where p_1 and p_2 represent the population proportions of applicants in each store whose incomes exceeded $10,000.

The sample data were

Store 1	Store 2
$\bar{p}_1 = 0.50$	$\bar{p}_2 = 0.18$
$\bar{q}_1 = 0.50$	$\bar{q}_2 = 0.82$
$n_1 = 150$	$n_2 = 160$

where these symbols have their conventional meanings. As in the preceding problem, we can start with a point estimate. The number $\bar{p}_1 - \bar{p}_2$ is the obvious point estimate of $p_1 - p_2$, and the sampling distribution of $\bar{p}_1 - \bar{p}_2$ may be assumed to be approximately normal with mean

$$\mu_{\bar{p}_1 - \bar{p}_2} = p_1 - p_2$$

and standard deviation

$$\sigma_{\bar{p}_1 - \bar{p}_2} = \sqrt{\frac{p_1 q_1}{n_1} + \frac{p_2 q_2}{n_2}}$$

Since the population proportions p_1 and p_2 are unknown and the sample sizes are large, $\bar{p}_1$ and $\bar{p}_2$ may be substituted for p_1 and p_2 to obtain the estimated standard error of the difference between percentages.

(6.12)
$$s_{\bar{p}_1 - \bar{p}_2} = \sqrt{\frac{\bar{p}_1 \bar{q}_1}{n_1} + \frac{\bar{p}_2 \bar{q}_2}{n_2}}$$

As in the procedure for differences between means, we use the Central Limit Theorem to argue that the sampling distribution of $\bar{p}_1 - \bar{p}_2$ is approximately normal, and we establish confidence limits of

(6.13)
$$(\bar{p}_1 - \bar{p}_2) \pm zs_{\bar{p}_1 - \bar{p}_2}$$

In this problem, the value of $s_{\bar{p}_1 - \bar{p}_2}$ is

$$s_{\bar{p}_1 - \bar{p}_2} = \sqrt{\frac{(0.50)(0.50)}{150} + \frac{(0.18)(0.82)}{160}} = 0.051$$

and since a 90% confidence coefficient is desired, $z = 1.65$. Therefore, the required confidence interval for the difference in the proportion of the applicants in the two stores whose incomes exceeded \$10,000 is

$$(0.50 - 0.18) \pm 1.65(0.051) = 0.32 \pm 0.084$$

The confidence limits are 0.236 and 0.404.

Exercises

1. Explain the following statement in detail: The standard deviation of $\bar{x}$ ($\sigma_{\bar{x}}$) is both a measure of statistical error in point estimation and a measure of dispersion of a distribution.

2. In a simple random sample of 400 firms within a very large industry, the arithmetic mean number of employees was 232 with a standard deviation of 40.
   a. Establish a 95.5% confidence interval for the population mean.
   b. In general, what is the meaning of a 95.5% confidence interval?

3. A random sample of 217 middle management personnel in a state was taken in order to determine the mean net income for these executives. A 95% confidence interval (\$12,000 to \$40,000) was established on the basis of the

sample results. Using only this information, state which of the following are valid conclusions.

a. If all possible samples of size 217 were drawn from this population, 95% of the sample means would fall in the interval $12,000 to $40,000.

b. If all possible samples of size 217 were drawn from this population, 95% of the universe means would fall in the interval.

c. If all possible samples of size 217 were drawn from this population, 95% of the confidence intervals established by the above method would contain the universe mean.

d. Of all similar executives in this state, 95% have mean incomes between $12,000 and 40,000.

e. Exactly 95% of the intervals established by the above method will contain the sample mean $\bar{X}$.

f. We do not know whether the universe mean is in the interval $12,000 to $40,000.

4. A simple random sample of 100 of the 820 students at a graduate school of business yielded an arithmetic mean entrance examination score of 550, a modal score of 600, and a standard deviation of 120. Suppose you desire to estimate the mean entrance examination score of all students of the business school with 97% confidence. What would your interval estimate be?

5. The mean life of 100 light bulbs selected at random from a shipment of 50,000 bulbs was 1000 hours with a standard deviation of 200 hours. What are the 95% confidence limits for the true mean life of all 50,000 bulbs?

6. Based on a simple random sample of 100 checking acccunts at a small suburban commercial bank, a 90% confidence interval of 2% to 4% is established for the percentage of overdrawn accounts. Is it correct to conclude that the probability is 90% that between 2% and 4% of all of the checking accounts at this bank are overdrawn?

7. In a simple random sample of 1000 stockholders of Atlas Credit Corporation, 600 were in favor of a new issue of bonds (with attached warrants for purchase of common stock), while 400 were against the bond issue. Construct a 95.5% confidence interval for the actual proportion of all stockholders who were in favor of the new issue. Assume that there is a very large number of stockholders.

8. In a random sample of 400 consumers in city A, an arithmetic mean expenditure on durable goods of $400 with a standard deviation of $120 was observed for 1976. In a random sample of 900 consumers in city B, an arithmetic mean of $410 with a standard deviation of $120 was observed for the same period.

a. Construct a 95% confidence interval for the mean expenditure on durable goods of all consumers in city A.

b. Construct a 95% confidence interval for the mean expenditure on durable goods of all consumers in city B.

c. Construct a 95% confidence interval for the difference in mean expenditures between the two cities.

9. An Internal Revenue Service agent is auditing the tax returns of a random

sample of executives, 40 in the oil industry and 40 in the chemical industry. The audits reveal the following information: (1) The average tax shelter write-off for the oil executive sample was $5000 with a standard deviation of $1000. (2) Twenty percent of the oil executives audited had a write-off of at least $6000. (3) The average tax shelter write-off for the chemical executive sample was $5500 with a standard deviation of $1000. (4) Fifteen percent of the chemical executives had a write-off of at least $6000.

a. Construct a 90% confidence interval for the difference between the average tax write-offs of oil industry executives and chemical industry executives.

b. Construct a 90% confidence interval for the true difference between the proportions of executives of the two industries with tax write-offs of at least $6000.

c. Interpret your results in (a) and (b).

10. The following results were obtained from a random sample of 400 shoppers in the Philadelphia area. (1) Sixty percent preferred sales help when purchasing clothing. (2) Of the 100 shoppers from New Jersey, 30 thought store A had the lowest prices in town. (3) Of the 300 shoppers from Philadelphia, 80 thought store A had the lowest prices in town.

a. What would be the 90% confidence interval for the true percentage of shoppers who prefer sales help when purchasing clothing?

b. Construct a 95% confidence interval for the true difference in opinion of the New Jersey and Philadelphia shoppers about store A's prices. Interpret the interval in light of the problem.

11. On the basis of a sample of 100 people, pollster A estimates that the percentage of people who are going to vote yes on a bond issue is 50%. Pollster B estimates the same percentage as 55% on the basis of a sample of 100 different randomly selected people. Is there reason to say that the difference between the pollsters' results is due to the different methods they use?

12. As part of the planning of capital expenditures on expansion for the next ten years, the finance division of a large corporation has asked the marketing research department to estimate the demand for the next ten years for one of the company's major products, homogenized milk. A random sample of 100 families from one large city is chosen in order to estimate the current average annual demand for the product by families in that city. The mean family demand in the sample is 150 gallons with a standard deviation of 40 gallons.

a. Construct a 95.5% confidence interval for the mean annual demand for the product by all families in the city.

b. If the range you obtained in (a) is larger than you are willing to accept, in what ways can you narrow it?

13. A random sample of 100 invoices was drawn from some 10,000 invoices of the Ace Company. For the sample, the mean dollar sales per invoice was $23.50 with a standard deviation of $6.00.

a. Construct a 99% confidence interval for the mean.

b. Give the verbal meaning of your result in part (a).

6.4 CONFIDENCE INTERVAL ESTIMATION (SMALL SAMPLES)

The estimation methods discussed thus far are appropriate when the sample size is large. The distinction between large and small sample sizes is important when the population standard deviation is *unknown* and therefore must be estimated from sample observations. The main point is as follows. We have seen that the ratio $z = (\bar{x} - \mu)/\sigma_{\bar{x}}$ (where $\sigma_{\bar{x}} = \sigma/\sqrt{n}$) is normally distributed for all sample sizes if the population is normal and approximately normally distributed for large samples if the population is not normally distributed. In words, this ratio is z = (sample mean − population mean)/*known* standard error. Furthermore, in Section 6.3 we observed that for *large samples*, even if an *estimated* standard error is used in the denominator of this ratio, the sampling distribution may be assumed to be a standard normal distribution for practical purposes. However, the ratio (sample mean − population mean)/*estimated* standard error is not approximately normally distributed for *all* sample sizes. Since this ratio is not approximately normally distributed for small samples, the theoretically correct distribution, known as the *t* distribution, must be used instead. Although the underlying mathematics involved in the derivation of the *t* distribution is complex and beyond the scope of our book, we can get an intuitive understanding of the nature of that distribution and its relationship to the normal curve.

The ratio $\dfrac{(\bar{x} - \mu)}{s/\sqrt{n}}$ is referred to as the *t* statistic. That is,

(6.14)
$$t = \frac{\bar{x} - \mu}{s/\sqrt{n}}$$

where, as defined in Equation 1.9, the sample standard deviation $s = \sqrt{\Sigma(x - \bar{x})^2/(n - 1)}$ is an estimator of the unknown population standard deviation σ.

Let us examine the *t* statistic and its relationship to the standard normal statistic, $z = (\bar{x} - \mu)/\sigma_{\bar{x}}$. As noted in the preceding paragraph, the denominator of the z ratio represents a *known* standard error, because it is based on a known population standard deviation. On the other hand, the denominator of the *t* statistic represents an *estimated* standard error, because s is an estimator of the population standard deviation.

The number $n - 1$ in the formula for s is referred to as the *number of degrees of freedom*, which we will denote ν (pronounced *nu*). It is not feasible to give a single simple verbal explanation of this concept. From a purely mathematical point of view, the number of degrees of freedom, ν, is simply a parameter that appears in the formula of the *t* distribution. However, in the present discussion, in which s is used as an estimator of the population standard deviation σ, the $n - 1$ may be interpreted as the number of independent deviations of the form $x - \bar{x}$ present in the calculation of s. Since the total of the deviations $\Sigma\ (x - \bar{x})$ for n observations equals 0, only $n - 1$ of them are independent. This means that if we were free to specify the deviations $x - \bar{x}$, we could designate only $n - 1$ of them independently. The nth one would be determined by the condition that the n deviations must add up to 0. Therefore, in the estimation of a population standard deviation or a population variance, if the divisor $n - 1$ is used in the estimator, the terminology is that there are $n - 1$ degrees of freedom present.

The t distribution has been derived mathematically under the assumption of a normally distributed population.[5] Just as is true of the standard normal distribution, the t distribution is symmetrical and has a mean of 0. However, the standard deviation of the t distribution is greater than that of the normal distribution but approaches the latter figure as the number of degrees of freedom (and, therefore, the sample size) becomes large. It can be demonstrated mathematically that for an infinite number of degrees of freedom, the t distribution and normal distribution are exactly equal. The approach to this limit is quite rapid. Hence, there is a widely applied rule of thumb that samples of size $n > 30$ may be considered "large" and that for such samples, the standard normal distribution may appropriately be used as an approximation to the t distribution, even though the latter is the theoretically correct functional form. Figure 6-5 shows the graphs of several t curves for different numbers of degrees of freedom. As can be seen from these graphs, the t curves are lower at the mean and higher in the tails than is the standard normal distribution. As the number of degrees of freedom increases, the t distribution rises at the mean and lowers at the tails until, for an infinite number of degrees of freedom, it coincides with the normal distribution. The use of tables of areas for the t distribution is explained in Example 6-5.

---

[5] The t distribution has the form

$$f(t) = c \left(1 + \frac{t^2}{\nu}\right)^{-(\nu+1)/2}$$

where $t = \dfrac{\bar{x} - \mu}{s_{\bar{x}}}$ (as previously defined)

    c = a constant required to make the area under the curve equal to 1
    $\nu = n - 1$, the number of degrees of freedom

The variable t ranges from minus infinity to plus infinity. The constant c is a function of ν, so that for a particular value of ν, the distribution of $f(t)$ is completely specified. Thus, $f(t)$ is a family of functions, one for each value of ν.

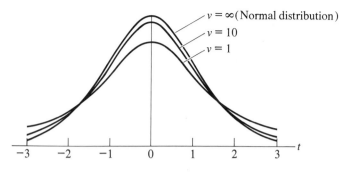

FIGURE 6-5
The t distributions for $\nu = 1$ and $\nu = 10$ compared to the normal distribution $(\nu = \infty)$.

Further Remarks

Early work on the t distribution was carried out by W. S. Gossett in the early 1900s. Gossett was an employee of Guinness Brewery in Dublin. Since the brewery did not permit publication of research findings by its employees under their own names, Gossett adopted "Student" as a pen name. Consequently, in addition to the term "t distribution" used here, the distribution has come to be known as "Student's distribution" or "Student's t distribution" and is so referred to in many books and journals.

EXAMPLE 6-5

Assume that a simple random sample of nine automobile tires was drawn from a large production run of a certain brand of tires. The mean life time of the tires in the sample was $\bar{x} = 22{,}010$ miles. This sample mean is the best single estimate of the corresponding population mean. The population standard deviation is unknown. Hence, an estimate of the population standard deviation was calculated by the formula $s = \sqrt{\Sigma (x - \bar{x})^2 / n - 1}$. The result was $s = 2520$ miles. What is the interval estimate for the population mean at a 95% level of confidence?

SOLUTION

By the same type of reasoning we used in the case of the normal sampling distribution for means, we find that confidence limits for the population mean, using the t distribution, are given by

(6.15)
$$\bar{x} \pm t \, \frac{s}{\sqrt{n}}$$

where t is determined for $n - 1$ degrees of freedom. The number of degrees of freedom is one less than the sample size, that is, $\nu = n - 1 = 9 - 1 = 8$.

Just as the z values in Examples 6-1 and 6-2 represented multiples of standard errors, the t value in Equation 6.15 represents a multiple of estimated standard errors. We find the t value in Table A-6 of Appendix A. A brief explanation of this table is required. In the table of areas under the normal curve, areas lying between the mean and specified z values were given. However, in the case of the t distribution, since there is a different t curve for each sample size, no single table of areas can be given for all these distributions. Therefore, for compactness, a t table shows the relationship between areas and t values for only a few "percentage points" in different t distributions. Specifically, the entries in the body of the table are t values for areas of 0.01, 0.02, 0.05, and 0.10 in the two tails of the distribution combined. In this problem, the number of degrees of freedom is 8. Referring to Table A-6 of Appendix A under column 0.05 for 8 degrees of freedom, we find $t = 2.306$. This means that, as shown in Figure 6-6, for 8 degrees of freedom, a total of 0.05 of the area in the t distribution lies below a t value of -2.306 or above $t = 2.306$. Correspondingly, the probability is 0.95 that for 8 degrees of freedom, the t value lies between -2.306 and 2.306.

In this problem, substituting $t = 2.306$ into Equation (6.15), we obtain the following 95% confidence limits:

$$22{,}010 \pm 2.306(840) = 22{,}010 \pm 1937.04 \text{ miles}$$

Hence, to the nearest mile, the confidence limits for the estimate of the mean life time of all tires in the production run are 20,073 and 23,947 miles.

The interpretation of this interval and the associated confidence coefficient is the same as in the case of large samples and a normal distribution. Comparing this procedure to the corresponding method for large samples for 95% confidence limits, we note that the t value of 2.306

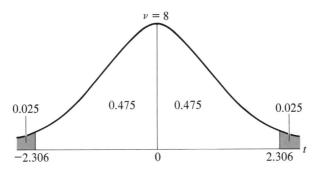

FIGURE 6-6
The relationship between t values and areas in the t distribution for 8 degrees of freedom.

replaces the 1.96 figure that is appropriate for the normal curve. Thus, we see that with small samples we have a wider confidence interval, leading to a vaguer result. This is to be expected, because σ is estimated by s using a smaller sample size n.

Note that since Table A-6 shows areas in the combined tails of the t distribution, we had to look under the column headed 0.05 for the t value corresponding to a 95% confidence interval. Correspondingly, we would find t values for 90%, 98%, and 99% confidence intervals under the columns headed 0.10, 0.02, and 0.01, respectively.

Exercises

1. Are the following statements true or false? Explain your answers.
   a. For small samples, if one is estimating the population mean, one uses $\bar{x} \pm ts_{\bar{x}}$.
   b. The more efficient of two unbiased estimators has the narrower confidence interval.
   c. If a 95% interval for the average strength of a certain type of wood beam is 12 to 15 pounds per square inch, then one can conclude that in a sample of 100 wood beams, 95 would have a strength between 12 and 15 pounds per square inch.

2. A sample of 16 high-technology firms selected at random yielded a mean annual expenditure on research and development of $600,000 with a standard deviation of $20,000.
   a. Construct a 98% confidence interval for the true mean annual research and development expenditures of all such firms.
   b. What would be the effect, if any, on the width of the interval in (a) if a higher level of confidence were used? If the interval had been based on a sample of 40 firms, assuming the same level of confidence and the same standard deviation?

3. A manufacturer of electronic components has a contract with NASA to provide components for weather satellites. The contract specifies that the components must have a mean life of two months when operating under extreme temperature conditions. To determine whether its components meet the requirements specified, the company decides to test several of the components, selected randomly. Since the testing process consists of subjecting the components (which are very costly) to extreme conditions that destroy the components, the company uses a small random sample. The results obtained from this sample are $n = 9$, $x = 9$ weeks, and $s = 4$ weeks.
   a. State a 95% confidence interval for μ, the population mean life of the components.
   b. Can the company be sure that there is no more than a 5% chance that the population mean, μ, lies outside this interval? Explain.

4. A simple random sample of 25 stocks from those listed on the New York Stock Exchange was selected by a security analyst in order to estimate the mean quarterly dividend payment of common stocks of the Exchange. The sample

yielded a mean dividend payment of $.50 per share per quarter with a standard deviation of $.12.

    a. Construct a 99% confidence interval for the true mean quarterly dividend payment per share of all New York Stock Exchange common stocks.

    b. Repeat part (a) for a 95% confidence coefficient and for a 90% confidence coefficient.

5. A company ran a test to determine the length of time required to complete service calls. The following times, in minutes, were obtained for a simple random sample of nine service calls: 48, 51, 28, 66, 81, 36, 40, 59, 50.

    a. Construct a 99% confidence interval for the mean time for completion of service calls.

    b. If times to complete service calls followed a highly skewed distribution, would the range you set up in part (a) really be a 99% confidence interval?

6.5 DETERMINATION OF SAMPLE SIZE

In all of the examples thus far, the sample size n was given. However, an important question is, How large should a sample be in a specific situation? If a larger sample than necessary is used, resources are wasted; if the sample is too small, the objectives of the analysis may not be achieved.

Sample Size for Estimation of a Proportion

Statistical inference provides the following type of answer to the question of sample size. Let us assume an investigator desires to estimate a certain population parameter and wants to know how large a simple random sample is required. We assume that the population is very large relative to the prospective sample size. The investigator must answer two questions in order to specify the required sample size. (1) What degree of precision is desired? and (2) How probable do you want it to be that the desired precision will be obtained? Clearly, the greater the degree of desired precision, the larger will be the necessary sample size; similarly, the greater the probability specified for obtaining the desired precision, the larger will be the required sample size. We will use examples to indicate the technique of determining sample size for estimation of a population proportion and a population mean.

EXAMPLE 6-6

Suppose we would like to conduct a poll among eligible voters in a city in order to determine the percentage who intend to vote for the Democratic candidate in an upcoming election. We specify that we want the probability to be 95.5% that we will estimate the percentage that will vote Demo-cratic within ±1 percentage point. What is the required sample size?

SOLUTION

We answer the question by first indicating the rationale of the procedure and then condensing this rationale into a simple summary formula. The statement of the question gives a relationship between sampling error that we are willing to tolerate and the probability of obtaining this level of pre-cision. In this problem, we have required that $2\sigma_{\bar{p}}$ be equal to 0.01. As can be seen in Figure 6-7, this means that we are willing to have a probabil-ity of 95.5% that our sample percentage $\bar{p}$ will fall within 0.01 of the true but unknown population proportion p. We may now write

$$2\sigma_{\bar{p}} = 0.01$$

or

$$2\sqrt{\frac{pq}{n}} = 0.01$$

and

$$\sqrt{\frac{pq}{n}} = 0.005$$

In all our previous problems, the sample size n was known, but here n is the unknown for which we must solve. However, it appears that there are too many unknowns, namely, the population parameters p and q as well as n. What we must do is estimate or guess values for p and q, and then we can solve for n. Suppose we wanted to make a very conservative estimate for n. What should we guess as a value for $\bar{p}$? In this context, by a conser-vative estimate we mean an estimate made in such a way as to ensure that the sample size will be large enough to deliver the precision desired. In

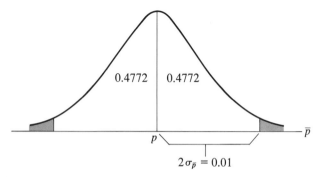

0.4772 | 0.4772

p $\bar{p}$

$$2\sigma_{\bar{p}} = 0.01$$

FIGURE 6-7
Sampling distribution of a proportion showing the relationship between error and probability of obtaining this degree of precision.

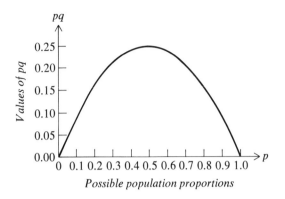

Possible population proportions

FIGURE 6-8
Relationship between possible values of p and the corresponding products pq (where $q = 1 - p$).

this problem, the "most conservative" estimate for n is given by assuming $p = 0.50$ and $q = 0.50$. This follows from the fact that the product $pq = 0.25$ is larger for $p = 0.50$ and $q = 0.50$ than for any other possible values of p and q where $p + q = 1$. For example, if $p = 0.70$ and $q = 0.30$, then $pq = 0.21$, which is less than 0.25. The relationships between possible values of p and the corresponding values of pq are shown in Figure 6-8. Thus, the largest, or "most conservative," value of n is determined by substituting $p = q = 0.50$ as follows:

$$0.005 = \sqrt{\frac{0.50 \times 0.50}{n}}$$

Squaring both sides gives

$$0.000025 = \frac{0.50 \times 0.50}{n}$$

and

$$n = \frac{0.50 \times 0.50}{0.000025} = 10,000$$

Hence, to achieve the desired degree of precision, a simple random sample of 10,000 eligible voters would be required. Of course, the large size of this sample is attributable to the high degree of precision specified. If $\sigma_{\bar{p}}$ were doubled from 0.005 to 0.01, the required sample size would be cut down to one fourth of 10,000, or 2500. This follows from the fact that the standard error varies inversely with the square root of sample size.

In this election problem, we assumed $p = q = 0.50$, although less conservative estimates are possible if it is believed that $p \neq 0.50$. In problems involving proportions, we would use whatever past knowl-

edge we have to estimate p. For example, suppose we wanted to determine the sample size to estimate an unemployment rate and we knew
from past experience that for the community of interest the proportion of
the labor force that was unemployed was somewhere between 0.05 and
0.10. We then would assume $p = 0.10$, since this would give us a more
conservative estimate (larger sample size) than assuming $p = 0.05$ or any
value between 0.05 and 0.10. (For example, if we assumed $p = 0.07$ and
in fact the true value of p was 0.09, the sample size we determined from a
calculation involving $p = 0.07$ would not be large enough to give us the
specified precision.) Assuming $p = 0.10$ assures us of obtaining the desired degree of precision regardless of what the true value of p is, as long
as it is in the range 0.05 to 0.10.

We can summarize this calculation for sample size by noting that we
started with the statement that

$$e = z\sigma_{\bar{p}}$$

where e is the specified (tolerable) sampling error and z is the multiple
of standard errors corresponding to the specified probability of obtaining
this precision. Then for infinite populations or large populations relative to sample size, we have

(6.16)
$$n = \frac{z^2(p)(1 - p)}{e^2}$$

Hence, in the preceding voting problem, application of Equation 6.16
yields

$$n = \frac{(2)^2(0.50)(0.50)}{(0.01)^2} = 10,000$$

Sample Size for Estimation of a Mean

The required sample size for estimation of a mean can be determined by
an analogous calculation. Suppose we wanted to estimate the arithmetic mean hourly wage rate for a group of skilled workers in a certain
industry. Let us further assume that from prior studies we estimate that
the population standard deviation of the hourly wage rates of these
workers is about $.15. How large a sample size would be required to
yield a probability of 99.7% that we will estimate the mean wage rate of
these workers within $\pm\$.03$?

Since the 99.7% probability corresponds to a level of three standard
errors, we can write

$$3\sigma_{\bar{x}} = \$.03$$

or

$$\frac{3\sigma}{\sqrt{n}} = \$.03$$

and

$$\frac{\sigma}{\sqrt{n}} = \$.01$$

The population standard deviation σ is known from past experience. Hence, substituting $\sigma = \$.15$ gives

$$\frac{\$.15}{\sqrt{n}} = \$.01$$

and

$$\sqrt{n} = \frac{\$.15}{\$.01} = 15$$

Squaring both sides yields the solution

$$n = (15)^2 = 225$$

Therefore, a simple random sample of 225 of these workers would be required. In summary, if we calculate $\bar{x}$ for the hourly wage rates of a simple random sample of 225 of these workers, we can estimate the mean wage rate of all skilled workers in this industry within $.03 with a probability of 99.7%.

In this problem, it was assumed that an estimate of the population standard deviation was available from prior studies; this type of situation may exist in governmental agencies that conduct repeated surveys of wage rates, population, and the like. If the population standard deviations (or estimated population standard deviations) in these past studies were not erratic or excessively unstable, they would provide useful bases for estimating σ values in the above procedures for computing sample size.

Of course, an estimate of the population standard deviation may not be available from past experience. It may be possible, however, to get a rough estimate of σ if there is at least some knowledge of the total range of the basic random variable in the population. For example, suppose we know that the difference between the wages of the highest- and lowest-paid workers is about $1.20. In a normal distribution, a range of three standard deviations on either side of the mean includes virtually the entire distribution. Thus, a range of 6σ includes almost all the frequencies, and we may state

$$6\sigma \approx \$1.20$$

or

$$\sigma \approx \$.20$$

Of course, the population distribution is probably nonnormal and $1.20 may not be exact. Consequently, the estimate of σ may be quite rough. Nevertheless, we may be able to obtain a reasonably good approximate estimate of the required sample size in a situation where, in the absence of this "guestimating" procedure, we may be at a loss for any notion of a suitable number of elements for the sample.

If we express in the form of an equation the technique for calculating a required sample size when estimating a population mean, we have

(6.17)
$$n = \frac{z^2\sigma^2}{e^2}$$

For the problem given earlier, this gives

$$n = \frac{(3)^2(\$.15)^2}{(\$.03)^2} = 225$$

In Equations 6.16 and 6.17, it is assumed that the n value determined is sufficiently large for the assumption of a normal sampling distribution to be appropriate and that the populations are large relative to this sample size. If the populations are not large relative to sample size, appropriate formulas may be derived for n by taking into account the finite population correction. The formulas for the sample sizes required for estimation of a proportion and mean, respectively, are

(6.18)
$$n = \frac{p(1-p)}{\dfrac{e^2}{z^2} + \dfrac{p(1-p)}{N}}$$

(6.19)
$$n = \frac{\sigma^2}{\dfrac{e^2}{z^2} + \dfrac{\sigma^2}{N}}$$

In Equations 6.18 and 6.19, N is the number of elements in the population (population size), and all other symbols are as previously defined. Equations 6.18 and 6.19 correspond respectively to Equations 6.16 and 6.17, in which infinite populations were assumed.

For example, in the voting problem discussed earlier in this section, if the population size had been 50,000, then the required sample size computed from Equation 6.18 would be

$$n = \frac{(0.50)(0.50)}{\dfrac{(0.01)^2}{(2)^2} + \dfrac{(0.50)(0.50)}{50,000}} = 8333$$

This result may be compared to the sample size of 10,000 computed earlier from Equation 6.16 assuming an infinite population size.

Analogously, in the problem involving the required sample size for estimating the population mean hourly wage for a group of skilled workers in a certain industry, let us assume that the population size was

1000 such skilled workers. Then the required sample size computed from Equation 6.19 would be

$$n = \frac{(0.15)^2}{\frac{(\$.03)^2}{(3)^2} + \frac{(0.15)^2}{1000}} = 184$$

This figure may be compared to the sample size of 225 computed previously from Equation 6.17 assuming an infinite population size. It would be instructive for the reader to verify that if a population size of 1,000,000 had been assumed, the computed sample size from Equation 6.19 would be 224.9, which rounds off to 225, the same figure obtained with the assumption of an infinite sample size.

Exercises

1. A random sample is to be selected to estimate the proportion of citizens of a large Texan city who favor federal price controls on natural gas that is transported interstate. The range of the estimate is to be kept within 4% with a confidence level of 96%. How large a simple random sample is required?

2. An efficiency expert made 100 random observations on a typist. In 56 of these observations, she found that the typist was "idle" (that is, without work to do).
   a. Construct a 98% confidence interval for the proportion of time the typist is idle.
   b. It was decided that the confidence interval in (a) was too large. If a new sample were drawn, how large would it have to be in order to estimate the true proportion of idle time within three percentage points, with a confidence coefficient of 98%?

3. A consumer research group is surveying the consumer population of the New England region to estimate the proportion of consumers who use biodegradable laundry detergent.
   a. How large a random sample should be drawn if the objective is to estimate the proportion of consumers using biodegradable detergent within four percentage points with 98% confidence? Assume that the proportion of consumers using this type of detergent is estimated from past experiences to be 0.20 or less.
   b. Assume, without regard to your answer in (a), that the research group drew a random sample of 400 consumers and found that 60 used biodegradable laundry detergent. Construct a 99% confidence interval for the proportion of consumers using the specified type of detergent.

4. A statistician wishes to determine the average hourly earnings for employees in a given occupation in a particular state. He runs a pilot study and obtains point estimates of $6.20 for the mean and $.50 for the standard deviation. He then specifies that when he takes his random sample, he wants to be 95.5% confident that the maximum error of estimate will not exceed $.05.
   a. Discuss the sense in which he will have 95.5% confidence in his estimate.
   b. What size sample should he take?

5. A public accountant wishes to estimate the percentage of companies in the United States that use the LIFO method of pricing inventory. He intends to do this on the basis of a random sample and wishes to be 95% confident that his estimate lies within three percentage points of the true percentage of companies using LIFO. He is quite certain that no more than 25% of the companies in the United States use this method of pricing inventory. How large a simple random sample should the accountant take?

6. One can always decrease the width of a confidence interval by increasing the sample size. Why then does one not always determine the desired width and then sample accordingly?

7. What are two ways of decreasing the width of a confidence interval for μ, given that the best point estimator of a sample is x?

8. The Chicago Board of Trade wishes to select a random sample of all wheat futures contracts in force at the present time to estimate the average premium being paid for an additional month in the term until maturity. For example, in a normal market, the price paid per bushel for immediate delivery of the commodity will be less than the selling price per bushel in a futures contract setting the delivery (and payment) date at one month from now. The price of a bushel of wheat might be $2.10 in a cash contract (immediate delivery), while the price of a bushel of wheat in a futures contract with a delivery date one month later might be $2.25. The premium in this case would be $2.25 − $2.10 = $.15. On the basis of past experience, it is estimated that the standard deviation of this premium is about $.04. How large a sample is needed to estimate the mean premium within $.005, with a 95.5% confidence coefficient?

9. An estimate of p, the percentage of unemployed executives in the United States, is desired. How large a sample should be drawn if it is desired to estimate p within 0.005 with 98% confidence? It may be assumed that $0.01 \leqslant p \leqslant 0.05$.

10. A popular U.S. senator who strongly encourages constituents to express their views on issues through letters and telegrams chairs a subcommittee that is attempting to formulate legislation to limit the amount organizations can contribute to the campaign of a presidential candidate. As a guideline, the senator has decided to estimate the mean of the limitations (in dollars) suggested in the thousands of letters received on the issue from the voters favoring such a limit. Instead of actually calculating the mean, the senator's staff will take a random sample of the letters and telegrams that have mentioned dollar amounts for the limitation.

   a. Estimate how large the sample should be in order for the appropriate interval estimate of the mean limitation to be no wider than $100. A 90% confidence coefficient is desired, and the standard deviation is taken to be $500.

   b. If the confidence coefficient was set at 99%, how large would the sample have to be?

7
Hypothesis Testing

In Chapter 6, we saw how statistical estimation techniques may be used to obtain and report information about population parameters. We now turn to the second basic subdivision of statistical inference, *hypothesis testing*, which deals with methods for testing hypotheses about population parameters. To put it another way, hypothesis testing addresses the important question of how to choose among alternative propositions or courses of action, while controlling and minimizing the risks of wrong decisions. We now briefly and informally summarize the rationale involved in testing hypotheses. Then we explain the details of these testing procedures by means of examples.

7.1 THE RATIONALE OF HYPOTHESIS TESTING

To gain some insight into the reasoning involved in statistical hypothesis testing, let us consider a nonstatistical hypothesis-testing procedure with which we are all familiar. As it turns out, the basic process of inference involved is strikingly similar to that employed in statistical methodology.

Consider the process by which an accused individual is judged in a court of law under our legal system. Under Anglo-Saxon law, the person before the bar is assumed innocent; the burden of proof of guilt rests on the prosecution. Using the language of hypothesis testing, let us say that we want to test a hypothesis, which we denote H_0, that the person before the bar is innocent. This means that there exists an alternative hypothesis, H_1, that the defendant is guilty. The jury examines the evidence to determine whether the prosecution has demonstrated that this evidence is inconsistent with the basic hypothesis, H_0, of innocence. If the jurors decide the evidence is inconsistent with H_0, they reject that hypothesis, and therefore accept its alternative, H_1, that the defendant is guilty.

If we analyze the situation that results when the jury makes its decision, we find that four possibilities exist. The first two possibilities pertain to the case in which the basic hypothesis H_0 is true, and the second two to the case in which the basic hypothesis H_0 is false.

1. The defendant is innocent (H_0 is true), and the jury finds that he is innocent (accepts H_0); hence, the correct decision has been made.
2. The defendant is innocent (H_0 is true), but the jury finds him guilty (rejects H_0); hence, an error has been made.
3. The defendant is guilty (H_0 is false), and the jury finds that he is guilty (rejects H_0); hence, the correct decision has been made.
4. The defendant is guilty (H_0 is false), but the jury finds him innocent (accepts H_0); hence, an error has been made.

In cases (1) and (3), the jury reaches the correct decision; in cases (2) and (4), it makes an error. Let us consider these errors in terms of conventional statistical terminology. The basic hypothesis, H_0, tested for possible rejection is generally referred to as the *null hypothesis*, and hypothesis H_1 is designated the *alternative hypothesis*. In case (2), hypothesis H_0 is erroneously rejected. To reject the null hypothesis when in fact it is true is referred to as a *Type I error*. In case (4), hypothesis H_0 is accepted in error. To accept the null hypothesis when it is false is termed a *Type II error*. It may be noted that under our legal system, a Type I error is considered far more serious than a Type II error; we feel that it is worse to convict an innocent person than to let a guilty one go

TABLE 7-1
The Relationship between Actions Concerning a Null Hypothesis
and the Truth or Falsity of the Hypothesis

Action *Concerning* *Hypothesis H_0*	*State of Nature*	
	H_0 Is True *(Innocent)*	*H_0 Is False* *(Guilty)*
Accept H_0	Correct decision	Type II error
Reject H_0	Type I error	Correct decision

free. Had we made H_0 the hypothesis that the defendant is guilty, the meaning of Type I and Type II errors would have been reversed. In the statistical formulation of hypotheses, how we choose to exercise control over the two types of errors is a basic guide in stating the hypotheses to be treated. We will see in this chapter how this error control is carried out in hypothesis testing. The cases listed above are summarized in Table 7-1; the headings in this table are in the terminology of modern decision theory and require a brief explanation. When hypothesis testing is viewed as a problem in decision making, two alternative actions can be taken: "accept H_0," and "reject H_0." The two alternatives, truth and falsity of hypothesis H_0, are viewed as "states of nature," or "states of the world," that affect the consequences, or "payoff," of the decision. The payoffs are listed in the table, and in the schematic presentation they are stated in terms of the correctness of the decision or the type of error made. We can see from the framework of the hypothesis-testing problem that what we need is some criterion for the decision either to accept or to reject the null hypothesis, H_0. Classical hypothesis testing attacks this problem by establishing decision rules based on data derived from simple random samples. The sample data are analogous to the evidence investigated by the jury. The decision procedure attempts to assess the risks of making incorrect decisions and, in a sense that we shall examine, to minimize them.

The Hypothesis-Testing Procedure

Two basic types of decision problems can be attacked by hypothesis-testing procedures. In the first type of problem, we want to know whether a population parameter has changed from or differs from a particular value. Here, we are interested in detecting whether the population parameter is *either* larger than or smaller than a particular value.

For example, suppose that the mean family income in a certain city was determined from a census to be $14,500 for a particular year and two years later we want to discover whether the mean income has *changed*. If it is not feasible to take another census, we may draw a simple random sample of families and try to reach a conclusion based on this sample. As in the illustration of the court of law, we can set up two competing hypotheses and choose between them. The null hypothesis H_0 would simply be an assertion that the mean family income was unchanged from the $14,500 figure; in statistical language, we write this hypothesis as $H_0: \mu = \$14,500$, where μ denotes the mean family income in the city. The alternative hypothesis is then that the mean family income *has* changed, or in statistical terminology, $H_1: \mu \neq \$14,500$. In this example, we would observe the mean family income in the simple random sample of, say, 1000 families. If the value of the sample mean $\bar{x}$ differs from the population mean $\mu = \$14,500$ by more than we would be willing to attribute to chance sampling error, we reject the null hypothesis H_0 and accept its alternative H_1. On the other hand, if the difference between the sample mean and the population mean assumed under H_0 is small enough to be attributed to chance sampling error, we accept H_0. How do we know for what values of the sample statistic to reject H_0 and for what values to accept H_0? The answer to this question is the essence of hypothesis testing.

The hypothesis testing procedure is simply a decision rule that specifies, for every possible value of a statistic observable in a simple random sample of size n, whether the null hypothesis H_0 should be accepted or rejected. The set of possible values of the sample statistic is referred to as the *sample space*. Therefore, the test procedure divides the sample space into mutually exclusive parts called the *acceptance region* and the *rejection* (or *critical*) *region*.

The nature of the division of the sample space for the example we have been discussing is illustrated in Figure 7-1. From the sampling

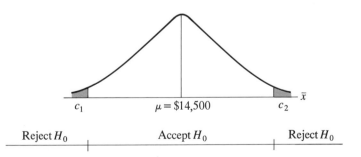

FIGURE 7-1
Two-tailed test: sampling distribution of the mean with acceptance and rejection regions for a null hypothesis.

theory developed in Chapter 5, we know that given a population with a mean of $14,500 and a known standard deviation, there would be sampling variation among the means of samples of the same size drawn from that population. According to the Central Limit Theorem, regardless of the shape of the population distribution, the sampling distribution of the mean for large sample sizes may be assumed to be normal. In order to decide whether family mean income has changed from $14,500, we determine two values, denoted c_1 and c_2 in Figure 7-1, which set limits on the amount of sampling variation we feel is consistent with the null hypothesis. Hence, as depicted in the figure, the decision rule in this case would be: If the mean income of the sample of 1000 families lies below c_1 or above c_2, reject the null hypothesis and conclude that the mean family income of the city has changed from $14,500. On the other hand, if the sample mean lies between c_1 and c_2, we cannot reject H_0. Hence, we cannot conclude that the city's mean family income has changed. A test in which we want to determine whether a population parameter has *changed* regardless of the direction of change is referred to as a *two-tailed test*, since the null hypothesis can be rejected by observing a statistic that falls in either of the two tails of the appropriate sampling distribution, as shown in Figure 7-1.

The second type of hypothesis test is one in which we wish to find out whether the sample came (1) from a population that has a parameter *less* than a hypothesized value or (2) from a population that has a parameter *more* than a hypothesized value. These situations, in which attention is focused upon the direction of change give rise to *one-tailed tests*. We will illustrate such a test with a quality control example. Suppose a construction company specified in a purchasing order that acceptable rivets must have a mean tensile strength of at least 40,000 pounds. The company decides to test the quality of an incoming shipment by taking a simple random sample of 100 rivets. The null hypothesis H_0 to be tested is "the true mean tensile strength of all rivets in the shipment is at least 40,000 pounds," and the alternative hypothesis H_1 is "the true mean tensile strength of all rivets in the shipment is less than 40,000 pounds." These two hypotheses may be expressed mathematically as follows:

$$H_0: \mu \geq 40,000 \text{ pounds}$$
$$H_1: \mu < 40,000 \text{ pounds}$$

where μ denotes the mean tensile strength of the shipment.

In this case, the decision rule for accepting or rejecting H_0 would be as follows: If the mean of the sample, $\bar{x}$, is less than some appropriate number c, reject H_0, and therefore reject the shipment. Otherwise, accept H_0 and the shipment. This decision rule is represented diagrammatically in Figure 7-2. Again, the number c represents a limit on the

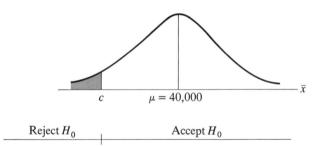

FIGURE 7-2
One-tailed test: sampling distribution of the mean with acceptance and rejection regions for a null hypothesis.

amount of sampling variation that we feel is consistent with the null hypothesis.

Concept of the Null Hypothesis

As can be seen from the preceding illustrations, a null hypothesis is a statement about a population parameter, as in the first example, H_0: $\mu = \$14,500$, and in the second, H_0: $\mu \geq 40,000$ pounds. Note that the equality sign is present in both hypotheses. It is standard procedure to include the equality sign in the null hypothesis. By having the null hypothesis assert that the population parameter is equal to some specific value, we are then able to decide for which values of the observed sample statistic we will reject or accept the null hypothesis. For example, in the two illustrations, when we hypothesized that $\mu = \$14,500$ and *included* in the second null hypothesis that $\mu = 40,000$ pounds, we were then able to establish sampling distributions of the sample mean $\bar{x}$ and to decide which values of $\bar{x}$ would lead us to reject the null hypothesis and which values would cause us to accept it. Furthermore, we can specify how much risk of making Type I errors we are willing to tolerate. For example, if we hypothesize that $\mu = 40,000$ pounds, we can compute the probability of making a Type I error, that is, the probability of erroneously rejecting that hypothesis. The application of this idea will be seen later in this chapter. The above points explain why null hypotheses are set up in the form stated earlier, rather than, say, in the form H_0: $\mu \neq \$14,500$ or H_0: $\mu < 40,000$ pounds.

We now turn to the application of some of these ideas. First, we will consider one-sample tests, which are tests of hypotheses based on data contained in a single sample. Tests involving means and proportions will be studied in that order.

Exercises

1. Distinguish between a parameter and a statistic.
2. Support or criticize the following statement: "A statistician will make very few errors by setting a very low significance level, say 0.001; hence, the statistician should always do so."
3. Distinguish briefly between
   a. a one-tailed test and a two-tailed test
   b. Type I and Type II errors

7.2 ONE-SAMPLE TESTS (LARGE SAMPLES)

It is assumed in the one-sample tests discussed in this section that

1. *the sample size is large ($n > 30$)*
2. *the sample size is decided upon before the test is conducted*
3. *the population standard deviation is known*

The assumption of large sample sizes ($n > 30$) is made so we can assume that the sampling distributions used in the testing procedure are normal. This is a useful simplification as compared to the small-sample case, as we shall see at the end of this chapter.

As a more detailed illustration of how the hypothesis-testing procedure we have discussed can be used, we consider first an example of an *acceptance sampling procedure.* The Morgan Company, a manufacturer of components for space vehicles regularly purchases a part required to withstand high temperatures. Specifications of mean heat resistance of at least 2250 degrees Fahrenheit (2250°F) are applied by the company to shipments sent by the supplier. In the past, all shipments have met specifications. We also assume that it is *known* from experience that the standard deviation of heat resistance of parts in a shipment is 300°F. In a simple random sample of 100 parts drawn from a particular shipment, a mean heat resistance of 2110°F was observed. Should the shipment be accepted, if it is desired that the risk of erroneously rejecting a shipment that actually meets the specifications for heat resistance be no more than 0.05? The shipment may be assumed to be very large relative to the sample size of 100.

We now convert this verbal problem to a hypothesis testing framework. As we shall see, this is an example of a one-tailed test, or one-sided alternative.

Test Involving a Mean: One-Tailed Test

We begin by stating in statistical terms the null hypothesis to be tested. Returning to our definition of a hypothesis as a quantitative statement about a population, we see that in this case the statement is about the mean of the population (shipment) sampled. (Note that an inference is to be drawn about the shipment and that the sample was randomly drawn from the shipment. Thus, the shipment is the population in this problem. The production process of the supplier may be viewed as a super population that gives rise to the shipment populations.) Letting μ denote the mean heat resistance of the parts in this particular shipment population, we state the null and alternative hypotheses as follows:

$$H_0: \mu \geqslant 2250°\text{F}$$
$$H_1: \mu < 2250°\text{F}$$

where the standard deviation σ of the population is assumed to be 300°F. The null hypothesis H_0 states that the population mean heat resistance is at least 2250°F. If our decision procedure leads us to accept this hypothesis, our action will be to accept the shipment. On the other hand, if we reject H_0, we will accept the alternative hypothesis H_1 and therefore conclude that the shipment mean heat resistance is less than 2250°F; then we will reject the shipment. Our decision will be based on the data observed in the sample of 100 parts. Therefore, the question is simply, Are the sample data so inconsistent with the null hypothesis that we shall be forced to reject that hypothesis?

Before we examine this question more closely, we must discuss an important technical point. The null hypothesis must be stated in such a way that the probability of a Type I error can be calculated. This was the reason for including the equality sign in the statement of the null hypothesis, so that a particular value of μ was specified. Since the mean heat resistance of the shipment can exceed 2250°F in an infinite number of ways, one for each possible value of μ, we cannot refer unambiguously to the probability of a Type I error as *the* probability of rejection of the null hypothesis when it is true. However, if we concentrate our attention on the single value $\mu = 2250°\text{F}$, we can refer to a Type I error for that particular value, and we will be able to compute the probability of such an error once we have settled on a decision procedure. Hence, for the moment, we focus attention on the particular value $\mu = 2250°\text{F}$, for which the null hypothesis is true.

We now return to our question regarding the consistency of the sample data with the null hypothesis. In terms of this particular problem, we can word the question as follows: Is the deviation between a mean of 2110°F observed in a simple random sample of size 100 and a hypothesized population mean of 2250°F so great that we would be unwilling to attribute such a difference to chance errors of sampling? If we conclude the difference is so large that it is improbable under the given hypothesis, then we say that a *significant difference* has been observed. An observed significant difference between a statistic and a parameter rejects the null hypothesis. Where the dividing line is set up between a significant difference and a nonsignificant difference depends on the risk we are willing to run of making a Type I error, that is, of rejecting the null hypothesis when it is true. For instance, we may set up the dividing line at a point such that the probability is 0.05 of erroneously rejecting the null hypothesis; then the test is said to have been conducted *at the 5% significance level.*

Significance levels such as 0.05 and 0.01 are very frequently used in classical hypothesis testing, because of the desire to maintain a low probability of rejecting the null hypothesis H_0 when it is in fact true. The level of significance is denoted by the Greek letter α *(alpha)*. As we shall see, in one-tailed test situations, α represents the *maximum* probability of a Type I error. In two-tailed test situations, in which the null hypothesis consists of only one value of a population parameter, α represents *the* probability of a Type I error.

Table 7-2 summarizes the two hypotheses tested in the present acceptance sampling problem and the possible actions concerning these hypotheses. After a brief discussion of the meaning of Type I and Type II errors in this problem, we will examine how the hypothesis-testing procedure is actually carried out quantitatively.

As we observe from Table 7-2, a Type I error in this problem (incorrect rejection of the null hypothesis) takes the form of the rejection of a "good" shipment. ("Good" and "bad" shipments in this context are those that meet specifications and those that do not meet specifications,

TABLE 7-2
Acceptance Sampling Problem: Relationship between Possible Actions and Hypotheses Concerning the Quality of a Shipment

Action Concerning Hypothesis H_0	State of Nature	
	$H_0: \mu \geq 2250°F$ (Shipment Meets Specifications)	$H_1: \mu < 2250°F$ (Shipment Does Not Meet Specifications)
Accept H_0 (Accept shipment)	No error	Type II Error: Acceptance of a shipment that does not meet specifications
Reject H_0 (Reject shipment)	Type I error: Rejection of a shipment that meets specifications	No error

respectively.) In quality control work, a Type I error is termed the "producer's risk," since the producer runs the risk of having a good shipment rejected because the data in a sample drawn from the shipment were misleading due to chance sampling error. A Type II error in this problem takes the form of the acceptance of a bad shipment and is referred to as the "consumer's risk;" that is, the consumer runs the risk of accepting a bad shipment. A realistic industrial sampling plan would require an equitable balancing of these types of risks, taking into account the costs involved to the producer and consumer. However, let us proceed with the present simplified problem, in order to study the classical approach to hypothesis testing.

Decision Rules

We now turn to the question of how to establish a *decision rule* on which to base our acceptance or rejection of the shipment. As indicated earlier, a hypothesis testing decision rule is simply a procedure that specifies the action to be taken for each possible sample outcome. Thus, we are interested in partitioning the sample space into a region in which we will reject the null hypothesis and a region in which we will accept it. The partitioning of the sample space is accomplished by considering the sampling distribution that results from assuming the null hypothesis is true and by specifying the probability of making a Type I error. In this particular problem, since the question concerns *mean* heat resistance, the sampling distribution of means is the relevant distribution, and we want to perform a test at the 5% significance level. Since our sample is large ($n = 100$), the Central Limit Theorem allows us to assume that the sampling distribution of means is normal. This distribution, for samples of size 100 from a population where $\mu = 2250°F$ (just in conformity with specifications) and $\sigma = 300°F$, is shown in Figure 7-3. The shaded region in this graph represents 5% of the area under the normal curve. Referring to Table A-5, we see that 5% of the area in a normal distribution lies to the right of $z = +1.65$, and correspondingly the same percentage of area lies to the left of $z = -1.65$. Thus, we would reject a shipment with a sample mean less than $\mu - 1.65\sigma_{\bar{x}}$. The standard error of the mean, $\sigma_{\bar{x}}$, is equal to

$$\sigma_{\bar{x}} = \frac{\sigma}{\sqrt{n}} = \frac{300°F}{\sqrt{100}} = 30°F$$

Thus, the critical value below which we would reject H_0 is

$$\mu - 1.65\sigma_{\bar{x}} = 2250°F - 1.65(30°F) \approx 2200°F$$

We can now see why this type of hypothesis testing situation is referred to as a one-tailed test or a one-sided alternative. Rejection of the null hypothesis takes place in only one tail of the sampling distribution.

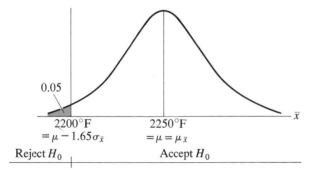

FIGURE 7-3
Sampling distribution of the mean showing regions of acceptance and rejection of H_0; population parameters are $\mu = 2250°F$ and $\sigma = 300°F$, and sample size is $n = 100$.

This is reasonable because in this example the other tail corresponds to shipments that significantly exceed specifications.

In summary, the Morgan Company should proceed as follows. Upon drawing a single random sample of 100 parts from the incoming shipment and observing $\bar{x}$, the sample mean heat resistance, the Morgan Company should apply this decision rule:

DECISION RULE

1. If $\bar{x} < 2200°F$, reject H_0 (reject the shipment)
2. If $\bar{x} \geqslant 2200°F$, accept H_0 (accept the shipment)[1]

We can now answer the original question. Since the sample yielded a mean heat resistance of $2110°F$, the null hypothesis H_0 should be rejected, and therefore the shipment should be rejected.

It is instructive to examine an alternative method of stating the decision rule. Instead of doing our work in terms of the original units, that is, in degrees, we could have calculated the z value in a standard normal distribution corresponding to an $\bar{x}$ value of $2110°F$. If the z value lies to the left of the critical value of -1.65, H_0 is rejected; if to the right of -1.65, H_0 is accepted. Thus, the decision rule can be rephrased as follows:

DECISION RULE

1. If $z < -1.65$, reject H_0 (reject the shipment)
2. If $z \geqslant -1.65$, accept H_0 (accept the shipment)

[1] There is some ambiguity whether the equality sign should appear in the rejection or acceptance part of the decision rule. From the theoretical point of view, it is inconsequential. This follows from the fact that the normal curve is a continuous probability distribution, so the probability of observing exactly $\bar{x} = 2200°F$ is 0.

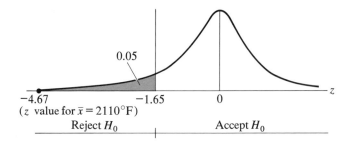

FIGURE 7-4
Standard normal curve for the acceptance sampling problem.

The arithmetic for the z value corresponding to $2110°F$ is

$$z = \frac{\bar{x} - \mu}{\sigma_{\bar{x}}} = \frac{2110°F - 2250°F}{30°F} = -\frac{140°F}{30°F} = -4.67$$

Therefore, since a mean of $2110°F$ falls 4.67 standard error units below the mean of the sampling distribution and the dividing line is at 1.65 standard error units, the null hypothesis H_0 is rejected. This situation is displayed in Figure 7-4.

The form of the decision rule we choose is inconsequential. However, as a practical matter, if the Morgan Company repeatedly applied the same sampling procedure to incoming shipments, the first form would be simpler. Once the rule is established for a particular sample size, no further arithmetic is required. Nevertheless, it is instructive in considering the rationale of the test to observe the implications of the computed z value. For example, in this case an $\bar{x}$ value of $2110°F$ corresponded to a z value of -4.67. Table A-5 of Appendix A does not give areas for z values above 4.0. Less than 0.0001 of the area in a normal distribution lies to the right of a z value of 4.0 or to the left of a z value of -4.0. Thus, if we interpret the z value of -4.67 in terms of the acceptance sampling problem, we see that if a sample of 100 parts was drawn at random from a shipment whose parts had a mean heat resistance of $2250°F$ and a standard deviation of $300°F$, the probability of observing a sample mean of $2110°F$ or lower was less than 0.0001. Since this sample result is so unlikely under the hypothesis of a shipment mean of $2250°F$, we reject that hypothesis.

For values of μ greater than $2250°F$, the probability of a Type I error, if we reject the null hypothesis whenever $\bar{x} < 2200°F$, is less than 0.05. We can see this from Figure 7-3. If $\mu > 2250°F$ (that is, if the sampling distribution shifts to the right), then less than 5% of the area will lie in the rejection region below $2200°F$. The larger the value of μ, the lower is the probability of a Type I error. This makes sense in terms of the acceptance sampling problem. The greater the actual mean heat resistance in the incoming shipment, the lower is the probability that the shipment will be erroneously rejected using the decision rule derived above. For $\mu = 2250°F$, the probability of a Type I error is 0.05. Now we can see the meaning of $\alpha = 0.05$, the significance level in this problem. It is the maximum probability of committing a Type I error. This sort of interpretation is typical for one-tailed tests.

Summary of Procedure

In the discussion so far, we have considered the situation in which the acceptance and rejection of the null hypothesis result in only two possible conclusions or actions. Furthermore, we have concentrated on the determination of decision rules stemming from control of Type I errors without reference to the corresponding implications for Type II errors. We shall deal with these matters subsequently, since it is advisable not to clutter the present discussion with too many details. We summarize the hypothesis-testing procedure discussed thus far as follows:

1. A null hypothesis and its alternative are drawn up. The null hypothesis is framed in such a way that we can compute the probability of a Type I error.
2. A level of significance, α, is decided upon. This controls the risk of a Type I error.
3. A decision rule is established by partitioning the relevant sample space into regions of acceptance and rejection of the null hypothesis. This partition is accomplished by consideration of the relevant sampling distribution. The nature of the null hypothesis and the choice of α determine the partition.
4. The decision rule is applied to the sample of size n. The null hypothesis is accepted or rejected. Rejection of the null hypothesis implies acceptance of the alternative.

Further Remarks

Several points should be made concerning the statistical theory involved in the acceptance sampling problem.

1. The normal curve was used as the appropriate sampling distribution of the mean. As we have seen, if the population distribution of mean heat resistances is normal, the normal curve is the theoretically correct sampling distribution; however, no statement about the population distribution was made in the acceptance sampling problem. The normal curve was used for the sampling distribution of $\bar{x}$ because the Central Limit Theorem tells us that no matter what the shape of the

population, the sampling distribution of x is approximately normal for a sample size as large as $n = 100$.

2. No finite population correction was used in the calculation of the standard error of the mean despite the fact that the sample of 100 parts was drawn without replacement from a finite population. However, it was stated in the problem that the population size was very large relative to the sample size. Therefore, the finite correction factor may be assumed to be approximately equal to 1. If the population size N is not large relative to the sample size, n, the standard error should be multiplied by $\sqrt{(N - n)/(N - 1)}$.

3. The population standard deviation, σ, was assumed to be known. If the population standard deviation is unknown and the sample size is large (say, $n > 30$), then the sample standard deviation s may be substituted for σ. Hence, instead of calculating $\sigma_{\bar{x}} = \sigma/\sqrt{n}$, an estimated standard error of the mean would be computed as $s/\sqrt{n}$. In all other respects, the decision procedure remains the same. In Section 7.4, we discuss the situation in which the sample size is small and the population standard deviation is unknown.

4. The nature of the z value computed in the problem is worth noting. In Chapter 5, when the idea of a standard score was discussed, it was in the context of a normally distributed *population*. In that case, $z = (x - \mu)/\sigma$ represented a deviation of the value of an individual item from the mean of the population expressed as a multiple of the population standard deviation. In our hypothesis-testing problem, the z values were of the form $z = (\bar{x} - \mu)/\sigma_{\bar{x}}$. Such a z value represents a deviation of a sample mean from the mean of the sampling distribution of $\bar{x}$, stated in multiples of the standard deviation of that distribution, $\sigma_{\bar{x}}$. (As we noted previously, the mean of the sampling distribution of $\bar{x}$, $\mu_{\bar{x}}$, is equal to the population mean, μ. Thus, we use μ and $\mu_{\bar{x}}$ interchangeably.) In hypothesis testing problems, z values generally take the form

$$z = \frac{\text{statistic} - \text{parameter}}{\text{standard error}}$$

For example, in the hypothesis testing problem just discussed, the sample value $\bar{x}$ is the statistic; μ, the population mean, is the parameter; and $\sigma_{\bar{x}}$, the standard error of the mean, is the appropriate standard error.

5. The size of the sample, $n = 100$, was predetermined in our illustration. Thus, the case we discussed was one in which the sample size was large, it was predetermined, and the construction of the decision rule was based on the control of only one type of incorrect decision, namely, Type I errors. The next section, on the power curve, discusses the measurement of Type II errors for such a test.

6. It is important to realize that we did not *prove* that the null hypothesis was false, nor could sample evidence have proved the null hypothesis true. All we can do is discredit a null hypothesis or fail to discredit it on the basis of sample data. Actually, a single sample statistic such as $\bar{x}$ is consistent with an infinite number of hypotheses concerning μ.[2] From the standpoint of decision making and subsequent actions, if sample data do not discredit a null hypothesis, we act as though that hypothesis is true. However, in scientific experimentation, the truth of a hypothesis or theory is always considered an ongoing question, subject to further tests.

The Power Curve

The hypothesis-testing procedure outlined thus far has concentrated on the control of Type I errors. The question of how well this test controls Type II errors naturally arises. That is, when the null hypothesis is false, how frequently does the decision rule lead us to accept the null hypothesis erroneously? This question is answered by means of the *power curve*, also called the *power function*, which can be computed from the information of the problem and the decision rule. The Greek letter β (*beta*) is used to denote the probability of a Type II error; thus, β represents the probability of accepting the null hypothesis when it is false. In the acceptance sampling problem, the null hypothesis H_0 is false for each possible value of μ satisfying the alternative hypothesis H_1: $\mu <$ 2250°F. Therefore, for each particular value of μ less than 2250°F, we can determine a β value. Actually, by convention, the power curve gives the complementary probability to β, that is, $1 - \beta$, for each such value of μ. Thus, it shows the probability of rejecting the null hypothesis for each value for which the null hypothesis is false, which, of course, represents in each case the probability of selecting the correct course of action. The number $1 - \beta$ is referred to as the *power of the test* for each particular value of μ satisfying the alternative hypothesis. For completeness, in a power curve, the probabilities of rejection are also shown for each value for which the null hypothesis is true. In summary, a power curve is a function that gives the probabilities of rejecting the null hypothesis H_0 for all possible values of the parameter tested.[3]

---

[2] Doubtless we have all had the disconcerting experience of observing a number of experts in disagreement after observing ostensibly the same basic set of data. Perhaps you share the experience of finding it easiest to accept and reject hypotheses when there are no data available at all.

[3] It is usual in industrial quality control to use *operating-characteristic curves*, briefly referred to as O-C curves, rather than power curves to evaluate the discriminating power of a test. The O-C curve is simply the complement of the power curve. That is, the probability of acceptance rather than the probability of rejection of the null hypothesis is plotted on the vertical axis.

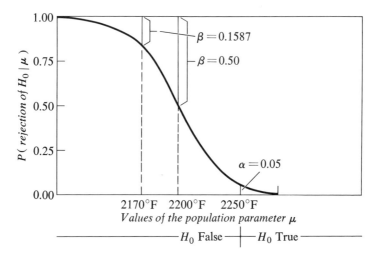

FIGURE 7-5
Power curve for the Morgan Company acceptance sampling problem.

Therefore, it shows the ability of the decision rule to discriminate between true and false hypotheses. It is useful in assessing the risks of making both Type I and Type II errors when using a decision rule.

The power curve for the acceptance sampling problem is shown in Figure 7-5. It has the typical reverse-S shape of a power curve for a one-tailed test with the rejection region in the left tail. For a one-tailed test with the rejection region in the right tail, the curve would be S-shaped, or dropping from upper right to lower left on the graph. Rejection probabilities for the null hypothesis are shown on the vertical axis and possible values of the population parameter μ on the horizontal axis. Specifically, the figures plotted on the vertical axis are conditional probabilities of the form $P(\text{rejection of } H_0|\mu) = P(\bar{x} < 2200°F|\mu)$.

The nature of the power curve in Figure 7-5 can be understood by considering a few of the plotted values. The value of $\alpha = 0.05$ is shown for $\mu = 2250°F$, indicating the significance level of the test. We can see that this is the maximum probability of erroneously rejecting the null hypothesis H_0, because in the region where H_0 is true, the ordinates of the curve drop off to the right as μ increases. The heights of the ordinates of the power curve to the right of $\mu = 2250°F$ represent the probabilities of making Type I errors. The heights of the ordinates to the left of $\mu = 2250°F$ give the values of $1 - \beta$, which are the probabilities of rejecting the null hypothesis when it is false. Therefore, the complementary distances from the curve to 1.0 are values of β, or probabilities of making Type II errors. Two such β values, for $\mu = 2200°F$ and for $\mu = 2170°F$, are displayed in the graph. Recall that our decision rule required the rejection of H_0 if the sample mean $\bar{x}$ was less than 2200°F. Thus, if the

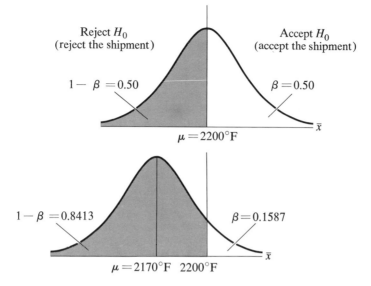

FIGURE 7-6
Graphs illustrating Type II error probabilities for $\mu = 2200°F$ and $\mu = 2170°F$.

shipment or population mean is 2200°F, obviously the probability of observing $\bar{x}$ values less than 2200°F (and, therefore, of rejecting H_0) is 0.50. This situation is shown graphically in Figure 7-6.

On the other hand, a bit of computation is required to obtain the β value for $\mu = 2170°F$. To compute this figure, we must refer to the sampling distribution of $\bar{x}$, given that $\mu = 2170°F$, and calculate the probability that a sample mean would lie in the rejection region, $\bar{x} < 2200°F$. The z value for 2200°F is

$$z = \frac{2200°F - 2170°F}{30°F} = +1.0$$

Thus, an $\bar{x}$ value of 2200°F lies one standard error of the mean to the right of 2170°F. Referring to Table A-5 of Appendix A, we find a figure of 0.3413, which when added to 0.5000 (the area to the left of $\mu = 2170°F$) gives 0.8413 as the probability of rejecting H_0. This probability is shown as the shaded region in Figure 7-6. Hence, the value of β for $\mu = 2170°F$ is $1.000 - 0.8413 = 0.1587$, or about 16%. Of course, this figure could have been obtained directly by subtracting 0.3413 from 0.5000, the area to the right of $\mu = 2170°F$.

For values of μ slightly less than 2250°F, the probability of a Type II error is very high. In fact, as can be seen from Figure 7-5, these probabilities exceed 0.50 for μ values between 2200°F and 2250°F. This simply indicates that the power of the test is low when the value of μ satisfies the alternative hypothesis but is close to values of μ satisfying the null hypothesis. For a fixed sample of size n, β can be decreased only by increasing α, and vice versa. If α is fixed, then as the sample size is increased, β is reduced for all values of μ in the region where H_0 is false. In this type of one-tailed test, the ideal power curve would be ⌐ shaped with the vertical line occurring at $\mu = 2250°F$; thus, the probability of rejecting H_0 would always be 1.0 when H_0 is false and 0.0 when H_0 is true. However, clearly this ideal curve is unattainable when sample data are used to test hypotheses, since sampling error will always be present. We may note that there is a trade-off relationship between Type I and Type II errors for a sample of fixed size, and thus in classical hypothesis tests such as the one discussed here, the level of significance should be decided by consideration of the relative seriousness of the two types of error.

Making Use of the Power Curve

How can the decision maker use power curves in setting up an appropriate hypothesis testing procedure? In the preceding discussion, the decision rule was determined for a fixed sample size n; α was specified, and the critical value of $\bar{x}$, 2200°F, was computed. Suppose that in examining the resulting power curve, the decision maker felt that the β values were too high. For instance, in the present example, it was deter-

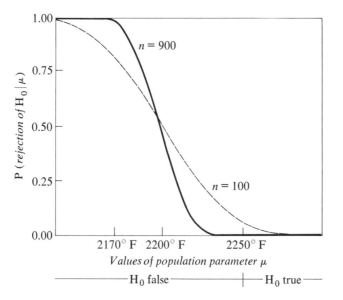

FIGURE 7-7
Comparison of power curves for two sample sizes, $n = 900$ and $n = 100$, for the Morgan Company acceptance sampling problem.

mined that if the shipment mean heat resistance was 2170°F, $\beta = 0.1587$ (see Figure 7-5). Since specifications called for a shipment mean of at least 2250°F, a shipment with $\mu = 2170$°F does not meet specifications, yet has a probability of 0.1587 of being accepted. If the Morgan Company wished to decrease this risk, it could increase the critical value of $\bar{x}$, thus raising the level of significance, which had been set at $\alpha = 0.05$. This would shift the entire power curve to the right. Thus, we see that if sample size n is unchanged, the only way to reduce β levels is to increase α. On the other hand, if the decision maker is unwilling to increase the significance level α, then the only way to reduce the β values is by increasing the sample size n. Of course, larger sample sizes involve increased costs.

To illustrate that larger sample sizes provide more discriminating tests, let us consider the effect of increasing the sample size in the Morgan Company example from $n = 100$ to $n = 900$, while retaining the critical value at $\bar{x} = 2200$°F. Figure 7-7 gives the corresponding power curves for the tests provided by these two sample sizes. Note in Figure 7-7 that the larger sample size provides a more discriminating test because its power curve implies lower probabilities both of accepting false hypotheses and of rejecting true hypotheses.

Test Involving a Proportion: Two-Tailed Test

The preceding discussion dealt with a test of a hypothesis about a mean. We now turn to hypothesis testing for a proportion. As an illustration, let us consider the case of an advertising agency that developed a general theme for the commercials on a certain TV show based on the assumption that 50% of the show's viewers were over 30 years of age. The agency was interested in determining whether the percentage had changed in either an upward or downward direction. If we use the symbol p to denote the proportion of all viewers of the show, we can state the null and alternative hypotheses as follows:

$$H_0: p = 0.50 \text{ viewers over 30 years of age}$$
$$H_1: p \neq 0.50 \text{ viewers over 30 years of age}$$

Let us assume that the agency wished to run a risk of 5% of erroneously rejecting the null hypothesis of "no change," i.e., $H_0: p = 0.50$. That is, the agency decided to test the null hypothesis at the 5% significance level ($\alpha = 0.05$).

In order to test the hypothesis, the agency conducted a survey of a simple random sample of 400 viewers of the TV show. Of the 400 viewers, 210 were over 30 years of age and 190 were 30 years of age or less. What conclusion should be reached?

In this problem, as contrasted with the previous one, the null hypothesis concerns a single value of p, which is a hypothetical population parameter of 0.50. The alternative hypothesis includes all other possible values of p. The reason for setting up the hypotheses this way can be seen by reflecting on how the test will be conducted. The hypothesized parameter under the null hypothesis is $p = 0.50$. We have observed in a sample a certain proportion, denoted $\bar{p}$, who were over 30 years of age. The testing procedure involves a comparison of $\bar{p}$ with the hypothesized value of p to determine whether a significant difference exists between them. If $\bar{p}$ does not differ significantly from p, and we accept the null hypothesis that $p = 0.50$, what we really mean is that the sample is consistent with a hypothesis that half the viewers of the TV show are over 30. On the other hand, if $\bar{p}$ is greater than 0.50 and a significant difference between $\bar{p}$ and p is observed, we will conclude that more than half the viewers are over 30. If the observed $\bar{p}$ is less than 0.50 and a significant difference from $p = 0.50$ is observed, we will conclude that fewer than half the viewers are over 30.

It is important to note that in hypothesis-testing procedures, the two hypotheses and the significance level of the test must be selected before the data are examined. We can easily see the difficulty with a procedure that would permit the investigator to select α after examination of the sample data. It would always be possible to accept a null hypothesis simply by choosing a sufficiently small significance level, thereby setting up a large enough region of acceptance. Thus, the first step in the present problem is the usual one of setting up the competing hypotheses, with the null hypothesis stated in such a way that the probability of a Type I error can be calculated. We have accomplished this by a single-valued null hypothesis, H_0: $p = 0.50$. Our next step is to set the significance level, which we have taken as $\alpha = 0.05$.

We proceed with the test. The simple random sample of size 400 is drawn, the statistic $\bar{p}$ is observed, and we can now establish the appropriate decision rule. Since the sample size is large, the theory developed in Chapter 5 allows us to use the normal curve as an appropriate approximation for the sampling distribution of the percentage $\bar{p}$. As in the preceding problem, for illustrative purposes, we will establish the decision rule in two different forms, first in terms of $\bar{p}$ values, then in terms of the corresponding z values in a standard normal distribution. Under the assumption that the null hypothesis is true, the sampling distribution of $\bar{p}$ has a mean of p and a standard deviation $\sigma_{\bar{p}} = \sqrt{pq/n}$. Again, we ignore the finite population correction, because the population is so large relative to the sample size. The sampling distribution of $\bar{p}$ for the present problem, in which $p = 0.50$, is shown in Figure 7-8. Also shown is the horizontal axis of the corresponding standard normal distribution, scaled in z values.

Since the null hypothesis will be rejected by an observation of a $\bar{p}$ value that lies significantly below or significantly above $p = 0.50$, we clearly must use a two-tailed test. The critical regions (rejection

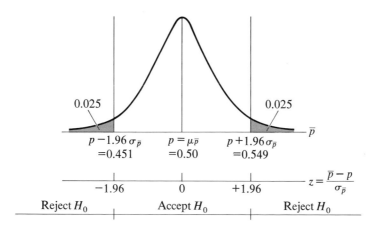

FIGURE 7-8
Sampling distribution of a proportion, with $p = 0.50$ and $n = 400$. This is a two-tailed test with $\alpha = 0.05$.

regions) are displayed in Figure 7-8. The arithmetic involved in establishing regions of acceptance and rejection of H_0 is as follows. The standard error of $\bar{p}$ is

$$\sigma_{\bar{p}} = \sqrt{\frac{(0.50)(0.50)}{400}} = 0.025$$

Referring to Table A-5 of Appendix A, we find that 2.5% of the area in a normal distribution lies to the right of $z = +1.96$, and therefore 2.5% also lies to the left of $z = -1.96$. Thus, we establish a significance level of 5% by marking off an acceptance range for H_0 of $p \pm 1.96\sigma_{\bar{p}}$. The calculation is

$$p + 1.96\sigma_{\bar{p}} = 0.50 + (1.96)(0.025) = 0.50 + 0.049 = 0.549$$
$$p - 1.96\sigma_{\bar{p}} = 0.50 - (1.96)(0.025) = 0.50 - 0.049 = 0.451$$

We can now state the decision rule. The agency draws a simple random sample of 400 viewers of the TV show, observes $\bar{p}$ (the proportion of persons in the sample who are over 30 years of age), and then applies the following decision rule:

DECISION RULE

1. If $\bar{p} < 0.451$ or $\bar{p} > 0.549$, reject H_0
2. If $0.451 \leqslant \bar{p} \leqslant 0.549$, accept H_0

Again, for illustrative purposes, we restate the decision rule in terms of z values.

DECISION RULE

1. If $z < -1.96$ or $z > +1.96$, reject H_0
2. If $-1.96 \leqslant z \leqslant +1.96$, accept H_0

Let us apply this decision rule to the present problem. The observed sample $\bar{p}$ was

$$\bar{p} = \frac{210}{400} = 0.525$$

and

$$z = \frac{\bar{p} - p}{\sigma_{\bar{p}}} = \frac{0.525 - 0.500}{0.025} = +1.0$$

Therefore, the null hypothesis H_0 is accepted. This leads us to a rather vague conclusion. If $\bar{p}$ fell in the rejection region in the right tail of the sampling distribution in Figure 7-8 (that is, if $\bar{p}$ were greater than 0.549), we would conclude that more than half the viewers of the TV show are over 30 years of age. If $\bar{p}$ had fallen in the left tail of the rejection region, we would conclude that less than half are over 30. However, if as in this case, $\bar{p}$ lies in the acceptance region, we *cannot* conclude that more than half of the viewers are over 30 or that less than half are over 30. The sample evidence is *consistent with the hypothesis of a 50–50 split*. Thus, acceptance of the null hypothesis means that on the basis of the available evidence, we simply are not in a position to conclude that more than half of the viewers are over 30 years of age or that less than half of them are. In some instances, the best course is to reserve judgment. Sequential decision procedures are often an appropriate technique to use where one of the alternatives is to delay the decision. A method of sequential decision making is discussed in Chapter 16.

Further Remarks

Power curves can be computed for two-tailed tests in an analogous way to that for one-tailed tests. However, such calculations will not be illustrated here. The power curve for the two-tailed test in the TV viewer problem is shown in Figure 7–9. The curve has the characteristic U-shape of a power function for a two-tailed test. The possible values of the parameter p are shown on the horizontal axis. The height of the ordinate at $p = 0.50$ is 0.05, which is the value of α. Since the only value of p for which the null hypothesis is true is 0.50, the ordinates at all other p values denote probabilities of rejecting the null hypothesis when it is false. The complements of these ordinates are equal to the values of β, the probabilities of making Type II errors.

The hypothesized proportion according to the null hypothesis in this problem was 0.50. This stemmed from the fact that the agency was

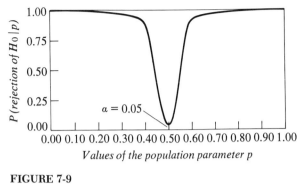

FIGURE 7-9
Power curve for the TV viewer problem.

interested in whether or not more than 50% of the viewers were over 30 years of age. On the other hand, if we wanted to test the assertion that the population proportion was 0.55, 0.60, or some other number, we would have used these figures as the respective hypothesized parameters. Also in this problem, the null hypothesis was single-valued ($p = 0.50$), whereas the alternative hypothesis was many-valued ($p \neq 0.50$). We saw that this resulted in a two-tailed test. Whether a test of a hypothesis is one-tailed or two-tailed depends on the question to be answered. For example, suppose an assertion had been made that at least 50% of the viewers of the TV show were over 30 and that we wish to run a maximum risk of 5% of erroneously rejecting this assertion. The null and alternative hypotheses in this instance would be

$$H_0: p \geqslant 0.50$$
$$H_1: p < 0.50$$

This would involve a one-tailed test with a rejection region lying in the left tail and containing 5% of the area under the normal curve. The critical region would be in the left tail, since only a significant difference for a $\bar{p}$ value lying *below* 0.50 could result in the rejection of the stated null hypothesis. The decision rule in terms of z values would be

DECISION RULE

1. If $z < -1.65$, reject H_0
2. If $z \geqslant -1.65$, accept H_0

On the other hand, if the assertion had been that 50% or fewer of the viewers were over 30 and if we wanted to run a maximum risk of 5% of erroneously rejecting this assertion, the rejection region would be the 5% area in the right tail. The corresponding hypotheses and decision rule would be

$$H_0: p \leqslant 0.50$$
$$H_1: p > 0.50$$

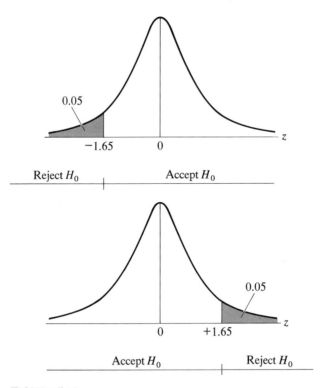

FIGURE 7-10
Standard normal distribution with decision rules for one-tailed tests in terms of z values; $\alpha = 0.05$.

DECISION RULE

1. If $z \geqslant + 1.65$, reject H_0
2. If $z < + 1.65$, accept H_0

Standard normal distributions with the decision rules for these one-tailed tests are depicted in Figure 7-10. When tests are conducted for means or other statistical measures, they may also be either one- or two-tailed, depending on the context of the problem.

Exercises

1. What is meant by the *power* of a test? How is the power related to the Type II error?

2. Agree with or criticize the following statements:
   a. It is more important to control Type I error than Type II error; hence, we should design our tests on the basis of controlling Type I error.

b. It is impossible to control both Type I and Type II errors, since to decrease one increases the other.

c. A Type I error occurs when we reject the null hypothesis incorrectly, and a Type II error occurs when we accept the null hypothesis incorrectly.

d. Once a decision has been made, we must then consider the possibility that Type I and Type II errors both may have occurred.

3. Agree with or criticize the following statements:

a. The power curve has the probability of rejecting the null hypothesis on one axis and the possible values of the parameter on the other, whereas the operating characteristic curve has the possible values of the statistics on one axis and the probability of rejecting the null hypothesis on the other.

b. Beta represents the probability that one will reject the null hypothesis incorrectly.

4. The Nevada Club, a nonprofit organization dedicated to the preservation of the environment in the Western region of the United States, claims that the average level of sulphur dioxide in the air is above the maximum permissible level of 0.15 parts per million in a certain area. Fifty random observations of the sulphur dioxide level were made, and the result obtained was $\bar{x} = 0.165$ parts per million. The population standard deviation is known to be 0.065 parts per million. Would you conclude that the average sulphur dioxide level in the area is greater than the maximum permissible level? Should the Club's claim be accepted? Use a 0.01 level of significance.

5. A specification calls for a drug to have a therapeutic effectiveness for a mean period of at least 40 hours. The standard deviation of the period of effectiveness is known to be 15 hours. A shipment of the drug is to be accepted or rejected on the basis of a simple random sample of 100 items drawn from the shipment.

a. What decision rule should be used if the maximum probability of erroneously rejecting the incoming shipment is to be 0.05? State clearly the null and alternative hypotheses. Indicate the decision to be reached if a simple random sample of 100 items shows a mean period of effectiveness of 38.5 hours. Show your reasoning.

b. Sketch the power curve for the test you designed in (a) and identify and label two points on the curve.

6. The Consumer Fraud Council claims that the Skimpy Foods Company does not put the required weight of tomato ketchup in its 20-ounce bottle. A simple random sample of 300 bottles is selected and weighed. From past experience it is known that the population standard deviation of weights of ketchup in bottles is 0.8 ounce. Set up a test such that if the firm, on the average, actually places the required weight in its 20-ounce bottle, one would accuse it unjustly only once in 200 times.

7. A financial analyst has determined that the net present value of a certain investment proposal is a random variable with mean $10,000 and standard deviation $750. An alternative investment is possible; a sample of 30 investments in the past similar to this alternative produced a sample mean value of $9800 and standard deviation of $1000. The financial analyst's assistant has noted, "Since the sample mean is not significantly less than $10,000 at

$\alpha = 0.05$, and the risk of this type of investment is less than that of the first proposal, we have strong evidence that we should choose the second alternative." Do you agree with the assistant's statement?

8. In a certain year, the arithmetic mean interest on loans to all large retailers (i.e., those with assets of $5,000,000 or more) was 6.0% and the standard deviation was 0.2 percentage points. Two years later, a simple random sample of 100 loans to large retailers yielded an arithmetic mean interest rate of 6.015%.

   a. Would you be willing to conclude that there has been a change in the average level of interest rates for large retailers? Assume that you are willing to run a 5% risk of concluding there has been a change when in fact there has been none.

   b. Using the same decision-making rule as in (a), find the probability of making a Type II error if the average interest rate for all large retailers was 6.01%.

9. It has been determined that the arithmetic mean length of life of safety tires produced by the Firerock process is 30,000 miles with a standard deviation of 4000 miles. A new process, Morelife, is developed and used on a simple random sample of 100 tires.

   a. How long must this sample of 100 tires last on the average in order for the president of Safety Tire Company to conclude that tires produced by the new process will have a higher mean than tires prepared by the old process? The president is willing to run a risk of no more than 0.02 of drawing such a conclusion in error.

   b. Suppose the universe mean for the new process is 30,200 miles and you employ the critical value(s) established in part (a).

      1. What is the probability of reaching an incorrect conclusion?
      2. If you reach an incorrect conclusion in this situation, would it be a Type I or Type II error? Why?

10. Assume that the Food and Drug Administration limits the caffein content of cola drinks to 1.2 grains per 12-ounce bottle and that the actual caffein content of cola drinks is normally distributed and varies from 0.55 to 0.85 grains per 12-ounce bottle. Let us estimate μ and σ as follows:

$$\mu = 0.70 \text{ grains} \qquad \sigma = 0.05 \text{ grains}$$

Suppose the FDA adopts an inspection plan that calls for rejection of a production lot if the population mean caffein content exceeds 0.7 grains per bottle. Assume that the sample size is 225 and that mistaken rejection of lots should not occur more than 1% of the time. Then what values of sample means will lead to rejection of a lot?

11. Over the past ten years, students receiving their MBA degrees from the Branford Business School received an average of six job offers, with a standard deviation of 2. In a random sample of 100 members of the most recent class, the mean number of job offers was 5.5. The school graduates 500 students each year. Assuming $\alpha = 0.05$, would you conclude that there has been a change in the average number of job offers received by Branford graduates? Use the finite population correction.

12. A standard departmental examination in accounting has been given for sev-

eral years to all second-year accounting majors at a prestigious Eastern graduate school of business. The average score has been 85 with a standard deviation of 6. A group of 100 students has attended special intensive classes in accounting principles. If the 100 students obtained a mean grade of 87 on the exam, is there reason to believe that the special classes increase students' results on the test? Use a 0.05 significance level.

13. Last year, a wholesale distributor found that the mean sales per invoice was $60, with standard deviation $20. This year, a random sample of 400 invoices is to be drawn in order to test the hypothesis that the mean sale per invoice has not changed. It is assumed that α will not change. The acceptance region of the test is agreed to be

$$\text{if } \$58.72 < \bar{x} < \$61.28, \text{ accept } H_0$$

a. Suppose that in fact this year $\mu = \$61$. What is the probability of accepting H_0?
b. Calculate and explain the meaning of the power of the test when $\mu = \$61$.
c. What level of significance was used for the test?

14. Suppose you are responsible for the quality control of a certain part bought from a supplier. Inspection tests destroy the part, so you must use sampling. A 5% defective rate is tolerable, but in your sample of 100 from a lot of 10,000, eight parts are defective. Is this sufficient evidence that the lot of parts has too many defectives?

15. Two hundred purchasers of a new product were selected at random and were asked whether they liked the product. The company conducting the survey felt that at least 20% of all purchasers must like the product in order for the firm to continue marketing it. Therefore, the firm decided that if 30 or fewer people responded favorably, it would stop marketing the product. State in words the nature of the Type I error involved here. How large is the risk of such an error?

16. A housing survey is to be undertaken to determine whether the proportion of substandard dwelling units in a certain large city has changed. At the time of the last housing census, 10% of the dwelling units were classified as substandard. It is desired to maintain a 0.01 risk of erroneously concluding that a change has occurred in the proportion of substandard dwelling units. A simple random sample of 900 dwelling units is contemplated.
a. Set up the decision rule for this test.
b. Calculate the probability of concluding that no change has taken place if the true population proportion of substandard dwelling units is 11%. What type of error is involved in this erroneous conclusion?

17. The major oil companies report that last year, 5% of all credit charges for gasoline, car repairs, and parts were never collected and must be written off as bad debts. Recently, the oil companies have installed a central computer credit check system. That is, for any credit purchase of over $10, the local gas station must call the central computer center, where after a computer search of the customer's payment record, the purchase is either given a credit acceptance number or refused credit. Any sales of over $10 not given a credit acceptance number will not be honored as a credit sale by the oil

company. To see whether the system is effective, 1000 credit charges that were accepted were selected at random, and it was found that 36 charges were uncollected and written off as bad debts. Do you think the system is effective? Use a 5% significance level.

18. Given the following hypothesis test:

H_0: The percentage p of losing stocks selected by a stockbroker is 20%
H_1: The percentage p of losing stocks selected by a stockbroker is greater than 20%

The sample is 50, and the significance level is 5%. The decision rule is

If the percentage of losers is greater than 29.34%, reject H_0 and switch to another stockbroker. Otherwise, accept H_0 and keep the present stockbroker.

Comment on the following statements:
a. If p is really greater than 20%, the probability that we will reject H_0 is 0.05.
b. If 32% of the sample stocks are losers, this proves that p is greater than 20%.
c. The above is a one-tailed test; the decision rule would be the same for H_0: $p \leqslant 20\%$.
d. If 33% of the stocks are losers, the probability that we would make a Type II error is

$$P\left(z < \frac{0.33 - 0.20}{\sqrt{\dfrac{(0.20)(0.80)}{50}}} \right)$$

19. A bank in a growing metropolitan area determined that a new office, located in the suburbs, should be opened if more than 30% of the depositors using the city office would do business at the new branch. From the bank's list of active accounts, 400 depositors were randomly selected. Of these, 144 indicated that they would patronize the new office if it were established. What is the probability of drawing a random sample of size 400 with 144 or more potential new office patrons, from a statistical universe that in reality contains exactly 30% potential patrons? The bank concluded that more than 30% of its depositors were potential users of the proposed new office. Should it have so concluded, in your opinion? Use $\alpha = 0.05$.

   In answering the above question, you had to locate a sample statistic (proportion) on a random sampling distribution. Draw a rough sketch of this distribution showing
a. the value and location of the hypothesized parameter
b. the value and location of the statistic
c. the vertical and horizontal scale descriptions
d. the portion of the distribution corresponding to the probability computed above

20. A random sample of 200 certified public accountants is polled on the issue of the desirability of price-level accounting. One hundred and ten favored inclusion of financial statements showing adjustments for price-level changes, and ninety indicated opposition to the proposal. Let p be the proportion of accountants supporting the proposal, and consider the hypotheses

$$H_0: p \leq 0.50$$
$$H_1: p > 0.50$$

The following test is proposed: Reject H_0 if $\bar{p}$, the sample proportion, is greater than 0.56.

a. What is the greatest probability of a Type I error with the above test?

b. Design a test such that α is at most 0.01, and test the hypothesis with the given sample data.

c. If you want to minimize the probability of a Type II error, which of the two tests would you use? Why?

21. The central government of a wealthy country has instituted a program to discourage investment in foreign countries by its citizens. Concern over balance of payments deficits has prompted these restrictions on capital outflows. It is known that in the past, 35% of the country's adult citizens, on average, have held investments in foreign countries. The government wishes to determine whether the current percentage of adult citizens who own foreign assets is greater than this long-term figure. A random sample of 800 adults is selected, and it is determined that 320 of these citizens hold foreign assets. Is this percentage significantly greater than 35%? Use a 10% significance level.

22. Experience has shown that at least 90% of Zimbel's Department Store's customers pay their bills on time. Last week the credit manager selected a simple random sample of 500 customers accounts and determined that 420 of the 500 customers had paid their outstanding bills on time.

a. The credit manager has designed a test that yields a maximum probability of 0.02 of erroneously concluding that the proportion of customers paying their bills on time is now less than it has been in the past. State the hypothesis being tested, give the decision rule, and state the conclusion that should be drawn.

b. Suppose that if a census had been taken, it would have been found that 85% of all Zimbel's customers had paid their bills on time. Given the sample information for the 500 customer accounts and the decision rule you designed in (a), what is your probability of making an error? What type of error is this?

23. A manufacturer claims that a customer will find that no more than 8% of parts in a given shipment are defective. The customer decides to test this claim against the alternative hypothesis $p > 0.08$, using a 0.05 probability of rejecting this claim when it is true.

a. The customer randomly selects 200 parts from the shipment and finds 28 defectives. What conclusion should be reached?

b. Might the decision reached in part (a) be in error? If yes, what type of error would this represent? If no, explain.

24. The J. T. McClay Company, a manufacturer of quality stereo receivers, recently held a meeting to review its marketing program. During the review, it was decided to confirm a previous study indicating that 30% of its customers were university students. If the percentage was different from 30%, the company would require a new marketing plan. The company wished to run a risk of 2% of erroneously rejecting the hypothesis that 30% of its customers were university students. The company conducted a simple random sample

of 600 potential customers. Of the 600, 140 were university students and 460 were not.

a. Should the J. T. McClay Company alter its marketing campaign on the basis of this sample? Roughly sketch the sampling distribution and label the horizontal axis properly. Indicate the acceptance region for the test and the location of the test statistic on which your conclusion was based.

b. Suppose the same company desired to test the hypothesis that at least 30% of its customers were university students. On the basis of the same sample data, should the company reject the hypothesis at the 2% significance level? Justify statistically.

c. In regard to your answer in part (b):
   1. Why is α considered the maximum probability of a Type I error?
   2. Calculate the power of this test when 22% of the customers are university students. Give a one-sentence interpretation of the number you calculated.
   3. Sketch the power curve and label both axes.

25. The Constitution of the United States requires a two-thirds majority of both House and Senate to override a presidential veto. In a given situation, the necessary majority was secured in the Senate, but the action of the House was uncertain. Prior to the actual vote in the House, newspaper reporters took a straw vote of 90 representatives, and 57 of them indicated their intention to vote to override the veto. If you wished to be wrong no more than 5% of the time, would you conclude (assuming that the sample was taken at random) that the veto would not be overridden? Justify your conclusion statistically. Use the finite population correction, assuming that there are 435 representatives in the House.

7.3 TWO-SAMPLE TESTS (LARGE SAMPLES)

The discussion thus far has involved testing of hypotheses using data from a single random sample. Another important class of problems involves the question of whether statistics observed in two simple random samples differ significantly. Recalling that all statistical hypotheses are statements concerning population parameters, we see that this question implies a corresponding question about the underlying parameters in the populations from which the samples were drawn. For example, if the statistics observed in the two samples are arithmetic means (say, $\bar{x}_1$ and $\bar{x}_2$), the question is whether we are willing to attribute the difference between these two sample means to chance errors of sampling or whether we will conclude that the populations from which the samples were drawn have *unequal means*. We illustrate these tests, first for differences between means and then for differences between proportions. In both cases, we assume as we did in Section 7.2 that we are using large samples.

Test for Difference between Means: Two-Tailed Test

A consulting firm conducting research for a client was asked to test whether the wage levels of unskilled workers in a certain industry were the same in two different geographical areas, referred to as Area A and Area B. The firm took simple random samples of the unskilled workers in the two areas and obtained the following sample data for weekly wages:

Area	Mean	Standard Deviation	Size of Sample
A	$\bar{x}_1 = \$100.01$	$s_1 = \$4.00$	$n_1 = 100$
B	$\bar{x}_2 = \quad 95.21$	$s_2 = \quad 4.50$	$n_2 = 200$

If the client wished to run a risk of 0.02 of incorrectly rejecting the hypothesis that the population means in these two areas were the same, what conclusion should be reached? Note that the samples need not be of the same size; different sample sizes have been assumed in this problem in order to keep the example completely general.

Let us refer to the means and standard deviations of *all* unskilled workers in this industry in Areas A and B, respectively, as μ_1 and μ_2, and σ_1 and σ_2. These are the population parameters corresponding to the sample statistics $\bar{x}_1$, $\bar{x}_2$, s_1, and s_2. The hypotheses to be tested are

$$H_0: \mu_1 - \mu_2 = 0$$
$$H_1: \mu_1 - \mu_2 \neq 0$$

That is, the null hypothesis asserts that the population parameters μ_1 and μ_2 are equal. We form the statistic $\bar{x}_1 - \bar{x}_2$, the difference between the sample means. If $\bar{x}_1 - \bar{x}_2$ differs significantly from 0, the hypothesized value for $\mu_1 - \mu_2$, we will reject the null hypothesis and conclude that the population parameters μ_1 and μ_2 are indeed different.

Since the risk of a Type I error has been set, we turn now to determining the decision rule based on the appropriate random sampling distribution. Let us examine some of the important characteristics of this distribution. The two random samples are independent; that is, the probabilities of selection of the elements in one sample are not affected by the selection of the other sample. Hence, $\bar{x}_1$ and $\bar{x}_2$ are independent random variables. It has been shown that the mean and standard deviation of the sampling distribution of $\bar{x}_1 - \bar{x}_2$ are, respectively,

(7.1) $$\mu_{\bar{x}_1 - \bar{x}_2} = 0$$

and

(7.2) $$\sigma_{\bar{x}_1 - \bar{x}_2} = \sqrt{\frac{\sigma_1^2}{n_1} + \frac{\sigma_2^2}{n_2}}$$

For large, independent samples, the sampling distribution of $\bar{x}_1 - \bar{x}_2$ is approximately normal by the Central Limit Theorem. *In summary, therefore, if $\bar{x}_1$, and $\bar{x}_2$ are the means of two large, independent samples from populations with means μ_1 and μ_2 and standard deviations σ_1 and σ_2, and if we hypothesize that μ_1 and μ_2 are equal, then the sampling distribution of $\bar{x}_1 - \bar{x}_2$ may be approximated by a normal curve with mean $\mu_{\bar{x}_1 - \bar{x}_2} = 0$ and standard deviation $\sigma_{\bar{x}_1 - \bar{x}_2} = \sqrt{\sigma_1^2/n_1 + \sigma_2^2/n_2}$.* It is helpful to think of this sampling distribution as the frequency distribution that would be obtained by grouping the $\bar{x}_1 - \bar{x}_2$ values observed in repeated pairs of samples drawn independently from two populations with the same means.

The standard deviation $\sigma_{\bar{x}_1 - \bar{x}_2}$ is referred to as the *standard error of the difference between two means.* We see from Equation 7.2 that we must know the population standard deviations in order to calculate this standard error. However, for *large samples,* we can approximate σ_1 and σ_2 using the sample standard deviations s_1 and s_2. The resulting estimated (or approximate) standard error is symbolized $s_{\bar{x}_1 - \bar{x}_2}$ and may be written

(7.3)
$$s_{\bar{x}_1 - \bar{x}_2} = \sqrt{\frac{s_1^2}{n_1} + \frac{s_2^2}{n_2}}$$

We can now proceed to establish the decision rule for the problem. The test is clearly two-tailed, because the hypothesis of equal population means would be rejected if $\bar{x}_1 - \bar{x}_2$ differed significantly from 0 in either the positive or the negative direction. The sampling distribution of $\bar{x}_1 - \bar{x}_2$ is shown in Figure 7-11. The horizontal scale of the distribution shows the difference between the sample means $\bar{x}_1 - \bar{x}_2$. As indicated, the mean of the distribution is equal to 0; in other words, under the null hypothesis, the expected value of $\bar{x}_1 - \bar{x}_2$ is 0. Another way of interpreting the 0 is that under the null hypothesis $H_0: \mu_1 - \mu_2 = 0$, we

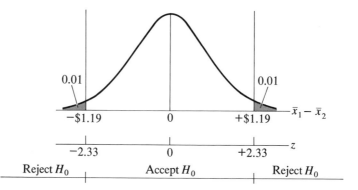

FIGURE 7-11
Sampling distribution of the difference between two means. This is a two-tailed test with $\alpha = 0.02$.

have assumed that the mean wages of the populations of unskilled workers are the same in Area A and Area B for the industry in question. Since the significance level is 0.02, 1% of the area under the normal curve is shown in each tail. From Table A-5 of Appendix A we find that in a normal distribution 1% of the area lies to the right of $z = +2.33$ and (by symmetry) 1% to the left of $z = -2.33$. Thus, we would reject the null hypothesis if the sample difference $\bar{x}_1 - \bar{x}_2$ fell more than 2.33 standard errors from the expected value of 0. The estimated standard error of the difference between means, $s_{\bar{x}_1-\bar{x}_2}$, is, by Equation 7.3,

$$s_{\bar{x}_1-\bar{x}_2} = \sqrt{\frac{(\$4.00)^2}{100} + \frac{(\$4.50)^2}{200}} = \$0.51$$

and

$$2.33 s_{\bar{x}_1-\bar{x}_2} = (2.33)(\$0.51) = \$1.19$$

Thus, the decision rule may be stated as follows:

DECISION RULE

1. If $\bar{x}_1 - \bar{x}_2 < -\$1.19$ or $\bar{x}_1 - \bar{x}_2 > \$1.19$, reject H_0
2. If $-\$1.19 \leq \bar{x}_1 - \bar{x}_2 \leq \$1.19$, accept H_0

In terms of z values, we have

DECISION RULE

1. If $z < -2.33$ or $z > +2.33$, reject H_0
2. If $-2.33 \leq z \leq +2.33$, accept H_0

where

$$z = \frac{(\bar{x}_1 - \bar{x}_2) = (\mu_1 - \mu_2)}{s_{\bar{x}_1-\bar{x}_2}} = \frac{\bar{x}_1 - \bar{x}_2}{s_{\bar{x}_1-\bar{x}_2}}$$

Note that this z value is in the usual form of the ratio (statistic − parameter)/standard error. The difference $\bar{x}_1 - \bar{x}_2$ is the statistic. The parameter under test is $\mu_1 - \mu_2$, which by the null hypothesis is 0 and thus need not be shown in the numerator of the ratio. As previously indicated, we have substituted an approximate standard error for the true standard error in the denominator.

Applying this decision rule to the present problem, we have

$$\bar{x}_1 - \bar{x}_2 = \$100.01 - \$95.21 = \$4.80$$

and

$$z = \frac{\bar{x}_1 - \bar{x}_2}{s_{\bar{x}_1-\bar{x}_2}} = \frac{\$4.80}{\$.51} = 9.4$$

Since \$4.80 far exceeds \$1.19 (and correspondingly, 9.4 far exceeds 2.33), the null hypothesis is rejected. Hence, it is extremely unlikely that these two samples were drawn from populations having the same mean. We conclude that the sample mean wages of unskilled workers in this industry *differed significantly* between Areas A and B and thus that the population means *differ* between Areas A and B. Note that it is incorrect to use the term "significant difference" when referring to the relationship between two population parameters (in this case, the population means). The student should also keep in mind that, in this as in all other hypothesis testing situations, we are assuming random sampling. Obviously, if the samples were not randomly drawn from the two populations, the foregoing procedure and conclusion would be invalid.

Test for Difference between Proportions: Two-Tailed Test

Another important case of two-sample hypothesis testing is one in which the observed statistics are proportions. The decision procedure is conceptually the same as when the sample statistics are means; only the computational details differ. In order to illustrate the technique, let us consider the following example. Workers in the Stanley Morgan Company and Rock Hayden Company, two firms in the same industry, were asked whether they preferred to receive a specified package of increased fringe benefits or a specified increase in base pay. For brevity in this problem, we will refer to the companies as the S. M. Company and the R. H. Company and the proposed increases as "increased fringe benefits" and "increase base pay." In a simple random sample of 150 workers in the S. M. Company, 75 indicated that they preferred increased base pay. In the R. H. Company, 103 out of a simple random sample of 200 preferred increased base pay. In each company, the sample was less than 5% of the total number of workers. It was desirable to have a very low probability of erroneously rejecting the hypothesis of equal proportions of workers in the two companies who preferred increased base pay. Therefore, a 1% level of significance was used for the test. Can it be concluded at the 1% level of significance that these two companies differed in the proportion of workers who preferred increased base pay?

Using the subscripts 1 and 2 to refer to the S. M. Company and R. H. Company, respectively, we can organize the sample data as follows:

S. M. Company	R. H. Company
$\bar{p}_1 = \dfrac{75}{150} = 0.50$	$\bar{p}_2 = \dfrac{103}{200} = 0.515$

(Continued)

S. M. Company	R. H. Company
$\bar{q}_1 = \dfrac{75}{150} = 0.50$	$\bar{q}_2 = \dfrac{97}{200} = 0.485$
$n_1 = 150$	$n_2 = 200$

where $\bar{p}_1$ and $\bar{q}_1$ refer to the sample proportions in the S. M. Company in favor of and opposed to increased base pay, respectively. The sample size in the S. M. Company is denoted n_1. Corresponding notation is used for the R. H. Company. If we designate the population proportions in favor of increased pay in the two companies as p_1 and p_2, then in a manner analogous to that of the preceding problem, we set up the two hypotheses

$$H_0: p_1 - p_2 = 0$$
$$H_1: p_1 - p_2 \neq 0$$

The underlying theory for the test is similar to that in the two-sample test for the difference between two means. If $\bar{p}_1$ and $\bar{p}_2$ are the observed sample proportions in large simple random samples drawn from populations with parameters p_1 and p_2, then the sampling distribution of the statistic $\bar{p}_1 - \bar{p}_2$ has a mean

(7.4) $$\mu_{\bar{p}_1 - \bar{p}_2} = p_1 - p_2$$

and a standard deviation

(7.5) $$\sigma_{\bar{p}_1 - \bar{p}_2} = \sqrt{\sigma_{\bar{p}_1}^2 + \sigma_{\bar{p}_2}^2}$$

where $\sigma_{\bar{p}_1}^2$ and $\sigma_{\bar{p}_2}^2$ are the variances of the sampling distributions of $\bar{p}_1$ and $\bar{p}_2$. Under assumptions of a binomial distribution, $\sigma_{\bar{p}_1}^2 = p_1 q_1 / n_1$ and $\sigma_{\bar{p}_2}^2 = p_2 q_2 / n_2$. Although the sampling was conducted without replacement, since each of the samples constituted only a small percentage of the corresponding population (less than 5%), the binomial distribution assumption appears reasonable. Thus, Equation 7.5 becomes

(7.6) $$\sigma_{\bar{p}_1 - \bar{p}_2} = \sqrt{\frac{p_1 q_1}{n_1} + \frac{p_2 q_2}{n_2}}$$

If we hypothesize that $p_1 = p_2$ (the null hypothesis in this problem) and refer to the common value of p_1 and p_2 as p, Equations 7.4 and 7.6 become

(7.7) $$\mu_{\bar{p}_1 - \bar{p}_2} = p - p = 0$$

and

(7.8) $$\sigma_{\bar{p}_1 - \bar{p}_2} = \sqrt{\frac{pq}{n_1} + \frac{pq}{n_2}} = \sqrt{pq \left(\frac{1}{n_1} + \frac{1}{n_2} \right)}$$

Since the common proportion p hypothesized under the null hypothesis is unknown, we estimate it for the hypothesis test by taking a weighted mean of the observed sample percentages. If we refer to this "pooled estimator" as $\bar{p}$, we have

(7.9)
$$\bar{p} = \frac{n_1\bar{p}_1 + n_2\bar{p}_2}{n_1 + n_2}$$

The numerator of Equation 7.9 is simply the total number of "successes" in the two samples combined, and the denominator is the total number of observations in the two samples. The standard deviation in Equation 7.8, $\sigma_{\bar{p}_1 - \bar{p}_2}$, is referred to as the *standard error of the difference between two proportions*. Substituting the "pooled estimator" $\bar{p}$ for p in (7.8), we have the following formula for the *estimated* or *approximate* standard error $s_{\bar{p}_1 - \bar{p}_2}$

(7.10)
$$s_{\bar{p}_1 - \bar{p}_2} = \sqrt{\bar{p}\bar{q}\left(\frac{1}{n_1} + \frac{1}{n_2}\right)}$$

We can now summarize these results. Let $\bar{p}_1$ and $\bar{p}_2$ be proportions of successes observed in two large, independent samples from populations with parameters p_1 and p_2. If we hypothesize that $p_1 = p_2 = p$, we obtain a pooled estimator $\bar{p}$ for p, where $\bar{p} = (n_1\bar{p}_1 + n_2\bar{p}_2)/(n_1 + n_2)$. *Then, the sampling distribution of $\bar{p}_1 - \bar{p}_2$ may be approximated by a normal curve with mean $\mu_{\bar{p}_1 - \bar{p}_2} = 0$ and estimated standard deviation*

$$s_{\bar{p}_1 - \bar{p}_2} = \sqrt{\bar{p}\bar{q}\left(\frac{1}{n_1} + \frac{1}{n_2}\right)}$$

We may think of this sampling distribution as the frequency distribution of $\bar{p}_1 - \bar{p}_2$ values observed in repeated pairs of samples drawn independently from two populations having the same proportions.

Proceeding to the decision rule, we see again that the test is two-tailed, because the hypothesis of equal population proportions would be rejected for $\bar{p}_1 - \bar{p}_2$ values significantly above or below 0. The sampling distribution of $\bar{p}_1 - \bar{p}_2$ for the present problem is shown in Figure 7-12. Since the significance level is $\alpha = 0.01$, the area under the normal distribution curve shown in each tail is $\alpha/2$ or $\frac{1}{2}(1\%)$. Referring to Table A-5 of Appendix A, we find that 0.005 of the area under a normal curve lies above a z value of $+2.58$, and thus the same percentage lies below $z = -2.58$. Hence, rejection of the null hypothesis H_0: $p_1 - p_2 = 0$ occurs if the sample difference $\bar{p}_1 - \bar{p}_2$ falls more than 2.58 standard error units from 0. By Equation 7.10, the standard error of the difference between proportions is

$$s_{\bar{p}_1 - \bar{p}_2} = \sqrt{\bar{p}\bar{q}\left(\frac{1}{n_1} + \frac{1}{n_2}\right)} = \sqrt{(0.51)(0.49)\left(\frac{1}{150} + \frac{1}{200}\right)}$$
$$= 0.054$$

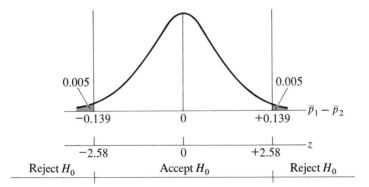

FIGURE 7-12
Sampling distribution of the difference between two proportions. This is a two-tailed test with $\alpha = 0.01$.

where

$$\bar{p} = \frac{n_1\bar{p}_1 + n_2\bar{p}_2}{n_1 + n_2} = \frac{(150)(0.50) + 200(0.515)}{150 + 200} = \frac{75 + 103}{150 + 200} = 0.51$$

Hence,

$$2.58 s_{\bar{p}_1 - \bar{p}_2} = (2.58)(0.054) = 0.139$$

Therefore, the decision rule is

DECISION RULE

1. If $\bar{p}_1 - \bar{p}_2 < -0.139$ or $\bar{p}_1 - \bar{p}_2 > 0.139$, reject H_0
2. If $-0.139 \leqslant \bar{p}_1 - \bar{p}_2 \leqslant 0.139$, accept H_0

In terms of z values, the rule is

DECISION RULE

1. If $z < -2.58$ or $z > +2.58$, reject H_0
2. If $-2.58 \leqslant z \leqslant +2.58$, accept H_0

where

$$z = \frac{\bar{p}_1 - \bar{p}_2}{s_{\bar{p}_1 - \bar{p}_2}}$$

Applying this decision rule yields

$$\bar{p}_1 - \bar{p}_2 = 0.500 - 0.515 = -0.015$$

and

$$z = \frac{\bar{p}_1 - \bar{p}_2}{s_{\bar{p}_1 - \bar{p}_2}} = \frac{-0.015}{0.054} = -0.28$$

Thus, the null hypothesis is accepted. In summary, the sample proportions $\bar{p}_1$ and $\bar{p}_2$ did not differ significantly, and therefore we cannot conclude that the two companies differed with respect to the proportion of workers who preferred increased base pay. This conclusion may be useful, for example, in testing a claim of a participant in labor negotiations that such a difference exists. Our reasoning is based on the finding that if the population proportions were equal, a difference between the sample proportions as large as the one observed could not at all be considered unusual.

Test for Differences between Proportions: One-Tailed Test

The two preceding examples illustrated two-tailed tests for cases where data are available for samples from two populations. Just as in the one-sample case, the question we wish to answer may give rise to a one-tailed test. In order to illustrate this point, let us examine the following problem.

Two competing drugs are available for treating a certain physical ailment. There are no apparent side-effects from administration of the first drug, whereas there are some definite side-effects (nausea and mild headaches) from use of the second. A group of medical researchers has decided that it would nevertheless be willing to recommend use of the second drug in preference to the first if the proportion of cures effected by the second were higher than those by the first drug. The group felt that the potential benefits of achieving increased cures of the ailment would far outweigh the disadvantages of the possible side-effects. On the other hand, if the proportion of cures effected by the second drug was equal to or less than that of the first drug, the group would recommend use of the first one. In terms of hypothesis testing, we can state the alternatives and consequent actions as

$$H_0: p_2 \leq p_1 \text{ (use the first drug)}$$
$$H_1: p_2 > p_1 \text{ (use the second drug)}$$

where p_1 and p_2 denote the population proportions of cures effected by the first and second drugs. Another way we may write these alternatives is

$$H_0: p_2 - p_1 \leq 0 \text{ (use the first drug)}$$
$$H_1: p_2 - p_1 > 0 \text{ (use the second drug)}$$

For purposes of comparison, note that the hypotheses in the preceding problem, a two-tailed testing situation, were

$$H_0: p_1 = p_2$$
$$H_1: p_1 \neq p_2$$

or in the alternative form (in terms of differences),

$$H_0: p_1 - p_2 = 0$$
$$H_1: p_1 - p_2 \neq 0$$

Clearly, the present problem involves a one-tailed test, in which we would reject the null hypothesis only if the sample difference $\bar{p}_2 - \bar{p}_1$ differed significantly from 0 and was a positive number.

The medical researchers used the drugs experimentally on two random samples of persons suffering from the ailment, administering the first drug to a group of 80 patients and the second drug to a group of 90 patients. By the end of the experimental period, 52 of those treated with the first drug were classified as "cured," whereas 63 of those treated with the second drug were so classified. The sample results may be summarized as follows:

First Drug	Second Drug
$\bar{p}_1 = \dfrac{52}{80} = 0.65$ cured	$\bar{p}_2 = \dfrac{63}{90} = 0.70$ cured
$\bar{q}_1 = \dfrac{28}{80} = 0.35$ not cured	$\bar{q}_2 = \dfrac{27}{90} = 0.30$ not cured
$n_1 = 80$	$n_2 = 90$

The pooled sample proportion cured is

$$\bar{p} = \frac{52 + 63}{80 + 90} = \frac{115}{170} = 0.676$$

and the estimated standard error of the difference between proportions is

$$s_{\bar{p}_2 - \bar{p}_1} = \sqrt{(0.676)(0.324)\left(\frac{1}{80} + \frac{1}{90}\right)} = 0.0719$$

Since the medical group wished to maintain a low probability of erroneously adopting the second drug, it selected a 1% significance level for the test. One percent of the area under the normal curve lies to the right of $z = +2.33$. Therefore, the null hypothesis would be rejected if $\bar{p}_2 - \bar{p}_1$ falls at least 2.33 standard error units above 0. In terms of proportions,

$$2.33 s_{\bar{p}_1 - \bar{p}_2} = 2.33(0.0719) = 0.168$$

Hence, the decision rule is

DECISION RULE

1. If $\bar{p}_2 - \bar{p}_1 > 0.168$, reject H_0
2. If $\bar{p}_2 - \bar{p}_1 \leq 0.168$, accept H_0

In terms of z values, the rule is

DECISION RULE

1. If $z > +2.33$, reject H_0
2. If $z \leq +2.33$, accept H_0

where

$$z = \frac{\bar{p}_2 - \bar{p}_1}{s_{\bar{p}_2 - \bar{p}_1}}$$

In the present problem,

$$\bar{p}_2 - \bar{p}_1 = 0.70 - 0.65 = 0.05$$

so

$$z = \frac{0.70 - 0.65}{0.0719} = 0.70$$

Thus, the null hypothesis is accepted. On the basis of the sample data, we cannot conclude that the second drug accomplishes a greater proportion of cures than the first. The sampling distribution of $\bar{p}_2 - \bar{p}_1$ is given in Figure 7-13. Note that it is immaterial whether the difference between proportions is stated as $\bar{p}_1 - \bar{p}_2$ or $\bar{p}_2 - \bar{p}_1$, but care must be ex-

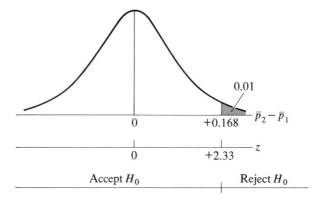

FIGURE 7-13
Sampling distribution of the difference between two proportions. This is a one-tailed test with $\alpha = 0.01$.

ercised concerning the correspondence between the way the hypothesis is stated, the sign of the difference between the sample proportions, and the tail of the sampling distribution in which rejection of the null hypothesis takes place.

Exercises

1. Match the correct test statistics with the following four null hypotheses.
   a. $H_0: \mu = \mu_0$ 1. $\bar{p}$
   b. $H_0: p = p_0$ 2. $\bar{x}_1 - \bar{x}_2$
   c. $H_0: \mu_1 \doteq \mu_2$ 3. $\bar{x}_1 + \bar{x}_2$
   d. $H_0: p_1 = p_2$ 4. $\bar{p}_1 - \bar{p}_2$
                             5. $\bar{x}$
                             6. $\bar{p}/\bar{x}$

2. A bank is considering opening a new branch in one of two neighborhoods. One of the factors considered by the bank is whether the average family incomes in the two neighborhoods differ. From census records, the bank draws two simple random samples of 200 families each. The following information is obtained:

$$x_1 = \$15,600 \qquad x_2 = \$15,490$$
$$s_1 = \$500 \qquad s_2 = \$700$$
$$n_1 = 200 \qquad n_2 = 200$$

   Formulate an appropriate null hypothesis and test it at the 5% significance level.

3. The Stillwater Brewing Company felt that two of its market areas exhibited equivalent sales patterns. The company wished to reduce its advertising budget, but it decided to test the hypothesis about sales patterns prior to any advertising change. The following random sample data were obtained for daily consumption:

Area	Mean Daily Consumption	Standard Deviation	Sample Size
1	1500	140	100
2	1450	120	150

   a. Would you conclude that areas 1 and 2 have equivalent mean daily consumption rates? Justify your answer using a 2% significance level.
   In answering question (a), you had to locate a sample statistic (the arithmetic mean) on a random sampling distribution. Draw a rough sketch of this distribution, showing
   b. the value and location of the hypothetical parameter
   c. the value and location of the statistic
   d. the horizontal scale description

e. the portion of the distribution corresponding to the probability $\alpha = 0.02$

4. Suppose you took a census of the incomes of all vice-presidents in two chemical firms and found the following:

Firm	Mean Income
Chemetrex	$65,000
Oxygen products	60,000

Can you conclude statistically that the average income of vice-presidents of Chemetrex exceeded that of Oxygen Products vice-presidents? Would you test a hypothesis? If yes, what hypothesis? If no, why not?

5. a. In a simple random sample of 225 foreign exchange traders, 135 expressed the opinion that a flexible exchange rate system would enhance the stability of the international monetary system. In a simple random sample of 200 professors of economics, 154 expressed the same opinion. Do you believe that there is a real difference in the attitude of the two groups with regard to the effect of flexible exchange rates on the stability of the international monetary system? Justify your answer statistically and indicate which significance level you use.

b. Explain specifically the meaning of a Type I error in this particular problem.

6. Workers in two different industries were asked what they considered the most important labor–management problem in their industry. In industry A, 200 out of a random sample of 400 workers felt that a fair adjustment of grievances was the most important problem. In Industry B, 60 out of a random sample of 100 workers felt that this was the most important problem.

a. Would you conclude that these two industries differed with respect to the proportion of workers who believed that a fair adjustment of grievances was the most important problem? Support your answer statistically, and give a brief statement of your reasoning. Use a 1% significance level.

b. Make a rough sketch (labeling the horizontal scale) of the random sampling distribution on which your answer to (a) depends. Show on the sketch the approximate positions of the values pertinent to the solution.

7. A company specializing in wood products owns a large tract of timber land. The trees on this tract are sprayed periodically in order to reduce damage to the trees by insects. A new spray has become available on the market, and company executives have asked their research department to determine whether the new spray is more effective than the old spray in reducing the number of damaged trees. The investigators have sprayed 200 trees with the old spray and 200 trees with the new spray. After a period of time, they observe that with the old spray, 86 trees are damaged, and with the new spray, 74 trees are damaged. At a 0.01 significance level, would you be willing to conclude that the new spray is more effective? Your answer should include a clear statement of the statistical hypotheses and of the decision rule.

8. Suppose that in a simple random sample of 400 people from one city, 188 pre-

ferred a particular brand of tea to all others, and in a similar sample of 500 people from another city, 210 preferred the same product. At the 0.05 significance level, is there reason to doubt the hypothesis that equal proportions of persons in the two cities prefer this brand of tea?

7.4 THE *t* DISTRIBUTION: SMALL SAMPLES WITH UNKNOWN POPULATION STANDARD DEVIATION(S)

The hypothesis-testing methods discussed in the preceding sections are appropriate for large samples. In this section, we concern ourselves with the case where the sample size is small. The underlying theory is exactly the same as that given in Section 6.3, where confidence interval estimation for small samples was discussed. In hypothesis testing, as in confidence interval estimation, the distinction between large and small sample tests becomes important when the population standard deviation is *unknown* and therefore must be *estimated* from the sample observations. The main principles will be reviewed here. The statistic $(\bar{x} - \mu)/s_{\bar{x}}$, where $s_{\bar{x}}$ denotes an estimated standard error, is not approximately normally distributed for all sample sizes. As we have noted earlier, $s_{\bar{x}}$ is computed by the formula $s_{\bar{x}} = s/\sqrt{n}$ where s represents an estimate of the true population standard deviation. For large samples, the ratio $(\bar{x} - \mu)/s_{\bar{x}}$ is approximately normally distributed, and we may use the methods discussed in Sections 7.2 and 7.3. However, since this statistic is not approximately normally distributed for small samples, the *t* distribution should be used instead. The use of the *t* distribution for testing a hypothesis concerning a population mean is demonstrated in Example 7-1. A small sample ($n \leq 30$) is assumed, and the population standard deviation is unknown.

EXAMPLE 7-1

One-Sample Test of a Hypothesis about the Mean: Two-Tailed Test.
    The personnel department of a company developed an aptitude test for a certain type of semiskilled worker. The individual test scores were assumed to be normally distributed. The developers of the test asserted a tentative hypothesis that the arithmetic mean grade obtained by this type of semiskilled worker would be 100. It was agreed that this hypothesis would be subjected to a two-tailed test at the 5% level of significance. The aptitude test was given to a simple random sample of 16 semiskilled workers with the following results:

$$\bar{x} = 94$$
$$s = 5$$
$$n = 16$$

The competing hypotheses are

$$H_0: \mu = 100$$
$$H_1: \mu \neq 100$$

SOLUTION

To carry out the test, the quantities

$$s_{\bar{x}} = \frac{s}{\sqrt{n}} = \frac{5}{\sqrt{16}} = 1.25$$

and

$$t = \frac{\bar{x} - \mu}{s_{\bar{x}}} = \frac{94 - 100}{1.25} = -4.80$$

were calculated.

The significance of this t value is judged from Table A-6 of Appendix A. The meaning of the table was discussed in Section 6.3. We will explain the use of the table for this hypothesis-testing problem. Let us set up the areas of acceptance and rejection of the hypothesis. Since the sample size is 16, the number of degrees of freedom is $\nu = 16 - 1 = 15$. Looking along the row of Table A-6 for 15 under the column 0.05, we find the t value, 2.131. This means that in a t distribution for $\nu = 15$, the probability is 5% that t is greater than 2.131 or is less than -2.131. Thus, in the present problem, at the 5% level of significance, the null hypothesis H_0: $\mu = 100$ is rejected if a t value exceeding 2.131 or less than -2.131 is observed. Since the computed t value in this problem is -4.80, we reject the null hypothesis. In other words, we are unwilling to attribute the difference between our sample mean of 94 and the hypothesized population mean of 100 merely to chance errors of sampling. The t distribution for this problem is shown in Figure 7-14.

A few remarks can be made about this problem. Since the computed t value of -4.80 was so much less than 0, the null hypothesis would have been rejected even at the 2% or 1% levels of significance (see Table A-6 of Appendix A). Had the test been one-tailed at the 5% level of significance,

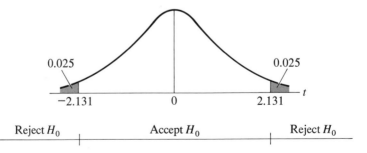

FIGURE 7-14
The t distribution for $\nu = 15$.

we would have had to obtain the critical *t* value by looking under 0.10 in Table A-6, since the 0.10 figure is the combined area in both tails. Thus, for a one-tailed test at the 5% level of significance and a rejection region in the lower tail, the critical *t* value would have been -1.753.

It is interesting to compare these critical *t* values with analogous critical *z* values for the normal curve. From Table A-5 of Appendix A, we find that the critical *z* values at the 5% level of significance are -1.96 and 1.96 for a two-tailed test and -1.65 for a one-tailed test with a rejection region in the lower tail. As we have just seen, the corresponding figures for the critical *t* values in a test involving 15 degrees of freedom are -2.131 and 2.131 for a two-tailed test and -1.753 for a one-tailed test.

An underlying assumption in applying the *t* test in this problem is that the population is closely approximated by a normal distribution. Since the population standard deviation, σ, is unknown, the *t* distribution is the theoretically correct sampling distribution. However, if the sample size had been large, even with an unknown population standard deviation, the normal curve could have been used as an approximation to the *t* distribution. As we saw in this problem, for $\nu = 15$, a total of 5% of the area in the *t* distribution falls to the right of $t = +2.131$ and to the left of $t = -2.131$. The corresponding *z* values in the normal distribution are $+1.96$ and -1.96. As can be seen from Table A-6, the *t* value for $\nu = 30$ is 2.042. The closeness of this figure to $+1.96$ gives rise to the usual rule of thumb, which uses $n > 30$ as the arbitrary dividing line between large-sample and small-sample methods. We have used this convenient rule in Chapters 6 and 7. However, what constitutes a suitable approximation really depends on the context of the particular problem. Furthermore, if the population is highly skewed, a sample size as large as 100 may be required for the assumption of a normal sampling distribution of $\bar{x}$ to be appropriate.

EXAMPLE 7-2

Two-Sample Test for Means: Two-Tailed Test.
   Small simple random samples of the freshmen and senior classes were drawn at a large university, and the amounts of money (excluding checks) that these individuals had on their persons were determined. The following statistics were calculated:

Freshmen	Seniors
$\bar{x}_1 = \$5.36$	$\bar{x}_2 = \$4.62$
$s_1 = \$0.51$	$s_2 = \$0.43$
$n_1 = 10$	$n_2 = 12$

Is the difference observed between the sample means significant at the 2% level?

SOLUTION

The alternative hypotheses are

$$H_0: \mu_1 - \mu_2 = 0$$
$$H_1: \mu_1 - \mu_2 \neq 0$$

To test the null hypothesis, we use the t statistic

$$t = \frac{(\bar{x}_1 - \bar{x}_2) - 0}{s_{\bar{x}_1 - \bar{x}_2}} = \frac{\bar{x}_1 - \bar{x}_2}{s_{\bar{x}_1 - \bar{x}_2}}$$

where $s_{\bar{x}_1 - \bar{x}_2}$ is the estimated standard error of the difference between two means.

Unlike the case with large samples, it is necessary here to assume equal population variances. An estimate of this common variance is obtained by pooling the two sample variances into a weighted average, using the numbers of degrees of freedom, $n_1 - 1$ and $n_2 - 1$, as weights. This pooled estimate of the common variance, which we denote as s^2, is given by

$$(7.11) \qquad s^2 = \frac{(n_1 - 1)s_1^2 + (n_2 - 1)s_2^2}{n_1 + n_2 - 2}$$

The estimated standard error of the difference between two means is then

$$(7.12) \qquad s_{\bar{x}_1 - \bar{x}_2} = \sqrt{\frac{s^2}{n_1} + \frac{s^2}{n_2}} = s \sqrt{\frac{1}{n_1} + \frac{1}{n_2}}$$

A number of alternative mathematical expressions are possible for Equation 7.12, but because of its similarity in appearance to previously used standard error formulas, we shall use it in this form.

We now work out the present problem. Substitution into (7.11) gives

$$s^2 = \frac{(10 - 1)(0.51)^2 + (12 - 1)(0.43)^2}{10 + 12 - 2} = 0.2187$$

and

$$s = \sqrt{0.2187} = \$0.47$$

Thus, the estimated standard error is

$$s_{\bar{x}_1 - \bar{x}_2} = (\$0.47) \sqrt{\frac{1}{10} + \frac{1}{12}} = \$0.20$$

and the t value is

$$t = \frac{\$5.36 - \$4.62}{\$0.20} = \frac{\$0.74}{\$0.20} = 3.70$$

The number of degrees of freedom in this problem is $n_1 + n_2 - 2$, that is, $10 + 12 - 2 = 20$. One way of explaining the number of degrees of freedom in this case is as follows. In the one-sample case, where the sample standard deviation was used as an estimate of the population standard deviation, there was a loss of 1 degree of freedom; hence, the number of degrees of freedom was $n - 1$. In the two-sample case, each of the sample variances was used in the pooled estimate of population variance; hence, 2 degrees of freedom were lost, and the number of degrees of freedom is $n_1 + n_2 - 2$.

The critical *t* value at the 2% significance level for 20 degrees of freedom is 2.528 (see Table A-6 of Appendix A). Since the observed *t* value, 3.70, exceeds this critical *t* value, the null hypothesis is rejected and we conclude on the basis of the sample data that the population means are indeed different. In terms of the problem, we are unwilling to attribute the difference in average amount of pocket money between freshmen and seniors at this university to chance errors of sampling. On the basis of this test, we conclude that freshmen at the university carry more pocket change than do seniors.

Exercises

1. In the following problem, a statistical procedure has been misused. Describe the incorrect procedure.

   NASA wants to test the reliability of a certain magneto relay used in the Surveyor spacecraft. According to specifications, these relays should have an average life of 180 hours before failure. A sample of 20 switches is tested, and the average switch life is 177.5 hours with a standard deviation of 4. Since the *z* statistic

$$z = \frac{177.5 - 180}{4/\sqrt{20}}$$

   is less than -1.65, it is concluded (on the basis of the normal distribution) at a 0.05 significance level that the switches do not meet specifications.

2. A simple random sample of ten delinquent charge accounts at a certain department store showed a mean of $79 and a standard deviation of $21.

   a. Are these data consistent with a claim that the average size of delinquent charge accounts in this department store is $70? Use a two-tailed test with a 0.05 level of significance.

   b. Now instead of the above sample size, assume a sample of size 100. Would you reach the same conclusion as in (a)? Comment on the results of tests you applied in (a) and (b).

3. In a study of use of bar soap by a simple random sample of 15 suburban families, the consumption of such soap was found to have an arithmetic mean of 50 ounces per family per month with a standard deviation of ten ounces. In another similar study of 11 urban families, consumption was found to average 55 ounces with a standard deviation of 12 ounces.

   a. At the 10% level of significance, would you conclude that there was a statistically significant difference in the sample averages of consumption of bar soap?

b. For this problem, state the null and alternative hypotheses and the decision rule employed.

4. A company seeking to place a large bond issue is trying to determine which investment banking firm should head the syndicate formed to handle the issue. The financial vice-president of the firm has instructed an assistant to take a small sample of past syndicates headed by F. E. Mutton and Company, a prominent New York firm, and determine the average percentage of the issues it has placed at the initial offering price (before any discount). Using a random sample of 18 issues, the assistant finds that on the average (unweighted mean), the firm has successfully placed 95% of each total issue in the initial offering with a standard deviation of 5%. With a random sample of 12, the assistant finds that the success rate for The Second Boston Corporation has averaged 90% with a standard deviation of 4%. Assuming a 2% significance level, should the assistant conclude that the sample average success rates of the two investment banking firms differ significantly? State the alternative hypotheses and the decision rule employed.

5. A successful tax advisor is analyzing a motion picture tax shelter for possible investment. The prospectus claims that movies in the same general category as the one under consideration have grossed an average of $10,000,000 each in the past. This is an optimal level of receipts for the current film (that is, gross receipts less than this would result in an inadequate return on investment or a loss of investment, and gross receipts more than this would result in taxable income, which is not the primary objective of a tax shelter). The advisor has taken a random sample of similar films and has compiled the following statistics to test the claim of the prospectus:

Film	Gross Receipts (in $Millions)
1	11.5
2	7
3	9
4	10
5	10
6	12
7	8.5
8	9
9	8
10	6
11	7.5
12	9.5

Formulate an appropriate test using $\alpha = 0.02$.

6. During a recession year, 15 government economists and 15 economists from private industry were polled on the question of what the appropriate budget

deficit for the United States should be to return the economy to full employment. The following results were obtained from these random samples:

Economist	Government Deficit Estimate (in $Billions)	Economist	Private Industry Deficit Estimate (in $Billions)
1	40	1	30
2	30	2	40
3	35	3	50
4	25	4	35
5	30	5	25
6	30	6	20
7	45	7	30
8	50	8	30
9	60	9	40
10	40	10	45
11	45	11	30
12	50	12	25
13	35	13	40
14	40	14	45
15	45	15	40

At a 0.05 significance level, would you conclude that there is a significant difference in the sample means?

7.5 THE *t* TEST FOR PAIRED OBSERVATIONS

In the two-sample tests considered so far, the two samples had to be considered independent. That is, it was necessary that the values of observations in one sample be *independent* of the values in the other. Situations arise in practice in which this condition does not hold. In fact, the
two samples may consist of pairs of observations made on the same individual, the same object, or more generally, the same selected population
elements. Clearly, the independence condition is violated in these
cases.

As a concrete example, let us consider a case in which a group of ten
men was given a special diet, and it was desired to test weight loss in
pounds at the end of a two-week period. The observed data are shown
in Table 7-3.

As indicated in Table 7-3, X_1 denotes the weight before the diet and
X_2 the weight after the diet. It would be incorrect to run a *t* test to deter-

TABLE 7-3
Weights before and after a Special Diet for a Simple Random Sample of Ten Men

Man	Weight before Diet X_1	Weight after Diet X_2	Difference in Weight $d = X_2 - X_1$	$d - \bar{d}$	$(d - \bar{d})^2$
1	181	178	−3	+1	1
2	172	172	0	+4	16
3	190	185	−5	−1	1
4	187	184	−3	+1	1
5	210	201	−9	−5	25
6	202	201	−1	+3	9
7	166	160	−6	−2	4
8	173	168	−5	−1	1
9	183	180	−3	+1	1
10	184	179	−5	−1	1
			−40		60

$$\bar{d} = \frac{-40}{10} = -4 \text{ pounds}$$

$$s_d = \sqrt{\frac{\Sigma(d - \bar{d})^2}{n - 1}} = \sqrt{\frac{60}{10 - 1}} = 2.58 \text{ pounds}$$

$$s_{\bar{d}} = \frac{s_d}{\sqrt{n}} = \frac{2.58}{\sqrt{10}} = 0.82 \text{ pounds}$$

mine whether there is a significant difference between the mean of the X_1 values and the mean of the X_2 values, because of the nonindependence of the two samples. An individual's weight after the test is certainly not independent of his weight before the test. Each of the d values, $d = X_2 - X_1$, represents a difference between two observations on the same individual. The assumption is made that the subtraction of one value from the other removes the effect of factors other than that of the diet; that is, we assume that these other factors affect each member of any pair of X_1 and X_2 values in the same way.

We can state the hypotheses to be tested as

$$H_0: \mu_2 - \mu_1 = 0$$
$$H_1: \mu_2 - \mu_1 < 0$$

where μ_1 and μ_2 are the population mean weights before and after the diet, respectively. Let us assume that the test is to be carried out with $\alpha = 0.05$. The null hypothesis states that there is no difference between mean weight after the diet and mean weight before the diet, whereas the alternative hypothesis says that mean weight after the diet is less than mean weight before. We can visualize the situation as one in which, if the null hypothesis is true, there is a population of numbers representing differences in the weight after the diet and before the diet, and the mean of these numbers is 0. Hence, we wish to test the hypothesis that our simple random sample of $d = X_2 - X_1$ values comes from this universe. The procedure used in these cases is to obtain the mean of the sample differences and to test whether this average, $\bar{d}$, differs significantly from 0. The estimated standard error of $\bar{d}$, denoted $s_{\bar{d}}$, is given by

$$s_{\bar{d}} = \frac{s_d}{\sqrt{n}}$$

where

$$s_d = \sqrt{\frac{\Sigma\,(d - \bar{d})^2}{n - 1}}$$

In this problem, $s_{\bar{d}} = 0.82$ pounds, as shown by the calculations in Table 7-3. Assuming that the population of differences (d values) is normally distributed, the ratio $(\bar{d} - 0)/s_{\bar{d}}$ is distributed according to the t distribution.

$$t = \frac{\bar{d} - 0}{s_{\bar{d}}} = \frac{-4}{0.82} = -4.88$$

The number of degrees of freedom is $n - 1$, where n is the number of d values. Hence, in this problem, $n - 1 = 10 - 1 = 9$. The test is one-tailed, because only a $\bar{d}$ value that is negative and significantly different from 0 could result in acceptance of the alternative hypothesis, $H_1: \mu_2 - \mu_1 < 0$. Since $\alpha = 0.05$ and the test is one-tailed, we look in Table A-6 under the heading 0.10. The critical t value for $\nu = 9$ is 1.833, which for our purposes is interpreted as -1.833. Since the observed t value of -4.88 is less than (lies to the left of) this critical point, the null hypothesis is rejected and we accept its alternative. Therefore, on the basis of this experiment, we conclude that the special diet does result in an average weight loss over a two-week period.

This method of pairing observations is also used to reduce the effect of extraneous factors that could cause a significant difference in means, whereas the factor whose effect we are really interested in may not have resulted in such a difference if the extraneous factors had not been present.

For example, if medical experimenters wanted to test two different treatments to judge which was better, they might administer one treatment to one group of persons and the other treatment to a second group. Suppose on the basis of the usual significance test for means, it is concluded that one treatment is better than the other. Let us also assume that the group receiving the supposedly better treatment was much younger and much healthier at the beginning of the experiment than the other group and that these factors could have an effect on the reaction to the treatments. Then clearly, the relative effectiveness of the two treatments would be obscured.

On the other hand, assume that individuals were selected in pairs in which both members were about the same age and in about the same health condition. If the first treatment is given to one member of a pair and the second treatment to the other, and then a difference measure is calculated for the effect of treatment, neither age nor health condition would affect this measurement. Ideally, we would like to select pairs that are identical in all characteristics other than the factor whose effects we are attempting to measure. Obviously, as a practical matter this is impossible, but the guiding principle is clear. Once differences are taken between members of each pair, the t test proceeds exactly as in the preceding example. It may be noted that in the weight example, the differences were measured on the same individual, whereas in the present illustration the differences are derived from the two members of each pair.

The method of paired observations is a very useful technique. As compared to the standard two-sample t test, in addition to the advantage that we do not have to assume that the two samples are independent, we also need not assume that the variances of the two samples are equal.

7.6 SUMMARY AND A LOOK AHEAD

In this chapter, we have considered some classical hypothesis-testing techniques. These tests represent only a few of the simplest methods. All the cases discussed thus far have involved only one or two samples, but methods are available for testing hypotheses concerning three or more samples. The cases we have dealt with also have tested only one parameter of a probability distribution. However, techniques are available for testing whether an entire frequency distribution is in conformity with a theoretical model, such as a specified probability distribution. Finally, the tests we have considered involved a final decision on the basis of the sample evidence; that is, a decision concerning acceptance or rejection of hypotheses was reached on the basis of the evidence contained in one or two samples. Sequential decision procedures are available to permit postponement of decision pending further sample evidence. Some of the broader decision procedures are discussed in subsequent chapters.

Although classical hypothesis testing techniques of the type discussed in this chapter have been widely applied in a great many fields, it would be incorrect to infer that their use is noncontroversial and that they can simply be employed in a mechanical way. At this point, it suffices to indicate that the methods discussed are admittedly incomplete and that Bayesian decision theory addresses itself to the required completion. Thus, for example, in hypothesis testing, the establishing of significance levels such as 0.05 or 0.01 inevitably appears to be a rather arbitrary procedure, despite the fact that the relative seriousness of Type I and Type II errors is supposed to be considered in designing a test. Although costs of Type I and Type II errors can theoretically be considered in the classical formulation, as a matter of actual practice, they are rarely included explicitly in the analysis. In Bayesian decision theory, the costs of Type I and Type II errors, as well as the payoffs of correct decisions, become an explicit part of the formal analysis.

Moreover, in classical hypothesis testing, decisions are reached solely on the basis of the present sample information without reference to any prior knowledge concerning the hypothesis under test. On the other hand, Bayesian decision theory provides a method for combining prior knowledge with current sample information for decision-making purposes. These Bayesian decision-theory methods are discussed in subsequent chapters.

Exercises

1. A research group ran the following experiment to determine whether monetary incentives would increase learning. It selected a random sample of students at a university. For some semesters, the students were given X dollars per week if their work met a certain standard, and no compensation otherwise. During other semesters, no rewards were given at all, regardless of performance. The results of eight students were as follows:

Student	Average during Terms with Payment	Average during Terms without Payment
1	3.10	3.00
2	2.95	2.45
3	2.00	2.25
4	1.95	1.95
5	3.80	3.75
6	2.43	2.65
7	2.65	2.55
8	2.40	2.20

Using a paired t test, would you conclude that monetary incentives increase grade average? A higher grade average represents a better average. Use $\alpha = 0.05$.

2. The marketing research department of the Pacific Petroleum Company wished to test the effect of increased radio advertising on the sales of one of

the firm's major products, a gasoline additive that increases the efficiency of automobile engines. The manager in charge of the project decided to increase substantially the radio advertising of the product in 13 metropolitan areas in the Western sales region for a five-week period. The average weekly sales (in units) of the product for the five-week period before the project began and for the five-week trial period in the 13 metropolitan areas are given below:

Metropolitan Area	Average Weekly Sales before Project Period	Average Weekly Sales during Project Period
	(in Thousands of Units)	
Seattle	55	60
Olympia	55	58
Portland	47	48
San Francisco	38	42
Sacramento	30	32
Los Angeles	68	65
San Diego	16	20
Reno	50	53
Las Vegas	23	28
Boise	36	34
Tucson	25	27
Phoenix	23	22
Salt Lake City	23	26

Using the paired t test, would you conclude that increased radio advertising increases sales? Assume that other factors affecting sales are unchanged over the ten-week period, and use a 5% significance level.

3. Ten middle managers from the Mammoth Manufacturing Company attended a three-week session of group meetings and workshops to increase their managerial effectiveness and sensitivity. The top management of the firm evaluates its middle managers using the "grid" methodology; each manager is given a rating consisting of two numbers between 1 (worst rating) and 9 (best rating). The first number indicates the evaluation of the manager's concern for production, and the second reflects the manager's concern for people. The following table shows the ratings given to each manager before and after attending the session:

Manager	Rating before Session	Rating after Session
1	7, 5	8, 6
2	4, 5	6, 8
3	8, 2	8, 4
4	8, 7	7, 7

(Continued)

Manager	Rating before Session	Rating after Session
5	6, 6	6, 9
6	4, 2	7, 4
7	7, 3	8, 7
8	3, 6	6, 7
9	4, 9	7, 8
10	6, 5	4, 5

Assuming consistency in the top management's rating determinations, would you conclude that the session has had a positive effect on the managers' ratings (a) on concern for production? (b) on concern for people? Use a 2.5% significance level.

8

Chi-Square Tests and Analysis of Variance

In Chapter 7, procedures were discussed for testing hypotheses with data obtained from a single simple random sample or from two such samples. For example, we considered tests of whether two population proportions or two population means were equal. Obvious generalizations of such techniques are tests for the equality of more than two proportions or more than two means. The two topics discussed in this chapter supply these generalizations. *Chi-square tests*[1] provide the basis for judging whether more than two population proportions may be considered to be equal; *analysis of variance* techniques provide ways to test whether more than two population means may be considered to be equal.

---

[1] The procedures are referred to as "χ^2 tests" or "chi-square tests," where the symbol χ is the Greek lowercase *chi* (pronounced kye).

We discuss χ^2 tests first in this chapter, considering the topics of (1) tests of goodness of fit and (2) tests of independence in that order. Tests of goodness of fit provide a means for deciding whether a particular theoretical probability distribution, such as the binomial distribution, is a close enough approximation to a sample frequency distribution for the population from which the sample was drawn to be described by the theoretical distribution. Tests of independence constitute a method for deciding whether the hypothesis of independence between different variables is tenable. It is this type of procedure that provides a test for the equality of more than two population proportions. Both types of χ^2 tests furnish a conclusion on whether a set of observed frequencies differs so greatly from a set of theoretical frequencies that the hypothesis under which the theoretical frequencies were derived should be rejected.

8.1   TESTS OF GOODNESS OF FIT

One of the major problems in the application of probability theory, statistics, and mathematical models in general is that the real-world phenomena to which they are applied usually depart somewhat from the assumptions embodied in the theory or models. For example, let us consider use of the binomial probability distribution in a particular problem. As indicated in Section 3.4, two of the assumptions involved in the derivation of the binomial distribution as the probability distribution for a Bernoulli process are

1. The probability of a success, p, remains constant from trial to trial

2. The trials are independent

Let us consider whether these assumptions are met in the following problem.

   A firm bills its accounts at a 2% discount for payment within ten days and for the full amount due for payment after ten days. In the past, 40% of all invoices have been paid within ten days. In a particular week, the firm sends out 20 invoices. Is the binomial distribution appropriate for computing the probabilities that 0, 1, 2, . . . , 20 firms will receive the discount for payment within ten days?

   Considering the possible use of the binomial distribution, we can let $p = 0.40$ represent the probability that a firm will receive the discount and $n = 20$ firms represent the number of trials. Does it seem reasonable to assume that p, the probability of receiving the discount, is 0.40 for each firm? Past relative frequency data for *each* firm could be brought to bear on the answer to this question. In most practical situations of this sort, we would probably find that the practices of individual firms vary widely, with some firms virtually always receiving discounts, some firms virtually never receiving discounts, and most firms falling somewhere between these two extremes.

Does the assumption of independence seem tenable in this problem? That is, does it seem reasonable that whether one firm receives the discount is independent of whether another firm does? Probably not, since general monetary conditions doubtless affect many firms in a similar way. For example, when money is "tight" and it is difficult to acquire adequate amounts of working capital, the fact that one firm does not receive the discount is related to, rather than *independent* of, whether other firms have done so. Moreover, there may be traditional practices in certain industries concerning whether or not discounts are taken, and other factors that would interfere with the independence assumption may be present.

How great a departure from the assumptions underlying a probability distribution or, more generally, from the assumptions embodied in any theory or mathematical model can be tolerated before we should conclude that the distribution, theory, or model is no longer applicable? This is a very complex question that cannot be readily answered by any simple universally applicable rule. The purpose of χ^2 "goodness of fit" tests is to provide one type of answer to this question by comparing *observed frequencies* with *theoretical*, or *expected*, *frequencies* derived under specified probability distributions or hypotheses.

The sequence of steps in performing goodness of fit tests is very similar to previously discussed hypothesis-testing procedures.

1. Null and alternative hypotheses are established, and a significance level is selected for rejection of the null hypothesis.

2. A random sample of observations is drawn from a relevant statistical population.

3. A set of expected, or theoretical, frequencies is derived under the assumption that the null hypothesis is true. This generally takes the form of the assumption that a particular probability distribution is applicable to the statistical population under consideration.

4. The observed frequencies are compared to the expected, or theoretical, frequencies.

5. If the aggregate discrepancy between the observed and theoretical frequencies is too great to attribute to chance fluctuations at the selected significance level, the null hypothesis is rejected.

We illustrate goodness of fit tests and discuss some of the underlying theory for an example involving a uniform probability distribution (see Section 3.3). Additional examples follow.

Suppose a consumer research firm wished to determine whether any of five brands of coffee was preferred by coffee drinkers in a certain metropolitan area. The firm took a simple random sample of 1000 coffee

TABLE 8-1
Number of Coffee Drinkers in a Certain
Metropolitan Area Who Most Preferred the
Specified Brand of Coffee

Brand Preference	Number of Consumers
A	210
B	312
C	170
D	85
E	223
	1000

drinkers in the area and conducted the following experiment. Each consumer was given five cups of coffee, one of each brand $(A, B, C, D,$ and $E)$ without identification of the individual brands. The cups were presented to each consumer in a random order determined by sequential selection from five paper slips, each containing one of the letters A, B, C, D, and E. In Table 8-1 are shown the numbers of coffee drinkers who stated that they liked the indicated brands best.

Denoting the true proportions of preference for each brand as p_A, p_B, p_C, p_D, and p_E, we can state the null and alternative hypotheses as follows:

$$H_0: p_A = p_B = p_C = p_D = p_E = 0.20$$
$$H_1: \text{The } p \text{ values are not all equal}$$

That is, if in the population from which the sample was drawn there were no differences in preference among the five brands, 20% of coffee drinkers would prefer each brand. An equivalent way of stating these hypotheses is

$$H_0: \text{The probability distribution is uniform}$$
$$H_1: \text{The probability distribution is not uniform}$$

In other words, we want to know whether the sample of 1000 coffee drinkers can be considered a random sample from a population in which the proportions who prefer each of the five brands are equal. Of course, this hypothesis is only one of many that could conceivably be formulated. One of the strengths of the goodness of fit test discussed in this section is that it permits a variety of different hypotheses to be raised and tested.

If the null hypothesis of no difference in preference were true, the *expected* or *theoretical* number of the 1000 coffee drinkers in the sample who would prefer each brand would be $0.20 \times 1000 = 200$. Hence, the expected frequency corresponding to each of the observed frequencies in Table 8-1 is 200. We can now compare the set of observed frequencies with the set of theoretical frequencies derived under the assumption that the null hypothesis is true. The test statistic we compute to make this comparison is known as chi-square, denoted χ^2. The computed value of χ^2 is

(8.1)
$$\chi^2 = \sum \frac{(f_0 - f_t)^2}{f_t}$$

where f_0 = an observed frequency
f_t = a theoretical (or expected) frequency

As we can see from Equation 8.1, if every observed frequency is exactly equal to the corresponding theoretical frequency, the computed value of χ^2 is 0. This is the smallest value χ^2 can have. The larger the discrepancies between the observed and theoretical frequencies, the larger is χ^2.

The computed value of χ^2 is a random variable that takes on different values from sample to sample. That is, χ^2 has a sampling distribution just as do the test statistics discussed in Chapter 7. We wish to answer the question, Is the computed value of χ^2 so large that we are required to reject the null hypothesis? In other words, are the aggregate discrepancies between the observed frequencies, f_0, and theoretical frequencies, f_t, so large that we are unwilling to attribute them to chance, and have to reject the null hypothesis? The calculation of χ^2 for the present problem is shown in Table 8-2.

TABLE 8-2
Calculation of the χ^2 Statistic for the Coffee-Tasting Problem

	(1)	(2)	(3)	(4)	(5)
Brand Preference	Observed Frequency f_0	Theoretical (Expected) Frequency f_t	$(f_0 - f_t)$	$(f_0 - f_t)^2$	Column (4) / Column (2) $\frac{(f_0 - f_t)^2}{f_t}$
A	210	200	10	100	0.5
B	312	200	112	12,544	62.7
C	170	200	−30	900	4.5
D	85	200	−115	13,225	66.1
E	223	200	23	529	2.6
Total	1000	1000			136.4

$$\chi^2 = \sum \frac{(f_0 - f_t)^2}{f_t} = 136.4$$

The χ^2 Distribution

Before we can answer the questions raised in the preceding paragraph, we must digress for a discussion of the appropriate sampling distribution. Then we will complete the solution to the coffee-tasting problem. It can be shown that for large sample sizes the sampling (probability) distribution of χ^2 can be closely approximated by the χ^2 distribution whose probability function is

(8.2)
$$f(\chi^2) = c(\chi^2)^{(\nu/2)-1}e^{-\chi^2/2}$$

where $e = 2.71828 \ldots$
ν = number of degrees of freedom
c = a constant depending only on ν

The χ^2 distribution has only one parameter, ν, the number of degrees of freedom. This is similar to the case of the t distribution, discussed in Section 6.4. Hence, $f(\chi^2)$ is a family of distributions, one for each value of ν. χ^2 is a continuous random variable greater than or equal to 0. For small values of ν, the distribution is skewed to the right. As ν increases, the distribution rapidly becomes symmetrical. In fact, for large values of ν, the χ^2 distribution is closely approximated by the normal curve. Figure 8-1 depicts the χ^2 distributions for $\nu = 1, 5$, and 10.

Since the χ^2 distribution is a probability distribution, the area under the curve for each value of ν equals 1. Because there is a separate distribution for each value of ν, it is not practical to construct a detailed table of areas. Therefore, for compactness, what is generally shown in a χ^2 table is the relationship between areas and values of χ^2 for only a few levels of significance in different χ^2 distributions.

In Table A-7 of Appendix A are shown χ^2 values corresponding to selected areas in the right tail of the χ^2 distribution. These tabulations are shown separately for the number of degrees of freedom listed in the left

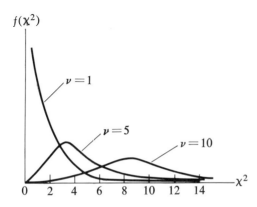

FIGURE 8-1
The χ^2 distributions for $\nu = 1, 5$, and 10.

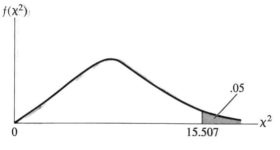

FIGURE 8-2
χ^2 distribution for 8 degrees of freedom.

column. The χ^2 values are shown in the body of the table, and the corresponding areas are shown in the column headings.

As an illustration of the use of the χ^2 table, let us assume a random variable having a χ^2 distribution with $\nu = 8$. In Table A-7, we find a χ^2 value of 15.507 corresponding to an area of 0.05 in the right tail. The relationships described in this illustrative problem are shown in Figure 8-2. Hence, if the random variable has a χ^2 distribution with $\nu = 8$, the probability that χ^2 is greater than 15.507 is 0.05. Let us give the corresponding interpretation in a hypothesis-testing context. If the null hypothesis being tested is true, the probability of observing a χ^2 figure greater than 15.507 because of chance variation is equal to 0.05. Therefore, for example, if the null hypothesis were tested at the 0.05 level of significance and we calculated $\chi^2 = 16$, we would reject the null hypothesis, because so large a χ^2 value would occur less than 5% of the time if the null hypothesis is true. We now turn to a discussion of the rules for determining the number of degrees of freedom involved in a χ^2 test.

Number of Degrees of Freedom

As we have seen earlier, a χ^2 goodness of fit test involves a comparison of a set of observed frequencies, denoted f_0, with a set of theoretical frequencies, denoted f_t. Let k be the number of classes for which these comparisons are made. For example, in the coffee-tasting problem $k = 5$, because there are five classes for which we computed relative deviations of the form $(f_0 - f_t)^2/f_t$. To determine the number of degrees of freedom, we must reduce k by 1 for each restriction imposed. In the coffee-tasting example, the number of degrees of freedom is equal to $\nu = k - 1 = 5 - 1 = 4$. The rationale for this computation follows.

In the calculations shown in Table 8-2, there are five classes for which f_0 and f_t values are to be compared. Hence, we start with $k = 5$ degrees of freedom. However, we have forced the total of the theoretical frequencies, Σf_t, to be equal to the total of the observed frequencies, Σf_0; that is, 1000. Therefore, we have reduced the number of degrees of freedom by 1 and there are now only 4 degrees of freedom. That is, once the total of the theoretical frequencies is fixed, only four of the f_t values may be freely assigned to the classes; when these four have been assigned, the fifth class is immediately determined, because the theoretical frequencies must total 1000.

If any further restrictions are imposed in the calculation of the theoretical frequencies, the number of degrees of freedom is reduced by 1 for each such restriction. Hence, for example, for each parameter value that has to be estimated from the sample, an additional degree of freedom is lost. For instance, if the mean μ of a Poisson distribution must be estimated from the mean $\bar{x}$ of a sample, there would be a reduction of 1 degree of freedom. In summary, we can state the following rules for determining ν, the number of degrees of freedom in a χ^2 test in which k classes of observed and theoretical frequencies are compared.

1. If the only restriction is $\Sigma f_t = \Sigma f_0$, the number of degrees of freedom is $\nu = k - 1$.
2. If in addition to the above restriction, m parameters are replaced by sample estimates, the number of degrees of freedom is $\nu = k - 1 - m$.

Decision Procedure

We can now return to the coffee-tasting example to perform the goodness of fit test. Let us assume we wish to test the null hypothesis at the 0.05 level of significance. Since the number of degrees of freedom is 4, we find the critical value of χ^2, which we denote as $\chi^2_{0.05}$, to be 9.488 (Table A-7 of Appendix A). This means that if the null hypothesis is true, the probability of observing a χ^2 value greater than 9.488 is 0.05. Specifically in terms of the problem, this means that if there were no difference in preference among brands, an aggregate discrepancy between the observed and theoretical frequencies larger than a $\chi^2 = 9.488$ would occur only 5% of the time. We can state the decision rule for this problem in which $\chi^2_{0.05} = 9.488$ as follows:

DECISION RULE

1. If $\chi^2 > 9.488$, reject H_0
2. If $\chi^2 \leqslant 9.488$, accept H_0

Since the computed χ^2 value in this problem is 136.4 and is thus very much larger than the critical $\chi^2_{0.05}$ value of 9.488, we reject the null hypothesis. Therefore, our conclusion is that there exist real differences in consumer preference among the brands of coffee involved in the experiment. In statistical terms, we cannot consider the 1000 coffee drinkers in the experiment to be a simple random sample from a population whose members prefer each of the five brands in equal proportions. In terms of goodness of fit, we reject the null hypothesis that the probability distribution is uniform. Hence, we conclude that the uniform distribution is decidedly not a "good fit" to the sample data.

We now turn to a number of other examples of χ^2 goodness of fit tests.

EXAMPLE 8-1

Jonathan Falk, a management scientist, was developing an inventory control system for a manufacturer of a diversified product line. He wanted to determine whether the Poisson distribution was an appropriate model for the demand for a particular product. He obtained the frequency distribution of the number of units of this product demanded per day for the past 200 business days. That distribution is shown in Columns (1) and (2) of Table 8-3. The mean number of units per day is shown at the bottom of the table; it is obtained by dividing the total of Column (3) by the total of Column (2).

TABLE 8-3
Number of Units of a Particular Product Demanded per Day for the Past 200 Business Days

(1) Number of Units Demanded per Day x	(2) Observed Number of Days f_0	(3) Column (1) × Column (2) $f_0 x$
0	11	0
1	28	28
2	43	86
3	47	141
4	32	128
5	28	140
6	7	42
7	0	0
8	2	16
9	1	9
10	1	10
Total	200	600

$$\bar{x} = \frac{600}{200} = 3 \text{ units per day}$$

TABLE 8-4
Theoretical Distribution of Demand Assuming a Poisson Distribution

(1) Number of Units Demand per Day x	(2) Probability $f(x)$	(3) Column (2) × 200 Expected Number of Days f_t
0	0.050	10.0
1	0.149	29.8
2	0.224	44.8
3	0.224	44.8
4	0.168	33.6
5	0.101	20.2
6	0.050	10.0
7	0.022	4.4
8	0.008	1.6
9	0.003	0.6
10	0.001	0.2
Total	1.000	200.0

Using the mean of this sample of observations, $\bar{x} = 3$ units of demand per day, as an estimate of the parameter μ of the corresponding theoretical Poisson distribution, Mr. Falk calculated the Poisson probability distribution shown in the first two columns of Table 8-4. Multiplying the probabilities in Column (2) by 200 days, he obtained the theoretical, or expected, frequencies if the demand were distributed according to the Poisson distribution. These theoretical frequencies are shown in Column (3). For example, if the *probability* that 0 units would be demanded is 0.050, then in 200 days the *expected number* of days in which 0 units would be demanded is $0.050 \times 200 = 10.0$ days. This figure of 10.0 days is the first entry in Column (3). The analyst was now able to apply a χ^2 goodness of fit test using the actual number of days in Column (2) of Table 8-3 as the observed frequencies, f_0, and the expected number of days in Column (3) of Table 8-4 as the theoretical frequencies, f_t. These two sets of frequencies are shown in Columns (2) and (3) of Table 8-5, where the calculation of the χ^2 value is carried out. The hypothesis under test in this problem may be stated as follows:

H_0: The population probability distribution is Poisson with $\mu = 3$

Assume that we wish to test the null hypothesis at the 0.05 level of significance.

As can be seen in Table 8-5, the last four classes of Tables 8-3 and 8-4 for 7, 8, 9, and 10 units of demand have been combined into one class entitled "7 or more." Both f_0 and f_t values have been cumulated for the four classes, and a single relative deviation of the form $(f_0 - f_t)^2/f_t$ has been cal-

TABLE 8-5
Calculation of the χ^2 Statistic for the Demand Distribution Problem

(1) *Number of* *Units Demanded* *per Day*	(2) *Observed* *Number of Days* f_0	(3) *Theoretical* *Number of Days* f_t	(4) $f_0 - f_t$	(5) $(f_0 - f_t)^2$	(6) $\dfrac{(f_0 - f_t)^2}{f_t}$
0	11	10.0	1.0	1.00	0.10
1	28	29.8	−1.8	3.24	0.11
2	43	44.8	−1.8	3.24	0.07
3	47	44.8	2.2	4.84	0.11
4	32	33.6	−1.6	2.56	0.08
5	28	20.2	7.8	60.84	3.01
6	7	10.0	−3.0	9.00	0.90
7 or more	4	6.8	−2.8	7.84	1.15
Total	200	200.0	0		5.53

$$\chi^2 = \sum \frac{(f_0 - f_t)^2}{f_t} = 5.53$$

culated for the combined class. There are now eight classes, $k = 8$, in Table 8-5 for which the χ^2 value has been computed and from which the number of degrees of freedom will be determined. The reason for this combination of classes will be explained at the completion of the problem.

Let us now compute the number of degrees of freedom in the test. As indicated in the earlier discussion, the number of degrees of freedom is given by $v = k - 1 - m$, where m is the number of parameters that have been replaced by sample estimates. Since the sample mean $\bar{x}$ was used as the estimate of the parameter μ in the Poisson distribution, $m = 1$. Hence, the number of degrees of freedom is $v = 8 - 1 - 1 = 6$. For 6 degrees of freedom, the critical value of χ^2 at the 0.05 level of significance is $\chi^2_{0.05} = 12.592$ (Table A-7 of Appendix A). Therefore, since the observed χ^2 value of 5.53 is less than 12.592, we accept the null hypothesis.

In other words, the aggregate discrepancy between the observed and theoretical frequencies is sufficiently small for us to *conclude that the Poisson distribution with $\mu = 3$ is a good fit.* Based on this result, the Poisson distribution can reasonably be used as a model for demand for the product.

In the foregoing example, the hypothesis that the population probability distribution is Poisson was tested and accepted. The same procedure could be used to test whether a shift away from a previously established Poisson distribution has occurred. The number of degrees of freedom would be $k - 1$ if μ was estimated from previous results or $k - 1 - 1$ if μ was estimated from the mean, $\bar{x}$, of the present sample data.

RULE CONCERNING SIZE OF THEORETICAL FREQUENCIES

As indicated earlier, for large sample sizes, the probability function of the computed χ^2 values can be closely approximated by the χ^2 distribution given in Equation (8.2), which is the distribution of a *continuous* random variable. However, there are only a finite number of possible combinations of f_t values, and hence only a finite number of computed χ^2 values. Thus, a computed χ^2 figure is one value of a *discrete* random variable. If the sample size is large, the approximation of the probability distribution of this discrete random variable by the continuous chi-square distribution will be a good one. This is analogous to approximating the binomial distribution, which is discrete, by the normal curve, which is continuous (see Section 5.3).

When the expected frequencies (the f_t values) are small, the approximation discussed in the preceding paragraph is inadequate. A frequently used rule is that each f_t value should be equal to or greater than 5. This is why the classes for 7, 8, 9, and 10 units of demand were combined in Example 8-1 in the computation of the χ^2 value. As shown in Table 8-5, the computed f_t value for the combined class is equal to 6.8, which satisfies the commonly used rule of thumb for a minimum expected frequency.

EXAMPLE 8-2

Researcher Lana Mauro hypothesized that human births could be considered a Bernoulli process. However, she suspected that male and female births were not equally likely. Specifically, she believed that large families tended to have more male children than female. She had data for a

TABLE 8-6
Number of Male Children in a Sample of Families
with Five Children Each

(1) *Number of Male Children* x	(2) *Observed Number of Families* f_0
0	12
1	42
2	92
3	108
4	46
5	20
	320

TABLE 8-7
Calculation of $\bar{p}$, the Proportion of Male Children in the
Sample of 320 Families

(1) Number of Male Children x	(2) Observed Number of Families f_0	(3) Column (1) $\times$ Column (2) $f_0 x$
0	12	0
1	42	42
2	92	184
3	108	324
4	46	184
5	20	100
Total	320	834

$$\bar{x} = \frac{834}{320} = 2.61 \text{ male children per family}$$

$$\bar{p} = \frac{2.61}{5} = 0.522$$

TABLE 8-8
Calculation of Expected Frequencies of Number of Male Children:
Assumed Binomial Distribution with $p = 0.522$ and $n = 5$

(1) Number of Male Children x	(2) Probability $f(x)$	(3) Column (2) $\times$ 320 Expected Number of Families f_t
0	0.025	8.00
1	0.136	43.52
2	0.298	95.36
3	0.324	103.68
4	0.178	56.96
5	0.039	12.48
	1.000	320.00

simple random sample of 320 families with five children each, which had been drawn for another purpose, so she decided to conduct a partial test of her theory using these data. The frequency distribution of male children in these 320 families is shown in Table 8-6.

Since Ms. Mauro hypothesized that human births could be treated as a Bernoulli process, she decided to fit a binomial distribution to the data in Table 8-6. She further decided to estimate p, the probability of a male birth, by using $\bar{p}$, the proportion of male births in the sample. As indicated in Table 8-7, she calculated $\bar{x}$, the mean number of male children per family, and divided that figure by 5. These computations are shown in Columns (1), (2), and (3) and at the bottom of Table 8-7. As shown, the sample proportion of male children was $\bar{p} = 0.522$ for the sample of 320 families. Therefore, the investigator stated the null hypothesis to be tested as follows:

H_0: The population probability distribution is binomial with $p = 0.522$

Since 0.05 was the conventional level of significance ordinarily used by other researchers in her field, Ms. Mauro decided to use that level of significance in testing the hypothesis. Letting

$$p = \text{the probability of a male birth} = 0.522$$
$$q = \text{the probability of a female birth} = 0.478$$

and

$$n = \text{number of children per family} = 5$$

she computed the binomial probability distribution given in Columns (1) and (2) of Table 8-8. Multiplying the probabilities in Column (2) by 320 families, she determined the expected number of families with zero, one, two, three, four, and five male children each. These theoretical frequencies are shown in Column (3) of Table 8-8. She then proceeded with the χ^2 test of goodness of fit using the observed frequencies in Column (2) of Table 8-6 as the f_0 values and the expected frequencies in Column (3) of Table 8-8 as the f_t. Table 8-9 shows the calculation of the χ^2 value. As indicated at the bottom of Table 8-9, the computed χ^2 value was 8.99.

The number of degrees of freedom is $\nu = k - 1 - m = 6 - 1 - 1 = 4$. The number of classes is $k = 6$, and $m = 1$ because one sample estimate $(\bar{p})$ was used to replace a population parameter (p). For $\nu = 4$, the critical χ^2 value is $\chi^2_{0.05} = 9.488$. Since the computed χ^2 value was only 8.99, the null hypothesis that the population probability distribution is binomial with $p = 0.522$ was accepted. Hence, Ms. Mauro concluded that the evidence represented by the observed frequency distribution on the number of male children in the sample families was consistent with the hypothesis of the operation of a Bernoulli process with $p = 0.522$ for births.

As a further test of her belief that large families have more male than female children, the investigator carried out a one-tailed test of the null hypothesis H_0: $p = 0.50$ males against the alternative hypothesis H_1: $p > 0.50$ males at the 0.05 level of significance. The sample statistic $p = 0.522$

TABLE 8-9
Calculation of the χ^2 Statistic for the Number of Male Children Problem: Fit of
Binomial Distribution with $p = 0.522$ and $n = 5$

(1) Number of Male Children	(2) Observed Number of Families f_0	(3) Theoretical Number of Families f_t	(4) $f_0 - f_t$	(5) $(f_0 - f_t)^2$	(6) $(f_0 - f_t)^2/f_t$
0	12	8.00	4.00	16.00	2.00
1	42	43.52	−1.52	2.31	0.05
2	92	95.36	−3.36	11.29	0.12
3	108	103.68	4.32	18.66	0.18
4	46	56.96	−10.96	120.12	2.11
5	20	12.48	7.52	56.55	4.53
Total	320	320.00	0		8.99

$$\chi^2 = \sum \frac{(f_0 - f_t)^2}{f_t} = 8.99$$

males for a sample of 1600 births (five children per family times 320 families) differed significantly from $p = 0.50$. Hence, she was pleased to observe this additional evidence in favor of her belief that male births tended to occur more frequently than female births in large families.

In the preceding example, the investigator tested a hypothesis using the χ^2 goodness of fit procedure, and she tested another related hypothesis using a standard one-tailed test for a proportion. Her conclusions from these two tests were consistent. However, this example can be used to illustrate the tentative nature of conclusions drawn from hypothesis-testing procedures. It is conceivable that a different sample when used in the χ^2 test could have led to the acceptance of an hypothesis involving $p = 0.522$, but on the other hand, when used in the hypothesis test of a proportion, could also have led to the acceptance of the hypothesis that $p = 0.50$, even at the same significance level.

There is a subtle point involved in the example. Under classical hypothesis-testing procedures, the hypothesis should be set up before the data are gathered. However, note that in the null hypothesis of the χ^2 test (H_0: the population probability distribution is binomial with $p = 0.522$), the hypothesized value of p was derived from the sample statistic, $\bar{p}$. In actual practice, many hypotheses are tested after sample evidence is obtained, sometimes by persons who had control over collection and tabulation of the data, and sometimes by others.

Classical statistics does not provide separate techniques for testing hypotheses before and after sample evidence has been collected. However, Bayesian decision theory, discussed in Chapters 13 through 17, provides different techniques for decision making prior to obtaining sample data and for the incorporation of sample information with prior knowledge.

Exercises

1. In an investigation of trust departments of commercial banks in the United States, a simple random sample of portfolios containing investments in six different stocks were examined to determine how many of the stocks declined in price during the past year. The results were as follows:

Number of Stocks Declining	Number of Portfolios
0	25
1	205
2	483
3	583
4	500
5	168
6	36
Total	2000

Determine whether the binomial distribution is a good fit to these data. Assume that the probability of a particular stock declining in price during the past year is 0.50.

2. In a particular region of the United States the number of small businesses that failed per day over a 100-day period during the last recession was as follows:

Number of Failures per Day: 0   1   2   3   4   5   6   7   8
Number of Days on Which Specified Number of Failures Occurred:
             3   5   27   34   23   4   2   1   1

Fit a Poisson distribution to these data, and use a χ^2 test to determine the goodness of fit. Use a 0.01 significance level.

3. A consumer research firm wished to determine whether drinkers of iced tea in a certain metropolitan area had a real difference in taste preference among five brands of iced tea. The firm took a simple random sample of 100 iced tea drinkers in the area and conducted the following experiment: Five glasses of iced tea marked A, B, C, D, and E, each containing one of the brands, were given to each consumer. The glasses were presented to each consumer in a random order determined by sequential selection from five paper slips, each containing one of the letters A, B, C, D, and E. In the table are shown the numbers of iced tea customers who stated that they most preferred the indicated brands.

Brand Preferred	Number of Drinkers
A	27
B	16
C	22
D	18
E	17

Use a χ^2 test to determine whether the null hypothesis

H_0: the probability distribution is uniform

should be rejected. Do the test using both a 0.05 and a 0.01 level of significance.

4. A Hartford-based life insurance company was considering a change in the premium charges on its policies. One of its actuaries, who was reformulating life expectancy tables, gathered some data from the records of 20 Boston-area hospitals. The number of deaths resulting from childbirth from 1970 to 1975, tabulated by number of hospital-years, was as follows:

Number of Deaths:	0	1	2	3	4	5	6	7	8	9
Number of Hospital-Years:	12	22	30	15	7	6	4	2	1	1

Use a χ^2 test to determine whether the Poisson distribution is a good fit.

5. A leading New York commercial bank employs a large number of MBA graduates but limits its recruiting and hiring to graduates of four top business schools, which we call A, B, C, and D. A simple random sample of 24 MBA holders drawn from the bank's employee population reveals that four have degrees from school A, nine from school B, eight from school C, and three from school D. Test the hypothesis that the bank's employees are equally distributed among the four schools.

8.2 TESTS OF INDEPENDENCE

Another important application of the χ^2 distribution is in testing for the independence of two variables on the basis of sample data. The general nature of the test is best explained with a specific example.

In connection with an investigation of the socioeconomic characteristics of the families in a certain city, a market research firm wished to determine whether the number of telephones owned was independent of the number of automobiles owned. The firm obtained this ownership information from a simple random sample of 10,000 families who lived in the city. The results are shown in Table 8-10. This type of table, which has one basis of classification across the columns (in this case, number of automobiles owned) and another across the rows (in this case,

TABLE 8-10
A Simple Random Sample of 10,000 Families Classified by Number of Automobiles and Telephones Owned

Number of Telephones	Number of Automobiles Owned (A_1) Zero	(A_2) One	(A_3) Two	Total
(B_1) Zero	1000	900	100	2000
(B_2) One	1500	2600	500	4600
(B_3) Two or more	500	2500	400	3400
Total	3000	6000	1000	10,000

number of telephones owned), is known as a *contingency table*. If the table has three rows and three columns, as Table 8-10 has, it is referred to as a *three-by-three* (often written 3 × 3) *contingency table*. In general, in an $r \times c$ contingency table, where r denotes the number of rows and c the number of columns, there are $r \times c$ cells. For example, in the 3 × 3 table under discussion, there are 3 × 3 = 9 cells with observed frequencies. In a 3 × 2 table, there are 3 × 2 = 6 cells, and so on. The χ^2 test consists in calculating expected frequencies under the hypothesis of independence and comparing the observed and expected frequencies.

The competing hypotheses under test in this problem may be stated as follows:

H_0: The number of automobiles owned is independent of the number of telephones owned
H_1: The number of automobiles owned is not independent of the number of telephones owned

Calculation of Theoretical (Expected) Frequencies

Since we are interested in determining whether the hypothesis of independence is tenable, we calculate the theoretical, or expected, frequencies by assuming that the null hypothesis is true. We observe from the marginal totals in the last column of Table 8-10 that 2000/10,000, or 20%, of the families do not own telephones. If the null hypothesis H_0 is true, that is, if ownership of automobiles is independent of ownership of telephones, then 20% of the 3000 families owning no automobiles, 20% of the 6000 families owning one automobile, and 20% of the 1000 families owning two automobiles would be expected to have no telephones.

Thus, the expected number of "no car" families who do not own telephones is

$$\frac{2000}{10,000} \times 3000 = 600$$

This is the *expected* frequency that corresponds to 1000, the *observed* number of "no car" families who do not own telephones.

Similarly, the expected number of "one car" families who do not own telephones is

$$\frac{2000}{10,000} \times 6000 = 1200$$

This figure corresponds to the 900 shown in the first row.

In general, the theoretical or expected frequency for a cell in the ith row and jth column is calculated as follows:

(8.3)
$$(f_t)_{ij} = \frac{(\Sigma \text{ row } i)(\Sigma \text{ column } j)}{\text{grand total}}$$

where $(f_t)_{ij}$ = the theoretical (expected) frequency for a cell in the ith row and jth column
Σ row i = the total of the frequencies in the ith row
Σ column j = the total of the frequencies in the jth column
grand total = the total of all of the frequencies in the table

For example, the theoretical frequency in the first row and first column of Table 8-11, whose rationale of calculation was just explained, is computed by Equation 8.3 as

$$(f_t)_{11} = \frac{(2000)(3000)}{10,000} = 600$$

In order to keep the notation uncluttered, we will drop the subscripts denoting rows and columns for f_t values in the subsequent discussion.

The expected frequencies for the present problem are shown in Table 8-11. Because of the method of calculating the expected frequencies, the totals in the margins of the table are the same as the totals

TABLE 8-11
Expected Frequencies for the Problem on the Relationship between Telephone and Automobile Ownership

Number of Telephones	Number of Automobiles Owned			
	Zero	One	Two	Total
Zero	600	1200	200	2000
One	1380	2760	460	4600
Two or more	1020	2040	340	3400
Total	3000	6000	1000	10,000

in the margins of the table of observed frequencies (Table 8-10). Note that the method of computing the expected frequencies under the null hypothesis of independence is simply an application of the multiplication rule for independent events given in Equation 1.8. For example, in Table 8-10, the "zero car" and "zero telephone" categories have been denoted A_1 and B_1, respectively. Under independence, $P(A_1 \text{ and } B_1) = P(A_1)P(B_1)$. The marginal probabilities $P(A_1)$ and $P(B_1)$ are given by

$$P(A_1) = \frac{3000}{10,000} = 0.30$$

$$P(B_1) = \frac{2000}{10,000} = 0.20$$

$$P(A_1 \text{ and } B_1) = P(A_1)P(B_1) = (0.30)(0.20) = 0.06$$

Multiplying this joint probability by the total frequency (10,000), we obtain the expected frequency previously derived for the upper left cell.

$$0.06 \times 10,000 = 600$$

The χ^2 Test

How great a departure from the theoretical frequencies under the assumption of independence can be tolerated before we reject the hypothesis of independence? The purpose of the χ^2 test is to provide an answer to this question by comparing observed frequencies with the theoretical, or expected, frequencies derived under the hypothesis of independence. The test statistic used to make this comparison is the chisquare statistic, $\chi^2 = \Sigma (f_0 - f_t)^2/f_t$, as defined in Equation 8.1, where f_0 is an observed frequency and f_t is a theoretical frequency.

Number of Degrees of Freedom

The number of degrees of freedom in the contingency table must be determined in order to apply the χ^2 test. The number of degrees of freedom in a 3×3 contingency table is 4. This can be rationalized intuitively as follows. In determining the expected frequencies, we used the marginal row and column totals. With three rows, only two row totals are "free," since the row totals must sum to Σf_0, which is 10,000 in the present illustration. Correspondingly, with three columns, by the same reasoning, two column totals are "free." This gives us the freedom to specify four cell totals, where the "free" columns and "free" rows intersect. *In general, in a contingency table containing r rows and c columns, there are (r − 1)(c − 1) degrees of freedom.* Thus, in a 2×2 table, ν is $(2 − 1)(2 − 1) = 1$; in a 3×2 table, ν is $(3 − 1)(2 − 1) = 2$; in a 3×3 table, ν is $(3 − 1)(3 − 1) = 4$.

TABLE 8-12
Calculation of the χ^2 Statistic for the Telephone and Automobile Ownership Problem

Observed Number of Families f_0	Expected Number of Families f_t	$f_0 - f_t$	$(f_0 - f_t)^2$	$(f_0 - f_t)^2/f_t$
1000	600	400	160,000	266.7
1500	1380	120	14,400	10.4
500	1020	−520	270,400	265.1
900	1200	−300	90,000	75.0
2600	2760	−160	25,600	9.3
2500	2040	460	211,600	103.7
100	200	−100	10,000	50.0
500	460	40	1600	3.5
400	340	60	3600	10.6
Total 10,000	10,000	0		$\chi^2 = 794.3$

We now return to our example to perform the χ^2 test of independence. Again denoting the observed frequencies as f_0 and the expected frequencies as f_t, we have shown the calculation of the χ^2 statistic in Table 8-12. No cell designations are indicated, but of course every f_0 value is compared to the corresponding f_t figure. As shown at the bottom of the table, the computed value of χ^2 is equal to 794.3. The number of degrees of freedom is $(r - 1)(c - 1)$ or $(3 - 1)(3 - 1) = 4$. In Table A-7 of Appendix A, we find a critical value at the 0.01 level of significance of $\chi^2_{0.01} = 13.277$. This means that if the null hypothesis is true, the probability of observing a χ^2 value greater than 13.277 is 0.01. Specifically in terms of the problem, this means that if ownership of telephones was independent of ownership of automobiles, an aggregate discrepancy between the observed and theoretical frequencies larger than a χ^2 value of 13.277 would occur only 1% of the time. We can state the decision rule for this problem as follows:

DECISION RULE

1. If $\chi^2 > 13.277$, reject H_0
2. If $\chi^2 \leq 13.277$, accept H_0

Since the computed χ^2 value of 794.3 so greatly exceeds this critical value, the null hypothesis of independence between telephone and automobile ownership is rejected.

Further Comments

We have seen how the χ^2 test for independence in contingency tables is a means of determining whether a relationship exists between two bases of classification, or, in other words, whether a relationship exists between two variables. Although this type of tabulation provides a basis for testing whether there is a dependence between the two classificatory variables, it does not yield a method for estimating the values of one variable from known values or assumed values of the other. In the next chapter, which deals with *regression* and *correlation analysis*, methods for providing such estimates are discussed. For example, regression analysis provides a method for obtaining estimates or predictions of the number of telephones owned by a family with a specific number of automobiles. Regression analysis, in particular, provides a very powerful tool for stating in explicit mathematical form the nature of the relationship that exists between two or more variables.

However, at least some indication may be obtained of the nature of the relationship between the two variables in a contingency table. Equivalently to the null hypothesis of independence rejected in our example, we have rejected the null hypothesis $H_0: p_1 = p_2 = p_3$, where p_1, p_2, and p_3 denote the population proportions of zero-, one-, and two-car families who do not have telephones. Reference to Table 8-11 makes it obvious why the null hypothesis was rejected. Of the 3000 families who did not own automobiles, 1000/3000 = 0.33 did not own a telephone. Let $\bar{p}_1 = 0.33$. The corresponding proportions of one- and two-car families who did not own telephones were $\bar{p}_2 = 900/6000 = 0.15$ and $\bar{p}_3 = 100/1000 = 0.10$. Hence, we have concluded that it is highly unlikely that these three statistics represent samples drawn from populations that have the same proportions ($p_1 = p_2 = p_3$). Clearly, the proportion of no-telephone families declines as automobile ownership increases. The data suggest a strong relationship between the ownership of telephones and automobiles for the families studied.

A powerful generalization develops from the preceding discussion. It can be shown that a χ^2 test applied to a 2×2 contingency table is algebraically identical to the two-sample test for difference between proportions by the methods of Section 7.3 using Equation 7.10 to calculate the estimated standard error of the difference. This means that the test of the hypothesis of independence carried out in a χ^2 test for a 2×2 contingency table is identical to the testing of the hypotheses

$$H_0: p_1 = p_2$$
$$H_1: p_1 \neq p_2$$

As we have seen, in our illustrative problem involving a 3×3 contingency table, we tested the null hypothesis

$$H_0: p_1 = p_2 = p_3$$

against the alternative that the p values were not all equal. The analogous test can be applied in general to c categories, where $c \geqslant 2$.

Additional Comments

Since the sampling distribution of the χ^2 statistic, $\chi^2 = \Sigma \ (f_0 - f_t)^2/f_t$, is only an approximation to the theoretical distribution defined in Equation 8.2, the sample size must be large for a good approximation to be obtained. As in the goodness of fit tests, in contingency tables, cells with frequencies of less than 5 should be combined.

Furthermore, in 2 × 2 tables (that is, when there is 1 degree of freedom), an adjustment known as Yates' correction for continuity may be used. This correction is introduced because the theoretical χ^2 distribution is continuous, whereas the tabulated values in Table A-7 of Appendix A are based on the distribution of the discrete χ^2 statistic of Equation 8.1. The correction is applied by computing the following χ^2 statistic:

(8.4)
$$\chi^2 = \Sigma \ \frac{(|f_0 - f_t| - \frac{1}{2})^2}{f_t}$$

In this correction, $\frac{1}{2}$ is subtracted from the absolute value of the difference between f_0 and f_t before squaring. The effect is to reduce the calculated value of χ^2 as compared to the corresponding calculation by Equation 8.1 without the correction.[2] In an example such as the one just discussed, where the expected frequencies are large, the effect of this correction is clearly unimportant, but it may be of greater significance for smaller samples.

As we have seen, in both χ^2 goodness of fit tests and tests of independence, the null hypothesis is rejected when large enough values of χ^2 are observed. Some investigators have raised the question whether the null hypothesis should also be rejected when the computed value of χ^2 is too low, that is, too close to 0. This is a situation in which the observed frequencies, f_0, all appear to *agree too well* with the theoretical frequencies, f_t. The recommended course of action is to examine the data very closely to see whether errors have been made in recording them. Perhaps the data rather than the null hypothesis should be rejected. An experience of one researcher is relevant to this point. He was analyzing some data on oral temperatures and found that a disturbingly large number of the recorded temperatures were equal to the "normal" figure

---

[2] For a more complete discussion of Yates' correction, see F. Yates, "Contingency Tables Involving Small Numbers and the χ^2 Test." Suppl. *J. Royal Stat. Soc.*, 1, 1934, 217–235 and Snedecor, George W. and William G. Cochran, *Statistical Methods*, 6th ed., Iowa State University Press, 1967.

of 98.6°F. He suspected that these data were "too good to be true." Upon investigation, he found that the temperatures were recorded by relatively untrained nurse's aides. Several of them had misread temperatures by recording the number to which the arrow on the thermometer pointed, namely, 98.6°! Clearly, this was a case where the data rather than an investigator's null hypothesis should be rejected.

Exercises

1. An investigator from the Justice Department has been told by the personnel department of a large corporation that the firm does not discriminate against women or minorities in its hiring practices. Using a random sample from the firm's personnel records, the investigator compiled the following data. The figures show the number of applicants for employment during the past year who were hired and rejected in each classification shown.

	Caucasian Males	Females	Minority Males	Total
Number hired	76	23	26	125
Number rejected	164	93	118	375
Total	240	116	144	500

Do these data support the hypothesis that the proportions hired were the same for the three categories of applicants? Use a 0.05 significance level.

2. A subscription service stated that preferences for different national magazines were independent of geographical location. A survey was taken in which 300 persons randomly chosen from three areas were given a choice among three different magazines. Each person expressed his or her favorite. The following results were obtained:

Region	Magazine X	Magazine Y	Magazine Z	
New England	75	50	175	300
Northeastern	120	85	95	300
Southern	105	110	85	300
Total	300	245	355	900

Would you agree with the subscription service's assertion? Use a 0.05 significance level.

3. The admissions office of a large Canadian university wishes to know whether course grades are independent of the scores achieved on the entrance examinations required of all applicants for admission. The admissions office obtained the grade averages and entrance examination scores of a simple

random sample of the members of the classes of 1973, 1974, and 1975 and set up the following table. Using a 0.01 significance level, test the hypothesis that grade performance is independent of entrance examination scores. Scores on the entrance examinations range from 300 to 800.

		Score Ranges		
Grade Average	300–449	450–649	650–800	Total
A	8	35	22	65
B	10	83	27	120
C	7	17	11	35
Total	25	135	60	220

4. A men's cosmetic manufacturer has test-marketed a new men's deodorant. Simple random samples of men in three cities were asked to try the product for two weeks and then were questioned carefully. Results for the question "Would you purchase this product?" are shown in the following table.

Test Results	Birmingham	Detroit	Philadelphia	Total
Will not purchase	13	12	25	50
Will purchase	87	88	275	450
Total	100	100	300	500

Would you conclude that respondents' preferences are independent of location? Use a 0.05 significance level.

5. In doing research to determine whether social class has any effect on newspaper buying, the editor of *The Star* took a poll of 150 people from each of three social classes and ascertained whether they read *The Star* or its competitor, *The Press.* The results were

Social Class	The Star	The Press	Total
Lower	80	70	150
Middle	90	60	150
Upper	50	100	150
Total	220	230	450

Test the hypothesis that choice of newspaper and social class are independent. Use a 0.05 significance level.

6. The following table describes two characteristics of a simple random sample of 200 registered voters in an affluent suburban community. A lobbyist for a

large oil firm wants to determine whether one characteristic, political viewpoint, is independent of the other, opinion on what should be done with the investment tax credit (ITC). Perform the test, using a 0.05 significance level.

Opinion on Issue	Political Viewpoint				
	Radical	Liberal	Moderate	Conservative	Total
Increase ITC	5	30	20	25	80
Leave ITC unchanged	5	15	15	15	50
Decrease or eliminate ITC	10	35	15	10	70
Total	20	80	50	50	200

7. Components are supplied to a television manufacturer by two subcontractors. Each component is tested with respect to five characteristics before it is accepted by the manufacturer. Records have been kept for one month on the number of different types of defects for each contractor, and the following table has been constructed from these records.

Supplier	Type of Defect					
	A	B	C	D	E	Total
1	70	10	10	30	0	120
2	10	10	20	20	20	80
Total	80	20	30	50	20	200

Would you conclude that type of defect and supplier are independent? Use a 0.01 significance level.

8.3 ANALYSIS OF VARIANCE: TESTS FOR EQUALITY OF SEVERAL MEANS

In Section 8.2, we saw that the χ^2 *test is a generalization of the two-sample test for proportions* and enables us to test for the significance of the difference among c $(c > 2)$ sample proportions. Conceptually, this represents a test of whether the c samples can be treated as having been drawn from the same population, or in other words, from populations having the same proportions. Similarly, in this section we consider a very ingenious technique known as the *analysis of variance*, which is a *generalization of the two-sample test for means* and enables us to test for the significance of the difference among c $(c > 2)$ sample means. Analogously to the case of the χ^2 test, this represents a test of whether the c samples can be treated as having been drawn from the same population, or more precisely, from populations having the same means.

A central point is that although the analysis of variance is literally a technique that analyzes or tests variances, by doing so, it provides us with a test for the significance of the difference among *means*. The rationale by which a test of variances is in fact a test for means will be explained shortly.

As an example, suppose we wish to test whether three methods of teaching a basic statistics course differ in effectiveness. It has been agreed that student grades on a final examination covering the work of the entire course will be used as the measure of effectiveness. The three methods of teaching are

Method 1: The lecturer does not work out nor assign problems.
Method 2: The lecturer works out and assigns problems.
Method 3: The lecturer works out and assigns problems. Students are also required to construct and solve their own problems.

The same professor taught three different sections of students, using one of the three methods in each class. All the students were sophomores at the same university and were randomly assigned to the three sections. In order to explain the principles of the analysis without cumbersome computational detail, we assume that there were only 12 students in the experiment, four in each of the three different sections. Of course, in

TABLE 8-13
Final Examination Grades of 12 Students Taught
by Three Different Methods

Student	*Teaching Method* 1	2	3
1	16	19	24
2	21	20	21
3	18	21	22
4	13	20	25
Total	68	80	92
Mean	17	20	23

$$\text{Grand mean} = \frac{17 + 20 + 23}{3} = 20$$

actual practice, a substantially larger number of observations would be required to furnish convincing results. The final examination was graded on the basis of 25 as the maximum score and 0 as the minimum score. The final examination grades of the 12 students in the three sections are given in Table 8-13. As shown in the table, the mean grades for students taught by methods 1, 2, and 3 were 17, 20, and 23, respectively, and the overall average of the 12 students, referred to as the "grand mean," was 20. (Note that the grand mean of 20 is the same figure that would be obtained by adding up all 12 grades and dividing by 12.)

Notation

At this point, we introduce some useful notation. In Table 8-13, there are four rows and three columns. As in the discussion of χ^2 tests for contingency tests, let r represent the number of rows and c the number of columns. Hence, there is a total of $r \times c$ observations in the table, in this case $4 \times 3 = 12$. Let X_{ij} be the score of the ith student taught by the jth method, where $i = 1, 2, 3, 4$ and $j = 1, 2, 3$. (Thus, for example, X_{12} denotes the score of student 1 taught by method 2 and is equal to 19; $X_{23} = 21$, and so on). In this problem, the different methods of instruction are indicated in the columns of the table, and interest centers on the differences among the scores in the three columns. This is typical of the so-called *one factor* (or *one-way*) *analysis of variance*, in which an attempt is made to assess the effect of only one factor (in this case, instructional method) on the observations. In the present problem, there are three columns. Hence, we denote the values in the columns as X_{i1}, X_{i2}, and X_{i3}, and the totals of these columns as $\sum_i X_{i1}, \sum_i X_{i2}$, and $\sum_i X_{i3}$. The

TABLE 8-14
Notation Corresponding to the Data of Table 8-13

i	X_{i1}	X_{i2}	X_{i3}
1	X_{11}	X_{12}	X_{12}
2	X_{21}	X_{22}	X_{23}
3	X_{31}	X_{32}	X_{33}
4	X_{41}	X_{42}	X_{43}
Total	$\sum_i X_{i1}$	$\sum_i X_{i2}$	$\sum_i X_{i3}$
Mean	$\bar{X}_1$	$\bar{X}_2$	$\bar{X}_3$

$$\bar{\bar{X}} = \frac{\bar{X}_1 + \bar{X}_2 + \bar{X}_3}{3}$$

subscript i under the summation signs indicates that the total of each of the columns is obtained by summing the entries over the rows. Adopting a simplified notation, we will refer to the means of the three columns as $\bar{X}_1, \bar{X}_2$, and $\bar{X}_3$, or in general, $\bar{X}_j$. Finally, we denote the grand mean as $\bar{\bar{X}}$ (pronounced "X double-bar"), where $\bar{\bar{X}}$ is the mean of all $r \times c$ observations. Since each column in our example contains the same number of observations, $\bar{\bar{X}}$ can be obtained by taking the mean of the three sample means $\bar{X}_1, \bar{X}_2$, and $\bar{X}_3$. This notation is summarized in Table 8-14. It is suggested that you study Table 8-14 carefully and compare the notation with the corresponding entries in Table 8-13.

The Hypothesis to Be Tested

As indicated earlier, we want to test whether the three methods of teaching a basic statistics course differ in effectiveness. We have calculated the following mean final examination scores of students taught by the three methods $\bar{X}_1 = 17, \bar{X}_2, = 20$, and $\bar{X}_3 = 23$. The statistical question is, Can the three samples represented by these three means be considered as having been drawn from populations having the same mean? Denoting the population means corresponding to $\bar{X}_1, \bar{X}_2$, and $\bar{X}_3$ as μ_1, μ_2, and μ_3, we can state the null hypothesis as

$$H_0: \mu_1 = \mu_2 = \mu_3$$

This hypothesis is to be tested against the alternative,

$$H_1: \text{The means } \mu_1, \mu_2, \text{ and } \mu_3 \text{ are not all equal}$$

Hence, what we wish to determine is whether the differences among the sample means $\bar{X}_1, \bar{X}_2$, and $\bar{X}_3$ are too great to be attributed to the chance errors of drawing samples from populations having the same means. If we do decide that the sample means differ significantly, our substantive conclusion is that the teaching methods differ in effectiveness.

Although we will specify the assumptions underlying the test procedure at the end of the problem, we indicate one of them at this point, namely, the assumption that the variances of the three populations are all equal. Therefore, rewording the hypothesis slightly, we can say that we wish to test whether our three samples were drawn from populations having the same means and variances.

Decomposition of Total Variation

Before discussing the procedures involved in the analysis of variance, we consider the general rationale underlying the test. If the null hypothesis that the three population means (μ_1, μ_2, and μ_3) are equal is true, then both the variation among the sample means ($\bar{X}_1, \bar{X}_2$, and $\bar{X}_3$) and the variation within the three groups reflect chance errors of the sampling process. The first of these types of variation is conventionally referred to as *variation between the c means, between-group variation,* or *between-column variation* (despite the English barbarism involved in using the word "between" rather than "among" when there are more than 2 groups present). The second type is referred to as *within-group variation* or simply the *between-row variation.* We will use the latter term in this chapter. Between-column variation is variation of the sample means, $\bar{X}_1$, $\bar{X}_2$, and $\bar{X}_3$, around the grand mean, $\bar{\bar{X}}$. On the other hand, between-row variation is variation of the individual observations within each column from their respective means, $\bar{X}_1$, $\bar{X}_2$, and $\bar{X}_3$.

Under the null hypothesis that the population means are equal, the between-column variation and the between-row variation would be expected not to differ significantly from one another, since they both reflect the same type of chance sampling errors. On the other hand, if the null hypothesis is false and the population column means are indeed different, then the between-column variation should significantly exceed the between-row variation. This follows from the fact that the between-column variation would now be produced by the inherent differences among the column means as well as by chance sampling error. On the other hand, the between-row variation would still reflect chance sampling errors only. *Hence, a comparison of between-column variation and between-row variation yields information concerning differences among the column means.* This is the central insight provided by the analysis of variance technique.

The term "variation" is used in statistics in a very specific way to refer to a sum of squared deviations and is often referred to simply as a *sum of squares.* When a measure of variation is divided by an appropriate number of degrees of freedom, as we have seen earlier in this text, it is referred to as a *variance.* In the analysis of variance, such a variance is referred to as a *mean square.* For example, the variation of a set of sample observations, denoted X, around their mean $\bar{X}$ is $\Sigma\,(X - \bar{X})^2$. Dividing this sum of squares by the number of degrees of freedom $n - 1$ (where n is the number of observations), we obtain $\Sigma\,(X - \bar{X})^2/(n - 1)$, the sample variance, which as indicated in Section 1.14 is an unbiased estimator of the variance of an infinite population. This sample variance can also be referred to as a mean square.

We now proceed with the analysis of variance by calculating the between-column variation and between-row variation for our problem.

Between-Column Variation

As indicated earlier, the between-column variation, or between-column sum of squares, measures the variation among the sample column means. It is calculated as follows:

(8.5) $\qquad \begin{array}{c}\text{Between-column}\\[-2pt]\text{sum of squares}\end{array} = \sum_{j} r(\bar{X}_j - \bar{\bar{X}})^2$

where r = number of rows (sample size involved in the calculation of each column mean)[3]

$\bar{X}_j$ = the mean of the jth column

$\bar{\bar{X}}$ = the grand mean

$\displaystyle\sum_j$ = summation is taken over all columns

As indicated in Equation 8.5, the between-column sum of squares is calculated by the following steps:

1. Compute the deviation of each column mean from the grand mean.
2. Square the deviations obtained in Step 1.
3. Weight each deviation by the sample size involved in calculating the respective mean. In the illustrative example, all sample sizes are the same and are equal to the number of rows, $r = 4$.
4. Sum over all columns the products obtained in Step 3.

The calculation of the between-column sum of squares for the numerical example involving three different teaching methods is given in Table 8-15. As indicated in the table, the between-column variation is 72.

TABLE 8-15
Calculation of the Between-Column Sum of Squares for the Teaching Methods Problem

$$(\bar{X}_1 - \bar{\bar{X}})^2 = (17 - 20)^2 = 9$$
$$(\bar{X}_2 - \bar{\bar{X}})^2 = (20 - 20)^2 = 0$$
$$(\bar{X}_3 - \bar{\bar{X}})^2 = (23 - 20)^2 = 9$$
$$\sum_j r(\bar{X}_j - \bar{\bar{X}})^2 = 4(9) + 4(0) + 4(9) = 72$$

Between-Row Variation

The between-row sum of squares is a summary measure of the random errors of the individual observations around their column means. The formula for its computation is

$$(8.6) \qquad \begin{array}{c} \text{Between-row} \\ \text{sum of squares} \end{array} = \sum_j \sum_i (X_{ij} - \bar{X}_j)^2$$

where X_{ij} = the value of the observation in the ith row and jth column

$\bar{X}_j$ = the mean of the jth column

---

[3] In this example, equal sample sizes (equal numbers of rows) are assumed. This treatment can be generalized to allow for different sample sizes.

$\sum_j \sum_i$ means that the squared deviations are first summed over all sample observations within a given column, then summed over all columns

As indicated in Equation 8.6, the between-row sum of squares is calculated as follows:

1. Calculate the deviation of each observation from its column mean.
2. Square the deviations obtained in Step 1.
3. Add the squared deviations within each column.
4. Sum over all columns the figures obtained in Step 3.

The computation of the between-row variation for the teaching methods problem is given in Table 8-16.

TABLE 8-16
Calculation of the Between-Row Sum of Squares for the Teaching Methods Problem

i	$(X_{i1} - \bar{X}_1)$	$(X_{i1} - \bar{X}_1)^2$	$(X_{i2} - \bar{X}_2)$	$(X_{i2} - \bar{X}_2)^2$	$(X_{i3} - \bar{X}_3)$	$(X_{i3} - \bar{X}_3)^2$
1	$(16 - 17) = -1$	1	$(19 - 20) = -1$	1	$(24 - 23) = 1$	1
2	$(21 - 17) = 4$	16	$(20 - 20) = 0$	0	$(21 - 23) = -2$	4
3	$(18 - 17) = 1$	1	$(21 - 20) = 1$	1	$(22 - 23) = -1$	1
4	$(13 - 17) = -4$	16	$(20 - 20) = 0$	0	$(25 - 23) = 2$	4
		$\overline{34}$		$\overline{2}$		$\overline{10}$

$$\sum_j \sum_i (X_{ij} - \bar{X}_j)^2 = 34 + 2 + 10 = 46$$

Total Variation

The between-column variation and between-row variation represent the two components of the total variation in the overall set of experimental data. The total variation, or total sum of squares, is calculated by adding the squared deviations of all of the individual observations from the grand mean $\bar{\bar{X}}$. Hence, the formula for the total sum of squares is

(8.7) $\qquad$ Total sum of squares $= \sum_j \sum_i (X_{ij} - \bar{\bar{X}})^2$

The total sum of squares is computed by the following steps:

1. Calculate the deviation of each observation from the grand mean.
2. Square the deviations obtained in Step 1.
3. Add the squared deviations over all rows and columns.

The total sum of squares, or total variation of the 12 observations in the teaching methods problem, is $(16 - 20)^2 + (21 - 20)^2 + \cdots + (25 - 20)^2 = 118$. Referring to the results obtained in Tables 8-15 and 8-16, we see that the total sum of squares, 118; is equal to the sum of the between-column sum of squares, 72, and the between-row sum of squares, 46. In general, the following relationship holds:

(8.8)   Total variation = Between-column variation
                         + Between-row variation

Although, as we have indicated earlier, the test of the null hypothesis in a one-factor analysis of variance involves only the between-column variation and the between-row variation, it is useful to calculate the total variation as well. This computation is helpful as a check procedure and is instructive in indicating the relationship between total variation and its components.

Shortcut Computational Formulas

The formulas we have given for calculating the between-column sum of squares (8.5), the between-row sum of squares (8.6), and the total sum of squares (8.7) are the clearest ones for revealing the rationale of the analysis of variance procedure. However, the following shortcut computational formulas are often used to calculate these sums of squares.

(8.9) $\begin{array}{c} \text{Between-column} \\ \text{sum of squares} \end{array} = \dfrac{\sum\limits_{j} T_j^2}{r} - C$

(8.10) $\begin{array}{c} \text{Between-row} \\ \text{sum of squares} \end{array} = \sum\limits_{j} \sum\limits_{i} X_{ij}^2 - \sum\limits_{j} \dfrac{T_j^2}{r}$

(8.11) $\begin{array}{c} \text{Total sum of} \\ \text{squares} \end{array} = \sum\limits_{j} \sum\limits_{i} X_{ij}^2 - C$

where C, the so-called *correction term*, is given by

(8.12) $$C = \dfrac{T^2}{rc}$$

and where T_j is the total of the r observations in the jth column and T is the grand total of all rc observations, that is,

(8.13) $$T = \sum\limits_{j} \sum\limits_{i} X_{ij}$$

All other terms are as previously defined.

These formulas are especially useful when the column means and grand mean are not integers. The shortcut formulas not only save time and computational labor, but also are more accurate because of avoidance of rounding problems, which usually occur with the use of Equations 8.5, 8.6, and 8.7.

The shortcut computations for the present example are as follows:

$$C = \frac{(240)^2}{(3)(4)} = 4800$$

$$\text{Between-column sum of squares} = \frac{(68)^2 + (80)^2 + (92)^2}{4} - 4800 = 72$$

$$\text{Between-row sum of squares} = (16)^2 + (21)^2 + \cdots + (25)^2$$

$$- \frac{(68)^2 + (80)^2 + (92)^2}{4}$$

$$= 46$$

$$\text{Total sum of squares} = (16)^2 + (21)^2 + \cdots + (25)^2 - 4800 = 118$$

It is recommended that the shortcut formulas be used, particularly when carrying out computations by hand.

Number of Degrees of Freedom

Although the preceding discussion was in terms of *variation* or *sums of squares* rather than *variance*, the actual test of the null hypothesis in the analysis of variance involves a comparison of the *between-column variance* with the *between-row variance*, or in equivalent terminology, a comparison of the *between-column mean square* with the *between-row mean square*. Hence, the next step in our procedure is to determine the number of degrees of freedom associated with each of the measures of variation. As stated earlier in this section, if a measure of variation, that is, a sum of squares, is divided by the appropriate number of degrees of freedom, the resulting measure is a variance, that is, a mean square.

The number of degrees of freedom associated with the between-column sum of squares is $c - 1$. We can see the reason for this by applying the same general principles indicated earlier for determining number of degrees of freedom in t tests and χ^2 tests. Since there are c columns, or c group means, there are c sums of squares involved in measuring the variation of these column means around the grand mean. Because the sample grand mean is only an estimate of the unknown population mean, we lose one degree of freedom. Hence, there are $c - 1$ degrees of freedom present.

The number of degrees of freedom in our example, which has three different teaching methods (that is, three columns), is $c - 1 = 3 - 1 = 2$.

The number of degrees of freedom associated with the between-row variation is $rc - c = c(r - 1)$. This may be reasoned as follows. There are a total of rc observations. In determining the between-row variation, the squared deviations within each column were taken around the column mean. There are c column means, each of which is an estimate of the true unknown population column mean. Hence, there is a loss of c degrees of freedom, and c must be subtracted from rc, the total number of observations.

Alternatively, there are r squared deviations in each column taken around the column mean and a total sum of squares for the column. $r - 1$ of the sums of squares can be assigned arbitrarily, and the last becomes fixed in order for the sum to equal the column sum. Since there are c columns, we have $c(r - 1)$ degrees of freedom.

In the present problem, the number of degrees of freedom associated with the between-row sum of squares is $c(r - 1) = 3(4 - 1) = 9$.

The number of degrees of freedom associated with the total variation is $rc - 1$. There are rc squared deviations taken from the sample grand mean, $\bar{\bar{X}}$. Since $\bar{\bar{X}}$ is an estimate of the true but unknown population mean, there is a loss of one degree of freedom. Alternatively, in the determination of the total sum of squares, there are rc squared deviations. $rc - 1$ of which may be arbitrarily assigned, but the last one is then constrained in order for the sum to be equal to the total sum of squares.

In the example, the number of degrees of freedom associated with the total variation is $rc - 1 = (4)(3) - 1 = 11$.

Just as the between-column and the between-row variation sum to the total variation, the numbers of degrees of freedom associated with the between-column and between-row variations add to the number associated with the total variation. In symbols,

$$(8.14) \qquad rc - 1 = (c - 1) + (rc - c)$$

In the teaching methods problem, the numerical values corresponding to Equation 8.14 are $11 = 2 + 9$.

The Analysis of Variance Table

An analysis of variance table for the teaching methods problem is given in Table 8-17. The calculations at the bottom of the table will be described presently. The table is in the standard form ordinarily employed to summarize the results of an analysis of variance. In Columns (1), (2), and (3) are listed the possible sources of variation, the sum of squares for each of these sources, and the number of degrees of freedom associated with each of the sums of squares. We again note that both sums of squares and numbers of degrees of freedom are additive;

TABLE 8-17
Analysis of Variance Table for the Teaching Methods Problem

(1) Source of Variation	(2) Sum of Squares	(3) Degrees of Freedom	(4) Mean Square
Between columns	72	2	36
Between rows	46	9	5.11
Total	118	11	

$$F(2, 9) = \frac{36}{5.11} = 7.05$$

$$F_{0.05}(2, 9) = 4.26$$

Since $7.05 > 4.26$, reject H_0

that is, these figures for between-column and between-row sources of variation add to the corresponding figure for total variation. Dividing the sums of squares in Column (2) by the numbers of degrees of freedom in Column (3) yields the between-column and between-row variances shown in Column (4). As indicated earlier, another name for a sum of squares divided by the appropriate number of degrees of freedom is a "mean square," and it is conventional to use this term in an analysis of variance table. Thus, in our problem, the between-column mean square, denoted MS_c, is equal to $72/2 = 36$. The between-row mean square, denoted MS_r, is equal to $46/9 = 5.11$. The test of the null hypothesis that the population column means are equal is carried out by a comparison of MS_c to MS_r.

TABLE 8-18
General Format of a One-Factor Analysis of Variance Table

(1) Source of Variation	(2) Sum of Squares	(3) Degrees of Freedom	(4) Mean Square
Between columns	SS_c	$\nu_1 = c - 1$	$MS_c = SS_c/(c - 1)$
Between rows	SS_r	$\nu_2 = c(r - 1)$	$MS_r = SS_r/c(r - 1)$
Total	SS_t	$rc - 1$	

$$F(\nu_1, \nu_2) = \frac{MS_c}{MS_r}$$

Table 8-18 gives the general format of a one-factor analysis of variance table. The sums of squares are denoted as follows:

$$SS_c = \text{between-column sum of squares}$$
$$SS_r = \text{between-row sum of squares}$$
$$SS_t = \text{total sum of squares}$$

Other notation is as previously defined or as given in the next subsection.

The *F* Test and *F* Distribution

The comparison of the between-column mean square to the between-row mean square is made by computing their ratio, referred to as F. Hence, F is given by

(8.15) $$F = \frac{MS_c}{MS_r}$$

In the F ratio, the between-column variance is always placed in the numerator and the between-row variance in the denominator. Under the null hypothesis that the population column means are equal, the F ratio would tend to be equal to 1. On the other hand, if the population column means do indeed differ, then the between-column mean square, MS_c, will tend to exceed the between-row mean square, MS_r, and the F ratio will then be greater than 1. In terms of our illustrative problem concerning different teaching methods, if F is large, we will reject the null hypothesis that the population mean examination scores are all equal; that is, we will reject $H_0: \mu_1 = \mu_2 = \mu_3$. On the other hand, if F is close to 1, we will accept the null hypothesis. The answer to how large the test statistic F must be in order to reject the null hypothesis is given by reference to the probability distribution of the F random variable. This distribution is complex, and its mathematical expression will be given here for reference only. Fortunately, critical values of the F ratio have been tabulated for frequently used significance levels analogous to the case of the χ^2 distribution. The probability density function of F is

(8.16) $$f(F) = cF^{(\nu_1/2)-1}\left(1 + \frac{\nu_1 F}{\nu_2}\right)^{-(\nu_1 + \nu_2)/2}$$

where $\nu_1 =$ the number of degrees of freedom of the numerator of F
$\nu_2 =$ the number of degrees of freedom of the denominator of F
$c =$ a constant depending only on ν_1 and ν_2.

The underlying assumptions are that two random samples are drawn from normally distributed populations with equal variances σ_1^2 and σ_2^2.[4]

---

[4] Research in recent years on "robustness" has shown that minor departures from these assumptions, particularly that of normality, do not materially affect the results of the analysis.

The term *homoscedasticity* is used in this and other statistical tests for the assumption of equal variances. Unbiased estimators $\hat{\sigma}_1^2$ and $\hat{\sigma}_2^2$ of the population variances are constructed from the sample, and

$$F = \frac{\hat{\sigma}_1^2}{\hat{\sigma}_2^2}$$

The F distribution is similar to the distributions of t and χ^2 in that it is actually a family of distributions. Each pair of values of ν_1 and ν_2 specifies a different distribution. F is a continuous random variable that ranges from 0 to infinity. Since the variances in both the numerator and denominator of the F ratio are squared quantities, F cannot take on negative values. The F distribution has a single mode, and although the specific distribution depends on the value of ν_1 and ν_2, its shape is generally asymmetrical and skewed to the right. The distribution tends towards symmetry as ν_1 and ν_2 increase. We will use the notation $F(\nu_1, \nu_2)$ to denote the F ratio defined in Equation 8.15, that is, $F = MS_c/MS_r$, where the numerator and denominator are between-column mean squares and between-row mean squares with ν_1 and ν_2 degrees of freedom, respectively. Table A-8 of Appendix A presents the critical values of the F distribution for two selected significance levels, $\alpha = 0.05$ and $\alpha = 0.01$. In this table, ν_1 values are listed across the columns and ν_2 down the rows. There are two entries in the table corresponding to every pair of ν_1 and ν_2 values. The upper figure in light-face type is an F value that corresponds to an area of 0.05 in the right tail of the F distribution with ν_1 and ν_2 degrees of freedom. That is, it is an F value that would be exceeded only five times in 100 if the null hypothesis under consideration were true. The lower figure in boldface type is an F value corresponding to a 0.01 area in the right tail.

We will illustrate the use of the F table in terms of the teaching methods problem. Assuming that we wish to test the null hypothesis $H_0: \mu_1 = \mu_2 = \mu_3$ at the 0.05 level of significance, we find in Table A-8 of Appendix A that for $\nu_1 = 2$ and $\nu_2 = 9$ degrees of freedom, an F value of 4.26 would be exceeded 5% of the time, if the null hypothesis were true. As indicated at the bottom of Table 8-17, we denote this critical value as $F_{0.05}(2, 9) = 4.26$. This relationship is depicted in Figure 8-3. Again referring to Table 8-17, since the computed value of the F ratio (the between-column mean square over the between-row mean square) is 7.05, and therefore greater than the critical value of 4.26, we reject the null hypothesis. Hence, we conclude that the column means (that is, the sample means of final examination scores in classes taught by the three teaching methods) differ significantly. The inference about the corresponding population means is that they are not all the same. Referring to Table 8-13, we see that average grades under Method 3 exceed those under Method 2, which are higher than those under Method 1. Hence, based on these data, our inference is that the three teaching methods are not equally effective, and there is evidence that Method 3 is

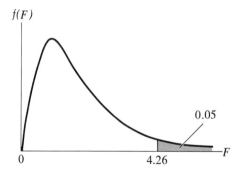

FIGURE 8-3
The F distribution for the teaching methods problem indicating the critical F value at the 5% level of significance.

the most effective and Method 1 the least effective. If we are interested in specific comparisons (as for example, between Method 1 and Method 2), we can use the usual two-sample hypothesis-testing procedures for means; that is, we could test the null hypothesis H_0: $\mu_1 - \mu_2 = 0$ by the procedures discussed in Sections 7.3 and 7.4.

The foregoing example was used to illustrate the rationale involved in the analysis of variance, the statistical technique employed, and the nature of the conclusions that can be drawn. However, we repeat the caveat that the sample sizes in this illustration are doubtless too small for safe conclusions to be drawn about differences in effectiveness of different teaching methods. After all, only four observations were made under each teaching method. Although the risk of a Type I error was controlled at .05 in this problem, the risk of Type II errors may be intolerably high. Suffice it to say that larger sample sizes are generally required. In the next subsection, we discuss some of the more complex experimental designs.

Other Experimental Designs

In the teaching methods example in this chapter, the samples were all of the same size; the same number of students (four) were taught by each method. Although it is often simplest in the collection of data to work with samples of the same size, this is not always feasible. The analysis of variance computational procedure given earlier is easily adjusted to allow for differing sample sizes. Let r_j denote the number of observations in the jth column. For example, if five students were taught by Method 1, then $r_1 = 5$. The general format for a one-factor analysis of variance for unequal sample sizes is given in Table 8-19. In this modification of Table 8-18, all symbols retain their original meanings.

We turn now to another characteristic of the one-factor illustration

TABLE 8-19
General Format of a One-Factor Analysis of Variance for Unequal Sample Sizes

(1) Source of Variation	(2) Sum of Squares	(3) Degrees of Freedom	(4) Mean Square
Between columns	SS_c	$\nu_1 = c - 1$	$MS_c = SS_c/(c-1)$
Between rows	SS_r	$\nu_2 = \sum_{j=1}^{c} (r_j - 1)$	$MS_r = SS_r/\sum_{j=1}^{c} (r_j - 1)$

$$F = \frac{MS_c}{MS_r}$$

(on teaching methods) used in this chapter; this type of example is conventionally designated as a *completely randomized experimental design*. As indicated earlier in the statement of underlying assumptions, this type of design involves the random selection of independent samples from c (the number of columns) normal populations. This implies that the students taught by each of the three methods were randomly selected from the sophomore class. We performed a one-factor analysis of variance, rejected the null hypothesis, and concluded that the three teaching methods are not equally effective.

However, let us be somewhat more critical and look at the possible interpretations of our findings more closely. We assumed that the same teacher taught three different sections of the basic statistics course by the three specified methods, 1, 2, and 3. Method 3 appeared to be most effective and Method 1 least effective. Suppose the class taught by Method 3 is given early in the morning. Hence, the teacher (and students too?) may be fresh, wide awake, and enthusiastic. On the other hand, suppose the classes taught by Methods 2 and 1 are in the middle of the day and the late afternoon, respectively. Let us assume that by late afternoon, the instructor is tired, sleepy, and rather unenthusiastic. Then it is possible that the differences in teaching effectiveness are not really attributable solely to the different teaching methods, but rather to some unknown mixture of the difference in teaching methods and the aforementioned factors associated with time of day. Time of day may be thought of as an extraneous factor that influences the variable of interest, teaching methods. A *two-factor analysis of variance* is required to isolate the influence of this extraneous factor in order that the effects of the main variable may be more accurately judged.

There are two basic types of two-factor analysis of variance. The first, known as a *randomized block design*, is appropriate for the situation just described, in which we are interested in the influence of a certain factor (for example, teaching methods) but we also wish to isolate the effects of a second extraneous factor, such as time of day. The term "block" derives from experimental design work in agriculture, in which parcels of land are referred to as blocks. In a randomized block design, treatments are randomly assigned to units within each block. This insures, for example, in testing the yield of different types of fertilizer, that the best fertilizer would not be applied only to the best soil. In the teaching methods illustration, times of day could be treated as the blocks, and the teaching methods (treatments) would be randomly assigned within the blocks. Thus, Methods 1, 2, and 3 would each be equally represented in the early morning, middle of the day, and late afternoon. The null hypothesis tested is the same as in the one-factor analysis of variance, namely, H_0: $\mu_1 = \mu_2 = \mu_3$. However, the design enables the variation in the time of day to be removed from the comparison of the three teaching methods. No inference is attempted about the effect of time of day, since as noted earlier, time of day is viewed as an extraneous factor. The extraneous blocking factor usually represents time, location, or experimental material. Just as the one-factor analysis of variance represents a generalization of the t test for means of two independent samples, the randomized block design represents a generalization of the t test for paired observations, discussed in Section 7.5.

The second type of two-factor analysis of variance is used when inferences about both factors are desired. For example, suppose that four different teachers were involved in the statistics course and that inferences were desired about differences in effectiveness among methods of teaching *and* among the different teachers. Of course, the sample sizes in this experiment and in the randomized block design would have to be much larger than the twelve students earlier assumed for computational convenience. The two-factor study here would use a *completely randomized design,* rather than the aforementioned randomized block design. In the completely randomized design, the sample units (students, in this case) would be randomly assigned to each factor combination. For example, teacher 1 using Method 1, teacher 1 using Method 2, etc., would represent factor combinations. Since there are three teaching methods and four teachers, there would be twelve possible factor combinations. This experimental design would then permit the testing of two null hypotheses: (1) no difference in population mean final examination scores among the different teaching methods, and (2) no difference in population mean final examination scores among the different teachers.

Two-factor analyses of variance have some distinct advantages. It may be noted from the above description that we were able to test two separate null hypotheses from the same set of experimental data. We did not need to run two one-factor experiments to get information about two factors. Furthermore, certain types of questions can be answered by two-factor designs but cannot be treated in one-factor analyses. For example, the interaction, or joint effects, of the two factors may be examined as well as their separate effects.

Only a very brief introduction to the analysis of variance has been given in this chapter. More elaborate designs than those considered in this book are available; they attempt to control and test for the effects of more factors, both qualitative and quantitative.

Further Remarks

As we saw in the one-factor teaching methods example, it is possible that observed differences in teaching effectiveness may not have been attributable solely to the different teaching methods, but rather to some unknown mixture of teaching methods and factors associated with time of day, and perhaps other factors as well. Hence, unless careful thought is given to the experimental design from which data are to be collected, erroneous inferences may be drawn. This applies equally well to the hypothesis-testing methods considered earlier, as for example in Chapter 7, because we might have had only two teaching methods to compare rather than three. Thus, we must guard against mechanical or rote application of statistical techniques such as hypothesis-testing methods. In this book, we consider the general principles involved in some of the simpler, basic procedures. More refined and sophisticated techniques may very well be required in particular instances.

One of the points we have attempted to convey in the preceding discussion is that statistical results are virtually always consistent with more than one interpretation. The researcher must avoid naïvely leaping to conclusions and must give careful consideration to alternative interpretations and explanations. We conclude this chapter with two anonymous humorous stories that are relevant to the point that alternative interpretations and explanations of experimental results are often possible.

An investigator wished to determine the differential effects involved in the intake of various types of mixed drinks. Therefore, he had subjects drink substantial quantities of scotch and water, bourbon and water, and rye and water. All of the subjects became intoxicated. The investigator concluded that since water was the one factor common to all of these drinks, the imbibing of water makes people drunk.

The heroine of our second story is a grammar school teacher, who wished to explain the harmful effects of drinking liquor to her class of eight-year-olds. She placed two glass jars of worms on her desk. Into the first jar, she poured some water. The worms continued to move about, and did not appear to have been adversely affected at all by the contact with the water. Then she poured a bottle of whiskey into the second jar. The worms became still and appeared to have been mortally stricken.

The teacher then called upon a student and asked, "Johnny, what is the lesson to be learned from this experiment?" Johnny, looking very thoughtful, replied, "I guess it proves that it is good to drink whiskey, because it will kill any worms you may have in your body."

Exercises

1. Seven samples, each of size ten, are drawn from a normally distributed population. The means and variances of the seven samples are shown in the following table:

Sample	Mean	Variance
A	50	5
B	47	4
C	52	5
D	51	6
E	49	5
F	52	4
G	49	6

Would you conclude that the seven samples were drawn *randomly* from the same population? Is your conclusion the same for both the 0.05 and 0.01 levels of significance?

2. A manufacturer wished to select the best advertising display for a new product. Because there was a choice of five different displays, the manufacturer randomly selected 25 different stores and placed each type of display in five stores. For the first six months, the following are the average amount sold and the variance for each display, expressed in dozens.

Type of Display	Mean	Variance
1	78	9
2	76	7
3	77	8
4	74	8
5	76	10

Can the manufacturer assume that it doesn't matter which display is used? Use a 0.01 significance level.

3. As a department head in a consumer research organization, you are responsible for testing and comparing life times of four different brands of a particular product. Suppose you test the life time of four items of each of the four brands. Your test data are as follows, each entry representing the life time of an item, measured in hundreds of hours.

	Brand		
A	B	C	D
44	47	41	43
41	45	40	43
45	46	45	41
42	42	42	41

Can you conclude that the mean life times of the four brands are equal?

4. The supervisor of the credit department in Racy's Department Store is evaluating the performance of three collectors, whose responsibilities include the collection of delinquent accounts. The following figures show the number of accounts each collector successfully removed from delinquency during each week of a five-week period. (The accounts to be investigated were equally divided among the collectors each week.)

Week	A	B	C
1	25	45	45
2	35	45	40
3	30	25	35
4	25	30	35
5	20	30	45

Should the supervisor conclude that there is no real difference among these collectors in the average number of successful investigations per week? Use a 0.05 significance level.

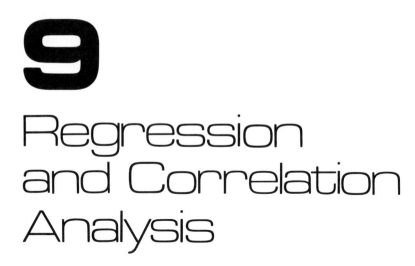

9

Regression and Correlation Analysis

9.1 INTRODUCTION

Prediction is required in virtually every aspect of the management of enterprises. Indeed, business planning and decision making are inseparable from prediction. Forecasting of sales, earnings, costs, production, personnel requirements, inventories, purchases, and capital requirements is the basic foundation of company planning and control. Many other illustrations could be given of the need for prediction in the management of private and public organizations.

In this chapter, we concern ourselves with a broad class of techniques for prediction, namely, *regression* and *correlation analysis*. This type of analysis is undoubtedly one of the most widely used statistical methods. Regression analysis provides the basis for predicting the values of a variable from the values of one or more other variables; correlation analysis enables us to assess the strength of the relationships (correlations) among the variables.

Equations are used in mathematics to express the relationships among variables. In fields such as geometry or trigonometry, these mathematical equations, or functions, express the *deterministic (exact) relationships* among the variables of interest. Thus, the equation $A = s^2$ describes the relationship between s, the length of the side of a square, and A, the area of the square. The equation $A = ab/2$ expresses the relationship between b, the length of any side of a triangle, a, the altitude or perpendicular distance to that side from the angle opposite it, and A, the area of the triangle. By substituting numerical values for the variables on the right sides of these equations, we can *determine* the *exact values* of the quantities on the left-hand sides. In the social sciences and in fields such as business and government administration, exact relationships are not generally observed among variables, but rather *statistical relationships* prevail. That is, certain average relationships may be observed among variables, but these average relationships do not provide a basis for perfect predictions. For example, if we know how much money a corporation spends on television advertising, we cannot make an exact prediction of the amount of sales this promotional expenditure will generate. If we know a family's net income, we cannot make an exact forecast of the amount of money that family saves. On the other hand, we can measure statistically how sales vary, on the average, with differences in television advertising, or how family savings vary, on the average, with differences in income. We can also determine to what extent actual figures vary from these average relationships. On the basis of these relationships, we may be able to estimate the values of the variables of interest closely enough for decision-making purposes. The techniques of regression and correlation analysis are important statistical tools in this measurement and estimation process.

The term *regression analysis* refers to the methods by which estimates are made of the values of a variable from a knowledge of the values of one or more other variables, and to the measurement of the errors involved in this estimation process. The term *correlation analysis* refers to methods for measuring the degree of association among these variables.

We begin by discussing *two-variable linear regression and correlation analysis*. The term "linear" means that an equation of a straight line of the form $Y = A + BX$, where A and B are fixed numbers, is used to describe the average relationship between the two variables and to carry out the estimation process. The factor whose values we wish to estimate is referred to as the *dependent variable* and is denoted by the symbol Y. The factor from which these estimates are made is called the *independent variable* and is denoted by X.

In the formulation of a regression analysis, the investigator should use prior knowledge and the results of past research to select independent variables that are potentially helpful in predicting the values of the dependent variable. Hence, the investigator should use any available information concerning the direction of cause and effect in the selection of relevant variables. As indicated earlier, correlation analysis can be used to measure the strength or degree of correlation among the variables of interest.

Let us consider some illustrative cases of variables that it is reasonable to assume are related to one another, that is, correlated. If suitable data were available, we might attempt to construct an equation that would permit us to estimate the values of one variable from the values of the other. We shall assume that the first named factor in each pair is the variable to be estimated, that is, the dependent variable, and the second one is independent. Consumption expenditures might be estimated from a knowledge of income; investment in telephone equipment from expenditures on new construction; personal net savings from disposable income; commercial bank discount rates from Federal Reserve Bank discount rates; and success in college from Scholastic Aptitude Test scores.

Of course, additional definitions are required to attach meaning to the estimation problems listed above. Thus, in the illustration of consumption expenditures and income, we would have to specify whose expenditures and whose income are involved. If we wanted to estimate family consumption expenditures from family income, the family would be said to be the "unit of association." The estimating equation would be constructed from data representing observations of these two variables for individual families. We would have to define the variables more specifically. For example, we might be interested in estimates of annual family consumption expenditures from annual family net income, where again these terms would require precise definitions.

In each of these examples, it is possible to specify other independent variables that might be included to aid in obtaining good estimates of the dependent variable. Hence, in estimating a family's consumption expenditures, we might wish to use knowledge of the size of the family in addition to information on the family's income. This would be an illustration of *multiple regression analysis,* where two independent variables (family income and family size) are used to obtain estimates of a dependent variable. As indicated earlier in this chapter, we first consider two-variable problems involving a dependent factor and only one independent factor and then turn to a brief discussion of multiple regression problems involving more than one independent variable.

The Simple Two-Variable Linear Regression Model

The use of a variable to predict the values of another variable may be viewed as a problem of statistical inference. The population consists of all relevant pairs of observations of the dependent and independent vari-

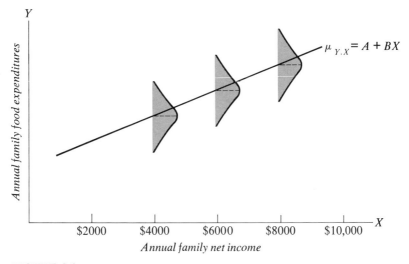

FIGURE 9-1
The linear regression model and conditional probability distributions of annual family food expenditures for selected annual family net incomes.

able. Generally, estimates or predictions must be made from only a sample of that population. To take a concrete problem, suppose we are interested in estimating a family's annual food expenditures (the dependent variable) from a knowledge of the family's annual net income (the independent variable). Furthermore, let us assume that we are interested in these estimates for 1976 for families with $10,000 or less of annual net income in a particular metropolitan area. Figure 9-1 shows what the population relationship between these two variables might look like for families in the given income range if we assume that this relationship is linear (that is, in the form of a straight line). The fact that the line runs from the lower left to the upper right side of the graph indicates that as family net annual income increases, family food expenditures also increase. We denote this population regression line as

(9.1) $$\mu_{Y.X} = A + BX$$

where the meaning of $\mu_{Y.X}$ is explained in the next paragraph.

In order to understand the assumptions involved in the linear regression model, let us consider families with certain specified incomes (say, $4000, $6000, and $8000), as indicated in Figure 9-1. If we now focus on families at a given income level (say, $6000), the probability distribution of the Y variable, food expenditures, is a *conditional probability distribution* of Y given $X = \$6000$. Such a conditional probability distribution may be symbolized in the usual way as $f(Y|X = \$6000)$ or, in general, $f(Y|X)$. This distribution has a mean, which may be denoted $\mu_{Y.X}$, and a standard deviation, which may be denoted $\sigma_{Y.X}$. In the linear regression model, the assumed relationship between $\mu_{Y.X}$ and X can be graphed as a straight line. That is, the graphs of the means of the conditional probability distributions are assumed to lie along a straight line. In this model of a linear regression function, A and B are population parameters that must be *estimated* from sample data.

The conditional probability distributions describe the variability of the Y values. Thus, for families with \$6000 annual net income in the given metropolitan area, the conditional probability distribution indicates the variability of family annual food expenditures around the conditional mean food expenditures, $\mu_{Y.X}$ (in this case $\mu_{Y.\,\$6000}$). The symbol $\mu_{Y.X}$ represents the mean of the Y variable for a given X value.

In addition to the assumption of a linear relationship, the following assumptions are involved in the use of the linear regression model:

1. The Y values are independent of one another
2. The conditional probability distributions of Y given X are normal
3. The conditional standard deviations, $\sigma_{Y.X}$, are equal for all values of X

A few comments about these assumptions are in order. The first assumption means that there is independence between successive observations. This means, for example, that a low value for Y on the first observation does not tend to imply that the second Y value will also be low. In certain situations, such an assumption may not be particularly valid. This is seen clearly in time-series data that move in cycles around a fitted trend line. Observations in the expansion phase of a business cycle will tend to lie above the trend line and to have relatively high values; the opposite is true for observations in a contraction or recession phase. Hence, successive observations in this case tend to be related rather than independent.

The second assumption means that for each value of X, we are assuming that the Y values are normally distributed around $\mu_{Y.X}$. As we will see, this assumption is useful for making probability statements about estimates of the dependent variable Y.

The third assumption implies that there is the same amount of variability around the regression line at each value of the independent variable, X. This characteristic is referred to as *homoscedasticity*. Note that in regression analysis, according to the second assumption, only Y is considered a random variable. X is considered fixed. Hence, if we attempt to predict a Y value (for example, a family's food expenditures) from a knowledge of X (for example, a family's income), the predicted Y value is subject to error. X is assumed to be known without error. On the other hand, in correlation analysis, both X and Y are treated as normally distributed random variables.

Clearly the three assumptions given above are never perfectly met in the real world. In many situations, however, these assumptions are approximately true and the model is useful. In this chapter, we develop the standard linear regression model under these assumptions.

Before turning to the methodology of regression and correlation analysis beginning in the next section, it is important to emphasize that the *formulation of the problem* is of critical importance. In the formulation stage, the analyst must identify the dependent and independent variables, must specify the relationships among these variables (e.g., linear or nonlinear), must decide what the relevant population is (e.g., low-income families in a particular metropolitan area), and must supply the data on the variables for a sample or for the entire relevant population.

9.2 SCATTER DIAGRAMS

A useful aid in studying the relationship between two variables is to plot the data on a graph. This allows visual examination of the extent to which the variables are related and aids in choosing the appropriate type of model for estimation. The chart used for this purpose is known as a *scatter diagram,* which is a graph on which each plotted point represents an observed pair of values of the dependent and independent variables. We will illustrate this by plotting a scatter diagram for the data given in Table 9-1. These figures represent observations for a sample of ten families of annual expenditures on food,which we shall treat as the

TABLE 9-1
Annual Food Expenditures and Annual Net Income of a Sample of Ten Families in a Metropolitan Area in 1976

Family	Annual Food Expenditures ($Hundreds) Y	Annual Net Income ($Thousands) X
A	22	8
B	23	10
C	18	7
D	9	2
E	14	4
F	20	6
G	21	7
H	18	6
I	16	4
J	19	6

dependent variable, Y (the factor to be estimated), and annual net income, X, which is the independent variable (the factor from which the estimates are to be made). As in the illustration given in Section 9.1, we assume that the ten families constitute a simple random sample of families with \$10,000 or less of annual net income in a metropolitan area in 1976. Although only ten families are too small a sample from which to draw very useful conclusions that would apply to all such families in a metropolitan area, we shall use such a small sample in order to limit the amount of arithmetic needed. Furthermore, we have assumed relatively low incomes in order to simplify the numerical work.

Figure 9-2 presents the data of Table 9-1 plotted as a scatter diagram. On the Y axis are plotted the figures on food expenditures and on the X axis annual net income. This follows the standard convention of plotting the dependent variable along the Y axis and the independent variable along the X axis. The pair of observations for each family determines one point on the scatter diagram. For example, for family A, a point is plotted corresponding to $X = 8$ along the horizontal axis and $Y = 22$ along the vertical axis; for family B, a point is plotted corresponding to $X = 10$ and $Y = 23$. An examination of the scatter diagram gives some useful indications of the nature and strength of the relationship between the two variables. For example, depending upon whether the Y values tend to increase or to decrease as the values of X increase, there is said to be a *direct* or *inverse* relationship, respectively, between the two variables. The configuration in Figure 9-2 indicates a general tendency for the points to run from the lower left to the upper right side

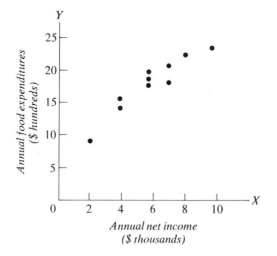

FIGURE 9-2

A scatter diagram of annual food expenditures and annual net income of a sample of ten families in a metropolitan area in 1976.

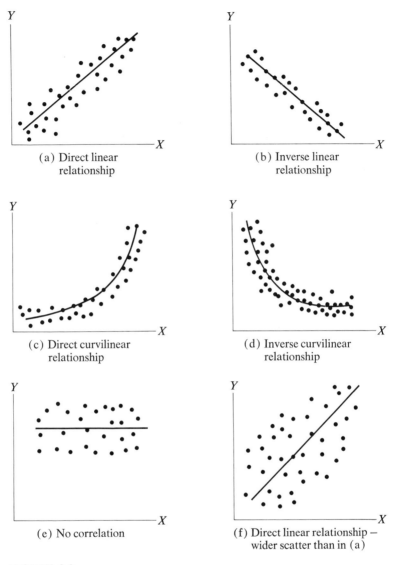

FIGURE 9-3
Scatter diagrams.

of the graph. Hence, as noted in Section 9.1, as we move from lower- to higher-income families, food expenditures tend to increase. This is an example of a *direct relationship* between the two variables. On the other hand, if the scatter of points runs from the upper left to the lower right (that is, if the Y variable tends to decrease as X increases), there is said to be an *inverse relationship* between the variables. Also, an examination of the scatter diagram gives an indication of whether a straight line appears to be an adequate description of the average relationship between the two variables. If a straight line is used to describe the average relationship between Y and X, a *linear relationship* is said to be present. On the other hand, if the points on the scatter diagram appear to fall along a curved line rather than a straight line, a *curvilinear relationship* is said to exist. Figure 9-3 presents illustrative combinations of the foregoing types of relationships. Parts (a), (b), (c), and (d) of Figure 9-3 show, respectively, direct linear, inverse linear, direct curvilinear, and inverse curvilinear relationships. As can be seen, the points tend to follow a straight line sloping upward in (a), a straight line sloping downward in (b), a curved line sloping upward in (c), and a curved line sloping downward in (d). Of course, the relationships are not always so obvious. In (e), the points appear to follow a horizontal straight line. Such a case depicts a situation of "no correlation" between the X and Y variables, or no evident relationship, since the horizontal line implies no change in Y, on the average, as X increases. In (f), the points follow a straight line sloping upward as in (a), but there is a much wider scatter of points around the line than in (a).

In our present problem, we assume that our prior expectation was for a direct linear relationship between the two variables and that a visual examination of Figure 9-2 confirmed this expectation. In the next section, we discuss the procedures used in regression and correlation analysis. We begin by considering what we hope to accomplish through the use of such techniques.

9.3 PURPOSES OF REGRESSION AND CORRELATION ANALYSIS

What does a regression and correlation analysis attempt to accomplish in studying the relationship between two variables, such as expenditures on food and net annual income of families? We will concentrate on these basic goals, which emphasize the relationships contained in the particular sample under study. Later, we will consider other objectives involving statistical inference, that is, inferences concerning the population from which the sample was drawn.

The first two objectives and the statistical procedures involved in their accomplishment fall under the heading of *regression analysis,* whereas the third objective and related procedures are classified as *correlation analysis.* Below, the objectives are stated and the statistical measures used to achieve these objectives are named. However, the mathematical definitions of these measures are postponed until the discussion of their use in the problem involving family expenditures on food and family income.

1. *The first purpose of regression analysis is to provide estimates of values of the dependent variable from values of the independent variable.* The device used to accomplish this estimation procedure is the *regression line,* which is a line fitted to the data by a method to be subsequently described. The regression line describes the average relationship existing between the X and Y variables. Somewhat more precisely, it is a line that displays mean values of Y for given values of X. The equation of this line, known as the *regression equation,* provides estimates of the dependent variable when values of the independent variable are inserted into the equation.

2. *A second goal of regression analysis is to obtain measures of the error involved in using the regression line as a basis of estimation.* For this purpose, the *standard error of estimate* and related measures are calculated. The standard error of estimate measures the scatter, or spread, of the observed values of Y around the corresponding values estimated from the fitted regression line. Measures of forecast error take into account this scatter as well as the probable difference between the regression line fitted to a sample of data and the true but unknown population regression line.

3. *The third objective, which we have classified as correlation analysis, is to obtain a measure of the degree of association or correlation between the two variables.* The *coefficient of correlation* and the *coefficient of determination,* calculated for this purpose, measure the strength of the relationship between the two variables.

9.4 ESTIMATION USING THE REGRESSION LINE

As indicated in the preceding section, to accomplish the first objective of a regression analysis, we must obtain the mathematical equation of a line that describes the average relationship between the dependent and independent variable. We can then use this line to estimate values of the dependent variable. Since the present discussion is limited to *linear* regression analysis, the line we are referring to is a straight line. Ideally, what we would like to obtain is the equation of the straight line that best fits the data. Let us defer for the moment what we mean by "best fits" and review the concept of the equation of a straight line.

The equation of a straight line is $Y = a + bX$, where a is the so-called "Y intercept," or the computed value of Y when $X = 0$, and b is the slope of the line, or the amount by which the computed value of Y changes with each unit change in X. In regression analysis, we use the notation

(9.2) $$Y_C = a + bX$$

for the equation of the regression line. It is useful to use the different symbols Y_C, which denotes a *computed* value of the dependent variable, and Y, which denotes an *observed* value. For example, in the data given in Table 9-1, the observed value of Y for family A is 22, or \$2200 annual food expenditures. The observed X value is 8, indicating that this family's net income is \$8000. When we obtain a regression equation, we may wish to estimate annual food expenditures for a family with an annual net income of \$8000. By substituting $X = 8$ into the regression equation, we can obtain the required estimate. Since the computed figure will in general be different from the observed value $Y = 22$, it is useful to have a separate symbol, such as Y_C, to denote this estimated or computed value of the dependent variable.

Let us review by means of a simple illustration the relationship between the equation $Y_C = a + bX$ and the straight line that represents the graph of the equation. Suppose the equation is

(9.3) $$Y_C = 2 + 3X$$

Thus, $a = 2$ and $b = 3$. If we substitute a value of X into this equation, we can obtain the corresponding computed value of Y_C. Each pair of X and Y_C values represents a single point. Although only two points are required to determine a straight line, several pairs of X and Y_C values for the line $Y_C = 2 + 3X$ are shown in Table 9-2. The graph corresponding to this line is shown in Figure 9-4. As can be seen on the graph, since the a value in the equation of the line is 2, the line intersects the Y axis at a height of 2 units. Also, since the b value, or slope of the line, is 3, we note in Table 9-2 that the Y_C values increase by 3 units each time X increases by 1 unit. This is shown graphically in Figure 9-4 as a rise of 3 units in the line when X increases by 1 unit.

The terms "regression line" and "regression equation" for the estimating line and equation stem from the pioneer work in regression and correlation analysis of the British biologist Sir Francis Galton in the

TABLE 9-2
Calculation of Pairs of X
and Y_C Values for the Line
$Y_C = 2 + 3X$

X	Y_C
0	2
1	5
2	8
3	11
4	14

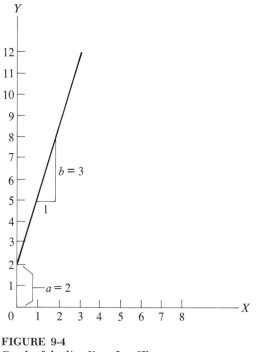

FIGURE 9-4
Graph of the line $Y_c = 2 + 3X$.

nineteenth century. The lines that he fitted to scatter diagrams of data on heights of fathers and sons in this early work came to be known as "regression lines" and the equations of these lines as "regression equations," because Galton found that the heights of the sons "regressed" toward an average height. Unfortunately, the terminology has persisted. Thus, these terms for the estimating line and estimating equation are used in the wide variety of fields in which regression analysis is applied, despite the fact that the original implication of a regression toward an average is not necessarily present for the phenomena under investigation.

We now turn to the question of obtaining a best fitting line to the data plotted on a scatter diagram in a two-variable linear regression problem. The fitting procedure to be discussed is the *method of least squares,* undoubtedly the most widely applied curve-fitting technique in statistics.

The Method of Least Squares

In order to establish a best fitting line to a set of data on a scatter diagram, we must have criteria concerning what constitutes *goodness of fit.* A number of criteria that might at first seem reasonable turn out to be unsuitable. For example, we might entertain the idea of fitting a straight line to the data in such a way that half of the points fall above the line and half below. However, this requirement can be easily dismissed, since such a line may represent a quite poor fit to the data if, say, the points that fall above the line lie very close to it whereas the points below deviate considerably from it.

Let us now consider the most generally applied curve-fitting technique in regression analysis, namely, the *method of least squares*. This method imposes the requirement that the *sum of the squares* of the deviations of the observed values of the dependent variable from the corresponding computed values on the regression line must be a minimum. Thus, if a straight line is fitted to a set of data by the method of least squares, it is a "best fit" in the sense that the sum of the squared deviations, $\Sigma (Y - Y_c)^2$, is less than it would be for any other possible straight line. Another useful characteristic of the least squares straight line is that it passes through the point of means, $(\bar{X}, \bar{Y})$, and therefore makes the total of the positive and negative deviations equal to 0. In summary, the least squares straight line possesses the following mathematical properties:

(9.4) $$\Sigma (Y - Y_c)^2 \text{ is a minimum}$$

(9.5) $$\Sigma (Y - Y_c) = 0$$

Figure 9-5 presents graphically the nature of the least squares property. A scatter plot is shown around an estimated least squares regression line, denoted $Y_c = a + bX$. Let us consider the first point, whose coordinates are $X = 2$ and $Y = 10$. The vertical distance of the point from the

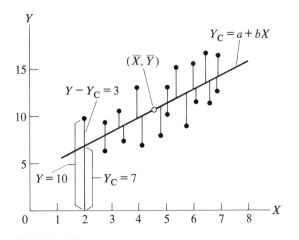

FIGURE 9-5
Scatter diagram and graphic representation of deviations from a fitted least squares regression line.

X axis is its Y value, in this case, 10. The Y_C value for the same point is the vertical distance from the X axis to the regression line. In this case, $Y_C = 7$. The vertical distance of the point from the corresponding value on the regression line is the deviation, or residual, $Y - Y_C$. In this illustration, $Y - Y_C = 10 - 7 = 3$. Since the Y value is above the regression line, the deviation is positive; for Y values below the line, the deviations are negative. The square of the deviation for the first point is $3^2 = 9$. As mentioned above, the line fitted by this method has the property that the total of all squared deviations is less than the corresponding totals for any other straight line that could have been fitted to the same data.

The regression line is depicted as passing through the point established by the means of the X and Y variables, $(\bar{X}, \bar{Y})$, which in this case is (4.5, 11). It can be shown that a least squares straight line must include this point of means and that this property makes the algebraic sum of the deviations above and below the line equal to 0. That is, the sum of the deviations of the points lying above the regression line, which are positive, and the deviations of the points lying below the regression line, which are negative, is 0.

Computational Procedure

By using calculus methods to apply the condition that the sum of the squared deviations from a straight line must be a minimum, two equations, known as the *normal equations*, are derived,[1] which can be solved for the values of a and b in the regression equation $Y_C = a + bX$.

---

[1] The derivation of the equations is as follows. We denote the sum of squared deviations that must be minimized as some function of the unknown quantities a and b. Thus, let

$$F(a, b) = \Sigma (Y - Y_C)^2$$

Substituting $a + bX$ for Y_C into the above equation gives

$$F(a, b) = \Sigma (Y - a - bX)^2$$

We impose the condition of a minimum value for $F(a, b)$ by obtaining its partial derivatives with respect to a and b and setting them equal to zero. Thus,

$$\frac{\partial F(a, b)}{\partial a} = -2\Sigma (Y - a - bX) = 0$$

$$\frac{\partial F(a, b)}{\partial b} = -2\Sigma (Y - a - bX)(X) = 0$$

Solving these equations yields

$$\Sigma Y = na + b\Sigma X$$
$$\Sigma XY = a\Sigma X + b\Sigma X^2$$

from which Equations 9.6 and 9.7 can be derived. A check reveals that the second derivatives of $F(a, b)$ are positive, so a minimum has been found.

TABLE 9-3
Computations for a Regression and Correlation Analysis for the Data
Shown in Table 9-1

Family	Y	X	XY	X²	Y²
A	22	8	176	64	484
B	23	10	230	100	529
C	18	7	126	49	324
D	9	2	18	4	81
E	14	4	56	16	196
F	20	6	120	36	400
G	21	7	147	49	441
H	18	6	108	36	324
I	16	4	64	16	256
J	19	6	114	36	361
	180	60	1159	406	3396

The values of a and b can then be determined from the following general solution for a and b:

$$(9.6) \qquad a = \bar{Y} - b\bar{X}$$

$$(9.7) \qquad b = \frac{\Sigma\, XY - n\bar{X}\bar{Y}}{\Sigma\, X^2 - n\bar{X}^2}$$

where $\bar{X}$ and $\bar{Y}$ are the arithmetic means of the X and Y variables.

Let us return to the problem involving the sample of ten families and assume that we have decided to fit a *straight line* to the data. From the original observations, we can determine the various quantities $(n, \Sigma\, Y, \Sigma\, X, \Sigma\, XY,$ and $\Sigma\, X^2)$ required in Equations 9.6 and 9.7, where n is the number of pairs of X and Y values (in this case, ten). For our illustration, the computation of the required totals is shown in Table 9-3. Although $\Sigma\, Y^2$ is not needed for the calculation of a and b, its computation is also shown. This figure is useful for calculating the standard error of estimate, to be discussed shortly.

From Table 9-3, we compute the means of X and Y as

$$\bar{X} = \frac{\Sigma\, X}{n} = \frac{60}{10} = 6 \ (\$ \text{ thousands})$$

$$\bar{Y} = \frac{\Sigma\, Y}{n} = \frac{180}{10} = 18 \ (\$ \text{ hundreds})$$

Substituting the additional quantities $n = 10$, $\Sigma\, XY = 1159$, and $\Sigma\, X^2 = 406$ from Table 9-3 into Equations 9.6 and 9.7, we obtain the following values for a and b:

$$b = \frac{1159 - (10)(6)(18)}{406 - 10(6)^2} = \frac{79}{46} = 1.717$$

$$a = 18 - 1.717(6) = 18 - 10.302 = 7.698$$

Hence, the least squares regression line is

(9.8) $$Y_c = 7.698 + 1.717X$$

If a family drawn from the same population had an annual net income of \$8000 in 1976, its estimated annual food expenditures from Equation 9.8 would be

$$Y_c = 7.698 + 1.717(8) = 21.434 \text{ (\$ hundreds)}$$

By plotting the point thus determined ($X = 8, Y_c = 21.434$) and one other point, or by plotting any two points derived from the regression equation, we can graph the regression line. The line is shown in Figure 9-6, along with the original data. Hence, in this case $a = 7.698$ means that the estimated annual food expenditures for a family whose income is zero dollars in 1976 is 7.698 (in hundreds), or \$769.80. Since no families in the original sample had incomes less than \$2000, it would be extremely hazardous to make predictions for families with incomes less than the \$2000 figure. Prediction outside the range of the original observations is discussed in a later section of this chapter.

The b value in the regression equation is often referred to as the *regression coefficient* or *slope coefficient*. The figure of $b = 1.717$ indi-

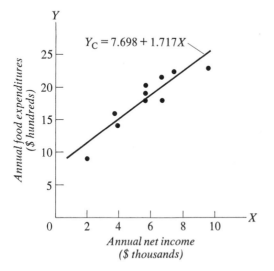

FIGURE 9-6
Least squares regression line for food expenditures and annual net income of a sample of ten families in a metropolitan area in 1976.

cates first of all that the slope of the regression line is positive. Thus, as income increases, estimated food expenditures increase. Taking into account the units in which the X and Y variables are stated, $b = 1.717$ means that for two families whose annual net income differs by $1000, the *estimated difference* in their annual food expenditures is $171.70. This is an interpretation in terms of the *regression line*. If we think of the figure $b = 1.717$ in terms of the *sample studied*, we can say that for two families whose annual net incomes differed by $1000, their annual food expenditures differed, *on the average*, by $171.70.

Exercises

1. Below are a scatter diagram and regression line showing the relationship between earnings during 1976 and price per share at the end of 1976 for selected common stocks.

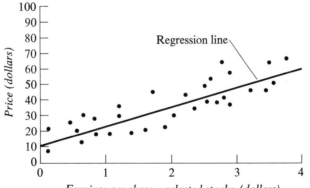

Earnings per share — selected stocks (dollars)

On the basis of the above chart, estimate the values of a and b in the equation of the line of regression, $Y_c = a + bX$. Use specific numbers.

2. Martin Breit, the financial vice-president of Lemming Lubricating Liquids, Ltd., wished to estimate the interest rate that his firm would have to pay on a bond issue to be floated soon. Because of past experience, he believes that the size of the issue has a significant effect on the rate and that an estimated relationship between these two variables would be of value in prediction. The following data on bond issues of Lemming, Ltd. have been collected:

Issue Size ($Millions)	Rate (%)	Issue Size ($Millions)	Rate (%)
$ 5	5.5	$25	7.5
5	6.5	30	7.0
10	6.0	30	8.0
15	6.5	30	8.5

(Continued)

Issue Size ($Millions$)	Rate (%)	Issue Size ($Millions$)	Rate (%)
15	6.0	40	9.0
20	7.0	45	9.5
20	7.5	60	10.5

a. Determine the linear regression equation by the method of least squares, using issue size as the independent variable.
b. If the size of the upcoming bond issue will be $35 million, what is the estimated interest rate?
c. Describe the main assumptions of the linear regression model.

3. An insurance company wished to examine the relationship between income and amount of life insurance held by heads of families. The company drew a simple random sample of twelve family heads and obtained the following results:

Family	Amount of Life Insurance ($Thousands$)	Income ($Thousands$)
A	$32	$14
B	40	19
C	50	23
D	20	12
E	22	9
F	35	15
G	55	22
H	45	25
I	28	15
J	22	10
K	24	12
L	35	16

a. Determine the linear regression equation using the method of least squares with income as the independent variable. Calculate Y_c for each income level given above.
b. What is the meaning of the regression coefficient, b, in this case?
c. What is your estimate of the amount of life insurance carried by a family head from the same population whose income is $20,000?

4. A national education organization is trying to build a strong case for increased expenditures on public education below the college level. The organization has taken a random sample of towns and cities across the United States and has collected the following information:

X = average annual expenditure per student in town or city (in dollars)
Y = average performance of college-oriented students on standard college entrance examination
X ranges from \$475 to \$700
$n = 200$
$\Sigma X = 125{,}000$
$\Sigma Y = 120{,}000$
$\Sigma XY = 77{,}625{,}000$
$\Sigma X^2 = 80{,}000{,}000$

a. Using the least squares method, estimate an appropriate regression equation.
b. Give two interpretations of the regression coefficient, b.
c. The range of possible scores on the standard examinations is 200 to 800. Calculate the estimated average performance for college-oriented students in a city that spends an average of \$800 per student on public education. Is there any problem with this estimate?

5. What are the three main objectives of regression and correlation analysis? What statistical measures are used to achieve these?

6. A research analyst for a federal reserve bank has compiled information on loan activity of a simple random sample of 45 banks in the reserve bank's district. The analyst has calculated the following statistics on net return on loans as a percentage of total loans from each bank (Y) and total deposits, in tens of millions of dollars, for each bank (X):

X = total deposits in tens of millions of dollars
Y = net return on loans as a percentage of total loans
$n = 45$
$\Sigma X = 787.5$
$\Sigma Y = 225$
$\Sigma XY = 3800$
$\Sigma X^2 = 15{,}150$

a. Estimate a least squares regression equation, obtaining estimates for a and b.
b. The Naugatuck Valley National Bank has total deposits of approximately \$225 million. Estimate the net return on loans, using the above equation.

9.5 CONFIDENCE INTERVALS AND PREDICTION INTERVALS IN REGRESSION ANALYSIS

Standard Error of Estimate

Now that we have seen that the regression equation is used for estimation, we can turn to the second objective referred to in Section 9.3, that of obtaining measures of the error involved in using the regression line for estimation. If there is a great deal of scatter of the observed Y values around the fitted line, estimates of Y values based on computed values from the regression line will not be very close to the observed Y values. On the other hand, if every point falls on the regression line, insofar as the *sample observations* are concerned, perfect estimates of the Y values can be made from the fitted regression line. Just as the standard deviation was used as a measure of scatter of a set of observations about the mean, it seems that an analogous measure of dispersion of observed Y values around the regression line is desirable. The measure of dispersion, referred to as the *standard error of estimate* (or the *standard deviation of the residuals*), is obtained by solving the equation

$$(9.9) \qquad s_{Y.X} = \sqrt{\frac{\Sigma\,(Y - Y_c)^2}{n - 2}}$$

where, as before, n is the size of sample.

Hence, although $s_{Y.X}$ is called the *standard error* of estimate, it is not a standard error in the sense that the term was used in Chapters 6 and 7, that is, as a measure of sampling error. The standard error of estimate simply measures the scatter of the observed values of Y around the corresponding computed Y_C values on the regression line. The sum of squared deviations is divided by $n - 2$ because this divisor makes $s^2_{Y.X}$ an unbiased estimator of the conditional variance around the population regression line,[2] $\sigma^2_{Y.X}$. The $n - 2$ represents the number of degrees of freedom around the fitted regression line. In general, the denominator is $n - k$ where k is the number of constants in the regression equation. In the case of a straight line, the denominator is $n - 2$ because two degrees of freedom are lost when a and b are used as estimates of the corresponding constants in the population regression line.

It is useful to consider the nature of the notation for the standard error of estimate, $s_{Y.X}$. In the discussion of dispersion in Section 1.14, the symbol s was used to denote the standard deviation of a sample of observations. The use of the letter s in $s_{Y.X}$ is analogous, since, as explained in the preceding paragraph, $s_{Y.X}$ is also a measure of dispersion computed from a sample. However, since both the variables Y and X are present in a two-variable regression and correlation analysis, subscript notation is required to distinguish among the various possible dispersion measures. Hence, as we have seen, the notation for the standard error of estimate, where Y and X are, respectively, the dependent and independent variables, is $s_{Y.X}$. The letter to the left of the period in the subscript is the dependent variable, and the letter to the right denotes the inde-

---

[2] In using $s_{Y.X}$ to estimate $\sigma_{Y.X}$, the deviation of each Y value is taken around its own estimated conditional mean, Y_c. We can combine the squares of these deviations, which are obtained from different conditional probability distributions because of the assumption of the linear regression model that the conditional standard deviations, $\sigma_{Y.X}$, are all equal.

pendent variable. Subscripts are also required to distinguish standard deviations around the means of the two variables. Thus, s_Y denotes the standard deviation of the Y values of a sample around the mean, $\bar{Y}$, and s_X denotes the standard deviation of the X values around their mean, $\bar{X}$.

In a problem containing large numbers of observations, the computation of the standard error of estimate using Equation 9.9 clearly involves a great deal of arithmetic. Calculation of Y_C for each X value in the sample is required, and then the arithmetic implied by the formula must be carried out. It is useful to have a shortcut method involving only quantities already computed. A convenient form of such a shortcut formula is given by

$$(9.10) \qquad s_{Y.X} = \sqrt{\frac{\Sigma\, Y^2 - a\Sigma\, Y - b\Sigma\, XY}{n - 2}}$$

All quantities required by Equation 9.10 were calculated for our illustrative problem in Table 9-3, or were computed in obtaining the constants of the regression line. Hence, the standard error of estimate for these data is

$$s_{Y.X} = \sqrt{\frac{3396 - (7.698)(180) - (1.717)(1159)}{10 - 2}} = 1.595 \text{ (\$ hundreds)}$$

Since the standard error of estimate, $s_{Y.X}$, is an estimate of $\sigma_{Y.X}$ (the standard deviation around the true but unknown population regression line), $s_{Y.X}$ may be used and interpreted as a standard deviation. If every sample point falls on the regression line, i.e., if there is no scatter around the line, $s_{Y.X}$ is 0. This indicates that the regression line is a perfect fit to the sample data. The larger the value of $s_{Y.X}$, the greater is the scatter around the regression line. As indicated earlier, in the case of a perfect linear relationship between X and Y for a *sample* of data, given a value of X, we could estimate or predict the corresponding *sample* Y value perfectly. However, to make predictions about values in the population not included in our sample, we would have to take account of the fact that the sample regression line we have fit to the data plotted on a scatter diagram may differ from the unknown regression line because of chance errors of sampling. That is, because of chance sampling errors, the a and b values in the sample regression line, $Y_C = a + bX$, may differ from the A and B values in the true population regression line, $\mu_{Y.X} = A + BX$. This means that the height and slope of the sample regression line may differ from the height and slope of the population regression line. Therefore, in making predictions about items in the population that were not included in the sample, we cannot simply use $s_{Y.X}$ in establishing confidence intervals, but must use appropriate standard errors that take the aforementioned chance errors into account. We now turn to the discussion of such confidence intervals.

Types of Interval Estimates

There are two types of estimates, or predictions, of values of the dependent variable that are ordinarily made in regression analysis. The first is an estimate of a *conditional mean*. Thus, for example, in our food expenditures illustration, we may be interested in estimating *average* food expenditures for families with annual net incomes of $8000; that is, we may want an estimate of the mean of the conditional probability distribution of food expenditures for families with annual net incomes of $8000.

The second type of estimate is the case of predicting an *individual value* of the dependent variable Y. In this situation, we wish to predict a *single value* of the dependent variable Y rather than a *conditional average value,* as in the first type of estimate. Thus, for example, we may wish to predict food expenditures for a *particular family* whose income is $8000. Such single values cannot be predicted with as much precision as for conditional means. *Interval estimates* are generally made for both types of estimates, with the width of the intervals indicating the precision of the estimation procedure. Interval estimates for a conditional mean are usually referred to as *confidence intervals*, those for an individual value of Y as *prediction intervals.* We consider the two types of estimates in the following subsections.

Confidence Interval
Estimate of a Conditional Mean

In Chapter 6, we discussed methods for estimating a population parameter from a statistic observed in a simple random sample. Thus, for example, we can establish a *confidence interval* for the population mean of food expenditures for the sample of ten families for whom data were shown in Table 9.1. Assuming that the dependent variable Y is normally distributed, and using Equation 6.14, we can calculate the confidence limits for population mean food expenditures as

$$\bar{Y} \pm t s_{\bar{Y}} = \bar{Y} \pm t \frac{s_Y}{\sqrt{n}}$$

where $\bar{Y}$ is the sample mean for the desired confidence level, t is determined for $n - 1$ degrees of freedom for the desired confidence level, $s_{\bar{Y}}$ is the estimated standard error of the mean, s_Y is the sample standard deviation of the Y data, and n is the sample size (in this case, 10). The symbol Y is used for the variable of interest here rather than the X used in Equation 6.14. Carrying out the arithmetic in this illustration for a 95% confidence interval, we have:

$$\bar{Y} = 18 \text{ (\$ hundreds)}$$

$$s_Y = \sqrt{\frac{\Sigma (Y - \bar{Y})^2}{n - 1}} = \sqrt{\frac{156}{9}} = 4.163 \text{ (\$ hundreds)}$$

We have that $t = 2.262$ for $\nu = 10 - 1 = 9$ degrees of freedom (obtained from Table A-6 of Appendix A for a 95% confidence coefficient, that is, 5% area in both tails, or 2.5% in each tail), and

$$s_{\bar{Y}} = \frac{s_Y}{\sqrt{n}} = \frac{4.163}{\sqrt{10}} = 1.317$$

Therefore, $\bar{Y} \pm 2.262 s_{\bar{Y}} = 18 \pm 2.262(1.317)$ ($ hundreds).

Hence, the 95% confidence limits are 15.021 and 20.979 ($ hundreds). That is, population mean food expenditures would be included in 95% of the intervals so constructed.

Interval estimates can be determined from a regression equation in an essentially similar way. Let us consider the construction of an interval estimate for a conditional mean $\mu_{Y.X}$. Returning to our illustration, suppose we wanted to estimate the *mean* food expenditures for families with annual net incomes of $8000. As we have seen, another way of stating this is that we wish to estimate the mean of the conditional probability distribution of food expenditures for $X = 8$ ($ thousands). An unbiased estimate of this conditional mean, denoted $\mu_{Y.8}$, is given by Y_C from our sample regression line for $X = 8$ ($ thousands). We previously calculated Y_C in Section 9.4 as

$$Y_C = 7.698 + 1.717(8) = 21.434 \text{ ($ hundreds)}$$

The confidence interval estimate for the conditional mean, also called the *confidence interval for the regression line*, is given by

(9.11) $$Y_C \pm ts_{Y_C}$$

where s_{Y_C} is the estimated standard error of the conditional mean and t is the t-multiple determined for $n - 2$ degrees of freedom for the desired confidence level. The number of degrees of freedom is $n - 2$ because 2 degrees of freedom are lost by using a and b from the sample regression equation, $Y_C = a + bX$, to estimate A and B, in the population regression line, $\mu_{Y.X} = A + BX$.

The estimated standard error of the conditional mean is given by[3]

---

[3] The formula for s_{Y_C} may be derived as follows. The sample regression equation can be expressed as

$$Y_C = \bar{Y} + b(X - \bar{X})$$

Using Rule 9 of Appendix C, we have

$$\sigma^2(Y_C) = \sigma^2[\bar{Y} + b(X - \bar{X})]$$
$$= \sigma^2(\bar{Y}) + \sigma^2[b(X - \bar{X})]$$
$$= \sigma^2(\bar{Y}) + (X - \bar{X})^2\sigma^2(b)$$

Hence,

$$\sigma^2(Y_C) = \frac{\sigma^2_{Y.X}}{n} + (X - \bar{X})^2 \frac{\sigma^2_{Y.X}}{\Sigma (X - \bar{X})^2}$$

(9.12)
$$s_{Y_C} = s_{Y.X} \sqrt{\frac{1}{n} + \frac{(X - \bar{X})^2}{\Sigma\,(X - \bar{X})^2}}$$

This expression for s_{Y_C} can be derived by taking the variance (and then the standard deviation) of the sample regression equation, which can be expressed for this purpose in the form

(9.13)
$$Y_C = \bar{Y} + b(X - \bar{X})$$

This regression equation expresses Y_C, the estimated value of the dependent variable, as a function of $X - \bar{X}$, a deviation of an observation X from the mean, $\bar{X}$. For a particular X value, the same estimate Y_C is obtained from $Y_C = a + bX$ as from the regression equation given in (9.13). When we view the problem of estimating a conditional mean according to (9.13), we note that the variance of Y_C is composed of two *independent* parts, the variance in the average height of the regression line ($\bar{Y}$) and the variance in the slope of the line (b). As is evident from Equation 9.12, the farther an X value lies from the mean, the larger is $(X - \bar{X})^2$, and the wider is the confidence interval at that X. Note that in Equation 9.13, if $X = \bar{X}$, then $Y_C = \bar{Y}$. That is, we obtain the familiar result that the estimated value of the dependent variable for an observation at the mean of the X variable, $\bar{X}$, is $\bar{Y}$, the mean of the dependent variable.

These situations are depicted in Figure 9-7.

Continuing with our illustration of obtaining an interval estimate of the conditional mean food expenditures for families with annual net incomes of $8000, we compute the estimated standard error of the conditional mean. A more convenient form of Equation 9.12 for computation purposes is

(9.14)
$$s_{Y_C} = s_{Y.X} \sqrt{\frac{1}{n} + \frac{(X - \bar{X})^2}{\Sigma\,X^2 - \dfrac{(\Sigma X)^2}{n}}}$$

Substituting into Equation 9.14 for $X = 8$ ($ thousands), we get

---

Substituting the estimator $s^2_{Y.X}$ for $\sigma^2_{Y.X}$ on the right side, and using the notation $s^2_{Y_C}$ for the estimator of $\sigma^2(Y_C)$ on the left side, we obtain

$$s^2_{Y_C} = \frac{s^2_{Y.X}}{n} + (X - \bar{X})^2\,\frac{s^2_{Y.X}}{\Sigma\,(X - \bar{X})^2}$$

Taking the square root of this last expression leads to the formula for the estimated standard error of the conditional mean:

$$s_{Y_C} = s_{Y.X} \sqrt{\frac{1}{n} + \frac{(X - \bar{X})^2}{\Sigma\,(X - \bar{X})^2}}$$

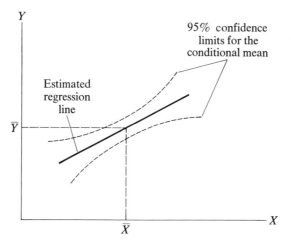

FIGURE 9-7
95% confidence interval for the conditional mean.

$$ s_{Y_c} = 1.595 \sqrt{\frac{1}{10} + \frac{(8-6)^2}{406 - \frac{(60)^2}{10}}} $$

$$ s_{Y_c} = 0.6897 \ (\$ \text{ hundreds}) $$

Hence, a 95% confidence interval for the conditional mean obtained by substitution into Equation 9.11 is

$$ 21.434 \pm (2.306)(0.6897) \quad \text{or} \quad (19.844, 23.024) \text{ in } \$ \text{ hundreds} $$

where the t value of 2.306 for a 95% confidence coefficient was obtained from Table A-6 of Appendix A for $n - 2 = 10 - 2 = 8$ degrees of freedom. Therefore, the desired 95% confidence limits are about $1980 to $2300 for mean food expenditures for families with annual net incomes of $8000.

Prediction Interval for an Individual Value of Y

Just as in the case of estimating a conditional mean for a given X value, an unbiased estimate in predicting an *individual Y value* for a given X is obtained from the point on the estimated regression line corresponding to X. That is, the best point prediction for the individual Y value is obtained from the regression $Y_c = a + bX$, by substituting the given X value.

However, for a given X value, we would not be able to predict an individual value of Y with as much precision as we can estimate a conditional mean, which is a more stable value. The reason is that the sampling variability of a predicted individual Y value consists of two components, namely, the sampling error in using Y_C to estimate $\mu_{Y.X}$ and the variation in the conditional probability distribution of Y values. The first component is measured by the sampling variance of a conditional mean (the square of Equation 9.12); the second component is measured by the estimated conditional variance of an individual Y observation. Hence, using the symbol s^2_{IND} to denote the variance of a predicted individual Y value, which is the total of the two components, we have

$$(9.15) \qquad s^2_{\text{IND}} = s^2_{Y.X}\left(\frac{1}{n} + \frac{(X - \bar{X})^2}{\Sigma\,(X - \bar{X})^2}\right) + s^2_{Y.X}$$

Factoring out $s^2_{Y.X}$ on the right side of (9.15) and rearranging, we get

$$(9.16) \qquad s^2_{\text{IND}} = s^2_{Y.X}\left(1 + \frac{1}{n} + \frac{(X - \bar{X})^2}{\Sigma\,(X - \bar{X})^2}\right)$$

Taking the square root of formula (9.16), we have

$$(9.17) \qquad s_{\text{IND}} = s_{Y.X}\sqrt{1 + \frac{1}{n} + \frac{(X - \bar{X})^2}{\Sigma\,(X - \bar{X})^2}}$$

Just as in the case of the standard error of the conditional mean, a more convenient form of Equation 9.17 for purposes of calculation is

$$(9.18) \qquad s_{\text{IND}} = s_{Y.X}\sqrt{1 + \frac{1}{n} + \frac{(X - \bar{X})^2}{\Sigma\,X^2 - \dfrac{(\Sigma\,X)^2}{n}}}$$

The prediction of an individual value of Y is usually referred to as an *individual forecast*, and s_{IND} is called the *standard error of forecast*.[4] The standard error of forecast can be used to set up a prediction interval for an individual Y value in a manner analogous to setting up confidence intervals for the conditional mean.

Returning to the food expenditures example, assume that we want to predict food expenditures for a particular family with an annual net income of $8000. The prediction interval for an individual Y value is given by

$$(9.19) \qquad Y_C \pm t s_{\text{IND}}$$

where Y_C is the individual forecast as determined from the sample regression line, s_{IND} is the standard error of forecast, and t is the t multiple

---

[4] Actually, s_{IND} should be referred to as the "estimated standard error of forecast" just as s_{Y_C} is the "estimated standard error of the conditional mean." In both of these cases, "estimated" is appropriate because $s_{Y.X}$ is an estimate of the population $\sigma_{Y.X}$. However, it has become conventional to refer to the "standard error of forecast" without the use of the word "estimated."

determined for $n - 2$ degrees of freedom for the desired confidence level.

Substituting into formula (9.18) for $X = 8$ ($ thousands), we obtain for the standard error of forecast

$$s_{Y_{IND}} = 1.595 \sqrt{1 + \frac{1}{10} + \frac{(8 - 6)^2}{406 - \frac{(60)^2}{10}}}$$

$$s_{Y_{IND}} = 1.738 \text{ ($ hundreds)}$$

Therefore, by formula (9.19), a 95% prediction interval for the individual forecast of the Y value for a family with an $8000 annual net income is

$$21.434 \pm 2.306 \ (1.738)$$

where, as earlier, the t value of 2.306 for a 95% confidence coefficient was obtained from Table A-6 for $n - 2 = 10 - 2 = 8$ degrees of freedom.

Hence, the prediction interval is from 17.426 to 25.442 ($ hundreds), or approximately $1740 to $2540 of food expenditures. We note that, as expected, these limits are wider than those previously computed for the conditional mean. Also, as was true for confidence limits for the conditional mean, the further the X value is from $\bar{X}$, the wider is the prediction interval. The relationship between confidence intervals for the conditional mean and prediction intervals for individual forecasts is depicted in Figure 9-8.

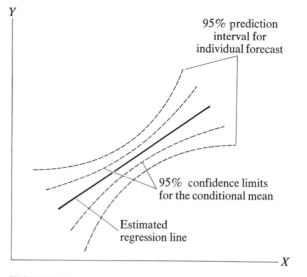

FIGURE 9-8
Confidence intervals for the conditional mean and prediction interval for an individual forecast.

Width of Confidence and Prediction Intervals

As can be seen from Equations 9.12 and 9.17, a number of factors affect the sizes of the estimated standard error of the conditional mean and the standard error of forecast and, hence, the width of confidence and prediction intervals. The larger the sample size, n, the smaller are these standard errors, and therefore the narrower are the widths of the intervals. This agrees with our intuitive notion that larger sample sizes provide greater precision of estimation.

Second, as noted earlier, the greater the deviation of X from $\bar{X}$, the greater is the standard error and the wider are the intervals for the given X value. This means that confidence and prediction intervals may be quite wide for very small or very large X values as compared to the analogous intervals for average-sized X values.

Third, the larger the estimated standard error of estimate, $s_{Y.X}$, the larger are the confidence and prediction intervals. This agrees with the intuitive idea that the more variable (less uniform) the data, the less precise would be predictions from these data.

Finally, we note that the more variability there is in the sample of X values (in our illustrative problem, the more variability there is in family incomes), the larger is $\Sigma (X - \bar{X})^2$, and hence the smaller are the standard errors and the narrower are the confidence and prediction intervals. This is consistent with the idea that, for example, if we observe a relationship for families with large variation in incomes, we would expect to be able to estimate food expenditures better than if the variation in incomes was very small.

Further Remarks

We have discussed estimation procedures for confidence intervals for conditional means and for prediction intervals for individual forecasts. Setting up confidence intervals for conditional means is appropriate whenever we are interested in estimating the *average value* of Y for a given X. In this chapter, we illustrated the estimation of *average food expenditures* for families with $8000 of income. The same procedure would be used if in other regression analyses we were interested, for example, in confidence intervals for *average* or *expected sales* for an expenditure of $1 million on advertising, for *average number of accidents* for a company with a certain number of workers exposed to some hazard, or *average weight* for black males in the United States who are 70 inches tall.

On the other hand, prediction intervals for individual forecasts are appropriate whenever we are interested in predicting an *individual Y value* for a given X. Conceptually, we wish to predict the Y value for a new observation drawn from the same population as the sample from which we have constructed the regression line. Hence, in this chapter, we illustrated the prediction interval estimation of *actual food expenditures* of a *particular* family with an $8000 income. Analogously, we would use this procedure if we were interested in prediction intervals for the *specific amount of sales* for *the next company* that spends $1 million on advertising, for the *specific number* of *accidents* for a *particular company* with 150 employees exposed to some hazard, or the *actual weight* of a *particular* black male in the United States who is 70 inches tall.

As we have noted, when we predict an individual Y value, the prediction interval is wider than confidence intervals for a conditional mean because in the former case we are concerned with the dispersion of the distribution within which *individual values* of Y are predicted to lie. This distribution of individual values is not subject to the dispersion-reducing effect of averaging that is present in setting up confidence intervals for a conditional mean.

The usefulness of confidence intervals or prediction intervals depends on the purposes for which they are used. For example, for long-range planning purposes, relatively wide limits may be appropriate and useful. On the other hand, for short-term operational decision making, narrower and therefore more precise intervals may be required. In a two-variable regression analysis, the standard error of estimate, $s_{Y.X}$, may be so large as to yield confidence and prediction intervals that are too wide for the investigator's purposes. In such a case, the introduction of additional independent variables may be required to obtain greater precision of estimation and prediction. We discuss the use of two or more independent variables in Section 9.9, which deals with *multiple* regression and correlation analysis.

Exercises

1. An accounting standards board investigating the treatment of research and development expenses by the nation's major electronics firms was interested in the relationship between a firm's research and development expenditures and its earnings. The board compiled the following data on 15 firms for the year 1976:

Firm	Earnings (1976) ($Millions)	Research and Development Expenditures for Previous Year (1975) ($Millions)
A	$221	$15.0
B	83	8.5
C	147	12.0
D	69	6.5

(Continued)

Firm	Earnings (1976) ($Millions)	Research and Development Expenditures for Previous Year (1975) ($Millions)
E	41	4.5
F	26	2.0
G	35	0.5
H	40	1.5
I	125	14.0
J	97	9.0
K	53	7.5
L	12	0.5
M	34	2.5
N	48	3.0
O	64	6.0

a. Estimate the linear regression equation, using the method of least squares, with research and development expenditures as the independent variable.
b. What are the estimated 1976 earnings for a firm that spent $10 million on research and development in 1975?
c. Calculate the standard error of estimate, $s_{Y.X}$. What does this measure?
d. Obtain a 95% confidence interval for the conditional mean 1976 earnings for the firm in part (b).

2. Texecon Products, Inc., a major producer of household products, has introduced 14 new products during the past two years. The marketing research unit needs an estimate of the relationship between first-year sales and an appropriate independent variable; this relationship will be used in future planning in marketing and advertising. The researchers have constructed a variable called "customer awareness," measured by the proportion of consumers who had heard of the product by the third month after its introduction. The data are as follows:

Product	First Year Sales ($Millions)	Customer Awareness (%)
E-Z-Kleen	$ 82	50
Sud-Z-Est	46	45
Alumofoil	17	15
Backscratcher	21	15
Pest Killer	112	70
Liquid Lush	105	75
Bubbly Bath	65	60
Wipe Away	55	40
Whirlwind	80	60
Magic Mop	43	25

(Continued)

Product	First Year Sales ($Millions)	Customer Awareness (%)
Cobweb Cure	79	50
Oven Eater	24	20
Dirt-B-Gone	30	30
Dustbowl	11	5

a. Find the least squares regression equation, with customer awareness as the independent variable.

b. Calculate the standard error of estimate, $s_{Y.X}$. What assumption is necessary in using $s_{Y.X}$ to estimate $\sigma_{Y.X}$, the conditional standard deviation around the population regression line?

c. Calculate a 95% confidence interval estimate for conditional mean sales for a customer awareness level of 35%.

d. Obtain a 95% prediction interval for new product sales for the same customer awareness level as in part (c).

e. Which interval was wider: that obtained in part (c) or that obtained in part (d)? Why?

3. The marketing research department of Texecon Products, Inc. decided that the customer awareness variable in Exercise 2 suffered from too many shortcomings. You have been instructed to perform the same analysis as in Exercise 2, using advertising expenditures as the independent variable.

Product	Advertising Expenditures ($Millions)
E-Z-Kleen	$1.8
Sud-Z-Est	1.2
Alumofoil	0.4
Backscratcher	0.5
Pest Killer	2.5
Liquid Lush	2.5
Bubbly Bath	1.5
Wipe Away	1.2
Whirlwind	1.6
Magic Mop	1.0
Cobweb Cure	1.5
Oven Eater	0.7
Dirt-B-Gone	1.0
Dustbowl	0.8

a. Estimate the linear regression equation, using the method of least squares.

b. Calculate the standard error of estimate, $s_{Y.X}$.

   c. Obtain a 95% confidence interval estimate for conditional mean sales for an advertising expenditure level of $1 million.

   d. Obtain a 95% prediction interval for new product sales for the same advertising expenditure level as in part (c).

   e. What factors affect the width of confidence and prediction interval estimates?

4. A congressional committee studying federal tax reform is taking a close look at the taxes that large corporations pay on reported income. A random sample of 12 firms yielded the following statistics:

Firm	Reported Income in 1976 ($Millions)	Income Taxes Paid in 1976 ($Millions)
Excelsior	$300	$120
Diamond International	250	110
General Nuts and Bolts	425	200
Scolding, Inc.	210	100
American Abacus	170	75
Stronghold Safes, Inc.	125	65
Dooley Brothers	100	40
Marathon Motors	280	125
Slick Oil Company	375	175
United Pickle	115	50
Agony Airlines	80	40
Stalwart Steel, Inc.	210	100

   a. Estimate the linear regression equation, using the method of least squares, with reported 1976 income as the independent variable.

   b. Calculate the standard error of estimate, $s_{Y.X}$.

   c. Interpret the estimated regression coefficient, b, in terms of this problem. Calculate a weighted average tax rate. How close are the two estimates?

   d. Obtain 95% confidence intervals for conditional mean income taxes paid for the following levels of reported income: $100 million, $220 million, and $340 million. Calculate the width of these intervals. What do you observe?

   e. Follow the instructions in part (d) for estimating prediction intervals for income taxes paid for each income level. Calculate and comment upon the interval widths.

5. A federal agency is completing a study on the feasibility of national health insurance and has requested information from its research department on family medical expenditures. The researchers have collected data on 122 families and present these results:

$$X = \text{annual family income (thousands of dollars)}$$
$$Y = \text{annual family medical expenditures (hundreds of dollars)}$$

$$n = 122$$
$$\Sigma X = 1464$$
$$\Sigma X^2 = 18{,}000$$
$$\Sigma Y = 1220$$
$$\Sigma Y^2 = 12{,}475$$
$$\Sigma XY = 14{,}900$$

a. Estimate the linear regression equation, using the method of least squares, with annual family income as the independent variable.
b. Calculate the standard error of estimate, $s_{Y \cdot X}$.
c. Estimate the annual family medical expenditures for a family with an annual income of $8000. Do the same for a family with an income of $25,000. Compare the two results as percentages of annual family income.
d. Obtain 95% confidence interval estimates for the conditional mean annual family medical expenditures for the two families in part (c).
e. Describe the two components of the variance of Y_c.

9.6 CORRELATION ANALYSIS: MEASURES OF ASSOCIATION

In the preceding two sections, regression analysis was discussed, with emphasis on estimation and measures of error in the estimation process. We now turn to correlation analysis, in which the basic objective is to obtain a measure of the degree of association between two variables. In this analysis, interest centers on the strength of the relationship between the variables, that is, on how well the variables are correlated. The assumptions of the two-variable correlation model are as follows:

1. Both X and Y are random variables

2. Both X and Y are normally distributed. The two distributions need not be independent

3. The standard deviations of the Ys are assumed to be equal for all values of X, and the standard deviations of the Xs are assumed to be equal for all values of Y

Note that in the correlation model, both X and Y are assumed to be random variables. On the other hand, in the regression model, only Y is a random variable, and the Y observations are treated as a random sample from the conditional distribution of Y for a given X.

The Coefficient of Determination

A measure of the amount of correlation between Y and X can be explained in terms of the relative variation of the Y values around the regression line and the corresponding variation around the mean of the Y

variable. The term "variation," as used in statistics, conventionally refers to a sum of squared deviations.

The variation of Y values around the regression line is measured by

(9.20) $$\Sigma\,(Y - Y_c)^2$$

The variation of Y values around the mean of the Y variable is measured by

(9.21) $$\Sigma\,(Y - \bar{Y})^2$$

The first of these expressions (9.20) is the sum of the squared vertical deviations of the Y values from the regression line. The second expression is the sum of the squared vertical deviations from the horizontal line $Y = \bar{Y}$. The relationship between the variations around the regression line and the mean can be summarized in a single measure to indicate the degree of association between X and Y. The measure used for this purpose is the *sample coefficient of determination*, defined as follows:

(9.22) $$r^2 = 1 - \frac{\Sigma\,(Y - Y_c)^2}{\Sigma\,(Y - \bar{Y})^2}$$

As we shall see from the subsequent discussion, r^2 may be interpreted as the proportion of variation in the dependent variable Y that has been accounted for, or "explained," by the relationship between Y and X expressed in the regression line. Hence, it is a measure of the degree of association or correlation between Y and X.

In order to present the rationale of this measure of strength of the relationship between Y and X, we will consider two extreme cases, zero linear correlation and perfect direct linear correlation. The term

TABLE 9-4
Two Sets of Data Displaying (a) Zero Linear Correlation and (b) Perfect Direct Linear Correlation

Observation	(a) X	(a) Y	Observation	(b) X	(b) Y
A	1	4	A	1	2
B	1	6	B	2	4
C	2	4	C	3	6
D	2	6	D	4	8
E	3	4	E	5	10
F	3	6	F	6	12
G	4	4	G	7	14
H	4	6	H	8	16
		$\bar{Y} = 5$			$\bar{Y} = 9$

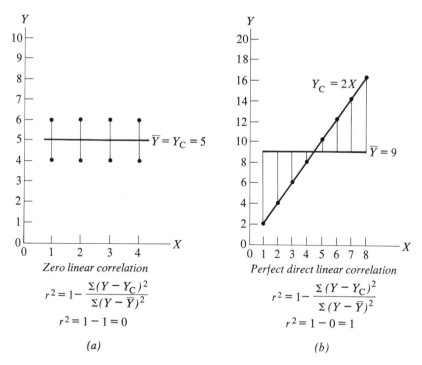

FIGURE 9-9
Scatter diagrams representing zero linear correlation and perfect direct linear correlation.

"linear" indicates that a straight line has been fitted to the X and Y values, and the term "direct" indicates that the line has a positive slope.

Two sets of data are presented in Table 9-4 labeled (a) and (b). They are shown in the form of scatter diagrams in Figure 9-9. As we shall see, the data in (a) and (b) illustrate the case of zero linear correlation and perfect direct linear correlation, respectively. In the discussion that follows, we will assume that the observations shown in (a) and (b) represent simple random samples from their respective universes. Therefore, we employ notation appropriate to samples. An analogous argument can be presented assuming that the observations represent population data, and the notation would change correspondingly. The calculations given below the scatter diagrams in Figure 9-9 will be explained in terms of the data displayed in the charts.

Case (a) represents a situation in which $\bar{Y}$, the mean of the Y values, coincides with a least squares regression line fitted to these data. Even without doing the arithmetic, we can see why this is so. The slope of the regression line is 0, because the same Y values are observed for $X = 1, 2, 3,$ and 4. Thus, the regression line would coincide with the mean of the Y values, balancing deviations above and below the regression line. Another way of observing this relationship is in terms of the first of the two equations used to solve for a and b. In Equation 9.6, i.e., $a = \bar{Y} - b\bar{X}$, since $b = 0$, $a = \bar{Y}$. Hence, the regression line has a Y intercept equal to $\bar{Y}$. Since it is also a horizontal line, the regression line coincides with $\bar{Y}$. From the point of view of estimation of the Y variable, the regression line represents no improvement over the mean of the Y values. This can be shown by a comparison of $\Sigma (Y - Y_c)^2$, the variation around the regression line, and $\Sigma (Y - \bar{Y})^2$, the variation around the mean of the Y values. In this case, the two variations are equal. These variations may be interpreted graphically as the sum of the squares of the vertical distances between the points on the scatter diagram and $\bar{Y}$, shown in Figure 9-9(A).

Now, let us consider case (b). This is a situation in which the regression line is a perfect fit to the data. The regression equation is a very simple one, which can be determined by inspection. The Y intercept is 0, since the line passes through the origin (0, 0). The slope is 2, because for every unit increase in X, Y increases by 2 units. Hence, the regression equation is $Y_C = 2X$, and all the data points lie on the regression line. As far as the data in the sample are concerned, perfect predictions are provided by this regression line. Given a value of X, the corresponding value of Y can be correctly estimated from the regression equation, indicating a perfect linear relationship between the two variables. Again, a comparison can be made of $\Sigma (Y - Y_c)^2$ and $\Sigma (Y - \bar{Y})^2$. Since all points lie on the regression line, the variation around the line, $\Sigma (Y - Y_c)^2$, is 0. On the other hand, the variation around the mean, $\Sigma (Y - \bar{Y})^2$, is some positive number, in this case, 168.

As indicated in Figure 9-9(a), when there is no linear correlation between X and Y, the sample coefficient of determination, r^2, is 0. This follows from the fact that since $\Sigma (Y - Y_c)^2$ and $\Sigma (Y - \bar{Y})^2$ are equal, the ratio $\Sigma (Y - Y_c)^2/\Sigma (Y - \bar{Y})^2$ equals 1. Hence, from Equation 9.22, r^2 equals 0, because the computation of the coefficient of determination requires subtraction of this ratio from 1.

On the other hand, as indicated in Figure 9-9(b), when there is perfect linear correlation between X and Y, the sample coefficient of determination, r^2, is 1. In this case, the variation around the regression line is 0, while the variation around the mean is some positive number. Thus, the ratio $\Sigma (Y - Y_c)^2/\Sigma (Y - \bar{Y})^2$ is 0. Hence, from Equation 9.22, r^2 equals 1 when the value of this ratio is subtracted from 1.

In realistic problems, r^2 falls somewhere between the two limits 0 and 1. A value close to 0 suggests that there is not much linear correlation between X and Y; a value close to 1 connotes a strong linear relationship between X and Y.

Population Coefficient of Determination

The measure r^2 has been referred to as the *sample* coefficient of determination. That is, it pertains only to the sample of n observations studied. The regression line computed from the sample may be viewed as an estimate of the true population regression line, which may be denoted

(9.23)
$$\mu_{Y.X} = A + BX$$

The corresponding population coefficient of determination is defined as

(9.24)
$$\rho^2 = 1 - \frac{\sigma^2_{Y.X}}{\sigma^2_Y}$$

The use of the symbol ρ^2 (rho squared) adheres to the usual convention of employing a Greek letter for a population parameter corresponding to the same letter in our alphabet that denotes a sample statistic. In the definition of ρ^2, $\sigma^2_{Y.X}$ is the variance around the population regression line $\mu_{Y.X} = A + BX$ and σ^2_Y is the variance around the population mean of the Ys, denoted μ_Y. Hence, both the sample and population coefficients of determination are equal to 1 minus a ratio of the variability around the regression line to the variability around the mean of the Y values.

A slightly different form of the *sample* coefficient of determination, which is directly parallel to Equation 9.24, is

(9.25)
$$r^2_C = 1 - \frac{s^2_{Y.X}}{s^2_Y} = 1 - \frac{\Sigma (Y - Y_C)^2/(n - 2)}{\Sigma (Y - \bar{Y})^2/(n - 1)}$$

The quantity r^2_C is referred to as the *corrected* or *adjusted* sample coefficient of determination. This terminology is used because $s^2_{Y.X}$ and s^2_Y are estimators of $\sigma^2_{Y.X}$ and σ^2_Y that make the appropriate corrections or adjustments for degrees of freedom.[5] Since $s^2_{Y.X}$ and s^2_Y are unbiased estimators of $\sigma^2_{Y.X}$ and σ^2_Y, the adjusted sample coefficient of determination, r^2_C, rather than the unadjusted coefficient, r^2, is ordinarily used when interest is focused on estimating the population coefficient of determination, ρ^2. However, in the discussion that follows we use only the unadjusted value, r^2, because of the complication associated with the divisors $n - 2$ and $n - 1$ in the adjusted measure.

Interpretation of the Coefficient of Determination

It is useful to consider in more detail the specific interpretations that may be made of coefficients of determination. For convenience, only the sample coefficient r^2 will be discussed, but the corresponding meanings for ρ^2 are obvious.

---

[5] The relationship between r^2_C and r^2 is given by $r^2_C = 1 - (1 - r^2)\left(\dfrac{n - 1}{n - 2}\right)$. For large sample sizes $\left(\dfrac{n - 1}{n - 2}\right)$ is close to 1 and r^2_C and r^2 are approximately equal.

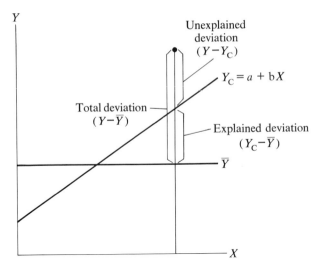

FIGURE 9-10
Graphical representation of total, explained, and unexplained variation.

An important interpretation of r^2 may be made in terms of variation in the dependent variable Y, which has been explained by the regression line. The problem of estimation is conceived of in terms of "explaining" or accounting for the variation in the dependent variable Y. Figure 9-10, on which a single point is shown, gives a graphical interpretation of the situation. In this context, if $\bar{Y}$, the mean of the Y values, were used to estimate the value of Y, the total deviation would be $Y - \bar{Y}$. We can think of this deviation as being decomposed as follows:

$$\text{total deviation} = \text{explained deviation} + \text{unexplained deviation}$$
$$(Y - \bar{Y}) \quad = \quad (Y_c - \bar{Y}) \quad + \quad (Y - Y_c)$$

If the regression line were used to estimate the value of Y, we would now have a closer estimate. As shown in the figure, there is still an "unexplained deviation" of $(Y - Y_c)$, but we have explained (or accounted for) $(Y_c - \bar{Y})$ out of the total deviation by assuming that the relationship between X and Y is given by the regression line.

In an analogous manner, we can partition the total variation of the dependent variable (or total sum of squares), $\Sigma (Y - \bar{Y})^2$, as follows:

$$\text{total variation} = \text{explained variation} + \text{unexplained variation}$$
$$\Sigma (Y - \bar{Y})^2 \quad = \quad \Sigma (Y_c - \bar{Y})^2 \quad + \quad \Sigma (Y - Y_c)^2$$

The ratio $\Sigma (Y - Y_c)^2 / \Sigma (Y - \bar{Y})^2$ is the proportion of total variation that remains unexplained by the regression equation; correspondingly, $1 - [\Sigma (Y - Y_c)^2 / \Sigma (Y - \bar{Y})^2]$ represents the *proportion of total variation in Y that has been explained* by the regression equation. These ideas may be summarized as follows:

(9.26) $$r^2 = 1 - \frac{\Sigma\,(Y - Y_c)^2}{\Sigma\,(Y - \bar{Y})^2} = 1 - \frac{\text{unexplained variation}}{\text{total variation}}$$

$$r^2 = \frac{\text{explained variation}}{\text{total variation}}$$

A simple numerical example helps to clarify these relationships. Let $\Sigma\,(Y - \bar{Y})^2 = 10$ and $\Sigma\,(Y - Y_c)^2 = 4$. Thus, $r^2 = 1 - \frac{4}{10} = \frac{6}{10} = 60\%$. In this problem, 10 units of total variation in Y have to be accounted for. After we fit the regression line, the residual variation or unexplained variation amounts to 4 units. Hence, 60% of the total variation in the dependent variable is explained by the relationship between Y and X expressed in the regression line.

Calculation of the Sample Coefficient of Determination

The computation of r^2 from the definitional formula, Equation 9.22, becomes quite tedious, particularly when there are a large number of observations in the sample. Just as in the case of the standard error of estimate, shorter methods of calculation are ordinarily used. These shortcut formulas are particularly helpful when computations are carried out by hand or on a calculator, but even when computers are used, they represent more efficient methods of computation. Such a formula, which only involves quantities already calculated, is

(9.27) $$r^2 = \frac{a\Sigma\,Y + b\Sigma\,XY - n\bar{Y}^2}{\Sigma\,Y^2 - n\bar{Y}^2}$$

Substituting into Equation 9.27, we obtain

$$r^2 = \frac{(7.698)(180) + (1.717)(1159) - 10(18)^2}{(3396) - 10(18)^2} = 0.870$$

Thus, for our sample of ten families, about 87% of the variation in annual food expenditures was explained by the regression equation, which related such expenditures to annual net income.

The Coefficient of Correlation

A widely used measure of the degree of association between two variables is the coefficient of correlation, which is simply the square root of the coefficient of determination. Thus, the population and sample coefficients of correlation are

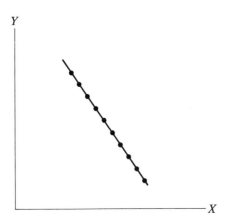

FIGURE 9-11
Scatter diagram representing perfect inverse linear correlation, $r = -1$.

(9.28) $$\rho = \sqrt{\rho^2}$$

and

(9.29) $$r = \sqrt{r^2}$$

Again, for convenience, our discussion will relate only to the sample value.

The algebraic sign attached to r is the same as that of the regression coefficient, b. Thus, if the slope of the regression line, b, is positive, then r is also positive; if b is negative, r is negative.[6] Hence, r ranges in value from -1 to $+1$. A figure of $r = -1$ indicates a perfect inverse linear relationship, $r = +1$ indicates a perfect direct linear relationship, and $r = 0$ indicates no linear relationship.

Illustrative scatter diagrams for the cases of $r = +1$ ($r^2 = 1$) and $r = 0$ ($r^2 = 0$) were given in Figure 9-9. A corresponding scatter diagram for $r = -1$, the case of perfect inverse linear correlation, is shown in Figure 9-11. As indicated in the graph, the slope of the regression line is negative and every point falls on the line. Thus, for example, if the slope of the regression line, b, were equal to -2, this would mean that with each increase of 1 unit in X, Y would decrease by 2 units. Since all points fall on the regression line in the case of perfect inverse correlation, $\Sigma (Y - Y_c)^2 = 0$. Therefore, substituting into Equation 9.22 to compute the sample coefficient of determination, we have $r^2 = 1 - 0 = 1$. Taking the square root, we obtain $r = \sqrt{1} = \pm 1$.

---

[6] An interesting relationship between r and b is given by $b = r(\Sigma (Y - \bar{Y})^2 / \Sigma (X - \bar{X})^2)$. Since $\Sigma (Y - \bar{Y})^2$ and $\Sigma (X - \bar{X})^2$ are positive numbers, b has the same sign as r. We also note that when the regression line is horizontal, $b = 0$ and also $r = 0$.

However, since the b value is negative (that is, X and Y are inversely correlated), we assign the negative sign to r, and $r = -1$.

In our problem,

$$r = \sqrt{0.870} = 0.933$$

The sign is positive because b was positive, indicating a direct relationship between food expenditures and net income.

Despite the rather common use of the coefficient of correlation, it is preferable for interpretation purposes to use the coefficient of determination. As we have seen, r^2 can be interpreted as a proportion or a percentage figure. When the square root of a percentage is taken, the specific meaning becomes obscure. Furthermore, since r^2 is a decimal value (unless it is equal to 0 or 1), its square root, or r, is a larger number. Thus, the use of r values to indicate the degree of correlation between two variables tends to give the impression of a stronger relationship than is actually present. For example, an r value of $+0.7$ or -0.7 seems to represent a reasonably high degree of association. However, since $r^2 = 0.49$, less than half of the total variance in Y has been explained by the regression equation.

It is useful to observe that the values of r and r^2 do not depend on the units in which X and Y are stated nor on which of these variables is selected as the dependent or independent variable. Whether a value of r or r^2 may be considered high depends somewhat upon the specific field of application. With some types of data, it is relatively unusual to find r values in excess of about 0.80. On the other hand, particularly in the case of time series data, r values in excess of 0.90 are quite common. In the following section, we consider the matter of determining whether the observed degree of correlation in a sample is sufficiently large to justify a conclusion that correlation between X and Y actually exists in the population.

9.7 INFERENCE ABOUT POPULATION PARAMETERS IN REGRESSION AND CORRELATION

In the procedures discussed to this point, computation and interpretation of *sample* measures have been emphasized. However, as we know from our study of statistical inference, sample statistics ordinarily differ from corresponding population parameters because of chance errors of sampling. Therefore, it is useful to have a protective procedure against the possible error of concluding from a sample that an association exists between two variables, while actually no such relationship exists in the population from which the sample was drawn. Hypothesis-testing techniques, such as those discussed in Chapter 7, can be employed for this purpose.

Inference about the Population Correlation Coefficient, ρ

Let us assume a situation in which we take a simple random sample of n units from a population and make paired observations of X and Y for each unit. The sample correlation coefficient, r, as defined in Equation 9.29, is calculated. The procedure involves a test of the hypothesis that the population correlation coefficient, ρ, is 0 in the universe from which the sample was drawn. In keeping with the language used in Chapter 7, we wish to test the null hypothesis that $\rho = 0$ versus the alternative that $\rho \neq 0$. Symbolically, we may write

$$H_0: \rho = 0$$
$$H_1: \rho \neq 0$$

If the computed r values in successive samples of the same size from the hypothesized population were distributed normally around $\rho = 0$, we would only have to know the standard error of r, σ_r, to perform the usual test involving the normal distribution. Although r values are not normally distributed, a similar procedure is provided by the statistic

(9.30) $$t = \frac{r - \rho}{s_r} = \frac{r}{\sqrt{(1 - r^2)/(n - 2)}}$$

which has a t distribution for $n - 2$ degrees of freedom. $s_r = \sqrt{(1 - r^2)/(n - 2)}$ is the estimated standard error of r, and we have $\rho = 0$ by the null hypothesis. It may be noted that despite the previous explanation that r^2 is easier to interpret than r, the hypothesis-testing procedure is in terms of r rather than r^2 values. The reason is that under the null hypothesis, $H_0: \rho = 0$, the sampling distribution of r leads to the t statistic, which is a well-known distribution and is relatively easy to work with. On the other hand, under the same hypothesis of no correlation in the universe, r^2 values, which range from 0 to 1, would not even be symmetrically distributed, and the sampling distribution would be more difficult to deal with. Suppose we wish to test the hypothesis that $\rho = 0$ at the 5% level of significance for our problem involving ten families. Since $r = 0.933$ and $n = 10$, substitution into (9.30) yields

$$t = \frac{0.933}{\sqrt{\dfrac{1 - 0.870}{10 - 2}}} = 7.32$$

Referring to Table A-6, we find a critical t value of 2.306 at the 5% level of significance, for eight degrees of freedom. Therefore, the decision rule is

DECISION RULE

1. If $-2.306 \leq t \leq 2.306$, accept H_0
2. If $t < -2.306$ or $t > 2.306$, reject H_0

Since our computed t value is 7.32, far in excess of the critical value, we conclude that the sample r value differs significantly from 0. We reject the hypothesis that $\rho = 0$ and conclude that there is a positive relationship between annual food expenditures and annual net income in the population from which our sample was drawn. Since the critical t value is 3.355 at the 1% level of significance (the smallest level shown in Table A-6 of Appendix A), it is extremely unlikely that an r value as high as 0.93 would have been observed in a sample of ten items drawn from a population in which X and Y were uncorrelated.

A few comments may be made concerning this hypothesis-testing procedure. First of all, this technique is valid only for a hypothesized universe value of $\rho = 0$. Other procedures must be used for assumed universe correlation coefficients other than 0.[7]

Second, only Type I errors are controlled by this testing procedure. That is, when the significance level is set at, say, 5%, the test provides a 5% risk of incorrectly rejecting the null hypothesis of no correlation. No attempt is made to fix the risks of Type II errors (i.e., the risk of accepting H_0: $\rho = 0$ when $\rho \neq 0$) at specific levels.

Third, even though the sample r value is significant according to this test, in some instances the amount of correlation may not be considered substantively important. For example, in a large sample, a quite low r value may be found to differ significantly from 0. However, since relatively little correlation has been found between the two variables, we may be unwilling to use the relationship observed between X and Y for decision-making purposes. Furthermore, prediction intervals based on the use of the applicable standard errors of estimate may be too wide to be of practical use.

Fourth, the distributions of t values computed by Equation 9.30, as in previously discussed cases, approach the normal distribution as sample size increases. Hence, for large sample sizes, the t value is approximately equal to z in the standard normal distribution, and critical values applicable to the normal distribution may be used instead. For example, in the preceding illustration, in which the critical t value was

---

[7] Fisher's z transformation may be used when ρ is hypothesized to be nonzero. In this procedure, a change of variable is made from the sample r to a statistic z, defined as $z = \frac{1}{2} \log_e[(1 + r)/(1 - r)]$. This statistic is approximately normally distributed with mean $\mu_z = \frac{1}{2} \log_e[(1 + \rho)/(1 - \rho)]$ and standard deviation $\sigma_z = 1/\sqrt{n - 3}$.

2.306 for 8 degrees of freedom at the 5% level of significance, the corresponding critical z value would be 1.96 at the same significance level. For large sample sizes, these values would be much closer.

Fifth, as noted earlier, in correlation analysis both X and Y are assumed to be normally distributed *random variables*. Hence, in order to use a hypothesis-testing procedure about the value of ρ, such as the one illustrated above, X and Y should both be random variables. On the other hand, in the regression model, the independent variable X is not a random variable. In a particular analysis, if the X values are indeed fixed or predetermined, it is not proper to use the sample correlation coefficient r to test hypotheses about ρ. However, in that case, as we have seen earlier, r or r^2 may be used to measure the effectiveness of the regression equation in explaining variation in the dependent variable Y.

Inference about the Population Regression Coefficient, B

In many cases, a great deal of interest is centered on the value of b, the slope of the regression line computed from a sample. Statistical inference procedures involving either hypothesis testing or confidence interval estimation are often useful for answering questions concerning the size of the population regression coefficient, B, in the population regression equation, $\mu_{Y \cdot X} = A + BX$.

In order to illustrate the hypothesis-testing procedure for a regression coefficient, let us return to the data in our problem, in which $b = 1.717$. We recall that the interpretation of this figure was that the estimated difference in annual food expenditures for two families whose annual net income in 1976 differed by \$1000 was \$172. Suppose that on the basis of similar studies in the same metropolitan area, it had been concluded that in previous years, a valid assumption for the true population regression coefficient was $B = 2$. Can we conclude that the population regression coefficient had changed?

To answer this question, we use a familiar hypothesis-testing procedure. We establish the following null and alternative hypotheses:

$$H_0: B = 2$$
$$H_1: B \neq 2$$

Assume that we were willing to run a 5% risk of erroneously rejecting the null hypothesis that $B = 2$. The procedure involves a two-tailed t test in which the estimated standard error of the regression coefficient, denoted s_b, is given by

(9.31)
$$s_b = \frac{s_{Y \cdot X}}{\sqrt{\Sigma (X - \bar{X})^2}}$$

Hence, s_b, the estimated standard deviation of the sampling distribution of b values, is a function of the scatter of points around the regression line and the dispersion of the X values around their mean. The t statistic, computed in the usual way, is given by

(9.32)
$$t = \frac{b - B}{s_b}$$

We calculate s_b according to (9.31) as follows:

$$s_b = \frac{1.595}{\sqrt{46}} = 0.235$$

Substituting this value for s_b into (9.32) gives

$$t = \frac{1.717 - 2}{0.235} = -1.204$$

Since the same level of significance and the same number of degrees of freedom $(n - 2)$ are involved as in the preceding test for the significance of r, the decision rule is identical. With critical t values of ± 2.306 at the 5% level of significance, we cannot reject the null hypothesis that $B = 2$. Hence, we cannot conclude that the regression coefficient has changed from $B = 2$ for families in the given metropolitan area. That is, we retain the hypothesis that if a family had an annual net income $1000 higher than that of a second family, then on the average, the first family's food expenditures would be $200 higher.

The corresponding confidence interval procedure involves setting up the interval

(9.33)
$$b \pm t s_b$$

In this problem, the 95% confidence interval for B is $1.717 \pm (2.306)(0.235) = 1.717 \pm 0.542$. Therefore, we can assert that the population B figure is included in the interval 1.175 to 2.259 with an associated confidence coefficient of 95%.

It may be noted that we used the t distribution in both the hypothesis-testing and confidence interval procedures just discussed. As in previous examples, we may note that normal curve procedures can be used for large sample sizes. Hence, for two-tailed hypothesis testing at the 5% level of significance and for 95% confidence interval estimation, the 2.306 t value given in the preceding examples would be replaced by a normal curve z value of 1.96.

A frequently used hypothesis test concerning the parameter B is for the null hypothesis, $H_0: B = 0$. This is a test to determine whether the slope of the sample regression line differs significantly from a hypothesized value of 0. A slope of 0 for the population regression coefficient B

implies that there is no linear relationship between the variables X and Y and that the population regression line is horizontal. Another way of stating this is that a B value of 0 in the linear model implies that all of the conditional probability distributions have the same mean. Hence, regardless of the value of X, all conditional probability distributions of Y are identical, so the same value of Y would be predicted for all values of X.

If B is assumed to be 0, Equation 9.32 reduces to

(9.34)
$$t = \frac{b}{s_b}$$

Substitution into Equation 9.34 in the present example yields

$$t = \frac{1.717}{0.235} = 7.31$$

Of course, the same number of degrees of freedom, $n - 2 = 10 - 2 = 8$, is involved in this test as in the preceding test that $B = 2$, and at the 5% level of significance, the critical t values are again ± 2.306. Hence, with a t value of 7.31, we reject the hypothesis that $B = 0$. We conclude that the slope of the population regression line is not 0. Based on our simple random sample, our best estimate of the slope of the population regression line is 1.717. It may be noted that the null hypothesis $H_0: \rho = 0$ and $H_0: B = 0$ are essentially equivalent assumptions. In packaged computer programs for regression and correlation analysis, the printouts generally do not show the t test for $H_0: \rho = 0$. However, these printouts for simple two-variable and multiple regression analyses either display the t values for all regression coefficients (b values) or display all regression coefficients and their corresponding standard errors, so that t values may easily be computed.

A final caution is appropriate. The tests discussed here pertain to the simple two-variable linear regression model. Although we might conclude on the basis of such tests that there is no linear relationship between X and Y, there may be some other statistically significant relationship (such as curvilinear or logarithmic); moreover, if additional variables are present but unnoticed, they may obscure the relationship (linear or otherwise) between X and Y.

9.8 CAVEATS AND LIMITATIONS

Regression analysis and correlation analysis are very useful and widely applied techniques. However, it is important to understand the limitations of these methods and to interpret the results with care.

Cause and Effect Relationships

In correlation analysis, the value of the coefficient of determination, r^2, is calculated. This statistic measures the degree of association between two variables. Neither this quantity nor any other statistical technique that measures or expresses the relationship among variables can prove that one variable is the *cause* and one or more other variables are the *effects*. Indeed, through the centuries there has been philosophical speculation and debate about the meaning of cause and effect and whether such a relationship can ever be demonstrated by experimental methods. In any event, a measure such as r^2 does not prove the existence of a cause-and-effect relationship between two variables X and Y.

In a situation in which a high value of r^2 is obtained, X may be producing variations in Y, or third and fourth (etc.) variables W and Z may be producing variations in both X and Y. Numerous examples, frequently humorous in nature, have been given to demonstrate the pitfalls in attempting to draw conclusions about cause and effect in such cases. For example, if the average salaries of ministers are associated with the average price of a bottle of scotch whiskey over time, a high degree of correlation between these two variables will probably be observed. Doubtless, we would be reluctant to conclude that the fluctuations in ministers' salaries cause the variations in the price of a bottle of scotch, or vice versa. This is a case where a third variable, which we may conveniently designate as the general level of economic activity, operates to produce variations in both of the aforementioned variables. From the economic standpoint, salaries of ministers represent the price paid for a particular type of labor; the cost of a bottle of scotch is also a price. When the general level of economic activity is high, both of these prices tend to be high. When the general level of economic activity is low, as in periods of recession or depression, both of these prices tend to be lower than during more prosperous times. Thus, the high degree of correlation between the two variables of interest is produced by a third variable (and possibly others); certainly, neither variable is *causing* the variations in the other.

Furthermore, it is important to keep in mind the problem of sampling error. As we have seen, it is conceivable that in a particular sample a high degree of correlation, either direct or inverse, may be observed, when in fact there is no correlation (or very little correlation) between the two variables in the population.

Finally, in applying critical judgment to the evaluation of observed relationships, one must be on guard against "nonsense correlations" where no meaningful unit of association is present. For example, suppose we record in a column labeled X the distance from the ground of the skirt hemlines of the first 100 women who pass a particular street corner. In a column labeled Y, we record 100 observations of the heights of the Himalaya mountains along a certain latitude at five-mile intervals. It is possible that a high r^2 value might be obtained for these data. Clearly, the result is nonsensical, because there is no meaningful unit or entity through which these data are related. In the illustrative example used in this chapter, expenditures and income were observed for the same family. Hence, the family may be referred to as the *unit of association*. The unit of association might be a time period or some other entity, but it must provide a reasonable link between the variables studied.

Extrapolation beyond
the Range of Observed Data

In regression analysis, an estimating equation is established on the basis of a particular set of observations. A great deal of care must be exercised in making predictions of values of the dependent variable based on values of the independent variable outside the range of the observed data. Such predictions are referred to as *extrapolations*. For example, in the problem considered in this chapter, a regression line was computed for families whose annual net income ranged from $2000 to $10,000. It would be extremely unwise to make a prediction of food expenditures for a family with an annual net income of $25,000 using the computed regression line. To do so would imply that the straight-line relationship could be projected up to a value of $25,000 for the independent variable. Clearly, in the absence of other information, we simply do not know whether the same functional form of the estimating equation is valid outside the range of the observed data. In fact, in certain cases, unreasonable or even impossible values may result from such extrapolations. For example, suppose a regression line with a negative slope had been computed relating the percentage of defective articles produced, Y, with the number of weeks of on-the-job training received, X, by a group of workers. An extrapolation for a large enough number of weeks of training would produce a negative value for the percentage of articles produced, which is an impossible result. Clearly in this case, although the computed estimating equation may be a good description of the relation between X and Y within the range of the observed data, an equation with different parameters or even a completely different functional form is required outside this range. Without a specific investigation or a pertinent theory, one simply does not know what the appropriate estimating device is outside the range of observed data.

However, sometimes the exigencies of a situation require an estimate, even though it is impractical or impossible to obtain additional data. Extrapolations and alternative methods of prediction have to be used, but the limitations and risks involved must be kept constantly in mind.

Other Regression Models

So far, we have considered only one particular form of the regression model, namely, that of a straight-line equation relating the dependent variable, Y, to the independent variable, X. Sometimes, theoretical considerations indicate that this is the form of model required. On the other hand, a linear model is often used because either the theoretical form of the relationship is unknown and a linear equation appears to be adequate, or the theoretical form is known but rather complex and a linear equation may provide a sufficiently good approximation. In all cases, the determination of the most appropriate regression model should be the result of a combination of theoretical reasoning, practical considerations, and careful scrutiny of the available data.

Often, the straight-line model $Y_c = a + bX$ is not an adequate description of the relationship between the two variables. In some situations, models involving transformations of one or both of the variables may provide better fits to the data. For example, if the dependent variable Y is transformed to a new variable, $\log Y$, a regression equation of the form

$$(9.35) \qquad \log Y_c = a + bX$$

may yield a better fit. Insofar as arithmetic is concerned, $\log Y$ is substituted for Y everywhere that Y appeared previously in the formulas. However, care must be used in the interpretation. The antilogarithm of $\log Y_c$ must be taken to provide an estimate of the dependent variable Y for a given value of X. It must also be recognized that different assumptions are involved in this model than in the model $Y_c = a + bX$. It is now assumed that $\log Y$ rather than Y is a normally distributed random variable. Furthermore, the logarithmic model implies that there is constant *percentage* change in Y_c per unit change in X, whereas the $Y_c = a + bX$ model implies a constant *amount* of change in Y_c per unit change in X.

Possible transformations include the use of square roots, reciprocals, and logarithms of one or both of the variables. As an example of one such useful transformation, if a straight-line equation is fitted to the logarithms of both variables, the model is of the form

$$(9.36) \qquad \log Y_c = a + b \log X$$

The regression coefficient, b, in this model has an interesting interpretation, if as in the illustrative example used in this chapter, Y is a consumption variable and X is income. For such variables, the regression coefficient b in the model $Y_c = a + bX$ can be interpreted as a marginal propensity-to-consume coefficient; that is, it estimates the dollar change in consumption per dollar change in income. Analogously, in the model $\log Y_c = a + b \log X$, the coefficient b can be interpreted as an income elasticity of consumption coefficient; that is, it estimates the *percentage change* in consumption per *one percent change* in income. Of course, fitting an equation of the form $\log Y_c = a + b \log X$ in the illustration under discussion implies that the income elasticity of consumption is constant over the range of income observed. Similarly, a model of the form $Y_c = a + bX$ implies that the marginal propensity to consume is constant over the range of observed income. In fact, according to Keynesian economic theory, the marginal propensity to consume (for total consumption expenditures) decreases with increasing income. The fitting of such regression models clearly cannot be merely a mechanistic procedure, but must involve a combination of knowledge of the field of application, good judgment, and experimentation.

In some applications, a curvilinear regression function may be more appropriate than a linear one. Polynomial functions are particularly convenient to fit by the method of least squares. The straight-line regression equation of the form $Y_c = a + bX$ is a polynomial of the first degree, since X is raised to the first power. A second-degree polynomial would involve a regression function of the form

$$(9.37) \qquad Y_c = a + bX + cX^2$$

in which the highest power to which X is raised is 2. This is the equation of a second-degree parabola, which has the characteristic that one change in direction can take place in Y as X increases, whereas in the case of a straight line, no changes in direction can take place. A third-degree polynomial permits two changes in direction, and so on. In the straight-line function, the amount of change in Y_c is constant per unit change in X. In the second-degree parabola, the amounts of change in Y_c may decrease or increase per unit change in X, depending upon the shape of the function. Figure 9-12 shows two scatter diagrams for situations in which a second-degree regression function of the form of (9.37) may provide a good fit. The probable shape of the regression function has been indicated. Analogous situations could be portrayed for cases of inverse relationships between X and Y.

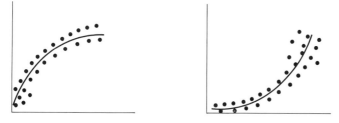

FIGURE 9-12
Scatter diagrams for two situations in which a second-degree polynomial regression function might be appropriate.

In the case of the straight-line regression function, the application of the method of least squares leads to two normal equations that must be solved for a and b. Analogously, to obtain the values of a, b, and c in the second-degree polynomial function, the following three simultaneous equations must be solved:

(9.38)
$$\Sigma\, Y = na + b\Sigma\, X + c\Sigma\, X^2$$
$$\Sigma\, XY = a\Sigma\, X + b\Sigma\, X^2 + c\Sigma\, X^3$$
$$\Sigma\, X^2 Y = a\Sigma\, X^2 + b\Sigma\, X^3 + c\Sigma\, X^4$$

Various computer programs have been developed to solve such normal equation systems, provide for transformations of the variables, and calculate all the regression and correlation measures we have discussed (as well as others). In applied problems in which large quantities of data are present and considerable experimentation with the form of the regression model is required, or in which more complex models than those so far discussed are appropriate, the use of modern computers may be the only feasible method of implementation.

Thus far, the discussion in this chapter has been limited to two-variable regression and correlation analysis. In many problems, the inclusion of more than one independent variable in a regression model may be required to provide useful estimates of the dependent variable. Suppose, for example, that in the illustration involving family food expenditures and family income, poor predictions were made based on the single independent variable "income." Other factors such as family size, age of the head of the family, and number of employed persons in the family might be considered as possible additional independent variables to aid in the estimation of food expenditures. When two or more independent variables are utilized, the problem is referred to as a *multiple regression and correlation analysis*. A brief description of the technique is included in the next section. Following that description, a case study is given to illustrate the use of such a model in a marketing managerial problem.

Exercises

1. A regression of company profits on company asset size for a random sample of 42 firms yielded the following results:

$$Y = \text{company profits (in millions of dollars)}$$
$$X = \text{company asset size (in millions of dollars)}$$
$$Y_C = 2.0 + 0.15X$$
$$\Sigma\,(Y - Y_C)^2 = 2000$$
$$\Sigma\,(Y - \bar{Y})^2 = 17{,}000$$
$$\Sigma\,(X - \bar{X})^2 = 500{,}000$$

a. Calculate and interpret the standard error of estimate, $s_{Y \cdot X}$.

b. Calculate the unadjusted and adjusted coefficients of determination. What does the coefficient of determination measure?

c. Describe and carry out a test of the significance of the estimated relationship between the two variables.

2. The Big Business School of Finance placement office has initiated a study of the determinants of a graduate's starting salary level. The independent variable selected for consideration in one part of the study was a rating of all graduates developed by the school based on quality of undergraduate school, academic performance, work experience, and achievement in nonacademic activities. The following statistics were computed:

$$Y = \text{starting salary (thousands of dollars)}$$
$$X = \text{rating}$$
$$n = 62$$
$$0 \leqslant X \leqslant 100$$
$$Y_C = 1.0 + 0.3X$$
$$s_{Y.X} = 5$$
$$s_Y = 12$$
$$\Sigma (X - \bar{X})^2 = 10{,}000$$

a. Calculate the adjusted coefficient of determination and the adjusted coefficient of correlation. Why is it preferable for interpretation purposes to use r_C^2 rather than r_C?

b. Is the estimated regression coefficient, b, significantly different from 0? Use a 1% significance level.

3. One of the models used in security analysis to evaluate the expected return for a particular stock is $E(R_i) = R_F + \beta_i[E(R_m) - R_F]$, where R_i is the return on the stock, R_F is the risk-free rate (return on a riskless asset), R_m is the return on a market portfolio, and β_i is the stock's beta coefficient (a measure of the market risk of the stock). For regression purposes, this equation is expressed as

$$R_i = (1 - \beta_i)R_F + \beta_i R_m \quad \text{or} \quad R_i = a + bR_m$$

where $a = (1 - \beta_i)R_F$
$b = \beta_i$

The following data have been collected on Strumming Consolidated, a multinational firm:

Year	R_i	R_m
1957	0.15	0.18
1958	0.11	0.10
1959	0.15	0.13
1960	0.10	0.08
1961	0.05	0.07
1962	0.13	0.11
1963	0.17	0.15
1964	0.08	0.09

(Continued)

Year	R_i	R_m
1965	0.14	0.10
1966	0.16	0.15
1967	0.18	0.20
1968	0.24	0.22
1969	0.09	0.11
1970	0.04	0.06
1971	0.06	0.09
1972	0.14	0.17
1973	0.19	0.25
1974	0.08	0.08
1975	0.02	0.05
1976	0.22	0.21

a. Estimate a and b in the equation $R_i = a + bR_m$, by the method of least squares. Assuming that the above model is correct, derive an estimate of R_F (this assumes that R_F has remained constant over time).
b. Calculate the adjusted and unadjusted coefficients of determination.
c. Theoretically, a market portfolio would have a beta coefficient of 1. Is Strumming Consolidated's beta coefficient significantly different from 1? Use a 1% level of significance.
d. Obtain an interval estimate of the conditional mean return on Strumming Consolidated stock if the market return (the average annual return on common stocks during the past 40 years) is 0.093. Use a 95% level of confidence.

4. The Warlord Motorcycle Company is considering the use of a model to predict company sales. As a first step in the determination of such a model, the researchers have decided to test the usefulness of disposable personal income as a reliable indicator for predicting sales. The data collected are shown below:

Year	Disposable Personal Income ($Billions)	Company Sales ($Millions)
1965	473	60
1966	512	71
1967	546	76
1968	591	92
1969	634	101
1970	692	115
1971	746	130
1972	803	148
1973	904	193
1974	980	204

a. Estimate the linear regression equation, using the method of least squares. Which variable did you treat as the independent variable?

b. Test the null hypothesis that the population regression coefficient is 0. Use a 1% level of significance.

c. Calculate the coefficient of determination and interpret the result. Calculate the adjusted coefficient of determination.

d. Obtain an interval estimate of the conditional mean company sales if the level of disposable personal income is $650 billion. Use a 95% confidence level.

5. As a second step in their study, the researchers at the Warlord Motorcycle Company performed a regression of company sales on industry sales. They used the following data:

Year	Industry Sales ($Millions)
1965	615
1966	760
1967	800
1968	885
1969	950
1970	1020
1971	1255
1972	1600
1973	1855
1974	2075

a. Estimate the linear regression equation, using the method of least squares.

b. Calculate the adjusted and unadjusted coefficients of determination.

c. Obtain an interval estimate of the conditional mean company sales for a level of industry sales of $1070 million. Use a 95% confidence level.

d. Obtain a prediction interval estimate of company sales for the same level of industry sales as in part (c). Use a 95% confidence level.

6. A plant manager computed the following regression equation between weekly production and cost per unit:

$$Y_c = 12.5 - 0.1X$$
$$X = \text{weekly production (in thousands of units)}$$
$$Y = \text{cost per unit (in dollars)}$$
$$s_Y = 1.5$$
$$s_{Y.X} = 0.8$$
$$n = 102 \text{ weeks}$$

Weekly production range: 0 to 20,000 units

$$\bar{X} = 12.5$$
$$\Sigma (X - \bar{X})^2 = 1650$$

a. Interpret the regression coefficient, b, specifically in terms of this problem.

    b. Calculate and interpret the adjusted coefficient of determination.

    c. An executive asserted that there is no linear relation between unit cost and production level. Do you agree?

    d. Of what utility is the Y intercept $(a = 12.5)$?

    e. Obtain a 95% prediction interval for the cost per unit for a weekly production level of 8000 units.

7. A least squares linear regression and correlation analysis was conducted on data obtained for a simple random sample of 62 sales representatives of the International Conglomerate Company, with the following results:

$$Y_C = -12 + X$$
$$X = \text{age in years}$$
$$Y = \text{annual commissions (in thousands of dollars)}$$

X ranges between 25 and 55 years

$$n = 62$$
$$s_Y = 7.5 \text{ (thousand dollars)}$$
$$s_{Y.X} = 3.5 \text{ (thousand dollars)}$$
$$\bar{X} = 38$$
$$\Sigma (X - \bar{X})^2 = 625$$

    a. The sales manager objected to the results of the equation because the value of the intercept (-12) did not seem reasonable. How would you reply?

    b. Interpret the regression coefficient, $b = 1$, specifically in terms of this problem. Is it significantly different from 0 at the 1% level?

    c. Calculate the adjusted coefficient of determination.

    d. Would a 45-year-old sales representative who received $20,000 in annual sales commissions be considered a "poor performer" or just an average performer? Why?

8. Discuss four important areas of error in the use of regression and correlation analysis.

9.9 MULTIPLE REGRESSION AND CORRELATION ANALYSIS

Multiple regression analysis represents a logical extension of two-variable regression analysis. Instead of a single independent variable, two or more independent variables are used to estimate the values of a dependent variable. However, the fundamental concepts in the analysis remain the same. Thus, just as in the analysis involving the dependent and only one independent variable, there are the following three general purposes of multiple regression and correlation analysis:

1. To derive an equation that provides estimates of the dependent variable from values of the two or more independent variables

2. To obtain measures of the error involved in using this regression equation as a basis for estimation
3. To obtain a measure of the proportion of variance in the dependent variable accounted for, or "explained by," the independent variables

The first purpose is accomplished by deriving an appropriate regression equation by the method of least squares. The second is achieved through the calculation of a standard error of estimate and related measures. The third purpose is accomplished by computing the multiple coefficient of determination, which is analogous to the coefficient of determination in the two-variable case and, as indicated in (3) above, measures the proportion of variation in the dependent variable explained by the independent variables.

As an example, let us return to our problem of estimating family food expenditures from family net income, both variables being stated on an annual basis. As indicated in the preceding section, the use of additional variables to income might be expected to improve prediction of the dependent variable. Let us assume that family size is selected as a second independent variable. Estimates of food expenditures may now be made from the following multiple regression equation:

$$(9.39) \qquad Y_C = a + b_1 X_1 + b_2 X_2$$

TABLE 9-5
Computations for Linear Multiple Regression Analysis: Family Food Expenditures (Y), Family Income (X_1), and Family Size (X_2)

Family	Annual Food Expenditures ($Hundreds) Y	Annual Net Income ($Thousands) X_1	Family Size (Number of Persons in Family) X_2	X_1Y	X_2Y	X_1X_2	Y^2	X_1^2	X_2^2
A	22	8	6	176	132	48	484	64	36
B	23	10	7	230	161	70	529	100	49
C	18	7	5	126	90	35	324	49	25
D	9	2	2	18	18	4	81	4	4
E	14	4	3	56	42	12	196	16	9
F	20	6	4	120	80	24	400	36	16
G	21	7	4	147	84	28	441	49	16
H	18	6	3	108	54	18	324	36	9
I	16	4	3	64	48	12	256	16	9
J	19	6	3	114	57	18	361	36	9
Total	180	60	40	1159	766	269	3396	406	182
Mean	18	6	4						

where Y_C = family food expenditures (estimated)

X_1 = family net income

X_2 = family size

and a, b_1, and b_2 are numerical constants that must be determined from the data in a manner analogous to that of the two-variable case. For simplicity, we have assumed a linear regression function.

As an example, we carry out a multiple regression and correlation analysis, fitting the linear regression Equation 9.39 to data for the indicated variables. The basic data for family food expenditures, family income, and family size are shown in the first three columns of Table 9-5. The data for the first two of these variables are the same as those given in Table 9-1 for the two-variable problem previously solved. The data on family size represent the total number of persons in each of the families in the sample.

The Multiple Regression Equation

We begin the analysis by using the method of least squares to obtain the best fitting three-variable linear regression equation of the form given in (9.39). In the two-variable regression problem, the method of least squares was used to obtain the best fitting straight line. In the present problem, the analogous geometric interpretation is that the method of least squares is used to obtain the best fitting plane. In a three-variable regression problem, the points can be plotted in three dimensions, along the X_1, X_2, and Y axes (analogous to the case of a two-variable problem, in which the points are plotted in two dimensions along an X and Y axis). The best fitting plane would pass through the points as shown in Figure 9-13, with some falling above and some below the plane in such a way that $\Sigma (Y - Y_C)^2$ is a minimum. Whereas in our previous illustration (involving two variables), two normal equations resulted from the minimization procedure, now three normal equations must be solved to determine the values of a, b_1, and b_2:[8]

(9.40)
$$\Sigma Y = na + b_1 \Sigma X_1 + b_2 \Sigma X_2$$
$$\Sigma X_1 Y = a\Sigma X_1 + b_1 \Sigma X_1^2 + b_2 \Sigma X_1 X_2$$
$$\Sigma X_2 Y = a\Sigma X_2 + b_1 \Sigma X_1 X_2 + b_2 \Sigma X_2^2$$

---

[8] In a manner similar to that of the two-variable case, a function of the form

$$F(a, b_1, b_2) = \Sigma (Y - Y_C)^2 = \Sigma (Y - a - b_1 X_1 - b_2 X_2)^2$$

is set up. This function is minimized by the standard calculus method of taking its partial derivatives with respect to a, b_1, and b_2 and equating these derivatives to 0. This procedure results in the three normal equations (9.40).

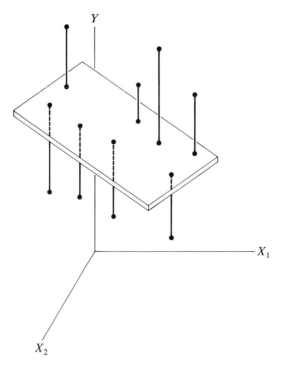

FIGURE 9-13
Graph of a multiple regression plane for data on the variables Y, X_1, and X_2.

The calculations of the required sums are shown in Table 9-5. Substitution into the normal equations (9.40) yields

$$180 = 10a + 60b_1 + 40b_2$$
$$1159 = 60a + 406b_1 + 269b_2$$
$$766 = 40a + 269b_1 + 182b_2$$

Solving these three equations simultaneously, we obtain the following values for a, b_1, and b_2:

$$a = 7.918$$
$$b_1 = 2.363$$
$$b_2 = -1.024$$

The calculations to obtain the values of a, b_1, and b_2 can be simplified somewhat by transforming (or coding) the original data into deviations from the mean of each of the variables. This procedure reduces by 1 the number of simultaneous equations to be solved. Denoting the deviations from the means of Y, X_1, and X_2 by lower case letters y, x_1, and x_2, respectively, we have

TABLE 9-6

Calculations in Terms of Deviations from Means for the Multiple Regression between Food Expenditures (Y), Family Income (X_1), and Family Size (X_2)

$$\Sigma\ y^2 = \Sigma\ Y^2 - (\bar{Y})(\Sigma\ Y) = 3396 - (18)(180) = 156$$
$$\Sigma\ x_1^2 = \Sigma\ X_1^2 - (\bar{X}_1)(\Sigma\ X_1) = 406 - (6)(60) = 46$$
$$\Sigma\ x_2^2 = \Sigma\ X_2^2 - (\bar{X}_2)(\Sigma\ X_2) = 182 - (4)(40) = 22$$
$$\Sigma\ x_1 y = \Sigma\ X_1 Y - (\bar{X}_1)(\Sigma\ Y) = 1159 - (6)(180) = 79$$
$$\Sigma\ x_2 y = \Sigma\ X_2 Y - (\bar{X}_2)(\Sigma\ Y) = 766 - (4)(180) = 46$$
$$\Sigma\ x_1 x_2 = \Sigma\ X_1 X_2 - (\bar{X}_1)(\Sigma\ X_2) = 269 - (6)(40) = 29$$

(9.41)
$$y = Y - \bar{Y}$$
$$x_1 = X_1 - \bar{X}_1$$
$$x_2 = X_2 - \bar{X}_2$$

When expressed in terms of deviations, the second and third normal equations of (9.40) become

(9.42)
$$\Sigma\ x_1 y = b_1 \Sigma\ x_1^2 + b_2 \Sigma\ x_1 x_2$$
$$\Sigma\ x_2 y = b_1 \Sigma\ x_1 x_2 + b_2 \Sigma\ x_2^2$$

These equations contain fewer terms than their counterparts in (9.40), since $\Sigma\ x_1$ and $\Sigma\ x_2$ are equal to 0. This is an example of the property that the sum of the deviations of any set of observations from their mean is equal to 0.

The required computations for the lower case normal equations of (9.42) are given in Table 9-6. Shortcut methods have been utilized to calculate the sums of squares and cross products. Substitution of the numerical results of Table 9-6 into the normal equations (9.42) yields

$$79 = 46b_1 + 29b_2$$
$$46 = 29b_1 + 22b_2$$

Solving these simultaneous equations for b_1 and b_2, we again find

$$b_1 = 2.363$$
$$b_2 = -1.024$$

The value of the constant a is obtained from

(9.43)
$$a = \bar{Y} - b_1 \bar{X}_1 - b_2 \bar{X}_2$$

Substituting the values of $\bar{Y}$, $\bar{X}_1$, and $\bar{X}_2$ from Table 9-5, we find that a is the same value previously computed:

$$a = 18 - (2.363)(6) - (-1.024)(4) = 7.918$$

When large numbers of variables and observations are present, even calculations involving deviations from means are likely to be too laborious. The use of electronic computing equipment in such cases is usually the only feasible alternative.

The multiple regression equation may now be written as

(9.44) $$Y_C = 7.918 + 2.363X_1 - 1.024X_2$$

It may be noted that more decimal values are usually shown for the constants in the multiple regression equation than are justified by the usual rules of rounding according to the number of significant digits. This is a standard procedure in regression analysis, because many of the measures computed in that type of analysis are particularly sensitive to rounding off. Hence, one rule of thumb often followed is to carry large numbers of digits and to round off only when final results are presented for measures such as regression coefficients, standard errors, and coefficients of determination.

Let us illustrate the use of the regression equation for estimation. Suppose we want to estimate food expenditures for a family from the same population as the sample studied. The family's income is \$6000, and there are four persons in the family. Substituting $X_1 = 6$ and $X_2 = 4$ yields the following estimated expenditures on food:

$$
\begin{aligned}
Y_C &= 7.918 + 2.363(6) - 1.024(4) \\
&= 18.00 \text{ (hundreds of dollars)} \\
&= 1800 \text{ (dollars)}
\end{aligned}
$$

In two-variable analysis, we discussed the interpretation of the constants a and b in the regression equation. Let us consider the analogous interpretation of the constants a, b_1, and b_2 in the multiple regression equation. The constant a is again the Y intercept. However, now it is interpreted as the value of Y_C when X_1 and X_2 are both 0. The b values are referred to in multiple regression analysis as *net regression coefficients*. The b_1 coefficient measures the change in Y_C per unit change in X_1 when X_2 is held fixed, and b_2 measures the change in Y_C per unit change in X_2 when X_1 is held fixed.[9]

Hence, in the present problem, the b_1 value of 2.363 indicates that if a family has an income \$1000 greater than another's (a unit change in X_1) and *the families are of the same size* (X_2 is held constant), then the estimated expenditures on food of the higher-income family exceed those of the other by 2.363 hundreds of dollars, or by about \$236. Similarly, the b_2 value of -1.024 means that if a family has one person more than another (a unit change in X_2) and *the families have the same income* (X_1 is

---

[9] In the language of calculus, b_1 and b_2 are the partial derivatives of Y_C with respect to X_1 and X_2, respectively; that is,

$$\frac{\partial Y_C}{\partial X_1} = b_1 \quad \text{and} \quad \frac{\partial Y_C}{\partial X_2} = b_2$$

held constant), then the estimated food expenditures of the larger family are less than those of the smaller by about $102.

Two properties of these net regression coefficients are worth noting. The b_1 value of 2.363 hundreds of dollars implies that an increment of one unit in X_1, or a $1000 increment in income, occasions an increase of $236.3 in Y_C, estimated food expenditures, regardless of the size of the family (for families of the sizes studied). Hence, an increase of $1000 in income adds $236.3 to estimated food expenditures, regardless of whether there are two or six people in the family. An analogous interpretation holds for b_2. These interpretations follow from the fact that a *linear* multiple regression equation was used, and are embodied in the assumption of linearity.

A second property of regression coefficients is apparent from a comparison of the b value of 1.72 in the simple regression equation (9.7), $Y_C = 7.70 + 1.72X$, previously obtained when family income, X, was the only independent variable, with the b_1 value of 2.36, the net regression coefficient of income in the multiple regression equation $Y_C = 7.92 + 2.36X_1 - 1.02X_2$, when the family size variable is included in the regression equation. (All coefficients have been rounded to two decimal places for convenience of comparison.) The coefficient $b = 1.72$ in the simple two-variable regression equation makes no explicit allowance for family size. The net regression coefficient $b_1 = 2.36$, on the other hand, "nets out" the effect of family size. A net regression coefficient may in general be greater or less than the corresponding regression coefficient in a two-variable analysis.

In this problem, the families with larger incomes were also the larger families. The positive correlation between income and family size is indicated by the correlation coefficient $r = 0.912$, as shown in Table 9-7. The foregoing pattern exemplifies an important characteristic of regression coefficients, regardless of the number of independent variables included in the study: A regression coefficient for any specific independent variable (for example, income) measures not only the effect on the dependent variable of income, but also the effect attributable to any other independent variables that happen to be correlated with it but have not been explicitly included in the analysis. This is true for both two-variable and multiple regression analyses.

When independent variables are highly correlated, rather odd results may be obtained in a multiple regression analysis. For instance, a regression coefficient that is positive (negative) in sign in a two-variable regression equation may change to a negative (positive) sign for the same independent variable in a multiple regression equation containing other independent variables that are highly intercorrelated with the one in question. For example, in this problem, the dependent variable, food expenditures (Y), is positively correlated with family size (X_2), as indicated by the correlation coefficient of $+0.785$ (Table 9-7). Hence, the regression coefficient for family size would also be positive in sign. However, as we have seen, the net regression coefficient for family size (b_2) in the three-variable regression equation is -1.02 and is thus negative in sign.

It will be shown in the discussion of statistical inference in multiple regression that the net regression coefficients for highly intercorrelated independent variables tend to be unreliable. The importance of this is that when independent variables are highly intercorrelated, it is extremely difficult to separate out the individual influences of each variable. This can be seen by considering an extreme case. Suppose a two-variable regression and correlation analysis is carried out between a dependent variable, denoted Y, and an independent variable, denoted X_1. Furthermore, let us assume that we introduce another independent variable, X_2, which has perfect positive correlation with X_1, that is, the correlation coefficient between X_1 and X_2 is $+1$. We now conduct a three-variable regression and correlation analysis. It is clear that X_2 cannot account for or explain any additional variance in the dependent variable Y after X_1 has been taken into account. The same argument could be made if X_1 were introduced after X_2. As indicated in the ensuing discussion of statistical inference in multiple regression, the net regression coefficients, b_1 and b_2, in cases of high intercorrelation between X_1 and X_2 will tend not to differ significantly from 0. Yet, if separate two-variable analyses had been run between Y and X_1 and Y and X_2, the individual regression coefficients might have differed significantly from 0. There is a great deal of concern in fields such as econometrics and applied statistics with this problem of intercorrelation among independent variables, often referred to as *multicollinearity*. One of the simplest solutions to the problem of two highly correlated independent variables is merely to discard one of them, but often more sophisticated procedures are required.

The illustration in this section used only two independent variables. The general form of the linear multiple regression function for $k-1$ independent variables $X_1, X_2, \ldots, X_{k-1}$ is

(9.45) $$Y_c = a + b_1X_1 + b_2X_2 + \cdots + b_{k-1}X_{k-1}$$

A linear function that is fitted to data for two variables is referred to as a straight line, for three variables as a plane, and for four or more variables, as a *hyperplane*. Although we cannot visualize a hyperplane, its linear characteristics are analogous to those of the linear functions of two or three variables. With the use of electronic computers, it is possible to test and include large numbers of independent variables in a multiple regression analysis. However, good judgment and knowledge of the logical relationships involved must always be the main guides to deciding which variables to include in the construction of a regression equation.

Standard Error of Estimate

As in simple two-variable regression analysis, a measure of dispersion, or scatter, around the regression plane or hyperplane can be used as an indicator of the error of estimation. Again, probability assumptions similar in principle to those of the simple regression model must be introduced. The following are the usual assumptions made in a linear multiple regression analysis, illustrated for the case of two independent variables:

1. The Y values are assumed to be independent of one another.
2. The conditional distributions of Y for given X_1 and X_2 are assumed to be normal.
3. These conditional distributions for each independent variable are assumed to have equal standard deviations.

The variance around a regression hyperplane involving k variables, of which one is dependent and $k - 1$ are independent, is

(9.46) $$S^2_{Y.12\ldots(k-1)} = \frac{\Sigma\,(Y - Y_c)^2}{n - k}$$

where n is the number of observations and k is the number of constants in the regression equation. The divisor $n - k$ represents the number of degrees of freedom, and its use provides an unbiased estimator of the population variance. The subscript notation to S^2 lists the dependent variable to the left of the period and the $k - 1$ independent variables to the right. The subscripts $1, 2, \ldots, k - 1$ denote the variables X_1, $X_2, \ldots, X_{k-1}$, respectively. Hence, in our example involving the three variables Y_1, X_1, and X_2, the variance around the regression plane $Y_c = 7.92 + 2.36X_1 - 1.02X_2$ is given by

(9.47) $$S^2_{Y.12} = \frac{\Sigma\,(Y - Y_c)^2}{n - 3}$$

The standard error of estimate, which is the square root of this variance, is

(9.48) $$S_{Y.12} = \sqrt{\frac{\Sigma\,(Y - Y_c)^2}{n - 3}}$$

A shortcut formula for calculating this standard error of estimate in terms of deviations from means is the following:

(9.49) $$S_{Y.12} = \sqrt{\frac{\Sigma\,y^2 - b_1\Sigma\,x_1y - b_2\Sigma\,x_2y}{n - 3}}$$

Substituting the values previously obtained in our illustrative problem into (9.49), we find

$$S_{Y.12} = \sqrt{\frac{156 - (2.362573)(79) - (-1.023392)(46)}{7}}$$

$$= 1.53 \ (\$hundreds)$$

As noted earlier, a large number of digits was used in calculating the standard error of estimate in order to prevent rounding error.

As in two-variable analysis, the standard error of estimate is the standard deviation of the observed values of Y around the fitted regression equation. It is possible to obtain interval estimates for the conditional mean and for individual forecasts using standard errors of estimate from multiple regression analysis, but the calculations are much more tedious than in the simple two-variable case and will not be presented here.

Coefficient of Multiple Determination

In two-variable correlation analysis, the degree of association between the two variables may be stated in terms of the adjusted coefficient of determination, r_C^2, which was defined in (9.25) as

$$r_C^2 = 1 - \frac{s_{Y.X}^2}{s_Y^2}$$

In this form, r_C^2 measures the proportion of variance in the dependent variable explained by the regression equation relating Y to X; that is, r_C^2 measures the proportion of variance in the Y variable accounted for by, or associated with, the independent variable X.

An analogous measure, the *coefficient of multiple determination*, denoted R^2, with appropriate subscripts, quantifies the degree of association that exists when more than two variables are present. For the case of one dependent and two independent variables, the coefficient of multiple determination, corrected for degrees of freedom, is defined as

(9.50)
$$R_{Y.12}^2 = 1 - \frac{S_{Y.12}^2}{s_Y^2}$$

where

(9.51)
$$S_{Y.12}^2 = \frac{\Sigma \ (Y - Y_C)^2}{n - 3}$$

and

(9.52)
$$s_Y^2 = \frac{\Sigma \ (Y - \bar{Y})^2}{n - 1}$$

In keeping with convention, the subscript notation in $R_{Y.12}^2$ lists the dependent variable to the left of the period and the independent variables to the right. As can be seen from these definitions, $S_{Y.12}^2$ is the variance of Y values around the regression plane and s_Y^2 is the variance of the Y values around their mean. Just as $\Sigma (Y - Y_C)^2$ in Equation 9.51 was earlier referred to as "unexplained variation" and $\Sigma (Y - \bar{Y})^2$ in Equation 9.52 as the "total variation" of the Y variable, $S_{Y.12}^2$ is referred to as the "unexplained variance" and s_Y^2 as the "total variance." Hence, in a manner similar to the interpretation of r_C^2, we may interpret $R_{Y.12}^2$ as the proportion of variance in the dependent variable explained by the regression equation relating Y to X_1 and X_2. Alternatively, it measures the proportion of variance in the Y variable accounted for by all the independent variables combined.

We illustrate the calculation of $R_{Y.12}^2$ for the example with which we have been working. As calculated by Equation 9.49 to six decimal places, $S_{Y.12}$ was found to be equal to 1.532168 ($ hundreds). Hence,

$$S_{Y.12}^2 = (1.532168)^2 = 2.347539$$

The standard deviation, s_Y, was computed to be 4.163 ($ hundreds). The total variance, or variance around $\bar{Y}$, is

$$s_Y^2 = (4.163)^2 = 17.330569$$

Therefore, substituting into (9.50), the coefficient of multiple determination is

$$R_{Y.12}^2 = 1 - \frac{2.347539}{17.330569} = 0.865$$

Thus, we have found that 86.5% of the variance in food expenditures has been explained by the linear regression equation that related that variable to family income and family size. Comparing this figure to the corresponding two-variable r_C^2 value of 0.853 obtained in Section 9.6 for the correlation between expenditures on food and family income, we find that the value of $R_{Y.12}^2$ is only 0.012, or about one percentage point, higher than the figure for r_C^2. This means that the addition of the second independent variable, family size, has explained very little of the variance in food expenditures, Y, beyond that which was already accounted for by family income alone. As we have noted earlier, one reason for this is the high correlation between the independent variables. Once family income has been taken into account, since family size moves together with that variable, family size can do very little to explain residual variation in food expenditures.

Two-Variable Correlation Coefficients

From the preceding discussion of the difficulties encountered in multiple correlation analysis when independent variables are intercorrelated, it is evident that it is good practice to compute coefficients of correlation or determination between each pair of independent variables that the analyst plans to enter into the regression equation. It is standard procedure in most multiple regression and correlation analysis computer programs to present a table of correlation coefficients for every pair of variables, including the dependent as well as all independent variables.

A convenient formula for calculating the correlation coefficient in terms of deviations from means, illustrated for the variables X_1 and X_2, is

(9.53)
$$r_{12} = \sqrt{\frac{(\Sigma\, x_1 x_2)^2}{(\Sigma\, x_1^2)(\Sigma\, x_2^2)}}$$

where $x_1 = X_1 - \bar{X}_1$
and $x_2 = X_2 - \bar{X}_2$

Substituting figures from Table 9-6 for the illustrative problem, we find

$$r_{12} = \sqrt{\frac{(29)^2}{(46)(22)}} = 0.912$$

Similar calculations for the other two pairs of variables give the following results:

$$r_{Y_1} = \sqrt{\frac{(79)^2}{(46)(156)}} = 0.933$$

$$r_{Y_2} = \sqrt{\frac{(46)^2}{(22)(156)}} = 0.785$$

In the printout of computer programs, the correlation coefficients are often presented in the form of a triangular table, such as is shown in Table 9-7. The 1.00s along the diagonal of this table indicate that the correlation coefficient of each variable with itself is 1.00, that is, each variable is perfectly and directly correlated with itself.

TABLE 9-7
Correlation Coefficients for Each Pair of the Three Variables: Expenditures on Food (Y), Family Income (X_1), and Family Size (X_2)

	Y	X_1	X_2
Y	1.00		
X_1	0.933	1.00	
X_2	0.785	0.912	1.00

Inferences about
Population Net Regression Coefficients

In the preceding discussion of correlation and regression analysis, the various equations and measures were all stated in terms of sample values, rather than in terms of the corresponding population equations and characteristics. If the assumptions given at the beginning of the discussion of the standard error of estimate are met, then appropriate inferences and probability statements can be made concerning population parameters. In multiple regression analysis, a great deal of interest is centered on the reliability of the observed net regression coefficients. Just as in the two-variable case referred to in Section 9.7, where statistical inference about the population regression coefficient B was discussed, analogous hypothesis-testing and estimation techniques are available for regression coefficients, where three or more variables are involved.

In the two-variable problem, the regression coefficient b in the equation $Y_C = a + bX$ is an estimate of the population parameter B in the population relationship $\mu_{Y.X} = A + BX$. Correspondingly, the regression coefficients in a three-variable problem, b_1 and b_2 in the equation $Y_C = a + b_1X_1 + b_2X_2$, are estimates of the parameters B_1 and B_2 in a population relationship denoted $\mu_{Y.12} = A + B_1X_1 + B_2X_2$. The standard errors of the net regression coefficients, which represent the estimated standard deviations of the sampling distributions of b_1 and b_2 values, are given by

(9.54)
$$s_{b_1} = \frac{S_{Y.12}}{\sqrt{\Sigma\, x_1^2(1 - r_{12}^2)}}$$

and

(9.55)
$$s_{b_2} = \frac{S_{Y.12}}{\sqrt{\Sigma\, x_2^2(1 - r_{12}^2)}}$$

where all terms in (9.54) and (9.55) have the definitions stated above. Substituting the required numerical values, we find

$$s_{b_1} = \frac{1.532168}{\sqrt{46(1 - (0.912)^2)}}$$
$$= 0.551$$

and

$$s_{b_2} = \frac{1.532168}{\sqrt{22(1 - (0.912)^2)}}$$
$$= 0.796$$

We can test hypotheses concerning B_1 and B_2 by computing t statistics in the usual way:

(9.56)
$$t_1 = \frac{b_1 - B_1}{s_{b_1}}$$

and

$$t_2 = \frac{b_2 - B_2}{s_{b_2}}$$

These t statistics approach normality as the sample size and number of degrees of freedom become large.

Hence, to test the hypotheses that the net regression coefficients are equal to 0, that is, that family income and family size have no effect on food expenditures, or

$$II_0: B_1 = 0$$
$$H_1: B_1 \neq 0$$

and

$$H_0: B_2 = 0$$
$$H_1: B_2 \neq 0$$

we calculate

(9.57)
$$t_1 = \frac{b_1 - 0}{s_{b_1}} = \frac{b_1}{s_{b_1}}$$

and

$$t_2 = \frac{b_2 - 0}{s_{b_2}} = \frac{b_2}{s_{b_2}}$$

In the illustrative problem, we find

$$t_1 = \frac{2.363}{0.551} = 4.289$$

and

$$t_2 = \frac{-1.023}{0.796} = -1.285$$

The number of degrees of freedom used to look up the critical t values for this test is $n - k$, which in this case is equal to $10 - 3 = 7$. This is the number of degrees of freedom used to estimate $S_{Y.12}$ in the calculation of s_{b_1} and s_{b_2}. The two-tailed critical t values at the 5% and 1% level of significance are ± 2.365 and ± 3.499, respectively (Table A-6 of Appendix A). Thus, since for b_1, the computed t_1 value of 4.289 exceeds the positive critical values, we conclude that b_1 differs significantly from 0 at both the 5% and 1% levels of significance. Therefore, we reject the null hypothesis that $B_1 = 0$. The computed t_2 value of -1.285 for b_2 means that the b_2 value lies 1.285 estimated standard errors below 0. Comparing this figure of -1.285 with the critical values of -2.365 and -3.499, we conclude that b_2 does not differ significantly from 0 at either the 5% or 1% level of significance. Hence, we accept the null hypothesis that $B_2 = 0$.

In summary, we conclude that family income, X_1, has a statistically significant effect on food expenditures, Y, but that after this income effect has been accounted for, family size, X_2, does not have a statistically significant influence. This result is consistent with the previous discussion of the difficulty of measuring the separate effects of two intercorrelated independent variables.

An important point concerning the interpretation of the results of a multiple regression analysis follows from the above discussion. If the basic purpose of computing a regression equation is to make predictions of values of the dependent variable, then the reliability of the individual net regression coefficients is not of great consequence. On the other hand, if the purpose of the analysis is to measure accurately the separate effects of each of the independent variables on the dependent variable, then the reliability of the individual net regression coefficients is clearly of importance.

Other Measures in Multiple Regression Analysis

A number of other measures are sometimes calculated in a multiple regression and correlation analysis. Only brief reference will be made to them here.

The coefficient of multiple determination, R^2, has been described as a measure of the effect of all the independent variables combined on the dependent variable. More specifically, this coefficient measures the percentage of variance in the dependent variable that has been accounted for by all the independent variables combined. The square root of the coefficient of multiple determination, $R = \sqrt{R^2}$, is referred to as the *coefficient of multiple correlation*. It is always assigned a plus sign. Since some of the individual independent variables may be positively correlated with the dependent variable and others negatively correlated, there would be no meaning in distinguishing between a positive and negative value for R. As in the case of r^2 and r in two-variable analysis, R^2 is easier to interpret, since R^2 is a percentage figure whereas R is not.

It is possible to calculate measures that indicate the separate effect of each of the independent variables on the dependent variable, if the influence of all the other independent variables has been accounted for.

For this purpose, it is conventional to compute so-called *coefficients of partial correlation*. For example, the partial correlation coefficient for family income in our illustrative problem, designated $r_{Y1.2}$, would show the partial correlation between Y and X_1 after the effect of X_2 on Y had been removed. The square of this coefficient, $r^2_{Y1.2}$, measures the reduction in variance brought about by introducing variable X_1 after X_2 has already been accounted for.

Sometimes, it is difficult to compare the differences in net regression coefficients because the independent variables are stated in different units. For example, in the illustrative example, b_1 indicates the average difference in food expenditures, Y, *per unit difference in family income* X_1, whereas b_2 indicates the average difference in food expenditures, Y, *per unit difference in family size*. (In both cases, the other independent variable is held constant.) However, unit differences in X_1 and X_2 are in different units ($1000 and one person, respectively). For improved comparability, the regression equation can be stated in a different form, where each of the variables is given in units of its own standard deviation. The transformed net regression coefficients are termed "beta coefficients." For example, in terms of beta coefficients, the linear regression equation for three variables would be

(9.58)
$$\frac{Y}{s_Y} = \alpha + \beta_1 \frac{X_1}{s_{X_1}} + \beta_2 \frac{X_2}{s_{X_2}}$$

Thus, the beta coefficients are equal to

(9.59)
$$\beta_1 = b_1 \frac{s_{X_1}}{s_Y}$$

and

$$\beta_2 = b_2 \frac{s_{X_2}}{s_Y}$$

As an illustration of the meaning of the beta coefficients, β_1 measures the number of standard deviations that Y_C changes with each change of one standard deviation in X_1.[10]

Selected General Considerations

A great deal of care must be exercised in the use of multiple regression and correlation techniques. In the development of the model, theoretical analysis, knowledge of the field of application, and logical judgment

---

[10] The reader is alerted that the battle of notation must be continually fought. In Chapter 7, β denoted the probability of a Type II error. The specific meaning of this symbol must be determined from the context of the particular discussion.

should aid in the selection of variables to be used in the study. Frequently, in business and economic applications, some of the relevant variables may not be easily quantifiable. Sometimes variables are not readily available and must be constructed from different sets of data.

In the discussion of multiple regression and correlation analysis, we have confined ourselves to the case of a linear model. The underlying assumptions should be checked for their validity. Simple graphic checks involve the examination of graphs of the dependent variable against each of the independent variables at the outset of the analysis and of plots of the $Y - Y_c$ deviations against each of the independent variables after fitting the regression equation. Sometimes transformations, such as taking logarithms, reciprocals, or square roots of original observations, may provide better adherence to original assumptions and better fits of regression equations to the data. Of course, when a linear regression equation is used, it simply represents a convenient approximation to the unknown "true" relationship. Where linear relationships provide inadequate fits, curvilinear regression equations may be required.

The quest for a good fit of the regression equation to the data leads to adding more and more independent variables. However, cost considerations, difficulties of providing data in the implementation and monitoring of the model, and the search for a reasonably simple model ("parsimony") point toward the use of as few independent variables as possible. Since no mechanistic statistical procedure exists to resolve this dilemma and many other problems of multiple regression and correlation analysis, subjective judgment inevitably plays a large role. It is not clear from statistical theory alone which variables should be included in a regression analysis. Prior knowledge of the field of application is important in the initial selection of independent variables and in choices of variables to include or exclude based on the statistical analysis.

The dangers of extrapolation must be carefully guarded against. There are subtle difficulties in multiple analysis as compared to two-variable analysis. Even within the range of the data, certain combinations of values of the independent variables may not have been observed. This means that statistically valid estimates of the dependent variable cannot be made for these combinations of values.

The Use of Computers in
Multiple Regression Analysis

The use of high-speed electronic computers has greatly simplified the testing and analysis of statistical relationships among variables. The libraries of most computer centers contain programs for various types of multivariate analysis including multiple regression analysis (for brevity, we will use that term rather than the longer "multiple regression and correlation analysis"). In the past, the cost and tedious labor involved in multiple regression analyses involving more than two or three independent variables severely restricted the analyst's ability to test and experiment. With the use of modern computer programs, the analyst now has a much wider range of choice in selection of variables, in options for performing transformations, in adding and deleting variables at various stages of the analysis, and in testing curvilinear as well as linear relationships.

Stepwise regression analysis is a versatile form of multiple regression analysis for which computers are particularly useful and for which there are a number of available computer programs. In this type of analysis, at the first stage, the computer determines which of the independent variables (as many as about 30 may be included in some programs) is most highly correlated with the dependent variable. The computer printout then displays all the usual statistical measures for the two-variable relationship. At the next stage, the program selects the independent variable that accomplishes the greatest reduction in the unexplained variance remaining after the two-variable analysis. As in the previous stage, the computer printout then displays all the usual statistical measures for the three-variable relationship. The program continues in this stepwise fashion, at each stage entering the "best" independent variable in terms of ability to reduce the remaining unexplained variance. Analysis of variance tables and lists of residuals $(Y - Y_c)$ are provided at each stage. It is evident that without the use of computers and "canned" programs, the time needed to perform such an analysis by hand would be prohibitive.

A number of other types of multiple regression analysis programs are available in which the analyst initially includes a certain number of independent variables and then can delete and add variables as desired. As an illustration, we have reproduced in Tables 9-8, 9-9, and 9-10 selected computer output for the food expenditures vs income and family

TABLE 9-8
Computer Output for the Variables in the Regression Equation to Predict Family Expenditures for Food

	VARIABLES IN THE EQUATION		
VARIABLE	B	BETA	STD ERROR B
INCOME	2.36257	1.28293	0.54957
FAMSIZE	−1.02339	−0.38432	0.79467
(CONSTANT)	7.91813		

TABLE 9-9
**Computer Output for Coefficients of Multiple
Correlation, Multiple Determination, and the
Standard Error of Estimate**

MULTIPLE R	0.94587
R SQUARE	0.89466
STANDARD ERROR	1.53217

size problem discussed earlier in this section (Table 9-5). The data were run on the DEC system 1070 computer at the Harvard University Graduate School of Business Administration using the Statistical Package for the Social Sciences (SPSS) written at Stanford University.

In Table 9-8 are shown

1. Names of the variables; (CONSTANT) is the Y intercept or a value
2. Estimated b_1, b_2, and a values
3. Beta coefficients
4. Standard errors of the regression coefficients (s_{b_1} and s_{b_2})

The minor differences between the figures shown in the computer printout and those calculated earlier in this chapter are due to rounding.

In Table 9-9 are shown

1. The multiple coefficient of correlation
2. The multiple coefficient of determination
3. The standard error of estimate

The R and R^2 values shown in the computer printout are unadjusted for degrees of freedom. Hence, the $R^2 = 0.89466$ value given in Table 9-9 exceeds the $R^2_{Y.12} = 0.865$ figure computed earlier in this section. Denoting the R^2 value adjusted for degrees of freedom as R^2_C and the unadjusted figure as R^2, we have the following relationship between the two coefficients:

$$(9.60) \qquad R^2_C = 1 - (1 - R^2)\left(\frac{n-1}{n-k}\right)$$

TABLE 9-10
Computer Output for Analysis of Variance of the Regression Problem to Predict Family Expenditures for Food

ANALYSIS OF VARIANCE	DF	SUM OF SQUARES	MEAN SQUARE	F
REGRESSION	2.	139.56725	69.78363	29.72633
RESIDUAL	7.	16.43275	2.34754	

where n = the number of observations

k = the number of constants in the regression equation

The adjustment factor $(n - 1)/(n - k)$ is approximately equal to 1 for large sample sizes. Therefore, R^2 and R_C^2 are approximately equal for large samples.

In Table 9-10 is given an analysis of variance for the regression problem. The analysis of variance provides an appraisal of the overall significance of the regression as opposed to the significance of each of the regression coefficients. It gives a test of the null hypothesis of no relationship between the dependent variable and all of the independent variables considered collectively. Another way of wording the hypotheses is

H_0: the regression is not significant

H_1: the regression is significant

The null hypothesis is accepted or rejected on the basis of the usual F test. In a simple two-variable regression analysis, the F test gives exactly the same result as the t test for the null hypothesis H_0: $\rho = 0$.

In the analysis of variance discussed in Chapter 8, we made two estimates of variance, referred to as "between columns" and "between rows." In regression analysis, the corresponding variance estimates are called "due to regression" and "residual variance" (also often referred to as "deviation about regression"). The variance due to regression is the explained variance, and the residual variance is the unexplained variance. The F ratio is

$$(9.61) \quad F = \frac{\text{explained variance}}{\text{unexplained variance}} = \frac{\text{variance due to regression}}{\text{residual variance}}$$

$$= \frac{\Sigma (Y_C - \bar{Y})^2/(k - 1)}{\Sigma (Y - Y_C)^2/(n - k)}$$

where k = number of constants in the regression equation

n = number of observations

$k - 1$ = number of degrees of freedom in the numerator of the F ratio

$n - k$ = number of degrees of freedom in the denominator of the F ratio

The terms usually used in computer printouts that correspond to terminology used in this chapter are as follows:

Computer Printout	Terminology of This Chapter
Regression sum of squares	Explained variation

(Continued)

Computer Printout	Terminology of This Chapter
Residual sum of squares	Unexplained variation
Total sum of squares (not shown in Table 9-10)	Total variation

The regression sum of squares, $\Sigma (Y_c - \bar{Y})^2$, and the residual sum of squares, $\Sigma (Y - Y_c)^2$, are shown in Table 9-10. It is necessary to divide these sums of squares by the appropriate numbers of degrees of freedom to obtain the corresponding mean squares or variances. In this problem, the number of degrees of freedom for the regression sum of squares is $k - 1 = 3 - 1 = 2$. The number of degrees of freedom for the residual sum of squares is $n - k = 10 - 3 = 7$. If the variance due to regression (explained variance) is significantly larger than the residual variance, the F ratio will significantly exceed 1 and the regression is significant. This is the situation in which R^2 is relatively high and there is a high degree of association between this dependent variable and the independent variables. The degree to which the explained variance must exceed the unexplained variance is determined by Appendix Table A-8 of Appendix A.

As indicated by Equation 9.61, the F ratio is calculated by dividing the mean square (variance) due to regression by the residual mean square (variance). In the food expenditures problem, as shown in Table 9-10, we have

$$F = \frac{139.56725/2}{16.43275/7} = \frac{69.78363}{2.34754} = 29.72633$$

We turn to Table A-8 of Appendix A, where degrees of freedom for the numerator are read across the top of the table and degrees of freedom for the denominator are read down the side. There are 2 and 7 degrees of freedom, respectively, in the food expenditures problem, which yields a critical F value of 4.74 at the 5% level of significance and a critical F value of 9.55 at the 1% level of significance. Since the computed F value shown for the computer output in Table 9-10 is 29.72633, the hypothesis that the regression is not significant is rejected at both the 5% and 1% levels. These relationships are shown in Figure 9-14.

One final comment is in order. The advent of modern electronic computers has opened up a greater choice than was formerly available in selection of variables, the inclusion of larger numbers of variables, more options in performing transformations of variables, and more testing and experimentation with different types of statistical relationships. However, because of these increased possibilities, it becomes even more im-

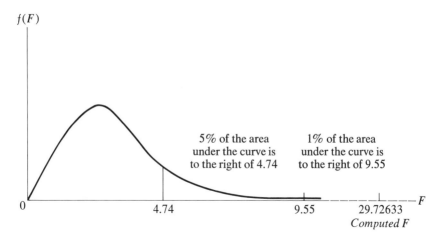

FIGURE 9-14
The critical and computed values of F for the food expenditures problem.

portant that care and good judgment be exercised to avoid misuse of methods and misinterpretation of findings.

A case study of the use of a series of multiple regression models, developed with the aid of a computer for a marketing management decision problem in the area of forecasting, is given in the next section.

Exercises

1. In connection with fiscal policy planning, a government economic research group attempted to analyze the impact of cash tax rebates on stimulating the economy. A regression was run using the following data:

Y = percentage of cash rebate spent within a year
X_1 = total income of family (in thousands of dollars)
X_2 = size of family

A random sample of 123 families was selected. The computer output yielded the following data:

Variable	Mean	Regression Coefficient	Standard Error of Regression Coefficient
X_1	12.476	-3.875 (b_1)	0.655
X_2	4.389	7.955 (b_2)	4.118

Intercept (a): 94.210
Standard error of estimate ($S_{Y.12}$): 22.042
Standard deviation of Y (s_Y): 40.035

Analysis of variance table:

Source	Sum of Squares	Degrees of Freedom
Due to regression	137237.891	2
Residual	58302.667	120
Total sum of squares	195540.558	

a. Construct and carry out a test of the significance of the net regression coefficients. Use a 1% significance level.
b. State exactly the conclusions to be drawn from the results of part (a).
c. Describe two properties of net regression coefficients in a multiple regression analysis.
d. Calculate and interpret the multiple coefficient of determination.
e. Estimate the percentage of cash rebate spent by a family of six with an income of $12,000. Assume that this situation has been observed.
f. Test the overall significance of the regression. State the terms given in the analysis of variance table in another way.

2. The personnel department of a large manufacturing firm drew a random sample of 23 workers. The workers were carefully interviewed and given several tests. On the basis of the tests, the following quantities were obtained for each worker:

$$X_1 = \text{manual dexterity score}$$
$$X_2 = \text{mental aptitude score}$$
$$X_3 = \text{personnel assessment score}$$

Subsequently, the workers were observed in order to determine the average number of units of work completed (Y) in a given time period for each worker. Regression analysis yielded these results:

$$Y_C = -212 + 1.90X_1 + 2.00X_2 - 0.25X_3$$
$$\quad\quad\quad (0.50)\quad\quad (0.60)\quad\quad (0.20)$$

The quantities in parentheses are the standard errors of the net regression coefficients. The standard error of estimate, $S_{Y.123}$, was 25. The standard deviation of the dependent variable, s_Y, was 50.

a. Give an interpretation of the net regression coefficient, b_1.
b. Are each of the net regression coefficients significantly different from 0? Test at the 1% level of significance.
c. What is the expected effect when highly correlated independent variables are included in a multiple regression equation?
d. Calculate the coefficient of multiple correlation and the multiple coefficient of determination.
e. Estimate the average number of units of work completed by a worker in the given time period with a manual dexterity score of 100, a mental aptitude score of 90, and a personnel assessment score of 15. Assume that such a worker has been observed.

f. Assume the following standard deviations:

$$s_{X_1} = 15, \quad s_{X_2} = 10, \quad s_{X_3} = 5$$

Calculate the beta coefficients, β_1, β_2, and β_3. Give an interpretation of β_1.

3. A study undertaken by the American Society of Monetarists inquired into the relationship between the unemployment rate in percent (Y), per capita money supply in dollars (X_1), and the logarithm of federal expenditures (X_2). The last named variable was used as an indicator of the yearly percentage change in federal expenditures. The following information refers to an attempt to establish a least squares linear relationship using annual data for a 21-year period. The relation estimated was

$$Y_C = a + b_1X_1 + b_2X_2$$

The statistics computed were

$n = 21$	$a = 18$
$\Sigma (Y - Y_C)^2 = 25$	$b_1 = -0.04$
$\Sigma (Y - \bar{Y})^2 = 100$	$b_2 = -3$
$r_{12} = 0.8$	$\Sigma X_1^2 = 900{,}000$
$r_{Y_1} = -0.4$	$\Sigma X_1 = 4200$
$r_{Y_2} = -0.2$	$\Sigma X_2^2 = 100$
$\Sigma (Y_C - \bar{Y})^2 = 75$	$\Sigma X_2 = 42$

a. Calculate and interpret the standard error of estimate. What assumptions are necessary in using this measure? How many degrees of freedom are there?
b. Calculate and interpret the multiple coefficient of determination.
c. Calculate the standard errors of the net regression coefficients. Carry out appropriate tests of the statistical significance of the net regression coefficients. For each case, state the null and alternative hypotheses.
d. Estimate the percentage unemployed if the per capita money supply is $200 and the logarithm of federal expenditures is 2.00. Assume that these values of X_1 and X_2 have been observed.
e. Carry out an appropriate test of the overall significance of the regression. State the null and alternative hypotheses. Use a 1% level of significance.

4. The research department of the Inexperienced Investment Company has been providing the firm's account executives with information on options for use in their management of clients' investment portfolios. An analysis of a valuation formula for a random sample of 43 options yielded the following results:

Y = value of option (in dollars)
X_1 = volatility of underlying stock
X_2 = time remaining before expiration of the option (in months)
$Y_C = a + b_1X_1 + b_2X_2$

The statistics computed were

$a = 2.0$	$\Sigma X_2^2 = 1225$
$b_1 = 0.2$	$\Sigma X_1X_2 = 5620$

$$b_2 = 0.8 \qquad \Sigma (Y - Y_C)^2 = 160$$
$$\Sigma X_1 = 1075 \qquad \Sigma (Y - \bar{Y})^2 = 760$$
$$\Sigma X_2 = 215 \qquad \Sigma (Y_C - \bar{Y})^2 = 600$$
$$\Sigma X_1^2 = 28{,}875$$

a. Interpret the net regression coefficient, b_2, in the above multiple regression.

b. Under what circumstances is the reliability of the individual net regression coefficients of particular importance?

c. Calculate the standard error of estimate.

d. Calculate the coefficient of multiple determination.

e. Calculate the correlation coefficient between X_1 and X_2 and the standard errors of the net regression coefficients. Are both net regression coefficients significantly different from 0?

f. Estimate the value of an option whose underlying stock has a price volatility of 20 and whose expiration date is six months away. Assume that such an option is included in the sample.

g. Using a 1% level of significance, perform an appropriate test of the overall significance of the regression.

5. A random sample of 16 sales of homes in a certain town was analyzed in an attempt to establish a relationship between selling price (Y) and assessed value (X_1). Y and X_1 are expressed in thousands of dollars. The analysis resulted in the following estimated equation:

$$Y_C = 4 + 0.8X_1$$

The statistics computed were

$$s_{Y.X} = 5$$
$$s_Y = 12$$
$$s_{b_1} = 0.15$$

In a second analysis, a second independent variable, age in years (X_2), was added in an attempt to improve the explanatory power of the regression. The results were

$$Y_C = 4 + 0.9X_1 - 0.15X_2$$

The statistics computed were

$$S_{Y.12} = 4.8$$
$$s_Y = 12$$
$$s_{b_1} = 0.25$$
$$s_{b_2} = 0.1$$

a. Compute the coefficient of determination for the first regression.

b. Test the significance of b_1 in the first regression. How many degrees of freedom are there?

c. Compute the multiple coefficient of determination in the second regression. Has the inclusion of the second independent variable explained very much of the variance in Y beyond that already accounted for by X_1 alone?

d. Interpret the coefficient b_1 in the second regression.

e. Test the significance of the regression coefficients in the second regression. Interpret the results carefully. What could account for these results?

6. The president of Channel WTV divides advertising (in minutes per hour) into two classes: advertising of sponsors (X_1) and advertising of future network programs (X_2). You, as a consultant to the president, collect data from 33 periods and run a regression with the number of viewers in tens of thousands (Y) as the dependent variable and X_1 and X_2 as the independent variables. The results are summarized below:

Variable	Regression Coefficient	Standard Error of the Regression Coefficient
X_1	−0.24438	0.04838
X_2	−0.02865	0.00816

Intercept (a): 6.30567

Analysis of variance table:

Source	Sum of Squares	Degrees of Freedom
Due to regression	3.6786	2
Residual	2.9418	30
Total	6.6204	

a. Are the regression coefficients significantly different from 0? Test at the 1% level of significance.

b. Calculate the multiple coefficient of determination and the coefficient of multiple correlation. Which is easier to interpret?

c. Test the overall significance of the regression. Express the F statistic in mathematical terms. Describe in words the two quantities that are compared in the calculation of this statistic.

9.10 A CASE STUDY

The setting of this case study involves a large pharmaceutical company that has developed a new ethical drug product and has made the critical decision to launch it on the market. (An ethical drug is one that can only be purchased upon prescription from a physician.) A heavy investment has already been made in the development of the product, through the various stages of preliminary screening, more detailed evaluation, and final preparation for commercialization. Once the decision to market the product has been made, the marketing executive must decide on such matters as the amount of promotional spending, the allocation of the budget over a period of time, the amount of sales effort to be invested, and the duration of any special support for the project. Many other important decisions are initiated in other departments of the company, such as production, purchasing, and finance, as operations are geared to adjust to the new requirements. In this type of situation, it is crucial that good short-term demand forecasts be provided in the initial months of marketing (say, for the first year of the product's life). Furthermore, it is important to know the expected demand levels at various points during this first year in order to plan the distribution of the promotion budget, the allocation of sales effort, and the timing of various types of special support for the product. This case study deals with the development of a sales forecasting model based on the methodology of multiple regression analysis. The purpose of this discussion is to indicate the general nature of project development. Thus, greater insight is obtained into some of the broader problems of the use of quantitative models in managerial applications than is possible when attention is directed solely to the specific mathematical or statistical techniques employed.

Let us consider first of all why any formal forecasting model was desired in the above situation. Typically, in the past, during the first few months a new product was on the market, marketing management had found itself confronted with uncertainties and a flood of puzzling and often conflicting information. There was a flow of data, primarily from pharmaceutical market research firms, concerning the product's sales during these early months, the reactions of prescribing physicians, the success of competitive products, and the comparative promotional efforts exerted by the company and its competitors. However, some of this information was of dubious utility, and some gave conflicting evidence. How should the company sift this information, extract the important items, and make the correct decisions concerning the potential future success of the product?

A considerable amount of effort went into the problem definition phase and the determination of management objectives. It was decided that relatively short-term forecasts were desired for short-run decision-making purposes as opposed to long-term forecasts for long-run, broader planning purposes. Marketing management had to decide whether it was interested in predictions in dollars or in number of physical units, whether it required forecasts for periods of time (as for example, the last quarter of the first year after introduction) or at "time points" (as for example, the monthly rate of sales during the twelfth month) or both. These basic definitional problems were resolved primarily from the viewpoint of the informational needs of the marketing vice-president. In the first months the new product was on the market, this executive needed good short-term forecasts of its demand, say for the last quarter of the first year of the product's life and for the entire first year. Furthermore, it was important to know the expected levels of demand at various points during this first year in order to make decisions on the distribution of the promotional budget, the allocation of sales effort, and so on.

It was recognized that the crucial point in these early forecasts was to recognize an unsuccessful product or a successful high-volume product quickly. The critical matter for decision-making purposes is to place the product in its proper size class (for example, large volume, medium volume, or low volume) rather than to estimate dollar sales with extreme precision. Then, as time progresses during the year following introduction and as more current information becomes available, the marketing executive needs revised estimates of the new product's market position.

The first stage of the project involved a detailed investigation to determine the *feasibility* of constructing a sales forecasting model to operate during the early months after product introduction—a model that would provide early projections of future demand. A number of alternative methodological approaches were reviewed and rejected. Due to the nature of the case, traditional time-series methods, such as those discussed in Chapter 10, were not feasible, because there was no past sales history of the new product that might be analyzed and projected. The technique finally selected was a series of least squares multiple regression equations, with the model using source data available during the early months of the new product's life.

In Chapter 4, we referred to the idea that in any statistical inquiry it is useful for the investigators to consider at the design stage what the ideal experiment might be. In the project under discussion, it was fruitful to consider the ideal regression equation for predicting new product sales. On the left side of the equation would be the dependent variable (in this case, new product sales, say, for the fourth quarter of the first year after introduction). On the right side would be variables representing the determinants of this demand, such as the underlying medical and economic factors and market variables that measure the relative preference for this product over its competitors. Probably, if a group of knowledgeable medical and marketing persons were assembled, they would not be able to agree on what these determinants were. However, even if agreement could be reached, the relevant information would simply not be available in a form suitable for inclusion in the type of model under discussion. Hence, as is invariably the case in such investigations, compromises with the ideal situation had to be made because of practical considerations.

It was decided that the methodology, similar to other forecasting tools, would use historical information. In this case, the information was the data for other new pharmaceutical products, for the variables included, over a reasonable prior period of time. By measuring various factors related to sales of all new products, the existing relationships could be measured statistically. The specific variables for which data were obtainable could not all be considered actual determinants of sales. However, some of the variables, such as sales of the product in the first few months of the product's life, reflected the operation of the underlying determinants of demand and thus might represent good predictive factors.

The basic rationale of the regression equation approach was that if for all new products introduced over a substantial period of time there was a strong relationship between demand levels and the specific variables considered during the early months of a product's life, and if this relationship persisted, then the probable market activity for the new product currently under consideration could be estimated from the corresponding early-month variables.

Another comment on practical expedients may be noted. Why use data pertaining to all new products, that is, all types of pharmaceutical products for all companies? It would seem more reasonable to use only information about other new products that were in some way similar to this one as the basic data from which to construct the model. For example, in the pharmaceutical industry, products are classified into more or less homogeneous groupings known as "therapeutic classes." Hence, it might appear reasonable to use as a data base only figures relating to new products in the same therapeutic class as this one that were placed on the market during the past few years. This procedure proved not to be feasible, because there simply were not enough new products in the same therapeutic class to provide an adequate data base for the multiple regression analysis. The possibility of using only data concerning all new products produced by this particular company had to be rejected on the same grounds of insufficient data. Therefore, it was decided that the historical data base would consist of information on all new pharmaceutical products placed on the market during the past few years. Intensive testing was carried out to evaluate the feasibility of constructing a workable forecasting system from this type of data base.

It was important in the initial phase of the study to define and evaluate the specific variables to be included in the model. The first step in this connection was an itemization of all candidate variables, dependent and independent, based on the subjective judgments of marketing research and other personnel. Knowledge of the marketing environment was essential for this task. The unit of association for the proposed regression analyses was the new product. A review of the sources of the necessary historical information revealed that considerable pertinent data were available for new products for the prior few years.

Approximately 30 factors were defined for possible inclusion in the model. Five different dependent variables were defined, such as

"Fourth Quarter Sales" (sales of the product during the fourth quarter of the first year of the product's life) and "First Year Sales." The 25 candidate independent variables included current market activity for new products, changes in market activity during the early months, attitudinal factors, and promotional expenditures. For example, among the factors considered were purchases of new products by drug stores, prescription activity, market-share measures, stocking and repurchase patterns of drugstores, measures of physician awareness and use, attitudes of physicians toward future use, expenditures on different types of advertising, and a variable that isolated differences in characteristics of certain product groups. In this as in other studies, the figures for a number of variables were not directly available, but had to be constructed from available data. For example, as indicated above, it was felt that promising independent variables for predictive purposes would include market-share measures. Such measures were not directly available and had to be constructed. The *market-share* of a new product was defined basically as the proportion of total sales in its therapeutic class accounted for by that product. In some cases, it was necessary to redefine the therapeutic class to include other products that were competitive with the new product but were not encompassed in the standard definition of the class. It was also hypothesized that month-to-month change in market-share during the early months of the new product's life might be a good predictive factor, so this variable had to be calculated.

In this study, the data search and collection phase was time consuming and costly. Hence, numerous practical decisions were required. If two similar variables were available but the data for one were costly to obtain relative to the second, figures were compiled only for the second. Careful checks had to be made of the consistency in definition and reporting of each variable over time and between different sources. In many cases, computational adjustments were required to obtain the necessary consistency. Data had to be checked for completeness; for example, certain variables had to be dropped from consideration because available data were incomplete. Clerical forms had to be devised for recording and checking the data. Subsequently, the data were transferred onto computer tapes.

The preliminary experimentation and evaluation phase consisted of a series of computer runs in which a large number of multiple regression equations were calculated. The purposes of these test runs were to determine (a) the overall predictability of future product demand and the relative accuracy of estimates and (b) the relative importance of the individual variables considered. The tests consisted of experiments with various combinations of dependent and independent variables for different time periods. Specifically, a series of multiple regression equations were calculated from the data; that is, a set of little models was constructed on a test basis.

A computer program for stepwise multiple regression analysis was used to derive the prediction equations. Scores of computer runs were made during the test and evaluation stage. This is a type of model-construction procedure that simply would not have been feasible before the availability of modern electronic data processing equipment. The evaluation of the test results was very encouraging. The degree of prediction accuracy appeared to be very high, as reflected by the standard errors of estimate and correlation measures. There was indication of decided month-to-month improvement in forecasting ability during the early months. Better predictions were made for aggregate sales over a time period than for levels of sales at points in time. The reason for this was that the former type of variable is more stable and therefore easier to predict; seasonal and random factors tend to be offset over a longer time span. It was possible to draw certain conclusions about the individual variables under consideration. For example, it became apparent that the factors reflecting actual market activity in the early months were better predictors that those reflecting attitudes and promotional effort. On the basis of the results obtained in the initial feasibility study, the decision was made to proceed with the next phase, the implementation of the model.

The implementation stage represented an extension of the initial effort, culminating in the development of a final working model. A basic goal was a more detailed analysis of the variables under consideration in order to select the combinations that furnished the highest degree of predictability and provided relatively simple and manageable model equations. The latter criteria were established so that the final model would be relatively easy to maintain and would be less vulnerable to potential future data inadequacies. As a result of considerable experimentation and analysis, an "initial forecasting system" was developed comprising five individual forecasts of aggregate demand for time periods and levels of demand at points in time. This system was then tested with prior new products, to see how it would have performed had it been available. On the basis of these past relationships, there appeared to be good predictability of future sales of new pharmaceutical products from factors available early in the lives of these products.

The following is an example of the type of model that was developed:

Month 3 Model for Estimating the Dollar Volume of Fourth Quarter New Prescriptions

Y = fourth-quarter product new prescriptions ($ thousands)
X_1 = cumulative first three-month direct mail advertising ($ thousands)
X_2 = cumulative first three-month product new prescriptions ($ thousands)
X_3 = market share of new prescriptions in third month (percentage)

X_4 = change in market share of new prescriptions (ratio of third-month to first-month new prescriptions)

$Y_C = -30.731 + 3.224X_1 + 0.561X_2 + 4.593X_3 + 2.970X_4$

$R^2 = 0.88$

Therefore, if the following values were observed for the third month after the product was introduced

$$X_1 = 20$$
$$X_2 = 240$$
$$X_3 = 18$$
$$X_4 = 1.19$$

then the forecast of fourth-quarter new prescription sales would be

$Y_C = -30.731 + 3.224(20) + 0.561(240) + 4.593(18)$
$$+ 2.970(1.19) = 254.597 \text{ (\$ thousands)}$$

Thus, based on information available from the first three months of sales of the new product, a prediction of about \$255,000 would be made for fourth quarter new prescription sales.

The next phase of the implementation study consisted of the construction of a system to produce revised forecasts as the time since product introduction increased. Essentially, this system consisted of a series of regression equations based upon more current information than was included in the initial forecast. Separate forecasting models were constructed for predicting fourth-quarter product new prescriptions at months 5, 7, and 9. In each case, the same basic methodology was used as in the development of the month 3 models. This forecasting system was considered necessary because of the volatile nature of markets for new products. Important changes in market conditions frequently occur between the time at which the forecast is made and the time or period for which it is made.

A maintenance and control system should be established as part of the overall construction of any model system. In this case, this entailed periodic updating of the model by inclusion of new data and recalculation of coefficients in the forecasting system. A continuous review of the basic data was required. Source data should be monitored for the presence of any possible changes in definition, inadequacies, or atypical situations. If new factors become predominant, new source data become available, or other similar conditions evolve, it is appropriate to test for the usefulness of inclusion of these new variables in the system.

The forecasting model system proved its worth by providing accurate and useful predictions, and it made a number of valuable contributions to marketing management. The use of a formal model for forecasting purposes removed the elements of emotion and vested-interest influences from the forecast and based it on the most appropriate available current and past information. Furthermore, since the model provided a quantitative statement of the relationship of future sales to a variety of factors, it was possible to estimate, by manipulating the model, the probable effects on sales of changes in these factors and the interactions among them. Despite the possibilities of error in the model, the marketing executive is doubtless in a better position when it is possible to combine judgment with the insights generated by this type of model than when one must rely on unaided judgment alone.

10
Time Series

10.1 INTRODUCTION

Decisions in private and public sector enterprises depend on perceptions of future outcomes that will affect the benefits and costs of possible alternative courses of action. Since these outcomes occur in the future, they must be forecast. Not only must managers forecast, but they must also plan and think through the nature of the activities that will permit them to accomplish their objectives. However, it is clear that managerial planning and decision making are inseparable from forecasting.

Methods of forecasting vary considerably. For example, in forecasting customer demand, the range of prevailing methods includes informal "seat of the pants" estimating, executive panels and composite opinions, consensus of sales force opinions, combined user responses, statistical techniques, and various combinations of these methods.

Because of such factors as the increased complexity of business operations, the need for greater accuracy and timeliness, the depen-

443

dence of outcomes on so many different variables, and the demonstrated utility of the techniques, management is increasingly turning to formal models, such as those provided by statistical methods, for assistance in the difficult task of peering into the future. A very widely applied and extremely useful set of procedures is *time series analysis*. A time series is a set of statistical observations arranged in chronological order. Examples include a weekly series of end-of-week stock prices, a monthly series of amounts of steel production, and an annual series of national income. Such time series are essentially historical series, whose values at any given time result from the interplay of large numbers of diverse economic, political, social, and other factors.

A first step in the prediction of any series involves an examination of past observations. Time series analysis deals with the methods of analyzing these past data and of projecting them to obtain estimates of future values. The traditional, or "classical," methods of time series analysis, which will be our primary subject of discussion, are descriptive in nature and do not provide for probability statements concerning future events. These time series models, although admittedly only approximate and not highly refined, have proven their worth when cautiously and sensibly applied. It is very important to realize that these methods cannot simply be used mechanistically but must at all times be supplemented by sound subjective judgment.

Although the preceding discussion has referred to the use of time series analysis for the purposes of forecasting, planning, and control, these procedures also are often used for the simple purpose of historical description. Hence, for example, they may be usefully employed in an analysis in which interest centers upon the differences in the nature of variations in different time series. The general nature of the classical time series model is described in the next section.

10.2 THE CLASSICAL TIME SERIES MODEL

If we wished to construct an ideally satisfying mathematical model of an economic time series, we might seek to define and measure the many factors determining the variations in the time series and then proceed to state the mathematical relationships between these and the particular series in question. However, the determinants of change in an economic time series are multitudinous, including such factors as changes in population, consumer tastes, technology, investment or capital-goods formation, weather, customs, and numerous other variables both of an economic and of a noneconomic nature. The enormity and impracticability of the task of measuring all these factors and then relating them mathematically to an economic time series militates against the use of this direct approach to time series analysis. Hence, it is not surprising that a more indirect and practical approach has come into use. Classical time series analysis is essentially a descriptive method that attempts to break down an economic time series into distinct components representing the effects of groups of explanatory factors such as those given earlier. These component variations are

1. Trend
2. Cyclical fluctuations
3. Seasonal variations
4. Irregular movements

Trend refers to a smooth upward or downward movement of a time series over a long period of time. Such movements are thought of as requiring a minimum of about 15 or 20 years to describe, and as being attributable to factors such as population change, technological progress, and large-scale shifts in consumer tastes.

Cyclical fluctuations, or business cycle movements, are recurrent upward and downward movements around trend levels that have a duration of anywhere from about 2–15 years. There is no single simple explanation of business cycle activity, and there are different types of cycles of varying length and size. Therefore, it is not surprising that no generally satisfactory mathematical model has been constructed for either describing or forecasting these cycles, and perhaps none will ever be.

Seasonal variations are cycles that complete themselves within the period of a calendar year and then continue in a repetition of this basic pattern. The major factors in producing these annually repetitive patterns of seasonal variations are weather and customs, where the latter term is broadly interpreted to include observance of various holidays such as Easter and Christmas. Series of monthly and quarterly data are ordinarily used to examine these seasonal variations. Hence, regardless of trend or cyclical levels, one can observe in the United States that each year more ice cream is sold during the summer months than during the winter, whereas more fuel oil for home heating is consumed in the winter than during the summer months. Both of these cases illustrate the effect of weather or climatic factors in determining seasonal patterns. Also, department store sales generally reveal a minor peak during the months in which Easter occurs and a larger peak in December, when Christmas occurs, reflecting the shopping customs of consumers. The techniques of measurement of seasonal variations that we will discuss are particularly well suited to the measurement of relatively stable patterns of seasonal variations, but they can be adapted to cases of changing seasonal movements as well.

Irregular movements are fluctuations in time series that are short in

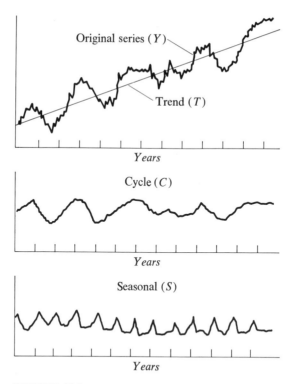

FIGURE 10-1
The components of a time series.

duration, erratic in nature, and follow no regularly recurrent or other discernible pattern. These movements are sometimes referred to as *residual variations*, since, by definition, they represent what is left over in an economic time series after trend, cyclical, and seasonal elements have been accounted for. These irregular fluctuations result from sporadic, unsystematic occurrences such as erratic shifts in purchasing habits, accidents, strikes, and the like. Whereas in the classical time series model, the elements of trend, cyclical fluctuations, and seasonal variations are viewed as resulting from systematic influences leading to either gradual growth, decline, or recurrent movements, irregular movements are considered to be so erratic that it would be fruitless to attempt to describe them in terms of a formal model.

Figure 10-1 presents the typical pattern of the trend, cyclical, and seasonal components of a time series. Time is plotted on the horizontal axes of the graphs, and the values of the particular series (which might be retail sales, ice cream production, or airline revenues) are plotted on the vertical axes.

10.3 DESCRIPTION OF TREND

As pointed out in the preceding section, the classical model involves the separate statistical treatment of the component elements of a time series. We begin our discussion by indicating how the description of the underlying trend is accomplished.

Before the trend of a particular time series can be determined, it is generally necessary to subject the data to some preliminary treatment. The amount of such adjustment depends to some extent on the time period for which the data are stated. For example, if the time series is in monthly form, it may be necessary to revise the monthly data to take account of the differing number of days per month. This may be accomplished by dividing the monthly figures by the number of days in the respective months, or by the number of working days per month, to state the data for each month on a per day basis.

Even when the original data are in annual form, which is often the case when primary interest is centered on the long-term trend of the series, the data may require a considerable amount of preliminary treatment before a meaningful analysis can be carried out. Adjustments for changes in population size are often made by dividing the original series by population figures to state the series in per capita form. Frequently, comparisons of trends in these per capita figures are far more meaningful than corresponding comparisons in the unadjusted figures.

It is particularly important to scrutinize a time series and adjust it for differences in definitions of statistical units, the consistency and coverage of the reported data, and similar items. It is important to realize that one cannot simply proceed in a mechanical fashion to analyze a time series. Careful and critical preliminary treatment of such data is required to ensure meaningful results.

Purposes for Fitting Trend Lines

The trend in a time series can be measured by the free-hand drawing of a line or curve that seems to fit the data, by fitting appropriate mathematical functions, or by the use of "moving average" methods. Moving averages are discussed later in this chapter in connection with seasonal variations.

A free-hand curve may be fitted to a time series by visual inspection. When this type of characterization of a trend is employed, the investigator is usually interested in a quick description of the underlying growth or decline in a series, without any careful further analysis. In many instances, this rapid graphic method may suffice. However, it clearly has certain disadvantages. Different investigators would surely obtain different results for the same time series. Indeed, even the same analyst would probably not sketch in exactly the same trend line in two different attempts on the same series. This excessive amount of subjectivity in choice of a trend line is especially problematic if further quantitative analysis is planned. In the ensuing discussion, we will concentrate on mathematically fitted trend lines.

Even in the case of the mathematical measurement of trend, the *purpose* of the analysis is of considerable importance in the selection of the appropriate trend line. Several different types of purposes can be specified.

1. Trend lines may be fitted for the purpose of historical description. If so, any line that fits well will suffice. The line need not have logical implications for forecasting purposes, nor should it be evaluated primarily by characteristics that might be desirable for other purposes.

2. A second purpose is that of prediction or projection into the future. In this case, particularly if long-term projection is desired, the selected line should have logical implications when it is extended into the future. The analyst, when engaging in prediction, must always carefully weigh the implications of the models being projected into the future as regards their reasonableness for the phenomena being described and predicted. For example, constant amounts of growth per unit time are implied if one projects a *straight-line* trend into the future. This may not be a reasonable long-term projection for many series.

3. A third purpose for which trend lines are fitted to economic data is to describe and eliminate trend movements from the series in order to study nontrend elements. Thus, if the analyst's primary interest is to study cyclical fluctuations, freeing the original data of trend makes it possible to examine cyclical movements without the presence of the trend factor. For this purpose, any type of trend line that does a reasonably good job of bisecting the individual business cycles in the data would be appropriate.

Types of Trend Movements

There have been considerable variations in the trend movements of different economic and business series. Over long periods of time, some companies and industries have experienced periods of growth and then have gone into steep declines when more modern competitive processes and products have emerged in other companies and industries. Real GNP in the United States has exhibited a relatively constant rate of growth of about 3% per year since the early part of this century. Since real GNP is a measure of overall economic activity, it represents a type of average of all series for production of goods and services. Although many series have shown more or less similar trends to that of real GNP, sharp divergences have also occurred. One example of these differential movements is that in recent years, the service sector of the economy has been growing relative to the agricultural sector. In fact, in the post-World War II period, while employment in the service industries generally was increasing, the number of persons employed in agriculture was decreasing.

Numerous studies have shown, for a large number of American industries, a trend that may be characterized as increasing at a decreasing percentage rate. Indeed, some investigators adapted growth curves originally used to describe biological growth to depict the past change of many industrial series. These growth curves, of which the *Gompertz* and *logistic* are the best known, are S-shaped for increasing series when plotted on graph paper with an arithmetic vertical scale, and are concave downward on a semilogarithmic chart. (A semilogarithmic chart has a logarithmic vertical scale and an arithmetic horizontal scale.) It is conventional in graphing time series on either arithmetic or semilogarithmic paper to plot the variable of interest on the vertical axis and time on the horizontal axis.[1] The general shape of such a growth curve, plotted on arithmetic graph paper and on semilogarithmic graph paper, is depicted in Figure 10-2. The so-called "law of growth" has been used to describe this type of change over time in an industry. As can be seen on the arithmetic chart, in the early stages of the industry, the growth is slow at first and then becomes rapid, with the series increasing by increasing amounts. Then the industry moves through a point beyond which it increases by decreasing amounts, and finally there is a tapering off into a period of "maturity." Throughout all stages, as seen on the semilogarithmic chart, although the industry is growing, the increases are at a decreasing *percentage rate*. Various reasons for this type of industrial growth were propounded by the investigators who discerned analogous changes in biological and industrial time series data. Since the equations of these growth curves are exponential in character, it is extremely difficult to fit such curves by the method of least squares described later in this chapter.[2]

---

[1] The familiar arithmetic graph has an arithmetically ruled vertical scale. Hence, equal vertical distances represent equal *amounts* of change. Semilogarithmic graphs have a logarithmically ruled vertical scale. Here, equal vertical distances represent equal *percentage rates* of change. For example, on an arithmetic graph, a straight line inclined upward depicts a series that is increasing with constant amounts of change. On a semilogarithmic graph, a straight line inclined upward depicts a series that is increasing at a constant percentage rate.

[2] A special technique known as "the method of selected points" is ordinarily used for this purpose.

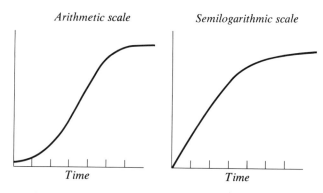

FIGURE 10-2
A growth curve plotted on arithmetic and on semilogarithmic paper.

As was indicated earlier, there are a great variety of types of trend movements in economic time series, and many of these trends cannot be adequately described or projected by means of growth curves. The growth curves have a number of desirable characteristics. For example, they have finite lower and upper limits, which are determined by the data to which the curves are fitted. However, no one family of curves is apt to be generally satisfactory for trend fitting purposes. In fact, the growth curves have been found to be quite inadequate for industrial growth prediction. The most commonly used polynomial-type trend lines, which are fitted by the method of least squares, are discussed in this chapter. The purpose of fitting, the goodness of fit obtained, knowledge of the growth and decay processes involved, and trial-and-error experimentation are all essential ingredients in the selection of the appropriate trend line.

10.4 FITTING TREND LINES BY THE METHOD OF LEAST SQUARES

For situations in which it is desirable to have a mathematical equation to describe the trend of a time series, the most widely used method is to fit some type of polynomial function to the data. In this section, we illustrate the general method by means of very simple examples, fitting a straight line and a second-degree parabola to time series data by the method of least squares.

The Method of Least Squares

The method of least squares, when used to fit trend lines to time series data, is employed mainly because it is a simple, practical method that provides best fits according to a reasonable criterion. However, it should be recognized that the method of least squares does not have the same type of theoretical underpinning when applied to fitting trend lines as when used in regression and correlation analysis, as described in Chapter 9. The major difficulty is that the usual probabilistic assumptions made in regression and correlation analysis are simply not met in the case of time series data. For example, in the illustrative problem in Chapter 9 involving the relationship between food expenditures and income for a sample of families, there were two possible theoretical models. In the first, both food expenditures and income were random variables; in the second, income was a controlled variable, that is, families of prespecified incomes were selected and food expenditures was a random variable. Hence, in each model, the dependent variable was a random variable, and the model assumed conditional probability distributions of this random variable around the computed values of the dependent variable, which fell along the regression line. These computed Y values were the means of the conditional probability distributions. A number of assumptions are implicit in this type of model: Deviations from the regression line are considered to be random errors describable by a probability distribution. The successive observations of the dependent variable are assumed to be independent. For example, Family B's expenditures were assumed to be independent of Family A's, and so on.

Clearly, in the fitting of trend lines to time series data, the probabilistic assumptions of the method of least squares are not met. If a trend line is fitted, for example, to an annual time series of department store sales, time is treated as the independent variable X and department store sales as the dependent variable Y. It is not reasonable to think of the deviation of actual sales in a given year from the computed trend value as a random error. Indeed, if the original data are annual, then deviations from trend would be considered to represent the operation of cyclical and irregular factors. (Seasonal factors would not be present in annual data, because by definition they complete themselves within a year.) Finally, the assumption of independence is not met in the case of time series data. A department store's sales in a given year surely are not independent of what they were in the preceding year. In summary, returning to the point made at the outset of this discussion, the method of least squares when used to fit trend lines is employed primarily because of its practicality, simplicity, and good fit characteristics rather than because of its justification from a theoretical viewpoint.

Fitting an Arithmetic Straight-Line Trend

As an example, we will fit a straight line by the method of least squares to an annual series on the number of industrial and commercial consumers of natural gas in the United States from 1956 to 1974. Although we wrote the equation of a straight line in the discussion of regression analysis in Chapter 9 as $Y_c = a + bX$, in time series analysis we will use the equation

(10.1) $Y_t = a + bx$

The computed trend value is denoted Y_t, with the subscript t standing for trend. That is, Y_t is the computed trend figure for the time period x. In time series analysis, the computations can be simplified by transforming the X variable, which is the independent variable "time," to a simpler variable with fewer digits. This is accomplished by stating the time variable in terms of deviations from the arithmetic mean time period, which is simply the middle time period.

The transformed time variable is denoted by lower case x. Hence in the illustrative example in Table 10-1, $x = 0$ in 1965, the middle year in the time series that runs from 1956 through 1974. The x values (or $X - \bar{X}$ figures) for years before and after 1965 are, respectively, $-1, -2, -3, \ldots$, and $1, 2, 3, \ldots$. For example, the x value for 1966 is equal to 1 because $X - \bar{X} = 1966 - 1965 = 1$. The constants in the trend equation are interpreted in a similar way to those in the straight line discussed in regression analysis; a is the computed trend figure for the period when $x = 0$, in this case, 1965; b is the slope of the trend line, or the amount of change in Y_t per unit change in x (per year in the present example). Because the sum of the deviations of a set of observations from their mean is equal to 0, $\Sigma x = 0$. This property makes the computation of the constants for the trend line simpler than in the corresponding case of the straight-line regression equation. In Chapter 9, the equations for fitting a straight line were given as

(10.2) $a = \bar{Y} - b\bar{X}$

(10.3) $b = \dfrac{\Sigma XY - n\bar{X}\bar{Y}}{\Sigma X^2 - n\bar{X}^2}$

In the least squares fitting of a straight-line trend equation, x is substituted for X. Since $\Sigma x = 0$, and therefore, $\bar{x} = 0$, the equations become

(10.4) $a = \bar{Y} = \dfrac{\Sigma Y}{n}$

(10.5) $b = \dfrac{\Sigma xY}{\Sigma x^2}$

Hence, the constant a is simply the mean of the Y values, and b is the quotient of two numbers easily determined from the original data. The calculations for fitting a straight-line trend to the time series on the number of industrial and commercial consumers of natural gas are given in Table 10-1. Columns (2) through (5) contain the basic computations for determining the values of a and b. As indicated in the calculation of

TABLE 10-1
Straight-Line Trend Fitted by the Method of Least Squares to the Number of Industrial
and Commercial Consumers of Natural Gas in the United States, 1956–1974

Year (1)	x (2)	Industrial and Commercial Consumers (Thousands) Y (3)	xY (4)	x^2 (5)	Y_t (6)	Percentage of Trend $\dfrac{Y}{Y_t} \cdot 100$ (7)
1956	−9	2107	−18963	81	2165.7	97.3
1957	−8	2193	−17544	64	2252.0	97.4
1958	−7	2293	−16051	49	2338.3	98.1
1959	−6	2412	−14472	36	2424.6	99.5
1960	−5	2525	−12625	25	2510.9	100.6
1961	−4	2621	−10484	16	2597.2	100.9
1962	−3	2716	− 8148	9	2683.5	101.2
1963	−2	2807	− 5614	4	2769.8	101.3
1964	−1	2908	− 2908	1	2856.1	101.8
1965	0	2997	0	0	2942.4	101.9
1966	1	3082	3082	1	3028.7	101.8
1967	2	3152	6304	4	3115.0	101.2
1968	3	3231	9693	9	3201.3	100.9
1969	4	3320	13280	16	3287.6	101.0
1970	5	3326	16630	25	3373.9	98.6
1971	6	3424	20544	36	3460.2	99.0
1972	7	3546	24822	49	3546.5	100.0
1973	8	3594	28752	64	3632.8	98.9
1974	9	3652	32868	81	3719.1	98.2
Total	0	55906	49166	570		

$$a = \frac{55906}{19} = 2942.4$$

$$b = \frac{49166}{570} = 86.3$$

$$Y_t = 2942.4 + 86.3x$$
$$x = 0 \text{ in } 1965$$
x is in one-year intervals
Y is in thousands of consumers

SOURCES: *Business Statistics*, 1971 Edition (18th Biennial Edition), U.S. Dept. of Commerce (Washington, D.C.: U.S. Government Printing Office). *Survey of Current Business*, U.S. Dept. of Commerce: Vol. 53, No. 1 (January 1973); Vol. 55, No. 1 (January 1975); Vol. 56, No. 2 (February 1976).

these constants at the bottom of the table, $a = 2942.4$ and $b = 86.3$. The trend equation is $Y_t = 2942.4 + 86.3x$. An identification statement such as the one given below the trend equation that $x = 0$ in 1965 and Y is in thousands of consumers should always accompany the equation, since it is not possible to interpret the meaning of the trend line fully without it. The trend figures are determined by substituting the appropriate values of x into the trend equation. Hence, for example, the trend figure for 1956 is

$$Y_{t,1956} = 2942.4 + 86.3(-9) = 2165.7 \text{ (thousands)}$$

Since the b value measures the change in Y_t per year, it can be added to each trend value to obtain the following year's figure. The trend figures are given in Column (6) of Table 10-1.

The trend line is graphed in Figure 10-3. Any two points can be plotted to determine the line. Interpreting the values $a = 2942.4$ and $b = 86.3$, we have a computed trend figure of 2942.4 thousands for the number of industrial and commercial consumers of natural gas in 1965 and an increase in trend of 86.3 thousands per year. As can be seen from the graph, the trend line fits the data rather well. Since the line was fitted by the method of least squares, the sum of the squared deviations of the actual data from the trend line is less than from any other straight line that could have been fitted to the data, and the total of the deviations above the line is equal to the total below the line.

A couple of technical points concerning the fitting procedure may be noted. Since the present illustration contained an odd number of

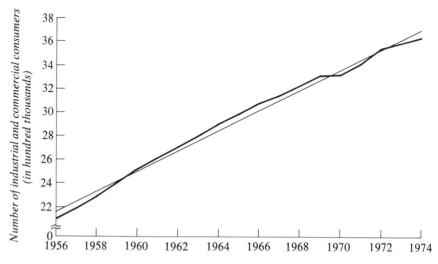

FIGURE 10-3
Straight-line trend fitted to industrial and commercial consumers of natural gas in the United States, 1956–1974.

years, the time period at which $x = 0$, or the x origin, coincided with one of the years of data, and the x values were stated as $1, 2, 3, \ldots$ for years after the x origin and $-1, -2, -3, \ldots$ for years before the origin. On the other hand, if there had been an even number of years, then the mean time period, at which $x = 0$, would fall midway between the two central years. For example, suppose there had been one less year of data and the figures on consumers of natural gas were available only for 1956–1973. Then there would be 18 annual figures and $x = 0$ at $1964\frac{1}{2}$. The two central years, 1964 and 1965, would deviate from this origin by $-\frac{1}{2}$ and $+\frac{1}{2}$, respectively. To avoid the use of fractions, it is usual to state the deviations in terms of half-year intervals rather than a year. Hence, the x values for 1965, 1966, 1967, $\ldots$, would be $1, 3, 5, \ldots$, and for 1964, 1963, 1962, $\ldots$, they would be $-1, -3, -5, \ldots$. The computation of the constants a and b would proceed in the usual way. However, now a would be interpreted as the computed trend figure for a time point midway between the two central years, and the b value would be the amount of change in trend per half year.

If the time intervals of the original data were not annual, the transformed time variable x would have to be appropriately interpreted. For example, if the data were stated in the form of five-year averages and there were an odd number of such figures, then x would be in five-year intervals. If there were an even number of figures and the nonfractional method of stating x referred to above were used, then x would be in $2\frac{1}{2}$-year intervals.

Projection of the Trend Line

Projections of the computed trend line can be obtained by substituting the appropriate values of x into the trend equation. For example, if a projected trend figure for 1982 were desired for the number of industrial and commercial consumers of natural gas, it would be computed by substituting $x = 17$ in the previously determined trend equation. Hence,

$$Y_{t,1982} = 2942.4 + 86.3(17) = 4409.5 \text{ (thousands)}$$

A rougher estimate of this trend figure would be obtained by extending the straight line graphically in Figure 10-3 to the year 1982. It must be remembered that these projections are estimates of only the trend level in 1982 and not of the actual figure for the number of industrial and commercial consumers of natural gas in that year. If a prediction of the latter figure were desired, estimates of the nontrend factors would have to be combined with the trend estimate. This means that a prediction of cyclical fluctuations would have to be made and incorporated with the trend figure. Accurate forecasts of this type are difficult to make over extended time periods. However, insofar as managerial applications of trend analysis are concerned, for long-range planning purposes often all that is desired is a projection of the trend level of the economic variable of interest. For example, a good estimate of the trend of demand would be adequate for a business firm planning a plant expansion to anticipate demand many years into the future. Accompanying predictions of business cycle standings many years into the future would not be required; nor, for that matter, would they be realistically feasible.

Cyclical Fluctuations

As was previously indicated, when a time series consists of annual data, it contains trend, cyclical, and irregular elements. The seasonal variations are absent, since they occur within a year. Hence, deviations of the actual annual data from a computed trend line are attributable to cyclical and irregular factors. Since the cyclical element is the dominant factor, a study of these deviations from trend essentially represents an examination of business cycle fluctuations. The deviations from trend are most easily observed by dividing the original data by the corresponding trend figures for the same time period. By convention, the result of this division of an original figure by a trend value is multiplied by 100 to express the figure as a percentage of trend. Hence, if the original figure is exactly equal to the trend figure, the percentage of trend is 100; if the original figure exceeds the trend value, the percentage of trend is above 100; and if the original figure is less than the trend value, the percentage of trend is below 100.

The formula for percentage of trend figures is

(**10.6**) $$\text{Percentage of trend} = \frac{Y}{Y_t} \cdot 100$$

where Y = annual time series data
Y_t = trend values

In summary, the original annual data contain trend, cyclical, and irregular factors. Since the data are annual, the seasonal component is not included. When converted to percentage of trend, these numbers contain only cyclical and irregular movements, since the division by trend eliminates that factor. The rationale of this procedure is easily seen by assuming a so-called multiplicative model for the analysis. That is, the original annual figures are viewed as representing the combined effect of trend, cyclical, and irregular factors. In symbols, let T, C, and I represent trend, cyclical, and irregular factors, respectively, and let Y and Y_t mean the same as in Equation 10.6. Then dividing the original time series by the corresponding trend values yields

(**10.7**) $$\frac{Y}{Y_t} = \frac{T \times C \times I}{T} = C \times I$$

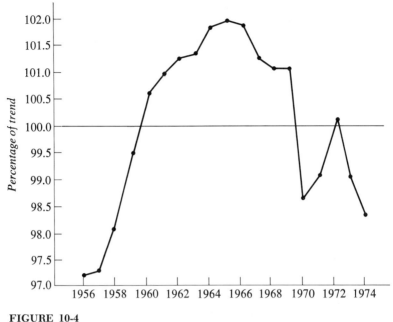

Percentage of trend

FIGURE 10-4
Percentages of trend for industrial and commercial consumers of natural gas in the
United States, 1956–1974.

The percentages of trend for the series on the number of industrial
and commercial consumers of natural gas are given in Column (7) of
Table 10-1 and are plotted in Figure 10-4. As may be seen from the
chart, the underlying upward trend movement is no longer present. Instead, the percentage of trend series fluctuates about the line labeled
100, which is the trend level. These percentages of trend are sometimes
referred to as *cyclical relatives;* that is, the original data are stated relative to the trend figure. (Of course, strictly speaking, Y/Y_t is the cyclical
relative, and the multiplication by 100 converts the relative to a percentage figure.) Another way of depicting cyclical fluctuations is in terms of
relative cyclical residuals, which are percentage deviations from trend
and are computed by the formula

(10.8)   Relative cyclical residual $= \dfrac{Y - Y_t}{Y_t} \cdot 100$

Hence, for example, if we refer to the consumers of natural gas data in
Table 10-1 for 1969, the actual figure is 3320, the computed trend value
is 3287.6, and the percentage of trend is 101.0. The relative cyclical
residual in this case is $+1.0\%$, indicating that the actual number of consumers is 1.0% above the trend figure because of cyclical and irregular
factors. These residuals are positive or negative depending on whether
the actual time series figures fall above or below the computed trend val-

ues. The graph of relative cyclical residuals is visually identical to that of the percentage of trend values except that relative cyclical residuals are shown as fluctuations around a zero base line rather than around a base line of 100%.

The familiar charts of business cycle fluctuations that often appear in publications such as the financial pages of newspapers and business periodicals are usually graphs of either percentages of trend or relative cyclical residuals. These charts may be studied for timing of peaks and troughs of cyclical activity, for amplitude of fluctuations, for duration of periods of expansion and contraction, and for other items of interest to the business cycle analyst.

Fitting a Second-Degree Trend Line

The preceding discussion on the fitting of a straight line pertains to the case in which the trend of the time series can be characterized as increasing or decreasing by constant amounts per time period. Actually very few economic time series exhibit this type of constant change over a long period of time (say, over a period of several business cycles). Therefore, it generally is necessary to fit other types of lines or curves to the given time series. Polynomial functions are particularly convenient to fit by the method of least squares. Frequently, a second-degree parabola provides a good description of the trend of a time series. In this type of curve, the amounts of change in the trend figures, Y_t, may increase or decrease per time period. Hence, a second-degree parabola may provide a good fit to a series whose trend is increasing by increasing amounts, increasing by decreasing amounts, etc. The procedure of fitting a parabola by the method of least squares involves the same general principles as fitting a straight line, but it entails somewhat more arithmetic.

EXAMPLE 10-1

We illustrate the method of fitting a second-degree parabola to a time series in terms of a very simple illustration. The reader is warned that the time period in this example is too short to permit a valid description of trend. However, the illustration is given for expository purposes only to indicate the procedure involved. In Table 10-2 is given a time series on the number of housing starts in a certain region from 1970 to 1976. This series is graphed in Figure 10-5. The trend of these data may be described as decreasing by decreasing amounts. The general form of a second-degree parabola is $Y_t = a + bX + cX^2$. Analogous to the method of stating the equation for a straight-line trend, the trend line for a second-degree parabola may be written

(10.9) $$Y_t = a + bx + cx^2$$

TABLE 10-2
Second-Degree Parabola Fitted by the Method of Least Squares to the Number of Housing Starts in a Certain Region, 1970–1976

Year (1)	x (2)	Y (3)	xY (4)	x^2Y (5)	x^2 (6)	x^4 (7)	Y_t (8)
			Number of Housing Starts (in Thousands)				
1970	−3	83	−249	747	9	81	84
1971	−2	60	−120	240	4	16	62
1972	−1	54	−54	54	1	1	44
1973	0	21	0	0	0	0	30
1974	1	22	22	22	1	1	20
1975	2	13	26	52	4	16	14
1976	3	13	39	117	9	81	12
Total	0	266	−336	1232	28	196	

SOURCE: Hypothetical data.

where $\quad Y_t$ = the trend values
$\qquad a, b, c$ = constants to be determined
$\qquad x$ = deviations from the middle time period

If the transformed variable x, representing deviations from the mean time period, is used, the equations for fitting a second-degree parabola are

(10.10) $\qquad\qquad \Sigma Y = na + c\Sigma x^2$

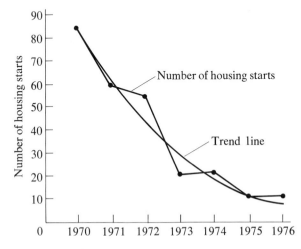

FIGURE 10-5
Second-degree parabola fitted to the number of housing starts in a certain region, 1970–1976.

(10.11) $$\Sigma\, x^2 Y = a\Sigma\, x^2 + c\Sigma\, x^4$$

(10.12) $$b = \frac{\Sigma\, xY}{\Sigma\, x^2}$$

Hence, the constant b is determined by the same equation as in fitting the straight line. The constants a and c are found by solving simultaneously Equations 10.10 and 10.11.

In the present problem, since there are an odd number of years, $x = 0$ in the middle year, 1973. Solving for b by substituting the appropriate totals from Table 10-2, we have

$$b = \frac{-336}{28} = -12$$

Substituting into Equations 10.10 and 10.11 gives

$$266 = 7a + 28c$$
$$1232 = 28a + 196c$$

Dividing the second equation by 4 to equate the coefficients of a, we obtain

$$266 = 7a + 28c$$
$$308 = 7a + 49c$$

Subtracting the first equation from the second, we have

$$42 = 21c$$

and

$$c = \frac{42}{21} = 2$$

Substituting this value for c into the first equation yields

$$266 = 7a + 28(2)$$
$$a = 30$$

Therefore, the equation of the second-degree parabola fitted to the time series on housing starts is

(10.13) $$Y_t = 30 - 12x + 2x^2$$

where $x = 0$ in 1973
$\quad\quad$ x is in one-year intervals
$\quad\quad$ Y is in thousands of housing starts

The trend figures, Y_t, shown in Column (8) of Table 10-2 are obtained by substituting the appropriate values of x into Equation 10.13. The constants $a, b,$ and c may be interpreted as follows: a is the computed trend figure at the time origin, that is, when $x = 0$; b is the slope of the parabola at the time origin; and c indicates the amount of acceleration or deceleration in the curve, or the amount by which the slope changes per time period.[3]

Although the second-degree parabola appears from Figure 10-5 to provide a reasonably good fit to the data in this example, the dangers of a mechanistic projection of a trend line are clearly illustrated. The parabola would begin to turn upward after 1976, and the projected trend figure for each year would be higher than the preceding year's figure. Therefore, only if an analysis of all the underlying factors determining the trend of this series revealed reasons for a reversal of the observed decline should one be willing to entertain the notion of extending the trend line into the future for forecasts, even for relatively short periods.

Fitting Logarithmic Trend Lines

As discussed earlier, the equations of trend lines embody assumptions concerning the type of change that takes place over time. Hence, the arithmetic straight line assumes a trend that increases or decreases by constant amounts, whereas the second-degree parabola assumes that the change in these amounts of change is constant per unit time. It is often useful to describe the trend of an economic time series in terms of the percentage rates of change that are taking place. Logarithmic trend lines are useful for this purpose.

If a time series increases or decreases at exactly a constant percentage rate, a straight line fitted to the logarithms of the data constitutes a perfect fit. For example, suppose a time series has successive values of 10, 100, 1000, 10,000. This series may be characterized as increasing at a constant percentage rate of 900%. The logarithms of these values are log 10 = 1, log 100 = 2, log 1000 = 3, and log 10,000 = 4. Hence, the logarithms are increasing by a constant amount, namely 1 unit, and a straight line could be drawn through the numbers 1, 2, 3, and 4. Some economic time series in the United States (for example, GNP), although not changing exactly at a constant rate, have exhibited trends of approximately constant percentage increases over substantial periods of time. The equation of the logarithmic straight line that would describe the trend of such series is

(10.14) $\log Y_t = a + bx$

---

[3] The derivative of the second-degree parabola trend equation is

$$\frac{dY_t}{dx} = b + 2cx$$

Hence, the slope of the curve differs at each time period x. When $x = 0$, $\frac{dY_t}{dx} = b$. Therefore, the slope at the time origin is b. The second derivative is $\frac{d^2Y_t}{dx^2} = 2c$. Thus the acceleration, or rate of change in the slope, is $2c$ per time period.

The method of fitting this line is the same as for the arithmetic straight line, $Y_t = a + bx$, except that wherever Y appeared before, $\log Y$ now appears. Hence, the values of the constants a and b are computed as follows.

$$(10.15) \qquad\qquad a = \frac{\Sigma \log Y}{n}$$

$$(10.16) \qquad\qquad b = \frac{\Sigma x \log Y}{\Sigma x^2}$$

After a and b have been calculated, trend figures are determined by substituting values of x into the trend equation, computing $\log Y_t$, and taking the antilogarithm to obtain Y_t. Although we will not present another example to indicate the fitting process, since there really are no new principles involved, we will illustrate the calculation of a trend figure for the type of trend line under discussion.

Suppose the logarithmic trend line for a particular series had been determined as

$$\log Y_t = 2.3657 + 0.0170x$$

Then the logarithm of the trend figure for the year in which $x = 2$ would be given by substituting this value of x into the trend equation to obtain

$$\log Y_t = 2.3657 + 0.0170(2)$$
$$\log Y_t = 2.3997$$

Taking the antilog of this value yields the trend figure

$$Y_t = \text{antilog } 2.3997 = 251.0$$

The rate of change implied by this trend line can be obtained by calculating antilog $b - 1$. For example, in the above illustration the antilog of the slope coefficient b is

$$\text{antilog } 0.0170 = 1.040$$

This figure is the ratio of each trend figure to the preceding one. Subtracting 1.00 from this figure yields $1.04 - 1.00 = 0.04$. Hence, the trend figures increase by 4% per time period. If the series had been a declining one and the result of the above calculation, for example, was -0.04, this would mean that the trend figures decrease by 4% per time period. Even though a time series may not exhibit a trend with constant rates of change throughout, sometimes it can be broken down into segments during which the rate of change has been approximately constant. It is often useful in such cases to make comparisons of the rates of change similarly determined from different economic time series of interest.

Logarithmic second-degree parabolas can also be fitted to time series in which the trend is increasing at an increasing percentage rate, increasing at a decreasing percentage rate, etc. However, ordinarily polynomials of third or higher degree are not fitted to time series in either arithmetic or logarithmic form. The reason is that such curves permit too many changes in direction and tend to follow the cyclical fluctuations in the data as well as the trend. Therefore, these curves often do not have the required characteristic of a trend line of depicting the smooth, continuous movement underlying the cyclical swings in a time series.

In the attempt to find an appropriate trend line, the analyst should always plot the time series on both arithmetic and semilogarithmic graph paper. These two types of graphs may help in determining whether an arithmetic or logarithmic line would provide a better description of the trend.

Exercises

1. Discuss the nature and causes of the component variations of an economic time series.

2. The following table gives the total output of services for the United States for the period 1964–1974 in billions of 1958 constant dollars.

Year	GNP – Services (Billions of 1958 Dollars)	Year	GNP – Services (Billions of 1958 Dollars)
1964	$210.8	1970	273.3
1965	221.9	1971	279.7
1966	236.3	1972	291.4
1967	249.1	1973	304.5
1968	259.7	1974	310.5
1969	268.2		

   a. Fit a straight line by the method of least squares to this annual series.
   b. Compute the percentage of trend for each year. What is accomplished in the conversion of the annual series into a percentage of trend series?
   c. Graph the trend line against the actual data.

3. Follow the instructions given in (a) and (b) of Exercise 2 for the following economic series. The series shown is the same as that in Exercise 2, except that the data are expressed in current dollars rather than constant dollars.

Year	GNP – Services (Current Dollars)	Year	GNP – Services (Current Dollars)
1964	$244.2	1970	$410.3
1965	262.9	1971	446.0
1966	289.1	1972	488.1

(Continued)

Year	GNP–Services (Current Dollars)	Year	GNP–Services (Current Dollars)
1967	316.5	1973	534.4
1968	346.6	1974	589.1
1969	377.9		

Graph this trend line against the actual data. Which trend line appears to fit the data more closely, the trend line for the constant dollar series or that for the current dollar series?

4. Obtain trend figures for the 1976–1980 period by projecting a straight-line trend for U.S. population (age 16 and over), using the end-of-year data in the following table:

Year	U.S. Population Age 16 and over (in Thousands)	Year	U.S. Population Age 16 and over (in Thousands)
1961	121,343	1968	135,562
1962	122,981	1969	137,841
1963	125,154	1970	140,182
1964	127,224	1971	142,596
1965	129,236	1972	145,775
1966	131,180	1973	148,263
1967	133,319	1974	150,827

5. The following series shows the total national income in $billions from 1933 to 1937:

Year	National Income ($Billions)
1933	35.0
1934	40.2
1935	44.0
1936	49.9
1937	55.0

a. For each year, compute the Y_t values for the equation

$$Y_t = a + bx$$

b. Are these Y_t values a good description of the trend of national income? Why or why not?

c. Compute the relative cyclical residual for 1935 and explain what it means.

6. The following table presents the consumption of electric power in a certain region of the United States:

Year	Consumption (Billion Kilowatt Hours)
1925	25
1935	60
1945	135
1955	330
1965	780
1975	1855

a. What was the average amount of increase per decade in the above series between 1925 and 1975? (Do not use trend line.)
b. Fit a linear trend line by the method of least squares to (1) the original data and (2) the logarithms of the data.
c. In terms of this problem, interpret the meaning of the constants of the trend equation found for the original data in part (b).
d. Is the answer to (a) consistent with the slope obtained for the original data in (b)? Why or why not?
e. In 1947, consumption of electric power was 200 billion kilowatt hours. Compute and interpret the absolute cyclical residual and the relative cyclical residual for that year, using the straight-line trend fitted to the original data in (b).

7. The following trend equation resulted from the fitting of a least squares second-degree parabola to labor force data in a Southern county:

$$Y_t = 74.62 + 6.83x - 0.31x^2$$

where $x = 0$ in 1955
      x is in $2\frac{1}{2}$ year intervals
      Y is the size of the labor force in thousands

a. Assume that this trend line is "a good fit." What generalizations can you make concerning the way in which the labor force of this county has grown in absolute amounts? Concerning the percentage rate at which it has grown?
b. In part (a), you were instructed to assume that the trend line was a good fit. However, the actual size of the labor force in 1975 was 126,430. This is rather striking evidence that the equation is not a good fit. Do you agree? Discuss.

8. The following three series were obtained from the *Wall Street Journal* of October 20, 1975. The figures for the first series are rates recorded in the last week of each year. The figures for the other two series are monthly averages for the years shown.

Year	(1)	(2)	(3)
1950	1.83 %	2.75 %	4.08 %
1951	2.26	3.22	4.26
1952	2.44	3.04	4.32
1953	2.55	3.23	4.78
1954	2.35	2.87	4.56
1955	2.59	3.27	4.73
1956	3.25	4.26	5.76
1957	2.98	4.04	5.61
1958	3.33	4.44	5.60
1959	3.83	5.27	6.23
1960	3.37	4.94	6.04
1961	3.50	4.58	5.69
1962	3.14	4.28	5.53
1963	3.32	4.49	5.45
1964	3.17	4.49	5.45
1965	3.59	4.92	5.62
1966	3.74	5.98	6.77
1967	4.52	6.93	6.81
1968	4.84	7.28	7.50
1969	6.85	9.22	8.62
1970	5.53	8.13	8.40
1971	4.97	7.54	7.59
1972	5.11	7.50	7.56
1973	5.19	8.09	8.78
1974	7.15	9.47	9.51

(1) Interest rates on municipal bonds
(2) Interest rates on new issues of
     high-grade corporate bonds
(3) Mortgage interest rates (secondary
     market yields on FHA mortgages)

    a. Fit a straight line by the method of least squares to each of these series.
    b. Calculate projected trend figures for the 1976–1978 period for each series.
9. A research analyst was studying federal, state, and local government expenditures over the recent past. As part of the study, the analyst wished to establish a trend line for the following data:

Year	State and Local Government Expenditures ($Billions)	Year	State and Local Government Expenditures ($Billions)
		1968	100.8
1962	53.7	1969	111.2

(Continued)

Year	State and Local Government Expenditures ($Billions)	Year	State and Local Government Expenditures ($Billions)
1963	58.2	1970	123.3
1964	63.5	1971	136.6
1965	70.1	1972	150.8
1966	79.0	1973	169.8
1967	89.4	1974	192.4

    a. Fit a straight line to this series by the method of least squares.
    b. Fit a second-degree parabola trend line to this series by the method of least squares.
    c. Which trend line seems to reflect more accurately the trend inherent in this time series?

10. The marketing department of the Century Corporation has initiated a research project on demand for the company's principal products. Because the products are consumed mainly in household operation, the following series has been selected as an appropriate one for study:

Year	Personal Consumption Expenditures – Household Operation (Billions of Current Dollars)	Year	Personal Consumption Expenditures – Household Operation (Billions of Current Dollars)
1960	$20.0	1968	31.2
1961	20.8	1969	33.8
1962	22.0	1970	36.4
1963	23.1	1971	39.4
1964	24.3	1972	43.3
1965	25.6	1973	47.3
1966	27.1	1974	52.9
1967	29.1		

    a. Fit a second-degree line by the method of least squares to this series and calculate the trend values for the period 1960–1974.
    b. Calculate and graph the percentage of trend figures for the period 1960–1974. Is there any evidence of cyclical fluctuations?

11. You have been assigned the task of ascertaining the trend movement of total farm employment in the United States. The series shown below has been provided for use in your analysis.

Year	Total U.S. Farm Employment (in Thousands)	Year	Total U.S. Farm Employment (in Thousands)
1960	7057	1968	4749
1961	6919	1969	4596
1962	6700	1970	4523
1963	6518	1971	4436
1964	6110	1972	4373
1965	5610	1973	4337
1966	5214	1974	4294
1967	4903		

Fit a logarithmic straight line to this series by the method of least squares and calculate the trend values for the period 1960–1974. Calculate the percentage of trend figures for this same period. What are the estimated trend values for 1975, 1976, and 1977?

NOTE: The data for many of the Exercises in Section 10-4 were obtained from *The Economic Report of the President, Transmitted to the Congress, February 1975* (U.S. Government Printing Office, Washington, 1975).

10.5 MEASUREMENT OF SEASONAL VARIATIONS

For long-range planning and decision making, in terms of time series components, executives of a business or government enterprise concentrate primarily on forecasts of trend movements. For intermediate planning periods, say from about two to five years, business cycle fluctuations are of critical importance, too. For shorter range planning, and for purposes of operational decisions and control, seasonal variations must also be taken into account.

Seasonal movements, as indicated in Section 10.2, are periodic patterns of variation in a time series. Strictly speaking, the terms "seasonal movements" and "seasonal variations" can be applied to any regularly repetitive movements that occur in a time series where the interval of time for completion of a cycle is one year or less. Hence, under this classification are subsumed movements such as daily cycles in utilization of electrical energy and weekly cycles in the use of public transportation vehicles. However, these terms generally refer to the annual repetitive patterns of economic activity associated with climate and custom. As noted earlier, these movements are generally examined by using series of monthly or quarterly data.

Purpose of Analyzing Seasonal Variations

Just as was true in the case of the study of trend movements, seasonal variations may be studied because *interest is primarily centered upon these movements,* or they may be measured merely *in order to eliminate them,* so that business cycle fluctuations can be more clearly revealed. As an illustration of the first purpose, a company might be interested in analyzing the seasonal variations in sales of a product it produces in order to iron out variations in production, scheduling, and personnel requirements. Another reason a company's interest may be primarily focused on seasonal variations is to budget a predicted annual sales figure by monthly or quarterly periods based on seasonal patterns observed in the past.

On the other hand, as an illustration of the second purpose, an economist may wish to eliminate the usual month-to-month variations in series such as personal income, unemployment rates, and housing starts in order to study the underlying business cycle fluctuations present in these data.

Rationale of the Ratio to Moving Average Method

There are a number of techniques by which seasonal variations can be measured, but only the most widely used one, the so-called *ratio to moving average method,* will be discussed here. It is most frequently applied to monthly data, but we will illustrate its use for a series of quarterly figures, thus reducing substantially the required number of computations.

In acquiring an understanding of the rationale of the measurement of seasonal fluctuations, it is helpful to begin with the final product, the seasonal indices. The object of the calculations when the raw data are for quarterly periods and a stable or regular seasonal pattern is present is to obtain four seasonal indices, each one indicating the seasonal importance of a quarter of the year. The arithmetic mean of these four indices is 100.0. Hence, if the seasonal index for, say, the first quarter is 105, this means that the first quarter averages 5% higher than the average for the year as a whole. If the original data had been monthly, there would be twelve seasonal indices, which average 100.0, and each index would indicate the seasonal importance of a particular month. These indices are descriptive of the recurrent seasonal pattern in the original series.

As an example of how these seasonal indices might be used, we can refer to budgeting a predicted annual sales figure, say, by quarterly periods. Suppose that $40,000,000 of sales of particular products was budgeted for the next year, or an average of $10,000,000 per quarter. If the quarterly seasonal indices based on an observed stable seasonal pattern in the past were 97.0, 110.0, 85.0, and 108.0, then the amounts of sales budgeted for each quarter would be

$$
\begin{aligned}
&\text{First quarter:} &&0.97 \times \$10 \text{ million} = \$\ 9.7 \text{ million} \\
&\text{Second quarter:} &&1.10 \times \quad 10 \text{ million} = \quad 11.0 \text{ million} \\
&\text{Third quarter:} &&0.85 \times \quad 10 \text{ million} = \quad\ \ 8.5 \text{ million} \\
&\text{Fourth quarter:} &&1.08 \times \quad 10 \text{ million} = \quad 10.8 \text{ million}
\end{aligned}
$$

The essential problem in the measurement of seasonal variations is eliminating from the original data the nonseasonal elements in order to isolate the stable seasonal component. In trend analysis, when annual data were used and it was desired to arrive at cyclical fluctuations, a similar problem existed. It was solved by obtaining measures of trend and using them as base line or reference figures. Deviations from trend were then measures of cyclical (and irregular) movements. Analogously, when we have monthly or quarterly original data, which consist of all of the components of trend, cycle, seasonal, and irregular movements, ideally we would like to obtain a series of base line figures that contain all the nonseasonal elements. Then deviations from the base line would represent the pattern of seasonal variations. Unsurprisingly, this ideal method of measurement is not feasible. However, the practical method is to obtain a series of moving averages that roughly include the trend and cycle components. Dividing the original data by these moving average figures eliminates the trend and cyclical elements and yields a series of figures that contain seasonal and irregular movements. These data are then averaged by months or by quarters to eliminate the irregular disturbances and to isolate the seasonal factor. This method of describing a pattern of stable seasonal movements is explained below.

Ratio to Moving Average Method

In order to derive a set of seasonal indices from a series characterized by a stable seasonal pattern, about five to eight years of monthly or quarterly data are required. A stable seasonal pattern means that the peaks and troughs generally occur in the same months or quarters of each year.

The ratio to moving average method of computing seasonal indices for quarterly data may be summarized as consisting of the following steps:

1. Derive a four-quarter moving average that contains the trend and cyclical components present in the original quarterly series. A four-quarter moving average is simply an annual average of the original quarterly data successively advanced one quarter at a time. For example, the first moving average figure contains the first four quarters. Then the first quarter is dropped, and the second through fifth quarterly figures are averaged. The computation proceeds this way until the last moving average is calculated, containing the last four quarters of the original series. In the actual calculation, an adjustment is made (as explained below) in order to center the moving average figures so their timing corresponds to that of the original data.

The reason these moving averages include the trend and cyclical components may perhaps be most easily understood by considering what these averages do *not* contain. Since they are annual averages, they do not contain seasonal movements (since such fluctuations, by definition, average out over a one-year period). Furthermore, the irregular movements that tend to raise the figures for certain months or quarters and to lower them in others tend to cancel out when averaged over the year. Thus, only the trend and cyclical elements tend to be present in the moving averages.

2. Divide the original data for each quarter by the corresponding moving average figure. These "ratio to moving average" numbers contain only the seasonal and irregular movements, since the trend and cyclical components were eliminated in the division by the moving average.

3. Arrange the ratio to moving average figures by quarters, that is, all the first quarters in one group, all the second quarters in another, and so forth. Average these ratio to moving average figures for each quarter in an attempt to eliminate the irregular movements, and thus to isolate the stable seasonal component. One type of average used for this purpose is referred to as a "modified mean." This is an arithmetic mean of the ratio to moving average figures taken after dropping the highest and lowest extreme values.

4. Make an adjustment to force the four modified means to total 400, and thus average out to 100.0. The resulting four figures, one for each quarter of the year, constitute the seasonal indices for the series in question.

In symbols, this procedure may be summarized as follows. Let Y be the original quarterly observations; MA the moving average figures; and T, C, S, and I the trend, cyclical, seasonal, and irregular components, respectively. Then dividing the original data by the moving average values gives

$$(10.17) \qquad \frac{Y}{MA} = \frac{T \times C \times S \times I}{T \times C} = S \times I$$

Averaging these ratio to moving average figures (Y/MA) eliminates the irregular movements that tend to make the Y/MA values too high in certain years and too low in others. Hence, if the eliminations of the nonseasonal elements were perfect, the final seasonal indices would reflect only the effect of seasonal variations. Of course, since the entire method is a rather rough and approximate procedure, the nonseasonal elements are generally not completely eliminated. The moving average usually contains the trend and *most* of the cyclical fluctuations. Therefore, the cyclical component is usually not completely absent in the Y/MA values. Moreover, the modified means do not ordinarily remove all of the erratic disturbances attributed to the irregular component. Nevertheless, in the case of series with a stable seasonal pattern, the computed seasonal indices generally isolate the underlying seasonal pattern quite well.

In Table 10-3 is given a quarterly series of feed grain price index numbers of average prices received by farmers from 1959 to 1966. The base period of the index number series is 1957–1959. As is indicated in the section on index numbers in Chapter 11, this means that the average level of prices during this period is designated as 100. Index figures above and below 100 represent price levels that are higher and lower, respectively, than during the base period. Examination of this series reveals that feed grain prices tend to be highest during the second and third quarters, that is, during the spring and summer months, and lowest during the first and fourth quarters, or during the fall and winter. The calculation of quarterly seasonal indices will be illustrated in terms of this series.

The feed grain price indices have been listed in Column (2) of Table 10-3 from the first quarter of 1959 through the second quarter of 1966. The inclusion of the first two quarters of 1966 permits the computation of the moving averages for all four quarters of 1965. Our first task is the calculation of the four-quarter moving average. This moving average would simply be calculated as indicated above by averaging four quarters at a time, continually moving the average up by a quarter. However, because of a problem of centering of dates, a slightly different type of average, a so-called "two-of-a-four-quarter moving average" is calculated. The problem is as follows. An average of four quarterly figures would be centered halfway between the dating of the second and third figures and would thus not correspond to the date of either of those figures. For example, the average of the four quarters of 1959, the first figures shown in Column (2) of Table 10-3, would be centered midway between the second and third quarter dates, or at the center of the year, July 1, 1959. The original quarterly figures are centered at the middle of their respective time periods, or, for simplicity, say, February 15, May 15, August 15, and November 15. Hence, the dates of a simple four-quarter moving average would not correspond to those of the original data. This problem is easily solved by averaging the moving averages two at a time. For example, as we have seen, the first moving average obtainable from Table 10-3 is centered at July 1, 1959. The second moving average, which contains the last three quarters of 1959 and the first quarter of 1960, is centered at October 1, 1959. Averaging these two figures yields a figure centered at August 15, the same as the dating of the third quarter.

The easiest way to calculate this properly centered moving average is given in Columns (3) through (5) of Table 10-3. In Column (3) is given a four-quarter moving total. The first figure, 391, is the total of the first four quarterly figures, 96, 103, 100, and 92. This figure is listed opposite the third quarter, 1959, although actually it is centered at July 1. The next four-quarter moving total is obtained by dropping the figure for the first quarter of 1959 and including the figure for the first quarter of 1960. Hence, 388 is the total of 103, 100, 92, and 93. The total of 391

TABLE 10-3
Feed Grain Index Numbers of Average Prices Received by Farmers by Quarters,
1959–1966: Computations for Seasonal Indices and Deseasonalizing of Original Data

Quar-ter	Feed Grain Price Index Numbers (1957–1959 = 100)	Four-Quarter Moving Total	Two-of-a-Four-Quarter Moving Total	Moving Average Col (5) = Col (4) × $\frac{1}{8}$	Original Data as Percentage of Moving Average [Col (2) ÷ Col (5)] × 100	Sea-sonal Index	Deseason-alized Feed Grain Price Index Numbers [Col(2) ÷ Col (7)] × 100
(1)	(2)	(3)	(4)	(5)	(6)	(7)	(8)
1959							
I	96					98.8	97.2
II	103					102.9	100.1
III	100	391	779	97.38	102.69	102.6	97.5
IV	92	388	771	96.38	95.46	95.7	96.1
1960							
I	93	383	762	95.25	97.64	98.8	94.1
II	98	379	752	94.00	104.26	102.9	95.2
III	96	373	744	93.00	103.23	102.6	93.6
IV	86	371	737	92.13	93.35	95.7	89.9
1961							
I	91	366	733	91.63	99.31	98.8	92.1
II	93	367	741	92.63	100.40	102.9	90.4
III	97	374	750	93.75	103.47	102.6	94.5
IV	93	376	756	94.50	98.41	95.7	97.2
1962							
I	93	380	759	94.88	98.02	98.8	94.1
II	97	379	758	94.75	102.37	102.9	94.3
III	96	379	762	95.25	100.79	102.6	93.6
IV	93	383	771	96.38	96.49	95.7	97.2
1963							
I	97	388	786	98.25	98.73	98.8	98.2
II	102	398	801	100.13	101.87	102.9	99.1
III	106	403	809	101.13	104.82	102.6	103.3
IV	98	406	814	101.75	96.31	95.7	102.4
1964							
I	100	408	813	101.63	98.40	98.8	101.2
II	104	405	813	101.63	102.33	102.9	101.1
III	103	408	823	102.88	100.12	102.6	100.4
IV	101	415	838	104.75	96.42	95.7	105.5
1965							
I	107	423	851	106.38	100.58	98.8	108.3
II	112	428	853	106.63	105.04	102.9	108.8
III	108	425	848	106.00	101.89	102.6	105.3
IV	98	423	842	105.25	93.11	95.7	102.4
1966							
I	105	419				98.8	106.3
II	108					102.9	105.0

SOURCE: Agricultural Handbook No. 325, U.S. Department of Agriculture, 1966.

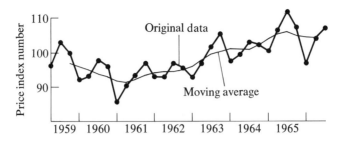

FIGURE 10-6
Feed grain index numbers of average prices received by farmers by quarters, 1959–1966.

and 388, or 779, is the first entry in Column (4). This represents the total for the eight months that would be present in the averaging of the first two simple four-quarter moving averages. Dividing this total by 8 yields the first two-of-a-four-quarter moving average figure of 97.38, properly centered at the middle of the third quarter, 1959.

The moving averages given in Column (5) of Table 10-3 are shown in Figure 10-6 along with the original data. It is very useful to examine graphs in the calculation of seasonal indices, because we can observe visually what is accomplished in each major step of the procedure. We have noted earlier that the original data, if stated in monthly or quarterly form, contain all of the components of trend, cycle, seasonal, and irregular movements. Although the time period is too short for trend to be revealed, we can observe in the series of feed grain price index numbers some effects of cyclical fluctuations as the data move into a trough at the end of 1960 and continue in an expansion swing thereafter. The repetitive annual rhythm of the seasonal movements is clearly discernible. Irregular movements are also present. The moving average, which runs smoothly through the original data, can be observed to follow the cyclical fluctuations rather closely, and if the series were long enough, we would be able to see how the moving average describes trend movements as well. Another way to view this point is to note that the seasonal variations and (to a large degree) the irregular movements are absent from the smooth line that traces the path of the moving average. It should be noted that there are no moving average figures corresponding to the first two and the last two quarters of original data. Correspondingly, if the original data were in monthly form and a twelve-month moving average were computed, there would be no moving averages to correspond to the first six months of data or to the last six months of data.

The "ratio to moving average" figures [that is, or original data in

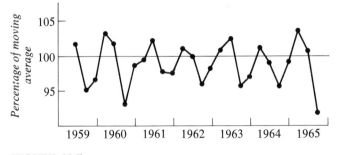

FIGURE 10-7
Percentage of moving averages for feed grain price index numbers, 1959–1966.

Column (2) divided by the moving averages in Column (5)] are given in Column (6) of Table 10-3. As is customary, these figures have been multiplied by 100 to express them in percentage form. They are often referred to as "percentage of moving average" values, and may be represented symbolically as $(Y/MA) \times 100$. These values are graphed in Figure 10-7. As can be seen in the graph, the trend and cyclical movements are no longer present in these figures. The 100-base line represents the level of the moving average, or the trend-cycle base. The fluctuations above and below this base line clearly reveal the repetitive seasonal movement of feed grain prices. As noted earlier, the irregular component is also present in these figures.

The next step in the procedure is the attempt to remove the effect of irregular movements from the $(Y/MA) \times 100$ values. This is accomplished by averaging the percentages of moving average figures for the same quarter. That is, the first-quarter $(Y/MA) \times 100$ values are averaged, the second-quarter values are averaged, and so on. The average customarily used in this procedure is a modified mean, which is simply the arithmetic mean of the percentages of moving average figures for each quarter over the different years, after eliminating the lowest and highest figures. It is desirable to make these deletions particularly when the highest and lowest figures tend to be atypical because of erratic or irregular factors such as strikes, work stoppages, or other unusual occurrences.

The percentage of moving average figures for each quarter are listed in Table 10-4. The highest and lowest figures have been deleted by a line drawn through them, and the modified means of the remaining values are shown for each quarter. These means are 98.6, 102.7, 102.4, and 95.6, respectively, for the first through fourth quarters. The total of these modified means is 399.3. Since it is desirable that the four indices total 400, in order that they average 100%, each of them is multiplied by the adjustment factor of 400/399.3. This adjustment has the effect of forcing a total of 400 by raising each of the unadjusted figures by the

TABLE 10-4
Feed Grain Price Index Numbers: Calculation of Quarterly Seasonal
Indices from Percentage of Moving Average Figures

| | *Percentage of Moving Averages Quarter* | | | |
	I	II	III	IV
1959			102.69	95.46
1960	97.64	104.26	103.23	93.35
1961	99.31	100.40	103.47	98.41
1962	98.02	102.37	100.79	96.49
1963	98.73	101.87	104.82	96.31
1964	98.40	102.33	100.12	96.42
1965	100.58	105.04	101.89	93.11
Modified				
Means	98.6	102.7	102.4	95.6

Total of modified means = 399.3
Adjustment factor　　= 400/399.3 = 1.0018
　　Seasonal indices

I	II	III	IV
98.8	102.9	102.6	95.7

same percentage. The final quarterly seasonal indices are shown on the bottom line of Table 10-4. In the case of monthly seasonal indices, a similar adjustment is made in order for the twelve monthly indices to total 1200; thus, the average monthly index equals 100%.

As indicated earlier, if interest centers on the pattern of seasonal variations itself, the four quarterly indices represent the final product of the analysis. On the other hand, sometimes the purpose of measuring seasonal variations is to eliminate them from the original data in order to examine, for example, the cyclical movements. The method of "deseasonalizing" the original data or adjusting these figures for seasonal movements is simply to divide them by the appropriate seasonal indices. This adjustment is shown in Table 10-3 for the feed grain price data by the division of the original figures in Column (2) by the seasonal indices in Column (7). The result is multiplied by 100, since the seasonal index is stated as a percentage rather than as a relative.

Let us illustrate the meaning of a deseasonalized figure by reference to the first line of figures in Table 10-3. The feed grain price index in the first quarter of 1959 was 96. Dividing this figure by the seasonal index for the first quarter, 98.8, and multiplying by 100 yields 97.2. This

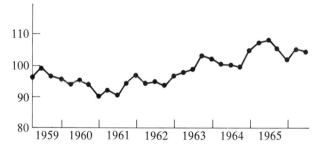

FIGURE 10-8
Deseasonalized figures for feed grain index numbers of average prices received by
farmers by quarters, 1959–1966.

is the feed grain price index for the first quarter of 1959 adjusted for seasonal variations. *That is, it represents the level that feed prices would
have attained if there had not been the depressing effect of seasonality
in the first quarter of the year.* All time series components other than
seasonal variations are present in these deseasonalized figures. This
idea can be expressed symbolically as follows in terms of the aforementioned multiplicative model of the time series analysis:

$$(10.18) \qquad \frac{Y}{SI} = \frac{T \times C \times S \times I}{S} = T \times C \times I$$

The figures for the feed grain price index numbers adjusted for seasonal
movements are graphed in Figure 10-8. It can be seen that the underlying cyclical movement is present in these data, irregular movements
are indicated, and if a sufficiently long period had been used, say, at
least a couple of business cycles, the trend would also be apparent. It
may be noted that as compared to the plot of the original data in Figure
10-6, most of the repetitive seasonal movements are no longer present in
the decentralized figures. However, ordinarily, as in this case too, the
adjustment for seasonality is not perfect. To the extent that seasonal
indices do not completely portray the effect of seasonality, division of
original data by seasonal indices will not entirely remove these influences.

   Seasonal indices are often used for the purpose just discussed. Economic time series adjusted for seasonal variations are often charted
in the *Federal Reserve Bulletin,* the *Survey of Current Business,* and
other publications. Quarterly gross national product figures are often
given as "seasonally adjusted at annual rates." These are simply deseasonalized quarterly figures multiplied by 4 to state the result in annual terms.

Exercises

1. Given the following data from the Fidelity Fence and Gate Company:

Y = actual sales, December 1976 = \$132,000
Y_t = trend value, December 1976 = \$125,000
SI = December seasonal index = 1.08

a. Express seasonally adjusted sales as a percentage of trend. What general factors account for the difference between your calculated value and 100%?
b. What is the meaning of the trend value?

2. A certain department store experiences marked seasonal variation in sales. The July seasonal index is 75, and the trend value for sales in July 1975 was \$42,600. Do you think sales in July 1975 were closer to \$42,600 or \$31,950 (\$42,600 × 0.75)? Discuss.

3. a. In the percentage of moving average method (ratio to moving average method) of determining seasonal indices, how is each of the nonseasonal elements removed? Explain.
b. Given the following data on milk production in a section of the United States during June, July, and August of 1975 and relevant seasonal indices on milk production:

Month	Milk Production (Thousands of Pounds)	Seasonal Index
June 1975	21,233	126.54
July 1975	19,942	117.90
August 1975	18,110	105.99

1. Would you attribute the decline in milk production during this period merely to seasonal variations? Specify the calculations you would make to answer this question, but do not perform these calculations.
2. Explain specifically how you would determine the effect of business cycle fluctuations on milk production. Specify any information that would be required in addition to that given above.

4. The trend equation for sales of the Expo Corporation is as follows:

$$Y_t = 214 + 0.32x$$

where $x = 0$ in July 1971
x is in monthly intervals
Y is monthly sales in \$millions

Figures for certain months in 1975 are given below:

Month	Sales (\$Millions)	Seasonal Index
September	190	86

(*Continued*)

Month	Sales ($Millions)	Seasonal Index
October	236	101
November	268	110
December	392	165

a. What does the seasonal index of 101 for October mean?

b. Isolate for September and November of 1975 the effect of each component of a time series (trend, cyclical and random, seasonal).

5. The following data pertain to the number of automobiles sold by the Valley Rudentino Corporation:

Year	Quarter	Number of Automobiles Sold
1970	1st	207
	2nd	231
	3rd	282
	4th	272
1971	1st	250
	2nd	278
	3rd	315
	4th	288
1972	1st	247
	2nd	265
	3rd	301
	4th	285
1973	1st	261
	2nd	285
	3rd	353
	4th	337
1974	1st	300
	2nd	325
	3rd	370
	4th	343
1975	1st	281
	2nd	317
	3rd	381
	4th	374

a. Using the ratio to moving average method, determine constant seasonal indices for each of the four quarters.

b. Do you think constant seasonal indices should be employed in this problem? Why or why not?

c. Assuming that constant seasonal indices are appropriate, adjust the quarterly sales figures between 1970 and 1975 for seasonal variations.

d. Assume that the trend in sales for the Valley Rudentino Corporation can be described by the equation

$$Y_t = 250 + 20x$$

where $x = 0$ in the second quarter of 1970
x is in one-year intervals
Y is the number of automobiles sold

Do you see any evidence of cycles in the data between 1970 and 1975? What is the basis for your answer?

6. Given the following values for a particular series:

Actual value, December 1975 = 430 units
Seasonal index, December = 180
$$Y_t = 230 + 4x - 0.1x^2$$
$$x = 0 \text{ at June 15, 1971}$$
x is in one-year units

a. What is the meaning of the seasonal index, 180?

b. What is the relative cyclical residual for December 1975 adjusted for seasonal variation?

c. What does the relative cyclical residual mean?

7. The following data represent the sales of a major winter sports equipment manufacturer from 1968 to 1974:

Year	Quarter	Sales (in $Millions)
1968	1st	$622
	2nd	576
	3rd	542
	4th	700
1969	1st	646
	2nd	600
	3rd	555
	4th	712
1970	1st	651
	2nd	624
	3rd	576
	4th	728
1971	1st	692
	2nd	649
	3rd	611
	4th	751

(Continued)

Year	Quarter	Sales (in $Millions)
1972	1st	727
	2nd	673
	3rd	634
	4th	779
1973	1st	745
	2nd	698
	3rd	663
	4th	807
1974	1st	770
	2nd	715
	3rd	688
	4th	829

Using the ratio to moving average method, determine constant seasonal indices for each of the four quarters. Adjust the quarterly sales figures for seasonal variations.

8. Given the following information for the production of the Petrocelli Pasta Corporation:

Month and Year	Production (in Thousands of Pounds)	SI
January 1975	64	115
February 1975	60	100
March 1975	55	85

The trend equation is

$$\log Y_t = 1.7351 + 0.0142x - 0.0005x^2$$

where $x = 0$ in December 1974
  x is in one-month intervals
  Y is production in thousands of pounds

a. Summarize in words how the trend of the above series is changing (based on the equation given).
b. Would it be proper to conclude that production of the Petrocelli Pasta Corporation declined cyclically between January 1975 and March 1975? Show all relevant calculations.
c. Explain the meaning of the January seasonal index, 115.
d. Would you be willing to use the above equation for forecasting? Discuss.

9. The following sentences refer to the ratio to moving method of measuring seasonal variation when applied to monthly gasoline sales in the United States from 1956 to 1975. Insert in the blank space of each sentence the number corresponding to the phrase that will complete the sentence most appropriately.

a. A twelve-month moving total was computed because _____.
   1. trend is thus eliminated
   2. this will give column totals equal to 1200
   3. seasonal variation cancels out over a period of twelve months
b. A two-item total is then taken of the twelve-month totals in order to _____.
   1. obtain moving average figures for the first and last six months
   2. center the moving average properly
   3. eliminate the rest of the random movements
c. This two-item total of a twelve-month moving total is divided by _____.
   1. 14
   2. 2
   3. 12
   4. 24
d. This moving average contains _____.
   1. all of the trend, most of the cyclical, all of the seasonal variation, and some irregular (random) variation
   2. all of the trend, most of the cyclical, and possibly some irregular (random) variation
   3. most of the trend, most of the cyclical, and all of the irregular (random) variation
e. The original data are then divided by the moving average figures. These specific seasonal relatives (Y/MA values) contain _____.
   1. all of the seasonal, possibly some of the cyclical, and almost all of the irregular
   2. seasonal only
   3. all of the trend, most of the cyclical, and none of the irregular
f. Modified means are taken of the specific seasonal relatives (Y/MA) in order to _____.
   1. eliminate from the specific seasonal relatives (Y/MA values) the nonseasonal elements
   2. get rid of seasonal elements
   3. eliminate the trend in the specific seasonal relatives (Y/MA)
   4. compensate for a changing seasonal pattern
g. To adjust the original data for seasonal variation, one computes _____.
   1. original data times seasonal index
   2. seasonal index divided by original data
   3. original data divided by seasonal index

10.6 METHODS OF FORECASTING

We have seen how classical methods are used in analyzing the separate components of an economic time series. These methods involve an implicit assumption that the various components act independently of one another. For example, there were no specific procedures established for taking into account cyclical influences on seasonal variations

or long-run changes in the structure of business cycles. Special procedures can be established to gauge some of these interactions, but basically the model used in classical time series analysis assumes that there
are independent sources of variation in economic time series and measures these sources separately. This decomposition or separation
process, although often very useful for descriptive or analytical purposes, is nevertheless artificial. Therefore, it is not surprising that for a
complex problem such as economic forecasting, it virtually never suffices simply to make mechanistic extrapolations based on classical time
series analysis alone. However, time series analysis frequently is a very
helpful starting point and an extremely useful supplement to other analytical and judgmental methods of forecasting.

   In short-term forecasting, a combined trend–seasonal projection
often provides a convenient first step. For example, as a first approximation in a company's forecast of next year's sales by months, a projection of a trend figure for annual sales might be obtained. Then this
figure might be allocated among months based on an appropriate set of
seasonal indices. Of course, the basic underlying assumption in this procedure is the persistence of the historical pattern of trend and seasonal
variations of the sales of this company into the next year. A more complete forecast might involve superimposing a cyclical prediction as well.
Thus, for example, the first step may again involve a projection of trend
to obtain an annual sales figure. Then an adjustment of this estimate
may be made based on judgment of recent cyclical growth rates. Suppose that the past few years represented the expansion phase of a business cycle and the cyclical growth rate for the economy during the next
year was predicted at about 4% by a group of economists. Assume further that the company in question had found that these forecasts in the
past were quite accurate and applicable to the company's own cyclical
growth rate—over and above its own forecast of trend levels. Then the
company might increase its trend forecast by this 4% figure to obtain a
trend–cycle prediction. Again, if predictions by months were required,
a monthly average could be obtained from the trend–cycle forecast and
seasonal indices could be applied to yield the monthly allocation. Ordinarily, no attempt would be made to predict the irregular movements.

Cyclical Forecasting and Business Indicators

Cyclical movements are more difficult to forecast than trend and seasonal elements. These cyclical fluctuations in a specific time series are
strongly influenced by the general business cycle movements characteristic of large sectors of the overall economy. However, since there is
considerable variability in the timing and amplitude with which many
individual economic series trace out their cyclical swings, there is no
simple mechanical method of projecting these movements.

Relatively "naive" methods such as the extension of the same percentage rate of increase or decrease in, say, sales as occurred last year or during the past few years are often used. These may be quite accurate, particularly if the period for which the forecast is made occurs during the same phase of the business cycle as the time periods from which the projections are made. However, the most difficult and most important items to forecast are the cyclical turning points at which reversals in direction occur. Obviously, managerial planning and implementation that presuppose a continuation of a cyclical expansion phase can give rise to serious problems if an unpredicted cyclical downturn occurs during the planning period.

Many statistical series produced by government and private sources have been extensively used as business indicators. Some of these series represent activity in specific areas of the economy such as employment in nonagricultural establishments or average hours worked per week in manufacturing. Others are very broad measures of aggregate activity pertaining to the economy as a whole, as for example, gross national product and personal income. We have noted earlier that economic series, while exhibiting a certain amount of resemblance in business cycle fluctuations, nevertheless display differences in timing and amplitude. The National Bureau of Economic Research has studied these differences carefully and has specified a number of time series as statistical indicators of cyclical revivals and recessions.

These time series have been classified into three groups. The first group consists of the so-called *leading series*. These series have usually reached their cyclical turning points prior to the analogous turns in general economic activity. The group includes series such as the layoff rate in manufacturing; the average workweek of production workers in manufacturing; the index of net business formation; the Standard & Poor's index of prices of 500 common stocks; and the index of homebuilding permits. The second group are series whose cyclical turns have roughly *coincided* with those of the general business cycle. Included are such series as the unemployment rate, the industrial production index, gross national product, and dollar sales of retail stores. Finally, the third group consists of the *lagging series,* those whose arrivals at cyclical peaks and troughs usually lag behind those of the general business cycle. This group includes series such as plant and equipment expenditures, consumer installment debt, and bank interest on short-term business loans. It is worth noting that rational explanations stemming from economic theory can be given for the logic of the placement of the various series into the respective groups, in addition to the empirical observations themselves. These statistical indicators are adjusted for seasonal movements. They are published monthly in *Business Conditions Digest* by the U.S. Department of Commerce. Another publication that carries the National Bureau statistical indicators, as well as other time series with accompanying analyses, is *Economic Indicators,* published by the Council of Economic Advisors.

Probably the most widespread application of these cycle indicators is as an aid in the prediction of the timing of *cyclical turning points.* If, for example, most of the leading indicators move in an opposite direction from the prevailing phase of the cyclical activity, this is taken to be a possible harbinger of a cyclical turning point. A subsequent similar movement by a majority of the roughly coincident indices would be considered a confirmation that a cyclical turn was in progress. These cyclical indicators, like all other statistical tools, have their limitations and must be used carefully. They are not completely consistent in their timing, and leading indicators sometimes give incorrect signals of forthcoming turning points because of erratic fluctuations in individual series. Furthermore, it is not possible to predict with any high degree of assurance the length of time between a signal given by the leading series group of an impending cyclical turning point and the turning point itself. There has been considerable variation in this lead time during past cycles of business activity.

Other Methods of Forecasting

Most individuals and companies engaged in forecasting do not depend on any single method, but rather utilize a variety of different approaches. It stands to reason that if there is substantial agreement among a number of forecasts arrived at by relatively independent methods, greater reliance would be placed on this consensus than would have been placed on the results of any single technique.

Other methods of prediction range from very informal judgmental techniques to highly sophisticated mathematical models. At the informal end of this scale, for example, sales forecasts are sometimes derived from the combined outlooks of the sales force of a company, from panels of executive opinion, or from a composite of both of these. At the other end of the scale are the more formal mathematical models such as regression equations or complex econometric models. There is widespread use of various types of regression equations, by which firms attempt to predict the movements of their own company's or industry's activity on the basis of relationships to other economic and demographic factors. Often, for example, a company's sales are predicted on the basis of relationships with other series whose movements precede those of the sales series to be forecasted.

Among the most formal and mathematically sophisticated methods of forecasting in current use are econometric models. An *econometric model* is a set of two or more simultaneous mathematical equations that describe the interrelationships among the variables in the system.

Some of the more complex models in current use for prediction of movements in overall economic activity include dozens of individual equations. Special methods of solution for the parameters of these equation systems have been developed, since in many instances ordinary least squares techniques are not appropriate. These econometric models have been used primarily for prediction at the level of the economy as a whole, and for industries, but they are coming into increasing use for prediction at the company level as well.

Management uses forecasts as an important ingredient of its planning, operational, and control functions. Invariably, no single method is relied upon; judgment is applied to the results of various forecasting methods. Often, formal prediction techniques make a useful contribution by narrowing considerably the area within which intuitive judgment is applied.

Exponential Smoothing. All methods of forecasting involve processing past information in order to extract insights for prediction into the future. Some methods, such as the classical time series analysis discussed in this chapter, require extensive manipulation of past data. Other techniques, such as regression models or econometric models, also require not only the use of considerable amounts of past data but also the fitting of formal mathematical models to these data. However, in some situations, the need is for a fast, simple, and rather mechanical way of forecasting for large numbers of items, without the computational burden of working with large quantities of data for each item for which forecasts are required. For example, in connection with inventory control, predictions of demand for thousands of items may be required for the determination of ordering points and quantities. *Exponential smoothing* is a forecasting method that meets the aforementioned requirements and is particularly appropriate for application on high-speed electronic computers. It is a technique that simply weights together the current (most recent) actual and the current forecast figures to obtain the new forecast value. The basic formula for exponential smoothing is

(10.19) $$F_t = wA_{t-1} + (1 - w)F_{t-1}$$

where F_t is the forecast for time period t

A_{t-1} is the actual figure for time period $t - 1$
F_{t-1} is the forecast for time period $t - 1$
w is a constant whose value is between 0 and 1

The time period $t - 1$ is the *current period,* and the prediction F_t is the *new forecast* for the next period. Therefore, Equation 10.19 may be written in words as

(10.20) New forecast $= w$(Current actual) $+ (1 - w)$(Current forecast)

Note that the new forecast F_t is made in the current period $t - 1$; the current forecast F_{t-1} was made in the preceding period $t - 2$. The new forecast can be seen to be a weighted average of the current actual figure A_{t-1} and the forecast for the current period F_{t-1}. Another way of writing (10.19) is

(10.21) $$F_t = A_{t-1} + (1 - w)(F_{t-1} - A_{t-1})$$

In this form, the new forecast may be viewed as the current actual figure A_{t-1} plus some fraction of the most recent forecast error, $F_{t-1} - A_{t-1}$, that is, the difference between the current forecast and current actual figures.

An important advantage of this method of forecasting is that the only quantity that must be stored until the next period is the new forecast, F_t. This modest storage requirement is an extremely useful feature for computer-based inventory management systems where, as mentioned earlier, there may be thousands of items for which forecasts are required. Despite the fact that exponential smoothing requires the storage of only one piece of data, that is, the new forecast, a bit of algebra demonstrates that each new forecast takes into account all past actual figures.

For example, we may write

$$F_2 = wA_1 + (1 - w)F_1$$
$$F_3 = wA_2 + (1 - w)F_2$$
$$F_4 = wA_3 + (1 - w)F_3$$

Substituting the first equation value for F_2 into the second equation and then the resulting value of F_3 into the third equation yields

(10.22) $$F_4 = wA_3 + (1 - w)wA_2 + (1 - w)^2 wA_1 + (1 - w)^3 F_1$$

We note that the last term of Equation 10.22 is the only one that is not expressed in terms of actual figures (A_i). However, F_1 can be written to summarize the results of the actual figures A_0, A_{-1}, A_{-2}, etc. Therefore, we can think of the forecast for any period as a weighted sum of all past actual figures, with the heaviest weight assigned to the most recent actual figure, the next highest to the second most recent one, and so on. We now can see why the method is referred to as *exponential smoothing*, since the weighting factor $(1 - w)$ has successively higher exponents the farther the data recede into the future. From a management point of view, the fact that the most recent experience gets the greatest weight, with earlier periods getting respectively less weight, makes a great deal of sense.

One choice that the analyst must make in exponential smoothing is the value of the weighting factor, w. As can be seen from Equations 10.19 and 10.22, if w is assigned a value close to 0, the most recent actual figure for, say, demand or sales will receive relatively little weight. Thus, forecasts made with the use of a small w factor tend to have little variation from period to period. Hence, low values for w would tend to minimize the reactions of forecasts to relatively brief or erratic fluctuations in demand. This points up the basic tradeoff involved in the

TABLE 10-5
Exponential Smoothing Forecasts for Weekly Demand Using Weight Factors of
$w = 0.1$ and $w = 0.8$

(1)	(2)	(3)	(4)
		Next	Next
	Actual	Forecast	Forecast
Week	Demand	$w = 0.1$	$w = 0.8$
1	$100 = A_1$	$100 = F_2$	$100 = F_2$
2	$100 = A_2$	$100 = F_3$	$100 = F_3$
3	$100 = A_3$	$100 = F_4$	$100 = F_4$
4	$100 = A_4$	$100 = F_5$	$100 = F_5$
5	$150 = A_5$	$105 = F_6$	$140 = F_6$
6	$100 = A_6$	$104.5 = F_7$	$108 = F_7$
7	$100 = A_7$	$104.05 = F_8$	$101.6 = F_8$
8	$100 = A_8$	$103.645 = F_9$	$100.32 = F_9$
9	$100 = A_9$	$103.2805 = F_{10}$	$100.064 = F_{10}$
10	$100 = A_{10}$	$102.95245 = F_{11}$	$100.0128 = F_{11}$

choice of w. If w is too low, the forecasts respond too slowly to basic
long-lived changes in the demand pattern. On the other hand, if w is too
high, the forecasts tend to react too sharply to random shifts in demand.
The values of w used in practice are relatively low, usually 0.3 or less,
with a value of 0.1 often proving to be quite effective in a variety of forecasting situations. When policy decisions that are expected to change
the pattern of demand are made, appropriate changes can be made in w
values.

As an example of the exponential smoothing procedure, let us examine Table 10-5. In Columns (1) and (2) are given the data for the weekly
number of units of demand for a certain product. This series is characterized by a constant demand of 100 units per week except for a sudden
sharp jump to 150 units in week 5. Columns (3) and (4) give the exponential smoothing forecasts for weight factors of $w = 0.1$ and $w = 0.8$,
respectively. Note that the first actual demand figure, A_1, is assumed to
be the first forecast figure, F_2. This is often the way an exponential
smoothing set of forecasts is begun. Note too that the forecasts on each
line of the table are the forecasts made in the given week for the following week.

As an illustration of the calculations, for example, the value of F_6 in
Column (3), which is 105, is derived as follows:

$$F_6 = wA_5 + (1 - w)F_5$$
$$= (0.1)(150) + (0.9)(100)$$
$$= 105$$

The effects of low and high values for the weighting factor w can be seen in Table 10-5. With the low value of $w = 0.1$, the forecasts show a sluggish response to the sudden spurt in demand to 150 units in week 5, with F_6 rising only to a value of 105. Furthermore, the effect of this spurt in actual demand lasts longer in the forecast series with $w = 0.1$ than with $w = 0.8$. With the higher value of $w = 0.8$, the forecast for F_6 jumps to 140 from the previous base level of 100, and the effect dies out more rapidly than in the case of the lower weighting factor, as may be seen in Columns (3) and (4). All the decimal places in the smoothing calculations have been shown in Table 10-5 for checking purposes. In actual practice, if demand can occur only in whole numbers of units, then only such integral values would be used as forecasts.

Summary. The method described above is the simplest and most basic form of exponential smoothing. It is particularly appropriate for economic time series that are relatively stable and do not have pronounced trend or seasonal movements. More complex forms of exponential smoothing are used, sometimes combined with spectral analysis (a method for determining which of various component cycles are most influential in a time series), to handle the more complicated series. In summary, exponential smoothing techniques are often very useful in situations in which a large number of economic time series must be forecast frequently. It is a relatively simple method characterized by economy of storage and computational time.

Exercises

1. Assume that the following figures represent the daily number of units of demand for a product sold by a retail store on ten successive business days:

   10, 10, 10, 20, 10, 10, 10, 10, 10, 10

   Use the exponential smoothing procedure to prepare next-day forecasts using weighting factors of $w = 0.1$ and $w = 0.7$. Assume that the first actual demand figure is used as the first next-day forecast.

2. Assume that the daily number of units demanded of the product discussed in Exercise (1) had been 10, 10, 10, 20, 20, 20, 20, 20, 20, 20. Again use the exponential smoothing procedure to prepare next-day forecasts using weighting factors of $w = 0.1$ and $w = 0.7$. As in Exercise (1), assume that the first actual figure is used to "initialize" the set of forecasts.

3. Discuss the effects of using low versus high values of the weighting factor for the demand series given in Exercises (1) and (2).

4. Using a weighting factor of 0.5, derive a forecast for 1975 for the mortgage interest rate series in Exercise (8) of Section 10.4. Does this procedure seem to be appropriate for forecasting this series? Assume that the first actual interest rate figure is used as the first next-year forecast.

5. Exponential smoothing is used in security analysis as a mechanical means of forecasting company earnings; the accuracy of analysts' forecasts can then be compared to the accuracy of these mechanical forecasts in order to evaluate the analysts' performances. The actual earnings per share for Atlantic Richfield Corporation for the period 1960–1974 are given below:

Year	Actual Earnings per Share	Year	Actual Earnings per Share
1960	$2.50	1968	$3.69
1961	2.46	1969	4.15
1962	2.445	1970	3.64
1963	2.19	1971	3.73
1964	2.345	1972	3.40
1965	2.82	1973	4.76
1966	3.605	1974	8.36
1967	4.28		

Use the exponential smoothing procedure to prepare next-year forecasts using a weighting factor of 0.4. Forecast earnings per share for 1975 and compare the forecasted value with the actual value of $6.16 per share. Assume that the first actual figure is used to initialize the set of forecasts.

11

Index
Numbers

11.1 THE NEED FOR AND USE OF INDEX NUMBERS

In our daily lives, we often make judgments that involve summarizing how an economic variable changes with time or place. As an example of variation over time, a family's income may have increased by 20% over a five-year period. Suppose this was a period of inflation, or generally rising prices. Has the family's "real income" increased? That is, can the family purchase more goods and services with its income than it could five years ago? If the general price level of items has increased more than the 20% figure, the family clearly cannot purchase as much. On the other hand, if prices have increased less than 20%, the family's real income has increased.

As an example of variation with changes in place, let us consider a company that wishes to transfer an executive from St. Louis to New York City. What should be the executive's minimum salary increase to allow for the cost of living in New York?

Both these cases require measurements of general price levels. The prices of the numerous items purchased by the family have doubtless increased at different rates; a few may even have decreased. Similarly, although some prices in New York City are much higher than in St. Louis, some may be lower. How can we summarize in a single composite figure the average differences between the two time periods or the two cities? Index numbers serve to answer questions of this type. An *index number* is a summary measure that states a relative comparison between groups of related items.

In its simplest form, an index number is nothing more than a percentage figure that expresses the relationship between two numbers with one of the numbers used as the base. For example, in a time series of prices of a particular commodity, the prices may be expressed as percentages by dividing each figure by the price in the base period. These percentages are referred to as *price relatives*. In the calculation of economic indices, it is conventional to state these relatives as numbers lying above and below 100, where the base period is 100%. For example, suppose the price of one pound of a certain brand of coffee was $1.20 in 1972, $1.38 in 1974, and $1.60 in 1976. Then the price relatives of the three figures, with 1972 as a base (written, 1972 = 100), are

Year	Price Relatives (1972 = 100)
1972	($1.20/$1.20)(100) = 100
1974	($1.38/$1.20)(100) = 115
1976	($1.60/$1.20)(100) = 133

As indicated, the price relative for any given year is obtained by dividing the price for that year by the figure for the base period. The resulting figure is multiplied by 100 to express the price relative in percentage form.

The price relatives may be interpreted as follows. In 1974, it would have cost 115% of the price in 1972 to purchase a pound of this brand of coffee; that is, there was a 15% increase in the price from 1972 to 1974. Similarly, in 1976, the price was 133% of the 1972 price, so the price had risen 33% from 1972 to 1976. Of course, we are usually interested in price changes for more than one item. For example, in the cases of the family and the transferred executive cited earlier, we were interested in the prices of all commodities and services included in the cost of living; this means combining the price relatives for many different items into a single summary figure for each time period. Similarly, we may wish to compute a food price index, a clothing price index, or an index of medical costs. Such summary figures constitute *composite* index number series. Since our discussion will pertain solely to composite indices, we will ordinarily use the term "index number" to mean "composite index number."

Series of index numbers are extremely useful in the study and analysis of economic activity. Every economy, regardless of the political and social structure within which it operates, is engaged in the production, distribution, and consumption of goods and services. Convenient methods of aggregation, averaging, and approximation are required to summarize the myriad individual activities and transactions. Index numbers have proved to be very useful tools in this connection. Thus, we find indices of industrial production, agricultural production, stock market prices, wholesale prices, consumer prices, prices of exports and imports, incomes of various types, and so on, in common use. Economic indices can be conveniently classified as indices of price, quantity, or value. The present discussion concentrates primarily on price indices, because most problems of construction, interpretation, and use of indices may be illustrated in terms of such measures. First, we deal with methods of construction of index numbers, using the illustrative data of a simple example. Then we consider some general problems of index number construction.

11.2 AGGREGATIVE PRICE INDICES

In order to illustrate the construction and interpretation of price indices, we consider the artificial problem of constructing a price index for a list of only four food commodities. A more realistic counterpart of this problem is represented by the Consumer Price Index (CPI) produced by the Bureau of Labor Statistics (BLS) of the U.S. Department of Labor. This index is used in many ways and provides the basis for many economic decisions. For example, fluctuations in the wages of over 5 million workers and the Social Security benefits of about 30 million people are partially based on changes that occur in the index figures. The series is also closely watched by monetary authorities as an indicator of inflationary price movements.

Unweighted Aggregates Index

In our simplified illustration, we use a base period of 1970, and we are interested in the change in these prices from 1970 to 1976 for a typical family of four that purchased these products at retail prices in a certain city. The universe and other basic elements of the problem should be very carefully defined. (For example, the price of a dozen eggs might be

TABLE 11-1
Calculation of the Unweighted Aggregates Index for Food
Prices in 1970 and 1976

	Unit Price	
	1970	*1976*
Food Commodity	P_{70}	P_{76}
Coffee (pound)	$1.00	$1.60
Bread (loaf)	0.35	0.53
Eggs (dozen)	0.65	0.99
Hamburger (pound)	1.00	1.50
	$3.00	$4.62

Unweighted Aggregates Index
for 1976, on 1970 Base

$$\frac{\Sigma P_{76}}{\Sigma P_{70}} \cdot 100 = \frac{\$4.62}{\$3.00} \times 100 = 154.0$$

the price of a dozen Grade A, large white eggs.) However, in this problem we purposely leave these matters very indefinite and concentrate on the construction and interpretation of the various indices. In Table 11-1 are shown the basic data of the problem and the calculation of the unweighted aggregates index. As indicated at the bottom of Table 11-1, the prices per unit are summed (or aggregated) for each year. Then one year (in this case, 1970) is selected as a base. The price index for any given year is obtained by dividing the sum of prices for that year by the sum for the base period. The resulting figure is multiplied by 100 to express the index in percentage form. (Hence, the index takes the value 100 in the base period.) If the symbol P_0 is used to denote the price in a base period and P_n the price in a nonbase period, the general formula for the unweighted aggregates index may be expressed as follows:

UNWEIGHTED AGGREGATES PRICE INDEX

(11.1) $$\frac{\Sigma P_n}{\Sigma P_0} \cdot 100$$

Let us interpret the index figure of 154.0 for 1976. In 1970, it would have cost $3.00 to purchase one pound of coffee, one loaf of bread, one dozen eggs, and one pound of hamburger. The corresponding cost in 1976 was $4.62. Expressing $4.62 as a percentage of $3.00, we find that in 1976 it would have cost 154% of the cost in 1970 to purchase one unit each of the specified commodities. In terms of percentage change, it would have cost 54% *more* in 1976 than in 1970 to purchase this "market basket" of goods.

The interpretation of the unweighted aggregates index is very straightforward. However, this type of index suffers from a serious limitation; it is unduly influenced by the price variations of high-priced commodities. The price totals in 1970 and 1976, respectively, were $3.00 and $4.62, an increase of $1.62. If we added to the list of commodities one that declined from $5.00 to $3.00 per unit from 1970 to 1976, the totals for 1970 and 1976 would then become $8.00 and $7.62. Hence, the price index figure for 1976 would be 95.3, indicating a decline in prices of 4.7%. Although the prices of four commodities increased and only one decreased, the overall index shows a decline, because of the dominance of the price change in one high-priced commodity. Furthermore, this high-priced commodity may be relatively unimportant in the consumption pattern of the group to which the index pertains. We conclude that this so-called "unweighted index" has an inherent haphazard weighting scheme.

Another deficiency of this type of index is the arbitrary nature of the units for which the prices are stated. For example, if the price of eggs were stated per half-dozen rather than per dozen, the calculated price index figure would change. However, even if all the prices were stated for the same unit of each commodity (say, per pound), the problem of the inherent haphazard weighting scheme would still remain; the index would be dominated by the commodities that happened to have high prices per pound. These may be the very commodities that are purchased least, because they are expensive. The difficulties of converting a simple aggregative index into an economically meaningful measure make the need to apply explicit weights apparent. We now turn to weighted aggregative price indices.

Weighted Aggregates Indices

In order to attribute the appropriate importance to each of the items included in an aggregative index, some reasonable weighting plan must be used. The weights to be used depend on the purposes of the index calculation, that is, on the economic question the index attempts to answer. In the case of a consumer food price index such as the one we have been discussing, reasonable weights would be given by the amounts of the individual food commodities purchased by the consumer units to whom the indices pertain. These would constitute *quantity weights*, since they represent quantities of commodities purchased. The specific types of quantities to be used in an aggregative index would depend, of course, on the economic nature of the index computed. Hence, an aggregative index of export prices would use quantities of commodities and services exported, whereas an index of import prices would use quantities imported.

TABLE 11-2
Calculation of the Weighted Aggregates Index for Food Prices, Using Base Period
Quantities Consumed as Weights (Laspeyres Method)

Food Commodity	1970 P_{70}	1976 P_{76}	Quantity 1970 Q_{70}	$P_{70}Q_{70}$	$P_{76}Q_{70}$
Coffee (pound)	$1.00	$1.60	1	$1.00	$1.60
Bread (loaf)	0.35	0.53	3	1.05	1.59
Eggs (dozen)	0.65	0.99	2	1.30	1.98
Hamburger (pound)	1.00	1.50	1	1.00	1.50
				$4.35	$6.67

Weighted Aggregates Index, with Base Period Weights:
for 1976, on 1970 Base

$$\frac{\Sigma \, P_{76}Q_{70}}{\Sigma \, P_{70}Q_{70}} \cdot 100 = \frac{\$6.67}{\$4.35} \times 100 = 153.3$$

Table 11-2 shows the prices of the same food commodities given in Table 11-1, but quantities consumed during the base period, 1970, are also shown. The figures in column Q_{70} (the symbol Q denotes quantity) represent average quantities consumed per week in 1970 by the consumer units to which the index pertains. Hence, they indicate an average consumption of one pound of coffee, three loaves of bread, and so on. The figures in the column labeled $P_{70}Q_{70}$ indicate the dollar expenditures for the quantities purchased in 1970. Correspondingly, the numbers in the column headed $P_{76}Q_{70}$ specify what it would cost to purchase these amounts of food in 1976. Hence, the sums, $\Sigma \, P_{70}Q_{70} = \$4.35$ and $\Sigma \, P_{76}Q_{70} = \$6.67$, indicate what it would have cost to purchase the specified quantities of food commodities in 1970 and 1976. The index number for 1976 on a 1970 base is given by expressing the figure for $\Sigma \, P_{76}Q_{70}$ as a percentage of the $\Sigma \, P_{70}Q_{70}$ figure, yielding in this case a figure of 153.3, as shown at the bottom of Table 11-2. Of course, the index number for the base period, 1970, would be 100.0.

What this type of index measures is the change in the total cost of a fixed market basket of goods. For example, in this case, the 153.3 figure indicates that in 1976 it would have cost 153.3% of what it cost in 1970 to purchase the weekly market basket of commodities representing a typical consumption pattern in 1970. Roughly speaking, this indicates an average price rise of 53.3% for this food market basket from 1970 to 1976. Referring to the corresponding index figure of 154% for the simple, or unweighted, index, we see that it is quite close to the 153.3 figure for the weighted index. The reason for this is that in our example we have assumed that the prices of all four commodities moved in the same direction with the percentage changes all falling between 10% and 25%. On the other hand, if there had been more dispersion in price movements (for example, if some prices increased while others decreased, as is often the case), the weighted index would have tended to differ more from the unweighted one.

The weighted aggregative index using base period weights is also known as the Laspeyres index. The general formula for this type of index may be expressed as follows:

**WEIGHTED AGGREGATES PRICE INDEX,
BASE PERIOD WEIGHTS (LASPEYRES METHOD)**

(11.2) $$\frac{\Sigma P_n Q_0}{\Sigma P_0 Q_0} \cdot 100$$

where P_0 = price in a base period
P_n = price in a nonbase period
Q_0 = quantity in a base period

The Laspeyres index clearly illustrates the basic dilemma posed by the use of any weighting system. Since an aggregative price index attempts to measure price changes and contains data on both prices and quantities, it appears logical to hold the quantity factor constant in order to isolate change attributable to price movements. If both prices and quantities were permitted to vary, their changes would be entangled and it would not be possible to ascertain what part of the movement was due to price changes. However, by keeping quantities fixed as of the base period, the Laspeyres index assumes a frozen consumption pattern. As time goes on, this assumption becomes more unrealistic and untenable. The consumption pattern of the current period would seem to represent a more realistic set of weights from the economic viewpoint.

However, let us consider the implications of the use of an aggregative index using current period (nonbase period) weights, which is known as the Paasche index. The general formula for a Paasche index is

**WEIGHTED AGGREGATES PRICE INDEX,
CURRENT PERIOD WEIGHTS (PAASCHE METHOD)**

(11.3) $$\frac{\Sigma P_n Q_n}{\Sigma P_0 Q_n} \cdot 100$$

If such an index is prepared on an annual basis, the weights would have to change each year, since they would consist of current year quantity figures. The 1971 Paasche index would be computed by the formula $\Sigma P_{71} Q_{71} / \Sigma P_{70} Q_{71}$, the 1972 index would be $\Sigma P_{72} Q_{72} / \Sigma P_{70} Q_{72}$, and so on. The interpretation of any one of the resulting figures (as price change from the base period, assuming the consumption pattern of the current period) is clear. However, the use of changing current period weights destroys the possibility of obtaining unequivocal measures of year-to-year price change. For example, if the Paasche formulas for the 1971 and 1972 indices on page 497 are compared, it will be noted that both prices and quantities have changed. Therefore, no clear statement can be made about price movements from 1971 to 1972. Thus, the use of current year weights makes year-to-year comparisons of price changes impossible.

Another practical disadvantage of using current period weights is the necessity of obtaining a new set of weights in each period. Let us consider the U.S. Bureau of Labor Statistics CPI to illustrate the difficulty of obtaining such weights. In order to obtain an appropriate set of weights for this index, the BLS conducts a massive sample survey of the expenditure patterns of families in a large number of cities. Such surveys have been carried out in 1917–19, 1934–36, 1950–51, 1960–61, and 1972–73. They are large-scale, expensive undertakings, and it would be infeasible for such surveys to be conducted, say, once a year or more frequently. Because of these disadvantages, the current period weighted aggregative method is not used in any well-known price index number series.

Because of these considerations and other factors, the most generally satisfactory type of price index is probably the weighted relative of aggregates index using a fixed set of weights. The term "fixed set of weights" rather than "base period weights" is used here, because the weights may pertain to a period somewhat different from the period that represents the base for measuring price changes. For example, one of the base periods for the CPI was 1957–1959, whereas the corresponding weights were derived from a 1960–1961 survey of consumer expenditures.[1] The base year for the present CPI is 1967 = 100. The BLS revises its weighting system about every ten years and also changes the reference base period for the measurement of price changes with about the same frequency. This procedure constitutes a workable solution to the dilemma of the need for both constant weights (in order to isolate price change) and up-to-date weights (in order to have a recent realistic description of consumption patterns).

The weighted [relative of] aggregates index using a fixed set of weights is referred to as the *fixed-weight aggregative index* and is defined by the formula

WEIGHTED AGGREGATES PRICE INDEX WITH FIXED WEIGHTS

(11.4)
$$\frac{\Sigma P_n Q_f}{\Sigma P_0 Q_f}$$

[1] An expenditures pattern survey was conducted in 1972–1973; the results of this survey, along with major changes in coverage and method of construction, are to be incorporated into a new CPI scheduled for publication in 1977.

where Q_f denotes a fixed set of quantity weights. The Laspeyres method may be viewed as a special case of this index, in which the period to which the weights refer is the same as the base period for prices. In order to clarify discussion of the two different time periods, the term "weight base" is used for the period to which the quantity weights pertain, whereas the term "reference base" is used to designate the time period from which the price changes are measured. Of course, a distinct advantage of a fixed-weight aggregative index is that the reference base period for measuring price changes may be changed without a corresponding change in the weight base. This is sometimes a useful and practical procedure, particularly in the case of some U.S. government indices that utilize data from censuses or large-scale sample surveys for changes in weights.

Exercises

1. The Jones Metal Company uses three raw materials in its business. Below are the average prices and quantities consumed of these three materials in 1970 and 1976.

	1970		1976	
Materials	Price	Quantity	Price	Quantity
A	$20	20	$25	30
B	1	100	2	120
C	5	50	6	70

a. Compute an appropriate weighted aggregates index for 1976 on a 1970 base.

b. A competitor reported a 1976 index for the same materials (1970 base year) of 120. Would you conclude that Jones paid more per unit in 1976 than this competitor? Why?

2. A security analyst wishes to construct an index to reflect the changes in the prices of five stocks in a certain portfolio. The following figures represent the average price and number of shares purchased for each stock during the year indicated.

	1974		1975		1976	
Stock	Average Price	Shares Purchased	Average Price	Shares Purchased	Average Price	Shares Purchased
A	20	100	30	200	25	200
B	40	200	35	150	40	100
C	50	150	70	250	60	100
D	35	300	40	250	45	200
E	10	200	15	400	25	300

     a. Construct an unweighted aggregates index using 1974 as a base year. What are the disadvantages of this type of index?

     b. Construct a Laspeyres index using 1974 as a base year. What are the disadvantages of this type of index?

     c. Construct a Paasche index using 1974 as a base year. What are the disadvantages of this type of index?

11.3   AVERAGE OF RELATIVES INDICES

A second basic method of construction of price indices is the *average of relatives* procedure. In an average of relatives index, the first step involves the calculation of a price relative for each commodity by dividing its price in a nonbase period by the price in a base period. Then an average of these price relatives is calculated. As in the case of aggregative indices, an average of relatives index may be either unweighted or weighted. We consider first the unweighted indices, using the same data on prices as in the preceding section.

Unweighted Arithmetic Mean of Relatives Index

The price data previously shown in Tables 11-1 and 11-2 are repeated in Table 11-3. The first step in the calculation of any average of relatives price index is the calculation of price relatives, which express the price of each commodity as a percentage of the price in the base period. These price relatives for 1976 on a 1970 base, denoted $(P_{76}/P_{70}) \times 100$, are shown in Column (4) of Table 11-3. Theoretically, once the price relatives are obtained, any average (including the arithmetic mean, median, and mode) could conceivably be used as a measure of their central tendency. The arithmetic mean has been most frequently used, doubtless because of its simplicity and familiarity. At the bottom of Table 11-3 is shown the calculation of the unweighted arithmetic mean of relatives for 1976 on a 1970 base. The formula is Equation 1.1, with the price relatives as the items to be averaged. Stated in general form, the formula for this unweighted arithmetic mean of relatives is

UNWEIGHTED ARITHMETIC MEAN OF RELATIVES INDEX

$$(11.5) \qquad \frac{\Sigma \left(\dfrac{P_n}{P_0} \cdot 100 \right)}{n}$$

where $\dfrac{P_n}{P_0} \cdot 100$ = the price relative for a commodity or service

          n = the number of commodities and services

Although this is an "unweighted" index, we find that as in the case of the unweighted aggregative index, an inherent weighting pattern is

TABLE 11-3
Calculation of the Unweighted Arithmetic Mean of Relatives Index of Food
Prices for 1976 on a 1970 Base

Food Commodity (1)	Unit Price 1970 P_{70} (2)	Unit Price 1976 P_{76} (3)	Price Relative $\dfrac{P_{76}}{P_{70}} = 100$ (4)
Coffee (pound)	$1.00	$1.60	160.0
Bread (loaf)	0.35	0.53	151.4
Eggs (dozen)	0.65	0.99	152.3
Hamburger (pound)	1.00	1.50	150.0
			613.7

Unweighted Arithmetic Mean of Relatives Index
for 1976, on a 1970 Base

$$\frac{\Sigma\left(\dfrac{P_{76}}{P_{70}} \cdot 100\right)}{4} = \frac{613.7}{4} = 153.4$$

present. It is useful to consider the implications of this inherent weighting system. In the unweighted arithmetic mean of relatives, percentage increases are balanced off against equal percentage decreases. For example, if we consider two commodities, one whose price increased by 50% and one whose price declined by 50% from 1970 to 1976, the respective price relatives for 1976 on a 1970 base would be 150 and 50. The unweighted arithmetic mean of these two figures is 100, indicating that, on the average, prices have remained unchanged. Thus, the unweighted arithmetic mean attaches the same weight to equal percentage changes in opposite directions. However, this method does not provide for explicit weighting in terms of the importance of the commodities whose prices have changed. Since it is widely recognized that explicit weighting is required to permit the individual items in an index to exert their proper influence, virtually none of the important government or private organization price indices are of the "unweighted" variety. We now turn to a consideration of weighted average of relatives indices.

Weighted Arithmetic Mean of Relatives Indices

Although several averages can theoretically be used for calculating weighted averages of relatives, only the weighted arithmetic mean is ordinarily employed.

The general formula for a weighted arithmetic mean of price relatives is

WEIGHTED ARITHMETIC MEAN OF RELATIVES, GENERAL FORM

(11.6)
$$\frac{\Sigma \left(\frac{P_n}{P_0} \cdot 100 \right) w}{\Sigma w}$$

where w = the weight applied to the price relatives.

Customarily, the weights used in this type of index are values, such as values consumed, produced, purchased, or sold. For example, in the food price index used as our illustrative problem, the weights would be values consumed, that is, dollar expenditures on the individual food commodities by the typical family to whom the index pertains. It seems reasonable that the importance attached to the price change for each commodity be determined by the amounts spent on these commodities. In index number construction, value = price × quantity. For example, if a commodity has a price of $.10 and the quantity consumed is three units, then the *value* of the commodity consumed is $.10 × 3 = $.30. Since prices and quantities can pertain to either a base period or a current period, the following systems of weights are all possibilities: $P_0 Q_0$, $P_0 Q_n$, $P_n Q_0$, and $P_n Q_n$. The weights $P_0 Q_0$ and $P_n Q_n$ are, respectively, base period values and current period values, the other two are mixtures of base and current period prices and quantities. Interestingly, the weighting systems $P_0 Q_0$ and $P_0 Q_n$, when used in the weighted arithmetic mean of relatives, result in indices that are algebraically identical to the Laspeyres and Paasche aggregative indices, respectively. This point is illustrated in (11.7), where base period weights $P_0 Q_0$ are used.

**WEIGHTED ARITHMETIC MEAN OF
RELATIVES, WITH BASE PERIOD VALUE WEIGHTS**

(11.7)
$$\frac{\Sigma \left(\frac{P_n}{P_0} \right) P_0 Q_0}{\Sigma P_0 Q_0} \cdot 100 = \frac{\Sigma P_n Q_0}{\Sigma P_0 Q_0} \cdot 100$$

As is clear from (11.7), the P_0s in the numerator cancel, yielding the Laspeyres index. The calculation of the weighted arithmetic mean of relatives using 1970 base period value weights is given in Table 11-4 for the data of our illustrative problem. The numerical value of the index is, of course, exactly the same as that obtained previously for the weighted aggregative index with base period quantity weights (Laspeyres method) in Table 11-2.

Since the two indices in (11.7) are algebraically identical, it would seem immaterial which is used, but there are instances when it is more feasible to compute one than the other. For example, it is more convenient to use the weighted average of relatives than the Laspeyres index

TABLE 11-4
Calculation of the Weighted Arithmetic Mean of Relatives Index of Food Prices for
1976 on a 1970 Base, Using Base Period Weights

| Food Commodity (1) | Prices | | Price Relatives $\frac{P_{76}}{P_{70}} \cdot 100$ (4) | Quan- tity 1970 Q_{70} (5) | $P_{70}Q_{70}$ (6) | Weighted Price Relatives Col. (4) × Col. (6) $\left(\frac{P_{76}}{P_{70}} \cdot 100\right)\left(P_{70}Q_{70}\right)$ |
	1970 P_{70} (2)	1976 P_{76} (3)				
Coffee (pound)	$1.00	$1.60	160.0	1	$1.00	$160.00
Bread (loaf)	0.35	0.53	151.4	3	1.05	158.97
Eggs (dozen)	0.65	0.99	152.3	2	1.30	197.99
Hamburger (pound)	1.00	1.50	150.0	1	1.00	150.00
					$4.35	$666.96

*Weighted Arithmetic Mean of Relatives for 1976, on 1970 Base,
Using Base Period Value Weights*

$$\frac{\Sigma \left(\frac{P_{76}}{P_{70}} \cdot 100\right) (P_{70}Q_{70})}{\Sigma\, P_{70}Q_{70}} = \frac{\$666.96}{4.35} = 153.3$$

when value weights are easier to obtain than quantity weights; when the
basic price data are more easily obtainable in the form of relatives than
absolute values; and when an overall index is broken down into a
number of component indices and we wish to compare the individual
components in the form of relatives. As an illustration of the first of
these situations, it is usually easier for manufacturing firms to furnish
value of production weights in the form of "value added by manufacturing" (sales minus cost of raw materials) than to provide detailed data
on quantities produced.

As indicated earlier, the Paasche index and the weighted arithmetic
mean of relatives with a P_0Q_n weighting system are algebraically identical. The reasons given for the wider use of the Laspeyres than the
Paasche index apply to a similarly wider usage of weighted means of relatives with P_0Q_0 than with P_0Q_n weights. The other two possible value
weighting systems, P_nQ_0 and P_nQ_n, create interpretational difficulties
and therefore are not utilized in any of the important indices.

Exercises

1. A small electrical company produces three models of household exhaust fans.
   Average unit selling prices and quantities sold in 1973 and 1976 were as
   follows.

Model	Price	1973 Quantity (in Thousands)	Price	1976 Quantity (in Thousands)
Economy	$26	20	$30	30
Model A	32	12	40	16
Model B	40	15	50	18

a. Calculate the index of fan prices for 1973 on a base year of 1976. Use the weighted arithmetic mean of relatives method with base period weights.

b. Explain, in words understandable to the lay person, precisely what the value of your index means.

2. a. Compute an index of Whipple Wine prices for the data below by the weighted arithmetic mean of relatives method with 1974 = 100, using base year weights.

b. Compute a price index by the weighted aggregate method, with 1974 = 100, using base year weights.

	Price (Dollars per Gallon) Blueberry	Strawberry	Production (Millions of Gallons) Blueberry	Strawberry
1974	$3.20	$3.05	2.45	2.60
1975	3.45	3.30	2.95	3.05
1976	3.55	3.50	2.80	3.30

3. A panel of economists studying the U.S. balance of payments with respect to a particular foreign country has assembled the following data. The average price and quantity imported by the United States of each of the five major export commodities of this foreign country are given for 1975 and 1976.

Commodity	Price	1975 Quantity (Millions of Units)	Price	1976 Quantity (Millions of Units)
1	$10	100	$16	200
2	6	200	14	150
3	25	50	30	100
4	20	175	15	150
5	10	125	15	250

Compute an index by the arithmetic mean of relatives method with 1975 as base year.

4. A price index of two commodities is to be constructed from the following data:

Commodity	Unit 1975	Price 1976	Quantities 1975	Consumed 1976
A	$1.00	$.50	3	2
B	.30	.60	7	8

A simple unweighted arithmetic mean of the two price relatives for 1976 on a 1975 base indicates that prices in 1976 were, on the average, 25% higher than in 1975 [(50 + 200)/2 = 125]. A simple unweighted arithmetic mean of the two price relatives for 1975 on a 1976 base indicates that prices in 1975 were, on the average, 25% higher than in 1976.

a. How do you explain these paradoxical results?
b. Compute what you consider the most generally satisfactory price index for 1976 using 1975 as a base year. You may use any of the above data you deem appropriate.
c. Explain precisely the meaning of the answer obtained from your calculation in part (b).

11.4 GENERAL PROBLEMS OF INDEX NUMBER CONSTRUCTION

In a brief treatment, it is not feasible to discuss all the problems of index number construction. However, many of the important matters are subsumed under the categories (1) selection of items to be included and (2) choice of a base period.

Selection of Items to Be Included

In the construction of price indices, as in other problems involving statistical methods, the definition of the problem and the statistical universe to be investigated are of paramount importance. Most of the widely used price index number series are produced by government agencies or sizable private organizations and are used in many ways. Hence, it is not feasible to state a simple purpose for each price index from which a clear definition of the problem and the statistical universe might follow. However, every index attempts to answer meaningful questions, and it is these general purposes of an index that determine the specific items to be included. For example, the CPI attempts to answer a question concerning the average movement of certain prices over time. The specific nature of this question about price movements determines the items to be included in the index. Similarly, many of the limitations of the use of the index stem from what the index does and does not attempt to measure.

Let us pursue the illustration of the CPI. Essentially, this index attempts to measure how much it would cost to purchase (at retail) a particular combination of goods and services compared to what it would have cost in a base period. More specifically, the combination of goods and services consists of items selected to represent a typical market basket of purchases by city wage earners and city clerical workers and their families. These families are considered to have "moderate" incomes. The relevant universe comprises about 40% of the U.S. population. Hence, the index does not attempt to describe changes in prices of purchases by low-income families, high-income families, farm families, or the families of business executives or professional people. By means of periodic consumer surveys, the BLS determines the goods and services purchased by the specified families and how these families spread their spending among these items. In summary, the general question the index purports to answer determines the items to be included. Obviously, indices constructed for other purposes, such as indices of export prices or agricultural prices, would be based on very different lists of items.

However, even when the general purpose of an index is clearly defined, many problems remain concerning the choice of items to be included. In the case of the CPI, the BLS has determined that there are about 2000 items purchased by moderate-income city wage earner families. However, the BLS includes only about 400 of these goods and services, having found that these few hundred accurately reflect the average change in the cost of the entire market basket. The choice of the commodities to be included in a price index is ordinarily not determined by usual sampling procedures. Each good or service cannot be considered a random sampling unit that is as representative as any other unit. Rather, an attempt is made to include practically all of the most important items and, by pricing these, to obtain a representative portrayal of the movement of the entire population of prices. If subgroup indices are required (for example, indices of food, housing, or medical care), as well as an overall consumers' price index, more items must be included than if only the overall index were desired. After the decisions have been made concerning the commodities to be included, sophisticated sampling procedures are often utilized to determine which specific prices will be included.

Choice of a Base Period

A second problem in the construction of a price index is the choice of a base period, that is, a period whose level of prices represents the base from which changes in prices are measured. As indicated earlier, the level of prices in the base period is taken as 100%. Price levels in non-base periods are stated as percentages of the base period level. The base period may be a conventional calendar time interval such as a month, a year, or even a period of years. It is usually considered advisable to use a time period with "normal" price levels. Of course, it is virtually impossible to devise a meaningful definition of what constitutes "normality" in almost any area of economic experience, but the time period selected should not be at or near the peaks or troughs of price fluctuations. Although there is nothing *mathematically* incorrect about using a base period with unusually low or high price levels, the use of such time intervals tends to produce distorted concepts, since comparisons are made with atypical periods.

The use of a period of years as a base provides an averaging effect on year-to-year variations. Any particular year may have unusual influences present, but if, say, a three- to five-year base period is used, these will tend to be evened out. Most of the U.S. government indices have used such time intervals (for example, 1935–1939, 1947–1949, and 1957–1959) as base periods.[2]

Another point to consider in choosing a base time interval is suggested by the three time periods just mentioned: The base period should not be too distant from the present. The farther away we move from the base period, the less we know about economic conditions prevailing at that time. Consequently, comparisons with these remote periods tend to lose significance and to become rather tenuous in meaning. This is why producers of index number series, such as U.S. government agencies, shift their base periods every decade or so, in order to make comparisons with a base time interval in the recent past. Furthermore, it is desirable to shift the base from time to time because a period previously thought of as normal or average may no longer be so considered after a long lapse of time.

Other considerations may also be involved in choosing a base period for an index. If a number of important existing indices have a certain base period, it is desirable for ease of comparison that newly constructed indices use the same time periods. Moreover, as new commodities are developed and indices are revised to include them, it becomes desirable to shift the base period to a time interval that reflects the newer economic environment.

11.5 QUANTITY INDICES

The discussion in the preceding sections referred to price indices. Another important group of summary measures of economic change is represented by *quantity indices,* which measure changes in physical *quantities* such as the volume of industrial production, physical volume

---

[2] A recent exception is the use of 1967 as a base. This was a year that was neither a peak nor a trough in business activity.

of imports and exports, quantities of goods and services consumed, and volume of stock transactions. In virtually all currently used *quantity indices*, what is actually measured is the change in the *value* of a set of goods from the base period to the current period attributed only to changes in *quantities*, prices being held constant. This corresponds to the interpretation of what is measured in weighted price indices as the change in the *value* of a set of goods attributed only to changes in *prices*, quantities being held constant. The same types of procedures used for the calculation of price indices are also employed to obtain quantity indices. Except for the case of the unweighted aggregate index, which would not be meaningful for a quantity index, corresponding quantity indices may be obtained by interchanging Ps and Qs in the formulas given earlier in this chapter.

An unweighted average of relatives quantity index can be determined by establishing quantity relatives $Q_n/Q_0 \cdot 100$ and calculating the arithmetic mean of these figures. As indicated in the preceding paragraph, an unweighted aggregative quantity index would not be meaningful, because it does not make sense to add up quantities stated in different units.

As was true for price indices, weighted quantity indices are preferable to unweighted ones. A weighted aggregative index of the Laspeyres type is given by the following formula:

**WEIGHTED RELATIVE OF AGGREGATES
QUANTITY INDEX, BASE PERIOD WEIGHTS (LASPEYRES METHOD)**

$$(11.8) \qquad \frac{\Sigma\, Q_n P_0}{\Sigma\, Q_0 P_0} \cdot 100$$

Just as the corresponding Laspeyres price index measures the change in price levels from a base period assuming a fixed set of quantities produced or consumed in the base period, etc., this quantity index measures change in quantities produced or consumed, etc., assuming a fixed set of prices that existed in the base period. As with price indices, a quantity index computed by the weighted average of relatives method, using the base period value weights given in (11.9), is algebraically identical to this Laspeyres index.

**WEIGHTED ARITHMETIC MEAN OF RELATIVES
QUANTITY INDEX, BASE PERIOD VALUE WEIGHTS**

$$(11.9) \qquad \frac{\Sigma\, \left(\dfrac{Q_n}{Q_0} \cdot 100\right) Q_0 P_0}{\Sigma\, Q_0 P_0}$$

Let us interpret these two equivalent weighted indices by considering the Laspeyres version, given in (11.9). We continue the assumption that the raw data refer to quantities of food items consumed (during one week) and prices paid by a typical family in an urban area. The numerator of the index shows the value of the specified food items consumed in the nonbase year at base year prices. The denominator refers to the value of the food items consumed in the base year. Suppose a figure of 125 resulted from such an index. Since prices were kept constant, the increase would be solely attributable to an average increase of 25% in the quantity of these items consumed.

FRB Index of Industrial Production

Probably the most widely used and best known quantity index in the United States is the Federal Reserve Board (FRB) Index of Industrial Production. This index measures changes in the physical volume of output of manufacturing, mining, and utilities. In addition to the overall index of industrial production, component indices are published by industry groupings such as Manufactures and Minerals, and by subcomponents such as Durable Manufactures and Nondurable Manufactures. Separate indices are reported for the output of consumer goods, output of equipment for business and government use, and output of materials. Based on the groupings used by the Standard Industrial Classification of the U.S. Bureau of the Budget, indices are also prepared for major industrial groups and subgroups. The indices are now issued monthly, with 1967 as both reference base and weight base. The Index of Industrial Production is closely watched by business executives, economists, and financial analysts as a major indicator of the physical output of the economy.

The method of construction is the weighted arithmetic mean of relatives, using the base periods mentioned above. Numerous problems have had to be resolved concerning both the quantity relatives and value weights. Since many industries cannot easily provide physical output data for the quantity relatives, related data that tend to move more or less parallel to output (such as shipments and man-hours worked) are sometimes used instead. The weights used are value-added data, which at the individual company level represent the sales of the firm minus all purchases of materials and services from other business firms. The reason for using value-added rather than value of final product weights is to avoid the problem of double counting. For example, if the value of final product were used for a steel company that sells its steel to an automobile company, and the value of the final product of the automobile company were also used, there would be double counting of the steel that went into making the automobile. Hence, the weights used follow the value-added approach, in which the values of so-called "intermediate products" produced at all stages prior to the final product are excluded. From the viewpoint of the economist, the value added of a firm is conceptually equivalent to the total of its factor of production payments—wages, interest, rent, and profits.

11.6 DEFLATION OF VALUE SERIES BY PRICE INDICES

One of the most useful applications of price indices is in adjusting series of dollar figures for changes in levels of prices. The result of this adjustment procedure, known as "deflation," is restatement of the original dollar figures in terms of so-called "constant dollars." The rationale of the procedure can be illustrated in terms of the simple example given in Table 11-5. In Column (2) are shown average (arithmetic mean) weekly wage figures for factory workers in a large city in 1970 and 1976. Such unadjusted dollar figures are usually said to be stated in "current dollars." In Column (3) is shown a CPI for the given city, with 1970 as reference base period. For simplicity of interpretation, let us assume that the CPI was computed by the Laspeyres method. As we note from Column (2), average weekly wages of the given workers have increased from \$110.00 in 1970 to \$170.00 in 1976, a gain of 54.5%. But can these workers purchase 54.5% more goods and services with this increased income? If prices of all these goods and services had remained unchanged between 1970 and 1976, then all other things being equal, the answer would be yes. But as can be seen in Column (3), prices rose 45% over this period. To determine what average weekly wages are in terms of 1970 constant dollars (dollars with 1970 purchasing power), we carry out the division \$170.00/1.45 = \$117.24. That is, we divide the 1976 weekly wage figure in current dollars by the 1976 CPI stated as a decimal figure (using a base of 1.00 rather than 100) to obtain the figure \$117.24 for average weekly wages in 1970 constant dollars. As indicated

TABLE 11-5
Calculation of Average Weekly Wages in 1970 Constant Dollars for Factory Workers in a Large City, 1970 and 1976

Year (1)	Average Weekly Wages (2)	Consumer Price Index (1970 = 100) (3)	"Real" Average Weekly Wages (1970 Constant Dollars) (4)
1970	\$110.00	100	\$110.00
1976	170.00	145	117.24

in the heading of Column (4), the result of this adjustment for price change is referred to as "real wages" (in this case, "real average weekly wages"). The implication of the term "real" is that a portion of the dollar increase in wages is absorbed by the increase in prices, so the adjustment attempts to isolate the "real change" in the volume of goods and services the weekly wages can purchase at base year prices. In summary, the dollar value figures in Column (2) are divided by the price index figures in Column (3) (stated on a base of 1.00) to obtain the real value figures in Column (4). The same procedure would have been followed if there had been a series of figures in Columns (2) and (3) (perhaps for several consecutive years) rather than just the current and base period figures. This division of a dollar figure by a price index is referred to as a *deflation* of the current dollar value series, whether a decrease or an increase occurs in going from figures in current dollars to constant dollars.

The rationale of the deflation procedure stems from the basic relationship value = price times quantity. The weekly wages in current dollars are value figures; they may be viewed as *value aggregates,* that is, as sums of prices of labor times quantities of such labor. By dividing such a figure by a price index, an attempt is made to isolate the change attributable to quantity or physical volume. Hence, we may think of the real average weekly wage figures as reflecting the changes in quantities of goods over which the wage figures have command.

This deflation procedure is very widely used in business and economics. One interesting application is in measuring economic well-being and growth. For example, in comparing growth rates among countries, one of the most important indicators used is per capita growth in real GNP. The division of GNP by population to yield per capita figures may be viewed as an adjustment for differences in population size. The division of the figures by a relevant price index to obtain real GNP is an adjustment for change in price levels. Per capita real GNP is an extremely useful measure of physical volume of production.

Of course, there are numerous limitations to the use of the deflation procedure. For example, in the weekly wages illustration, the market basket of commodities and services implicit in the consumer price index may not refer specifically to the factory workers to whom the weekly wages pertain. Even if the index had been constructed for this specific group of factory workers, it is, after all, only an average, subject to all the interpretational problems of any measure of central tendency. Furthermore, inferences from such data must be made with care. For instance, even if there has been an increase in real average weekly wages from one period to another, we cannot immediately infer an increase in economic welfare for the factory worker group in question; in the later period, there may be a less equitable distribution of this income, taxes may be higher (leading to lower disposable income), and so on. Nevertheless, despite such limitations and caveats, the deflation procedure is a very useful, practical, and widely utilized tool of business and economic analysis.

We have seen how a price index may be used to remove from a value aggregate the change attributable to price movements. Another way to view the deflation procedure is as a method of adjusting a value figure for changes in the purchasing power of money. In this connection, it is important to note that a *purchasing power index* is conceptually the *reciprocal of a price index*. For example, suppose you have $20 to purchase shoes in a certain year when a pair of shoes costs $10. The $20 enables you to purchase two pairs of shoes. But if in a later year the price of shoes rises to $20, you can then purchase only one pair of shoes. Let us imagine a price index composed solely of the price of this pair of shoes. If the earlier year is the base period, the base period price index is 100 and the later period figure is 200. On the other hand, if the price of these shoes has doubled, the purchasing power of the dollar relative to shoes has halved. Hence, a purchasing power index that was at a level of 100 in the earlier base year should stand at a level of 50 in the later period. If the indices are stated using 1.00 rather than 100 in the base period, the reciprocal relationship can be expressed as $2 \times \frac{1}{2} = 1$. That is, the doubling in the price index and the halving in the purchasing power index are reciprocals. It is this relationship between price and purchasing power indices that is implied in such popular statements as that a dollar today is worth only fifty cents in terms of the dollar in some earlier period.

Exercises

1. a. Using the data of Exercise 2, Section 11.3, compute an index of wine production by the weighted arithmetic mean of relatives method, with 1974 = 100, using base year weights.
   b. Compute an index by the weighted aggregates method, on the same base.
2. The Klondike Mining Company produces two types of coal: anthracite and bituminous. The price and quantities produced of each are shown in the table below:

Type	1974 Price (per Ton)	1974 Quantity (Thousands of Tons)	1975 Price (per Ton)	1975 Quantity (Thousands of Tons)	1976 Price (per Ton)	1976 Quantity (Thousands of Tons)
Bituminous	$20	1000	$25	1500	$30	2000
Anthracite	$25	500	$30	1000	$40	1250

Compute an index of coal production by the weighted aggregates method, with 1974 = 100, using base year weights.

3. The National Paper Company had gross sales of $20,000,000 in 1971; $32,000,000 in 1973; and $43,000,000 in 1975. It uses the following whole-sale price index for pulp, paper, and allied products as a price deflator.

	Wholesale Price Index (1967 = 100)
1971	110.1
1973	122.1
1975	170.4

By what percentage did "real gross sales" change from 1971 to 1975? From 1973 to 1975?

4. The figures given below represent the disposable personal income (DPI) and

Year	DPI ($Billions)	CPI (1967 = 100)
1960	349.4	88.7
1965	472.2	94.5
1970	685.9	116.3
1975	1076.8	161.2

consumer price index (CPI) for the United States for the years indicated. Adjust the above series on disposable income to 1970 constant dollars.

5. Average weekly earnings of manufacturing workers in the United States and values of the CPI for the United States for 1972 through 1975 are given in the following table:

Year	Average Weekly Earnings	CPI (1967 = 100)
1972	$154.69	125.3
1973	166.06	133.1
1974	176.40	147.7
1975	189.51	161.2

Calculate the year-to-year changes in average weekly earnings, deflated by the CPI.

6. GNP and personal consumption figures for the United States for the years 1971–1975 are given below. The appropriate implicit price deflators are also presented for each series. For the deflator series, 1972 = 100.

Year	GNP ($Billions)	GNP Implicit Deflator	Personal Consumption ($Billions)	Personal Consumption Implicit Deflator
1971	$1063.4	96.02	$668.2	96.60
1972	1171.1	100.00	733.0	100.00
1973	1306.3	105.92	808.5	105.50
1974	1406.9	116.20	885.9	116.60
1975	1499.0	126.35	963.2	125.60

Calculate the ratio of real personal consumption (in 1972 constant dollars) to real GNP (in 1972 constant dollars) for each year.

(NOTE: The source for data in Exercises 4, 5, and 6 is the *Economic Report of the President,* U.S. Government Printing Office, Washington, D.C., January 1976.)

11.7 SOME CONSIDERATIONS IN THE USE OF INDEX NUMBERS

Numerous problems arise in connection with the use of index numbers for analysis and decision purposes. A few of these are discussed below.

Shifting the Base

For a variety of reasons, it is often necessary to change the reference base of an index number series from one time period to another without returning to the original raw data and recomputing the entire series. This change of reference base period is usually referred to as *shifting* the base. For example, it may be desired to compare several index number series computed on different base periods. Particularly if the several series are to be shown on the same graph, it may be desirable for them to have the same base period. In other situations, the shifting of a base period may simply reflect the desire to state the series in terms of a more recent time period. The procedure for accomplishing the shift is simple and is illustrated in the following example, in which a housing price index with a reference base of 1969 is shifted to a new base period of 1975.

In Table 11-6, the original price index is shown in the first column stated on a base period of 1969. The shift to a 1975 base period is accomplished by dividing each figure in the original series by the index number for the desired new base period stated in decimal form. Hence, in this illustration, each index number on the old base of 1969 is divided by 1.237, the figure for 1975 stated as a decimal. Thus, the index number for 1968 shifted to the new base of 1975 is 97.3/1.237 = 78.7; for 1969, the new figure is 100.0/1.237 = 80.8.

TABLE 11-6

	Housing Price Index (1969 = 100)	Housing Price Index (1975 = 100)
1968	97.3	78.7
1969	100.0	80.8
1970	103.7	83.8
1971	106.9	86.4
1972	109.4	88.4
1973	113.7	91.9
1974	118.3	95.6
1975	123.7	100.0
1976	129.8	104.9

Note that the relationships among the new index figures after the base is shifted are the same as in the old series. For example, the index number for 1969 exceeds that of 1968 by the same percentage in both series. That is, $100.0/97.3 = 80.8/78.7 = 1.027$, and so on throughout the two series. However, a subtle problem arises with the weighting scheme. Let us suppose that the old series was computed using a Laspeyres type index. Hence, both the reference base and weight base periods are 1969. The procedure of dividing the series by the 1975 index number changes the reference period, but the weights still pertain to 1969. That is, the raw data originally collected for quantity weights pertained to 1969. The mere procedure of dividing the index numbers in the old series by one of its members does nothing to change these weights. Indeed, obtaining new weights for 1975 would involve a new data collection process. In summary, the new series has been shifted to a reference base period of 1975 for measuring price changes, but the weights are fixed at 1969. This point may be demonstrated algebraically as follows. Consider the Laspeyres price index for 1970 computed on the reference base of 1969. The formula for computing this figure may be written as $\Sigma P_{70}Q_{69}/\Sigma P_{69}Q_{69}$. Correspondingly, the formula for the 1975 price index on the 1969 base is $\Sigma P_{75}Q_{69}/\Sigma P_{69}Q_{69}$. (The multiplication by 100 has been dropped in these formulas to simplify the discussion.) To obtain the new price index figure for 1970 on a 1975 base, we divide the old 1970 figure by the old 1975 figure.

$$\frac{\Sigma P_{70}Q_{69}}{\Sigma P_{69}Q_{69}} \div \frac{\Sigma P_{75}Q_{69}}{\Sigma P_{69}Q_{69}} = \frac{\Sigma P_{70}Q_{69}}{\Sigma P_{75}Q_{69}}$$

Since the $\Sigma P_{69}Q_{69}$ values cancel, we see that the resulting index figure for 1970 is stated on a reference base of 1975, but the weights still pertain to 1969.

Despite the fact that weights are not changed by the procedure discussed here, this method of shifting reference bases is widely employed. It often represents the only practical way of shifting a base, and analysts ordinarily do not view as a matter of serious concern the fact that the weighting system remains unchanged.

Splicing

Sometimes an index number series is available for a period of time and then undergoes substantial revision, including a shift in the reference base period. In these cases, if it is desired to obtain a continuous series going back through the time period of the older series, the old and revised series must be *spliced* together. Splicing involves a similar arithmetic procedure to that for shifting a base. For example, suppose a price index number series was revised by inclusion of certain new products, exclusion of some old products, and change in definition of some other products. In Table 11-7 are shown such an old series on a reference base of 1970 and the revised series on a base of 1972. There must be an overlapping period of the old and revised series to provide for splicing them, and the period of overlap in this example is 1972. The splicing of the two series to obtain a continuous series on the new base of 1972 is accomplished by dividing each figure in the old series by the old index figure for 1972 stated in decimal form, that is, by 1.09. This restates the old series on the new base of 1972. The resulting spliced series is shown in the last column of Table 11-7. Had it been desired to state the continuous series on the old reference base of 1970, each figure in the revised index would be multiplied by 1.09.

Although the arithmetic procedure involved in splicing is very simple, the interpretation of the resulting continuous series may be extremely difficult, particularly if long time periods are involved. For

TABLE 11-7

	Old Price Index (1970 = 100)	Revised Price Index (1972 = 100)	Spliced Price Index (1972 = 100)
1969	96.9		88.9
1970	100.0		91.7
1971	105.2		96.5
1972	109.0	100.0	100.0
1973		104.3	104.3
1974		106.8	106.8
1975		108.1	108.1
1976		110.1	110.1

example, it is difficult to specify precisely what is measured if, say, a price index in the later period contains prices of frozen foods, clothing made from synthetic fibers, television, and similar recently developed products, whereas the spliced indices for the earlier period (before these products were on the market) did not contain these products. However, despite such conceptual difficulties, splicing frequently represents the only practical method of providing for comparability in similar phenomena measured by indices over different time periods.

Quality Changes

In the construction of an index such as the CPI, the basic data on prices are collected by trained investigators who price goods for which detailed specifications have been made. The same items are always priced in the same stores. However, over time, as a result of technological and other improvements, there often is a corresponding improvement in the quality of many commodities. It is very difficult, and in many cases impossible, to make suitable adjustments in a price index for quality changes. The artificial but practical procedure adopted by the Bureau of Labor Statistics is to consider a product's quality improved only if changes that increase the cost of producing the product have occurred. Hence, for example, an automobile tire is not considered improved if it delivers increased mileage at the same cost of production. Because of such actual improvements in product quality, many analysts feel that over reasonably long periods of time, indices such as the CPI that have shown steady rises in price levels overstate the actual price increases of a fixed market basket of goods.

Uses of Indices

Index number series are widely used in connection with decision making and analysis in business and government. One of the best known applications of a price index is the use of the CPI as an escalator in collective bargaining contracts. In this connection, in 1976 over five million workers were covered by contracts that specify periodic changes in wage rates depending on the amount by which the CPI moves up or down. The Bureau of Labor Statistics Wholesale Price Index is similarly used for escalation clauses in contracts between business firms.

Much use is made of index numbers by individual companies as well as at the levels of entire industries and the overall economy. In certain industries, it is standard practice to key changes in selling prices to changes in indices of prices of raw materials and wage earnings. Assessments of past trends and current status and projection of future economic activity are made on the basis of appropriate indices. Economists follow many of the various indices in order to appraise the performance of the economy and to analyze its structure and behavior.

12
Nonparametric Statistics

12.1 INTRODUCTION

Most of the methods discussed thus far have involved assumptions about the distributions of the populations sampled. For example, as we have seen, when certain hypothesis-testing techniques are used it is assumed that the observations are drawn from normally distributed populations. Recently, a number of very useful techniques that do not make these restrictive assumptions have been developed. Such procedures are referred to as *nonparametric* or *distribution-free* tests. Many writers prefer the latter term, because it emphasizes the fact that the techniques are free of assumptions concerning the underlying population distribution. However, the two terms are generally used synonymously.

In addition to making less restrictive assumptions than the corresponding so-called "parametric" methods, nonparametric procedures are generally easy to carry out and understand. Furthermore, as is implied by the lack of underlying assumptions, they are applicable under a very wide range of conditions. Many nonparametric tests are in terms of the *ranks* or *order* rather than the numerical values of the observations. Sometimes, even ordering is not required.

However, when distribution-free procedures are applied where parametric techniques are possible, the nonparametric methods have some disadvantages. Since these nonparametric procedures may use ordering or ranking as opposed to the actual numerical values of the observations, they are, in effect, guilty of ignoring a certain amount of information. As a result, nonparametric tests are somewhat less efficient than the corresponding standard tests. This means that in testing at a given level of significance, say $\alpha = 0.05$, the probability of a Type II error, β, would be greater for the nonparametric than for the parametric test. Advocates of nonparametric tests argue, however, that despite the lessened efficiency, the analyst can have more confidence in these tests than in the standard ones, because of the restrictive and somewhat unrealistic assumptions often required in the latter procedures.

In this chapter, we consider a few simple and widely applied nonparametric techniques.

12.2 THE SIGN TEST

As we have seen in Chapter 7, the solution to many problems in business and public administration and in social science research centers on a comparison between two different samples. In some of the hypothesis-testing techniques previously discussed, very restrictive assumptions about the populations sampled were necessary. For example, the t test for the difference between two sample means assumes that the populations are normally distributed and have equal variances. Sometimes, one or both of these assumptions may be unwarranted. Furthermore, situations often exist in which quantitative measurements are impossible. In such cases, it may be possible to assign ranks or scores to the observations in each sample. In these situations, the *sign test* can be used. The name of the test indicates that the signs of observed differences (that is, positive or negative signs) are used rather than quantitative magnitudes.

We will illustrate the sign test in terms of data obtained from a panel of 60 beer-drinking consumers. Let us assume a blindfold test in which the tasters were asked to rate a glass of each of two brands of beer, Wudbeiser and Diller, on a scale from 1 to 5 with 1 representing the best taste (excellent), 5 the worst taste (poor), and the other scores denoting the appropriate intermediates. Table 12-1 shows a partial listing of the scores assigned by the panel members in this taste test.

Column (4) shows the signs of the differences between the scores as-

TABLE 12-1
Ranking Scores Assigned to Taste of Two Brands of Beer by a Panel

Panel Member (1)	Score for Wudbeiser (2)	Score for Diller (3)	Sign of Difference (4)
A	3	2	+
B	4	1	+
C	2	4	−
D	3	3	0
E	1	2	−
⋮	⋮	⋮	⋮

NOTE: Best score = 1; worst score = 5. Hence, a plus sign means Diller is preferred; a minus sign means Wudbeiser is preferred.

signed by each participant in columns (2) and (3). As indicated, a plus sign means a higher numerical score was assigned to Wudbeiser than to Diller beer, a minus sign means the reverse, and a 0 denotes a tie score. Let us assume that the following results were obtained:

+ scores	35
− scores	15
ties	10
Total	60

Method

By means of the sign test, we can test the null hypothesis of no difference in rankings of the two brands of beer. That is, more specifically, we can test the hypothesis that plus and minus signs are equally likely for the differences in rankings. If this null hypothesis were true, we would expect about equal numbers of plus and minus signs. We would reject the null hypothesis if too many of one type occurred. If we use p to denote the probability of obtaining a plus sign, we can indicate the hypotheses as

$$H_0: p = 0.50$$
$$H_1: p \neq 0.50$$

Since tied cases are excluded in the sign test, the data used for the test consist of 35 pluses and 15 minuses. The problem is conceptually the same as one in which a coin has been tossed 50 times, yielding 35 heads and 15 tails, and we wish to test the hypothesis that the coin is fair.

The binomial distribution is the theoretically correct one. However, we can use the large-sample method of Section 7.2 consisting of the normal curve approximation to the binomial distribution. In terms of proportions, the mean and standard deviation of the sampling distribution are

$$\mu_{\bar{p}} = p = 0.50$$

$$\sigma_{\bar{p}} = \sqrt{\frac{pq}{n}} = \sqrt{\frac{(0.50)(0.50)}{50}} = 0.071$$

Assuming that the test is performed at the 5% level of significance ($\alpha = 0.05$), we would reject the null hypothesis if $z < -1.96$ or $z > 1.96$.

Since, in this problem, the observed proportion of plus signs is $\bar{p} = 35/50 = 0.70$, then

$$z = \frac{\bar{p} - p}{\sigma_{\bar{p}}} = \frac{0.70 - 0.50}{0.071} = 2.82$$

Hence, we reject the null hypothesis that plus and minus signs are equally likely. Since the plus signs exceeded minus signs in the observed data, our interpretation of the experimental data is that Diller beer is preferred to Wudbeiser, according to the rank scores given by the consumer panel.

We note that the arithmetic could have been carried out in terms of critical limits for $\bar{p}$, rather than for z values. The critical limits for $\bar{p}$ are

$$p + 1.96\sigma_{\bar{p}} = 0.50 + (1.96)(0.071) = 0.639$$
$$p - 1.96\sigma_{\bar{p}} = 0.50 - (1.96)(0.071) = 0.361$$

Since the observed $\bar{p}$ of 0.70 exceeds 0.639, we reach the same conclusion and reject H_0. The testing procedure is shown in the usual way in Figure 12-1.

Two points may be made concerning the techniques used in this

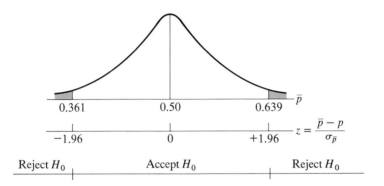

FIGURE 12-1
Sampling distribution of a proportion for beer tasting problem: $p = 0.50$, $n = 50$. This is a two-tailed test with $\alpha = 0.05$.

illustration of the sign test. First, although a two-tailed test was appropriate for the problem, the sign test can also be used in one-tailed test situations. Second, a normal curve approximation to the binomial distribution was used. For small samples, binomial probability calculations and tables should be used.

General Comments

As we have seen, the sign test is very simple to apply. In the example, it was not feasible to obtain quantitative data for beer tasting, so rank scores were used. Because of its simplicity, the sign test is sometimes used instead of a standard test even when quantitative data are available. For example, suppose we refer to the beer taste rankings for any particular individual as observations from a "matched pair." We might have matched-pair observations where the data are, say, weights before and after a diet, grades on a preliminary scholastic aptitude test and on the regular aptitude test, or other pairs of quantitative values. In applying the sign test, only the signs of the differences in the matched-pair observations would be used, rather than the actual magnitudes of the differences. Of course, as noted earlier, there would be some loss in the efficiency of the test as a result.

In addition to being simple to apply, the sign test is applicable in a wide variety of situations. As in the example given above, the two samples do not even have to be independent. Indeed, when matched pair observations are used, the elements in the first sample are usually matched as closely as possible with the corresponding elements in the second sample. Furthermore, the sign test may be used in cases of qualitative classifications where it may even be difficult to use a ranking scheme such as that used above. For example, an experimenter may classify subjects after a treatment has been applied as improved (+), worse (−), or the same (0), and then use the sign test.

Exercises

1. The management of a large manufacturing firm decided to test a new incentive program designed to increase the productivity of the firm's factory workers. The following results were reported for the new program as compared with a standard previous incentive program:

	Number of Workers
New program better	96
Previous program better	42
No difference	12
	150

Use the method developed in this section to test the null hypothesis of no difference in effectiveness of the two incentive programs. Use a one-tailed test at the 0.01 significance level to determine whether the new method is more effective than the old.

2. In a certain production process, the specified standard weight for a critical component was 14.5 pounds. In a production run of 200 of these components, 110 weighed in excess of 14.5 pounds and 90 weighed less. Replacing sample values in excess of specifications by a plus sign and those less than specifications by a minus sign, test the null hypothesis that p, the proportion of plus signs, equals 0.50. Use a two-tailed test at the 0.05 significance level.

3. A marketing executive wanted to determine the opinions of the sales force on a proposed promotional program for a certain product as compared to a current program. The executive asked a randomly selected sample of 35 sales representatives to assign ratings from 0 to 10 for the effectiveness of each program, with 10 being the highest possible score. The ratings are given below. For example, the first representative assigned a 9 to the new program and a 7 to the old.

New program: 9, 7, 10, 6, 9, 8, 5, 9, 7, 8, 6, 4, 10, 8, 9, 7, 8, 6, 9, 8, 9, 7, 6, 8, 9, 7, 6, 5, 8, 7, 8, 6, 4, 8, 8
Old program: 7, 8, 8, 7, 9, 7, 4, 5, 8, 6, 3, 7, 8, 5, 7, 4, 8, 7, 6, 9, 6, 7, 8, 7, 8, 8, 5, 6, 7, 9, 6, 5, 6, 7, 9

Match the scores, assigning a plus sign when the rating for the new program exceeds that of the old, etc. Since the executive wished to place the burden of proof on the new program, it was decided to use a one-tailed test of the following type:

$$H_0: p \leq 0.50$$
$$H_1: p > 0.50$$

where p is the probability of obtaining a plus sign. Carry out the appropriate sign test using a 0.05 significance level.

4. A research agency was interested in determining whether American citizens felt that there had been a decrease in the influence of the United States in world affairs over the past decade. In a simple random sample of 500 citizens, 296 felt that there had been a decrease, 150 felt there had been an increase, and 54 thought that there had been no change. Use a plus sign to represent a perceived increase, a minus sign for a perceived decrease, and a 0 for "no change," and apply a one-tailed test at the 0.02 significance level. Use the null hypothesis $H_0: p \leq 0.50$, where p is the probability of a perceived decrease in the influence of the United States.

12.3 COMPARISON OF TWO POPULATIONS USING PAIRED OBSERVATIONS (THE WILCOXON MATCHED-PAIRS SIGNED RANK TEST)

In Section 12.2, we discussed the sign test, which is useful in testing for significant differences between paired observations. As we have seen, the sign test is carried out in terms of the signs of the differences between matched pairs of observations, without regard to the magnitudes of these differences. We turn now to the Wilcoxon matched-pairs signed rank test, another nonparametric test for significant differences between paired observations, which *does* take account of the magnitudes of the differences. Therefore, when the differences between paired observations can be quantitatively measured rather than merely assigned rankings, the Wilcoxon matched-pairs signed rank test is preferable to the sign test.

Both the sign test and the Wilcoxon matched-pairs signed rank test may be considered substitutes for the analogous parametric *t* test for paired observations (see Section 7.5). However, while the parametric *t* test for paired observations requires the assumption that the underlying population of differences is normally distributed, the nonparametric tests have the advantage of making no assumption about the population distribution.

We illustrate the Wilcoxon matched-pairs signed rank test by an example consisting of the numerical grades on the midterm and final examinations in a certain course, for ten students selected at random from the course rolls. These data are given in Table 12-2. The Wilcoxon test is carried out in terms of the same paired differences as the parametric test, that is, $d = X_2 - X_1$, where the d values represent differences between two observations on the same individual or object. However, in the Wil-

TABLE 12-2
Calculations for the Wilcoxon Matched-Pairs Signed Rank Test of Significance of Differences in Grades on Two Examinations for a Simple Random Sample of Ten Students

Student	Grade on Midterm Examination X_1	Grade on Final Examination X_2	Difference $d = X_2 - X_1$	Rank of $|d|$	Signed Rank Rank (+)	Signed Rank Rank (−)
1	75	72	−3	3		3
2	87	94	+7	6.5	6.5	
3	72	92	+20	9	9	
4	65	67	+2	2	2	
5	93	86	−7	6.5		6.5
6	85	85	0			
7	59	58	−1	1		1
8	73	79	+6	5	5	
9	64	69	+5	4	4	
10	71	82	+11	8	8	
Total (Σ)					34.5	10.5

coxon test, the *absolute values* of the differences are obtained, pooled, and ranked from 1 to n, with the smallest difference being assigned the rank 1. These ranks are then given the sign (+ or −) of the corresponding value of d. If there are ties in rankings, the mean rank value is assigned to the tied items. (For example, if, as in the data of Table 12-2, the sixth and seventh ranked items are tied, a rank of $(6 + 7)/2 = 6.5$ is assigned to each item. The analogous procedure is used if more than two items are tied.) If the difference between the paired observations for an item is 0, as in the case of student 6 in Table 12-2, that item is dropped, and the number of differences is correspondingly reduced. Since there is one such item in the present example, the effective sample size is $n = 10 - 1 = 9$.

As indicated in the last two columns of Table 12-2, the sums of the ranks are obtained separately for the positive and negative differences. These sums, denoted Σ rank (+) and Σ rank (−), form the basis for the null hypothesis H_0: Σ rank (+) = Σ rank (−). The null hypothesis is often referred to as one of "identical population distributions." More specifically, the hypothesis is that the population positive and negative differences are symmetrically distributed about a mean of 0. The smaller of the two ranked sums, conventionally known as Wilcoxon's T statistic, is the test statistic. Hence, in Table 12-2, the test statistic is $T = \Sigma$ Rank (−) = 10.5.

It can be shown[1] that when n is large (at least 25), T is approximately normally distributed with mean and standard deviation

$$(12.1) \qquad \mu_T = \frac{n(n + 1)}{4}$$

$$(12.2) \qquad \sigma_T = \sqrt{\frac{n(n + 1)(2n + 1)}{24}}$$

Therefore, we can compute

$$z = \frac{T - \mu_T}{\sigma_T}$$

and carry out the test in the usual way.

The critical values of T are shown in Table A-10 of Appendix A, where, in keeping with the aforementioned convention, T is the smaller of the positive or negative rank sums. The table indicates that with $n = 9$ pairs, the null hypothesis of identical population distributions would be rejected at the 5% significance level using a two-tailed test at $T \leqslant 5$ ($T_{.05} = 5$). Note that Table A-10 of Appendix A presents the *maximum* values that T can have and still be considered significant at the stated significance levels. Thus, in this example, since the calculated

---

[1] See W. H. Kruskal and W. A. Wallis, "Use of Ranks in One-Criterion Variance Analysis," *Journal of the American Statistical Association*, Vol. 47, 1952, pp. 583–621.

value of T (10.5) exceeds 5, the evidence on the midterm and final examinations does not permit us to reject the null hypothesis of identical population distributions. Hence, we conclude that the performance of this sample of students on the midterm and final examinations did not differ significantly.

Exercises

1. A large number of high school seniors take college Scholastic Aptitude Tests (SAT) more than once in an effort to improve their scores. The following data represent the scores of a simple random sample of twelve students in a certain high school on the SAT mathematics test.

Student	Score on First Test	Score on Second Test
A	550	560
B	490	530
C	610	580
D	570	650
E	670	660
F	480	500
G	720	680
H	630	570
I	500	570
J	660	710
K	540	520
L	570	570

Use the Wilcoxon matched-pairs signed rank test to determine whether the scores on the second test are significantly higher than the scores on the first test. Use $\alpha = 0.05$ in a one-tailed test.

2. The following data represent the monthly numbers of sales by ten sales representatives for one-month periods before and after a special company promotional campaign.

Sales Representative	Sales before Campaign	Sales after Campaign
A	22	25
B	10	14
C	15	17
D	21	31
E	28	33
F	27	27
G	33	32

(Continued)

Sales Representative	Sales before Campaign	Sales after Campaign
H	18	19
I	17	28
J	16	22
K	15	22

Use the Wilcoxon matched-pairs signed rank test to determine whether the promotional campaign was accompanied by an increase in number of sales. Use $\alpha = 0.01$ in a one-tailed test.

3. A company made an improvement on a new product after experiencing what it felt were a large number of cutomer returns of the product. The following data present the numbers of returns for six-month periods before and after the product improvement, by cities in which sales took place.

City	Returns before Improvement	Returns after Improvement
Atlanta	110	42
Boston	122	70
Chicago	467	301
Detroit	206	325
Los Angeles	340	283
Miami	76	38
New Orleans	134	75
New York	643	397
Philadelphia	389	227
San Francisco	291	183

a. Would you conclude that the product improvement resulted in a decrease in the numbers of product returns? Use a Wilcoxon matched-pairs signed rank test in a one-tailed test at the 5% significance level.

b. Assuming that the normal distribution assumptions of the parametric t test for paired observations (discussed in Section 7.5) are satisfied, test the hypothesis of no difference between the average numbers of returns before and after the product improvement using this t test. Again, use a one-tailed test at the 5% significance level.

4. In a test involving 55 paired observations of portfolio performance before and after adoption of a new investment strategy, the lower of the two ranked sums was $T = \Sigma$ rank $(+) = 510$. Use the normal approximation to the Wilcoxon matched-pair signed rank test to determine whether the null hypothesis of identical population distribution should be rejected. Use $\alpha = 0.05$ in a two-tailed test.

12.4 MANN–WHITNEY *U* TEST (RANK SUM TEST)

Another very useful nonparametric technique involving a comparison of data from two samples is the Mann–Whitney U test, often referred to as the rank sum test. This procedure is used to test whether two independent samples have been drawn from populations having the same mean. Hence, the rank sum test may be viewed as a substitute for the parametric t test or the corresponding large-sample normal curve test for the difference between two means. Since the rank sum test explicitly takes into account the rankings of measurements in each sample, in that sense it uses more information than does the sign test.

As an illustration of the use of the rank sum test, consider the data shown in Table 12-3. These data represent the grades obtained on an aptitude test given by a large corporation to management training program applicants. The samples consist of graduates of two different universities, referred to as H and W.

Method

The first step in the rank sum test is to merge the two samples, arraying the individual scores in rank order as shown in Table 12-4. The test is

TABLE 12-3
Grades on an Aptitude Test Obtained by
Graduates of Two Universities

H University	W University
50	70
51	76
53	77
56	80
57	81
63	82
64	83
65	86
71	87
73	88
74	92
78	93
89	96
90	98
95	99

TABLE 12-4
Array of Grades on an Aptitude Test Obtained by Graduates of Two Universities

Rank	Grade	University
1	50	H
2	51	H
3	53	H
4	56	H
5	57	H
6	63	H
7	64	H
8	65	H
9	70	W
10	71	H
11	73	H
12	74	H
13	76	W
14	77	W
15	78	H
16	80	W
17	81	W
18	82	W
19	83	W
20	86	W
21	87	W
22	88	W
23	89	H
24	90	H
25	92	W
26	93	W
27	95	H
28	96	W
29	98	W
30	99	W

then carried out in terms of the sum of the ranks of the observations in either of the two samples. The following symbolism is used:

n_1 = number of observations in sample number one
n_2 = number of observations in sample number two
R_1 = sum of the ranks of the items in sample number one
R_2 = sum of the ranks of the items in sample number two

Treating the data for H University as sample number one, we find that R_1 is the sum of ranks 1, 2, 3, 4, 5, 6, 7, 8, 10, 11, 12, 15, 23, 24, and 27, which is 158. Correspondingly, $R_2 = 307$.

If the null hypothesis that the two samples were drawn from the same population were true, we would expect the totals of the ranks (or equivalently, the mean ranks) of the two samples to be about the same. In order to carry out the test, a new statistic, U, is calculated. This test statistic, which depends only on the number of items in the samples and the total of the ranks in one of the samples, is defined as follows:

(12.3)
$$U = n_1 n_2 + \frac{n_1(n_1 + 1)}{2} - R_1$$

The statistic U provides a measurement of the difference between the ranked observations of the two samples and yields evidence about the difference between the two population distributions. Very large or very small U values constitute evidence of the separation of the ordered observations of the two samples. Under the null hypothesis stated above, it can be shown that the sampling distribution of U has a mean equal to

(12.4)
$$\mu_U = \frac{n_1 n_2}{2}$$

and a standard deviation of

(12.5)
$$\sigma_U = \sqrt{\frac{n_1 n_2 (n_1 + n_2 + 1)}{12}}$$

Furthermore, it can be shown that the sampling distribution approaches normality very rapidly and may be considered approximately normal when both n_1 and n_2 are in excess of about ten items.

For the data in the present problem, substituting into Equations 12.3, 12.4, and 12.5, we have

$$U = (15)(15) + \frac{(15)(15 + 1)}{2} - 158 = 187$$

$$\mu_U = \frac{(15)(15)}{2} = 112.5$$

$$\sigma_U = \sqrt{\frac{(15)(15)(15 + 15 + 1)}{12}} = 24.1$$

Hence, proceeding in the usual manner, we calculate the standardized normal variate

$$z = \frac{U - \mu_U}{\sigma_U} = \frac{187 - 112.5}{24.1} = 3.09$$

Thus, if the test was originally set up as a two-tailed test at, say, the 1% significance level with a critical absolute value for z of 2.58, we would reject the null hypothesis that the samples were drawn from the same populations. On the other hand, if our alternative hypothesis predicted the direction of the difference, we would be dealing with a one-tailed test. For example, suppose our alternative hypothesis had been that applicants from W University had a higher average aptitude test score ranking than did applicants from H University. The calculated z value would be the same as previously ($z = 3.09$), but the critical absolute value for z for a one-tailed test at the 1% significance level is 2.33. Since $3.09 > 2.33$, again we would reject the null hypothesis that the samples were drawn from the same population. Referring to the original sample data in Tables 12-3 and 12-4, we observe that W University has the higher average ranking ($R_2/n_2 = 307/15 = 20.5$ as against $R_1/n_1 = 158/15 = 10.5$). We now accept the alternative hypothesis of the one-tailed test and conclude that the population of applicants from W University has a higher average aptitude test score than does the corresponding population from H University.

General Comments

The above test was carried out in terms of the sum of the ranks for sample number one. That is, the U statistic was defined in terms of R_1. It could similarly have been defined in terms of R_2 as

$$(12.6) \qquad U = n_1 n_2 + \frac{n_2(n_2 + 1)}{2} - R_2$$

The subsequent test would have yielded the same z value as the one previously calculated, except that the sign would change. Of course, the conclusion would be exactly the same.

There were no ties in rankings in the example given. However, if there are such ties, the average rank value is assigned to the tied items. A correction is available for the calculation of σ_U when ties occur, but the effect is generally negligible for large samples.

As mentioned earlier, the rank sum test may be viewed as a substitute for the t test for the difference between two means. The rank sum test may be particularly useful in this connection, because it does not require the restrictive assumptions of the t test. For example, the t test assumption of population normality may not be valid in the example given above.

Exercises

1. The following data represent the profits ($ thousands) generated by 28 foreign exchange traders during the past two months. The first group traded only European currencies, while the second group dealt with only Asian and South American currencies.

   Group 1: 10.4, 9.7, 9.6, 9.3, 8.9, 8.7, 8.2, 7.7, 7.5, 6.9, 6.2, 5.8, 5.5, 5.1
   Group 2: 9.8, 9.5, 8.8, 8.6, 8.4, 8.3, 7.9, 7.8, 7.6, 7.2, 7.1, 6.8, 5.4, 5.3

Use the rank sum procedure to test the hypothesis that the two samples were drawn from populations generating the same average profit. Use $\alpha = 0.01$.

2. A market research director drew simple random samples of 15 sales representatives from each of two sales regions in order to compare sales figures. When last year's dollar values of sales made by these sales representatives were arrayed for the two regions combined, the following rankings emerged:

Region A: 1, 2, 4, 7, 8, 10, 12, 13, 14, 17, 21, 24, 26, 27, 28
Region B: 3, 5, 6, 9, 11, 15, 16, 18, 19, 20, 22, 23, 25, 29, 30

Use the rank sum test at the 0.01 significance level to determine whether there is a significant difference in the average level of sales in the two samples.

3. A machinist was interested in comparing the times needed to complete a certain project using two different methods. The times were recorded under essentially similar situations for both methods. When the times were merged and ranked, the following results were observed:

Method 1: 1, 2, 3, 6, 10, 11, 13, 16, 17, 18, 21, 23, 27
Method 2: 4, 5, 7, 8, 9, 12, 14, 15, 19, 20, 22, 24, 25, 26, 28

Use the rank sum procedure to test the null hypothesis that there is no difference between the true average times for the two different methods. Use $\alpha = 0.05$.

4. A simple random sample of 15 companies was drawn from each of two highly competitive industries. The capital expenditures ($ hundreds of thousands) made by the firms last year are recorded below.

Industry A: 33.3, 18, 38.7, 48, 52, 30, 38.4, 42, 25, 44, 36, 51, 35, 26, 40
Industry B: 46, 17, 24.6, 24.3, 37.8, 39, 14, 23, 33.8, 37.1, 45, 13, 27, 21, 31

Use the rank sum procedure to test the null hypothesis that there is no difference between the average levels of capital expenditures in the two industries. Use $\alpha = 0.05$.

12.5 ONE-SAMPLE TESTS OF RUNS

As we have seen in Chapters 6, 7 and 8, estimation procedures and parametric tests of hypotheses are predicated on the assumption that the observed data have been obtained from random samples. Indeed, in many instances, evidence of nonrandomness can represent an important phenomenon. As an example, in the frontier days of the old Wild West, rather serious consequences were predictable if a card player suspected the randomness of the hands of cards dealt by another player. In many less exotic contexts as well, the randomness of selection of sampled items is of considerable import.

Let us consider a rather oversimplified situation. Suppose that in a certain city, there were about 50% whites and 50% blacks on the rolls of persons eligible for jury duty. Further, let us assume that the following sequence represents the order in which the first 48 persons were drawn from the rolls (B = Black, W = White):

BBBBBBBBBBBB WWWWWWWWWWWW BBBBBBBBBBBB
WWWWWWWWWWWW

Would you suspect, on an intuitive basis, that there was a real question concerning the randomness of selection of these persons? Undoubtedly your answer is in the affirmative, but why? Note that there are 24 Bs and 24 Ws. Hence, the observed proportion of blacks (or whites) is 50% which does not differ from the known population proportion. Your suspicions concerning nonrandomness doubtless stem from the *order* of the items listed, rather than from their frequency of occurrence. Similarly, we would find a perfectly alternating sequence, BWBWBWBW . . . , suspect with respect to randomness of order of occurrence. The *theory of runs* has been developed in order to test samples of data for randomness, with emphasis on the order in which these events occur.

A *run* is defined as a sequence of identical occurrences (symbols) that are followed and preceded by different occurrences (symbols) or by none at all. Hence, in the listing of 48 symbols, there are four runs, the first run consisting of the first 12 Bs, the second run consisting of 12Ws, and so on. Our intuitive feeling is that this represents too few runs. Analogously, in the perfectly alternating series, BWBWBWBW . . . , we would feel that there are too many runs to have occurred on the basis of chance alone.

Method

We illustrate the analytical procedure for the test of runs in terms of a somewhat less extreme illustration than that considered above. Let us assume that the following 42 symbols represent the successive occurrences of births of males (M) and females (F) in a certain hospital.

<u>MM</u> <u>F</u> <u>M</u> <u>FFF</u> <u>MM</u> <u>FF</u> <u>M</u> <u>F</u> <u>MMM</u> <u>FF</u> <u>M</u> <u>FFF</u> <u>MM</u> <u>FF</u> <u>MM</u> <u>FF</u>
<u>MMM</u> <u>FF</u> <u>MM</u> <u>FF</u> <u>MMM</u>

Using the symbol r to denote the number of runs, we have $r = 21$. The runs (which, of course, may be of differing lengths) have been indicated by separation of sequences and by underlining. As we have noted, if there are too few or too many runs, we have reason to doubt that their occurrences are random. The runs test is based on the idea that if there are n_1 symbols of one type and n_2 symbols of a second type, and r denotes the total number of runs, the sampling distribution of r has a mean of

(12.7)
$$\mu_r = \frac{2n_1 n_2}{n_1 + n_2} + 1$$

and a standard deviation of

(12.8)
$$\sigma_r = \sqrt{\frac{2n_1 n_2 (2n_1 n_2 - n_1 - n_2)}{(n_1 + n_2)^2 (n_1 + n_2 - 1)}}$$

If either n_1 or n_2 is larger than 20, the sampling distribution of r is closely approximated by the normal distribution. Hence, we can compute

$$z = \frac{r - \mu_r}{\sigma_r}$$

and proceed with the test in the usual manner.

In the present problem, where there are $n_1 = 22$ Ms, $n_2 = 20$ Fs, and $r = 21$, we have

$$\mu_r = \frac{(2)(22)(20)}{22 + 20} + 1 = 21.95$$

and

$$\sigma_r = \sqrt{\frac{(2)(22)(20)[(2)(22)(20) - 22 - 20]}{(22 + 20)^2 (22 + 20 - 1)}} = 3.19$$

Therefore,

$$z = \frac{21 - 21.95}{3.19} = -0.30$$

Testing at a significance level of, say, 5%, where a critical absolute value of 1.96 for z is required for rejection of the null hypothesis, we find that the randomness hypothesis cannot be rejected. In other words, the number of runs in this case is neither small enough nor large enough to lead us to conclude that the sequence of male and female births is nonrandom.

General Comments

The runs test has many applications, including cases where the sequential data are numerical rather than in the form of attributes such as the symbols of the above illustrations. Thus, for example, runs tests could be applied to sequences of random numbers, such as those in Table 4-1 on page 186. Such tests might be applied, say, to sequences of random numbers generated on computers. One form of the test might be in terms of runs of numbers above the median and runs of numbers below the median. For the digits 0, 1, 2, 3, 4, 5, 6, 7, 8, and 9, the median is 4.5. Hence, runs could be determined for digits that fall above and below the median. The test could also be applied in terms of runs of odd-numbered digits and runs of even-numbered ones. Another alternative is to group the numbers into pairs of digits, so that the possible occurrences are 00, 01, . . . , 99. Here the median is 49.5, and similar runs tests could be applied to those suggested for the one-digit case. It is clear that with a bit of imagination, an analyst can devise many useful and easily applied versions of the runs test.

Exercises

1. In the table of random numbers (Table 4-1 on page 186), consider the 50 digits in the first line. Label the even digits a and the odd digits b. Carry out a runs test at both the 0.05 and 0.01 levels of significance.

2. Consider the same 50 digits as in Exercise 1. Now label the digits a or b depending upon whether they are above or below the theoretical mean, 4.5. Carry out a runs test at both the 0.05 and 0.01 levels of significance.

3. The following figures represent the monthly numbers of on-the-job accidents occurring in a certain factory over a 24-month period:

$$2, 2, 1, 3, 4, 4, 1, 3, 1, 2, 4, 4$$
$$2, 3, 3, 2, 5, 6, 4, 5, 4, 6, 5, 5$$

Determine the median of this set of 24 figures. Label the numbers above and below the median a and b, respectively. Perform a runs test at the 0.01 level of significance on the series of as and bs. This type of test of runs above and below the median is particularly useful for determining the existence of trend patterns in data. If there is a trend, as will tend to appear in the early part of the series and bs in the later part, or vice versa.

4. The one-sample runs test is sometimes used to test series of stock price changes for evidence of randomness to support a random walk theory. A certain stock has shown the following behavior during the past 25 business days: $+ + + - - + - - - + + - + - + - - + + - + + - + -$ (+ indicates a price increase, and − indicates a price decrease). Carry out a runs test at both the 0.05 and 0.01 levels of significance. Would you consider the stock price changes to be random events?

12.6 ONE-FACTOR ANALYSIS OF VARIANCE BY RANKS (KRUSKAL–WALLIS TEST)

In Section 8.3, it was noted that the one-factor analysis of variance represents an extension of the two-sample test for means and provides a test for whether several independent samples can be considered to have been drawn from populations having the same mean. Analogously, the Kruskal–Wallis one-factor analysis of variance by ranks is a nonparametric test that represents a generalization of the two-sample Mann–Whitney U rank sum test. The parametric analysis of variance discussed in Section 8.3 involves the assumption that the populations are normally distributed; otherwise, the F-test procedure is invalid. The Kruskal–Wallis test makes no assumptions about the population distribution.

The Kruskal–Wallis test is based upon a test statistic calculated from ranks established by pooling the observations from c independent simple random samples, where $c > 2$. The null hypothesis is that the populations are identically distributed, or alternatively, that the samples were drawn from c identical populations. We illustrate the procedure for the test by the following example.

Simple random samples of corporate treasurers in a certain industry were drawn from firms classified into three size categories (large, medium, and small). These executives, after being assured of the confidentiality of their replies, were asked to rate the overall quality of performance of the Federal Reserve Board in discount rate policy during the past six-month period on a scale from 0 to 100, with 0 denoting the lowest quality rating and 100 the highest. The scores, classified by size of firm, and the rankings of the pooled sample scores are shown in Table 12-5. The result was the following pooled ranking, with the lowest score that was actually given being represented by rank 1 and the highest by rank $n = 20$ (where n denotes the total number of pooled sample observations):

Score: 50 60 61 62 65 68 70 72 73 75 77 78
       80 82 84 85 87 90 93 95
Rank: 1 2 3 4 5 6 7 8 9 10 11 12
       13 14 15 16 17 18 19 20

TABLE 12-5
Calculations for the Kruskal–Wallis One-Factor Analysis of Variance: Scores and Ranks Classified by Size of Firm

Large Firms (Group 1)		Medium-Sized Firms (Group 2)		Small Firms (Group 3)	
Score	Rank	Score	Rank	Score	Rank
78	12	68	6	82	14
95	20	77	11	65	5
85	16	84	15	50	1
87	17	61	3	93	19
75	10	62	4	70	7
90	18	72	8	60	2
80	13			73	9
$n_1 = 7$		$n_2 = 6$		$n_3 = 7$	
$R_1 = 106$		$R_2 = 47$		$R_3 = 57$	

The test statistic involves a comparison of the variation of the ranks of the sample groups. The Kruskal–Wallis test statistic is

$$(12.9) \qquad K = \frac{12}{n(n+1)} \left(\Sigma \frac{R_j^2}{n_j} \right) - 3(n+1)$$

where n_j = the number of observations in the jth sample

$n = n_1 + n_2 + \cdots + n_c$ = the total number of observations in the c samples

R_j = the sum of the ranks for the jth sample

The sample sizes and rank sums are shown in Table 12-5 for each sample group. Substituting into Equation 12.9, we compute the K statistic in the present example.

$$K = \frac{12}{20(20+1)} \left(\frac{106^2}{7} + \frac{47^2}{6} + \frac{57^2}{7} \right) - 3(20+1) = 6.641$$

It can be shown that the sampling distribution of K is approximately the same as the χ^2 distribution with $\nu = c - 1$ degrees of freedom (where c is the number of sample groups). In this example, where there are three sample groups, the number of degrees of freedom is $\nu = c - 1 = 3 - 1 = 2$. Testing the null hypothesis at the 5% level of significance ($\alpha = 0.05$) and using Table A-7 of Appendix A, we find the critical value of χ^2 to be $\chi^2_{0.05} = 5.991$. Hence, our decision rule is

If $K > 5.991$, reject the null hypothesis
If $K \leqslant 5.991$, accept the null hypothesis

Since $K = 6.641$ is greater than the critical value of 5.991, we reject the null hypothesis of identically distributed populations. Therefore, in terms of the example, we conclude that there are significant differences by size of firm in the scores assigned by these three samples of corporate treasurers. Looking back at the scores given in Table 12-5, we find that treasurers of large firms tended to assign higher scores than did their counterparts in medium-sized or small firms.

Further Remarks

As in other nonparametric tests, where there are ties, observations are assigned the mean of the tied ranks. In the case of ties, a corrected K value, K_c, should be computed as follows:

$$(12.10) \qquad K_c = \frac{K}{1 - \left[\dfrac{\Sigma\,(t_j{}^3 - t_j)}{(n^3 - n)} \right]}$$

where t_j is the number of tied scores in the jth sample.

Furthermore, for the χ^2 distribution to be applicable, the sample sizes (that is, the n_j values) should all be greater than 5.

Exercises

1. The numbers of units produced by simple random samples of workers in three different plants are given in the table below. The employees produced the same product in the same number of hours and under essentially the same conditions. Use the Kruskal–Wallis analysis of variance by ranks test to determine whether the three populations may be considered to be identically distributed. Use $\alpha = 0.05$.

Number of Units Produced		
Plant A	Plant B	Plant C
68	97	104
92	116	125
120	121	65
111	117	101
119	72	88
74	82	81
85	110	108
105		114

2. A manufacturer purchased large batches of a product from four subcontractors. There were the same number of articles in each batch. The table below gives the numbers of defective articles in each batch. Would you conclude that there was no difference among subcontractors in numbers of defectives per batch? Use the Kruskal–Wallis analysis of variance by ranks test at a 1% significance level.

Number of Defectives per Batch			
Subcontractor A	Subcontractor B	Subcontractor C	Subcontractor D
12	30	15	18
6	28	17	27
10	7	20	13
0	25	3	22
2	24	19	16
4	29	21	23
	31	8	
	14		

3. In a test, the marketing research department of a firm sent three differently designed advertisements to equal-sized simple random samples of potential customers in six cities. The numbers of units of sales that resulted from business reply cards attached to these advertisements are given on page 539.

| | Numbers of Units Sold | | |
City	Design A	Design B	Design C
1	38	64	55
2	59	75	82
3	30	36	80
4	52	77	66
5	61	69	73
6	43	67	47

a. Using the Kruskal–Wallis analysis of variance by ranks, test the hypothesis of no difference in effectiveness among the three advertisement designs in terms of numbers of sales that resulted. Use $\alpha = 0.01$.

b. Assuming that the normal distribution assumptions of analysis of variance are satisfied, use the parametric analysis of variance discussed in Section 8.3 to test the hypothesis of no difference in average numbers of sales resulting from the three advertisements. Use $\alpha = 0.01$.

12.7 RANK CORRELATION

Nonparametric procedures can be useful in correlation analysis where the basic data are not available in the form of numerical magnitudes but where rankings can be assigned. If two variables of interest can be ranked in separate ordered series, a *rank correlation coefficient* can be computed; this is a measure of the degree of correlation that exists between the two sets of ranks. We illustrate the method in terms of a simple random sample of individuals for whom such rankings have been established for two variables referring to ability in two different sports activities.

Method

For illustrative purposes, we consider two extreme cases, the first representing perfect *direct* correlation between two series, the second perfect *inverse* correlation. Table 12-6 displays data on the rankings of a simple random sample of ten individuals according to playing abilities in baseball and tennis. Clearly, this represents a case in which it would be extremely difficult, if not impossible, to obtain precise quantitative measures of these abilities, but in which rankings may be feasible. In rank correlation analysis, the rankings may be assigned in order from high to low, with 1 representing the highest rating, 2 next highest, etc., or 1 may

TABLE 12-6
Rank Correlation of Baseball Playing Ability with Tennis Playing Ability (Perfect Correlation Case)

Individual	Rank in Baseball Ability X	Rank in Tennis Ability Y	Difference in Ranks $d = X - Y$	$d^2 = (X - Y)^2$
A	1	1	0	0
B	2	2	0	0
C	3	3	0	0
D	4	4	0	0
E	5	5	0	0
F	6	6	0	0
G	7	7	0	0
H	8	8	0	0
I	9	9	0	0
J	10	10	0	0
Total				0

$$r_r = 1 - \frac{6\Sigma\ d^2}{n(n^2 - 1)} = 1 - \frac{6(0)}{10(10^2 - 1)} = 1$$

represent the lowest rank, 2 the next to lowest, etc. The computed rank correlation coefficient will be the same regardless of the rank ordering used. Let us assume in this case that 1 represents the highest or best rank, 2 the second highest, etc.

The rank correlation coefficient (also referred to as the Spearman rank correlation coefficient) can be derived mathematically from one of the formulas for r, the sample correlation coefficient discussed in Chapter 9, where ranks are used for the observations of X and Y. We will use the symbol r_r to denote the rank correlation coefficient. It is computed by the following formula:

(12.11)
$$r_r = 1 - \frac{6\Sigma\ d^2}{n(n^2 - 1)}$$

where d = difference between the ranks for the paired observations
n = number of paired observations

The calculations of the rank correlation coefficients for the two extreme cases mentioned earlier are shown in Tables 12-6 and 12-7. In Table 12-6, there is perfect direct correlation in the rankings. That is, the individual who ranks highest in baseball playing ability is also best in tennis, etc. On the other hand, in Table 12-7, there is perfect inverse correlation. That is, the individual who ranks highest in baseball

TABLE 12-7
Rank Correlation of Baseball Playing Ability with Tennis Playing Ability (Perfect Inverse Correlation Case)

Individual	Rank in Baseball Ability X	Rank in Tennis Ability Y	Difference in Ranks $d = X - Y$	$d^2 = (X - Y)^2$
A	1	10	-9	81
B	2	9	-7	49
C	3	8	-5	25
D	4	7	-3	9
E	5	6	-1	1
F	6	5	1	1
G	7	4	3	9
H	8	3	5	25
I	9	2	7	49
J	10	1	9	81
Total				330

$$r_r = 1 - \frac{6\Sigma\,d^2}{n(n^2 - 1)} = 1 - \frac{6(330)}{10(10^2 - 1)} = -1$$

playing ability is worst in tennis, etc.   It may be noted from the calculations shown in the two tables that in the case of perfect direct correlation between the ranks, $r_r = 1$; in the case of perfect inverse correlation, $r_r = -1$.   This is not surprising, because the rank correlation coefficient is derived mathematically from the sample correlation coefficient, r. Hence, the range of possible values of these coefficients is the same.   An r_r value of 0 would analogously indicate no correlation between the ranks.   Tied ranks are handled in the calculations by averaging in the usual way.

The significance of the rank correlation may be tested in the same way as for the sample correlation coefficient r.   That is, we compute the statistic

(12.12)
$$t = \frac{r_r}{\sqrt{(1 - r_r^2)/(n - 2)}}$$

which has a t distribution for $n - 2$ degrees of freedom.   Hence, for example, suppose in an example such as the one above, r_r had been computed to be 0.90.   Then substitution into (12.12) would yield

$$t = \frac{0.90}{\sqrt{(1 - 0.81)/(10 - 2)}} = 5.84$$

Let us assume that we are using a two-tailed test of the null hypothesis of zero correlation in the ranked data of the population. Then referring to Table A-6 of Appendix A, we find critical t values of 2.306 and 3.355 at the 5% and 1% levels of significance, respectively. Hence, we would reject the hypothesis of no rank correlation at both levels and conclude that there is a positive linear relationship between the rankings in baseball playing ability and tennis playing ability.

Exercises

1. The following were the performance rankings of ten executives of a commercial bank during an economic boom and an economic recession.

Officer	Rank during Boom	Rank during Recession
A	1	5
B	2	6
C	3	9
D	4	3
E	5	8
F	6	4
G	7	10
H	8	1
I	9	7
J	10	2

Calculate the rank correlation coefficient for the above set of data.

2. The data given below represent the rankings of a simple random sample of companies with regard to total sales and return on equity:

Company	Sales Rank	Return on Equity Rank
A	4	7
B	6	4
C	12	9
D	2	3
E	7	8
F	1	5
G	9	11
H	11	12
I	3	1
J	8	6
K	5	2
L	10	10

Calculate the rank correlation coefficient for the above set of data.

3. The following table shows the results of a survey taken by an insurance company to examine the relationship between income and amount of life insurance held by heads of families. A simple random sample of ten family heads was drawn.

Family	Amount of Insurance ($Thousands)	Income ($Thousands)
A	9	10
B	20	14
C	22	15
D	15	14
E	17	14
F	30	25
G	18	12
H	25	16
I	10	12
J	20	15

Calculate the rank correlation coefficient. Assign average rank values to tied items.

Decision Making Using Prior Information

13.1 INTRODUCTION

In recent years, there has been a shift of emphasis from classical, or traditional, statistical inference, to the problem of decision making under conditions of uncertainty. This modern formulation has come to be known as *statistical decision theory* or *Bayesian decision theory*. The latter term is often used to emphasize the role of Bayes' Theorem in this type of decision analysis. The two ways of referring to modern decision analysis have come to be used interchangeably and will be so used henceforth in this book. Statistical decision theory has developed into an important model for making rational selections among alternative courses of action when information is incomplete and uncertain. It is a prescriptive theory rather than a descriptive one. That is, it presents the principles and methods for making the best decisions under specified conditions, but it does not purport to describe how actual decisions are made in the real world.

13.2 STRUCTURE OF THE DECISION-MAKING PROBLEM

Managerial decision making has increased in complexity as the economy of the United States and the business units within it have grown larger and more intricate. However, Bayesian decision theory is based on the assumption that regardless of the type of decision (whether it involves long- or short-range consequences; whether it is in finance, production, marketing, or some other area; whether it is at a relatively high or low level of managerial responsibility), certain common characteristics of the decision problem can be discerned. These characteristics constitute the formal description of the problem and provide the structure for a solution. The decision problem under study may be represented by a model in terms of the following elements:

1. *The decision maker.* The agent charged with the responsibility for making the decision. The decision maker is viewed as an entity and may be a single individual, a corporation, a government agency, etc.

2. *Alternative courses of action.* The decision involves a selection among two or more alternative courses of action, referred to simply as "acts." The problem is to choose the best of these alternative acts. Sometimes the decision maker's problem is to choose the best of alternative "strategies," where each strategy is a decision rule indicating which act should be taken in response to a specific type of experimental or sample information.

3. *Events.* Occurrences that affect the achievement of the objectives. These are viewed as lying outside the control of the decision maker, who does not know for certain which event will occur. The events constitute a mutually exclusive and complete set of outcomes; hence, one and only one of them can occur. Events are also referred to as "states of nature," "states of the world," or simply "outcomes."

4. *Payoff.* A measure of net benefit to be received by the decision maker under particular circumstances. These payoffs are summarized in a *payoff table* or *payoff matrix*, which displays the consequences of each action selected and each event that occurs.

5. *Uncertainty.* The indefiniteness concerning which events or states of nature will occur. This uncertainty is indicated in terms of probabilities assigned to events. One of the distinguishing characteristics of Bayesian decision theory is the assignment of personalistic, or subjective, probabilities as well as other types of probabilities.

The payoff table, expressed symbolically in general terms, is given in Table 13-1. It is assumed that there are n alternative acts, denoted A_1, A_2, . . . , A_n. These different possible courses of action are listed as column headings in the table. There are m possible events or states of

TABLE 13-1
The Payoff Table

Event	A_1	A_2	Acts $\cdots$	A_n
θ_1	u_{11}	u_{12}	$\cdots$	u_{1n}
θ_2	u_{21}	u_{22}	$\cdots$	u_{2n}
.	.	.	$\cdots$	.
.	.	.	$\cdots$	.
.	.	.	$\cdots$	.
θ_m	u_{m1}	u_{m2}	$\cdots$	u_{mn}

nature, denoted θ_1, θ_2, . . . , θ_m. The payoffs resulting from each combination of an act and an event are designated by the symbol u with appropriate subscripts. The letter u has been used because it is the first letter of the word "utility." The net benefit, or payoff, of the selection of an act and the occurrence of a state of nature can be treated most generally in terms of the utility of this consequence to the decision maker. How these utilities are arrived at is a technical matter and is discussed later in this chapter. In summary, the utility of selecting act A_1 and having event θ_1 occur is denoted u_{11}; the utility of selecting act A_2 and having event θ_1 occur is u_{12}, and so on. Note that the first subscript in these utilities indicates the event that prevails and the second subscript denotes the act chosen. A convenient general notation is the symbol u_{ij}, which denotes the utility of selecting act A_j if subsequently event θ_i occurs. The rows of a table (or matrix) are commonly denoted by the letter i, where i can take on values 1, 2, . . . , m, and the columns by j, where j can take on values 1, 2, . . . , n.

If the event that will occur (for example, θ_3) were known with certainty beforehand, then the decision maker could simply look along row θ_3 in the payoff table and select the act that yields the greatest payoff. However, in the real world, since the states of nature lie beyond the control of the decision maker, one ordinarily does not know with certainty which specific event will occur. The choice of the best course of action in the face of this uncertainty is the crux of the decision maker's problem.

13.3 AN ILLUSTRATIVE EXAMPLE

In order to illustrate the ideas discussed in the preceding section, we will take a simplified business decision problem. (This problem will also be continued in later sections to exemplify other principles of deci-

TABLE 13-2
Payoff Table for the Inventor's Problem (in Units of $10,000 Profit)

Event	A_1 Manufacture Device Himself	A_2 Sell Patent Rights
θ_1: Strong sales	$80	$40
θ_2: Average sales	20	7
θ_3: Weak sales	−5	1

sion analysis.) A man has invented and patented a new device, and a bank is willing to lend him the money to manufacture the device himself. After some preliminary investigation, it is decided that the next five years is a suitable planning period for the comparison of payoffs from this invention. According to the inventor's analysis, if sales are strong, he anticipates profits of $800,000 over the next five years; if sales are average, he expects to make $200,000; and if they are weak, he expects to lose $50,000. A company, Nationwide Enterprises, Inc., has offered to purchase the patent rights from him. Based on the royalty arrangement offered to him, the inventor estimates that if he sells the patent rights and if sales are strong he can anticipate a net profit of $400,000; if sales are average, $70,000; and if weak, $10,000. The payoff table for the inventor's problem is given in Table 13-2.

In this problem, the alternative acts, denoted A_1 and A_2, respectively, are for the inventor to manufacture the device himself or to sell the patent rights. The events or states of nature (denoted θ_1, θ_2, and θ_3, respectively) are strong sales, average sales, and weak sales for the five-year planning period. The payoffs are in terms of the net profits that would accrue to the inventor under each act–event combination. In order to keep the numbers rather simple in this problem, the payoffs have been stated in units of $10,000; hence, a net profit of $800,000 has been recorded as $80, a net loss of $50,000 has been entered as −$5, etc.[1]

The types of events or states of nature used in the inventor's problem are, of course, very simplified. Generally, an unlimited number of possible events could occur in the future relating to such matters as the customers, technological change, competitors, etc., which

[1] It is good practice in the comparison of economic alternatives to compare the present values of discounted cash flows or, what amounts to the same thing, equivalent annual rates of return. Both of these methods take into account the time value of money; that is, the fact that a dollar received today is worth more than a dollar received in some future period. These are conceptually the types of monetary payoff values that should appear in the payoff table. This point is amply discussed in standard texts dealing with economy studies or investment analysis. To avoid a lengthy tangential discussion, we will not elaborate on the point here.

lie beyond the decision maker's control. These may all be considered states of nature that affect the potential payoffs of the alternative decisions to be made. However, in order to cut our way through the maze of complexities involved, and to construct a manageable framework of analysis for the problem, we can think of the variable "demand" as the resultant of all of these other underlying factors.

In the inventor's problem, three different levels of demand (strong, average, and weak) have been distinguished. It is helpful in this regard to think of "demand" as a variable. In the present problem, demand is a discrete random variable that can take on three possible values. Demand could have been conceived of as a discrete random variable taking on any finite or infinite number of values. (For example, it could have been stated in numbers of units demanded or, say, in hundreds of thousands of units demanded.) Demand can also be treated as a continuous rather than discrete variable. The conceptual framework of the solution to the decision problem remains the same, but the required mathematics differs somewhat from the case where the events are stated in the form of a discrete variable.

13.4 CRITERIA OF CHOICE

Assuming that the inventor in our illustrative problem has carried out the thinking, experiments, data collection, etc., required to construct the payoff matrix (Table 13-2), how should he now compare the alternative acts? Neither act is preferable to the other under all states of nature. For example, if event θ_1 occurs, that is, if sales are strong, the inventor would be better off to manufacture the device himself (act A_1), realizing a profit of $800,000, as compared to selling the patent rights (act A_2), which would yield a profit of only $400,000. On the other hand, if event θ_3 occurs, and sales are weak, the preferable course of action would be to sell the patent rights, thereby earning a profit of $10,000, as compared to a loss of $50,000. If the inventor knew with *certainty* which event was going to occur, his decision procedure would be very simple. He would merely have to look along the row represented by that event and select the act that yielded the highest payoff. However, it is the *uncertainty* with regard to which state of nature will prevail that makes the decision problem an interesting one.

Maximin Criterion

A number of different criteria for selecting the best act have been suggested. One of the earliest suggestions, made by the mathematical statistician Abraham Wald,[2] is known as the *maximin criterion*. Under

[2] Abraham Wald, *Statistical Decision Functions* (New York: John Wiley & Sons, 1950).

this method, the decision maker assumes that once a course of action has been chosen, nature might be malevolent and hence might select the state of nature that minimizes the decision maker's payoff. According to Wald, the decision maker should choose the act that maximizes the payoff under this pessimistic assumption concerning nature's activity. In other words, Wald suggested that a selection of the "best of the worst" is a reasonable form of protection. By this criterion, if the inventor chose act A_1, nature would cause event θ_3 to occur and the payoff would be a loss of $50,000. If the decision maker chose A_2, nature would again cause θ_3 to occur, since that would yield the worst payoff, in this case, a profit of $10,000. Comparing these worst, or minimum, payoffs, we have

	Action	
	A_1	A_2
Minimum payoffs (in units of $10,000)	−5	1

The decision maker is now supposed to do the best he can in the face of this sort of perverse nature, and select the act that yields the greatest minimum payoff, namely, act A_2. That is, he should sell the patent rights, for which the minimum payoff is $10,000. Thus, the proposed decision procedure is to choose the act that yields the maximum of the minimum payoffs—hence the use of the term "maximin."

Obviously, the maximin is a very pessimistic type of criterion. It is not reasonable to suppose that the usual executive would or should make decisions in this way. By following this decision procedure, the executive would always be concentrating on the worst things that could happen. In most situations, the maximin criterion would freeze the decision maker into complete inaction and would imply that it would be best to go out of business entirely. For example, let us consider an inventory stocking problem, where the events are possible levels of demand, the acts are possible stocking levels (that is, the numbers of items to be stocked), and the payoffs are in terms of profits. If no items are stocked, the payoffs will be 0 for every level of demand. For each of the other numbers of items stocked, we can assume that for some levels of demand, losses will occur. Since the worst that can happen if no items are stocked is that no profit will be made, whereas under all other courses of action the possibility of a loss exists, the maximin criterion would require the firm to carry no stock or, in effect, go out of business. Such a procedure is not necessarily irrational, and it might be consistent with some people's attitudes toward risk. However, for the person who is willing to take some risks, such an arbitrary decision rule would be completely unacceptable. A number of other decision criteria have been suggested by various writers, but, to avoid a lengthy digression, they will not be discussed here.[3]

It seems reasonable to argue that a decision maker should take into account the probabilities of occurrence of the different possible states of nature. As an extreme example, if the state of nature that results in the minimum payoff for a given act has (say) only one chance in a million of occurring, it would seem unwise to concentrate on the possibility of this occurrence. The decision procedures we will focus upon include the probabilities of states of nature as an important part of the problem.

Expected Profit under Uncertainty

In a realistic decision-making situation, it is reasonable to suppose that a decision maker would have some idea of the likelihood of occurrence of the various states of nature and that this knowledge would help in choosing a course of action. For example, in our illustrative problem, if the inventor felt very confident that sales would be strong, this would tend to move him toward manufacturing the device himself, since the payoff under that act would exceed that of selling the patent rights. By the same reasoning, if he were very confident that sales would be weak, he would be influenced to sell the patent rights. If there are many possible events and many possible courses of action, the problem becomes complex, and the decision maker clearly needs some orderly method of processing all the relevant information. Such a systematic procedure is provided by the computation of the *expected* monetary value of each course of action, and the selection of the act that yields the highest of these expected values. As we shall see, this procedure yields reasonable results in a wide class of decision problems. Furthermore, we will see how this method can be adjusted for the computation of expected utilities rather than expected monetary values in cases where the maximization of expected monetary values is not an appropriate criterion of choice.

We now return to the inventor's problem to illustrate the calculations for decision making by maximization of the expected monetary value criterion. In this case, the maximization takes the form of selecting the act that yields the largest expected profit. Let us assume that the inventor carries out the following probability assignment procedure. On the basis of extensive investigation of experience with similar devices in the past, and on the basis of interviews with experts, the inventor comes to the conclusion that the odds are 50:50 that sales will be average (that is, that the event we previously designated as θ_2 will occur). Furthermore, he concludes that it is somewhat less likely that sales will

---

[3] See, for example, Chapter 5 of D. W. Miller and M. K. Starr, *Executive Decisions and Operations Research* (Englewood Cliffs, N.J.: Prentice-Hall, 1960).

be strong (event θ_1) than that they will be weak (event θ_3). On this basis, the inventor assigns the following subjective probability distribution to the events in question:

Event	Probability
θ_1: Strong sales	0.2
θ_2: Average sales	0.5
θ_3: Weak sales	0.3
	1.0

In order to determine the basis for choice between the inventor's manufacturing the device himself (act A_1) and selling the patent rights (act A_2), we compute the expected profit for each of these courses of action. These calculations are shown in Table 13-3. As indicated in that table, profit is treated as a variable that takes on different values depending upon which event occurs. We compute its expected value in the usual way, according to Equation 3.11. The "expected value of an act" is the weighted average of the payoffs under that act, where the weights are the probabilities of the various events that can occur.

TABLE 13-3
Inventor's Expected Profits (in Units of $10,000 Profit)

Event	Act A_1: Manufacture Device Himself Probability	Profit	Weighted Profit
θ_1: Strong sales	0.2	$80	$16.0
θ_2: Average sales	0.5	20	10.0
θ_3: Weak sales	0.3	−5	−1.5
	1.0		$24.5

Expected profit = $24.5 ($ten thousands)
= $245,000

Event	Act A_2: Sell Patent Rights Probability	Profit	Weighted Profit
θ_1: Strong sales	0.2	$40	$ 8.0
θ_2: Average sales	0.5	7	3.5
θ_3: Weak sales	0.3	1	0.3
	1.0		$11.8

We see from Table 13-3 that the inventor's expected profit if he manufactures the device himself is $245,000, whereas if he sells the patent rights, the expected profit is only $118,000. If he acts to maximize his expected profit, the inventor will select A_1; that is, he will manufacture the device himself.

It is useful to have a brief term to refer to the expected benefit of choosing the optimal act under conditions of uncertainty. We shall refer to the expected value of the monetary payoff of the best act as the *expected profit under uncertainty*. Hence, in the foregoing problem, the expected profit under uncertainty is $245,000.

We can summarize the method of calculating the *expected profit under uncertainty* as follows:

1. Calculate the expected profit for each act as the weighted average of the profits under that act, where the weights are the probabilities of the various events that can occur.

2. The expected profit under uncertainty is the maximum of the expected profits calculated under Step 1.

Expected Opportunity Loss

A useful concept in the analysis of decisions under uncertainty is that of *opportunity loss*. An opportunity loss is the loss incurred because of failure to take the best possible action. Opportunity losses are calculated separately for each event that might occur. Given the occurrence of a specific event, we can determine the best possible act. For a given event, the opportunity loss of an act is the difference between the payoff of that act and the payoff for the best act that could have been selected. Thus, for example, in the inventor's problem, if event θ_1 (strong sales) occurs, the best act is A_1, for which the payoff is $80 (in units of $10,000). The opportunity loss of that act is $80 − $80 = $0. The payoff for Act A_2 is $40. The opportunity loss of Act A_2 is the amount by which the payoff of the best act, $80, exceeds the $40 payoff of Act A_2. Hence, the opportunity loss of Act A_2 is $80 − $40 = $40.

It is convenient to asterisk the payoff of the best act for each event in the original payoff table in order to denote that opportunity losses are measured from these figures. Both the original payoff table and the opportunity loss table are given in Table 13-4 for the inventor's problem.

We can now proceed with the calculation of expected opportunity loss in a manner completely analogous to the calculation of expected profits. Again, we use the probabilities of events as weights and determine the weighted average opportunity loss for each act. Our goal is to select the act that yields the *minimum* expected opportunity loss. The calculation of the expected opportunity losses for the two acts in the inventor's problem is given in Table 13-5. The symbol EOL will be used to represent expected opportunity loss. Hence, $EOL(A_1)$ and $EOL(A_2)$

TABLE 13-4
Payoff Table and Opportunity Loss Table for the Inventor's Problem (in Units of $10,000)

Event	Payoff Table Acts		Opportunity Loss Table Acts	
	A_1	A_2	A_1	A_2
θ_1: Strong sales	$80*	$40	$0	$40
θ_2: Average sales	20*	7	0	13
θ_3: Weak sales	−5	1*	6	0

denote the expected opportunity losses of acts A_1 and A_2, respectively. The inventor's EOL if he manufactures the device himself is $18,000, and if he sells the patent rights, his EOL is $145,000. If he selects the act that minimizes his EOL, he will choose A_1, that is, to manufacture the device himself. This is the same act that he selected under the criterion

TABLE 13-5
Expected Opportunity Losses in the Inventor's Problem (in Units of $10,000)

	Act A_1: Manufacture Device Himself		
Event	Probability	Opportunity Loss	Weighted Opportunity Loss
θ_1: Strong sales	0.2	0	0
θ_2: Average sales	0.5	0	0
θ_3: Weak sales	0.3	6	1.8
	1.0		1.8

$$EOL(A_1) = 1.8 \text{ (\$ten thousands)}$$
$$= \$18,000$$

	Act A_2: Sell Patent Rights		Weighted
Event	Probability	Opportunity Loss	Opportunity Loss
θ_1: Strong sales	0.2	40	8.0
θ_2: Average sales	0.5	13	6.5
θ_3: Weak sales	0.3	0	0
	1.0		14.5

$$EOL(A_2) = 14.5 \text{ (\$ten thousands)}$$
$$= \$145,000$$

of maximizing expected profit. It can be proved that the best act according to the criterion of maximizing expected profit is also best if the decision maker follows the criterion of minimizing expected opportunity loss. The relationship between the maximum expected profit and the minimum expected opportunity loss will be examined later. It should be noted that opportunity losses are not losses in the accountant's sense of profit and loss, because as we have seen, they even occur where only profits of different actions are compared for a given state. They represent opportunities foregone rather than monetary losses incurred.

Minimax Opportunity Loss

It was noted earlier that various criteria of choice have been suggested for the decision problem. One that has been advanced in terms of opportunity losses is that of *minimax opportunity loss*. Under this method, the decision maker selects the act that minimizes the worst possible opportunity loss that can be incurred among the various acts. As with the maximin criterion for payoffs, the minimax criterion for opportunity loss takes a pessimistic view toward which states of nature will occur. Once the opportunity loss table has been prepared as in Table 13-4, the decision maker determines for each act the largest opportunity loss that can be incurred. For example, in the inventor's problem, under act A_2, it is \$400,000. The decision maker then chooses that act for which these worst possible losses are the least, that is, the act that minimizes the maximum losses; hence, the use of the term "minimax." In the inventor's problem, the decision maker would choose act A_1, since the maximum possible opportunity loss (\$60,000) under this course of action is less than the corresponding worst loss under act A_2 (\$400,000). This criterion is also sometimes referred to as "minimax regret," where the opportunity losses are viewed as measures of regret for the taking of less than the best courses of action.[4]

In our illustrative problem, it happens that the minimax opportunity loss act, A_1, is the same decision as would be made under the criterion of maximizing expected profit. However, this is not necessarily so. The minimax loss criterion, like the maximin payoff rule, singles out for each course of action the worst consequence that can befall the decision maker, and then attempts to minimize this damage. As was true for the maximin payoff criterion, the minimax loss viewpoint yields results in many instances which imply that a business executive who faces risky ventures should simply go out of business.

Throughout the remainder of this book, we will use the Bayesian decision theory criterion of maximizing expected profit or its equivalent, minimizing expected opportunity loss.

[4] See L. J. Savage, "The Theory of Statistical Decision," *Journal of the American Statistical Association,* **46,** 55–57, 1951.

Exercises

1. If the possible states of nature are: the competitor will set the price
   a. higher than yours
   b. the same as yours
   c. lower than yours
   what is wrong with assessing prior probabilities as 0.6, 0.3, and 0.2, respectively?

2. A new appliance store finds that in its first week of business it sold five major appliances, ten home appliances, and 30 small appliances. Based solely on this past knowledge, what prior probability distribution would you formulate for the type of appliance to be sold?

3. As manager of a plant, you must decide to invest $25,000 in either a cost reduction program or a new advertising campaign. Assume that you know the cost reduction program will increase the profit-to-sales ratio from the present 10% to 11%. The sales campaign, if successful, is expected to increase the present $2 million of sales by 12%. The probability that the campaign will be successful is 0.8. Which is the better course of action?

4. R.B.A., Inc. is given the opportunity to submit a closed bid to the government to build certain electronic equipment. An examination of similar proposals made in the past revealed that the average profit per successful bid was $175,000 and that R.B.A., Inc. received the contract (that is, had the lowest bid) on 10% of its submitted bids. The cost of preparing a bid is, on the average, $10,000. Should R.B.A., Inc. prepare a bid?

5. In Exercise 4, suppose R.B.A., Inc. chose to prepare a bid. For this particular proposal, assume the company finds it can submit only the following four bids: $1,600,000, $1,700,000, $1,800,000, or $1,900,000. At $1,600,000, the expected profit is $160,000. Each successive bid yields an increase in profit equal to the increase in the bid. From an examination of past accounting records, R.B.A., Inc. assesses the probabilities that the bids will be the lowest ones as 0.4, 0.3, 0.2, and 0.1, respectively. Which bid should be submitted?

6. A retailer must decide how much inventory to carry, and the decision is dependent upon demand. Since the stock is perishable, a loss results from an overstock. Because of space limitations, the retailer can stock at most five items. The cost per item is $1, and the selling price is $5. The profit table and probabilities are

Probability	Number Demanded	Units Stocked					
		0	1	2	3	4	5
2/20	0	$0	$−1	$−2	$−3	$−4	$−5
3/20	1	0	4	3	2	1	0
5/20	2	0	4	8	7	6	5
5/20	3	0	4	8	12	11	10
4/20	4	0	4	8	12	16	15
1/20	5	0	4	8	12	16	20

What is the optimal stocking level and what is the retailer's expected profit?

7. An operations research team is trying to decide whether to put the predictions of the ten leading investment advice newsletters into an information system it is building. The cost of including the predictions is $4850 a year. It is estimated that in 20 decisions to be made in a year, the added information would result in a new decision only once. However, the change in decision would result on the average in a saving of $75,000. Should the team include the newsletter in its information system?

8. A company has $100,000 available to invest. The company can either build a new plant or put the money in the bank at 4% interest. If business conditions remain good, the company expects to make 10% on its investment in a new plant, but if there is a recession, the investment is expected to return only 2%. What probability must management assign to the occurrence of a recession to make the two investments equally attractive? Assume that the only two possible states for business conditions are "good" and "recession."

9. Farmer MacNeil is trying to decide whether to irrigate his 25 acres of cropland this year or next. If he irrigates this year, he will have to borrow $5000 for two years and pay a total of $600 in interest charges. If he waits until next year, he expects to have to borrow only $2500 for one year at 6% simple interest. An irrigation system is of value to him only if there is a drought. If there is a mild drought, he expects irrigation to result in an increased yield of one ton per acre compared to output without the system, and if there is a severe drought, an increase of three tons per acre. The farmer's crop sells for $18 a ton. If he feels that the probability of a mild drought each year is 0.40 and that of a severe drought 0.25, should he irrigate this year or wait until next year? (Assume that the irrigation system has an infinite lifetime.)

10. Given the following payoff matrix measured in utility units:

Projected Sales Level	Buy Specified Interest in Plywood, Inc.		
	100%	50%	25%
$30 Million	50	25	17
15 Million	35	15	5
10 Million	−6	−2	0
5 Million	−8	−4	−2

Construct the corresponding opportunity loss table.

13.5 EXPECTED VALUE OF PERFECT INFORMATION

Thus far in our discussion, we have considered situations in which the decision maker chooses among alternative courses of action on the basis of *prior information* without attempting to gather further information before making the decision. That is, the probabilities used in computing the expected value of each act, as shown in Table 13-3, are termed "prior probabilities" to indicate that they represent probabilities established prior to obtaining additional information through sampling. The procedure of calculating expected values of each act based on these prior probabilities and selecting the optimal act is referred to in Bayesian decision theory as *prior analysis*. In Chapter 14 we consider how courses of action may be compared after these prior probabilities are revised on the basis of sample information, experimental data, or information resulting from tests of any sort. However, the analysis carried out to this point provides a yardstick for measuring the value of perfect information concerning which events will occur. This yardstick will be referred to as the *expected value of perfect information*. In order to determine this value, we must calculate the *expected profit with perfect information*. Then, if we subtract the *expected profit under uncertainty*, whose calculation we previously examined, we will have the *expected value of perfect information*. These concepts will be explained in terms of the inventor's problem. We begin with the idea of expected profit with perfect information.

The calculation of the expected profit of acting with perfect information is based on the expected payoff if the decision maker has access to a perfect predictor. It is assumed that if this perfect predictor forecasts that a particular event will occur, then indeed that event will occur. The expected payoff under these conditions for the inventor's problem is given in Table 13-6. In order to understand the meaning of this calculation, it is necessary to adopt a long-run relative frequency point of view. If the forecaster says the event "strong sales" will prevail, the decision maker can look along that row in the payoff table and select the act that yields the highest profit. In the case of strong sales, the best act is A_1,

TABLE 13-6
Calculation of Expected Profit with Perfect Information for the Inventor's Problem (in Units of $10,000 Profit)

Predicted Event	Profit	Probability	Weighted Profit
θ_1: Strong sales	$80	0.2	$16.0
θ_2: Average sales	20	0.5	10.0
θ_3: Weak sales	1	0.3	.3
			$26.3

Expected profit with perfect information
$$= \$26.3 \ (\$\text{ten thousands})$$
$$= \$263,000$$

which yields a profit of $800,000. Hence, the figure $80 is entered under the profit column in Table 13-6. The same procedure is used to obtain the payoffs for each of the other possible events. The probabilities shown in the next column are the original probability assignments to the three states of nature. From a relative frequency viewpoint, these probabilities are now interpreted as the proportion of times the perfect predictor would forecast that each of the given states of nature would occur if the present situation were faced repetitively. Each time the predictor makes a forecast, the decision maker selects the optimal payoff. The expected profit with perfect information is then calculated as shown in Table 13-6 by weighting these best payoffs by the probabilities and totaling the products. The expected profit with perfect information in the inventor's problem is $263,000. This figure can be interpreted as the average profit the inventor could realize from this type of device if he were faced with this decision problem repeatedly under identical conditions, and if he always took the best action after receiving the forecast of the perfect indicator. Expected profit with perfect information has sometimes been called the "expected profit under certainty," but this term is clearly somewhat misleading. The inventor is not *certain* to earn any one profit figure. The expected profit with perfect information is to be interpreted as indicated in this discussion.

The expected value of perfect information, abbreviated as EVPI, is defined as *the expected profit with perfect information* minus *the expected profit under uncertainty.* Its calculation is shown in Table 13-7.

The interpretation of the EVPI is clear from its calculation. In the inventor's problem, his expected payoff if he selects the optimal act under conditions of uncertainty is $245,000. (See Table 13-3.) On the other hand, if the perfect predictor were available and the inventor acted according to those predictions, the expected payoff would be $263,000. (See Table 13-6.) The difference of $18,000 represents the increase in expected profit attributable to the use of the perfect forecaster. Hence, the expected value of perfect information may be interpreted as the most the inventor should be willing to pay to get perfect information on the level of sales for his device.

The expected opportunity loss of selecting the optimal act under

TABLE 13-7
Calculation of Expected Value of Perfect Information for the Inventor's Problem

Expected profit with perfect information	$263,000
Less: Expected profit under uncertainty	245,000
Expected value of perfect information (EVPI)	$ 18,000

EVPI = EOL of the optimal act under uncertainty = $18,000

conditions of uncertainty in the inventor's problem was also shown earlier to be $18,000. That is, this figure represented the minimum value among the expected opportunity losses associated with each act. As shown in Table 13-7, this figure is equal to the expected value of perfect information. It can be mathematically proved that this equality holds in general. Another term used for the expected opportunity loss of the optimal act under uncertainty is the *cost of uncertainty.* This term highlights the "cost" attached to the making of a decision under conditions of uncertainty. Expected profit would be larger if a perfect predictor were available and this uncertainty were removed. Hence, this cost of uncertainty is also equal to the expected value of perfect information. In summary, the following three quantities are equivalent:

Expected value of perfect information
Expected opportunity loss of the optimal act under uncertainty
Cost of uncertainty

13.6 REPRESENTATION BY A DECISION DIAGRAM

It is useful to represent the structure of a decision problem under uncertainty by a *decision tree diagram,* also called a *decision diagram* or, briefly, a *tree.* The problem can be depicted in terms of a series of choices made in alternating order by the decision maker and by "chance." Forks at which the decision maker is in control of choice are referred to as *decision forks;* those at which chance is in control are called *chance forks.* Decision forks will be represented by a little square, chance forks by a little circle. Forks may also be referred to as branching points or junctures.

A simplified decision diagram for the inventor's problem is given in Figure 13-1. After explaining this skeleton version, we will insert some additional information to obtain a completed diagram. As we can see from Figure 13-1, the first choice is the decision maker's at branching point 1. He can follow either branch A_1 or branch A_2; that is, he can choose either act A_1 or A_2. Assuming that he follows path A_1, he comes to another juncture, branching point 2, which is a chance fork. Chance now determines whether the event that will occur is θ_1, θ_2, or θ_3. If chance takes him down the θ_1 path, the terminal payoff is $800,000; the corresponding payoffs are indicated for the other paths. An analogous interpretation holds if he chooses to follow branch A_2. Thus, the decision diagram depicts the basic structure of the decision problem in schematic form. In Figure 13-2, additional information is superimposed on the diagram to represent the analysis and solution to the problem.

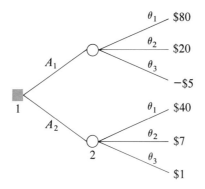

FIGURE 13-1
Simplified decision diagram for inventor's problem (payoffs are in units of $10,000 profit).

The decision analysis process represented by Figure 13-2 (and other decision diagrams to be considered at later points) is known as *backward induction*. We imagine ourselves as located at the right side of the tree diagram, where the monetary payoffs are. Let us consider first the upper three paths denoted θ_1, θ_2, and θ_3. Below these symbols, in parentheses, we enter the respective probability assignments (0.2, 0.5, and 0.3) as given in Table 13-3. These represent the probabilities assigned by chance to following these three paths, after the decision maker has selected act A_1. Moving back to the chance fork from which these three paths emanate, we can calculate the expected monetary value of being located at that fork. This expected monetary value is $24.5 (in units of $10,000, as are the other obvious corresponding numbers) and is calculated in the usual way:

$$\$24.5 = (0.2)(\$80) + (0.5)(\$20) + (0.3)(-\$5)$$

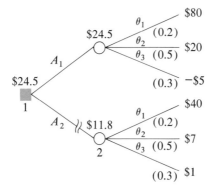

FIGURE 13-2
Decision diagram for inventor's problem (payoffs are in units of $10,000 profit).

This figure is entered at the chance fork under discussion. It represents the value of standing at that fork after choosing act A_1, as chance is about to select one of the three paths. The analogous figure entered at the lower chance fork is \$11.8. Therefore, imagining ourselves as being transferred back to branching point 1, where the little square represents a fork at which the decision maker can make a choice, we have the alternatives of selecting act A_1 or act A_2. Each of these acts leads us down a path at the end of which is a risky option whose expected profit has been indicated. Since following path A_1 yields a higher expected payoff than path A_2, we block off A_2 as a nonoptimal course of action. This is indicated on the diagram by the two wavy lines. Hence, A_1 is the optimal course of action, and it has the indicated expected payoff of \$24.5.

Thus, the decision tree diagram reproduces in compact schematic form the analysis given in Table 13-3. An analogous diagram could be constructed in terms of opportunity losses to reproduce the analysis of Table 13-5.

Exercises

1. Explain the meaning of expected value of perfect information.
2. Explain the difference between expected opportunity loss and expected value of perfect information.
3. Given an opportunity loss table, can you compute the corresponding payoff table? Explain why or why not.
4. The following is a payoff matrix in units of \$1000:

| | Price the Item at | | | |
| | A_1 | A_2 | A_3 | A_4 |
Competitor's Price	\$.90	\$.95	\$1.00	\$1.05
S_1: \$1.00	10	6	3	1
S_2: .95	5	8	4	6
S_3: .90	12	9	8	5
S_4: .85	8	10	12	14

The prior probabilities are

State of Nature	Probability
S_1	0.3
S_2	0.2
S_3	0.4
S_4	0.1

Compute the EVPI by two different methods.

5. The following is a payoff table in units of $1000:

	Action			
Demand Is	A_1	A_2	A_3	A_4
S_1: Above average	18	15	16	11
S_2: Average	8	12	12	10
S_3: Below average	2	5	3	8

where A_1 = keep store open weekdays, evenings, and Saturday
A_2 = keep store open weekdays plus Wednesday evening
A_3 = keep store open weekdays and Saturday
A_4 = keep store open only weekdays

The prior probability distribution of demand is

S_i	$P(S_i)$
S_1	0.5
S_2	0.3
S_3	0.2

a. Find the expected profit under certainty.
b. Find the expected profit under uncertainty.
c. How much would you pay for information that yields the true state of nature?

6. Trivia Press, Inc. has been offered an opportunity to publish a new novel. If the novel is a success, the firm can expect to earn $8 million over the next five years, and if it is a failure, to lose $4 million over the next five years. After reading the novel, the publisher assesses the probability of success as $\frac{1}{3}$. Should the firm publish the book? What is the expected value of perfect information?

7. An advertising firm submits for acceptance a campaign costing $55,000. The company's marketing manager estimates that if the campaign is received well by the public, profits will increase by $175,000; if it is received moderately well, profits will increase by $55,000; and if it is received poorly, profits will remain unchanged. Compute the appropriate opportunity loss table.

8. Suppose you are an executive considering taking some debatable deductions on your tax return for 1976. You will save $2000 in taxes if you take the deductions and the IRS does not audit your return. If the IRS does audit your return, you will have to pay a penalty of $1000 in addition to the tax payment of $2000. If the rest of your return is correct and would not be challenged in an audit, the payoff table would be:

State of Nature	Take Deductions	Do Not Take Deductions
IRS audits	−$1000	0
IRS does not audit	+$2000	0

Because of the complicated nature of your tax return, you estimate that there is a 65% chance that the IRS will audit it. Calculate the expected value of perfect information.

9. As personnel manager of Lemon Motors, you must decide whether to hire a new sales representative. Depending on the representative's performance, the payoff to the firm is

Sales	Payoff
High	$10,000
Average	3000
Low	−13,000

a. Judging from a particular candidate's application, you feel the probabilities of the three possible performance levels are as follows: high, 0.3; average, 0.4; and low, 0.3. Should you hire this candidate?

b. How much would you be willing to pay a perfect predictor to tell you what the representative's performance would be?

10. As marketing manager of a firm, you are trying to decide whether to open a new region for a product. Success of the product depends on demand in the new region. If demand is high, you expect to gain $100,000, if average to gain $10,000, and if low to lose $80,000. From your knowledge of the region and your product, you feel the chances are 0.4 that sales will be average, and equally likely that they will be high or low. Should you open the new region? How much would you be willing to pay to know the true state of nature?

11. In Exercise 7, if the president of the company, after examining the proposed campaign, feels that the probabilities that it would be received "well" and "moderately well" are 0.4 and 0.2, respectively, what is the expected opportunity loss for each action? What is the optimal decision?

12. A brewer who presently packages beer in old-style cans is debating whether to change the packaging of beer for next year. The choices are A_1, an easy-open aluminum can; A_2, a lift-top can; A_3, a new wide-mouth screw-top bottle; or A_4, the same old-style cans. Profits resulting from each move depend on what the brewer's competitor does for the next year. The payoff matrix and prior probabilities, measured in $10,000 units, are as follows:

Prior Probability	Competitor Uses	Action			
		A_1	A_2	A_3	A_4
0.5	Old-style bottles	15	14	13	16
0.2	Easy-open cans	12	11	10	8
0.1	Lift-top cans	6	9	8	6
0.2	Screw-top bottles	5	6	8	5

a. Find the expected opportunity loss for each act.
b. Determine EVPI.
c. Determine the optimal decision.

13.7 THE ASSESSMENT OF PROBABILITY DISTRIBUTIONS

As indicated in the preceding section, it is reasonable to assume that a decision maker would have some idea of the probabilities of the various values of the states of nature that are relevant to the decision to be made. In this section, we discuss how such probabilities may be assessed.[5]

If the random variable representing states of nature is discrete and there are only a small number of possible outcomes, then the decision maker will probably be able to assign probabilities directly to each possible outcome. For example, if the decision maker must assess the probability that a particular manufactured article is defective, he or she may be able to assign a probability of, say, 0.10 that the article is defective; hence, the complementary probability would be 0.90 that the article is not defective. If there were three possible outcomes, such as seriously defective, moderately defective, and not defective, again the decision maker can assign a probability to each of these three possible outcomes. As an example in a quite different context, the decision maker might have to assign probabilities to the following events: (1) the U.S. Congress *will* pass the bill on education currently before it or (2) the U.S. Congress *will not* pass the bill. In all of these cases, the decision maker may use a combination of past empirical information and a subjective evaluation of the effect of additional knowledge in making the probability assignments.

On the other hand, the random variable of interest may take on a very large number of possible values. For example, if the states of na-

---

[5] For a comprehensive discussion of the philosophy and practice of the assessment of subjective probabilities, see C. S. Spetzler and C. S. Staël Von Holstein, "Probability Encoding in Decision Analysis," *Management Science*, Vol. 22, No. 3 (November 1975).

ture are values for a company's sales next year for a certain product, the relevant values to be considered by the decision maker may range from, say, 100,000 to 500,000 units. It would be rather meaningless to attempt to assess a probability for each of the 400,001 possible outcomes for sales, that is, for 100,000, 100,001, . . . , 500,000. In this type of situation, the decision maker should treat the random variable as continuous and should set up a cumulative distribution function by making probability assessments for a number of selected ranges of the random variable. A distinction can be made between the following two situations.

1. The decision maker directly establishes a subjective cumulative probability distribution without formal processing of data.
2. There is a small quantity of past data, and the decision maker formally processes this information to set up a cumulative probability distribution.

We now discuss how these probability distributions may be constructed.

Direct Subjective Assessment

We illustrate the first situation by returning to a decision maker's forecasts of a company's sales of a certain product for next year. The decision maker feels that sales may range from 100,000 to 500,000 units and wishes to establish a subjective cumulative probability distribution without explicitly using any data. The basis of the procedure is to focus attention at a few key points in the distribution. These points are known as *fractiles*. For example, the 0.50 fractile is a value such that the decision maker believes the probability is 0.50 that the random variable is equal to or less than that value. (We note that the 0.50 fractile is simply another name for the median.) It is useful in this context to think in terms of hypothetical gambles. For example, one gamble might pay $100 if the random variable in question is less than or equal to some value selected by the decision maker, and a second gamble might pay $100 if the random variable turns out to be more than that value. The 0.50 fractile is the value such that the decision maker is indifferent to the choice between the two gambles. Let us assume that after some serious reflection on these gambles, the decision maker selects 350,000 units as the median or 0.50 fractile value.

We continue with the subjective assessment process. The decision maker should now be asked to select the 0.25 and 0.75 fractiles. For example, the 0.25 fractile is a figure such that the probability is 1/4 that the value of the random variable will lie below that figure and 3/4 that the value will lie above it. The analogous interpretation applies for the 0.75 fractile. We return to choices among gambles to determine the 0.25 fractile value. The decision maker might be asked to assume that sales next year will be less than 350,000 units and should then be presented with a pair of gambles: The first wager will pay $100 if sales fall between a value that the decision maker chooses and the 0.50 fractile; the second will pay $100 if sales are equal to or less than the chosen figure. The dividing value of the random variable determined by this "indifference point" between these two gambles is the 0.25 fractile. Obviously, if at first the decision maker selects a tentative dividing value but then finds one gamble more attractive than the other, the 0.25 fractile value has not been determined. The assessment procedure should continue until the indifference point has been found. Let us assume that the decision maker chooses 250,000 as the 0.25 fractile. The 0.75 fractile should then be determined by a similar procedure; let us assume that the decision maker determines 400,000 as the 0.75 fractile value.

Three selected points (the 0.25, 0.50, and 0.75 fractiles) have now been found in the cumulative distribution function. The usual procedure at this juncture is to determine two more values at the extremes of the distribution and to use the resulting five points as a basis for sketching the function. The extreme values often used are either the 0.01 and 0.99 or the 0.001 and 0.999 fractiles. In the present illustration, let us assume that the decision maker, when queried about the range of 100,000 to 500,000 units for next year's sales, stated that these were not absolute limits such that it was impossible for sales to fall outside them. After some further hard thinking about next year's sales, the decision maker might select these two values as the 0.01 and 0.99 fractiles. That is, the subjectively assigned odds are 99:1 that sales will exceed 100,000 (corresponding to a probability of 0.99). Similarly, the subjective odds are 99:1 that next year's sales would not exceed 500,000 units. Actually, these extreme fractiles are very difficult to assess. Empirical experiments have demonstrated that people tend to make the distributions too tight; that is, the 0.01 (or 0.001) and 0.99 (or 0.999) fractiles are generally not low enough and not high enough, respectively. Evidently, in formulating these assessments, one must make a very conscientious effort to spread the extreme values sufficiently.

To monitor further the probability assessments, the decision maker should carry out some tests for consistency of judgments. For example, since the 0.25 and 0.75 fractiles in this example are 250,000 and 400,000, the decision maker is asserting a 0.50 chance that next year's sales would fall within the range of these two figures (known, incidentally, as the *interquartile range*). Hence, these probability assessments imply that it is equally likely that next year's sales will fall inside this range or outside this range. Perhaps, upon reflection, the decision maker will choose to revise one or both of the two fractiles. After a few introspective checks of this sort, the five selected points should be plotted and a smooth curve drawn through them. Figure 13-3 shows a cumulative probability distribution that could be drawn through the 0.01, 0.25, 0.50, 0.75, and 0.99

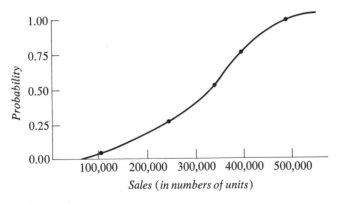

FIGURE 13-3
A cumulative probability distribution sketched through five selected fractiles.

fractiles referred to in our illustration. The curve is in the typical S-shaped form obtained for cumulative probability distributions for continuous random variables.

Even after the graph has been drawn, the checking or monitoring of the decision maker's probability assignments should continue. If the curve rises too slowly or too steeply in a certain portion of the random variable's range, the graph (and consequently the appropriate fractile values) should be adjusted. The reassessment procedure should continue until the decision maker feels that the curve appropriately describes the probability distribution.

Use of Past Data

The direct subjective assessment of probability distributions just described is appropriate in situations where there is little or no past historical data on which to base the construction of the function. That type of assessment is also appropriate when there *is* a substantial amount of past data but the decision maker is unwilling to use these data formally in the establishment of a probability distribution. This situation might arise whenever the decision maker decides that the underlying factors that gave rise to the past historical data have changed so greatly that the past data are not a reliable guide for the assessment of future probabilities. For example, suppose a company that purchases one of its raw materials from a supplier has a large quantity of past data on the percentages of defective product in past shipments from the supplier. Assuming that there have been no important changes in the supplier's manufacturing process and in policies governing the production of the product, the purchasing company may feel justified in using the past relative frequencies of occurrence of percentage defectives per shipment to assess a probability distribution for the percentage of defectives that might occur on the *next* shipment. On the other hand, if the purchasing company has just learned that the supplier has made an important change in its method of manufacture, or in the work force that produces the product, or in the method of manufacture, the purchaser may now feel that the past relative frequencies of occurrence are no longer relevant for assessing the afore-mentioned probability distribution. The purchaser may then appropriately proceed to make a direct subjective assessment of the probability distribution.

We turn now to the second situation listed at the beginning of this section, namely where there is a *small* quantity of data and the decision maker formally processes these data to construct a cumulative probability distribution for the relevant states of nature. In terms of the discussion in the previous paragraph, if there were a *large* quantity of relevant historical data in the form of relative frequencies of occurrence, the decision maker could simply use these figures to set up a discrete probability distribution. On the other hand, if there were only a *small* quantity of relevant data, how should these data be processed to construct the desired probability distribution? We will discuss here a systematic method for dealing with this type of situation.

Let us assume that as a consultant, you have been asked by a retail establishment to prepare a probabilistic forecast of the number of telephone orders that would be received next week for a certain product. The product has been sold for the past 20 weeks, and Table 13-8 gives a record of the numbers of weeks on which stated numbers of telephone orders were received. If you set up a relative frequency distribution of

TABLE 13-8
Numbers of Weeks on Which Specified Numbers of Telephone Orders Were Received

Number of Telephone Orders	Number of Weeks That Number of Orders Was Received	Number of Telephone Orders	Number of Weeks That Number of Orders Was Received
21	1	32	1
22	0	33	1
23	0	34	1
24	2	35	0
25	2	36	0
26	3	37	1
27	3	38	1
28	1	39	1
29	0	40	0
30	0	41	0
31	1	42	1

TABLE 13-9
Relative Frequency of Occurrence of Number of Telephone Orders

Number of Telephone Orders	Relative Frequency of Occurrence	Number of Telephone Orders	Relative Frequency of Occurrence
21	0.05	32	0.05
24	0.10	33	0.05
25	0.10	34	0.05
26	0.15	37	0.05
27	0.15	38	0.05
28	0.05	39	0.05
31	0.05	42	0.05

this discrete random variable, you would obtain the function shown in Table 13-9.

If you now think of the relative frequency distribution of Table 13-9 as a probability distribution, you find yourself involved in some rather odd interpretations. For example, starting at the beginning of the distribution, you would state that receiving 21 orders has a probability of 0.05, 22 and 23 orders have a probability of 0, 24 has a probability of 0.10, and so on. However, you really do not believe that it is impossible for 22 or 23 telephone orders to occur. You would recognize that these numbers of orders probably did not occur in the small sample of only twenty weeks of experience because of chance sampling fluctuations. It turns out that if you work in terms of a cumulative probability distribution rather than the probability mass function of Table 13-9, you can develop a more reasonable interpretation of the data. The cumulative relative frequencies of occurrence for the telephone order data are shown in Table 13-10. These cumulative frequencies represent the proportion of weeks in which the indicated numbers of telephone order *or fewer* were received. Interpreted as a probability, a cumulative frequency may be thought of as the probability that the random variable "number of telephone orders" is less than or equal to the specified value.

A graph of the cumulative frequency distribution of Table 13-10 is shown in Figure 13-4. In the present example, the raw data were ungrouped. As can be seen in Figure 13-4, the ungrouped data are graphed as a step function. The upper point plotted on each vertical line in the graph is the cumulative frequency corresponding to the value of the random variable plotted on the horizontal axis. As was noted concerning the original frequency distribution shown in Table 13-9, the irregularities of the step function graph of the cumulative relative fre-

TABLE 13-10
Cumulative Relative Frequency of Occurrence of Numbers of Telephone Orders

Number of Telephone Orders	Proportion of Weeks in Which the Specified Number or Fewer Orders Were Received	Number of Telephone Orders	Proportion of Weeks in Which the Specified Number or Fewer Orders Were Received
21	0.05	32	0.70
22	0.05	33	0.75
23	0.05	34	0.80
24	0.15	35	0.80
25	0.25	36	0.80
26	0.40	37	0.85
27	0.55	38	0.90
28	0.60	39	0.95
29	0.60	40	0.95
30	0.60	41	0.95
31	0.65	42	1.00

quency distribution of Figure 13-4 are doubtless the result of chance sampling variations. Therefore, as a way of smoothing out these sampling variations, it seems natural to try to fit a smooth curve to the graph representing a continuous cumulative probability function. Such a smooth curve, fitted to the step function data, is depicted in Figure

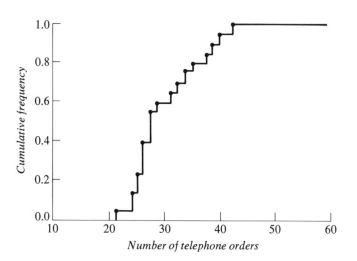

FIGURE 13-4
Step function graph of cumulative relative frequency distribution in Table 13-10.

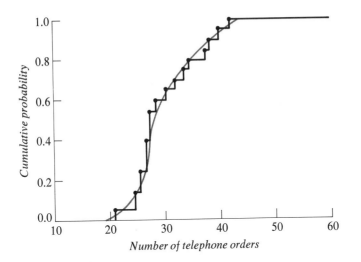

FIGURE 13-5
Continuous cumulative probability distribution fitted to the step function graph of
Figure 13-4.

13-5. In fitting this curve, you should attempt to draw it through the
centers of the flat portions ("flats") of the step function. This makes
sense because you would want to show a representative cumulative
probability for the range of values covered by these flats. Another point
to take into account in the fitting process is that the range of possible values shown on the graph should be wider than the analogous range observed in the small sample of data. Thus, for example, although the
original data had a lowest value of 21 and a highest value of 42, the horizontal axis of Figure 13-5 shows a range of from 10 to 60 telephone
orders. The lowest and highest values should reflect the decision
maker's best judgment concerning the range of possible outcomes. In
fact, the entire smooth cumulative probability distribution should represent the decision maker's judgment concerning the form of the curve
(guided, of course, by the rough form suggested by the step function
data).

Bracket Medians

In the preceding subsections, we have outlined the following ways of
constructing probability distributions for states of nature depending
upon the type and quantity of data available. If there are no historical
data available, or if the decision maker prefers not to process the available data formally because of doubts concerning the constancy of the
mechanisms producing these data, one may proceed to a direct subjec-

tive assessment of the cumulative probability distribution. As we have seen, by using selected fractiles, the decision maker can construct a continuous cumulative probability function.[6]

In the case in which there is a relatively small quantity of available data, we have noted how a cumulative relative frequency step function could be graphed and how a continuous cumulative probability distribution could be fitted to that function.

At the other end of the spectrum, if there were large quantities of relative frequency data available concerning states of nature, and if we were confident about the constancy of the causal systems producing the data, we could simply set up a discrete probability distribution from those data.

For the first two of the above situations (when there are no data or relatively small amounts of data), we have discussed how to construct continuous cumulative probability distributions for events or states of nature in decision making problems. However, in the earlier sections of this chapter, we discussed the structure of the decision making problem and decision diagram representation in terms of *discrete* probability distributions for states of nature. Because continuous probability distributions assign probability densities to every possible value of the states of nature, they are difficult to deal with in the analysis of many practical decision problems.[7] In this subsection, we discuss the bracket median technique, a method of approximating a continuous cumulative probability distribution by a discrete distribution. We can then go on to carry out decision analysis for states of nature represented by discrete random variables as outlined in this and subsequent chapters.

As an illustration of the bracket median technique, we consider the continuous cumulative probability distribution for next year's sales of a certain product, shown in Figure 13-6. The basic rationale of the method is to break up the probability distribution into groups having equal probabilities. Thus, in Figure 13-6, the vertical scale of cumulative probabilities is subdivided into five equal divisions, 0 to 0.2, 0.2 to 0.4, etc. This procedure correspondingly divides the random variable sales into five equally likely groups, each having a probability of 0.2.

Our next step is to determine a representative value, or *bracket median,* for each of the five groups of the random variable and assign a 0.2 probability to each of these representative values. These bracket medians are determined by bisecting each of the probability groups along the vertical axis, drawing the broken lines shown at 0.1, 0.3, 0.5,

---

[6] See R. S. Schlaifer, *Analysis of Decisions Under Uncertainty* (New York: McGraw-Hill, 1969) for a comprehensive discussion of the assessment of probability distributions based on the quantity and type of data available.

[7] More advanced texts in statistical decision theory deal with methods of analysis for continuous probability distributions whose formulas can be given algebraically (for example, the normal, gamma, and beta distributions).

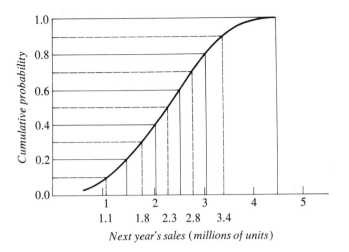

FIGURE 13-6
The representation of a continuous cumulative probability distribution by bracket medians.

0.7, and 0.9 over to the curve. We then read down on the horizontal axis to determine the bracket medians. In Figure 13-6, these bracket medians are at 1.1, 1.8, 2.3, 2.8, and 3.4 millions of units of sales. The resulting discrete probability distribution is given in Table 13-11.

In this illustration, five bracket medians were used. Of course, greater accuracy could be obtained by using larger numbers of groups. If ten groups were used, probabilities of 0.1 would be assigned to the equally likely groups, and so on. Computer programs are available to carry out the bracket median procedure for as many as 100 equally probable groupings. As noted previously, the resulting discrete probability

TABLE 13-11
Approximation of the Continuous Probability
Distribution of Figure 13-6 by a Discrete
Distribution Using Bracket Medians

Bracket Median Next Year's Sales (Millions of Units)	Probability
1.1	0.2
1.8	0.2
2.3	0.2
2.8	0.2
3.4	0.2
	1.0

distributions can be used in a decision analysis in the same way that the probability distribution of sales was used in the example given earlier in this chapter.

13.8 DECISION MAKING BASED ON EXPECTED UTILITY

In the decision analysis discussed up to this point, the criterion of choice was the maximization of expected monetary value. This criterion can be interpreted as a test of preferredness that selects as the optimal act the one that yields the greatest long-run average profit. That is, in a decision problem such as our example involving the inventor's choices, the optimal act is the one that would result in the largest long-run average profit if the same decision had to be made repeatedly under identical environmental conditions. In general, in such decision-making situations, as the number of repetitions becomes large, the observed average payoff approaches the theoretical expected payoff. Gamblers, baseball managers, insurance companies, and others who engage in what is colloquially called "playing the percentages," may often be characterized as using the aforementioned criterion. However, many of the most important personal and business decisions are made under unique sets of conditions, and in some of these occasions it may not be realistic to think in terms of many repetitions of the same decision situation. Indeed, in the business world, many of management's most important decisions are unique, high-risk, high-stake choice situations, whereas the less important, routine, repetitive decisions are ones customarily delegated to subordinates. Therefore, it is useful to have an apparatus for dealing with one-time decision making. Utility theory, which we discuss in this section, provides such an apparatus, as well as providing a logical method for repetitive decision making.

Whether an individual, a corporation, or other entity would be willing to make decisions on the basis of the expected monetary value criterion depends on the decision maker's attitude toward risk situations. Several simple choice situations are presented in Table 13-12 to illustrate that in choosing between two alternative acts we might select the one with the lower expected value. The reason we might make such a choice is our feeling that the increase in expected gain of the act with greater expected monetary value does not sufficiently reward us for the additional risk involved.

Table 13-12 gives three choice situations for alternative acts grouped in pairs. For each pair, a decision must be made between the two alternatives.

The illustrative choices are to be made once and only once. That is,

TABLE 13-12
Alternative Courses of Action with Different Expected Monetary Payoffs

Certainty Equivalent		Gamble
A_1: Receive $0 for certain. That is, you are certain to incur neither a gain nor a loss	or	A_2: Receive $.60 with probability 1/2 and lose $.40 with probability 1/2
B_1: Receive $0 for certain. That is, you are certain to incur neither a gain nor a loss	or	B_2: Receive $60,000 with probability 1/2 and lose $40,000 with probability 1/2
C_1: Receive a $1,000,000 gift for certain	or	C_2: Receive $2,100,000 with probability 1/2 and receive $0 with probability 1/2

there is to be no repetition of the decision experiment. We shall assume for simplicity that all monetary payoffs are tax free. Suppose you choose acts A_2, B_1, and C_1. In the case of the choice between A_1 and A_2, you might argue as follows. "The expected value of act A_1 is $0; the expected value of A_2 is $E(A_2) = (\frac{1}{2})(\$.60) + (\frac{1}{2})(-\$.40) = \$.10$. A_2 has the higher expected value, and since if I incur the loss of $.40, I can sustain such a loss with equanimity, I am willing to accept the risk involved in the selection of this course of action."

A useful way of viewing the choice between A_1 and A_2 is to think of A_2 as an option in which a fair coin is tossed. If it lands "heads" you receive a payment of $.60. If it lands "tails" you must pay $.40. If you select act A_1, you are choosing to receive $0 for certain rather than to play the game involved in flipping the coin. Hence, you neither lose nor gain anything. The $0 figure is referred to as the "certainty equivalent" of the indicated gamble.

On the other hand, you might very well choose act B_1 rather than B_2, even though the respective expected values are

$$E(B_1) = \$0$$

$$E(B_2) = \left(\frac{1}{2}\right)(\$60,000) + \left(\frac{1}{2}\right)(-\$40,000) = \$10,000$$

In this case, you might reason that even though act B_2 has the higher expected monetary value, a calamity of no mean proportions would occur if the coin landed tails, and you incurred a loss (say, a debt) of $40,000. Your present level of assets might cause you to view such a loss as intolerable. Hence, you would refuse to play the game. If you look at the difference between the two choices just discussed, A_1 versus A_2 and B_1 versus B_2, you will note that the only difference is that in A_2 we had the payoffs $.60 and $-\$.40$. In B_2 the decimal point has been moved five places to the right for each of these numbers, making the monetary gains and losses much larger than in A_2. In all other respects, the wording of the choice between A_1 and A_2 and between B_1 and B_2 is the same. Nevertheless, as we shall note after the ensuing discussion of the choice between acts C_1 and C_2, it is not necessarily irrational to select act A_2 over A_1, where A_1 has the greater expected monetary value, and B_1 over B_2, where B_1 has the smaller expected monetary value.

In the choice between acts C_1 and C_2, you would probably select act C_1, which has the lower expected monetary value. That is, most people would doubtless prefer a gift of $1,000,000 for certain to a 50:50 chance at $2,100,000 and $0, for which the expected payoff is

$$E(C_2) = \left(\frac{1}{2}\right) (\$2,100,000) + \left(\frac{1}{2}\right) (\$0) = \$1,050,000$$

In this case, you might argue that you would much prefer to have the $1,000,000 for certain, and go home to contemplate your good fortune in peace, than to play a game where on the flip of a coin you might receive nothing at all. You might also feel that there are relatively few things that you could do with $2,100,000 that you could not accomplish with $1,000,000. Hence, the increase in satisfaction to be derived even from winning on the toss of the coin in C_2 might not convince you to take the risk involved as compared to the "sure thing" of $1,000,000 in the selection of act C_1.

From the above discussion, we may conclude that it is reasonable to depart sometimes from the criterion of maximizing expected monetary values in making choices in risk situations. We cannot specify how a person *should* choose among alternative courses of action involving monetary payoffs, given only the type of information contained in Table 13-12. One's decisions will clearly depend upon one's *attitude toward risk*, which in turn will depend on a combination of factors such as one's level of assets, liking or distaste for gambling, and psychoemotional constitution. If we single out the factor of level of assets, for example, it is evident that a large corporation with a substantial level of assets may choose to undertake certain risky ventures that a smaller corporation with smaller assets would avoid. In the case of the larger corporation, an outcome of a loss of a certain number of dollars might represent an unfortunate occurrence but as a practical matter would not materially change the nature of operation of the business, whereas in the case of the smaller corporation, a loss of the same magnitude might constitute a catastrophe and might require the liquidation of the business. Hence, large and small corporations do and should have different attitudes toward risk. It may be noted that reverse attitudes toward risky ventures to those just indicated might be present in comparing a venturesome management of a small company with a highly conservative management of a large company.

To recapitulate, we can summarize the problem concerning decision making in problems involving payoffs that depend upon risky outcomes. Monetary payoffs are sometimes inappropriate as a measuring device, and it appears appropriate to substitute some other set of values or "numeraire," which reflects the decision maker's attitude toward risk. A clever approach to this problem has been furnished by Von Neumann and Morgenstern, who developed the so-called Von Neumann and Morgenstern utility measure.[8] In the next section, we consider how these utilities may be derived and the procedures for using them in decision analysis.

Construction of Utility Functions

We have seen that in certain risk situations we might prefer one course of action to another even though the first act has a lower expected monetary value. In the language of decision theory, the reason for preferring the first act is that it possesses greater expected utility than does the second act. The procedure used to establish the utility function of a decision maker requires a series of choices in each of which one receives with certainty an amount of money denoted M (for money) as opposed to a gamble in which one would receive an amount M_1 with probability p and an amount M_2 with probability $1 - p$. The question the decision maker must answer is, "What probability p for consequence M_1 would make one indifferent between receiving M for certain and partaking in the gamble involving the receipt of M_1 with probability p and M_2 with probability $1 - p$?" This probability assessment provides the assignment of a utility index to the monetary value M. The data obtained from the series of questions posed to the decision maker result in a set of utility-money pairs, which can be plotted on a graph and constitutes the decision maker's utility function for money. We will illustrate the procedure for constructing an individual's utility function by returning to one of the examples given in Table 13-12.

Suppose we ask the decision maker to choose between the following:

B_1: Receive \$0 for certain B_2: Receive \$60,000 with probability $1:2$
                                                  lose \$40,000 with probability $1:2$

[8] The term "utility" as used by Von Neumann and Morgenstern and as used in this text differs from the economist's use of the same word. In traditional economics, utility referred to the inherent satisfaction delivered by a commodity and was measured in terms of psychic gains and losses. On the other hand, Von Neumann and Morgenstern conceived of utility as a measure of value used in the assessment of situations involving risk, which provides a basis for choice making. The two concepts can give rise to widely differing numerical measures of utility.

Suppose the choice made is to receive $0 for certain. Our task then is to find out what probability for the receipt of the $60,000 would make the decision maker just indifferent between the gamble and the certain receipt of $0. This will enable us to determine the utility assigned to $0. The first step is the arbitrary assignment of "utilities" to the monetary consequences in the gamble, for example,

$$U(\$60,000) = 1$$
$$U(-\$40,000) = 0$$

where the symbol U denotes "utility" and $U(\$60,000) = 1$ is read "the utility of $60,000 is equal to one." It should be emphasized that the assignment of the numbers 0 and 1 as the utilities of the lowest and highest outcomes of the gamble is entirely arbitrary. Any other numbers could have been assigned, just so the utility assigned to the higher monetary outcome is greater than that assigned to the lower outcome. Thus, the utility scale has an arbitrary zero point, just like the $0°$ mark in temperature, which corresponds to different conditions depending upon whether the Centigrade or Fahrenheit scale is used.

The expected utility of the indicated gamble is

$$E[U(B_2)] = \frac{1}{2}[U(\$60,000)] + \frac{1}{2}[U(-\$40,000)] = \frac{1}{2}(1) + \frac{1}{2}(0) = \frac{1}{2}$$

Therefore, since the decision maker prefers $0 for certain to this gamble, it follows that the utility assigned to $0 is greater than $\frac{1}{2}$, or $U(\$0) > \frac{1}{2}$. In order to aid the decision maker in deciding how much greater than $\frac{1}{2}$ the utility assigned to the monetary outcome of $0 is, we introduce the concept of a hypothetical lottery for use in calibrating the decision maker's utility assessment.

Let us assume that we have a box with 100 balls in it, 50 black and 50 white. The balls are identical in all other respects. Furthermore, we assume that if a ball is drawn at random from the box and its color is black, the decision maker receives a payoff of $60,000. On the other hand, if the ball is white, the payoff is $-\$40,000$. We now have constructed a gamble denoted B_2. The question now is, If we retain the total number of balls in the calibrating box at 100 but vary the composition in terms of the number of black and white balls, how many black balls would be required for a decision maker to be indifferent between receiving $0 for certain and participating in the gamble? With 50 black balls in the box, the decision maker prefers $0 for certain. With 100 black balls (and no white balls), the decision maker would obviously prefer the gamble, since it would result in a payoff of $60,000 with certainty. For some number of black balls between 50 and 100, the decision maker should be indifferent, that is, at the threshold beyond which the gamble would become preferable to the certainty of the $0 payoff. Suppose we begin replacing white balls by black balls, and for some time the decision maker is still unwilling to participate in the gamble. Finally, when there are 70 black balls and 30 white balls, the decision maker announces that the point of indifference has been reached. We now can calculate the utility assigned to $0 as follows:

$$U(\$0) = 0.70[U(\$60,000)] + 0.30[U(-\$40,000)]$$
$$= 0.70(1) + 0.30(0) = 0.70$$

This utility calculation is a particular case of the general relationship

(13.1) $$U(M) = pU(M_1) + (1 - p)U(M_2)$$

where M is an amount of money received for certain and M_1 and M_2 are component prizes received in a gamble with respective probabilities p and $1 - p$.

We have now determined three money-utility pairs: $(-\$40,000, 0)$, $(\$60,000, 1)$, and $(\$0, 0.70)$. The first figure in each ordered pair represents a monetary payoff in dollars and the second figure the utility index assigned to this amount. The utility figures for other monetary payoffs between $-\$40,000$ and $\$60,000$ can be assessed in exactly the same way as for $0, assuming that the patience of our long-suffering decision maker holds out. A relatively small number of points could be determined and the rest of the function interpolated. Suppose the utility function shown in Figure 13-7 results from the indifference probabilities assigned by the decision maker in the set of gambles proposed. The one point whose determination was illustrated ($0, 0.70) is depicted on the graph. This utility function can now be used to evaluate risk alternatives that might be presented to the decision maker. The expected utility of an alternative can be calculated by reading off the utility figure corresponding to

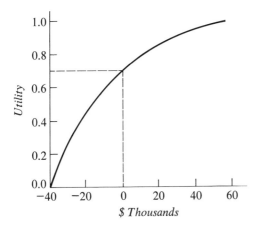

FIGURE 13-7
A utility function.

each monetary outcome and then weighting these utilities by the probabilities that pertain to the outcomes. In other words, the utility figures can now be used by the decision maker in place of the original monetary values, for calculation of expected utilities, whereas for a person with this type of utility function, calculation of expected monetary values is clearly an inadequate guide for decision making.

The preceding discussion indicated one particular method of constructing a decision maker's utility function. That procedure was explained in terms of a basic gamble involving a maximum payoff of $60,000 and a minimum payoff of −$40,000, to which respective utility values of 1 and 0 were assigned. Then other points on the utility curve were obtained by determining the probabilities required for the receipt of the $60,000 to make the decision maker indifferent between the gamble and the certainty equivalent of specified dollar amounts ($0 was illustrated).

Another somewhat more practical method for assessing utility functions will now be discussed. This method may be referred to as the *five-point procedure*,[9] because in its simplest form, five points on the utility curve are determined, namely, the points pertaining to utility values 1.00, 0.75, 0.50, 0.25, and 0. Since it is helpful in understanding this method to have a particular example in mind, suppose you are considering two alternative investment proposals and wish to formalize your attitude toward risk by determining your utility function. First of all you must determine the *criterion* by which you should investigate your attitude toward risk. A logical criterion would be your net asset position at the future time point at which the consequences of the investment can be evaluated.[10] Then you should estimate the best and worst possible consequences in terms of your criterion. Let us assume, for example, that you feel that under the best possible outcome of the two investment proposals, your net asset position two years hence (including the results of the proposed investments) would be $2,500,000; the worst possible outcome would be a net asset position of $0. You should then choose a pair of *reference consequences* for your criterion of net assets whose range is sufficiently wide to include the best and worst possible outcomes. Suppose you choose reference consequences of $3,000,000 and $0, because you feel that these limits are wide enough to include your possible net asset positions. We then arbitrarily assign utility values of 1 to $3,000,000 and 0 to $0. As in our previous example, we plot these as the two extreme points on our utility curve, shown in Figure 13-8. We then obtain your certainty equivalents for three 50:50 gambles in the

---

[9] See R. S. Schlaifer, *Analysis of Decisions Under Uncertainty* (New York: McGraw-Hill, 1969) for a detailed discussion of this method of assessing points on a utility curve.
[10] An alternative measure would be the *present value* of the net asset position at the future horizon date.

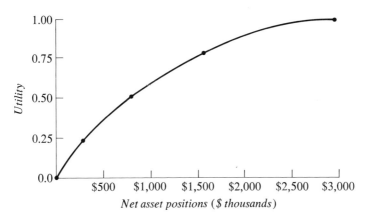

FIGURE 13-8
A utility function constructed by the five-point procedure.

range between the two reference consequences. Suppose that after considerable thought and discussion, you decide that you would take $800,000 for certain in exchange for 50:50 chances at the extreme outcomes of $3,000,000 and $0. Then the expected utility of the gamble would be

$$U(\$800,000) = \frac{1}{2}[U(\$3,000,000)] + \frac{1}{2}[U(\$0)]$$

$$= \frac{1}{2}(1) + \frac{1}{2}(0) = 0.50$$

We could then plot the point ($800,000, 0.50) on the graph. Now we can determine intermediate points between this 0.50 utility value and the two reference consequences. Hence, suppose that after suitable thought, you choose $1,600,000 as your certainty equivalent for a 50:50 gamble at consequences of $3,000,000 and $800,000. We now have

$$U(\$1,600,000) = \frac{1}{2}[U(\$3,000,000)] + \frac{1}{2}[U(\$800,000)]$$

$$= \frac{1}{2}(1) + \frac{1}{2}(0.50) = 0.75$$

This yields the point ($1,600,000, 0.75) that can be plotted on the graph. Finally, let us assume that in a 50:50 gamble between $800,000 and $0, you choose a certainty equivalent of $300,000. Then we have

$$U(300,000) = \frac{1}{2}[U(\$800,000)] + \frac{1}{2}[U(\$0)]$$

$$= \frac{1}{2}(0.50) + \frac{1}{2}(0) = 0.25$$

The five points we have now determined are

Utility	1.0	0.75	0.50	0.25	0
Net Asset Positions	$3,000,000	$1,600,000	$800,000	$300,000	$0

The smooth curve drawn through these five points is shown in Figure 13-8.

Of course, in a real-world situation, when the five points are plotted they may give an irregular appearance rather than lying along a smooth curve. The decision maker should be presented with the evidence and should be asked to resolve inconsistencies. Furthermore, additional gambles should be posed for other intermediate points as consistency checks. The purpose of this checking procedure is to make sure that the utility curve appropriately represents the decision maker's attitude toward risk for all outcomes in the range of the two reference consequences.

Characteristics and Types of Utility Functions

The utility functions depicted in Figures 13-7 and 13-8 rise consistently from the lower left to the upper right side of the chart. That is, the utility curves have positive slopes throughout their extent. This is a general characteristic of utility functions; it simply implies that people ordinarily attach greater utility to a larger sum of money than to a smaller sum.[11] Economists have noted this psychological trait in traditional demand theory and have referred to it as a "positive marginal utility for money." The concave downward shape shown in Figure 13-8 illustrates the utility curve of an individual who has a diminishing marginal utility for money, although the marginal utility is always positive. This type of utility curve is characteristic of a "risk avoider," and is so indicated in Figure 13-9(a). A person characterized by such a utility curve would prefer a small but certain monetary gain to a gamble whose expected monetary value is greater but may involve a large but unlikely gain, or a large and not unlikely loss. The linear function in Figure 13-9(b) depicts the behavior of a person who is "neutral" or "indifferent" to risk. For such a person every increase of, say, $1000 has an associated con-

---

[11] The almost infinite variety of types of human behavior is attested to by the fact that conduct that runs counter to a generalization of this sort is even occasionally observed. A few years ago, newspapers carried an account of an heir to a fortune of $30 million who committed suicide at the age of 23, indicating in a final letter that his great wealth prevented him from living a normal life.

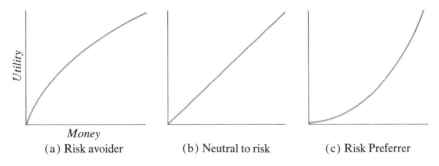

(a) Risk avoider   (b) Neutral to risk   (c) Risk Preferrer

FIGURE 13-9
Various types of utility functions.

stant increase in utility. This type of individual would use the criterion of *maximizing expected monetary value* in decision making, because this would also *maximize expected utility.*

In Figure 13-9(c) is shown the utility curve for a "risk preferrer." This type of person willingly accepts gambles that have a smaller expected monetary value than an alternative payoff received with certainty. For such an individual, the attractiveness of a possible large payoff in the gamble tends to outweigh the fact that the probability of such a payoff may indeed by very small.

Empirical research suggests that most individuals have utility functions in which for small changes in money amounts the slope does not change very much. Hence, over these ranges of money outcomes, the utility function may be considered approximately linear and constant in slope. However, in considering courses of action in which one of the consequences is very adverse or in which one of the payoffs is very large, individuals can be expected to depart from the maximization of expected monetary values as a guide to decision making. For many business decisions, where the monetary consequences may represent only a small fraction of the total assets of the business unit, the use of maximization of expected monetary payoff may constitute a reasonable approximation to the decision-making criterion of maximization of expected utility. In other words, in such cases, the utility function may often be treated as approximately linear over the range of monetary payoffs considered.

Assumptions Underlying Utility Theory

The utility measure we have discussed was derived by evoking the decision maker's preferences between sums of money obtainable with certainty and lotteries or gambles involving a set of basic alternative monetary outcomes. This procedure entails a number of assumptions.

It is assumed that an individual, when faced with the types of choices discussed, can determine whether an act, say A_1, is preferable to another act, A_2; whether these acts are indifferently regarded; or whether A_2 is preferred to A_1. If A_1 is preferred to A_2, then the utility assigned to A_1 should exceed the utility assigned to A_2.

Another behavioral assumption is that an individual who prefers A_1 to A_2 and A_2 to A_3, will also prefer A_1 to A_3. This is referred to as the principle of *transitivity*. The assumption extends also to indifference relationships. That is, the decision maker who is indifferent between A_1 and A_2 and between A_2 and A_3 should also be indifferent between A_1 and A_3.

Furthermore, it is assumed that if a payoff or consequence of an act is replaced by another, and one is indifferent between the former and new consequences, one should also be indifferent between the old and new acts. This is often referred to as the principle of *substitution*.

Finally, it is assumed that the utility function is bounded. This means that utility cannot increase or decrease without limit. As a practical matter, this simply means that the range of possible monetary values is limited. For example, at the lower end the range may be limited by a bankruptcy condition.

It may be argued that human beings do not always exhibit the type of consistency implied by these assumptions. However, the point is that if in the construction of one's utility function it is observed that one is behaving inconsistently and these incongruities are indicated to one, and if one is "reasonable" or "rational," one should adjust one's choices accordingly. If one insists on being irrational and refuses to adjust one's choices that violate the underlying assumptions of utility theory, then a utility function cannot be constructed for one and one cannot use maximization of expected utility as a criterion of rationality in choice making. It is important to keep in mind that the type of theory discussed here does not purport to describe the way people actually *do behave* in the real world, but rather specifies how they *should behave* if their decisions are to be consistent with their own expressed judgments as to preferences among consequences. Indeed, it may be argued that since human beings are fallible and do make mistakes, it is useful to have prescriptive procedures that police their behavior and provide ways in which it can be improved.

A Brief Note on Scales

Von Neumann–Morgenstern utility scales are examples of *interval scales*. Such scales have a constant unit of measurement but an arbitrary zero point. Differences between scale values can be expressed as multiples of one another, but individual values cannot.

EXAMPLE 13-1

The familiar scales for temperature are examples of interval scales. We cannot say that 100°C is twice as hot as 50°C. The corresponding Fahrenheit measures would not exhibit a ratio of 2:1. On the other hand, we can say that the intervals or differences between 100°C and 50°C and between 75°C and 50°C are in a 2:1 ratio. Thus, using the relationship $F = (9/5)C + 32°$, we have

$$C = 100°; \quad F = (9/5)(100°) + 32° = 212°$$
$$C = 75°; \quad F = (9/5)(75°) + 32° = 167°$$
$$C = 50°; \quad F = (9/5)(50°) + 32° = 122°$$

The difference between 100° and 50° = 50°; between 75° and 50° = 25°. The ratio of 50° to 25° is 2:1.

The difference between 212° and 122° = 90°; between 167° and 122° = 45°. The ratio of 90° to 45° is 2:1.

In decision making using utility measures, if a different zero point and a different scale are selected, the same choices will be made. A constant can be added to each utility value, and each utility value can be multiplied by a constant, without changing the properties of the utility function. Thus, if a is any constant, b is a positive constant, and x is an amount of money,

$$U_2(x) = a + bU_1(x)$$

and $U_2(x)$ is as legitimate a measure of utility as $U_1(x)$.

Exercises

1. a. The expected monetary return of the decision to buy life insurance is negative. Thus, it is irrational to buy life insurance. Do you agree or disagree? Explain.
   b. If a large company such as American Telephone and Telegraph does not carry automobile insurance, why do you think this is so?
2. If the following investment outcomes have the given utilities

Prospect	Units of Utility
A	10
B	8
C	5
D	3
E	2

would you prefer C for certain to

a. a chance of getting A with 0.4 probability and E with 0.6 probability?
b. a chance of getting B with 0.5 probability and E with 0.5 probability?
c. a chance of getting A with 0.3 probability and D with 0.7 probability?
d. a chance of getting B with 0.4 probability and E with 0.6 probability?

3. You have a choice of placing your money in the bank and receiving interest equal to ten units of utility or investing in Rerox stock. With a probability of 0.4, Rerox will yield gains equal to 45 units of utility and with a probability of 0.6, it will cause a loss of 15 units of utility.

a. What is the expected utility of the prospect "buy the stock"?
b. Should you buy Rerox or put the money in the bank?

4. Drillwell Oil Company is debating what it should do with an option on a parcel of land. If it takes the option, the firm can drill with 100% interest or with 50% interest (i.e., all costs and profits are split with another firm). It costs $50,000 to drill a well and $20,000 to operate a producing well until it is dry. The oil is worth $1 a barrel. Assume that the well is either dry or produces 200,000 or 500,000 barrels of oil. The firm assesses the probability of each outcome as 0.8, 0.1, and 0.1, respectively.

a. Based on expected monetary return, what is the best action?
b. Suppose Drillwell's management has the following utility function:

Dollars	Units of Utility
−50,000	−30
−25,000	−10
65,000	25
130,000	60
215,000	120
430,000	200

What is the best decision based on expected utility?

5. The IRS has audited your last year's income tax and has sent you a bill for $225 for back taxes. You now have the choice of paying the bill or disputing the audit. If you dispute it, it will cost you $20 for an accountant's fee to prepare your case. After preliminary talks with your accountant, you feel the chances of your winning the dispute are 0.05.

a. Should you dispute the case based on monetary expectations?
b. Assume that large losses of money are disastrous to you as a struggling student. This is reflected in your utility function, which indicates that $U(-\$20)$ is -4 units of utility, $U(-\$225)$ is -425 units of utility, and $U(-\$245)$ is -440 units of utility. Based on expected utility, what is your best course of action?

6. A drug manufacturer has developed a new drug named Curitol. Tests have shown it to be extremely effective with almost no side effects. However, it has only been tested for three years, and long-range side effects are really un-

known. The research department feels the probability the drug will have any serious long-range effects is 0.01. The Food and Drug Administration (FDA) must first clear the drug for sale. Assume that the FDA evaluates the loss to society because of serious long-range side effects as $-900,000$ units of utility, the gain to society because of the use of the drug as 8,000 units of utility, and the gain attributable to the economic advantages of production of a new drug as 1000 units of utility. If the FDA accepts the firm's appraisal of the probability of long-range side effects, should it "accept" the drug?

7. An investor who is considering buying a franchised furniture business estimates that the business will yield either a loss of $50,000 or a profit of $100,000, $200,000, or $500,000 every five years, with probabilities 0.5, 0.2, 0.1, 0.2, respectively. The investor's utility function is found to be

Dollars	Units of Utility	Dollars	Units of Utility	Dollars	Units of Utility
$-50,000$	-40	25,000	1.2	150,000	7.5
$-37,500$	-10	50,000	2.5	200,000	10.0
$-25,000$	-4.2	75,000	5.8	250,000	12.5
$-12,500$	-2	100,000	6.0	375,000	26.0
0	0	125,000	6.2	500,000	40.0

a. Graph the utility function and interpret the shape of the curve.
b. Based on expected utility value, should the investor buy or not?
c. Suppose the investor can buy either the whole franchise, three fourths, one half, or one fourth of it (i.e., a one-fourth interest means that the investor receives one fourth of all profits and pays one fourth of all losses). What is the best investment decision?

14

Decision Making Using Both Prior and Sample Information

14.1 INTRODUCTION

The discussion in Chapter 13 may be referred to as *prior analysis*, that is, decision making in which expected payoffs of acts are computed on the basis of prior probabilities. In this chapter, we discuss *posterior analysis*, in which expected payoffs are calculated with the use of *posterior probabilities*, which are revisions of prior probabilities on the basis of additional sample or experimental evidence. Bayes' Theorem is utilized to accomplish the revision of the prior probabilities. The terms "prior" and "posterior" in this context are relative ones. For example, subjective prior probabilities may be revised to incorporate the additional evidence of a particular sample. The revised probabilities then constitute posterior probabilities. If these probabilities are in turn revised on the basis of another sample, they represent prior probabilities relative to the new sample information, and the revised probabilities are "posteriors."

The basic purpose of attempting to incorporate more evidence through sampling is to reduce the expected cost of uncertainty. If the expected cost of uncertainty (or the expected opportunity loss of the optimal act) is high, then it will ordinarily be wise to engage in sampling. Sampling in this context is understood to include statistical sampling, experimentation, testing, and any other methods used to acquire additional information.

Here as in Chapters 13 through 17, for simplicity of presentation, we have carried out decision analyses in terms of *monetary payoffs* rather than *utility figures*. As noted in Section 13.8, this use of monetary values constitutes an assumption that the utility function is approximately linear over the range of monetary payoffs considered. Of course, when this assumption is not valid, the analyses should be performed with appropriate utility values substituted for monetary payoffs.

14.2 POSTERIOR ANALYSIS

The general method of incorporating sample evidence into the decision-making process can be illustrated in terms of two types of sample information, where (1) the reliability of the sample information is specified or (2) the sample size is specified. We will illustrate the first type in terms of the inventor problem of Chapter 13. Then an acceptance sampling problem will be considered in which the additional evidence is based on a sample of a given size.

Suppose in the problem discussed in Chapter 13, the inventor decided not to rely solely on prior probabilities concerning the demand for his new device, but to have a market research organization conduct a sample survey of potential consumers to gather additional evidence for the probable level of sales for his product. Let us assume that the survey can result in three types of sample results, denoted x_1, x_2, and x_3, corresponding to the three states of nature, sales levels θ_1, θ_2, and θ_3. Specifically, the possible results may be

x_1: sample indicates strong sales
x_2: sample indicates average sales
x_3: sample indicates weak sales

The survey is conducted, and the sample gives an indication of an average level of sales, that is, x_2 is observed. Assume that on the basis of previous surveys of this type, the market research organization can assess the reliability of the sample evidence in the following terms. In the past, when the actual sales level after a new device was placed on the market was average, sample surveys properly indicated an average level of demand 80% of the time. However, when the actual level was strong

TABLE 14-1
Computation of Posterior Probabilities in the Inventor's Problem for the Sample
Indication of an Average Level of Sales

Event θ_i	Prior Probability $P(\theta_i)$	Conditional Probability $P(x_2|\theta_i)$	Joint Probability $P(\theta_i)\,P(x_2|\theta_i)$	Posterior Probability $P(\theta_i|x_2)$
θ_1: Strong sales	0.2	0.1	0.02	0.042
θ_2: Average sales	0.5	0.8	0.40	0.833
θ_3: Weak sales	0.3	0.2	0.06	0.125
	1.0		0.48	1.000

sales, about 10% of the sample surveys incorrectly indicated demand as average; and when the actual level was weak sales, about 20% of the sample surveys gave an indication of average sales. These relative frequencies represent conditional probabilities that the sample evidence indicates "average sales," given the three possible actual events concerning sales level. They can be symbolized as follows:

$$P(x_2|\theta_1) = 0.1$$
$$P(x_2|\theta_2) = 0.8$$
$$P(x_2|\theta_3) = 0.2$$

The revision by means of Bayes' Theorem of the prior probabilities assigned to the three sales levels on the basis of the observed sample evidence x_2 (average sales) is given in Table 14-1. In terms of Equation 2.17 for Bayes' Theorem, x_2 plays the role of B, the sample observation; and θ_i replaces A_i, the possible events, or states of nature. In the usual way, after the joint probabilities are calculated, they are divided by their total (in this case, 0.48) to yield posterior, or revised, probabilities for the possible events. The effect of the weighting given to the sample evidence by Bayes' Theorem in the revision of the prior probabilities may be noted by comparing the posterior probabilities with the corresponding "priors" in Table 14-1. With a sample indication of average sales, the prior probability of the event "average sales," 0.5, was revised upward to 0.833. Correspondingly, the probabilities of events "strong sales" and "weak sales" declined from 0.2 to 0.042 and from 0.3 to 0.125, respectively.

Decision Making after the Observation of Sample Evidence

The revised probabilities calculated in Table 14-1 can now be used to compute the "posterior expected profits" of the inventor's alternative courses of action. In Table 13-3, expected payoffs were computed based on the subjective prior probabilities assigned to the possible events. These can now be referred to as *prior expected profits*. The calculation of the posterior expected profits (using the revised, or posterior, probabilities as weights) is displayed in Table 14-2. It is customary to denote prior probabilities as $P_0(\theta_i)$ and posterior probabilities as $P_1(\theta_i)$. That is, the subscript 0 is used to denote prior probabilities and the subscript 1 to signify posterior probabilities. A decision diagram of this posterior analysis is given in Figure 14-1. Note that the decision tree is essentially the same as the one for the prior analysis given in Figure 13-2, except that posterior probabilities have been substituted for prior probabilities.

Since the posterior expected profit of act A_1 exceeds that of A_2, the better of the two courses of action remains that of the inventor manufacturing the device himself. However, after the sample indication of "average sales," the expected profit of act A_1 has decreased from $245,000 based on the prior probabilities to $193,950 based on the revised probabilities. Moreover, the difference in the expected profits of the two acts has narrowed somewhat. The $245,000 and $193,950 fig-

TABLE 14-2
Calculation of Posterior Expected Profits in the Inventor's Problem Using Revised Probabilities of Events (in Units of $10,000)

Act A_1: *Manufacture Device Himself*

Event	Probability $P_1(\theta_i)$	Profit	Weighted Profit
θ_1: Strong sales	0.042	$80	3.360
θ_2: Average sales	0.833	20	16.660
θ_3: Weak sales	0.125	− 5	− .625
	1.000		19.395

Posterior expected profit $A_1 = \$19.395$ ($ten thousands) $= \$193,950$

Act A_2: *Sell Patent Rights*

Event	Probability $P_1(\theta_i)$	Profit	Weighted Profit
θ_1: Strong sales	0.042	$40	1.680
θ_2: Average sales	0.833	7	5.831
θ_3: Weak sales	0.125	1	0.125
	1.000		7.636

Posterior expected profit $A_2 = 7.636$ ($ten thousands) $= \$76,360$

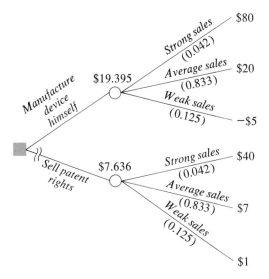

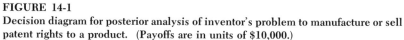

FIGURE 14-1
Decision diagram for posterior analysis of inventor's problem to manufacture or sell
patent rights to a product. (Payoffs are in units of $10,000.)

ures are, respectively, the *prior expected profit under uncertainty* and
the *posterior expected profit under uncertainty.* It is entirely possible
for the optimal course of action under a posterior analysis to change from
that of the prior analysis. In the present example, if the sample indication had been "weak sales," with appropriate conditional probabilities,
it would have been possible for the posterior expected profit of A_2 to
have exceeded that of A_1. (Assume some figures and demonstrate this
point.)

Insight can be gained into the cost of uncertainty and the value of
obtaining additional information by calculating the *posterior expected
value of perfect information,* which is simply the expected payoff, using
posterior probabilities, of decision making in conjunction with a perfect
predictor. We can now refer to the EVPI calculated in Chapter 13
(Table 13-7) as the prior EVPI. This prior EVPI was $18,000. There-
fore, it would have been worthwhile for the decision maker to pay up to
$18,000 for perfect information to eliminate his uncertainty concerning
states of nature. Since no sample could be expected to yield perfect information, the decision maker does not yet have a clear guide as to the
worth of obtaining additional information through sampling. Expected
value of sample information is discussed in Chapter 15. However, the
prior EVPI of $18,000 sets an upper limit to the worth of obtaining per-
fect information and thus eliminating uncertainty concerning events.
After the decision maker obtains additional information through
sampling, he can calculate the *posterior* EVPI. The change that occurs

TABLE 14-3
Calculation of the Posterior Expected Value of Perfect Information for the Inventor's
Problem (in Units of $10,000)

Event	Profit	Posterior Probability	Weighted Profit
θ_1: Strong sales	$80	0.042	3.360
θ_2: Average sales	20	0.833	16.660
θ_3: Weak sales	1	0.125	0.125
		1.000	20.145

Posterior expected profit with perfect information = 20.145 ($ten thousands)
= $201,450

Posterior expected profit with perfect information = $201,450
Less: posterior expected profit under uncertainty = 193,950
Posterior EVPI $ 7500

is useful in evaluating the decision to be made and the worth of attempting to obtain further information.

The posterior EVPI is computed to be $7500, as shown in Table
14-3. Analogously to prior analysis, the posterior EVPI is calculated by
subtracting the posterior expected profit under uncertainty from the posterior expected profit under certainty. The alternative determination of
the posterior EVPI as the expected opportunity loss of the optimal act
using posterior probabilities is given in Table 14-4. The only difference
between this calculation and the similar calculation in Table 13-5 for the

TABLE 14-4
Posterior Expected Opportunity Loss of the Optimal Act in the Investor's Problem
(in Units of $10,000)

Act A_1: *Manufacture Device Himself*

Event	Probability	Opportunity Loss	Weighted Opportunity Loss
θ_1: Strong sales	0.042	0	0
θ_2: Average sales	0.833	0	0
θ_3: Weak sales	0.125	6	0.75
			0.75

Posterior EOL of the optimal act = Posterior EVPI = 0.75 ($ten thousands)
= $7500

prior expected opportunity loss for act A_1 is the substitution of posterior probabilities for the prior probabilities.

In summary, the EVPI has been reduced from $18,000 to $7500 by the information obtained from the sample. Whereas the decision maker should have been willing to pay up to $18,000 for a perfect predictor prior to having the sample information, the expected value of perfect information is only $7500 after the sample indication of "average sales." In other words, the decision maker has reduced his cost of uncertainty; in view of the information already gained from the survey, the availability of a perfect forecaster is not as valuable as it was prior to the sample survey. Instead of a decrease occurring from the prior EVPI to the posterior EVPI as in this problem, there might very well have been an increase. This could occur if there were a marked difference between the posterior and prior probability distributions, and a reversal took place in the optimal act after the incorporation of sample information. Such an increase in the EVPI after inclusion of knowledge gained from sampling can be interpreted to mean that the doubt concerning the decision has been increased because of the additional evidence.

Of course, the determination of the prior and posterior EVPIs by alternative methods is not necessary in practice, but the computations have been shown here to indicate the relationships involved.

Exercises

1. Given

State of Nature	$P(\theta)$	$P(X|\theta)$	$P(\theta)P(X|\theta)$	$P(\theta|X)$
θ_1: Housing starts will increase next year	0.5	0.6		
θ_2: Housing starts will remain at the same level or will decline		0.4		

If X is the result of a survey of 100 construction companies, fill in the blanks and interpret the data.

2. A foreign exchange broker has ascertained the following probabilities for the spot rate of the Dutch guilder one month from now:

S	$P(S)$
$S_1 = \$.35$	0.15
$S_2 = .36$	0.25
$S_3 = .37$	0.30
$S_4 = .38$	0.20
$S_5 = .39$	0.10

A sample observation X with the following properties occurs:

$$P(X|S_1) = 0.90 \qquad P(X|S_2) = 0.80 \qquad P(X|S_3) = 0.45$$
$$P(X|S_4) = 0.20 \qquad P(X|S_5) = 0.10$$

What are the revised probabilities?

3. A firm is trying to decide whether to embark on a new advertising campaign. Management assigns the following prior probability distribution:

State of Nature	Probability
S_1: Successful	0.5
S_2: Unsuccessful	0.5

A sample result with the following probability of occurrence is observed:

$$0.3 \text{ if } S_1 \text{ is true}$$
$$0.6 \text{ if } S_2 \text{ is true}$$

Revise the prior probability distribution in the light of this new information.

4. In a certain situation, before sampling, the best act is A_1 and the EVPI is $100. After a sample was drawn, the best act was still A_1 and the revised EVPI was $50. The cost of sampling was $20. Can you conclude that the actual value of the sample information to the decision maker was $30? Explain your answer.

5. Given

State of Nature	$P(X|\theta)$	$P(\theta|X)$
θ_1	0.75	0.50
θ_2	0.25	0.50

where θ_1 is "product is superior to competitor's" and θ_2 is "product is as good as or inferior to competitor's." An experiment was run with the results given above. What was the prior probability distribution before sampling?

6. There are two possible actions to take, A_1 and A_2, and there are two states of nature, S_1 and S_2. A_1 is preferred if S_1 is true, and A_2 is preferred if S_2 is true. If the prior probabilities are $P(S_1) = 0.7$ and $P(S_2) = 0.3$ and you observe a sample observation S such that the $P(S|S_1) = 0.9$ and $P(S|S_2) = 0.2$, can you conclude that A_1 is the better act? Explain your answer.

7. Management's prior probability assessment of demand for a newly developed product is 0.55 for high demand, 0.25 for average demand, and 0.20 for low demand. A survey taken to help determine the true demand for the product indicates that demand is high. The reliability of the survey is such that it will indicate high demand 80% of the time when demand is actually high, 60% of the time when demand is actually average, and 15% of the time when demand

is actually low. In the light of this information, what would be the reassessed probabilities of the three states of nature?

8. A bond trader is considering the sale of her inventory holdings of New York City obligations. Her profit on such a transaction would be $5000 if undertaken immediately. If she waits six months before selling, her profit will depend on the direction of change in interest rates during that period. The potential profits and the trader's subjective probabilities for each possible event are shown below:

Event	Probability $P(\theta_i)$	Profit
θ_1: Higher interest rates	0.45	−$15,000
θ_2: No change	0.35	$ 5,000
θ_3: Lower interest rates	0.20	$18,000

What action should she take?

Now suppose the latest business statistics show that the number of companies planning to raise capital through debt during the next six months is very large. The trader knows that the probability of this situation given expectations of lower interest rates is 0.90, the probability given no expected change in interest rates is 0.25, and the probability given expectations of higher rates is 0.10. Compute the revised probabilities of the three events and the posterior expected profits of the two possible actions.

9. There are two states of nature and two alternative actions open to a firm. The payoff matrix in units of $ thousands and the initial probabilities are

Action	State of Nature S_1	S_2	
A_1	100	10	$P(S_1) = 0.3$
A_2	20	30	$P(S_2) = 0.7$

The firm is planning to construct an information system, that is, an organized system of data collection, storage, and analysis. It can construct two possible information structures, A and B. Both systems yield information either of type M_1 or of type M_2. The reliability (that is, $P(M_i|S_j)$) of the information from each system is as follows:

	System A M_1	M_2		System B M_1	M_2
S_1	0.9	0.1	S_1	0.6	0.4
S_2	0.2	0.8	S_2	0.4	0.6

a. What is the maximum amount the firm should pay for an information system?
b. Given that you receive information of type M_1 from system A, what are the revised prior probabilities, the best action, and the revised EVPI?
c. Given that you receive information of type M_1 from system B, what are the revised prior probabilities, the best action, and the revised EVPI?

An Acceptance Sampling Illustration

As the second illustration of posterior analysis, we consider a problem in acceptance sampling of a manufactured product. Let us assume that the Renny Corporation inspects incoming lots of articles produced by a supplier in order to determine whether to accept or reject these lots. In the past, incoming lots from this supplier have contained either 10%, 20%, or 30% defective articles. On a relative frequency basis, lots with these percentages of defectives have occurred 50%, 30%, and 20% of the time, respectively. The Renny Corporation feels justified in using these past percentages as prior probabilities for another lot just delivered by the supplier. A simple random sample of ten units is drawn with replacement from the incoming lot, and two defectives are found.

The Renny Corporation has found from past experience that it should accept lots that have 10% defectives and should reject those that have 20% or 30% defectives. That is, because of the costs of rework, it was not economical to accept lots with more than 10% defectives. On the basis of a careful analysis of past costs, the Renny Corporation constructed the payoff matrix in terms of opportunity losses shown in Table 14-5. The two possible courses of action are act A_1, to reject the incoming lot, and act A_2, to accept the incoming lot. The three states of nature are the possible proportions of defectives in the lot, namely, 0.10, 0.20, and 0.30. (It will be more convenient to consider the defectiveness of the lot in terms of decimals than percentages.) The proportion of defectives in the lot is denoted p and will be treated as a discrete random vari-

TABLE 14-5
Payoff Matrix Showing Opportunity Losses for Accepting and Rejecting Lots with Specified Proportions of Defectives

Event (p)	Act	
(Lot Proportion of Defectives)	A_1 Reject	A_2 Accept
0.10	$200	0
0.20	0	$100
0.30	0	$200

able that can take on only the three given values. Of course, it is rather unrealistic to assume that an incoming lot can be characterized only by a 0.10, 0.20, or 0.30 proportion of defectives. However, for convenience of exposition, that assumption will be made here. The same general principles would hold if more realistic assumptions were made, as, for example, that the proportion of defectives could take on values at intervals of one percentage point, namely, 0.00, 0.01, . . . , 1.00, or that there was a continuous probability distribution for proportion of defectives that was approximated by the bracket median method described in Section 13.7. We now proceed to apply some of the principles of decision analysis we have learned.

Suppose the Renny Corporation had to take action concerning acceptance of the present lot before drawing the sample of ten units from this lot. Assuming that the firm is willing to make its decision on the basis of prior information, what action should be taken? It seems reasonable that in the absence of additional information, the company should use the past relative frequencies of lot proportion of defectives as prior

TABLE 14-6
Prior Expected Opportunity Losses for the Renny Corporation Problem

Act A_1: Reject the Lot

Event p	Prior Probability $P_0(p)$	Opportunity Loss	Weighted Opportunity Loss
0.10	0.50	$200	$100
0.20	0.30	0	0
0.30	0.20	0	0
	1.00		$100

$$\text{EOL}(A_1) = \$100$$

Act A_2: Accept the Lot

Event p	Prior Probability $P_0(p)$	Opportunity Loss	Weighted Opportunity Loss
0.10	0.50	$ 0	$ 0
0.20	0.30	100	30
0.30	0.20	200	40
	1.00		$70

$$\text{EOL}(A_2) = \$70$$
$$\text{Prior EVPI} = \$70$$

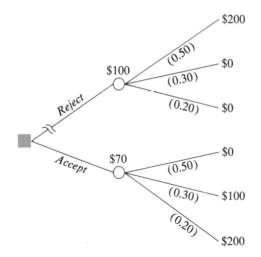

FIGURE 14-2
Decision diagram for prior analysis of the Renny Corporation problem of whether to accept or reject a lot produced by a supplier.

probability assignments. The prior analysis of the company's two courses of action based on expected opportunity losses is given in Table 14-6 and is depicted in Figure 14-2.[1] As shown in the table, where the past relative frequencies are used as prior probabilities, the expected opportunity loss of rejecting the lot is $100 and that of accepting the lot is $70. Hence, the optimal act is A_2, to accept the lot. As noted earlier, the $70 figure is the prior EVPI, since it represents the expected opportunity loss of the optimal act, using the prior probability distribution.

We turn now to the posterior analysis. Assume that the Renny Corporation proceeds to draw the simple random sample of ten units with replacement from the incoming lot and observes two defectives. If we take this sample evidence into account, what is the optimal course of action?

Following the same general procedure as in the inventor's problem, we can use the sample evidence to revise the prior probabilities assigned to the possible lot proportions of defectives. The application of Bayes' Theorem for this purpose is shown in Table 14-7. The condi-

---

[1] The notation $P_0(p)$ is used in Table 14-6 for the prior probability distribution of the random variable p, and $P_1(p)$ is used in Table 14-7 for the corresponding posterior probability distribution. Moreover, the random variable is referred to by lower case p in this and subsequent illustrations. This is a departure from the convention of using capital letters to denote random variables and lower case letters to represent the values taken on by the random variable. This is done to avoid confusion because of the common use of the capital P to mean "probability."

TABLE 14-7
Computation of Posterior Probabilities in the Renny Corporation Problem Incorporating Evidence Based on a Sample of Size 10

Event p	Prior Probability $P_0(p)$	Conditional Probability $P(X=2\mid n=10, p)$	Joint Probability $P_0(p)P(X=2\mid n=10, p)$	Posterior Probability $P_1(p)$
0.10	0.50	0.1937	0.09685	0.4136
0.20	0.30	0.3020	0.09060	0.3869
0.30	0.20	0.2335	0.04670	0.1995
	1.00		0.23415	1.0000

tional probabilities shown in the third column of Table 14-7 are often referred to as "likelihoods." That is, they represent the likelihoods of obtaining two defectives in ten units in a simple random sample drawn with replacement from the assumed incoming lots. When the basic random variable is the parameter p of a Bernoulli process, as in this problem, the likelihoods of the observed "number of successes" in the sample are computed by the binomial distribution. The likelihood figures in Table 14-7 were obtained from Table A-1 of Appendix A. The notation $P(X = 2\mid n = 10, p)$ means "the probability that the random variable "number of successes" is equal to 2 in ten trials of a Bernoulli process whose parameter is p." We will use this type of symbolism in this and other problems for likelihood calculations. With the evidence of two defectives in a sample of ten units, or 20% defectives in the sample, the prior probability that the lot contains 20% defectives is revised upward from 0.30 to 0.3869, as indicated in Table 14-7. Correspondingly, the probabilities that the lots contain 10% or 20% defectives are revised downward.

Expected payoffs of the two acts can now be recomputed using the revised, or posterior, probabilities. These computations are shown in Table 14-8 and are displayed in Figure 14-3. The optimal act is still A_2, to accept the lot. However, the posterior expected opportunity losses of the two acts are much closer together than were the prior ones. Furthermore, the posterior expected opportunity loss of the optimal act, A_2, is $78.59, which represents an increase from the prior expected opportunity loss of the optimal act, $70. In other words, the posterior EVPI now exceeds the prior EVPI, which indicates that the value of having a perfect predictor available has increased.

Effect of Sample Size

We can use this acceptance sampling problem to illustrate an important point in Bayesian decision analysis, namely, the effect of sample size on the posterior probability distribution. Suppose that instead of a simple

TABLE 14-8
Posterior Expected Opportunity Losses in the Renny Corporation Problem

Act A_1: Reject the Lot

Event p	Posterior Probability $P_1(p)$	Opportunity Loss	Weighted Opportunity Loss
0.10	0.4136	$200	$82.72
0.20	0.3869	0	0
0.30	0.1995	0	0
	1.0000		$82.72

$$EOL(A_1) = \$82.72$$

Act A_2: Accept the Lot

Event p	Posterior Probability $P_1(p)$	Opportunity Loss	Weighted Opportunity Loss
0.10	0.4136	$ 0	$ 0
0.20	0.3869	100	38.69
0.30	0.1995	200	39.90
	1.0000		$78.59

$$EOL(A_2) = \$78.59$$
$$\text{Posterior EVPI} = \$78.59$$

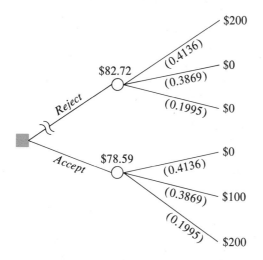

FIGURE 14-3
Decision diagram for posterior analysis of the Renny Corporation problem of whether to accept or reject a lot produced by a supplier.

TABLE 14-9
Computation of Posterior Probabilities in the Renny Corporation Problem
Incorporating Evidence Based on a Sample of Size 100

Event p	Prior Probability $P_0(p)$	Conditional Probability $P(X = 20\|n = 100, p)$	Joint Probability $P_0(p)P(X = 20\|n = 100, p)$	Posterior Probability $P_1(p)$
0.10	0.50	0.0012	0.00060	0.0188
0.20	0.30	0.0993	0.02979	0.9336
0.30	0.20	0.0076	0.00152	0.0476
	1.00		0.03191	1.0000

random sampe size of ten units being drawn from the incoming lot, a similar sample of 100 units was drawn. Furthermore, let us assume that 20 defectives were found in the sample; in other words, the fraction of defectives in this larger sample is 0.20, just as it was in the smaller sample of ten units. The computation of posterior probabilities using Bayes' Theorem and the information from the sample of size 100 is given in Table 14-9. The conditional probabilities in Table 14-9 were obtained from a table of binomial probabilities that includes $n = 100$.

Although 0.30 was the prior probability assigned to the state of nature that the incoming lot proportion of defectives was 0.20, as indicated in the last column of Table 14-9, the revised probability is 0.9336. Hence, because of the implicit weight given to the sample evidence by Bayes' Theorem, it is much more probable that the defective lot fraction is 0.20 according to the posterior distribution than according to the prior distribution. Furthermore, comparing the posterior probability distributions in Tables 14-7 and 14-9, a much higher probability (0.9336) is assigned to the event $p = 0.20$ after 20% defectives have been observed in a sample of size 100 than when that percentage of defectives is found in a sample of ten units (0.3869). A generalization of this result is that as sample size increases, the posterior distribution of the random variable "proportion of defectives" is influenced more and more by the sample evidence and less and less by the prior distribution.

Prior and Posterior Means

Let us consider a somewhat more formal explanation of the relationship between the prior distribution, the sample evidence, and the posterior distribution. This can be given in terms of the change that takes place between the mean of the prior distribution and the mean of the corresponding posterior distribution. For brevity, we will refer to the mean of a prior distribution of a random variable representing states of nature

TABLE 14-10
Calculation of the Prior Mean for the Defective
Proportion in the Renny Corporation Problem

Event p	Prior Probability $P_0(p)$	$pP_0(p)$
0.10	0.50	0.05
0.20	0.30	0.06
0.30	0.20	0.06
	1.00	0.17

Prior mean $= E_0(p) = 0.17$ defectives

as a "prior mean" or "prior expected value" and the corresponding mean
of a posterior distribution as a "posterior mean" or "posterior expected
value." The prior mean is obtained by the usual method for computing
the mean of any probability distribution. The calculation of the prior
mean for the acceptance sampling problem is given in Table 14-10.
Thus, the prior mean is 0.17 defective articles. The notation $E_0(p)$ is
used for the prior mean, the letter E denoting "expected value" and the
subscript 0 denoting "prior distribution." Analogously, the mean of the
posterior distribution is denoted $E_1(p)$. The corresponding computation
of the posterior means for the cases in which the posterior distributions
reflect sample evidence of two defectives in a sample of ten units and 20
defectives in a sample of 100 units are given in Table 14-11. The prior

TABLE 14-11
Calculation of Posterior Means for the Proportion of Defectives in the Renny
Corporation Problem

Posterior Distribution Incorporating Sample Evidence $X = 2$, $n = 10$			Posterior Distribution Incorporating Sample Evidence $X = 20$, $n = 100$		
Event p	Posterior Probability $P_1(p)$	$pP_1(p)$	Event p	Posterior Probability $P_1(p)$	$pP_1(p)$
0.10	0.4136	0.04136	0.10	0.0188	0.00188
0.20	0.3869	0.07738	0.20	0.9336	0.18672
0.30	0.1995	0.05985	0.30	0.0476	0.01428
	1.000	0.17859		1.0000	0.20288

Posterior mean $= E_1(p) = 0.17859$ Posterior mean $= E_1(p) = 0.20288$

mean is the expected proportion of defectives in the supplier's lot based on the Renny Corporation's prior assessment, that is, before the use of sample information. The posterior mean is the expected proportion of defectives in the supplier's lot based on the Renny Corporation's posterior assessment, that is, after incorporation of the sample information.

Rounding off the results obtained in Table 14-11, we observe posterior means of 0.179 defectives based on the sample evidence of two defectives in a sample of ten units, and .203 based on sample evidence of 20 defectives in a sample of 100. Hence, in the case of the smaller sample size, the posterior mean lies closer to the prior mean of 0.17 defectives than to the sample evidence of 0.20 defectives. On the other hand, when the larger sample is employed, the posterior mean falls closer to the sample evidence of 0.20 defectives than to the mean of the prior distribution.[2] This empirical finding is in keeping with our previous statement that as the sample size increases, the posterior distribution is progressively more influenced by the sample evidence and less by the prior distribution.

It is instructive to determine the optimal act using posterior probabilities that incorporate the evidence of 20 defectives in a sample of size 100. Table 14-12 gives the posterior expected opportunity losses based on the aforementioned probabilities. The posterior expected opportunity losses of act A_1 and act A_2 are further apart than in any of the preceding cases. The optimal act is A_1, to reject the lot, with a posterior expected opportunity loss of only $3.76 as compared to the corresponding figure of $102.88 for A_2. Comparing this decision in favor of act A_1 with the analogous choice of act A_2 based on prior expected opportunity losses, we see here an illustration of a situation where a reversal of decision takes place because of evidence observed in a sample. The very low figure of $3.76, which can be interpreted as the posterior EVPI, indicates that after the observation of 20 defectives in a sample of 100 units, it would not be wise to spend much money accumulating additional evidence before making the decision concerning acceptance or rejection of the lot.

---

[2] The posterior mean of 0.203 exceeds both the value of the prior mean, 0.17, and the sample value, 0.20 defectives. An examination of the results for $p = 0.10$ and $p = 0.30$ in Table 14-9 gives the reason. The likelihood of 20 defectives in a sample of 100 for $p = 0.30$ is more than six times the similar likelihood for $p = 0.10$. Despite a lower prior probability for $p = 0.30$ than for $p = 0.10$, the joint probability and, therefore, the posterior probability in the case of 0.30 far exceed those for $p = 0.10$. The net effect is to pull $E_1(p)$ somewhat closer to 0.30 than to 0.10.

TABLE 14-12
Posterior Expected Opportunity Losses in the Renny Corporation
Problem: The Posterior Probabilities Incorporate Sample Evidence $X = 20$,
$n = 100$

Act A_1: Reject the Lot

Event *p*	Posterior Probability $P_i(p)$	Opportunity Loss	Weighted Opportunity Loss
0.10	0.0188	$200	$3.76
0.20	0.9336	0	0
0.30	0.0476	0	0
	1.0000		$3.76

$$\text{EOL}(A_1) = \$3.76$$

Act A_2: Accept the Lot

Event *p*	Posterior Probability $P_1(p)$	Opportunity Loss	Weighted Opportunity Loss
0.10	0.0188	$ 0	$ 0
0.20	0.9336	100	93.36
0.30	0.0476	200	9.52
	1.0000		$102.88

$$\text{EOL}(A_2) = \$102.88$$
$$\text{Posterior EVPI} = \$3.76$$

Exercises

1. Let p be the true percentage of customers who will purchase a new product. Assume that p can take on the values given below with the respective prior probabilities.

p	$P_0(p)$
0.05	0.65
0.15	0.35

A simple random sample of 15 customers is drawn. The customers are asked whether they would purchase the product. What are the revised probabilities if

a. one customer would purchase?
b. two customers would purchase?
c. three customers would purchase?

2. Let μ equal the average number of pedestrians injured per month in a process for which the Poisson distribution is a suitable model. The state of nature, μ, can assume the values given below with the respective prior probabilities.

μ	$P_0(\mu)$
5	0.6
6	0.4

If during a particular month the number of pedestrians injured was six, what are the revised probabilities?

3. A small retail company is considering putting in a credit system. Let p be the proportion of new accounts that will be uncollectable if a credit system is installed. The states of nature and their respective prior probabilities are

p	$P_0(p)$
0.01	0.1
0.05	0.4
0.10	0.4
0.15	0.1

Assume that a credit system is installed. What are the revised probabilities if credit is extended to 100 customers and

a. four are uncollectable accounts?
b. 12 are uncollectable accounts?
Use a normal curve approximation.

4. Let p be the proportion of defective transistors in a lot offered to a radio manufacturer. Given:

p	$P_0(p)$	Opportunity Loss of Accepting Lot
0.10	0.40	0
0.20	0.30	100
0.25	0.20	300
0.30	0.10	600

A sample of 20 is taken, and four are found defective. What is the expected opportunity loss of the action "accept,"

a. if action is taken before sampling?

b. if action is taken after sampling?

5. In a previous problem describing an opportunity for Trivia Press, Inc. to publish a new novel, the probability of success for the novel was assessed as 1/3. The gain from a successful novel would be $8 million; the loss if the novel is unsuccessful would be $4 million. The publisher decides to send a copy of the book to ten critics for their opinions.

    The results are slightly unfavorable; six out of ten dislike the book. If the book should be a failure, the probability that a critic would dislike it is 0.8, and if the book should be a success, there is a 50:50 chance that a critic would like it. In view of this added information, would you recommend publication of the book?

6. The government is trying to decide whether to build a reservoir in Dodd County or in Todd County. Dodd County has a population of 800,000, and Todd has a population of 1,200,000. The cost of building the reservoir is the same in each location. The immediate expected benefit of the reservoir is computed as $10 per person in the county immediately *after* the reservoir is built. Building the reservoir will displace people, and the loss due to displacement is estimated at $50 per person. The reservoir in Dodd would displace 15,000 people, and in Todd it would displace 92,000 people. Of those displaced, the percentage of people who would move out of Todd County is between 10% and 20%, and out of Dodd County, between 5% and 15%. From economic studies and comparisons with previous reservoir sites, the prior probabilities of the proportion that would move out are estimated for each county as follows:

θ	Dodd $P_0(\theta)$	Todd $P_0(\theta)$
0.05	0.3	0.0
0.10	0.4	0.3
0.15	0.3	0.4
0.20	0.0	0.3

a. Write a mathematical expression for the benefit derived from each reservoir, letting x represent the proportion of displaced people who would move out of the county.

b. Based on prior beliefs, where should the reservoir be built?

c. In each county, a random sample of 20 persons was drawn. These people were asked whether they would move out of the county if displaced. Three people in Dodd and five in Todd said that they would move. Based on this new information, where should the reservoir be built?

7. A survey was taken by the Doody's Investor Service to determine the percentage of large corporations that use debt only as a last resort for financing, with

the aim of keeping their financial ratings as high as possible. The Investor Service's subjective probabilities associated with the possible percentage values, before considering the survey's results, were as follows:

p	$P_0(p)$
0.10	0.25
0.20	0.30
0.30	0.45

a. If the survey included 20 companies and 20% (four companies) maintained the conservative viewpoint on debt described above, what would be the revised probabilities after incorporating sample information?

b. If the survey included 100 companies and indicated the same percentage, what would be the revised probabilities? (See Table 14-9 for the conditional probabilities.)

8. Calculate the prior mean for the percentage of companies following the conservative financing practice described in Exercise 7. Calculate the posterior mean both for parts (a) and (b).

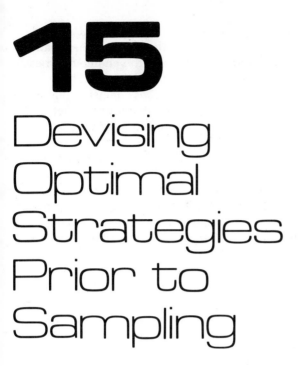

Devising
Optimal
Strategies
Prior to
Sampling

15.1 INTRODUCTION

In Chapter 13, we considered *prior analysis,* a method for decision making *prior* to the incorporation of additional information obtained through sampling or experimentation. In Chapter 14, we studied *posterior analysis,* a corresponding method for decision making *after* additional information has been obtained through sampling or experimentation. In posterior analysis, we draw a sample and choose the best act, taking into account both the sample data and prior probabilities of the events that affect payoffs. In this chapter, we are concerned with *preposterior analysis,* which answers the question whether it is worthwhile to collect sample or experimental data at all. If such collection of data is worthwhile, this method further specifies the best courses of action for each possible type of sample or experimental outcome, and how large a sample to take. This type of analysis leads to a more complex but more interesting view of decision making than those we have considered

609

so far. In both prior and posterior analysis, a final decision is made among the alternative courses of action based on the information at hand. Such a decision, which makes a final disposition of the choice of a best act, is referred to in Bayesian analysis as a *terminal decision*. The act itself is referred to as a *terminal act*. In many business decision-making situations, the wise course of action is not to choose a terminal act, but rather to delay making a terminal decision in order to obtain further information. Preposterior analysis, in addition to answering the questions whether additional information should be obtained and, if so, how much, also delineates the optimal decision rule to employ based on the possible types of evidence that can be produced by the additional information if it is gathered. An obvious difficulty in this type of decision procedure is that the specific outcome of a sample (or additional information) is unknown prior to taking the sample. Yet, in the nature of the case, the decision whether sample information will be worthwhile must be made prior to the actual drawing of the sample. Since there is a cost associated with acquiring additional knowledge through a sampling process, it pays to take a sample only if the anticipated worth of this sample information exceeds its cost. The methods discussed in this chapter provide a procedure for determining the *expected value of this sample information.*

The possibility of delaying a decision in order to acquire more information, as opposed to choosing a terminal act, opens up the further possibility of a multistage, or *sequential,* decision process. After the first sample is taken, the question again arises whether a terminal decision should be made now or whether another sample should be drawn. As long as the anticipated incremental worth of another sample exceeds its incremental cost, in keeping with general economic principles, it pays to continue sampling. Conceptually, we can envision the possibilities of a sample of one, two, or three items, and so on. It seems reasonable that there must be some optimal sample size, depending on the environmental circumstances beyond which further increases are uneconomical. However, we will begin by considering a simpler problem, involving only a single-stage sample of fixed size, before turning to the more complex problem of sequential decision making. In Section 15.2, we discuss two examples of preposterior analysis.

15.2 PREPOSTERIOR ANALYSIS

A preposterior analysis, as the name implies, is an investigation that must be carried out *before* sample information is obtained and therefore prior to the availability of posterior probabilities based on a particular sample outcome. However, this type of analysis takes into account all possible sample results and computes the expected worth (or expected opportunity loss) of a strategy that assumes that the best acts are selected depending on the type of sample information observed. We will illustrate preposterior analysis by an oversimplified example in order to convey the basic principles of the procedure without getting bogged down in computational detail.

A Marketing Example

The A. B. Westerhoff Company, a firm that manufactured consumer products, considered the marketing of a new product it had developed. However, the company wanted to appraise the advisability of engaging a market research firm to help determine whether sufficient consumer demand existed to warrant placing the product on the market. The market research firm offered to conduct a nationwide survey of consumers to obtain an appropriate indication of the market for the product. The fee for the survey was $15,000. The A. B. Westerhoff Company also wished to carry out an analysis that would specify whether it was better to act now (on the basis of prior betting odds as to the success or failure of the product and estimated payoffs) or to engage the market research firm and then act on the basis of the survey indication.

The company analyzed the situation as follows. There were two available actions:

a_1: Market the product
a_2: Do not market the product

The states of nature, or possible events, were defined as

θ_1: Successful product
θ_2: Unsuccessful product

Although only two states of nature for success of the product are used in this example, many states could be employed to indicate degree of success. Similarly, more than two courses of action might have been considered.

The company decided to view the problem in terms of opportunity losses of incorrect action. Based on appropriate estimates, the opportunity loss matrix (in $ thousands) shown in Table 15-1 was constructed.

The company, on the basis of its past experience with products of this type, assessed the odds that the product would be a failure at 3:1. That is, the company assigned prior probabilities to the success and failure of the product as follows:

$$P(\theta_1) = P(\text{successful product}) = \tfrac{1}{4} = 0.25$$
$$P(\theta_2) = P(\text{unsuccessful product}) = \tfrac{3}{4} = 0.75$$

TABLE 15-1
Payoff Table Showing Opportunity Losses for the A. B.
Westerhoff Company Problem (in $Thousands)

	Market a_1	Do Not Market a_2
θ_1: Product is a success	$ 0	$200
θ_2: Product is a failure	160	0

Hence, the prior analysis given in Table 15-2 was conducted to determine the optimal course of action if no additional information was obtained. As shown in Table 15-2, the optimal course of action was a_2 (do not market). The expected opportunity loss for this course of action was $50,000. A decision diagram of this prior analysis is given in Figure 15-1.

The A. B. Westerhoff Company then turned its attention to the problem of analyzing its expected opportunity losses if its action was

TABLE 15-2
Calculation of Prior Expected Opportunity Losses for the A. B. Westerhoff Company Problem (in $Thousands)

a_1: Market the Product

Event	Probability $P(\theta_i)$	Opportunity Loss	Weighted Opportunity Loss
θ_1: Successful product	0.25	$ 0	$ 0
θ_2: Unsuccessful product	0.75	160	120
	1.00		$120

$$\text{EOL}(a_1) = \$120 \text{ (thousand)}$$

a_2: Do Not Market the Product

Event	Probability $P(\theta_i)$	Opportunity Loss	Weighted Opportunity Loss
θ_1: Successful product	0.25	$200	$ 50
θ_2: Unsuccessful product	0.75	0	0
	1.00		$ 50

$$\text{EOL}(a_2) = \$50 \text{ (thousand)}$$
$$\text{EOL of the optimal act} = \text{EOL}(a_2) = \$50 \text{ (thousand)}$$

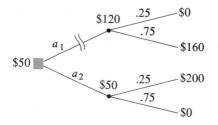

FIGURE 15-1
Decision diagram for action by the A. B. Westerhoff Company if no survey is conducted. (All losses are in $thousands.)

based on the survey results obtained by the market research firm. It requested the market research firm to indicate the nature and reliability of the results that would be supplied by the consumer survey. The market research firm replied that the survey would yield one of the following three types of indications:

X_1: Favorable
X_2: Intermediate
X_3: Unfavorable

That is, an X_1 indication meant an observed level of consumer demand in the survey that was favorable to the success of the product; an X_3 result was unfavorable, and an X_2 indication meant a situation falling between levels X_1 and X_3 and classified as "intermediate." Although we are using the verbal classifications "favorable," "intermediate," and "unfavorable" here for simplicity of reference, each of these indications will be understood to represent a specific numerical range of demand. How this type of sample is used in decision making is explained subsequently.

As a description of the anticipated *reliability* of the survey indications, the market research firm supplied the array of conditional probabilities shown in Table 15-3. The entries in this table are values of

TABLE 15-3
Conditional Probabilities of Three Types of Survey Evidence for the A. B. Westerhoff Company Problem of Deciding Whether to Conduct a Survey

Event	Conditional Probability $P(X_j \mid \theta_i)$			
	X_1	X_2	X_3	Total
θ_1: Successful product	0.72	0.16	0.12	1.00
θ_2: Unsuccessful product	0.08	0.12	0.80	1.00

$P(X_j|\theta_i)$ based on past relative frequencies in similar types of surveys. That is, they represent the conditional probabilities of each type of sample evidence given that the product actually was successful or unsuccessful. For example, the $P(X_1|\theta_1) = 0.72$ entry in the upper left corner of the table means the probability is 72% that the survey will yield a favorable indication, given a successful new product. The subscript j is used for the sample indication, denoted X_j, analogous to the use of the subscript i for the states of nature, denoted θ_i. It may not always be feasible to use past relative frequencies as the conditional probability assignments as we are doing here. The use of past relative frequencies as probabilities for future events always involves the assumption of a continuation of the same conditions as existed when the relative frequencies were established. As we have seen in other examples, probability functions such as the binomial distribution are often appropriate for computing these "likelihoods" or conditional probabilities of sample results.

The A. B. Westerhoff Company analysts decided to carry out a preposterior analysis in order to determine whether it was worthwhile to engage the market research firm to conduct the survey. For this purpose, it was necessary to compare the expected opportunity loss of purchasing the survey and then selecting a terminal act to the expected opportunity loss of terminal action without the survey. The latter expected loss figure was previously obtained from the prior analysis. As an intermediate step, the analysts computed the joint probability distribution of the sample evidence X_j and the events θ_i, as shown in Table 15-4. These joint probabilities were obtained by multiplying the prior (marginal) probabilities by the appropriate conditional probabilities. For example, 0.18, the upper left entry in the joint probability distribution in Table 15-4, was obtained by multiplying 0.25 by 0.72. In symbols, $P(X_1 \text{ and } \theta_1) = P(\theta_1)P(X_1|\theta_1)$, etc.

It is useful to consider the following interesting points about the

TABLE 15-4
Calculation of the Joint Probability Distribution of Survey Evidence and Events for the A. B. Westerhoff Company Problem of Deciding Whether to Conduct a Survey

Event θ_i	Prior Probability $P(\theta_i)$	Conditional Probability $P(X_j|\theta_i)$			Joint Probability $P(X_j \text{ and } \theta_i)$			Total
		X_1	X_2	X_3	X_1	X_2	X_3	
θ_1	0.25	0.72	0.16	0.12	0.18	0.04	0.03	0.25
θ_2	0.75	0.08	0.12	0.80	0.06	0.09	0.60	0.75
Total	1.00				0.24	0.13	0.63	1.00

joint probability distribution in order to understand the subsequent analysis. As in any joint bivariate probability distribution, the totals in the margins of the table are marginal probabilities. The row totals are the prior probabilities $P(\theta_1) = 0.25$ and $P(\theta_2) = 0.75$. The column totals are the marginal probabilities, of the survey evidence; that is, $P(X_1) = 0.24$, $P(X_2) = 0.13$, and $P(X_3) = 0.63$. Conditional probabilities of events given the survey evidence, that is, probabilities of the form $P(\theta_i|X_j)$, can be computed by dividing the joint probabilities by the appropriate column totals, $P(X_j)$. For example, the probability of a successful product, given a favorable survey indication, is given by

$$P(\theta_1|X_1) = \frac{P(\theta_1 \text{ and } X_1)}{P(X_1)} = \frac{0.18}{0.24}$$

and so on. These probabilities can be viewed as revised or posterior probability assignments to the events θ_1 and θ_2, given the survey evidence X_1, X_2, and X_3. The calculation of these posterior probabilities

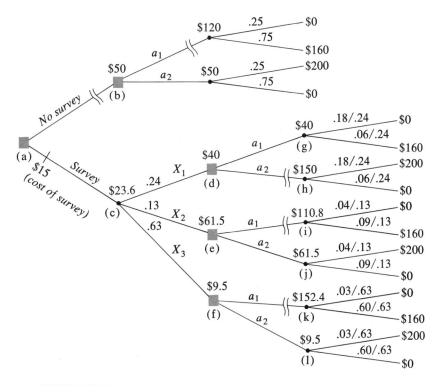

FIGURE 15-2
Decision diagram for preposterior analysis for the A. B. Westerhoff Company problem of deciding whether to conduct a survey (payoffs are in $thousands).

represent an application of Bayes' Theorem, Equations 2.12 and 2.16, as is shown in the following method of symbolizing the probability $P(\theta_1 | X_1)$ just calculated:

$$P(\theta_1 | X_1) = \frac{P(\theta_1 \text{ and } X_1)}{P(X_1)} = \frac{P(\theta_1)P(X_1 | \theta_1)}{P(\theta_1)P(X_1 | \theta_1) + P(\theta_2)P(X_1 | \theta_2)}$$

That is, the joint probability $P(\theta_1 \text{ and } X_1) = 0.18$ was calculated in Table 15-4 by multiplying the marginal probability $P(\theta_1) = 0.25$ by the conditional probability $P(X_1 | \theta_1) = 0.72$; the marginal probability $P(X_1) = 0.24$ was obtained by adding the joint probabilities $P(\theta_1)P(X_1 | \theta_1) = 0.18$ and $P(\theta_2)P(X_1 | \theta_2) = 0.06$.

The A. B. Westerhoff Company proceeded with its analysis and constructed the decision diagram shown in Figure 15-2. All monetary figures are in thousands of dollars. Returning to the beginning of the problem, the first choice is to purchase the survey or not to purchase it. Therefore, starting at node (a) at the left and following the "no survey" branch of the tree, we move along to node (b). From node (b) to the right is reproduced the decision tree depicted in Figure 15-1 for the prior analysis with no survey. As indicated earlier, the $50 entry at node (b) is the expected opportunity loss of choosing the optimal terminal act without conducting the survey.

Expected Value of Sample Information

On the other hand, suppose that at the outset [at node (a)], the decision is to conduct the survey. Making this decision, we move down to branch point (c). The results of the survey then determine which branch to follow. The three branches emanating from node (c), which represent X_1, X_2, and X_3 types of survey information, are marked with their respective marginal probabilities, 0.24, 0.13, and 0.63. Suppose type X_1 information (that is, a "favorable" indication) were observed. Moving ahead to branch point (d), we can choose either act a_1 or act a_2, that is, market or not market the product. If act a_1 is selected, we move to node (g); if act a_2 is selected, to node (h). At these points, the probability questions that must be answered are of the a posteriori or revised probability variety. For example if act a_1 is chosen, the probabilities shown on the two branches stemming from node (g), 0.18/0.24 and 0.06/0.24, are the conditional (posterior) probabilities, given type X_1 information, that the product is successful or unsuccessful, or in symbols, $P(\theta_1 | X_1)$ and $P(\theta_2 | X_1)$. Thus, they represent revised probabilities of these two events after observation of a particular type of sample evidence. These conditional probabilities are calculated from Table 15-4, as indicated earlier, by dividing joint probabilities by the appropriate marginal probabilities.

Now looking forward from node (g) and using the posterior probabilities (0.18/0.24 and 0.06/0.24) as weights applied to the losses attached to the events "successful product" and "unsuccessful product," we obtain an expected payoff of an opportunity loss of $40 for act a_1. This figure is entered at node (g). Comparing it with the corresponding figure of $150 for a_2, we block off act a_2 as being nonoptimal. Therefore, $40 is carried down to node (d), representing the payoff for the optimal act upon observation of type X_1 information. Similar calculations yield $61.5 and $9.5 at nodes (e) and (f) for X_2 and X_3 types of information. Weighting these three payoffs by the marginal probabilities of type X_1, X_2, and X_3 indications (0.24, 0.13, and 0.63, respectively), we obtain a loss of $23.6 as the expected payoff of conducting the survey and taking optimal action after the observation of the sample evidence.

Comparing the $23.6 figure with the $50 obtained under the "no survey" option, we see that it would be worthwhile to pay up to $50 − $23.6 = $26.4 (thousands) for the survey.

This difference represented by the $26.4 (thousands) is referred to as the *expected value of sample information*, and we will denote it as EVSI. Hence, if we are considering a choice between immediate terminal action without obtaining sample information and a decision to sample and then select a terminal act, EVSI is the *expected amount by which the terminal opportunity loss is reduced by the information to be derived from the sample*. This EVSI is a gross figure, since it has not taken into account the cost of obtaining the survey information. To calculate the *expected net gain of sample information*, which we denote as ENGS, the cost of obtaining the sample information must be subtracted from the expected value of this sample information. Hence in general,

(15.1)   ENGS = EVSI − Cost of sample information

and in this problem,

$$\text{ENGS} = \$26.4 - \$15 = \$11.4$$

In conclusion, since the expected value of sample information was $26,400 and the cost of the survey was $15,000, the ENGS was $11,400. Therefore, the A. B. Westerhoff Company decided that it was worthwhile to engage the market research firm to conduct the survey.

It is important to recognize that the EVSI and ENGS computations have been made with respect to the particular prior probability distribution used in the analysis. If different prior probabilities were used, the survey would have had correspondingly different EVSI and ENGS values. Sensitivity analysis, discussed in Section 15.5, can be used to test how sensitive the alternative actions are to the size of the prior probabilities.

Some Considerations in Preposterior Analysis

A few points concerning preposterior analysis arise from consideration of the foregoing example. First of all, we note that in this problem, the optimal action if no survey were conducted was a_2, not to market the product. On the other hand, as seen from Figure 15-2, if the survey were carried out and a favorable indication of demand, X_1, were obtained, the best course of action would be a_1, to market the product. The fact that for at least one of the survey outcomes it was possible for the decision maker to change the selected act is what gives the survey some value. Clearly, if a decision maker's course of action cannot be modified regardless of the experimental outcome, then the experiment is without value.

Second, the calculations in the preceding problem were carried out in terms of opportunity losses. If the payoffs had been expressed in terms of profits, the obvious equivalent analysis would have been required. For example, instead of the EVSI being the expected amount by which the terminal opportunity loss is reduced by the information to be derived from the sample, it would be the *expected amount by which the terminal profit is increased* by this information. Hence, expected profit of terminal action without sampling and expected profit of choosing a terminal action after sampling would be calculated. The first of these quantities would be subtracted from the second to yield the same EVSI figure obtained in the opportunity loss analysis.

Third, in this problem, the terms "sample" and "survey" were used interchangeably. Actually, even if the survey had represented a complete enumeration, the same analysis could have been carried out. Thus, the term "sample" in this context is used in a very general sense to include any sample from size one up to a complete census. Indeed, the sample outcomes X_1, X_2, and X_3 need not have arisen from a statistical sampling procedure but more generally could represent any set of experimental outcomes. Therefore, a preposterior analysis may be viewed as yielding the "expected value of experimental information" rather than the "expected value of sample information." However, since the latter term is conventionally used in Bayesian decision analysis, we have resisted the temptation to adopt a new term.

Finally, another level of generalization suggests itself. Only one survey of a fixed size was considered in this problem. Many surveys of different types and different sizes could possibly have been considered. The preposterior expected opportunity loss would then be calculated for each of these different surveys. The minimum of these figures would be subtracted from the prior expected opportunity loss to yield the EVSI. In order to obtain the ENGS, the total expected opportunity loss for each survey is calculated by adding the cost of the survey to the corresponding preposterior expected opportunity loss. Then these figures are subtracted from the prior expected loss. The maximum of these differences is the ENGS, since it represents the expected net gain of the survey with the lowest total expected opportunity loss. Although the theoretical number of possible surveys or experiments that might be considered is infinite, obviously there would ordinarily be a delimitation at the outset based on factors of practicability or feasibility. On the other hand, the use of computers increases considerably the number of possible alternatives that can practically be compared.

Exercises

1. Define and compare EVSI and ENGS.

2. Let θ_1 be the state of nature "the unemployment rate will decrease over the next six months" and θ_2 "the unemployment rate will stay the same or increase over the next six months." The chief economist for the Bank of Boston will predict either X, "the rate will decrease," or Y, "the rate will stay the same or increase." Fill in the missing entries of the following table:

θ	$P(\theta)$	$P(X|\theta)$	$P(Y|\theta)$	$P(\theta)P(X|\theta)$	$P(\theta)P(Y|\theta)$	$P(\theta|X)$	$P(\theta|Y)$
θ_1	0.2	0.7					
θ_2	0.8	0.2					

3. Let the two possible states of nature be θ_1: the average level of stock market prices will advance, and θ_2: the average level of stock market prices will stay the same or decrease. Your prior belief is that there is a 60% chance that market prices will advance. In a week the forecast of a well-known econometric model will be published. Either an advance or no advance will be indicated. The model is correct 80% of the time. Find the posterior probabilities for θ for each of these indications.

4. Given the illustration on page 620 of extensive analysis with payoffs in terms of opportunity losses:
   a. Fill in all the missing entries
   b. Interpret each part of the tree
   c. Find the EVPI before and after sampling
   d. Find the EVSI
   e. Find the expected net gain from sampling

5. A motel located near the site of a soon-to-be-opened world's fair is contemplating the construction of some temporary extra rooms. The rooms are of no value after the fair is closed, since the present size of the motel is more than adequate for normal demand. Let θ_1 be "demand for the temporary rooms is at least enough to cover the cost of building them," and θ_2 be, "demand for the temporary rooms is not sufficient to warrant construction." A survey can

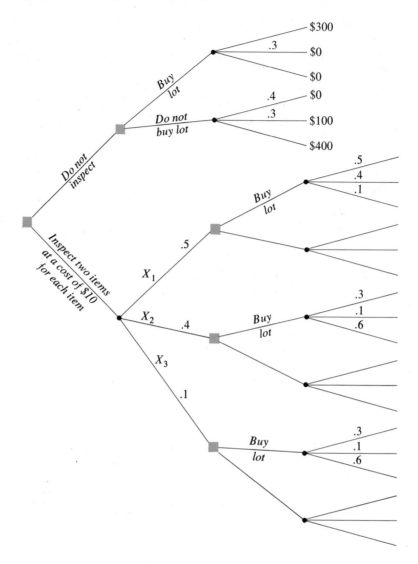

be taken that would yield two possible indications, X_1 or X_2. The following is known:

State of Nature	$P(\theta)$	$P(X_1\|\theta)$		Opportunity Loss Table Build	Do Not Build
θ_1	0.5	0.55	θ_1	0	6
θ_2	0.5	0.50	θ_2	5	0

a. Find
   1. the optimal act before sampling
   2. the optimal act if X_1 occurs
   3. the optimal act if X_2 occurs
   4. $P(X_1)$ and $P(X_2)$
   5. the EVSI
b. Should the survey be taken?

6. A mutual fund is contemplating the sale of a certain common stock. A study can be made that will yield two possible results, X_1 or X_2. Given the following information:

State of Nature	$P(\theta)$	$P(X_1|\theta)$	$P(X_2|\theta)$
θ_1: Stock price up	0.70	0.85	0.15
θ_2: Stock price down or same	0.30	0.60	0.40

	Opportunity Loss Table (Units of $10,000)	
	θ_1	θ_2
Sell	8	0
Do not sell	0	8

a. Find
   1. the optimal act before sampling
   2. the optimal act if X_1 results
   3. the optimal act if X_2 results
   4. $P(X_1)$ and $P(X_2)$
   5. the EVSI
b. If the cost of the study is $1000, what is the expected net gain from the study?

7. Assume that there are three possible states, S_1 (business will be a success), S_2 (business will have limited success), and S_3 (business will not be a success); with prior probabilities of occurrence 0.58, 0.25, and 0.17, respectively. The following is the opportunity loss table, in $thousands.

	Opportunity Loss Table		
	S_1	S_2	S_3
Invest in business	0	0	500
Do not invest	400	100	0

An investigation costing $10,000 is run for which there are three outcomes, X, Y, and Z. The probabilities of these outcomes are 0.6, 0.3, and 0.1, respectively. The posterior probabilities of S_i, given the respective outcomes, are

	X	Y	Z
S_1	0.8	0.3	0.1
S_2	0.1	0.5	0.4
S_3	0.1	0.2	0.5
	1.0	1.0	1.0

Draw a decision tree diagram and make all the necessary entries on the branches dealing with outcome X.

8. A pharmaceutical firm classifies its drugs as low-volume, medium-volume, and high-volume products. The firm introduces many new drugs each year. When a new drug is released, an initial marketing plan (that is, stocking of the drug, advertising, etc.) is put into effect. It is important that the firm initiate the proper market plan. For example, if a drug has a high-volume potential but a low-volume market plan is initiated, the drug will not realize its potential. The opportunity loss table for different marketing plans is

| Marketing | State of Nature | | |
Plan Used	Low Volume	Medium Volume	High Volume
Low	$ 0	$200,000	$700,000
Medium	200,000	0	500,000
High	400,000	200,000	0

A new drug, Novatol, is to be released. The marketing vice-president feels the probability that it will be high-volume is 0.2 and low-volume 0.1. A forecasting model has been developed that has the following reliability [that is, P(indication/state of nature)]

| Model | State of Nature | | |
Indicates	Low Volume	Medium Volume	High Volume
Low	0.85	0.10	0.05
Medium	0.10	0.80	0.10
High	0.05	0.10	0.85

What is the expected value of information from the forecasting model for the drug Novatol?

9. In Exercise 3 of Section 13.8, you are offered Stock, Inc. investment advice, which is correct 70% of the time. What is the maximum amount you should pay for Stock, Inc. advice on the predicted price movement of Rerox stock?

10. In Exercise 5 on p. 607, assume that the cost to the publisher was $200 to solicit the opinion of a professional critic.
    a. If the publisher solicits the opinion of just one critic, how much is the expected net gain from that critic's advice?
    b. What is the expected net gain of querying two critics?

11. A large manufacturing firm is planning to build a new plant in either Georgia or Pennsylvania. The cost of the required land would be $750,000 less in Georgia due to the inexpensiveness of the land and special tax incentives. However, in Pennsylvania a skilled labor force would be available, whereas that might not be the case in Georgia. It costs the company $5000 to train and pay an employee until the worker is productive. It would be cheaper to move a skilled labor force into the area, but it is against company policy to relocate labor. The firm needs 500 skilled workers. Assume that either 300, 350, or 400 skilled employees can be hired in Georgia and the rest would have to be trained. Management feels that the probabilities of hiring these numbers of employees are 0.3, 0.4, and 0.3, respectively. A survey of the labor situation can be undertaken. The reliability of the survey is

Survey	State of Nature		
Indicates	300	350	400
300	0.8	0.3	0.1
350	0.1	0.6	0.2
400	0.1	0.1	0.7

What is the maximum amount that the firm should pay for the survey?

12. In Exercise 6 above suppose the mutual fund felt that the prior probability the common stock price would increase was 0.95 instead of 0.70. What would be the expected net gain from the study? Compare it to the expected net gain if the prior probability is 0.70.

13. A mail order house with a fixed "market" of 100,000 people is deciding whether to sell a new line of goods. If more than 30% of its customers will purchase, the company should market the line, and if less than 30% purchase, it should not. For simplicity, assume that either 20% or 40% will purchase the new line. The manager of the firm believes the probability that only 20% of the market will buy is 0.6. The payoff table (in units of $10,000) is

	20% Will Purchase	40% Will Purchase
Market new line	−3	5
Do not market new line	0	0

a. One person is selected at random from the 100,000 and asked whether he will purchase. What is the expected value of this information?

b. Suppose two persons are selected at random from the 100,000 and asked whether they will purchase. What is the expected value of this information?

15.3 EXTENSIVE-FORM AND NORMAL-FORM ANALYSES

The type of preposterior investigation carried out in the A. B. Westerhoff Company problem is known as *extensive-form analysis*. It is perhaps easiest to characterize this type of analysis in terms of a decision tree diagram, such as is shown in Figure 15-2. In that diagram, as indicated previously, a prior analysis is given in the upper part of the tree with the resulting prior expected opportunity loss figure entered at node (b). Similarly, an extensive-form analysis is given in the lower part of the tree with the resulting preposterior expected opportunity loss entered at node (c).

For purposes of comparison, we summarize the procedures for prior, posterior, and extensive-form preposterior analysis. This comparison assumes that experimental (sample) information is collected by a single-stage procedure as in the A. B. Westerhoff Company problem. Each type of analysis may be thought of as starting at the right side of a decision tree diagram and then proceeding inward by the process referred to earlier as backward induction.

1. *Prior analysis*
   a. Sketch a tree and depict states of nature at the right tips
   b. Assign payoffs to each of the states for each possible action
   c. Assign prior probabilities to each state of nature
   d. Calculate expected terminal payoffs for each act
   e. Select the act with the highest expected terminal payoff

2. *Posterior analysis*
   a. Do steps (a) and (b) as in prior analysis
   b. Assign posterior probabilities to each state of nature based on a specific outcome of the experimental information
   c. Do steps (d) and (e) as in prior analysis

3. *Preposterior analysis—extensive-form*
   a. Do steps (a) and (b) as in prior analysis
   b. Assign marginal probabilities to each possible experimental (sample) outcome
   c. Assign posterior probabilities to each state of nature given specific experimental outcomes

d. For each experimental outcome, carry out the posterior analysis as in 2(c)
e. Weight the expected terminal payoffs of the best act for each type of experimental information by the marginal probability of occurrence of that type of information, and add these products to yield an overall preposterior expected payoff
f. The difference between the preposterior expected payoff and a prior expected payoff calculated as in 1 above is the EVSI

Concept of a Strategy

If there were many possible experiments, rather than one as assumed in the foregoing listing, the extensive-form analysis also specifies which of the possible experiments should be carried out. Moreover, the extensive-form analysis supplies a decision rule that selects an optimal act for each possible outcome of the chosen experiment. For example, in Figure 15-2, this decision rule was as follows:

$$X_1 \rightarrow a_1$$
$$X_2 \rightarrow a_2$$
$$X_3 \rightarrow a_2$$

where $X_1 \rightarrow a_1$ means that if experimental outcome X_1 is observed, select act a_1, etc. Such a decision rule is referred to as a *strategy*. Mathematically, a strategy can be defined as a function in which an act is assigned to each possible experimental outcome. We now turn to an alternative procedure to extensive-form analysis, known as *normal-form analysis*, in which the problem begins with a listing of all possible strategies that might be employed. Normal-form analysis then proceeds to make an explicit comparison of the worth of all of these strategies to arrive at the same optimal strategy as was derived in extensive-form analysis.

Normal-Form Analysis

In the preceding section, extensive-form analysis was applied to the A. B. Westerhoff Company problem. In order to indicate the nature of the normal-form procedure and the relationship between extensive and normal forms of analysis, the same problem will now be solved by the latter method. Then some factors relating to the choice between the two alternative procedures will be given.

All possible strategies or decision rules for the A. B. Westerhoff Company problem are listed in Table 15-5. The strategies, denoted s_1, s_2, . . . , s_8, indicate the acts taken in response to each sample outcome. Hence, for example, strategy s_3 is

TABLE 15-5
A Listing of All Possible Strategies in the A. B. Westerhoff Company Problem of
Deciding Whether to Conduct a Survey

Sample Outcome	Strategy							
	s_1	s_2	s_3	s_4	s_5	s_6	s_7	s_8
X_1	a_1	a_1	a_1	a_1	a_2	a_2	a_2	a_2
X_2	a_1	a_1	a_2	a_2	a_1	a_1	a_2	a_2
X_3	a_1	a_2	a_1	a_2	a_1	a_2	a_1	a_2

$$X_1 \rightarrow a_1$$
$$X_2 \rightarrow a_2$$
$$X_3 \rightarrow a_1$$

and s_4 is the one previously described. In this problem, there are eight possible strategies. There are three sample outcomes, X_1, X_2, and X_3, and two possible acts, a_1 and a_2. In general, the number of possible strategies is given by n^r, where n denotes the number of acts and r is the number of sample outcomes. Hence, in this case, there are $2^3 = 8$ possible strategies.

It is clear from an examination of Table 15-5 that some of the strategies are not very sensible. For example, strategies s_1 and s_8 select the same act regardless of the experimental outcome (survey indication). Strategy s_5 selects act a_2, not to market the product, if the "favorable" survey indication X_1 is obtained, but perversely it would market the product if the intermediate or "unfavorable" indications X_2 or X_3 are obtained. On the other hand, strategies s_2 and s_4 seem to be quite logical and would probably be the only ones a reasonable person would seriously consider on an intuitive basis if he or she contemplated using the experimental information at all.

Continuing with the normal-form analysis, we now compute the expected payoff of each possible strategy. The method consists in calculating for each strategy the expected opportunity loss conditional on the occurrence of each state of nature. These conditional expected losses are referred to in Bayesian decision analysis as *risk*. We will use that term or its equivalent, *"conditional expected losses."* The weighted average, or expected value, of these risks, using prior probabilities of states as weights, yields the *expected opportunity loss*, or *expected risk*, of the strategy. In Table 15-6 is shown the calculation of the risks associated with strategies s_2 and s_4. For example, let us consider the calculations for strategy s_2. In the left-hand portion of the table is given the payoff table in terms of opportunity losses. In the next section are shown "probabilities of action," which are probabilities of taking actions

TABLE 15-6

Calculation of Risks or Conditional Expected Opportunity Losses, for Strategies s_2 and s_4 (in \$Thousands)

State of Nature	Opportunity Loss		Strategy s_2 Probability of Action $s_2(a_1, a_1, a_2)$			Risk (Conditional Expected Loss) $R(s_2; \theta_i)$
	a_1	a_2	a_1	a_1	a_2	
θ_1	\$ 0	\$200	0.72	0.16	0.12	\$24.00
θ_2	160	0	0.08	0.12	0.80	32.00

$$R(s_2; \theta_1) = 0.72(\$0) + 0.16(\$0) + 0.12(\$200) = \$24.00$$
$$R(s_2; \theta_2) = 0.08(\$160) + 0.12(\$160) + 0.80(\$0) = \$32.00$$

State of Nature	Opportunity Loss		Strategy s_4 Probability of Action $s_1(a_1, a_2, a_2)$			Risk (Conditional Expected Loss) $R(s_4; \theta_i)$
	a_1	a_2	a_1	a_2	a_2	
θ_1	\$ 0	\$200	0.72	0.16	0.12	\$56.00
θ_2	160	0	0.08	0.12	0.80	12.80

$$R(s_4; \theta_1) = 0.72(\$0) + 0.16(\$200) + 0.12(\$200) = \$56.00$$
$$R(s_4; \theta_2) = 0.08(\$160) + 0.12(\$0) + 0.80(\$0) = \$12.80$$

specified by the given strategy based on the observations of the possible types of sample evidence. A convenient notation for designating a strategy is given in the caption of this section of the table. Hence, the notation $s_2(a_1, a_1, a_2)$ means strategy s_2 consists in taking act a_1 if the sample indication X_1 is observed; again selecting a_1 if X_2 is observed; and choosing a_2 if X_3 is observed. That is, the first element within the parentheses denotes the action to be taken upon observing the first sample indication, the second element specifies the action to be taken upon observing the second sample indication, and so on.

The risk, or expected opportunity loss, given the occurrence of a specific state of nature, is calculated by multiplying these probabilities of action by the respective losses incurred if these actions are taken and summing the products. Thus, the risk, or expected opportunity loss, associated with the use of strategy s_2, given that state of nature θ_1 occurs, denoted $R(s_2; \theta_1)$, is calculated to be \$24.00, as shown below the appropriate portion of Table 15-6. Correspondingly, the expected opportunity loss of strategy s_2, conditional on the occurrence of θ_2, denoted $R(s_2; \theta_2)$, is calculated to be $32.00. The analogous calculation of risks for strategy s_4 is given in the lower part of Table 15-6.

A revealing alternative way of thinking about these risks is that they are simply *calculations for each state of nature of the loss due to taking the wrong act times the probability that the wrong act will be taken.* For example, let us consider $R(s_2; \theta_1)$, the conditional expected loss of strategy s_2, given that θ_1 occurs. Now, if θ_1 occurs, the correct (best) course of action is a_1; the incorrect act is a_2. The $24.00 figure for $R(s_2; \theta_1)$ is merely $200, the loss of taking act a_2, times 0.12, the total probability of selecting a_2 under strategy s_2, given the occurrence of θ_1. Similarly, the $32.00 figure for $R(s_2; \theta_2)$ is equal to $160, the loss of taking act a_1, times 0.20, the total probability of selecting a_1 under strategy s_2, conditional on the occurrence of θ_2. The risk calculations of Table 15-6 are shown in Table 15-7, utilizing this alternative conception.

A summary of the risks, or conditional expected opportunity losses, for all eight strategies is given in Table 15-8. Let us review the interpretation of these figures, taking, as an example, strategy s_4. In a relative frequency sense (that is, in a large number of identical situations of successful new products, θ_1), if survey evidence X_1, X_2, and X_3 occur with the specified probabilities and if strategy s_4 is employed, the average opportunity loss per product would be $56.00. A similar interpretation holds for unsuccessful products, θ_2, with an average loss of $12.80. Now, we can weight these average losses by the prior probabilities of successful and unsuccessful products to obtain an overall expected opportunity loss for strategy s_4. The prior probabilities were given earlier as $P(\theta_1) = 0.25$

TABLE 15-7
Alternative Calculation of Risks, or Conditional Expected Opportunity Losses, for Strategies s_2 and s_4 (in $Thousands)

State of Nature	Opportunity Loss of Wrong Act	Strategy $s_2(a_1, a_1, a_2)$ Probability of Wrong Act Given θ_i	Conditional Expected Opportunity Loss $R(s_2; \theta_i)$
θ_1	$200	0.12	$24.00
θ_2	160	0.20	32.00

State of Nature	Opportunity Loss of Wrong Act	Strategy $s_4(a_1, a_2, a_2)$ Probability of Wrong Act Given θ_i	Conditional Expected Opportunity Loss $R(s_4; \theta_i)$
θ_1	$200	0.28	$56.00
θ_2	160	0.08	12.80

TABLE 15-8
Risks, or Conditional Expected Opportunity Losses, for the Eight Strategies in the A. B. Westerhoff Company Problem (in $Thousands)

State of Nature	Strategy							
	s_1	s_2	s_3	s_4	s_5	s_6	s_7	s_8
θ_1	$ 0	$24.00	$ 32.00	$56.00	$144.00	$168.00	$176.00	$200.00
θ_2	160.00	32.00	140.80	12.80	147.20	19.20	128.00	0

and $P(\theta_2) = 0.75$. Since θ_1 occurs with probability 0.25 and the conditional expected loss if θ_1 occurs is $56.00, and since θ_2 occurs with probability 0.75 and the conditional expected loss if θ_2 occurs is $12.80, the expected opportunity loss of using strategy s_4 is

$$\text{EOL}(s_4) = (0.25)(\$56.00) + (0.75)(\$12.80) = \$23.60$$

In symbols, we have

$$\text{EOL}(s_4) = P(\theta_1)R(s_4; \theta_1) + P(\theta_2)R(s_4; \theta_2)$$

To state this relationship in general form, the expected opportunity loss of the kth strategy, s_k, is

$$(15.2) \quad \text{EOL}(s_k) = P(\theta_1)R(s_k; \theta_1) + P(\theta_2)R(s_k; \theta_2) + \cdots + P(\theta_m)R(s_k; \theta_m)$$

$$= \sum_{i=1}^{m} P(\theta_i)R(s_k; \theta_i)$$

The decision rule for which this expected opportunity loss is a minimum is known as the "Bayes' strategy."

The expected opportunity losses of the eight strategies in the A. B. Westerhoff Company problem as calculated from Equation 15.2 are given in Table 15-9. The optimal strategy, or the one for which the expected opportunity loss is least, is s_4. An asterisk has been placed beside the $23.60 expected opportunity loss associated with decision rule s_4 to indicate that it is the minimum loss figure.

In summary, using normal-form analysis, the optimal strategy in this problem is $s_4(a_1, a_2, a_2)$; this strategy has an expected opportunity loss of $23.60 (thousands). If we refer to Figure 15-2, we see that this is exactly the same solution arrived at by extensive-form analysis. In that figure, the $23.6 is shown at node (c) and the optimal strategy (a_1, a_2, a_2) is arrived at by noting for the survey outcomes X_1, X_2, and X_3 the forks that have not been blocked off.

We can now summarize the steps involved in a normal-form analysis, in which sample evidence is obtained from a single sample (experiment) as in the foregoing example.

TABLE 15-9
Expected Opportunity Losses of the Eight
Strategies in the A. B. Westerhoff
Company Problem

Strategy	Expected Opportunity Loss
s_1	$120.00
s_2	30.00
s_3	113.60
s_4	23.60*
s_5	146.40
s_6	56.40
s_7	140.00
s_8	50.00

Normal-form analysis
1. List all possible strategies in terms of actions to be taken upon observation of sample outcomes
2. Calculate the conditional expected opportunity loss (risk) for each state of nature. The probabilities to be used in this calculation are conditional probabilities of sample outcomes, given states of nature
3. Compute the (unconditional) expected opportunity loss of each strategy by weighting the conditional expected opportunity losses by the prior probabilities of states of nature
4. Select the strategy that has the minimum expected opportunity loss

This summary of normal-form analysis has been given in terms of a single experiment. If more than one experiment were conducted, the decision maker should carry out steps (a) through (d) above and then select the experiment that yields the lowest expected opportunity loss. Furthermore, the summary has been expressed in terms of opportunity losses. If payoffs of utility or profits were used, the same procedures would be followed except that the decision maker would maximize expected utility or profit rather than minimize expected opportunity loss.

15.4 COMPARISON OF EXTENSIVE-FORM AND NORMAL-FORM ANALYSES

As we have seen, the extensive and normal forms of analysis are equivalent approaches. In both procedures, an expected opportunity loss (or expected profit) is calculated before experimental results are observed.

This expected payoff anticipates the selection of optimal acts after the observation of experimental outcomes. However, there are differences between the two types of analysis in the way in which the various components of the procedures are performed. These differences give rise to advantages and disadvantages in the two forms of analysis.

As was clear from the A. B. Westerhoff Company example, the extensive-form solution can be calculated more rapidly. This follows from the fact that *in the extensive-form approach, it is not necessary to carry out expected loss calculations for every possible decision rule.* Because of the blocking off of nonoptimal courses of action posterior to the observation of sample evidence, only the expected loss for the optimal strategy need be carried out. In some practical problems, the number of decision rules or strategies that must be evaluated in a normal-form analysis may be very large. For example, in the problem discussed in this chapter, the number of possible strategies was $2^3 = 8$, because there were two acts and three experimental outcomes. If there had been three acts $(n = 3)$ and four experimental outcomes $(r = 4)$, there would have been $n^r = 3^4 = 81$ strategies for which normal-form calculations would be required.

On the other hand, the normal form of analysis may appeal more to those who feel uneasy about making the subjective probability assessments involved in preposterior analysis. In many problems, the conditional probabilities of sample outcomes, given states of nature, may be based on relative frequencies, as in the A. B. Westerhoff Company case, or on an appropriate probability distribution, as in another problem discussed later in this chapter. However, the prior probabilities, $P(\theta_i)$, will most likely represent subjective or judgmental assignments in most real problems. In normal-form analysis, these prior probabilities are applied as the last step in the calculations. Therefore, it is possible to proceed all the way to this point without introducing subjective probability assessments. It is then possible to judge how sensitive the choice of the optimal strategy is to the prior probability assignments. That is, we can determine by how much the magnitudes of the prior probabilities may be permitted to change without a shift occurring in the optimal decision rule to be employed. This procedure is discussed below.

Exercises

1. Define and contrast the extensive-form and normal-form approaches.

2. For Exercise 4 of Section 15.2, construct a table showing all possible strategies and select the optimal strategy if the two items are inspected.

3. For Exercise 7 of Section 15.2, the following strategy table has been constructed. The two alternative courses of action are a_1: invest in business and a_2: do not invest. Fill in those strategies that are missing and select the optimal strategy.

Sample Outcome	s_1	s_2	s_3	s_4	s_5	s_6	s_7	s_8
X	a_1	a_1	a_1	a_1	a_2	a_2	a_2	a_2
Y		a_2	a_1	a_1		a_2	a_1	a_1
Z		a_1	a_2	a_1		a_1	a_2	a_1

4. In Exercise 8 of Section 15.2, there are 27 strategies. Assume that the optimal strategy is one of the following three:

Model Indication	s_5	s_{15}	s_{23}
Low	L	M	L
Medium	M	M	M
High	M	H	H

L, M, and H mean low, medium, and high market plans used, respectively. Compute conditional expected losses and expected opportunity losses to determine which of these three strategies is best.

5. The following table pertains to Exercise 11 of Section 15.2. Fill in the missing strategies. The two courses of action are a_1: build in Pennsylvania and a_2: build in Georgia.

Sample Outcome	s_1	s_2	s_3	s_4	s_5	s_6	s_7	s_8
300	a_1	a_1	a_1	a_1	a_2	a_2	a_2	a_2
350	a_1	a_1	a_2		a_1	a_1		a_2
400		a_1	a_2			a_1		a_1

Compute the conditional expected losses as well as the expected opportunity losses for strategies s_1, s_3, and s_4. Which strategy is optimal?

6. Suppose there are three states of nature, θ_1, θ_2, and θ_3; two actions, a_1 and a_2; and three outcomes from a forecast, X, Y, and Z. The outcome X indicates that θ_1 will occur, Y indicates θ_2 will occur, and Z indicates θ_3 will occur. The opportunity loss table is

	a_1	a_2
θ_1	0	1300
θ_2	300	0
θ_3	800	0

A forecasting model has been developed that has the following reliability:

| | Model Indicates | | |
State of Nature	X	Y	Z
θ_1	0.75	0.15	0.10
θ_2	0.09	0.50	0.41
θ_3	0.07	0.09	0.84

Of the eight strategies only two are plausible: $s_3(a_1, a_2, a_2)$ and $s_5(a_1, a_1, a_2)$. Use normal-form analysis to determine the optimal strategy. In computing the conditional expected losses for the two strategies, use the *loss* of taking the wrong act times the probability of taking the wrong act. The prior probabilities for θ_1, θ_2, and θ_3 are 0.35, 0.30, and 0.35, respectively.

15.5 SENSITIVITY ANALYSIS

In the new product decision problem discussed in this chapter, the point was made that only two decision rules appeared reasonable on an intuitive basis, namely, s_2 and s_4. As can be seen in Table 15-9, these strategies yielded the two lowest expected opportunity losses, with figures of $30.00 and $23.60, respectively, for s_2 and s_4. However, we can observe in Table 15-6 that although the conditional expected loss for s_2 is lower than s_4 if θ_1 occurs ($24.00 versus $56.00, respectively), the opposite is true if θ_2 occurs ($32.00 versus $12.80, respectively). Since these conditional expected losses were weighted by prior probabilities of states of nature, $P(\theta_1)$ and $P(\theta_2)$, it is clear that to a considerable extent, the choice between strategies s_2 and s_4 is dependent on the magnitudes of these prior probabilities. We can test how *sensitive* the choice between decision rules s_2 and s_4 is to the size of these prior probabilities. This test is an example of *sensitivity analysis*, that is, a study that tests how sensitive the solution to a decision problem is to changes in the data for the variables of the problem.

In the present problem, a sensitivity test can be accomplished by solving for a *break-even value for the prior probability of one of the two states of nature* such that the expected opportunity losses of the two strategies will be equal. If the prior probability rises above this break-even value, s_2 becomes the optimal act rather than s_4. We illustrate this procedure in the new product decision problem. As indicated earlier, the (unconditional) expected opportunity losses of the two strategies in question were computed as follows:

$$\text{EOL}(s_2) = (0.25)(\$24.00) + (0.75)(\$32.00) = \$30.00$$
$$\text{EOL}(s_4) = (0.25)(\$56.00) + (0.75)(\$12.80) = \$23.60$$

If we now substitute p for the 0.25 value of $P(\theta_1)$, and $1 - p$ for the 0.75 figure of $P(\theta_2)$, we can solve for the value of p for which $\text{EOL}(s_2) = \text{EOL}(s_4)$. Making this substitution and equating the resulting expressions for expected opportunity loss, we have

$$p(\$24.00) + (1 - p)(\$32.00) = p(\$56.00) + (1 - p)(\$12.80)$$

Carrying out the multiplications, we obtain

$$\$24.00p + \$32.00 - \$32.00p = \$56.00p + \$12.80 - \$12.80p$$

Collecting terms, we find the break-even value of p.

$$\$51.20p = \$19.20$$

$$p = \frac{19.20}{51.20} = 0.375$$

In summary, we conclude that if $p = P(\theta_1) = 0.375$, then the expected opportunity losses of strategies s_2 and s_4 would be equal, and we would have no preference between them. In the A. B. Westerhoff Company new product decision problem, $p = P(\theta_1) = 0.25$. Hence, we observe that this subjective prior probability assignment could have varied up to a value of 0.375 and s_4 would still have been a better strategy than s_2. However, for p values in excess of 0.375, s_2 has a lower expected opportunity loss and is therefore the better rule.

Of course, break-even values of p could be determined between s_4 and other strategies as well. In the case of a strategy such as s_5, whose conditional expected values given θ_1 and θ_2 are each higher than the corresponding figures for s_4 ($144.00 versus $56.00 and $147.20 versus $12.80, respectively), it is impossible for the weighted average or expected opportunity loss to be less than that of s_4, regardless of the weights p and $1 - p$. In a situation such as this, the strategy s_4 is said to *dominate* s_5. As a general definition, a strategy, say s_1, is said to be a *dominating strategy* with respect to s_2 if

$$R(\theta; s_1) \leq R(\theta; s_2) \qquad \text{for all values of } \theta$$

and

$$R(\theta; s_1) < R(\theta; s_2) \qquad \text{for at least one value of } \theta$$

In words, s_1 is said to dominate s_2 if the conditional expected losses (risks) for s_1 are *equal to or less than* the corresponding figures for s_2 for every state of nature, *and* the conditional expected loss for s_1 is *less than* the corresponding figure for s_2 for at least one value of θ.

The preceding discussion illustrated the use of sensitivity analysis to determine the effects of variations in prior probabilities on the selection of a best act. We could also determine the effects of changes in the entries in the payoff matrix on our choice of a best act. In general, in any decision analysis problem, it is very useful to test how sensitive the solution is to changes in all of the important variables of the problem. In problems involving numerous states of nature, experimental outcomes, and courses of action, many calculations may be required to carry out such an analysis. The use of computers in such situations is usually a practical necessity.

15.6 AN ACCEPTANCE SAMPLING EXAMPLE

As another illustration of preposterior analysis, we will return to the problem of acceptance sampling of a manufactured product discussed in Section 14.2. Since the problem was posed there as one in posterior analysis, we will make the necessary changes to convert it to a problem in preposterior analysis. We assume, as previously, a situation in which the Renny Corporation inspects incoming lots of articles produced by a supplier in order to determine whether to accept or reject these lots. We also retain the assumptions that in the past, incoming lots from this supplier have contained either 10%, 20%, or 30% defectives and that lots with these percentages of defectives have occurred 50%, 30%, and 20% of the time, respectively. The same payoff matrix as shown in Table 14-5 is assumed. Table 14-6 showed the calculation of prior expected opportunity losses for the two acts,

a_1: Reject the incoming lot[1]
a_2: Accept the incoming lot

if action were taken without sampling. For convenience, this information on events, prior probabilities, payoffs, and prior expected opportunity losses is summarized in Table 15-10. The preferable action without sampling is act a_2, "accept the incoming lot," which has a prior expected opportunity loss of $70 as opposed to act a_1, "reject the incoming lot," which has a prior expected opportunity loss of $100.

Let us now assume that we would like to know whether it is better to make the decision concerning acceptance or rejection of the incoming lot without sampling or to draw and inspect a random sample of items from the incoming shipment and then make the decision. We recognize this problem as one of preposterior analysis, specifically one that requires an evaluation of the *expected value of sample information* (EVSI) and the *expected net gain of sampling* (ENGS). We assume that a

---

[1] The alternative actions have been denoted here by lower case rather than capital letters to maintain consistency with the notation we have used in specifying strategies.

TABLE 15-10
Basic Data for Acceptance Sampling Problem for the Renny Corporation

Event p (Lot Proportion of Defectives)	Prior Probability $P_0(p)$	Opportunity Loss a_1 Reject	a_2 Accept
0.10	0.50	$200	0
0.20	0.30	0	100
0.30	0.20	0	200
	1.00		

Prior EOL$(a_1) = (0.50)(\$200) + (0.30)(\$0) + (0.20)(\$0) = \$100$
Prior EOL$(a_2) = (0.50)(\$0) + (0.30)(\$100) + (0.20)(\$200) = \$70$

sample of two articles drawn without replacement is contemplated and the cost of sampling and inspecting these two articles is $5. Hence, our problem is one of choice between a terminal decision without sampling and a decision to examine a fixed-size sample of two articles and then select a terminal act.

Of course, a sample of only two articles may be entirely too small to be realistic in many cases. On the other hand, in the case of certain manufactured complex assemblies of large unit cost where the testing procedure is destructive (for example, in missile testing or in space probing where the test vehicle is not retrievable), it may only be feasible to test a very small number of items. However, the principles illustrated in this problem are perfectly general. The assumption of a larger sample size would merely increase the computational burden. Furthermore, in Section 15.7 we will consider appropriate decision theory procedures for determining an optimal sample size.

Again as in the preceding example in this section, we will solve the problem in two different ways, first through the use of extensive-form analysis, and then by normal-form analysis.

Extensive-Form Analysis

The decision diagram for the preposterior analysis of the Renny Corporation problem is given in Figure 15-3. As indicated in the legend, the alternative actions in this problem have been denoted

a_1: Reject the lot
a_2: Accept the lot

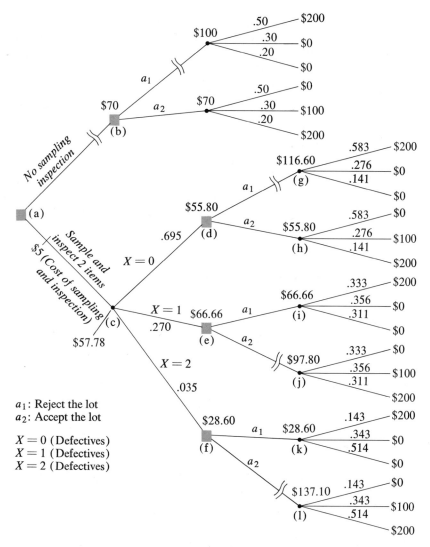

FIGURE 15-3
Decision diagram for preposterior analysis of the lot acceptance decision for the Renny Corporation problem.

The possible sample outcomes based on a random sample of two articles drawn from the incoming lot are denoted

$$X = 0 \text{ defectives}$$
$$X = 1 \text{ defective}$$
$$X = 2 \text{ defectives}$$

At node (a), the two basic choices are shown: not to engage in sampling inspection, or to sample and inspect two articles from the incoming lot. The prior analysis for the "no sampling" choice is shown in the usual way in the upper portion of the tree. As previously observed in Table 15-10, the prior expected opportunity loss of rejection, a_1, is $100 and that of acceptance, a_2, is $70. Hence, action a_1 has been blocked off on the branch emanating from node (b) as a nonoptimal act, and the lower figure of $70 for a_2 has been entered at node (b). This $70 figure is the expected opportunity loss of choosing the optimal act without sampling.

On the other hand, the alternative choice is to sample and inspect two items prior to making the terminal decision of acceptance or rejection of the lot. To carry out the calculations for this part of the analysis, we begin at the right tips of the tree and proceed inward by backward induction in the usual way. The expected values of $116.60 and $55.80 shown at nodes (g) and (h), for example, are the expected opportunity losses of taking acts a_1 and a_2, respectively, after having observed no defectives $(X = 0)$ in the sample of two articles. The probabilities used to calculate these expected values are the conditional (posterior) probabilities, given $X = 0$, that the lot proportion of defectives, p, is 0.10, 0.20, and 0.30, respectively. In symbols,

$$P(p = 0.10 \,|\, X = 0) = 0.583$$
$$P(p = 0.20 \,|\, X = 0) = 0.276$$
$$P(p = 0.30 \,|\, X = 0) = 0.141$$

The calculation of these posterior probabilities is shown in Table 15-12 and will be explained shortly. Since acceptance of the lot (act a_2) has a lower expected loss than does rejection (act a_1), given that no defectives have been observed in the sample, the lower figure, $55.80, is entered at node (d). Furthermore, act a_1 is blocked off as nonoptimal. Corresponding figures of $66.66 and $28.60 for expected losses of optimal acts after observing one and two defectives are shown at nodes (e) and (f), respectively. Rejection, act a_1, is shown to be optimal after observation of either one or two defectives. Weighting the three payoffs shown at nodes (d), (e), and (f) by the marginal probabilities of observing zero, one, and two defectives (0.695, 0.270, and 0.035, respectively), we find a figure of $57.78 as the expected opportunity loss of sampling and inspecting two articles and taking terminal action after observing the sample evidence. Before proceeding with the calculation of EVSI and ENGS, we pause to consider the method of calculating the marginal and posterior probabilities mentioned in this paragraph.

In order to calculate the marginal probabilities of the sample evidence, $P(X = 0)$, $P(X = 1)$, and $P(X = 2)$, and the posterior probabilities of the form $P(p \,|\, X)$, we must obtain the joint probability distribution of the sample evidence and events, that is, $P(X \text{ and } p)$, for $X = 0, 1,$ and 2, and $p = 0.10, p = 0.20,$ and $p = 0.30$. As in the A. B. Westerhoff Company problem, these joint probabilities are obtained by multiplying prior probabilities of events, in this case, $P_0(p)$, by the appropriate conditional probabilities of sample evidence given events, in this case, $P(X \mid p)$. In the A. B. Westerhoff Company problem, it was assumed that these probabilities of the type $P(\text{sample outcomes} \mid \text{events})$ were based on past relative frequencies of occurrence in surveys previously conducted by a market research firm. In the present problem, we must calculate these $P(X \mid p)$ values by using an appropriate probability distribution. Let us assume that the incoming lot from which the sample of two articles is drawn is very large relative to the size of the sample. Then, as previously indicated in Section 2.4, we can assume that even if the sample is selected without replacement, the drawings of the articles may be considered, for practical purposes, trials of a Bernoulli process, and the *binomial distribution* may be used to calculate probabilities of sample outcomes. That is, we may assume that since there is so little change in the population (lot) because of the drawing of the first article, the probability of obtaining a defective item on the second draw may be considered the same as on the first. On the other hand, if the incoming lot is not large relative to the sample size, that is, if it is not at least ten times the sample size, then the hypergeometric distribution is appropriate. The $P(X \mid p)$ values for the present problem (that is, the conditional probabilities of observing zero, one, or two defectives, given lot proportion defectives of 0.10, 0.20, and 0.30 calculated by the use of the binomial distribution) are shown in Table 15-11. As indicated in the table, the probabilities of zero, one, and two defectives in a sample of size two from a lot that contains 0.10 defectives are given by the respective terms of the binomial distribution whose parameters are $n = 2$, $p = 0.10$, and

TABLE 15-11
Conditional Probabilities of Specified Numbers of
Defectives in a Sample of Two Articles from an
Incoming Lot in the Renny Corporation Problem

$$P(X = 0 \mid p = 0.10) = \quad (0.90)^2 \qquad = 0.81$$
$$P(X = 1 \mid p = 0.10) = 2(0.90)(0.10) = 0.18$$
$$P(X = 2 \mid p = 0.10) = \quad (0.10)^2 \qquad = \underline{0.01}$$
$$1.00$$

$$P(X = 0 \mid p = 0.20) = \quad (0.80)^2 \qquad = 0.64$$
$$P(X = 1 \mid p = 0.20) = 2(0.80)(0.20) = 0.32$$
$$P(X = 2 \mid p = 0.20) = \quad (0.20)^2 \qquad = \underline{0.04}$$
$$1.00$$

$$P(X = 0 \mid p = 0.30) = \quad (0.70)^2 \qquad = 0.49$$
$$P(X = 1 \mid p = 0.30) = 2(0.70)(0.30) = 0.42$$
$$P(X = 2 \mid p = 0.30) = \quad (0.30)^2 \qquad = \underline{0.09}$$
$$1.00$$

corresponding calculations provide the probabilities for the cases in which $p = 0.20$ and 0.30.

Table 15-12 shows the computation of the joint probability distribution $P(X$ and $p)$. This joint distribution is obtained by multiplying the prior probabilities, $P_0(p)$, by the conditional probabilities, $P(X|p)$, derived in Table 15-11. For example, 0.405, the upper left entry in the joint probability distribution in Table 15-12, was obtained by multiplying 0.50 by 0.81. In symbols, $P(X = 0$ and $p = 0.10) = P_0(p = 0.10)$ $P(X = 0|p = 0.10)$. We can now obtain the marginal probabilities of the sample outcomes of zero, one, and two defectives. The column totals in the joint probability distribution are $P(X = 0) = 0.695$, $P(X = 1) = 0.270$, and $P(X = 2) = 0.035$, respectively. These are the marginal probabilities entered in the decision diagram in Figure 15-3, on the branches emanating from node (c).

The calculations of posterior, or revised, probabilities of lot proportion of defectives given the sample outcomes $X = 0$, 1, and 2 are shown in Table 15-13. These posterior probabilities were obtained by dividing the joint probabilities by the appropriate column totals, $P(X)$. For example, the probability that the incoming lot contains 0.10 defectives, given that no defectives were observed in the sample of two articles, is

$$P(p = 0.10 | X = 0) = \frac{P(X = 0 \text{ and } p = 0.10)}{P(X = 0)} = \frac{0.405}{0.695} = 0.583$$

As previously indicated in connection with Table 15-4, the calculation of these posterior probabilities represents an application of Bayes' Theorem. The posterior probabilities are shown in Figure 15-3 on the branches stemming from nodes (g), (h), (i), (j), (k), and (l).

We can now complete the extensive-form analysis. As we noted earlier, the $70 at node (b) in Figure 15-3 represents the expected loss of terminal action without sampling. Correspondingly, the $57.78 at node

TABLE 15-12
Calculation of the Joint Probability Distribution of Sample Outcomes and Events in the Renny Corporation Problem

Event p (Lot Proportion of Defectives)	Prior Probability $P_0(p)$	Conditional Probability $P(X\|p)$			Joint Probability $P(X$ and $p)$			
		$X = 0$	$X = 1$	$X = 2$	$X = 0$	$X = 1$	$X = 2$	Total
0.10	0.50	0.81	0.18	0.01	0.405	0.090	0.005	0.50
0.20	0.30	0.64	0.32	0.04	0.192	0.096	0.012	0.30
0.30	0.20	0.49	0.42	0.09	0.098	0.084	0.018	0.20
					0.695	0.270	0.035	1.00

TABLE 15-13
Calculation of the Posterior Probabilities of Lot
Proportion of Defectives in the Renny
Corporation Problem

$$P(p = 0.10|X = 0) = 0.405/0.695 = 0.583$$
$$P(p = 0.20|X = 0) = 0.192/0.695 = 0.276$$
$$P(p = 0.30|X = 0) = 0.098/0.695 = \underline{0.141}$$
$$1.000$$

$$P(p = 0.10|X = 1) = 0.090/0.270 = 0.333$$
$$P(p = 0.20|X = 1) = 0.096/0.270 = 0.356$$
$$P(p = 0.30|X = 1) = 0.084/0.270 = \underline{0.311}$$
$$1.000$$

$$P(p = 0.10|X = 2) = 0.005/0.035 = 0.143$$
$$P(p = 0.20|X = 2) = 0.012/0.035 = 0.343$$
$$P(p = 0.30|X = 2) = 0.018/0.035 = \underline{0.514}$$
$$1.000$$

(c) is the expected loss of sampling and inspecting two articles and then taking terminal action. Hence, the *expected value of sample information* is

$$\text{EVSI} = \$70 - \$57.78 = \$12.22$$

Thus, the expected amount by which the terminal opportunity loss of action without sampling is reduced by the procedure of sampling two articles and taking action after inspection of the sample outcomes is $12.22. Since the sampling and inspection of two articles costs $5.00, the *expected net gain of sample information* is

$$\text{ENGS} = \$12.22 - \$5.00 = \$7.22$$

Therefore, since the expected net gain of sample information is $7.22 for terminal action after sampling and inspecting two articles as compared to terminal action without sampling, it pays the Renny Corporation to follow the former course of action.

Normal-Form Analysis

We now turn to the normal-form analysis of the Renny Corporation problem, which we have just solved by extensive-form procedures. As in the A. B. Westerhoff problem, we begin the normal-form analysis by listing all possible strategies, or decision rules. These are shown in Table 15-14. We can see on an intuitive basis that some of the strategies are illogical and would therefore tend to have relatively large expected opportunity losses associated with them. For example, strategy $s_2(a_1, a_1, a_2)$ rejects the incoming lot if the sample of two articles yields zero or one

TABLE 15-14
A Listing of All Possible Strategies in the Renny Corporation Problem

Sample Outcome (Number of Defectives)	s_1	s_2	s_3	s_4	s_5	s_6	s_7	s_8
				Strategy				
$X = 0$	a_1	a_1	a_1	a_1	a_2	a_2	a_2	a_2
$X = 1$	a_1	a_1	a_2	a_2	a_1	a_1	a_2	a_2
$X = 2$	a_1	a_2	a_1	a_2	a_1	a_2	a_1	a_2

a_1: Reject the lot
a_2: Accept the lot

defectives, but accepts the lot if two defectives are observed. Strategy $s_4(a_1, a_2, a_2)$ also employs a perverse type of logic, since it rejects the lot on an observation of zero defectives but accepts it if one or two defectives are observed. Strategy $s_1(a_1, a_1, a_1)$ rejects the lot, and $s_8(a_2, a_2, a_2)$ accepts the lot, regardless of the sample observation. The two most logical decision rules appear to be strategy $s_5(a_2, a_1, a_1)$, which accepts the lot if zero defectives are observed in the sample and rejects the lot if one or two defectives are observed, and strategy $s_7(a_2, a_2, a_1)$, which accepts the lot if zero or one defectives are observed and rejects the lot if two defectives are found.

The next step is to calculate for each strategy the risks or conditional expected losses associated with the occurrence of each event (lot proportion of defectives). We will illustrate these calculations for strategies s_5 and s_7 by computing, for each proportion of defectives, the *loss* of taking the wrong act times the *probability* of taking the wrong act. These risks, denoted $R(s_5;p)$ and $R(s_7;p)$, are shown in Table 15-15. For example, let us consider strategy $s_5(a_2, a_1, a_1)$. The risk, or conditional expected loss, if the lot proportion of defectives is 0.10 is equal to $R(s_5;p = 0.10) = \$200 \times 0.19$ by the following reasoning. Referring back to the original payoff matrix in Table 15-10, we recall that the correct course of action if $p = 0.10$ is to accept, and if $p = 0.20$ or 0.30, to reject the lot. Strategy $s_5(a_2, a_1, a_1)$ accepts the lot on the observation of no defectives and rejects it otherwise. From Table 15-12 we find a conditional probability of 0.81 of observing no defectives, given that the lot proportion of defectives is $p = 0.10$. Hence, with strategy s_5, the probability of making the wrong decision, given a lot proportion of defectives $p = 0.10$, is $1 - 0.81 = 0.19$. The loss associated with rejection if $p = 0.10$ is \$200 (Table 15-10). Therefore, as shown

TABLE 15-15
Calculation of Risks or Conditional Expected Opportunity Losses for Strategies s_5 and s_7

Event p (Lot Proportion of Defectives)	Strategy $s_5(a_2, a_1, a_1)$		Risk (Conditional Expected Loss) $R(s_5; p)$
	Opportunity Loss of Wrong Act	Probability of Wrong Act Given p	
0.10	$200	0.19	$38
0.20	100	0.64	64
0.30	200	0.49	98

Event p (Lot Proportion of Defectives)	Strategy $s_7(a_2, a_2, a_1)$		Risk (Conditional Expected Loss) $R(s_7; p)$
	Opportunity Loss of Wrong Act	Probability of Wrong Act Given p	
0.10	$200	0.01	$ 2
0.20	100	0.96	96
0.30	200	0.91	182

in Table 15-15, the risk of strategy s_5, given $p = 0.10$, is $R(s_5; p = 0.10) = \$200 \times 0.19 = \$38$. Analogous calculations produce the other risks shown in Table 15-15. A summary of the risks for all eight strategies is presented in Table 15-16.

We can now calculate the expected loss of each strategy using Equation 15.2. Adapting the notation of that equation to that used in the

TABLE 15-16
Risks or Conditional Expected Opportunity Losses for the Eight Strategies in the Renny Corporation Problem

Event p (Lot Proportion of Defectives)	Strategy							
	s_1	s_2	s_3	s_4	s_5	s_6	s_7	s_8
0.10	$200	$198	$164	$162	$38	$ 36	$ 2	$ 0
0.20	0	4	32	36	64	68	96	100
0.30	0	18	84	102	98	116	182	200

TABLE 15-17
**Expected Opportunity Losses of the Eight
Strategies in the Renny Corporation Problem**

Strategy	Expected Opportunity Loss
s_1	\$100.00
s_2	103.80
s_3	108.40
s_4	112.20
s_5	57.80*
s_6	61.60
s_7	66.20
s_8	70.00

present problem, we obtain as the expected opportunity loss of the kth strategy

(15.3)
$$\text{EOL}(s_k) = \sum_p P_0(p)R(s_k;p)$$

That is, for each strategy, we weight the conditional expected losses associated with each lot proportion of defectives (Table 15-16) by the prior probabilities of such incoming lots (Table 15-12).

Hence, the expected opportunity losses (expected risks) of strategies s_5 and s_7 are

$$\text{EOL}(s_5) = (0.50)(\$38) + (0.30)(\$64) + (0.20)(\$98) = \$57.80$$
$$\text{EOL}(s_7) = (0.50)(\$2) + (0.30)(\$96) + (0.20)(\$182) = \$66.20$$

The expected opportunity losses of all eight strategies are given in Table 15-17. The optimal strategy is seen to be $s_5(a_2, a_1, a_1)$, since it has the minimum expected opportunity loss of \$57.80. This strategy, which accepts the lot if no defectives are observed in the sample and rejects the lot if one or two defectives are found, is seen to be the same as the one found in the preceding extensive-form analysis depicted in the decision diagram in Figure 15-3. The minor difference between the expected opportunity losses of the optimal strategy, \$57.78 in the extensive-form analysis and \$57.80 in the normal-form analysis, is attributable to rounding of decimal places.

Exercises

1. In Exercise 5 at the end of Section 14.2, at what value of the prior probability of success would the decision concerning publication change?

2. A certain manufacturing process produces lots of 500 units each. In each lot,

either 10% or 30% of the 500 items are defective. The quality control department inspects each lot before shipment. If accepted, the lot produces a profit of $500, but if rejected, the lot is sold for scrap at cost. If a 30% defective lot is sent out, confidence in the company is lost and the firm estimates its loss in good will, and hence in future orders, at $1500. In the past, 80% of the lots produced contained 50 defectives. It costs $3 to test an item, but the test is not destructive (that is, if the item was good, it still is good after the test and can be sold). What is the expected net gain from running a test on three items drawn at random and what is the best decision rule? Use the binomial probability distribution to compute conditional probabilities.

3. If the expected opportunity loss of strategy s_1 is $4500 and the expected opportunity loss of strategy s_2 is $6400, at what value of the prior probability of s_1 would one be indifferent between the two strategies?

15.7 OPTIMAL SAMPLE SIZE

In the above acceptance sampling problem, we indicated how the *expected* value of *sample information* can be derived prior to the actual drawing of a sample. This EVSI figure was obtained by subtracting the expected opportunity loss of the best terminal act without sampling from the expected loss of the optimal strategy with sampling. More precisely, the latter expected value is the expected opportunity loss of a decision to sample and then take optimal terminal action after observation of the sample outcome. In that problem, we assumed a fixed or predetermined sample size of two articles. However, as was previously indicated (p. 636), the method of analysis presented is a general one, and the only practical effect of an assumption of a larger sample size would have been an increase in the computational burden. Nevertheless, the question remains, Can an *optimal sample size* be derived in a problem such as the one presented? The answer is yes, and the general method for obtaining such an optimal value follows.

As might be suspected intuitively, an increase in sample size brings about an increase in the EVSI. However, an increase in sample size also results in an increase in the cost of sampling. (We are using the term "cost of sampling" here to mean the total cost of sampling and inspection.) Therefore, we would like to find the sample size for which the difference between the EVSI and the cost of sampling is the largest. An equivalent and more convenient approach is to minimize *total loss*, where the total loss associated with any sample size n is defined as

(15.4)   Total Loss = Cost of sampling + Expected opportunity loss
                            of the optimal strategy

Let us consider how the quantities in Equation 15.4 might be calculated. The cost of sampling would ordinarily not be difficult to calculate. In many instances, this cost may be entirely variable, that is, proportional to the number of articles sampled. In that case, the cost of sampling would be equal to

(15.5) $$C = vn$$

where C = cost of sampling
$\quad\quad v$ = cost of sampling each unit
$\quad\quad n$ = number of units in the sample

Hence, in the Renny Corporation example, where the cost of sampling was $5.00 for two articles, if costs were entirely variable, the cost of sampling each unit would be $2.50. Hence, v = $2.50 and n = 2. The cost of sampling for a sample of size ten would be C = $2.50 × 10 = $25.00, and so on.

On the other hand, in certain situations, a portion of the total cost of sampling might be fixed and the remaining part variable. In this case, the cost of sampling would be

(15.6) $$C = f + vn$$

where C = cost of sampling
$\quad\quad f$ = fixed cost
$\quad\quad v$ = cost of sampling each unit
$\quad\quad n$ = number of units in the sample

Thus, for example, in a case in which the fixed cost of sampling is $10 and the variable cost per unit is $2, the cost of sampling 20 units would be

$$C = \$10 + (\$2)(20) = \$50$$

Other formulas can be derived for more complex situations.

Turning now to the other term in Equation 15.4 for *total loss*, namely, the *expected opportunity loss of the optimal strategy*, we pause first to introduce more convenient notation. Let us assume we are dealing with an acceptance sampling problem in which the sample size is ten. Although in the normal-form analysis of the Renny Corporation problem, we considered every possible strategy for the acceptance of the incoming lot, it can be shown that the only strategies worth considering as potential optimal decision rules are those that accept the incoming lot if a certain number or less of defectives are observed and reject the lot otherwise. It is conventional to refer to this critical number as "the acceptance number," denoted c. Hence, in the present problem, in which the sample size is ten, the possible values for the acceptance number c are 0, 1, 2, . . . , 10. For example, if c = 0, the lot is accepted if no defectives are observed in a sample of size n = 10 and rejected otherwise. If c = 1, the lot is accepted if *one or fewer* defectives are observed and rejected otherwise. If c = 2, the lot is accepted if *two or fewer* defectives are observed and rejected otherwise, etc. Therefore, we can characterize the optimal strategy for each sample size by two figures: c, the acceptance number, and n, the sample size. For example, the c, n pair, denoted (c,n), for an optimal strategy with an acceptance number $c = 2$ and sample size $n = 10$ would be $(2, 10)$. In the Renny Corporation problem, the c, n pair was $(0, 2)$ for the optimal strategy s_5.

Theoretically, we could calculate for every possible sample size the expected opportunity loss of the optimal strategy (c, n). Adding this figure to the cost of sampling, we would obtain the total loss associated with each sample size. We would then select as the optimal sample size the one that yielded the *minimum total loss*.

At first, this might seem to be an impossible procedure, because if the population is infinite, the sample size could conceivably take on any positive integral value. However, n must be a finite value because of the cost of sampling involved. The expected value of sample information (EVSI) cannot exceed the expected value of perfect information (EVPI). As we saw in Section 13.5, EVPI is the expected opportunity loss of the optimal act prior to sampling. That is, it is the expected loss of the best terminal action without sampling. Since the value of sampling cannot exceed that EVPI figure, it would never be worthwhile to take a sample so large that the cost of sampling exceeded this EVPI figure. Hence, n for the optimal sample size will be a finite number.

The acceptance sampling type of problem we have been discussing is one that can be described as involving "binomial sampling." That is, the conditional probabilities of sample outcomes of the form $P(X \mid p)$ were calculated in the Renny Corporation problem by the binomial probability distribution. In this type of problem, the optimal strategies for the various sample sizes must be calculated by the methods we have indicated, which may be characterized as trial-and-error procedures. That is, there is no simple general formula that enables us to derive the optimal (c, n) pairs for every problem. Therefore, there may be a considerable amount of calculation involved to determine the optimal sample size. For sufficiently large and important problems, the use of computers may represent the only practical method of carrying out the computations.

15.8 GENERAL COMMENTS

In this chapter, we have discussed decision-making procedures for the selection of optimal strategies prior to obtaining experimental data or sample information. The method used, preposterior analysis, anticipates the adoption of best actions after the observation of the experimental or sample results. The general principles of this type of analysis have been discussed using two examples. The first illustration involved the introduction of a new product. In that case, experimental evidence in the form of results of a sample survey was obtainable for revising prior probabilities of occurrence of the states of nature "successful product" and "unsuccessful product." The second example involved a decision concerning acceptance or rejection of an incoming lot. Here, information in the form of results of a random sample drawn from the lot was obtainable for revising prior probabilities of the states of nature "lot proportion of defectives." In the first example, an empirical joint frequency distribution on the success or lack of it of past new products and survey results provided information for the calculation of probabilities required for the problem solution. In the second example, the decision depended on the proportion of defectives, denoted p. The observable sample data were represented by the random variable "number of defectives," denoted X, in a sample drawn from the incoming lot. This random variable was taken to be binomially distributed. In both of these problems, the states of nature and the experimental or sample evidence were discrete random variables. Another class of problems is comprised by situations in which the states of nature and sample evidence are represented by continuous random variables. For example, the decision may depend on a parameter μ, the mean of a population, and the observed sample evidence may be a sample mean, $\bar{X}$. The mathematics required for the solution of this class of problems is outside the scope of this book. However, the same basic principles discussed in this chapter for extensive-form and normal-form analysis are applicable, regardless of whether the random variables representing states of nature and experimental outcomes are discrete or continuous.

Sequential Decision-Making Procedures

16.1 INTRODUCTION

We have considered decision-making procedures based solely on prior knowledge, on an incorporation of experimental or sample information with this prior knowledge, and on the devising of optimal strategies prior to obtaining experimental or sample data. Whenever information from sampling was involved, we dealt with a *single sample of fixed size*. In this chapter, we discuss decision-making procedures in which a choice is made after some information has been obtained between taking terminal action and obtaining further information. If the better course of action is to obtain further information, and this is done, a decision has to be made again between taking terminal action and gathering even more information. These decisions proceed sequentially until finally, terminal action is taken. Such a multi-stage series of choices is referred to as a *sequential decision-making procedure*.

The additional information obtained at each stage may consist of a single observation or a group of observations. For example, an illustration of the single observation case occurs in quality control work, where a lot may be inspected by taking successive observations of one item each. After each item is drawn, a choice is made between (1) stopping sampling and making a terminal decision to reject or accept the lot, and (2) drawing an additional item before again deciding whether to take a terminal act or to continue sampling.

A corresponding illustration involving groups of observations also occurs in quality control work. Instead of successive observations of one item each, samples of n items are drawn at every stage. We will illustrate the procedure by assuming that $n = 10$. Upon the drawing of the first sample of ten items, if a very small number of defectives is present (say, no defectives), the lot is accepted. If a large number of defectives is present (say, six or more), the lot is rejected. If one through five defectives are observed, another sample of ten items is drawn. Now, if a small cumulative number of defectives is present in the two samples of ten items each, say a total of one defective, the lot may be accepted. On the other hand, a large cumulative number of defectives would result in rejection of the lot, and an intermediate number would require drawing another sample. Eventually, a stage would be reached at which only two decisions are possible, rejection or acceptance of the lot.

If the lot is of very good quality (that is, contains a small percentage of defectives), it will tend to be accepted very quickly, perhaps on the first or second sample. Similarly, if the lot is of very poor quality (that is, contains a large percentage of defectives), it will tend to be rejected very quickly. Greater numbers of samples would be required for lots of intermediate quality. Although theoretically an extremely large number of samples might be needed for a lot of intermediate quality, in actual practice the procedure is truncated and a decision for acceptance or rejection is *forced* after a specified number of samples.

A major advantage of these sequential sampling plans is that on the average, they require fewer observations than single sampling plans (fixed sample size) to achieve the same levels of Type I and Type II errors. In classical statistics, sequential sampling plans are designed to meet prescribed risks of making Type I and Type II errors. As was true of other procedures in classical statistical inference, discussed in Chapters 6, 7, and 8, costs or profits are not explicitly included in the analysis. On the other hand, Bayesian sequential decision procedures, which are the type discussed in this chapter, do explicitly incorporate costs and profits. Extensive-form analysis, as discussed in Chapter 15, is used in the Bayesian sequential decision procedure with calculations of new posterior probabilities as further information is acquired at each successive stage.

We will illustrate Bayesian sequential decision procedures by the following new product development problem.

16.2 A NEW PRODUCT DEVELOPMENT PROBLEM

A company was considering the development of a new product. Sufficient research had been carried out so that a possible course of action was to proceed to direct commercialization of the product. However, the company had had several unsuccessful products in the past. Therefore, deciding to analyze the development of the product quantitatively, the firm selected Bayesian sequential decision procedures for that purpose.

The company faced the full array of problems typical of the development of any new product. If the first stage of development was carried out, the firm could decide at that point, based on the information derived, to proceed directly to commercialization and to introduce the product, or it could terminate the project completely and decide not to introduce the product. On the other hand, it could choose to proceed to the second development stage, generating additional information concerning the product's chances for success. Again at the end of the second development stage, in the light of the new information, it would face the same set of decisions as at the end of the first stage. This sequential decision procedure would continue for as many development stages as the firm had under consideration. Often in fields such as marketing and production, the earliest development stages represent preliminary screening procedures, whereas later ones involve more costly detailed screening and intensive investigative efforts.

Assumptions

In order to keep the exposition simple, we will make a number of simplifying assumptions in our illustrative problem. First, we will disregard the time value of money in computing expected payoffs. That is, we will ignore the fact that a dollar received one year from today is worth less than a dollar today. In a more realistic analysis, this fact should be taken into account, and appropriate discount factors should be applied to both costs and payoffs pertaining to future time periods. Second, we will assume that only two stages of development are under consideration and that the company has decided to move ahead into the first stage. Third, as in some of our earlier problems, we will assume only two states of nature, namely, θ_1: the product is successful, and θ_2: the product is unsuccessful, and two alternative actions, A_1: introduce the product, and A_2: do not introduce the product. Fourth, we will assume that delays in marketing the product occasioned by the length of the development periods do not alter the payoffs. This will forestall possible confusion arising from changing payoff figures. However, in a real problem, the effects of competition and other factors that might bring about changes in payoff figures because of delays in introducing the product should be taken into account. Finally, as in previous problems, we are presupposing that a linear utility function for money over the range of payoffs considered is a suitable approximation. Therefore, it is meaningful to use the maximization of expected monetary payoffs as a criterion of preference among alternative courses of action.

Expected Value of Perfect Information

We will refer to the person who has responsibility for the decision process as the "decision maker." We begin the analysis by assuming that the decision maker has assessed payoffs in terms of net profit in dollars and has prior betting odds of 50:50 on the success of the product. These data are summarized in Table 16-1. The prior expected profits of the two courses of action have been calculated and are shown at the bottom of the table. As indicated in the table, the preferable course of action based on prior betting odds is to introduce the product (act A_1), yielding an expected payoff of $500,000 as opposed to $0, if the product is not introduced (act A_2). The *expected profit under uncertainty* is equal to $500,000, since that is the expected payoff of selecting the optimal act under conditions of uncertainty. Using methods discussed in Chapter 13, we now calculate the *expected value of perfect information* (EVPI), in order to evaluate the upper limit of the worth of obtaining additional information. Table 16-2 shows this computation by the two methods previously discussed. Since the expected value of perfect information is high ($500,000), it appears reasonable to investigate

TABLE 16-1
Calculation of Prior Expected Profits in the New Product Development Problem

Event θ_i	Prior Probability $P(\theta_i)$	Introduce Product A_1	Do Not Introduce Product A_2
θ_1: Successful product	0.50	$2,000,000	$0
θ_2: Unsuccessful product	0.50 ‾‾‾‾ 1.00	− 1,000,000	0

Prior expected profit $(A_1) = (0.50)(\$2,000,000) + (0.50)(-\$1,000,000) = \$500,000$
Prior expected profit $(A_2) = (0.50)(\$0) + (0.50)(\$0) = \$0$

Expected profit under uncertainty $= \$500,000$

TABLE 16-2
Calculation of the Expected Value of Perfect Information

Event	Prior Probability $P(\theta_i)$	Profit	Weighted profit
θ_1: Successful product	0.50	$2,000,000	$1,000,000
θ_2: Unsuccessful product	0.50	0	0
			$1,000,000

Expected profit with perfect information = $1,000,000
Less: Expected profit under uncertainty = 500,000
Expected value of perfect information = $ 500,000

Event	Prior Probability	Opportunity Loss	
		Introduce Product A_1	Do Not Introduce Product A_2
θ_1: Successful product	0.50	$ 0	$2,000,000
θ_2: Unsuccessful product	0.50	1,000,000	0

$\text{EOL}(A_1) = (.50)(\$0) + (.50)(\$1,000,000) = \$500,000$
$\text{EOL}(A_2) = (.50)(\$2,000,000) + (.50)(\$0) = \$1,000,000$

Expected value of perfect information = Minimum $\text{EOL}(A_j) = \$500,000$

whether the information-gathering process of sequential development is worthwhile.

Decision Tree Diagram

As indicated earlier, it was assumed in this problem that the company had decided to proceed with first stage development. At the end of the problem, we will return to the question of whether the company should undertake the development process at all or move directly to commercialization. However, assuming that the company enters first-stage development, the analysis must indicate whether it is better to take an optimal terminal action after gathering the information at this stage or to move on to second-stage development. The information-gathering costs are $20,000 for first-stage development and $100,000 for second-stage development. The first-stage information is 60% reliable; this means that if the product were successful, the probability is 0.60 that the first-stage information would indicate a successful product and 0.40 that it would indicate an unsuccessful product. Second-stage information is 70% reliable in the same sense.

Figures 16-1 and 16-2 show the tree diagram for the sequential decision procedure. The tree is shown in two sections to clarify the presentation. We will trace through Figure 16-1 to indicate the decision process. Two types of information can be obtained from first-stage development:

X_1: An indication that the product will be successful
X_2: An indication that the product will be unsuccessful

From Figure 16-1, we see that if type X_1 information is the outcome, the decision maker has two choices: to stop the development process and make a terminal decision, or to proceed with second-stage development. If development is stopped, the decision maker can choose terminal action A_1 or A_2. For A_2, the payoff is $0. If A_1 is chosen, state of nature θ_1 or θ_2 will eventuate, that is, the product will either be successful or unsuccessful with the payoffs indicated in Table 16-1.

On the other hand, if the company enters second-stage development, again two types of indications may be given, X_1' and X_2' (X_1 prime

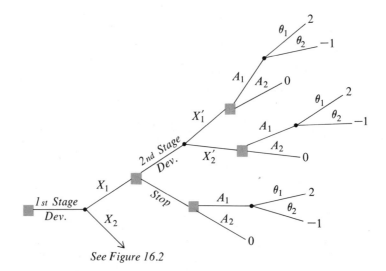

FIGURE 16-1
Decision tree diagram, assuming that a type X_1 indication is obtained in first-stage development (all payoffs are in $millions).

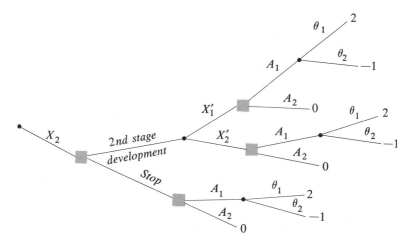

FIGURE 16-2
Decision tree diagram, assuming that a type X_2 indication is obtained in first-stage development (all payoffs are in $millions).

and X_2 prime). These symbols have the same meaning as X_1 and X_2, respectively, except that the prime signifies information generated at the second stage. At this point, we have assumed that the decision maker must take a terminal action. If a third stage of development were contemplated, the diagram would continue as at the earlier stages.

A similar description of the tree diagram applies to Figure 16-2, which pertains to the case where X_2 information is observed as a result of first-stage development.

Sequential Decision Analysis

We can now proceed with the sequential decision analysis, entering the appropriate numerical quantities in the decision tree. Since there is a possibility of information being generated at two development stages before a terminal choice is made, we must calculate posterior probabilities, which incorporate the information outcomes at each of the successive stages. The posterior probabilities of the events θ_1 and θ_2 after the information indication X_1 or X_2 has been observed become the prior probabilities of these events at the second stage. These prior probabilities are in turn revised to incorporate the information indications X_1' and X_2' generated from the second development stage. The revised figures represent the posterior probabilities after the second development stage. As usual, Bayes' Theorem is used to accomplish the revision of probabilities. These calculations are given in Tables 16-3 and 16-4. The conditional probabilities in these tables are based on the previously stated assumption that first-stage information is 60% reliable and second-stage information is 70% reliable.

At this point, we interrupt the explanation of the sequential decision problem in order to illustrate an important and very interesting aspect of Bayes' Theorem. In Tables 16-3 and 16-4, we used Bayes' Theorem to revise prior probabilities in a sequential manner. For example, there were subjective probabilities $P(\theta_i)$ assigned to the states of nature θ_1 and θ_2. If X_1 was observed at the first stage, posterior probabilities were calculated incorporating this information. These posterior probabilities became the prior probabilities for the second stage. If X_1' was observed at the second stage, posterior probabilities were again calculated to reflect this information. Now, an interesting question arises. Suppose, instead of revising prior probabilities one step at a time, we had revised our original subjective probabilities on the basis of all accumulated experimental information in one step. For example, suppose we were to revise the original subjective probabilities in one step in the light of the sequence of information X_1, X_1'; that is, we would compute $P(\theta_i|X_1$ and $X_1')$, which for simplicity, we denote $P(\theta_i|X_1X_1')$. How would these probabilities compare with the corresponding posterior probabilities resulting from the one-step-at-a-time revision? The answer is that the posterior probability distributions would be identical. The alternative one-step calculation of posterior probabilities on the basis of all

TABLE 16-3
Computation of Posterior Probabilities That Incorporate First-Stage Development Information

X_1 Observed at First Stage

Event θ_i	Prior Probability $P(\theta_i)$	Conditional Probability $P(X_1|\theta_i)$	Joint Probability $P(\theta_i)P(X_1|\theta_i)$	Posterior Probability $P(\theta_i|X_1)$
θ_1	0.50	0.60	0.30	0.60
θ_2	0.50	0.40	0.20	0.40
			$P(X_1) = 0.50$	1.00

X_2 Observed at Second Stage

θ_i	$P(\theta_i)$	$P(X_2|\theta_i)$	$P(\theta_i)P(X_2|\theta_i)$	$P(\theta_i|X_2)$
θ_1	0.50	0.40	0.20	0.40
θ_2	0.50	0.60	0.30	0.60
			$P(X_2) = 0.50$	1.00

TABLE 16-4
Computation of Posterior Probabilities That Incorporate First- and Second-Stage Development Information

Event θ_i	X_1 Observed at First Stage; X_1' at Second Stage			
	Prior Probability $P(\theta_i)$	Conditional Probability $P(X_1'\vert\theta_i)$	Joint Probability $P(\theta_i)P(X_1'\vert\theta_i)$	Posterior Probability $P(\theta_i\vert X_1')$
θ_1	0.60	0.70	0.42	0.78
θ_2	0.40	0.30	0.12	0.22
			$P(X_1') = 0.54$	1.00

θ_i	X_1 Observed at First Stage; X_2' at Second Stage			
	$P(\theta_i)$	$P(X_2'\vert\theta_i)$	$P(\theta_i)P(X_2'\vert\theta_i)$	$P(\theta_i\vert X_2')$
θ_1	0.60	0.30	0.18	0.39
θ_2	0.40	0.70	0.28	0.61
			$P(X_2') = 0.46$	1.00

θ_i	X_2 Observed at First Stage; X_1' at Second Stage			
	$P(\theta_i)$	$P(X_1'\vert\theta_i)$	$P(\theta_i)P(X_1'\vert\theta_i)$	$P(\theta_i\vert X_1')$
θ_1	0.40	0.70	0.28	0.61
θ_2	0.60	0.30	0.18	0.39
			$P(X_1') = 0.46$	1.00

θ_i	X_2 Observed at First Stage; X_2' at Second Stage			
	$P(\theta_i)$	$P(X_2'\vert\theta_i)$	$P(\theta_i)P(X_2'\vert\theta_i)$	$P(\theta_i\vert X_2')$
θ_1	0.40	0.30	0.12	0.22
θ_2	0.60	0.70	0.42	0.78
			$P(X_2') = 0.54$	1.00

cumulative experimental evidence is given in Table 16-5. These probabilities are seen to be identical with the corresponding posterior probabilities in the last column of Table 16-4.

Returning now to the sequential decision problem, we enter the probabilities derived in Tables 16-3 and 16-4 at the appropriate places in the decision tree diagrams of Figures 16-1 and 16-2. Then we use the method of *backward induction*, which, as we have seen in Section 13.6,

TABLE 16-5
Alternative One-Step Calculation of Posterior Probabilities

X_1 Observed at First Stage; X_1' at Second Stage

Event θ_i	Prior Probability $P(\theta_i)$	Conditional Probability $P(X_1 X_1' \vert \theta_i)$	Joint Probability $P(\theta_i) P(X_1 X_1' \vert \theta_i)$	Posterior Probability $P(\theta_i \vert X_1 X_1')$
θ_1	0.50	$(0.60)\,(0.70) = 0.42$	0.21	0.78
θ_2	0.50	$(0.40)\,(0.30) = 0.12$	0.06	0.22
			0.27	1.00

X_1 Observed at First Stage; X_2' at Second Stage

θ_i	$P(\theta_i)$	$P(X_1 X_2' \vert \theta_i)$	$P(\theta_i) P(X_1 X_2' \vert \theta_i)$	$P(\theta_i \vert X_1 X_2')$
θ_1	0.50	$(0.60)\,(0.30) = 0.18$	0.09	0.39
θ_2	0.50	$(0.40)\,(0.70) = 0.28$	0.14	0.61
			0.23	1.00

X_2 Observed at First Stage; X_1' at Second Stage

θ_i	$P(\theta_i)$	$P(X_2 X_1' \vert \theta_i)$	$P(\theta_i) P(X_2 X_1' \vert \theta_i)$	$P(\theta_i \vert X_2 X_1')$
θ_1	0.50	$(0.40)\,(0.70) = 0.28$	0.14	0.61
θ_2	0.50	$(0.60)\,(0.30) = 0.18$	0.09	0.39
			0.23	1.00

X_2 Observed at First Stage; X_2' at Second Stage

θ_i	$P(\theta_i)$	$P(X_2 X_2' \vert \theta_i)$	$P(\theta_i) P(X_2 X_2' \vert \theta_i)$	$P(\theta_i \vert X_2 X_2')$
θ_1	0.50	$(0.40)\,(0.30) = 0.12$	0.06	0.22
θ_2	0.50	$(0.60)\,(0.70) = 0.42$	0.21	0.78
			0.27	1.00

involves starting at the right side of the decision tree diagram and proceeding inward. We compute expected values and select optimal acts at
each stage, moving inward until we return to the beginning of the tree.
Thus, we evaluate the best courses of action at the later stages in order to
determine the best moves at earlier stages. Costs of obtaining information must be subtracted at each stage to ascertain whether to stop or continue development.

The evaluations of the portions of the decision tree diagram shown in Figures 16-1 and 16-2 are given in Figures 16-3 and 16-4, respectively. (All payoff figures are in millions of dollars.) The posterior probabilities assigned to the events θ_1 and θ_2 are the appropriate figures from Tables 16-3 and 16-4, depending on whether post-first-stage or post-second-stage probabilities are relevant. The marginal probabilities of the information indications also come from Tables 16-3 and 16-4. For example, the probabilities assigned to observing X_1' and X_2' information at the second stage after X_1 has been observed in first-stage development are $P(X_1') = 0.54$ and $P(X_2') = 0.46$, respectively. As can be seen from Table 16-4, these are marginal probabilities of the two types of information, computed after allowance has been made for the revision of prior probabilities in the light of information obtained at the first stage.

Expected values are computed in the usual way, using the appropriate marginal and posterior probabilities to weight payoffs. As can be seen from Figures 16-3 and 16-4, the decision maker would proceed as follows. If favorable information is obtained from first-stage development (that is, X_1 is observed), stop further development and take action A_1, that is, introduce the product. (See Figure 16-3.) The expected payoff of this course of action is $.80 million. On the other hand, if X_2 is observed on the first stage, the decision maker should proceed to second-stage development. If a favorable indication results from second-stage development (that is, X_1' is observed), the decision maker

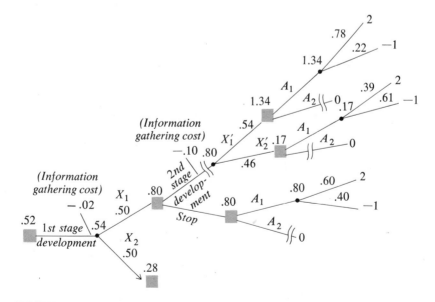

FIGURE 16-3
Evaluation of tree diagram depicted in Figure 16-1 (all payoffs are in $millions).

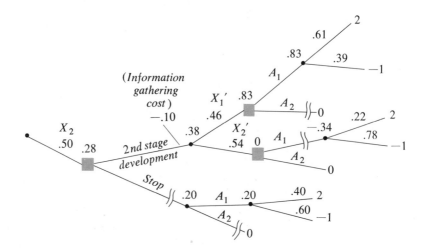

FIGURE 16-4
Evaluation of tree diagram depicted in Fig. 16-2 (all payoffs are in $millions).

should take action A_1 and introduce the product. However, if an unfavorable indication results (that is, X_2'), then the product should not be introduced. The expected payoff of proceeding into second-stage development, if X_2 is observed at the first stage, is $280,000. This figure is obtained by subtracting $100,000, the cost of gathering information by engaging in second-stage development, from $380,000, the gross expected payoff of proceeding into that development stage. (See Figure 16-4.)

Weighting the expected payoffs of taking the best action after observing type X_1 and X_2 information ($.80 million and $280,000, respectively) by the marginal probabilities of observing these types of information (0.50 and 0.50) yields $540,000 as the gross expected payoff of proceeding into first-stage development. Subtracting $20,000, the cost of gathering information by pursuing first-stage development, yields an expected payoff of $520,000 for entering first-stage development and selecting optimal courses of action thereafter. Since we assumed at the outset that the decision was made to enter first-stage development, the preceding analysis indicates the expected payoff associated with that course of action and the optimal moves in the sequential decision process thereafter.

However, suppose we wished to compare the advisability of entering the development process with the alternative of proceeding directly to commercialization of the product without any information gathering. Returning to Table 16-1, we find an expected payoff of $500,000 if action is taken solely on the basis of the decision maker's prior betting odds. Comparing this payoff with the aforementioned payoff of $520,000, the better course of action is to enter the development process, although the difference in payoffs is relatively small.

Truncating the Decision Tree

In the problem just described, the number of development stages was limited to two. If the sequential decision procedure involved many more information-gathering stages, the calculations would become very detailed and tedious. For example, in evaluating an item-by-item sequential sampling plan in quality control work, that is, a plan in which each successive sample observation consists of only one item, the tree can become extremely large. The general principle for determining when to end the sampling procedure is to stop wherever the cost of an additional observation exceeds the opportunity loss of optimal terminal action. Sometimes the use of computers may be required to carry out the computations. Whether or not computers are used, it is doubtless desirable to truncate or cut the tree. The probability of reaching some of the higher-level stages becomes very small, particularly if the number of stages is large. Most of the tree will already have been terminated at lower levels. Guesses can be made at the payoffs pertaining to the remaining high-level positions, and the tree may thus be truncated at arbitrary levels. It is also possible to use sensitivity analysis in which widely different payoffs (or opportunity losses) may be assumed for some of the remaining high-level positions. It will probably be observed that the effect of these widely different assumptions on calculated expected terminal payoffs is negligible. Hence, a substantial reduction in calculations may be effected at virtually no loss in accuracy.

General Remarks

The sequential decision procedures discussed in this chapter did not really introduce any new principles. However, they represent a powerful framework of analysis for determining what decisions should be made and when they should be made. This point is perhaps most clearly noted in the case of multistage business decision problems where a delay in decision is a definite alternative course of action. For example, if "delay one year" is one of the acts to be considered, it is evaluated in the light of all relevant future decisions. That is, the expected payoff assigned to "delay one year" takes into account the relationship between this decision and the future decisions stemming from it. Curiously, as we have seen, actions are planned in an optimal fashion from any point forward by a backward induction technique of problem solution.

Although in this chapter a new product development problem has been used as an example, Bayesian sequential decision procedures are clearly applicable to a wide variety of situations. Wherever sequential sampling or information gathering is involved, as in quality control inspection of product, market research surveys, or pilot investigations of various types, these techniques may be effectively employed.

Exercises

1. Explain the difference between a single-stage decision problem and a sequential decision problem.

2. Construct and solve a realistic example of a two-stage decision problem.

3. Let two possible states of nature be θ_1: stock market prices will advance and θ_2: stock market prices will stay the same or decline. Your prior belief is that there is a 75% chance of an advance. In a week the results of a Stiff Blarney research report will be available. Either an advance or no advance will be indicated. These research reports are correct 90% of the time. A Sherrill, Pinch, Fierce, Kenner, and Blithe report will be available one week after Stiff Blarney, and it is also correct 90% of the time. Suppose both reports indicate that stock prices will advance. Compute posterior probabilities that incorporate information from both research reports.

4. The capital expenditures committee of the Hoboken Needle Company is contemplating the addition of a plant extension. The total cost will be $700,000, but if demand for the company's products is higher than average, increased profit will be $1,800,000. If demand is not higher than average, revenues will remain unchanged. There is a 50% chance that demand will be higher than average, so the committee has determined that at least one sample test of the market should be taken. Each sample test costs $90,000. Information from the first test sample is 60% reliable. This means that if the demand is high, the probability is 0.60 that information from the first test sample will indicate a high demand. Information from a second test sample is 80% reliable in the same sense. Draw a decision tree for this problem, and make all necessary entries on the tree. There are only two states of nature; demand will be higher than average or not higher than average.

5. The "Prince of Pancakes" Corporation is a chain of restaurants across the northeastern United States. A vice-president of the chain is considering whether to locate a new restaurant on Route 66 just outside of West Buttermilk, Ohio. He feels that if this particular restaurant is successful, the company can make a profit over the planning period of $3,500,000, but if it is not successful, it will lose $1,200,000. The probabilities of these occurrences are 0.6 and 0.4, respectively. It is known that a sample of the demand in the area must be taken, but whether one sample or two sequential samples should be taken is still undecided. The reliability of the first sample is 0.75 and of the second sample 0.85. The cost of the first sample is $120,000, and the cost of the second is $160,000. Draw a tree diagram for this problem and make all necessary entries on the tree. Interpret the results of the diagram.

6. The Suboptimal Investment Corporation wishes to undertake a direct investment on the island of Krakatoa. In order to appraise the situation, the company feels that some research must be carried out. The research will yield one of two indications: X_1, favorable for investment, and X_2, unfavorable for investment. Prior probabilities had been determined as 0.4 that the volcano on the island would not erupt and 0.6 that the volcano would erupt and destroy everything on the island. If it invests, Suboptimal will either earn $10 million or lose $5 million depending on whether the volcano erupts. The reliability of the research is 0.60, and the cost of the research will be $500,000. The president of the corporation feels that more extensive research should be conducted after the results of the earlier research are deter-

mined. This extra extensive research will cost an additional $700,000, but its reliability is 0.90. Draw a decision tree for this problem.

7. A speculator in the commodities market is contemplating the sale of a presently held futures contract for wheat. A study can be purchased from the Commodities Investment Service that would reveal two possible results, X_1 or X_2, where X_1 indicates that the price of wheat (and the price of the futures contract, consequently) will increase over the period in question and X_2 indicates that the price will go down or remain the same (with identical effects on the contract price).

State of Nature	Prior Probability $P(\theta_i)$	Conditional Probability $P(X_1\|\theta_i)$	$P(X_2\|\theta_i)$
θ_1: Wheat price up	0.35	0.7	0.3
θ_2: Wheat price down or same	0.65	0.2	0.8

Opportunity Loss Table
(Units of $1,000)

	θ_1	θ_2
Sell	8	0
Do not sell	0	8

A second study can also be made that has reliability of 0.75. The speculator has decided to take the first survey, which costs $500, but as yet is undecided about the second study, which costs $850. Using a tree diagram, determine whether it is advisable to purchase the second study. Also, determine whether purchasing any kind of study is better than not purchasing any at all.

8. In Exercise 5 at the end of Section 15.2, suppose that a second survey is also possible. The reliability of the second survey is 0.7 and will cost 0.60 units of utility. If the cost of the first survey is 0.30 units of utility draw a decision tree diagram showing all possible occurrences with respective probabilities and expected opportunity losses.

9. Assume that in Exercise 7 at the end of Section 15.2, a second investigation costing $50,000 can also be conducted. There are three outcomes, X_1, Y_1, and Z_1. The probabilities of such outcomes given the respective states of nature are

	X_1	Y_1	Z_1
S_1	0.90	0.05	0.05
S_2	0.10	0.80	0.10
S_3	0.05	0.10	0.85

Draw a decision tree diagram and make all necessary entries.

10. A home appliance manufacturing firm is considering the marketing of a new product. Before marketing a new product, the firm mails out questionnaires to its panel members, consisting largely of randomly selected consumers. The survey results are then classified into three categories: favorable, uncertain, and unfavorable. After the first survey is completed, a second survey is undertaken a few months later using nonmembers selected at random. Based upon either the first or both survey results, the firm decides whether to market the new product in question. In detail, the situation is as follows. There are two possible states of nature: Q_1: New product is a success and Q_2: New product is a failure. There are two available actions: D_1: Market the new product and D_2: Do not market the new product. The firm's opportunity losses (in units of $10,000) due to incorrect action are assumed to be as follows:

	D_1 (*Market*)	D_2 (*Do Not Market*)
Q_1 (Success)	$ 0	$100
Q_2 (Failure)	120	0

The prior probabilities of success and failure are 0.25 and 0.75, respectively, and the conditional probabilities of observing favorable (X_1), uncertain (X_2), and unfavorable (X_3) survey results on the first sample, when Q is the true state of nature, are shown in the following table:

	X_1	X_2	X_3
Q_1	0.72	0.16	0.12
Q_2	0.08	0.12	0.80

A similar table for the second sample is

	X_1	X_2	X_3
Q_1	0.82	0.10	0.08
Q_2	0.04	0.06	0.90

Each sample costs $75,000. Draw a decision tree for the problem. Make all necessary entries and interpret the results.

17

Comparison
of Classical
and Bayesian
Statistics

17.1 INTRODUCTION

Topics in classical statistical inference and Bayesian decision theory were discussed in earlier chapters. In this chapter, we compare some aspects of *classical statistics* and *Bayesian statistics*. Classical statistics is a broad term that includes the two main topics of classical statistical inference (hypothesis testing and confidence interval estimation) as well as other topics, such as classical regression analysis, as discussed in Chapter 9. Bayesian statistics is also a broad term that analogously may be thought of as including Bayesian decision theory, Bayesian estimation, and other topics such as Bayesian regression analysis. Although the terminologies of classical and Bayesian statistics differ, there are many similarities in the structure of the problems to which they address themselves and in their methods of analysis. However, there are important differences, particularly in their methods of analysis, that are a matter of continuing discussion and debate. In order to compare these

665

two important types of statistical analysis, we consider in Section 17.2 an illustrative problem that presents a comparison of classical hypothesis-testing methods and Bayesian decision theory; in Section 17.3, we turn to a comparison of classical and Bayesian estimation procedures. In conclusion, Section 17.4 presents some general comments on the areas of common ground and differences between these two schools of thought.

To introduce the comparison, let us consider a standard hypothesis-testing problem. Suppose we wish to test the null hypothesis, H_0: $p \leqslant p_0$ (where p_0 is a known or hypothesized population proportion), against the alternative hypothesis, H_1: $p > p_0$. For example, we might test the hypothesis that p, the proportion of defectives in a shipment of a manufactured product, is less than or equal to 0.03 against the alternative hypothesis that $p > 0.03$. Using classical hypothesis-testing methods, we could design a decision rule, which would tell us whether to accept or to reject the null hypothesis on the basis of a random sample drawn from the shipment. We would fix α, the desired maximum probability of making a Type I error, and through the use of the power curve we could determine the risks of making Type II errors for values of p for which the alternative hypothesis, H_1, is true. Table 17-1 summarizes the relationship between actions concerning these hypotheses and the truth or falsity of the hypotheses. For convenience, the table is given in terms of the null hypothesis, H_0. However, it is understood that when H_0 is true, H_1 is false and when H_0 is false, H_1 is true. As earlier, we refer to the truth or falsity of H_0 as the prevailing "state of nature." As indicated in the column headings of the table, the symbols a_1 and a_2 denote the actions "accept H_0" and "reject H_0."

We see that the structure of this hypothesis-testing problem includes (1) states of nature representing the truth or falsity of the null hypothesis, (2) actions a_1 and a_2, which accept or reject the null hypothesis, and (3) sample or experimental data, which when examined in the light of a decision rule lead to one of the actions indicated under (2).

Let us rephrase the example in terms of Bayesian decision theory. We are dealing with a two-action problem involving acts a_1 and a_2, where

TABLE 17-1
Relationships between Actions Concerning a Null Hypothesis and the Truth or Falsity of the Hypothesis

State of Nature	Action Concerning the Null Hypothesis	
	a_1: Accept H_0	a_2: Reject H_0
H_0 is true	No error	Type I error
H_0 is false	Type II error	No error

the states of nature are the possible values of the proportion of defectives, p. Although p varies along a continuum, and hence may be considered a continuous random variable, we assume for comparative purposes that only two states of nature are distinguished, namely, θ_1: $p \leq 0.03$ and θ_2: $p > 0.03$. Hence, θ_1 and θ_2 correspond to truth and falsity of H_0, respectively, in the classical hypothesis-testing problem. Finally, a random sample can be drawn from the shipment, and the observed sample or experimental data can be used to help decide between a_1 and a_2 as the better action. Therefore, the same three components of the structures of the decision-theory problem are present as were discussed in the hypothesis-testing problem: (1) states of nature, (2) alternative actions, and (3) experimental data that aid in the choice of actions. Furthermore, Table 17-2 is a payoff table for this problem in terms of opportunity losses and is similar to Table 17-1. The symbols $L(a_2|\theta_1)$ and $L(a_1|\theta_2)$ denote the opportunity loss of action a_2 given that θ_1 is the true state of nature and a_1 given that θ_2 is the true state of nature. The zeros in the other two cells of the table indicate that there is no opportunity loss when the correct action is taken for the specified states of nature. Actually, payoffs would ordinarily be treated as a function of p and would vary with p. However, as indicated earlier, we have assumed in this discussion that only two states of nature are distinguished.

Now, let us consider the differences in the two approaches. In hypothesis testing, the choice of the significance level, α, establishes the decision rule and is thus the overriding feature in the choice between alternative actions. In symbols, $\alpha = P(a_2|H_0 \text{ is true})$. That is, α is the conditional probability of rejecting the null hypothesis given that it is true. Hence, a major criterion of choice among actions in hypothesis testing is the relative frequency of occurrence of this type of error. But how is α chosen? In many applications, conventional significance levels such as 0.05 and 0.01 are used uncritically with little or no thought given to underlying considerations. However, it would be unfair to criticize a methodological approach simply because there are misuses of it. In classical statistics, the investigator is supposed to consider the relative seriousness of both Type I and Type II errors in establishing alternative

TABLE 17-2
Payoff Table in Terms of Opportunity Losses for the
Two-Action Problem to Accept or Reject a Shipment

State of Nature	Action	
	a_1	a_2
θ_1	0	$L(a_2|\theta_1)$
θ_2	$L(a_1|\theta_2)$	0

hypotheses and significance levels at which these hypotheses are to be tested. Also, the investigator is aided by prior knowledge concerning the likelihood that H_0 and H_1 are true. For example, in the problem just discussed, why did the investigator not set up the null hypothesis as, say, $H_0: p \leq 0.001$ or $H_0: p \leq 0.60$? In this particular acceptance sampling problem, one may know that two hypotheses such as these would be utterly ridiculous because of the extremely low and extremely high proportion of defectives implied. Hence, one may be virtually certain that the first of these hypotheses is false and that the second is true. Consequently, it would not be useful to set up the hypotheses in these forms.

Prior knowledge concerning the likelihood of truth of the competing hypotheses also helps the investigator in establishing the significance level. Hence, if it is considered very likely that the null hypothesis is true, one will tend to set α at a very low figure, in order to maintain a low probability of erroneously rejecting that hypothesis.

However, advocates of Bayesian decision theory criticize these classical hypothesis-testing procedures for informality and for excessive reliance on unaided intuition and judgment. The Bayesians argue that the structure of their decision theory represents a logical extension of classical hypothesis testing, since it explicitly provides for the assignment of prior probability distributions to states of nature and incorporates losses into the formal structure of the problem. These decision theorists contend that losses are supposed to be considered in classical hypothesis testing in evaluating the relative seriousness of Type I and Type II errors, but they ask, How can they be considered if no explicit loss function is formulated?

We turn now to an acceptance sampling problem that contains a comparison of the Bayesian approach with classical hypothesis testing. The problem demonstrates that if tests of hypotheses are conducted in the usual manner of establishing decision rules of rejecting or failing to reject hypotheses at preset levels of significance, nonoptimal decisions may be made.

17.2 A COMPARATIVE PROBLEM

Let us assume a situation in which a company inspects incoming lots of articles produced by a particular supplier. Acceptance sampling inspection is carried out to decide whether to accept or reject these incoming lots by selecting a single random sample of n articles from each lot. As in previous problems of this type, we make the simplifying assumption that there are only a few possible levels of proportion of defectives, in this case, 0.02, 0.05, and 0.08. On the basis of an analysis of past costs, the company constructed the payoff table in terms of opportunity losses depicted in Table 17-3. From long experience, the firm has determined that lots containing 0.02 defectives are "good" and should be accepted. Hence, as indicated in Table 17-3, "accept" is the best act in the case of a 0.02 defective lot, and the opportunity loss in that case is $0. On the other hand, "reject" is the optimal act for lots containing 0.05 and 0.08 defectives, and correspondingly the opportunity loss is $0 for such correct action.

On the basis of past performance, it has been determined that 50% of the supplier's lots are 2% defective, 25% are 5% defective, and 25% are 8% defective. In the absence of any further information, these past relative frequencies are adopted as the prior probabilities that such lots will be submitted by the supplier for acceptance or rejection.

In order to compare the approaches of Bayesian decision theory and traditional hypothesis testing, we will first carry out a study of possible single sampling plans (see p. 649) by extensive- and normal-form preposterior analyses. The result of these analyses will be determination of the optimal sampling plan or strategy. Then a hypothesis-testing solution will be given, and a comparison will be made of the two approaches.

A decision tree diagram is given in Figure 17-1, beginning with the decision to sample and inspect n items. We move to branch point (b), where the results of the sampling inspection then determine which branch to follow. The possible results of sampling have been classified into three categories. The number of defectives, denoted X, may have been less than or equal to some number c_1, where $c_1 < n$. It may have been greater than c_1 but less than or equal to c_2, where $c_1 < c_2 < n$. Finally, the number of defectives may have been greater than c_2. These three types of results, for purposes of brevity, are referred to as Type L (low), Type M (middle), and Type H (high) information, respectively. In Table 17-4, a joint frequency distribution is given for sample results and states of nature. We will assume that these frequencies were derived from a large number of past observations and therefore may be

TABLE 17-3
Payoff Matrix Showing Opportunity Losses for Actions of Acceptance and Rejection

State of Nature (p = Lot Proportion of Defectives)	Prior Probability	Act a_1 Reject	Act a_2 Accept
0.02	0.50	$200	$ 0
0.05	0.25	0	300
0.08	0.25	0	500
	1.00		

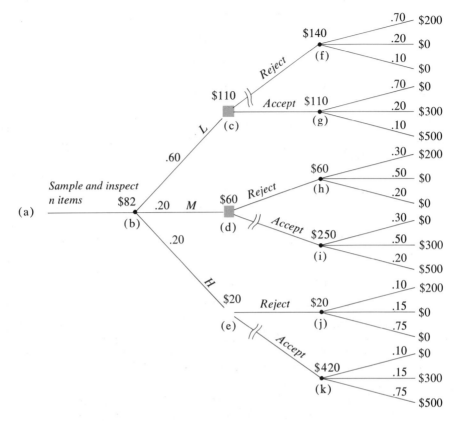

FIGURE 17-1
Decision tree diagram for the acceptance sampling problem.

taken to represent probability in the relative frequency sense.[1] For example, in the past, in 0.42 of the lots inspected, the number of defectives observed was c_1 or less, and the lots contained 0.02 defectives. In terms of marginal frequencies, Type L information ($X \leqslant c_1$) was observed in 0.60 of the lots, Type M ($c_1 < X \leqslant c_2$) was observed in 0.20 of the lots, and Type H ($X > c_2$) was observed in 0.20 of the lots.

Returning to the decision tree, we find the three branches representing L, M, and H types of information emanating from node (b) marked with their respective probabilities, 0.60, 0.20, and 0.20. We will give a brief explanation of the usual extensive-form analysis, using Type L information as an example. If Type L information is observed, we move to branch point (c), where we can either accept or reject the lot. If

---

[1] Alternatively, the conditional probabilities of sample results (likelihoods) derivable from this table may be thought of as having been calculated from an appropriate probability distribution such as the binomial or hypergeometric distribution. However, the basic methodological discussion remains unchanged.

TABLE 17-4
Joint Frequency Distribution of Sample Results and States of Nature

State of Nature ($p = $ Lot Proportion of Defectives)	Sample Result			
	Type L $X \leqslant c_1$	Type M $c_1 < X \leqslant c_2$	Type H $X > c_2$	Total
0.02	0.42	0.06	0.02	0.50
0.05	0.12	0.10	0.03	0.25
0.08	0.06	0.04	0.15	0.25
	0.60	0.20	0.20	1.00

we reject, we move to node (f); if we accept, to node (g). The probabilities shown on the three branches stemming from (f), which are 0.70, 0.20, and 0.10, are the posterior probabilities, given Type L information, that the lots contain 0.02, 0.05, and 0.08 defectives, respectively. The calculation of these posterior probabilities by Bayes' Theorem is given in Table 17-5. These probabilities can also be derived from Table 17-4 by dividing joint probabilities by the appropriate marginal probabilities, for example, $0.70 = 0.42/0.60$.

We now use the standard backward induction technique (see Section 13.6) to obtain the expected opportunity loss of the optimal strategy. Looking forward from node (f) and using the posterior probabilities 0.70, 0.20, and 0.10 as weights attached to the three states of nature (0.02, 0.05, and 0.08 defective lots), we obtain an expected opportunity loss of $140 for the act "reject." Comparing this figure with the corresponding one of $110 for "accept," we block off the action "reject" as nonoptimal. Therefore, $110 is carried down to node (c), representing the payoff for the optimal act upon observing Type L ($X \leqslant c_1$) information. Similar

TABLE 17-5
Calculation of Posterior Probabilities of States of Nature Given

State of Nature ($p = $ Lot Proportion of Defectives)	Type L Information ($X \leqslant c_1$)			
	Prior Probability $P_0(p)$	Conditional Probability $P(L\|p)$	Joint Probability $P_0(p)P(L\|p)$	Posterior Probability $P_1(p)$
0.02	0.50	0.84	0.42	0.70
0.05	0.25	0.48	0.12	0.20
0.08	0.25	0.24	0.06	0.10
			0.60	1.00

calculations yield \$60 and \$20 at (d) and (e) for Types M and H information. Weighting these three payoffs by the marginal probabilities of obtaining Types L, M, and H information, we obtain a loss of \$82 as the expected payoff of sampling and inspecting n items. The cost of sampling and inspection would then have to be added if, for example, we wished to make a comparison with the expected loss of terminal action without sampling. However, for our comparison of the Bayesian decision theory approach with hypothesis testing, we will focus attention on the \$82 figure, which has been entered at node (b). We note, in summary, that \$82 is the expected loss of the optimal strategy, which accepts the lot if Type L information is observed and rejects it otherwise.

We turn now to normal-form analysis, in which all possible decision rules or strategies will be considered as a means of commenting on traditional hypothesis-testing procedures. There are eight possible strategies implicit in the decision tree diagram shown in Figure 17-1. These are enumerated in Table 17-6. (An R denotes reject; an A denotes accept.) Therefore, for example, strategy s_3 means accept the lot if Type L or Type M information is observed, that is, if c_2 or fewer defectives are found. Strategy s_4 signifies acceptance if c_1 or fewer defectives are observed. Thus, a choice between strategies s_3 and s_4 means, in acceptance sampling terms, a selection between a single sampling plan with an acceptance number of c_2 and one with an acceptance number of c_1. The conclusion of the extensive-form analysis was that s_4 is the optimal strategy, that is, s_4 has the minimum expected opportunity loss.

As in previous problems, certain of the decision rules do not make much sense. For example, strategy s_1 would reject the lot and strategy s_2 would accept the lot, regardless of the type of information revealed by the sample. Strategy s_6 would reject a lot if a small number of defectives $(X \leq c_1)$ were observed in the sample, but would accept the lot for larger numbers of defectives $(X > c_1)$. The only strategies that appear to be at all logical are s_3 and s_4.

Now, let us suppose this problem had been approached from the standpoint of a hypothesis-testing procedure. The two alternative hypotheses would be

TABLE 17-6
Possible Decision Rules Based on Information Derived from Single Samples of Size n

Sample Information	s_1	s_2	s_3	s_4	s_5	s_6	s_7	s_8
Type $L(X \leq c_1)$	R	A	A	A	A	R	R	R
Type $M(c_1 < X \leq c_2)$	R	A	A	R	R	A	R	A
Type $H(X > c_2)$	R	A	R	R	A	A	A	R

$$H_0: p = 0.02$$
$$H_1: p = 0.05 \text{ or } 0.08$$

Acceptance or rejection of the null hypothesis, H_0, would mean acceptance or rejection of the lot, respectively. As indicated earlier, the company conducting the acceptance sampling wishes to accept lots that are 0.02 defective and to reject otherwise. Hence, the rejection of a good lot (one that contains 0.02 defectives) constitutes a Type I error. Let us assume that the company decides to test the null hypothesis at a preselected 0.05 significance level. That is, the company specifies that it wants to reject lots containing 0.02 defectives no more than 5% of the time. We will examine what this selection of $\alpha = 0.05$ implies concerning the choice of a decision rule.

Power curves may be plotted for each of the strategies or decision rules given in Table 17-6. However, the only ones we will show are for s_3 and s_4. As implied earlier, none of the other strategies are worthy of further consideration. The power curves are depicted in Figure 17-2.

Let us consider how the power curves are plotted by taking the points for $p = 0.02$ as an example. Strategy s_3 accepts H_0 (accepts the lot) if Type L or Type M information is observed. From Table 17-4, we find that the conditional probability of observing Type L or Type M information, given $p = 0.02$, is $0.42/0.50 + 0.06/0.50 = 0.96$. Therefore, the probability that H_0 will be accepted, given $p = 0.02$, is 0.96. Hence, the probability that H_0 will be rejected, given $p = 0.02$, is $1 - 0.96 =$

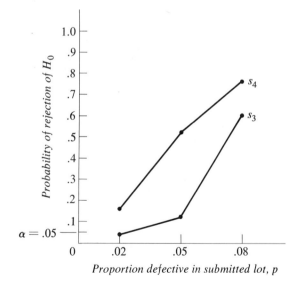

FIGURE 17-2
Power curves for strategies s_3 and s_4.

0.04. Symbolically, for strategy s_3, $P(\text{Rejection of } H_0 | p = 0.02) = 0.04$. Analogously, we find that for strategy s_4, $P(\text{Rejection of } H_0 | p = 0.02) = 1 - 0.42/0.50 = 0.16$.

Now, if we impose the condition of a 0.05 significance level, that is, lots containing 0.02 defectives should be rejected no more than 5% of the time, we find that strategy s_3 meets this criterion but strategy s_4 does not! Therefore, traditional hypothesis-testing procedures would require the use of strategy s_3, which has been shown to be nonoptimal. Looking at Figure 17-1, we can see why this is so. Under strategy s_3, if Type M information is observed, the lot must be accepted, incurring an expected loss of $250, whereas under strategy s_4, if Type M information is observed, the lot is rejected, with a loss of only $60. In summary, the expected opportunity losses of the two strategies are

$$\text{EOL}(s_3) = (0.60)(\$110) + (0.20)(\$250) + (0.20)(\$20) = \$120$$
$$\text{EOL}(s_4) = (0.60)(\$110) + (0.20)(\$60) + (0.20)(\$20) = \$82$$

The major criticism of traditional hypothesis-testing procedures implied by this example is that too much burden is placed on significance levels as a means of deciding between alternative acts. *Specifically, the inclusion of economic costs, or more generally, opportunity losses is not a standard procedure in the decision-making process.*

Another illustrative problem is given in the next section, which more thoroughly contrasts the two sets of procedures. This is followed by a general comparative discussion.

Exercises

1. On what major grounds do advocates of Bayesian decision theory criticize classical hypothesis-testing procedures?

2. Suppose the following table is a joint frequency distribution for a particular problem:

State of Nature	Sample Result			Total
	Type L	Type M	Type H	
θ_1	0.36	0.07	0.01	0.44
θ_2	0.10	0.20	0.02	0.32
θ_3	0.04	0.03	0.17	0.24
Total	0.50	0.30	0.20	1.00

The only strategies that make sense are $s_2(A, R, R)$ and $s_7(A, A, R)$, where A means accept H_0 and R means reject H_0. As an example, $s_2(A, R, R)$ means accept H_0 if type L information is observed, but reject H_0 if type M or H information is observed. Construct power curves for both of these strategies. If the probability of a Type I error has been set at 0.05, which strategy satisfies the requirement?

The alternative hypotheses are

$$H_0: \theta = \theta_1$$
$$H_1: \theta = \theta_2 \text{ or } \theta_3$$

3. Complete the following joint frequency distribution table. Determine which of the following strategies satisfy the requirement that the Type I error be no larger than 0.10: $s_4(R, R, A)$, $s_6(R, A, A)$. Draw power curves for both strategies. The hypotheses and the symbols have the same meaning as in Exercise 2.

State of Nature	Sample Result Type L	Type M	Type H	Total
θ_1		0.12		
θ_2	0.07		0.08	0.38
θ_3			0.17	0.29
Total	0.26	0.43		

4. The Stalwart Appliance Center regularly inspects incoming shipments of small appliances from its supplier to regulate the acceptance of defective items. From an analysis of past costs and past performance, the company has developed a payoff table showing prior probabilities and opportunity losses:

State of Nature ($p = $ Lot Proportion of Defectives)	Prior Probability	Act A_1 Reject	A_2 Accept
0.01	0.45	$400	$0
0.02	0.35	0	700
0.05	0.20	0	900

The company has set two limits, x_1 and x_2, for classification of the results of an inspection into one of three categories:

$$\text{Type } L: X \leqslant x_1$$
$$\text{Type } M: x_1 < X \leqslant x_2$$
$$\text{Type } H: X > x_2$$

where X is the number of defectives in a sample of n items.

From a large number of past samples, the following results are obtained:

State of Nature $(p = Lot$ Proportion of Defectives$)$	Type L $X \leq x_1$	Type M $x_1 < X \leq x_2$	Type H $X > x_2$	Total
0.01	0.30	0.11	0.04	0.45
0.02	0.15	0.13	0.07	0.35
0.05	0.05	0.06	0.09	0.20
	0.50	0.30	0.20	1.00

a. Draw a decision tree for this problem and make all necessary entries. What is the optimal decision (reject or accept shipment) if Type M information is obtained from a sample?

b. Construct power curves for the two strategies $s_3(A_2, A_2, A_1)$ and $s_4(A_2, A_1, A_1)$. If the probability of a Type I error has been set at 0.10, which strategy satisfies the requirement? The alternative hypotheses are

$$H_0: p = 0.01$$
$$H_1: p = 0.02 \text{ or } 0.05$$

17.3 CLASSICAL AND BAYESIAN ESTIMATION

In the preceding section, a comparison was made between hypothesis-testing procedures in classical statistical inference and the corresponding approaches in Bayesian decision theory. In this section, a comparison is made between the estimation techniques in the two approaches.

In Chapter 6, a brief description was given of classical point estimation techniques, that is, methods in which a population parameter value is estimated by a single statistic computed from the observations in a sample. For example, the mean of a sample may be used as the best single estimate of a population mean. In most practical problems, it is not sufficient to have merely a point estimate. If we were given two different point estimates of a population parameter and no further information, we could not distinguish the degree of reliability to be placed upon these estimates. Yet one estimate might be based on a sample of size 10,000 and the other on a sample of size ten. Clearly, these estimates differ greatly in reliability. As we have seen, traditional statistics handles the problem of indicating reliability by the use of the confidence interval procedure. In this section, we will compare this classical technique to the corresponding Bayesian approach. However, before making this comparison, we pause for a comment on point estimation techniques in the two approaches.

Point Estimation

In Section 6.2, criteria of goodness of estimation were discussed. We have become familiar with point estimators such as the observed sample proportion of successes, $\bar{p}$, in a Bernoulli process, which is used as the estimator of the population parameter p, and the observed sample mean, $\bar{x}$, in a process described by the normal distribution, which is used as the estimator of the population mean, μ.

Bayesian decision theory takes a different approach to the problem of point estimation. It views estimation as a straightforward problem of decision making. The estimator is the decision rule, the estimate is the action, and the possible values that the population parameter can assume are the states of nature. For example, the sample mean $\bar{x}$ might be the estimator (decision rule), 10.6 might be the estimate (action), and the possible values that the population mean μ can assume are the parameter values (states of nature). In this formulation, the unknown population parameter is treated as a random variable.

To clarify the method, we will introduce some notation. Let θ be the true value of the parameter we want to estimate and $\hat{\theta}$ the estimate or action. Then a loss is involved if the value of $\hat{\theta}$ differs from θ, and the amount of the loss is some function of the difference between $\hat{\theta}$ and θ. Hence, two possible loss functions might be

(17.1)
$$L(\hat{\theta}; \theta) = |\hat{\theta} - \theta|$$

and

(17.2)
$$L(\hat{\theta}; \theta) = (\hat{\theta} - \theta)^2$$

where $L(\hat{\theta}; \theta)$ is the loss involved in estimating (taking action) $\hat{\theta}$ when the parameter value (state of nature) is θ.

Somewhat more generally, the loss functions (17.1) and (17.2) may be written as

(17.3)
$$L(\hat{\theta}; \theta) = k(\theta)|\hat{\theta} - \theta|$$

and

(17.4)
$$L(\hat{\theta}; \theta) = k(\theta)(\hat{\theta} - \theta)^2$$

respectively, where $k(\theta)$ is a constant for a particular value of θ. This constant may be in money units, utility units, and so on. For simplicity, in the ensuing discussion, we will assume $k(\theta) = 1$ unit of utility, sometimes referred to as a *utile*. Therefore, we are dealing with functions of the form of (17.1) and (17.2), and the losses are given in units of utility.

Expression (17.1) is referred to as a *linear loss function;* (17.2) as a *quadratic loss function* (or *squared error loss function*). The nature of these functions can be illustrated by simple examples. Assume that the true value of the parameter θ is 10. Suppose we consider the losses involved if we estimate this parameter incorrectly as $\hat{\theta} = 11$ and $\hat{\theta} = 12$. For these two estimates, the respective linear loss functions (17.1) are

$$L(11; 10) = |11 - 10| = 1$$

and

$$L(12; 10) = |12 - 10| = 2$$

On the other hand, the quadratic loss function, (17.2), is equal to

$$L(11; 10) = (11 - 10)^2 = 1$$

and

$$L(12; 10) = (12 - 10)^2 = 4$$

In other words, in the linear case the loss in overestimating by two units is *twice* as much as in overestimating by one unit. In the quadratic case, the loss in overestimating by two units is *four* times as much as in overestimating by one unit. It may be noted that in both functions, an underestimate of a given size, say two units, is as serious as an overestimate of the same size. Such loss functions are said to be *symmetrical*.

The ideas of the two aforementioned loss functions were referred to earlier in Section 3.16 in somewhat different forms. There, we were concerned with guessing the value of an observation selected at random from a frequency distribution. The penalty of an incorrect guess or estimate was referred to as the "cost of error." That "cost" corresponds to "loss" in the present discussion. It was pointed out that if the cost of error varies directly with the size of error regardless of sign [the linear loss function, (17.1)], the median is the "best guess," since it minimizes average absolute deviations. On the other hand, if the cost of error varies according to the square of the error [the quadratic loss function, (17.2)], the mean should be the estimated value, since the average of the squared deviations around it is less than around any other figure. It is of interest to note that least squares methods of estimation in classical statistics assume a quadratic loss function, since they obtain estimates for which the average squared error is minimized. Whether or not this is an appropriate loss function for the particular problem involved is rarely investigated.

The Bayesian method of point estimation begins with setting up whatever loss function appears to be appropriate. Then these losses are used in the standard decision procedure. Risks (conditional expected losses) are computed for each decision rule, or estimator. Prior probabilities are assigned to states of nature, or parameter values. Expected risks are computed for each decision rule. Then the estimator for which the expected risk is the least is the one chosen.

No Bayesian point estimators will be derived here, but one result is of particular interest. If the parameter p of a Bernoulli process is estimated using a squared error loss function, and a uniform or rectangular (continuous) prior probability distribution for p is assumed (that is, all values between 0 and 1 are assumed to be equally likely), then the Bayesian estimator of p, denoted $\hat{p}$, is

(17.5)
$$\hat{p} = \frac{X + 1}{n + 2}$$

where X = the number of successes and
$\quad\quad n$ = the number of trials

It turns out that this value of $\hat{p}$ is also the mean or expected value of the posterior distribution of p if the prior distribution of p is assumed to be rectangular (and continuous) and the sample evidence is an observation of X successes in n trials. The estimate of p that we used earlier was simply the observed proportion of successes, X/n. If the sample size, n, is large, these two estimates are approximately equal. Furthermore, there are other prior probability distributions besides the rectangular (uniform) distribution for which the mean of the posterior distribution will have a difference of this order of magnitude when compared with an estimator derived by classical methods. This brings out a very interesting point. From the Bayesian point of view, the standard use of the X/n estimate in such situations carries with it assumptions concerning the nature of the prior distribution of p. The Bayesian decision analyst would argue that some of these prior distributions are very unreasonable in the context of particular problems. For example, a rectangular prior distribution implies that all values of the parameter (in its admissible range) are equally likely. Such an implication may be quite unrealistic based upon the prior knowledge of the individual carrying out the estimation.

Interval Estimation

We turn now to a consideration of interval estimation in classical and Bayesian statistics. We have seen that in estimating a population parameter in classical confidence interval estimation, an interval is set up on the basis of a sample of n observations and a so-called "confidence coefficient" is associated with this interval. Suppose, for example, we wanted to make a confidence interval estimate of p, the proportion of all potential customers on an importer's mailing list who would purchase special jars of cocktail onions if advertisements were sent to them, and let us assume that we want to make this estimate on the basis of the proportion $\bar{p}$ who purchased in a simple random sample of 100 drawn from the list. We could establish (say) a 95% confidence interval around $\bar{p}$ for the estimation of p in the usual way. Let us review the interpretation of this confidence interval. According to the classical school, it is definitely *incorrect* to say that the probability is 95% that the parameter is included in the interval. It is argued that the population parameter is a particular value and therefore cannot be considered a random variable. Indeed, in all of classical statistics, it is forbidden to make conditional probability statements about a *population parameter* given the value of a sample statistic, such as, $P(p \mid \bar{p})$. The permissible types of statements concern conditional probabilities of sample statistics given the value of a population parameter. For example, in a problem involving a Bernoulli process, we could compute probabilities of the type $P(\bar{p} \mid p)$.

Returning to the importer's problem, from the classical viewpoint, the confidence interval estimate of p cannot be interpreted as a probability statement about the proportion of all potential customers on the mailing list who would purchase the product. Since the interval is considered to be the random variable, the confidence coefficient refers to the concept that 95% of the intervals so constructed would bracket or include the true value of the population parameter. Thus, on a relative frequency basis, 95% of the statements made on the basis of such intervals would be correct. Furthermore, in keeping with the classical viewpoint, only the evidence of this particular sample can be used in establishing the confidence interval. Prior knowledge of any sort is not made a part of the estimation procedure. Finally, just as in hypothesis testing, the way in which the sample observations are to be used must be decided upon prior to the examination of these observations.

The Bayesian approach to this general problem stands in sharp contrast to the classical procedure. The Bayesian argues that if the value of the population parameter is unknown, then it can and should be treated as a random variable. In a setting such as the importer's problem, we would view the population parameter p as a basic random variable affecting a decision that must be made. Hence, we would be willing to compute conditional probabilities of the type $P(p \mid \bar{p})$. Furthermore, we would state that these conditional probabilities are the ones relevant to the decision maker, rather than those of the form $P(\bar{p} \mid p)$. For example, in problems similar to that of the importer's, we might be interested in the probability that at least a certain proportion of the population would purchase the product based on the sample evidence. We would not be interested in the reverse conditional probability concerning a proportion in the sample given some postulated value for the population. The Bayesian decision analyst would argue that the confidence interval information is not particularly relevant. The decision maker is not interested in the proportion of correct statements that would be made in the long run, but rather in making a correct decision in this particular case.

The Bayesian approach also maintains that it is not wise to restrict oneself to the evidence of the particular sample that has been drawn but that the sample evidence should be incorporated with prior information through the use of Bayes' Theorem to produce a posterior probability distribution. This leads to the Bayesian approach to the problem that classical inference solves by confidence interval estimation. The Bayesian procedure begins with the assignment of a prior probability distribution to the parameter being estimated. Then a sample is drawn, and the sample evidence is used to revise the prior probability distribution. This revision generates a posterior probability distribution. Then statements such as the following can be made: The probability is 0.90 that p lies between 0.04 and 0.06. The probability is 0.95 that the value of p is 0.07 or less, etc. We may have a large number of possible values for p if that random variable is discrete, or we may have a probability density function over p if the random variable is continuous. The principle remains the same. If the prior distribution was a subjective probability distribution, then the posterior probabilities similarly represent revised degrees of belief or betting odds.

An interesting result occurs that is analogous to a relationship indicated earlier between classical and Bayesian point estimation when a rectangular prior distribution was assumed. If a rectangular distribution is assumed for the random variable p, and if the sample size is large, then there is a close coincidence between the statements made under the two schools of thought. Specifically, for example, the posterior probability that p lies in a 0.95 confidence interval is approximately 0.95. A roughly rectangular or uniform prior distribution is often referred to by Bayesian decision theorists as a "diffuse" or "gentle" prior distribution. Such a distribution implies roughly equal likelihood of occurrence of all values of the random variable in its admissible range. This type of distribution is thought of as an appropriate subjective prior distribution when the decision maker has virtually complete ignorance of the value of the parameter being estimated. Doubtless, such states of almost complete lack of knowledge about parameter values are rare. Hence, Bayesian decision theorists argue that the uncritical use of confidence interval estimates may imply unreasonable assumptions about the investigator's prior knowledge concerning the parameter being estimated.

17.4 SOME REMARKS ON CLASSICAL AND BAYESIAN STATISTICS

As might be surmised from the material in this chapter, some controversy has arisen between those adhering to the classical, or orthodox, school of statistics and those advocating the Bayesian viewpoint. In this section, we will comment on some of the areas of common ground and some of the points of difference between the two schools of thought.

Despite differences in terminology, both schools conceptualize a problem of decision making in which there are states of nature and actions that must be taken in the light of sample or experimental evidence about these states of nature. Both schools use conditional probabilities of sample outcomes, given states of nature (population parameters) for the decision process. These conditional probabilities provide the error characteristic curve on the basis of which the classicist chooses the decision rule. Informally, one is supposed to take into account the relative seriousness of Type I and Type II errors by considering the entire error characteristic curve, but since losses virtually always vary with population parameter values, it is not clear how one can actually do this.

The Bayesian approach supplements or completes the classical analysis by formally providing a loss function that specifies the seriousness of errors in selecting acts and by assigning prior probabilities to states of nature on either an objective or a subjective basis. However, serious measurement problems are clearly present both in the establishment of loss functions and prior probability distributions.

It might be noted that some classicists have affirmed that hypothesis testing is not a decision problem, but rather one of drawing conclusions or inferences. However, other classical adherents have specifically formulated hypothesis testing as an action problem. In any event, it is not always clear whether a problem is one of inference or decision making.

An important area of disagreement between non-Bayesian and Bayesian analysts is the matter of subjective prior probability distributions. The non-Bayesians argue that the only legitimate types of probabilities are "objective" or relative frequency of occurrence probabilities. They find it difficult to accept the idea that subjective or personalistic probabilities should be processed together with relative frequencies, as in the Bayesian's use of Bayes' Theorem, to arrive at posterior probabilities. The Bayesian argues that in actual decision making we do exactly that type of analysis. We have prior betting odds on events that influence the payoffs of our actions. On the observation of sample or experimental information, we revise these prior betting odds. This is an argument centered upon "descriptive" behavior, that is, a purported description of how people actually behave. However, the Bayesian goes further, and says that Bayesian procedures are "normative" or "prescriptive," that is, they specify how a reasonable person *should* choose among alternatives to be consistent with one's own evaluations of payoffs and degrees of belief attached to uncertain events. The Bayesian also argues that if we rigidly maintain that only objective probabilities have meaning, we prevent ourselves from handling some of the most important uncertainties involved in problems of decision making. This latter point is surely a cogent one, particularly in areas such as business and economic decision making.

The problem of how to assign prior probabilities is troublesome, even to convinced Bayesians, and is a subject of ongoing research. There are unresolved problems involved in determining whether all events should be considered equally likely under ignorance, how to pose questions to a decision maker to derive that individual's distribution of betting odds, or more generally, how best to quantify judgments about uncertainty.

The Bayesian turns the tables on the orthodox school, which makes an accusation of excessive subjectivity, and directs a similar charge against classical statistics. The choices of hypotheses to test, probability distributions to use, significance and confidence levels, and what data to collect in order to obtain a relative frequency distribution are all inextricably interwoven with subjective judgments.

The preceding indication of some of the points of disagreement between the classical and Bayesian schools tends to emphasize a polarization of points of view. However, the fact is that even within each of these schools there are philosophical and methodological disagreements, as well. These diversities of viewpoint between and within schools of thought make statistical analysis for decision making a lively and growing field.

Exercises

1. Compare the characteristics of linear loss functions and quadratic loss functions.

2. Describe the steps in the Bayesian method of point estimation.

3. What is the major difference between the classical and the Bayesian viewpoints with regard to an unknown population parameter?

4. Discuss other similarities and differences between Bayesian decision theory and classical statistical inference.

Bibliography

1. Probability

Feller, W., *An Introduction to Probability Theory and Its Applications,* Vols. 1 and 2. New York: John Wiley & Sons, Inc., 1957, 1966.

Goldberg, S., *Probability: An Introduction.* Englewood Cliffs, N.J.: Prentice-Hall, Inc., 1960.

Hodges, J., and E. Lehman, *Basic Concepts of Probability and Statistics.* San Francisco: Holden-Day, Inc., 1964.

Hoel, P. G., S. C. Port, and C. J. Stone, *Introduction to Probability Theory.* Boston: Houghton Mifflin Co., 1972.

Kemeny, J. G., H. Mirkil, J. L. Snell, and G. L. Thompson, *Finite Mathematical Structures.* Englewood Cliffs, N.J.: Prentice-Hall, Inc., 1958.

Mosteller, F., R. Rourke, and G. Thomas, Jr., *Probability and Statistics.* Reading, Mass.: Addison-Wesley Publishing Co., Inc., 1961.

Parzen, E., *Modern Probability Theory and Its Applications.* New York: John Wiley & Sons, Inc., 1960.

2. General Statistics

Afifi, A. A., and S. P. Azen, *Statistical Analysis: A Computer Oriented Approach*, New York: Academic Press, Inc., 1972.

Anderson, T. W., *The Statistical Analysis of Time Series*. New York: John Wiley & Sons, Inc., 1971.

Anderson, R., and R. Bancroft, *Statistical Theory in Research*. New York: McGraw-Hill Book Co., 1952.

Anderson, T. W., and S. L. Sclove, *Introductory Statistical Analysis*. Boston: Houghton Mifflin Co., 1974.

Clelland, R. C., J. S. de Cani, and F. E. Brown, *Basic Statistics with Business Applications*, 2nd ed. New York: John Wiley & Sons, Inc., 1973.

Dixon, W., and F. Massey, Jr., *Introduction to Statistical Analysis*, 3rd ed. New York: McGraw-Hill Book Co., 1969.

Ehrenfeld, S., and S. Littauer, *Introduction to Statistical Method*. New York: McGraw-Hill Book Co., 1964.

Ezekiel, M., and K. Fox, *Methods of Correlation and Regression Analysis*, 3rd ed. New York: John Wiley & Sons, Inc., 1959.

Fox, K., *Intermediate Economic Statistics*. New York: John Wiley & Sons, Inc., 1968.

Freund, J., and F. Williams, *Modern Business Statistics*, 2nd ed. Englewood Cliffs, N.J.: Prentice-Hall, Inc., 1969.

Huff, D., *How to Lie with Statistics*. New York: W. W. Norton, 1954.

Lapin, L. L., *Statistics for Modern Business Decisions*. New York: Harcourt Brace Jovanovich, Inc., 1973.

Mood, A. M., F. A. Graybill, and D. C. Boes, *Introduction to the Theory of Statistics*, 3rd ed. New York: McGraw-Hill Book Co., 1974.

Neter, J., W. Wasserman, and G. Whitmore, *Fundamental Statistics for Business and Economics*, 4th ed. Boston: Allyn and Bacon, Inc., 1973.

Peters, William S., *Readings in Applied Statistics*. Englewood Cliffs, N.J.: Prentice-Hall, Inc., 1969.

Summers, G., and W. Peters, *Statistical Analysis for Decision Making*. Englewood Cliffs, N.J.: Prentice-Hall, Inc., 1968.

Robbins, H., and J. V. Ryzin, *Introduction to Statistics*. Chicago: Science Research Associates, Inc., 1975.

Tanur, J. M., et al. (eds.), *Statistics: A Guide to the Unknown*. San Francisco: Holden-Day, Inc., 1972.

Yule, G., and M. Kendall, *An Introduction to the Theory of Statistics*, 14th ed. New York: Hafner Press, 1950.

3. Decision Analysis

Brown, R. V., A. S. Kahr, and C. Peterson, *Decision Analysis for the Manager*. New York: Holt, Rinehart and Winston, Inc., 1974.

Chernoff, H., and L. Moses, *Elementary Decision Theory*. New York: John Wiley & Sons, Inc., 1959.

Edwards, W., and A. Tversky (eds.), *Decision Making: Selected Readings*. Harmondsworth, England: Penguin Books, Ltd., 1967.

Forester, J., *Statistical Selection of Business Strategies*. Homewood, Ill.: Richard D. Irwin, Inc., 1968.

Hadley, G., *Introduction to Probability and Statistical Decision Theory*. San Francisco: Holden-Day, Inc., 1967.

Harvard Business Review. *Statistical Decision Series* (Parts I–IV). Boston, 1951–70.

Lindley, D. V., *Introduction to Probability and Statistics from a Bayesian Viewpoint*, Part 2, *Inference*. New York: Cambridge University Press, 1965.

Lindley, D. V., *Making Decisions*. London: Wiley-Interscience, 1971.

Pessemier, Edgar A., *New Product Decisions—An Analytic Approach*. New York: McGraw-Hill Book Co., 1966.

Pratt, J., H. Raiffa, and R. Schlaifer, *Introduction to Statistical Decision Theory*. New York: McGraw-Hill Book Co., 1965.

Raiffa, H., *Decision Analysis, Introductory Lectures on Choices Under Uncertainty*. Reading, Mass.: Addison-Wesley Publishing Co., Inc., 1968.

Raiffa, H., and R. Schlaifer, *Applied Statistical Decision Theory*. Cambridge, Mass.: Division of Research, Graduate School of Business Administration, Harvard University, 1961.

Risk Analysis-Proceedings of the United States Army Operations Research Symposium, 15–18 May 1972. Washington, D.C.: Office of the Chief of Research and Development, Department of the Army, 1972.

Schlaifer, R., *Probability and Statistics for Business Decisions*. New York: McGraw-Hill Book Co., 1959.

Wald, A., *Statistical Decision Functions*. New York: John Wiley & Sons, Inc., 1950.

4. Sample Survey Methods

Cochran, W. G., *Sampling Techniques*, 2nd ed. New York: John Wiley & Sons, Inc., 1963.

Deming, W. E., *Sample Designs in Business Research*. New York: John Wiley & Sons, Inc., 1960.

Hansen, M. H., W. N. Hurwitz, and W. G. Madow, *Sample Survey Methods and Theory. Vol. I: Methods and Applications; Vol. II: Theory*. New York: John Wiley & Sons, Inc., 1953.

Kish, L., *Survey Sampling*. New York: John Wiley & Sons, Inc., 1965.

Namias, J., *Handbook of Selected Sample Surveys in the Federal Government with Annotated Bibliography*. New York: St. John's University Press, 1969.

5. Nonparametric Statistics

Bradley, J. V., *Distribution-Free Statistical Tests.* Englewood Cliffs, N.J.: Prentice-Hall, Inc., 1968.

Fisz, M., *Probability Theory and Mathematical Statistics,* 3rd ed. New York: John Wiley & Sons, Inc., 1963.

Gibbons, J. D., *Nonparametric Statistical Inference.* New York: McGraw-Hill Book Co., 1971.

Kraft, C. H., and C. van Eeden, *A Nonparametric Introduction to Statistics.* New York: Macmillan Publishing Co., 1968.

Noether, G. E., *Elements of Nonparametric Statistics.* New York: John Wiley & Sons, Inc., 1967.

Siegel, S., *Nonparametric Statistics for the Behavioral Sciences.* New York: McGraw-Hill Book Co., 1956.

6. Sources of Statistical Data

Daniells, L. M. *Business Information Sources.* Berkeley: University of California Press, 1976.

National Referral Center for Science and Technology, *A Directory of Information Resources in the United States, Social Sciences.* Washington, D.C.: U.S. Government Printing Office, 1967.

Silk, L. S., and M. L. Curley, *A Primer on Business Forecasting with a Guide to Sources of Business Data.* New York: Random House, Inc., 1970.

Wasserman, P., E. Allen, A. Kruzas, and C. Georgi, *Statistics Sources,* 4th ed. Detroit: Gale Research Co., 1971.

7. Statistical Tables

Beyer, W. H., *Handbook of Tables for Probability and Statistics,* 2nd ed. Cleveland: Chemical Rubber Company, 1968.

Burington, R. S., and D. C. May, *Handbook of Probability and Statistics with Tables,* 2nd ed. New York: McGraw-Hill Book Co., 1970.

Hald, A., *Statistical Tables and Formulas.* New York: John Wiley & Sons, Inc., 1952.

National Bureau of Standards, *Tables of the Binomial Probability Distribution.* Applied Mathematical Series 6. Washington, D.C.: U.S. Department of Commerce, 1950.

Owen, D., *Handbook of Statistical Tables.* Reading, Mass.: Addison-Wesley Publishing Co., Inc., 1962.

Pearson, E. S., and H. O. Hartley, *Biometrika Tables for Statisticians,* 2nd ed. Cambridge, England: Cambridge University Press, 1962.

RAND Corporation, *A Million Random Digits with 100,000 Normal Deviates.* New York: Free Press of Glencoe, 1955.

Zehna, P. W., and D. R. Barr, *Tables of the Common Probability Distributions.* Monterey: U.S. Naval Postgraduate School, 1970.

8. Dictionary of Statistical Terms

Freund, J., and F. Williams, *Dictionary/Outline of Basic Statistics.* New York: McGraw-Hill Book Co., 1966.

Kendall, Maurice, and W. R. Buckland, *Dictionary of Statistical Terms*, 3rd ed. New York: Hafner Press, 1971.

Statistical Tables

TABLE A-1
Selected Values of the Binomial Cumulative Distribution Function

$$F(c) = P(X \le c) = \sum_{x=0}^{c} \binom{n}{x}(1-p)^{n-x}p^{x}$$

Example If $p = 0.20$, $n = 7$, $c = 2$, then $F(2) = P(X \le 2) = 0.8520$.

n	c	0.05	0.10	0.15	0.20	0.25	*p* 0.30	0.35	0.40	0.45	0.50
2	0	0.9025	0.8100	0.7225	0.6400	0.5625	0.4900	0.4225	0.3600	0.3025	0.2500
	1	0.9975	0.9900	0.9775	0.9600	0.9375	0.9100	0.8775	0.8400	0.7975	0.7500
3	0	0.8574	0.7290	0.6141	0.5120	0.4219	0.3430	0.2746	0.2160	0.1664	0.1250
	1	0.9928	0.9720	0.9392	0.8960	0.8438	0.7840	0.7182	0.6480	0.5748	0.5000
	2	0.9999	0.9990	0.9966	0.9920	0.9844	0.9730	0.9571	0.9360	0.9089	0.8750
4	0	0.8145	0.6561	0.5220	0.4096	0.3164	0.2401	0.1785	0.1296	0.0915	0.0625
	1	0.9860	0.9477	0.8905	0.8192	0.7383	0.6517	0.5630	0.4752	0.3910	0.3125
	2	0.9995	0.9963	0.9880	0.9728	0.9492	0.9163	0.8735	0.8208	0.7585	0.6875
	3	1.0000	0.9999	0.9995	0.9984	0.9961	0.9919	0.9850	0.9744	0.9590	0.9375
5	0	0.7738	0.5905	0.4437	0.3277	0.2373	0.1681	0.1160	0.0778	0.0503	0.0312
	1	0.9774	0.9185	0.8352	0.7373	0.6328	0.5282	0.4284	0.3370	0.2562	0.1875
	2	0.9988	0.9914	0.9734	0.9421	0.8965	0.8369	0.7648	0.6826	0.5931	0.5000
	3	1.0000	0.9995	0.9978	0.9933	0.9844	0.9692	0.9460	0.9130	0.8688	0.8125
	4	1.0000	1.0000	0.9999	0.9997	0.9990	0.9976	0.9947	0.9898	0.9815	0.9688
6	0	0.7351	0.5314	0.3771	0.2621	0.1780	0.1176	0.0754	0.0467	0.0277	0.0156
	1	0.9672	0.8857	0.7765	0.6554	0.5339	0.4202	0.3191	0.2333	0.1636	0.1094
	2	0.9978	0.9842	0.9527	0.9011	0.8306	0.7443	0.6471	0.5443	0.4415	0.3438
	3	0.9999	0.9987	0.9941	0.9830	0.9624	0.9295	0.8826	0.8208	0.7447	0.6562
	4	1.0000	0.9999	0.9996	0.9984	0.9954	0.9891	0.9777	0.9590	0.9308	0.8906
	5	1.0000	1.0000	1.0000	0.9999	0.9998	0.9993	0.9982	0.9959	0.9917	0.9844
7	0	0.6983	0.4783	0.3206	0.2097	0.1335	0.0824	0.0490	0.0280	0.0152	0.0078
	1	0.9556	0.8503	0.7166	0.5767	0.4449	0.3294	0.2338	0.1586	0.1024	0.0625
	2	0.9962	0.9743	0.9262	0.8520	0.7564	0.6471	0.5323	0.4199	0.3164	0.2266
	3	0.9998	0.9973	0.9879	0.9667	0.9294	0.8740	0.8002	0.7102	0.6083	0.5000
	4	1.0000	0.9998	0.9988	0.9953	0.9871	0.9712	0.9444	0.9037	0.8471	0.7734
	5	1.0000	1.0000	0.9999	0.9996	0.9987	0.9962	0.9910	0.9812	0.9643	0.9375
	6	1.0000	1.0000	1.0000	1.0000	0.9999	0.9998	0.9994	0.9984	0.9963	0.9922
8	0	0.6634	0.4305	0.2725	0.1678	0.1001	0.0576	0.0319	0.0168	0.0084	0.0039
	1	0.9428	0.8131	0.6572	0.5033	0.3671	0.2553	0.1691	0.1064	0.0632	0.0352
	2	0.9942	0.9619	0.8948	0.7969	0.6785	0.5518	0.4278	0.3154	0.2201	0.1445
	3	0.9996	0.9950	0.9786	0.9437	0.8862	0.8059	0.7064	0.5941	0.4770	0.3633
	4	1.0000	0.9996	0.9971	0.9896	0.9727	0.9420	0.8939	0.8263	0.7396	0.6367
	5	1.0000	1.0000	0.9998	0.9988	0.9958	0.9887	0.9747	0.9502	0.9115	0.8555
	6	1.0000	1.0000	1.0000	0.9999	0.9996	0.9987	0.9964	0.9915	0.9819	0.9648
	7	1.0000	1.0000	1.0000	1.0000	1.0000	0.9999	0.9998	0.9993	0.9983	0.9961
9	0	0.6302	0.3874	0.2316	0.1342	0.0751	0.0404	0.0207	0.0101	0.0046	0.0020
	1	0.9288	0.7748	0.5995	0.4362	0.3003	0.1960	0.1211	0.0705	0.0385	0.0195
	2	0.9916	0.9470	0.8591	0.7382	0.6007	0.4628	0.3373	0.2318	0.1495	0.0898
	3	0.9994	0.9917	0.9661	0.9144	0.8343	0.7297	0.6089	0.4826	0.3614	0.2539
	4	1.0000	0.9991	0.9944	0.9804	0.9511	0.9012	0.8283	0.7334	0.6214	0.5000
	5	1.0000	0.9999	0.9994	0.9969	0.9900	0.9747	0.9464	0.9006	0.8342	0.7461
	6	1.0000	1.0000	1.0000	0.9997	0.9987	0.9957	0.9888	0.9750	0.9502	0.9102
	7	1.0000	1.0000	1.0000	1.0000	0.9999	0.9996	0.9986	0.9962	0.9909	0.9805
	8	1.0000	1.0000	1.0000	1.0000	1.0000	1.0000	0.9999	0.9997	0.9992	0.9980

Source: From Irwin Miller and John E. Freund, *Probability and Statistics for Engineers,* © 1965 by Prentice-Hall, Inc.

TABLE A-1 (continued)

n	c	0.05	0.10	0.15	0.20	0.25	*p* 0.30	0.35	0.40	0.45	0.50
10	0	0.5987	0.3487	0.1969	0.1074	0.0563	0.0282	0.0135	0.0060	0.0025	0.0010
	1	0.9139	0.7361	0.5443	0.3758	0.2440	0.1493	0.0860	0.0464	0.0232	0.0107
	2	0.9885	0.9298	0.8202	0.6778	0.5256	0.3828	0.2616	0.1673	0.0996	0.0547
	3	0.9990	0.9872	0.9500	0.8791	0.7759	0.6496	0.5138	0.3823	0.2660	0.1719
	4	0.9999	0.9984	0.9901	0.9672	0.9219	0.8497	0.7515	0.6331	0.5044	0.3770
	5	1.0000	0.9999	0.9986	0.9936	0.9803	0.9527	0.9051	0.8338	0.7384	0.6230
	6	1.0000	1.0000	0.9999	0.9991	0.9965	0.9894	0.9740	0.9452	0.8980	0.8281
	7	1.0000	1.0000	1.0000	0.9999	0.9996	0.9984	0.9952	0.9877	0.9726	0.9453
	8	1.0000	1.0000	1.0000	1.0000	1.0000	0.9999	0.9995	0.9983	0.9955	0.9893
	9	1.0000	1.0000	1.0000	1.0000	1.0000	1.0000	1.0000	0.9999	0.9997	0.9990
11	0	0.5688	0.3138	0.1673	0.0859	0.0422	0.0198	0.0088	0.0036	0.0014	0.0005
	1	0.8981	0.6974	0.4922	0.3221	0.1971	0.1130	0.0606	0.0302	0.0139	0.0059
	2	0.9848	0.9104	0.7788	0.6174	0.4552	0.3127	0.2001	0.1189	0.0652	0.0327
	3	0.9984	0.9815	0.9306	0.8389	0.7133	0.5696	0.4256	0.2963	0.1911	0.1133
	4	0.9999	0.9972	0.9841	0.9496	0.8854	0.7897	0.6683	0.5328	0.3971	0.2744
	5	1.0000	0.9997	0.9973	0.9883	0.9657	0.9218	0.8513	0.7535	0.6331	0.5000
	6	1.0000	1.0000	0.9997	0.9980	0.9924	0.9784	0.9499	0.9006	0.8262	0.7256
	7	1.0000	1.0000	1.0000	0.9998	0.9988	0.9957	0.9878	0.9707	0.9390	0.8867
	8	1.0000	1.0000	1.0000	1.0000	0.9999	0.9994	0.9980	0.9941	0.9852	0.9673
	9	1.0000	1.0000	1.0000	1.0000	1.0000	1.0000	0.9998	0.9993	0.9978	0.9941
	10	1.0000	1.0000	1.0000	1.0000	1.0000	1.0000	1.0000	1.0000	0.9998	0.9995
12	0	0.5404	0.2824	0.1422	0.0687	0.0317	0.0138	0.0057	0.0022	0.0008	0.0002
	1	0.8816	0.6590	0.4435	0.2749	0.1584	0.0850	0.0424	0.0196	0.0083	0.0032
	2	0.9804	0.8891	0.7358	0.5583	0.3907	0.2528	0.1513	0.0834	0.0421	0.0193
	3	0.9978	0.9744	0.9078	0.7946	0.6488	0.4925	0.3467	0.2253	0.1345	0.0730
	4	0.9998	0.9957	0.9761	0.9274	0.8424	0.7237	0.5833	0.4382	0.3044	0.1938
	5	1.0000	0.9995	0.9954	0.9806	0.9456	0.8822	0.7873	0.6652	0.5269	0.3872
	6	1.0000	0.9999	0.9993	0.9961	0.9857	0.9614	0.9154	0.8418	0.7393	0.6128
	7	1.0000	1.0000	0.9999	0.9994	0.9972	0.9905	0.9745	0.9427	0.8883	0.8062
	8	1.0000	1.0000	1.0000	0.9999	0.9996	0.9983	0.9944	0.9847	0.9644	0.9270
	9	1.0000	1.0000	1.0000	1.0000	1.0000	0.9998	0.9992	0.9972	0.9921	0.9807
	10	1.0000	1.0000	1.0000	1.0000	1.0000	1.0000	0.9999	0.9997	0.9989	0.9968
	11	1.0000	1.0000	1.0000	1.0000	1.0000	1.0000	1.0000	1.0000	0.9999	0.9998
13	0	0.5133	0.2542	0.1209	0.0550	0.0238	0.0097	0.0037	0.0013	0.0004	0.0001
	1	0.8646	0.6213	0.3983	0.2336	0.1267	0.0637	0.0296	0.0126	0.0049	0.0017
	2	0.9755	0.8661	0.6920	0.5017	0.3326	0.2025	0.1132	0.0579	0.0269	0.0112
	3	0.9969	0.9658	0.8820	0.7473	0.5843	0.4206	0.2783	0.1686	0.0929	0.0461
	4	0.9997	0.9935	0.9658	0.9009	0.7940	0.6543	0.5005	0.3530	0.2279	0.1334
	5	1.0000	0.9991	0.9925	0.9700	0.9198	0.8346	0.7159	0.5744	0.4268	0.2905
	6	1.0000	0.9999	0.9987	0.9930	0.9757	0.9376	0.8705	0.7712	0.6437	0.5000
	7	1.0000	1.0000	0.9998	0.9988	0.9944	0.9818	0.9538	0.9023	0.8212	0.7095
	8	1.0000	1.0000	1.0000	0.9998	0.9990	0.9960	0.9874	0.9679	0.9302	0.8666
	9	1.0000	1.0000	1.0000	1.0000	0.9999	0.9993	0.9975	0.9922	0.9797	0.9539
	10	1.0000	1.0000	1.0000	1.0000	1.0000	0.9999	0.9997	0.9987	0.9959	0.9888
	11	1.0000	1.0000	1.0000	1.0000	1.0000	1.0000	1.0000	0.9999	0.9995	0.9983
	12	1.0000	1.0000	1.0000	1.0000	1.0000	1.0000	1.0000	1.0000	1.0000	0.9999
14	0	0.4877	0.2288	0.1028	0.0440	0.0178	0.0068	0.0024	0.0008	0.0002	0.0001
	1	0.8470	0.5846	0.3567	0.1979	0.1010	0.0475	0.0205	0.0081	0.0029	0.0009

TABLE A-1 (continued)

n	c	0.05	0.10	0.15	0.20	0.25	0.30	0.35	0.40	0.45	0.50
14	2	0.9699	0.8416	0.6479	0.4481	0.2811	0.1608	0.0839	0.0398	0.0170	0.0065
	3	0.9958	0.9559	0.8535	0.6982	0.5213	0.3552	0.2205	0.1243	0.0632	0.0287
	4	0.9996	0.9908	0.9533	0.8702	0.7415	0.5842	0.4227	0.2793	0.1672	0.0898
	5	1.0000	0.9985	0.9885	0.9561	0.8883	0.7805	0.6405	0.4859	0.3373	0.2120
	6	1.0000	0.9998	0.9978	0.9884	0.9617	0.9067	0.8164	0.6925	0.5461	0.3953
	7	1.0000	1.0000	0.9997	0.9976	0.9897	0.9685	0.9247	0.8499	0.7414	0.6047
	8	1.0000	1.0000	1.0000	0.9996	0.9978	0.9917	0.9757	0.9417	0.8811	0.7880
	9	1.0000	1.0000	1.0000	1.0000	0.9997	0.9983	0.9940	0.9825	0.9574	0.9102
	10	1.0000	1.0000	1.0000	1.0000	1.0000	0.9998	0.9989	0.9961	0.9886	0.9713
	11	1.0000	1.0000	1.0000	1.0000	1.0000	1.0000	0.9999	0.9994	0.9978	0.9935
	12	1.0000	1.0000	1.0000	1.0000	1.0000	1.0000	1.0000	0.9999	0.9997	0.9991
	13	1.0000	1.0000	1.0000	1.0000	1.0000	1.0000	1.0000	1.0000	1.0000	0.9999
15	0	0.4633	0.2059	0.0874	0.0352	0.0134	0.0047	0.0016	0.0005	0.0001	0.0000
	1	0.8290	0.5490	0.3186	0.1671	0.0802	0.0353	0.0142	0.0052	0.0017	0.0005
	2	0.9638	0.8159	0.6042	0.3980	0.2361	0.1268	0.0617	0.0271	0.0107	0.0037
	3	0.9945	0.9444	0.8227	0.6482	0.4613	0.2969	0.1727	0.0905	0.0424	0.0176
	4	0.9994	0.9873	0.9383	0.8358	0.6865	0.5155	0.3519	0.2173	0.1204	0.0592
	5	0.9999	0.9978	0.9832	0.9389	0.8516	0.7216	0.5643	0.4032	0.2608	0.1509
	6	1.0000	0.9997	0.9964	0.9819	0.9434	0.8689	0.7548	0.6098	0.4522	0.3036
	7	1.0000	1.0000	0.9996	0.9958	0.9827	0.9500	0.8868	0.7869	0.6535	0.5000
	8	1.0000	1.0000	0.9999	0.9992	0.9958	0.9848	0.9578	0.9050	0.8182	0.6964
	9	1.0000	1.0000	1.0000	0.9999	0.9992	0.9963	0.9876	0.9662	0.9231	0.8491
	10	1.0000	1.0000	1.0000	1.0000	0.9999	0.9993	0.9972	0.9907	0.9745	0.9408
	11	1.0000	1.0000	1.0000	1.0000	1.0000	0.9999	0.9995	0.9981	0.9937	0.9824
	12	1.0000	1.0000	1.0000	1.0000	1.0000	1.0000	0.9999	0.9997	0.9989	0.9963
	13	1.0000	1.0000	1.0000	1.0000	1.0000	1.0000	1.0000	1.0000	0.9999	0.9995
	14	1.0000	1.0000	1.0000	1.0000	1.0000	1.0000	1.0000	1.0000	1.0000	1.0000
16	0	0.4401	0.1853	0.0743	0.0281	0.0100	0.0033	0.0010	0.0003	0.0001	0.0000
	1	0.8108	0.5147	0.2839	0.1407	0.0635	0.0261	0.0098	0.0033	0.0010	0.0003
	2	0.9571	0.7892	0.5614	0.3518	0.1971	0.0994	0.0451	0.0183	0.0066	0.0021
	3	0.9930	0.9316	0.7899	0.5981	0.4050	0.2459	0.1339	0.0651	0.0281	0.0106
	4	0.9991	0.9830	0.9209	0.7982	0.6302	0.4499	0.2892	0.1666	0.0853	0.0384
	5	0.9999	0.9967	0.9765	0.9183	0.8103	0.6598	0.4900	0.3288	0.1976	0.1051
	6	1.0000	0.9995	0.9944	0.9733	0.9204	0.8247	0.6881	0.5272	0.3660	0.2272
	7	1.0000	0.9999	0.9989	0.9930	0.9729	0.9256	0.8406	0.7161	0.5629	0.4018
	8	1.0000	1.0000	0.9998	0.9985	0.9925	0.9743	0.9329	0.8577	0.7441	0.5982
	9	1.0000	1.0000	1.0000	0.9998	0.9984	0.9929	0.9771	0.9417	0.8759	0.7728
	10	1.0000	1.0000	1.0000	1.0000	0.9997	0.9984	0.9938	0.9809	0.9514	0.8949
	11	1.0000	1.0000	1.0000	1.0000	1.0000	0.9997	0.9987	0.9951	0.9851	0.9616
	12	1.0000	1.0000	1.0000	1.0000	1.0000	1.0000	0.9998	0.9991	0.9965	0.9894
	13	1.0000	1.0000	1.0000	1.0000	1.0000	1.0000	1.0000	0.9999	0.9994	0.9979
	14	1.0000	1.0000	1.0000	1.0000	1.0000	1.0000	1.0000	1.0000	1.0000	0.9997
	15	1.0000	1.0000	1.0000	1.0000	1.0000	1.0000	1.0000	1.0000	1.0000	1.0000
17	0	0.4181	0.1668	0.0631	0.0225	0.0075	0.0023	0.0007	0.0002	0.0000	0.0000
	1	0.7922	0.4818	0.2525	0.1182	0.0501	0.0193	0.0067	0.0021	0.0006	0.0001
	2	0.9497	0.7618	0.5198	0.3096	0.1637	0.0774	0.0327	0.0123	0.0041	0.0012
	3	0.9912	0.9174	0.7556	0.5489	0.3530	0.2019	0.1028	0.0464	0.0184	0.0064
	4	0.9988	0.9779	0.9013	0.7582	0.5739	0.3887	0.2348	0.1260	0.0596	0.0245

TABLE A-1 (continued)

n	c	0.05	0.10	0.15	0.20	0.25	0.30	0.35	0.40	0.45	0.50
17	5	0.9999	0.9953	0.9681	0.8943	0.7653	0.5968	0.4197	0.2639	0.1471	0.0717
	6	1.0000	0.9992	0.9917	0.9623	0.8929	0.7752	0.6188	0.4478	0.2902	0.1662
	7	1.0000	0.9999	0.9983	0.9891	0.9598	0.8954	0.7872	0.6405	0.4743	0.3145
	8	1.0000	1.0000	0.9997	0.9974	0.9876	0.9597	0.9006	0.8011	0.6626	0.5000
	9	1.0000	1.0000	1.0000	0.9995	0.9969	0.9873	0.9617	0.9081	0.8166	0.6855
	10	1.0000	1.0000	1.0000	0.9999	0.9994	0.9968	0.9880	0.9652	0.9174	0.8338
	11	1.0000	1.0000	1.0000	1.0000	0.9999	0.9993	0.9970	0.9894	0.9699	0.9283
	12	1.0000	1.0000	1.0000	1.0000	1.0000	0.9999	0.9994	0.9975	0.9914	0.9755
	13	1.0000	1.0000	1.0000	1.0000	1.0000	1.0000	0.9999	0.9995	0.9981	0.9936
	14	1.0000	1.0000	1.0000	1.0000	1.0000	1.0000	1.0000	0.9999	0.9997	0.9988
	15	1.0000	1.0000	1.0000	1.0000	1.0000	1.0000	1.0000	1.0000	1.0000	0.9999
	16	1.0000	1.0000	1.0000	1.0000	1.0000	1.0000	1.0000	1.0000	1.0000	1.0000
18	0	0.3972	0.1501	0.0536	0.0180	0.0056	0.0016	0.0004	0.0001	0.0000	0.0000
	1	0.7735	0.4503	0.2241	0.0991	0.0395	0.0142	0.0046	0.0013	0.0003	0.0001
	2	0.9419	0.7338	0.4797	0.2713	0.1353	0.0600	0.0236	0.0082	0.0025	0.0007
	3	0.9891	0.9018	0.7202	0.5010	0.3057	0.1646	0.0783	0.0328	0.0120	0.0038
	4	0.9985	0.9718	0.8794	0.7164	0.5187	0.3327	0.1886	0.0942	0.0411	0.0154
	5	0.9998	0.9936	0.9581	0.8671	0.7175	0.5344	0.3550	0.2088	0.1077	0.0481
	6	1.0000	0.9988	0.9882	0.9487	0.8610	0.7217	0.5491	0.3743	0.2258	0.1189
	7	1.0000	0.9998	0.9973	0.9837	0.9431	0.8593	0.7283	0.5634	0.3915	0.2403
	8	1.0000	1.0000	0.9995	0.9957	0.9807	0.9404	0.8609	0.7368	0.5778	0.4073
	9	1.0000	1.0000-	0.9999	0.9991	0.9946	0.9790	0.9403	0.8653	0.7473	0.5927
	10	1.0000	1.0000	1.0000	0.9998	0.9988	0.9939	0.9788	0.9424	0.8720	0.7597
	11	1.0000	1.0000	1.0000	1.0000	0.9998	0.9986	0.9938	0.9797	0.9463	0.8811
	12	1.0000	1.0000	1.0000	1.0000	1.0000	0.9997	0.9986	0.9942	0.9817	0.9519
	13	1.0000	1.0000	1.0000	1.0000	1.0000	1.0000	0.9997	0.9987	0.9951	0.9846
	14	1.0000	1.0000	1.0000	1.0000	1.0000	1.0000	1.0000	0.9998	0.9990	0.9962
	15	1.0000	1.0000	1.0000	1.0000	1.0000	1.0000	1.0000	1.0000	0.9999	0.9993
	16	1.0000	1.0000	1.0000	1.0000	1.0000	1.0000	1.0000	1.0000	1.0000	0.9999
19	0	0.3774	0.1351	0.0456	0.0144	0.0042	0.0011	0.0003	0.0001	0.0000	0.0000
	1	0.7547	0.4203	0.1985	0.0829	0.0310	0.0104	0.0031	0.0008	0.0002	0.0000
	2	0.9335	0.7054	0.4413	0.2369	0.1113	0.0462	0.0170	0.0055	0.0015	0.0004
	3	0.9868	0.8850	0.6841	0.4551	0.2630	0.1332	0.0591	0.0230	0.0077	0.0022
	4	0.9980	0.9648	0.8556	0.6733	0.4654	0.2822	0.1500	0.0696	0.0280	0.0096
	5	0.9998	0.9914	0.9463	0.8369	0.6678	0.4739	0.2968	0.1629	0.0777	0.0318
	6	1.0000	0.9983	0.9837	0.9324	0.8251	0.6655	0.4812	0.3081	0.1727	0.0835
	7	1.0000	0.9997	0.9959	0.9767	0.9225	0.8180	0.6656	0.4878	0.3169	0.1796
	8	1.0000	1.0000	0.9992	0.9933	0.9713	0.9161	0.8145	0.6675	0.4940	0.3238
	9	1.0000	1.0000	0.9999	0.9984	0.9911	0.9674	0.9125	0.8139	0.6710	0.5000
	10	1.0000	1.0000	1.0000	0.9997	0.9977	0.9895	0.9653	0.9115	0.8159	0.6762
	11	1.0000	1.0000	1.0000	1.0000	0.9995	0.9972	0.9886	0.9648	0.9129	0.8204
	12	1.0000	1.0000	1.0000	1.0000	0.9999	0.9994	0.9969	0.9884	0.9658	0.9165
	13	1.0000	1.0000	1.0000	1.0000	1.0000	0.9999	0.9993	0.9969	0.9891	0.9682
	14	1.0000	1.0000	1.0000	1.0000	1.0000	1.0000	0.9999	0.9994	0.9972	0.9904
	15	1.0000	1.0000	1.0000	1.0000	1.0000	1.0000	1.0000	0.9999	0.9995	0.9978
	16	1.0000	1.0000	1.0000	1.0000	1.0000	1.0000	1.0000	1.0000	0.9999	0.9996
	17	1.0000	1.0000	1.0000	1.0000	1.0000	1.0000	1.0000	1.0000	1.0000	1.0000

TABLE A-1 (continued)

n	c	0.05	0.10	0.15	0.20	0.25	0.30	0.35	0.40	0.45	0.50
20	0	0.3585	0.1216	0.0388	0.0115	0.0032	0.0008	0.0002	0.0000	0.0000	0.0000
	1	0.7358	0.3917	0.1756	0.0692	0.0243	0.0076	0.0021	0.0005	0.0001	0.0000
	2	0.9245	0.6769	0.4049	0.2061	0.0913	0.0355	0.0121	0.0036	0.0009	0.0002
	3	0.9841	0.8670	0.6477	0.4114	0.2252	0.1071	0.0444	0.0160	0.0049	0.0013
	4	0.9974	0.9568	0.8298	0.6296	0.4148	0.2375	0.1182	0.0510	0.0189	0.0059
	5	0.9997	0.9887	0.9327	0.8042	0.6172	0.4164	0.2454	0.1256	0.0553	0.0207
	6	1.0000	0.9976	0.9781	0.9133	0.7858	0.6080	0.4166	0.2500	0.1299	0.0577
	7	1.0000	0.9996	0.9941	0.9679	0.8982	0.7723	0.6010	0.4159	0.2520	0.1316
	8	1.0000	0.9999	0.9987	0.9900	0.9591	0.8867	0.7624	0.5956	0.4143	0.2517
	9	1.0000	1.0000	0.9998	0.9974	0.9861	0.9520	0.8782	0.7553	0.5914	0.4119
	10	1.0000	1.0000	1.0000	0.9994	0.9961	0.9829	0.9468	0.8725	0.7507	0.5881
	11	1.0000	1.0000	1.0000	0.9999	0.9991	0.9949	0.9804	0.9435	0.8692	0.7483
	12	1.0000	1.0000	1.0000	1.0000	0.9998	0.9987	0.9940	0.9790	0.9420	0.8684
	13	1.0000	1.0000	1.0000	1.0000	1.0000	0.9997	0.9985	0.9935	0.9786	0.9423
	14	1.0000	1.0000	1.0000	1.0000	1.0000	1.0000	0.9997	0.9984	0.9936	0.9793
	15	1.0000	1.0000	1.0000	1.0000	1.0000	1.0000	1.0000	0.9997	0.9985	0.9941
	16	1.0000	1.0000	1.0000	1.0000	1.0000	1.0000	1.0000	1.0000	0.9997	0.9987
	17	1.0000	1.0000	1.0000	1.0000	1.0000	1.0000	1.0000	1.0000	1.0000	0.9998
	18	1.0000	1.0000	1.0000	1.0000	1.0000	1.0000	1.0000	1.0000	1.0000	1.0000

TABLE A-2
Coefficients of the Binomial Distribution

Example If $n = 8$ and $x = 6$, $\binom{8}{6} = 28$.

This table gives the value of $\binom{n}{x}$ in $\binom{n}{x}q^{n-x}p^{x}$, the general term of $(q + p)^n$.

n	$\binom{n}{0}$	$\binom{n}{1}$	$\binom{n}{2}$	$\binom{n}{3}$	$\binom{n}{4}$	$\binom{n}{5}$	$\binom{n}{6}$	$\binom{n}{7}$	$\binom{n}{8}$	$\binom{n}{9}$	$\binom{n}{10}$
0	1										
1	1	1									
2	1	2	1								
3	1	3	3	1							
4	1	4	6	4	1						
5	1	5	10	10	5	1					
6	1	6	15	20	15	6	1				
7	1	7	21	35	35	21	7	1			
8	1	8	28	56	70	56	28	8	1		
9	1	9	36	84	126	126	84	36	9	1	
10	1	10	45	120	210	252	210	120	45	10	1
11	1	11	55	165	330	462	462	330	165	55	11
12	1	12	66	220	495	792	924	792	495	220	66
13	1	13	78	286	715	1287	1716	1716	1287	715	286
14	1	14	91	364	1001	2002	3003	3432	3003	2002	1001
15	1	15	105	455	1365	3003	5005	6435	6435	5005	3003
16	1	16	120	560	1820	4368	8008	11440	12870	11440	8008
17	1	17	136	680	2380	6188	12376	19448	24310	24310	19448
18	1	18	153	816	3060	8568	18564	31824	43758	48620	43758
19	1	19	171	969	3876	11628	27132	50388	75582	92378	92378
20	1	20	190	1140	4845	15504	38760	77520	125970	167960	184756

TABLE A-3
Selected Values of the Poisson Cumulative Distribution

$$F(c) = P(X \le c) = \sum_{x=0}^{c} \frac{\mu^x e^{-\mu}}{x!}$$

Example If $\mu = 1.00$, then $F(2) = P(X \le 2) = 0.920$.

$\mu \backslash c$	0	1	2	3	4	5	6	7	8	9
0.02	0.980	1.000								
0.04	0.961	0.999	1.000							
0.06	0.942	0.998	1.000							
0.08	0.923	0.997	1.000							
0.10	0.905	0.995	1.000							
0.15	0.861	0.990	0.999	1.000						
0.20	0.819	0.982	0.999	1.000						
0.25	0.779	0.974	0.998	1.000						
0.30	0.741	0.963	0.996	1.000						
0.35	0.705	0.951	0.994	1.000						
0.40	0.670	0.938	0.992	0.999	1.000					
0.45	0.638	0.925	0.989	0.999	1.000					
0.50	0.607	0.910	0.986	0.998	1.000					
0.55	0.577	0.894	0.982	0.998	1.000					
0.60	0.549	0.878	0.977	0.997	1.000					
0.65	0.522	0.861	0.972	0.996	0.999	1.000				
0.70	0.497	0.844	0.966	0.994	0.999	1.000				
0.75	0.472	0.827	0.959	0.993	0.999	1.000				
0.80	0.449	0.809	0.953	0.991	0.999	1.000				
0.85	0.427	0.791	0.945	0.989	0.998	1.000				
0.90	0.407	0.772	0.937	0.987	0.998	1.000				
0.95	0.387	0.754	0.929	0.984	0.997	1.000				
1.00	0.368	0.736	0.920	0.981	0.996	0.999	1.000			
1.1	0.333	0.699	0.900	0.974	0.995	0.999	1.000			
1.2	0.301	0.663	0.879	0.966	0.992	0.998	1.000			
1.3	0.273	0.627	0.857	0.957	0.989	0.998	1.000			
1.4	0.247	0.592	0.833	0.946	0.986	0.997	0.999	1.000		
1.5	0.223	0.558	0.809	0.934	0.981	0.996	0.999	1.000		
1.6	0.202	0.525	0.783	0.921	0.976	0.994	0.999	1.000		
1.7	0.183	0.493	0.757	0.907	0.970	0.992	0.998	1.000		
1.8	0.165	0.463	0.731	0.891	0.964	0.990	0.997	0.999	1.000	
1.9	0.150	0.434	0.704	0.875	0.956	0.987	0.997	0.999	1.000	
2.0	0.135	0.406	0.677	0.857	0.947	0.983	0.995	0.999	1.000	
2.2	0.111	0.355	0.623	0.819	0.928	0.975	0.993	0.998	1.000	
2.4	0.091	0.308	0.570	0.779	0.904	0.964	0.988	0.997	0.999	1.000
2.6	0.074	0.267	0.518	0.736	0.877	0.951	0.983	0.995	0.999	1.000
2.8	0.061	0.231	0.469	0.692	0.848	0.935	0.976	0.992	0.998	0.999
3.0	0.050	0.199	0.423	0.647	0.815	0.916	0.966	0.988	0.996	0.999

Source: From *Statistical Quality Control*, by Eugene L. Grant. Copyright 1964 by McGraw-Hill Book Company. Used with permission of McGraw-Hill Book Company.

TABLE A-3 (continued)

μ\c	0	1	2	3	4	5	6	7	8	9
3.2	0.041	0.171	0.380	0.603	0.781	0.895	0.955	0.983	0.994	0.998
3.4	0.033	0.147	0.340	0.558	0.744	0.871	0.942	0.977	0.992	0.997
3.6	0.027	0.126	0.303	0.515	0.706	0.844	0.927	0.969	0.988	0.996
3.8	0.022	0.107	0.269	0.473	0.668	0.816	0.909	0.960	0.984	0.994
4.0	0.018	0.092	0.238	0.433	0.629	0.785	0.889	0.949	0.979	0.992
4.2	0.015	0.078	0.210	0.395	0.590	0.753	0.867	0.936	0.972	0.989
4.4	0.012	0.066	0.185	0.359	0.551	0.720	0.844	0.921	0.964	0.985
4.6	0.010	0.056	0.163	0.326	0.513	0.686	0.818	0.905	0.955	0.980
4.8	0.008	0.048	0.143	0.294	0.476	0.651	0.791	0.887	0.944	0.975
5.0	0.007	0.040	0.125	0.265	0.440	0.616	0.762	0.867	0.932	0.968
5.2	0.006	0.034	0.109	0.238	0.406	0.581	0.732	0.845	0.918	0.960
5.4	0.005	0.029	0.095	0.213	0.373	0.546	0.702	0.822	0.903	0.951
5.6	0.004	0.024	0.082	0.191	0.342	0.512	0.670	0.797	0.886	0.941
5.8	0.003	0.021	0.072	0.170	0.313	0.478	0.638	0.771	0.867	0.929
6.0	0.002	0.017	0.062	0.151	0.285	0.446	0.606	0.744	0.847	0.916

μ	10	11	12	13	14	15	16
2.8	1.000						
3.0	1.000						
3.2	1.000						
3.4	0.999	1.000					
3.6	0.999	1.000					
3.8	0.998	0.999	1.000				
4.0	0.997	0.999	1.000				
4.2	0.996	0.999	1.000				
4.4	0.994	0.998	0.999	1.000			
4.6	0.992	0.997	0.999	1.000			
4.8	0.990	0.996	0.999	1.000			
5.0	0.986	0.995	0.998	0.999	1.000		
5.2	0.982	0.993	0.997	0.999	1.000		
5.4	0.977	0.990	0.996	0.999	1.000		
5.6	0.972	0.988	0.995	0.998	0.999	1.000	
5.8	0.965	0.984	0.993	0.997	0.999	1.000	
6.0	0.957	0.980	0.991	0.996	0.999	0.999	1.000

μ\c	0	1	2	3	4	5	6	7	8	9
6.2	0.002	0.015	0.054	0.134	0.259	0.414	0.574	0.716	0.826	0.902
6.4	0.002	0.012	0.046	0.119	0.235	0.384	0.542	0.687	0.803	0.886
6.6	0.001	0.010	0.040	0.105	0.213	0.355	0.511	0.658	0.780	0.869
6.8	0.001	0.009	0.034	0.093	0.192	0.327	0.480	0.628	0.755	0.850
7.0	0.001	0.007	0.030	0.082	0.173	0.301	0.450	0.599	0.729	0.830
7.2	0.001	0.006	0.025	0.072	0.156	0.276	0.420	0.569	0.703	0.810
7.4	0.001	0.005	0.022	0.063	0.140	0.253	0.392	0.539	0.676	0.788
7.6	0.001	0.004	0.019	0.055	0.125	0.231	0.365	0.510	0.648	0.765
7.8	0.000	0.004	0.016	0.048	0.112	0.210	0.338	0.481	0.620	0.741

TABLE A-3 (continued)

$\mu\backslash c$	0	1	2	3	4	5	6	7	8	9
8.0	0.000	0.003	0.014	0.042	0.100	0.191	0.313	0.453	0.593	0.717
8.5	0.000	0.002	0.009	0.030	0.074	0.150	0.256	0.386	0.523	0.653
9.0	0.000	0.001	0.006	0.021	0.055	0.116	0.207	0.324	0.456	0.587
9.5	0.000	0.001	0.004	0.015	0.040	0.089	0.165	0.269	0.392	0.522
10.0	0.000	0.000	0.003	0.010	0.029	0.067	0.130	0.220	0.333	0.458

	10	11	12	13	14	15	16	17	18	19
6.2	0.949	0.975	0.989	0.995	0.998	0.999	1.000			
6.4	0.939	0.969	0.986	0.994	0.997	0.999	1.000			
6.6	0.927	0.963	0.982	0.992	0.997	0.999	0.999	1.000		
6.8	0.915	0.955	0.978	0.990	0.996	0.998	0.999	1.000		
7.0	0.901	0.947	0.973	0.987	0.994	0.998	0.999	1.000		
7.2	0.887	0.937	0.967	0.984	0.993	0.997	0.999	0.999	1.000	
7.4	0.871	0.926	0.961	0.980	0.991	0.996	0.998	0.999	1.000	
7.6	0.854	0.915	0.954	0.976	0.989	0.995	0.998	0.999	1.000	
7.8	0.835	0.902	0.945	0.971	0.986	0.993	0.997	0.999	1.000	
8.0	0.816	0.888	0.936	0.966	0.983	0.992	0.996	0.998	0.999	1.000
8.5	0.763	0.849	0.909	0.949	0.973	0.986	0.993	0.997	0.999	0.999
9.0	0.706	0.803	0.876	0.926	0.959	0.978	0.989	0.995	0.998	0.999
9.5	0.645	0.752	0.836	0.898	0.940	0.967	0.982	0.991	0.996	0.998
10.0	0.583	0.697	0.792	0.864	0.917	0.951	0.973	0.986	0.993	0.997

	20	21	22
8.5	1.000		
9.0	1.000		
9.5	0.999	1.000	
10.0	0.998	0.999	1.000

$\mu\backslash c$	0	1	2	3	4	5	6	7	8	9
10.5	0.000	0.000	0.002	0.007	0.021	0.050	0.102	0.179	0.279	0.397
11.0	0.000	0.000	0.001	0.005	0.015	0.038	0.079	0.143	0.232	0.341
11.5	0.000	0.000	0.001	0.003	0.011	0.028	0.060	0.114	0.191	0.289
12.0	0.000	0.000	0.001	0.002	0.008	0.020	0.046	0.090	0.155	0.242
12.5	0.000	0.000	0.000	0.002	0.005	0.015	0.035	0.070	0.125	0.201
13.0	0.000	0.000	0.000	0.001	0.004	0.011	0.026	0.054	0.100	0.166
13.5	0.000	0.000	0.000	0.001	0.003	0.008	0.019	0.041	0.079	0.135
14.0	0.000	0.000	0.000	0.000	0.002	0.006	0.014	0.032	0.062	0.109
14.5	0.000	0.000	0.000	0.000	0.001	0.004	0.010	0.024	0.048	0.088
15.0	0.000	0.000	0.000	0.000	0.001	0.003	0.008	0.018	0.037	0.070

	10	11	12	13	14	15	16	17	18	19
10.5	0.521	0.639	0.742	0.825	0.888	0.932	0.960	0.978	0.988	0.994
11.0	0.460	0.579	0.689	0.781	0.854	0.907	0.944	0.968	0.982	0.991
11.5	0.402	0.520	0.633	0.733	0.815	0.878	0.924	0.954	0.974	0.986
12.0	0.347	0.462	0.576	0.682	0.772	0.844	0.899	0.937	0.963	0.979
12.5	0.297	0.406	0.519	0.628	0.725	0.806	0.869	0.916	0.948	0.969

TABLE A-3 (continued)

$\mu\backslash c$	10	11	12	13	14	15	16	17	18	19
13.0	0.252	0.353	0.463	0.573	0.675	0.764	0.835	0.890	0.930	0.957
13.5	0.211	0.304	0.409	0.518	0.623	0.718	0.798	0.861	0.908	0.942
14.0	0.176	0.260	0.358	0.464	0.570	0.669	0.756	0.827	0.883	0.923
14.5	0.145	0.220	0.311	0.413	0.518	0.619	0.711	0.790	0.853	0.901
15.0	0.118	0.185	0.268	0.363	0.466	0.568	0.664	0.749	0.819	0.875

	20	21	22	23	24	25	26	27	28	29
10.5	0.997	0.999	0.999	1.000						
11.0	0.995	0.998	0.999	1.000						
11.5	0.992	0.996	0.998	0.999	1.000					
12.0	0.988	0.994	0.997	0.999	0.999	1.000				
12.5	0.983	0.991	0.995	0.998	0.999	0.999	1.000			
13.0	0.975	0.986	0.992	0.996	0.998	0.999	1.000			
13.5	0.965	0.980	0.989	0.994	0.997	0.998	0.999	1.000		
14.0	0.952	0.971	0.983	0.991	0.995	0.997	0.999	0.999	1.000	
14.5	0.936	0.960	0.976	0.986	0.992	0.996	0.998	0.999	0.999	1.000
15.0	0.917	0.947	0.967	0.981	0.989	0.994	0.997	0.998	0.999	1.000

$\mu\backslash c$	4	5	6	7	8	9	10	11	12	13
16	0.000	0.001	0.004	0.010	0.022	0.043	0.077	0.127	0.193	0.275
17	0.000	0.001	0.002	0.005	0.013	0.026	0.049	0.085	0.135	0.201
18	0.000	0.000	0.001	0.003	0.007	0.015	0.030	0.055	0.092	0.143
19	0.000	0.000	0.001	0.002	0.004	0.009	0.018	0.035	0.061	0.098
20	0.000	0.000	0.000	0.001	0.002	0.005	0.011	0.021	0.039	0.066
21	0.000	0.000	0.000	0.000	0.001	0.003	0.006	0.013	0.025	0.043
22	0.000	0.000	0.000	0.000	0.001	0.002	0.004	0.008	0.015	0.028
23	0.000	0.000	0.000	0.000	0.000	0.001	0.002	0.004	0.009	0.017
24	0.000	0.000	0.000	0.000	0.000	0.000	0.001	0.003	0.005	0.011
25	0.000	0.000	0.000	0.000	0.000	0.000	0.001	0.001	0.003	0.006

	14	15	16	17	18	19	20	21	22	23
16	0.368	0.467	0.566	0.659	0.742	0.812	0.868	0.911	0.942	0.963
17	0.281	0.371	0.468	0.564	0.655	0.736	0.805	0.861	0.905	0.937
18	0.208	0.287	0.375	0.469	0.562	0.651	0.731	0.799	0.855	0.899
19	0.150	0.215	0.292	0.378	0.469	0.561	0.647	0.725	0.793	0.849
20	0.105	0.157	0.221	0.297	0.381	0.470	0.559	0.644	0.721	0.787
21	0.072	0.111	0.163	0.227	0.302	0.384	0.471	0.558	0.640	0.716
22	0.048	0.077	0.117	0.169	0.232	0.306	0.387	0.472	0.556	0.637
23	0.031	0.052	0.082	0.123	0.175	0.238	0.310	0.389	0.472	0.555
24	0.020	0.034	0.056	0.087	0.128	0.180	0.243	0.314	0.392	0.473
25	0.012	0.022	0.038	0.060	0.092	0.134	0.185	0.247	0.318	0.394

TABLE A-3 (continued)

	24	25	26	27	28	29	30	31	32	33
16	0.978	0.987	0.993	0.996	0.998	0.999	0.999	1.000		
17	0.959	0.975	0.985	0.991	0.995	0.997	0.999	0.999	1.000	
18	0.932	0.955	0.972	0.983	0.990	0.994	0.997	0.998	0.999	1.000
19	0.893	0.927	0.951	0.969	0.980	0.988	0.993	0.996	0.998	0.999
20	0.843	0.888	0.922	0.948	0.966	0.978	0.987	0.992	0.995	0.997
21	0.782	0.838	0.883	0.917	0.944	0.963	0.976	0.985	0.991	0.994
22	0.712	0.777	0.832	0.877	0.913	0.940	0.959	0.973	0.983	0.989
23	0.635	0.708	0.772	0.827	0.873	0.908	0.936	0.956	0.971	0.981
24	0.554	0.632	0.704	0.768	0.823	0.868	0.904	0.932	0.953	0.969
25	0.473	0.553	0.629	0.700	0.763	0.818	0.863	0.900	0.929	0.950

	34	35	36	37	38	39	40	41	42	43
19	0.999	1.000								
20	0.999	0.999	1.000							
21	0.997	0.998	0.999	0.999	1.000					
22	0.994	0.996	0.998	0.999	0.999	1.000				
23	0.988	0.993	0.996	0.997	0.999	0.999	1.000			
24	0.979	0.987	0.992	0.995	0.997	0.998	0.999	0.999	1.000	
25	0.966	0.978	0.985	0.991	0.994	0.997	0.998	0.999	0.999	1.000

TABLE A-4
Four-Place Common Logarithms

N	0	1	2	3	4	5	6	7	8	9		1	2	3	4	5	6	7	8	9
														Proportional Parts						
10	0000	0043	0086	0128	0170	0212	0253	0294	0334	0374		4	8	12	17	21	25	29	33	37
11	0414	0453	0492	0531	0569	0607	0645	0682	0719	0755		4	8	11	15	19	23	26	30	34
12	0792	0828	0864	0899	0934	0969	1004	1038	1072	1106		3	7	10	14	17	21	24	28	31
13	1139	1173	1206	1239	1271	1303	1335	1367	1399	1430		3	6	10	13	16	19	23	26	29
14	1461	1492	1523	1553	1584	1614	1644	1673	1703	1732		3	6	9	12	15	18	21	24	27
15	1761	1790	1818	1847	1875	1903	1931	1959	1987	2014		3	6	8	11	14	17	20	22	25
16	2041	2068	2095	2122	2148	2175	2201	2227	2253	2279		3	5	8	11	13	16	18	21	24
17	2304	2330	2355	2380	2405	2430	2455	2480	2504	2529		2	5	7	10	12	15	17	20	22
18	2553	2577	2601	2625	2648	2672	2695	2718	2742	2765		2	5	7	9	12	14	16	19	21
19	2788	2810	2833	2856	2878	2900	2923	2945	2967	2989		2	4	7	9	11	13	16	18	20
20	3010	3032	3054	3075	3096	3118	3139	3160	3181	3201		2	4	6	8	11	13	15	17	19
21	3222	3243	3263	3284	3304	3324	3345	3365	3385	3404		2	4	6	8	10	12	14	16	18
22	3424	3444	3464	3483	3502	3522	3541	3560	3579	3598		2	4	6	8	10	12	14	15	17
23	3617	3636	3655	3674	3692	3711	3729	3747	3766	3784		2	4	6	7	9	11	13	15	17
24	3802	3820	3838	3856	3874	3892	3909	3927	3945	3962		2	4	5	7	9	11	12	14	16
25	3979	3997	4014	4031	4048	4065	4082	4099	4116	4133		2	3	5	7	9	10	12	14	15
26	4150	4166	4183	4200	4216	4232	4249	4265	4281	4298		2	3	5	7	8	10	11	13	15
27	4314	4330	4346	4362	4378	4393	4409	4425	4440	4456		2	3	5	6	8	9	11	13	14
28	4472	4487	4502	4518	4533	4548	4564	4579	4594	4609		2	3	5	6	8	9	11	12	14
29	4624	4639	4654	4669	4683	4698	4713	4728	4742	4757		1	3	4	6	7	9	10	12	13
30	4771	4786	4800	4814	4829	4843	4857	4871	4886	4900		1	3	4	6	7	9	10	11	13
31	4914	4928	4942	4955	4969	4983	4997	5011	5024	5038		1	3	4	6	7	8	10	11	12
32	5051	5065	5079	5092	5105	5119	5132	5145	5159	5172		1	3	4	5	7	8	9	11	12
33	5185	5198	5211	5224	5237	5250	5263	5276	5289	5302		1	3	4	5	6	8	9	10	12
34	5315	5328	5340	5353	5366	5378	5391	5403	5416	5428		1	3	4	5	6	8	9	10	11
35	5441	5453	5465	5478	5490	5502	5514	5527	5539	5551		1	2	4	5	6	7	9	10	11
36	5563	5575	5587	5599	5611	5623	5635	5647	5658	5670		1	2	4	5	6	7	8	10	11
37	5682	5694	5705	5717	5729	5740	5752	5763	5775	5786		1	2	3	5	6	7	8	9	10
38	5798	5809	5821	5832	5843	5855	5866	5877	5888	5899		1	2	3	5	6	7	8	9	10
39	5911	5922	5933	5944	5955	5966	5977	5988	5999	6010		1	2	3	4	5	7	8	9	10
40	6021	6031	6042	6053	6064	6075	6085	6096	6107	6117		1	2	3	4	5	6	8	9	10
41	6128	6138	6149	6160	6170	6180	6191	6201	6212	6222		1	2	3	4	5	6	7	8	9
42	6232	6243	6253	6263	6274	6284	6294	6304	6314	6325		1	2	3	4	5	6	7	8	9
43	6335	6345	6355	6365	6375	6385	6395	6405	6415	6425		1	2	3	4	5	6	7	8	9
44	6435	6444	6454	6464	6474	6484	6493	6503	6513	6522		1	2	3	4	5	6	7	8	9
45	6532	6542	6551	6561	6571	6580	6590	6599	6609	6618		1	2	3	4	5	6	7	8	9
46	6628	6637	6646	6656	6665	6675	6684	6693	6702	6712		1	2	3	4	5	6	7	7	8
47	6721	6730	6739	6749	6758	6767	6776	6785	6794	6803		1	2	3	4	5	5	6	7	8
48	6812	6821	6830	6839	6848	6857	6866	6875	6884	6893		1	2	3	4	4	5	6	7	8
49	6902	6911	6920	6928	6937	6946	6955	6964	6972	6981		1	2	3	4	4	5	6	7	8
50	6990	6998	7007	7016	7024	7033	7042	7050	7059	7067		1	2	3	3	4	5	6	7	8
51	7076	7084	7093	7101	7110	7118	7126	7135	7143	7152		1	2	3	3	4	5	6	7	8
52	7160	7168	7177	7185	7193	7202	7210	7218	7226	7235		1	2	2	3	4	5	6	7	7
53	7243	7251	7259	7267	7275	7284	7292	7300	7308	7316		1	2	2	3	4	5	6	6	7
54	7324	7332	7340	7348	7356	7364	7372	7380	7388	7396		1	2	2	3	4	5	6	6	7
N	0	1	2	3	4	5	6	7	8	9		1	2	3	4	5	6	7	8	9

TABLE A-4 (continued)

N	0	1	2	3	4	5	6	7	8	9	1	2	3	4	5	6	7	8	9
													Proportional Parts						
55	7404	7412	7419	7427	7435	7443	7451	7459	7466	7474	1	2	2	3	4	5	5	6	7
56	7482	7490	7497	7505	7513	7520	7528	7536	7543	7551	1	2	2	3	4	5	5	6	7
57	7559	7566	7574	7582	7589	7597	7604	7612	7619	7627	1	2	2	3	4	5	5	6	7
58	7634	7642	7649	7657	7664	7672	7679	7686	7694	7701	1	1	2	3	4	4	5	6	7
59	7709	7716	7723	7731	7738	7745	7752	7760	7767	7774	1	1	2	3	4	4	5	6	7
60	7782	7789	7796	7803	7810	7818	7825	7832	7839	7846	1	1	2	3	4	4	5	6	6
61	7853	7860	7868	7875	7882	7889	7896	7903	7910	7917	1	1	2	3	4	4	5	6	6
62	7924	7931	7938	7945	7952	7959	7966	7973	7980	7987	1	1	2	3	3	4	5	6	6
63	7993	8000	8007	8014	8021	8028	8035	8041	8048	8055	1	1	2	3	3	4	5	5	6
64	8062	8069	8075	8082	8089	8096	8102	8109	8116	8122	1	1	2	3	3	4	5	5	6
65	8129	8136	8142	8149	8156	8162	8169	8176	8182	8189	1	1	2	3	3	4	5	5	6
66	8195	8202	8209	8215	8222	8228	8235	8241	8248	8254	1	1	2	3	3	4	5	5	6
67	8261	8267	8274	8280	8287	8293	8299	8306	8312	8319	1	1	2	3	3	4	5	5	6
68	8325	8331	8338	8344	8351	8357	8363	8370	8376	8382	1	1	2	3	3	4	4	5	6
69	8388	8395	8401	8407	8414	8420	8426	8432	8439	8445	1	1	2	2	3	4	4	5	6
70	8451	8457	8463	8470	8476	8482	8488	8494	8500	8506	1	1	2	2	3	4	4	5	6
71	8513	8519	8525	8531	8537	8543	8549	8555	8561	8567	1	1	2	2	3	4	4	5	5
72	8573	8579	8585	8591	8597	8603	8609	8615	8621	8627	1	1	2	2	3	4	4	5	5
73	8633	8639	8645	8651	8657	8663	8669	8675	8681	8686	1	1	2	2	3	4	4	5	5
74	8692	8698	8704	8710	8716	8722	8727	8733	8739	8745	1	1	2	2	3	4	4	5	5
75	8751	8756	8762	8768	8774	8779	8785	8791	8797	8802	1	1	2	2	3	3	4	5	5
76	8808	8814	8820	8825	8831	8837	8842	8848	8854	8859	1	1	2	2	3	3	4	5	5
77	8865	8871	8876	8882	8887	8893	8899	8904	8910	8915	1	1	2	2	3	3	4	4	5
78	8921	8927	8932	8938	8943	8949	8954	8960	8965	8971	1	1	2	2	3	3	4	4	5
79	8976	8982	8987	8993	8998	9004	9009	9015	9020	9025	1	1	2	2	3	3	4	4	5
80	9031	9036	9042	9047	9053	9058	9063	9069	9074	9079	1	1	2	2	3	3	4	4	5
81	9085	9090	9096	9101	9106	9112	9117	9122	9128	9133	1	1	2	2	3	3	4	4	5
82	9138	9143	9149	9154	9159	9165	9170	9175	9180	9186	1	1	2	2	3	3	4	4	5
83	9191	9196	9201	9206	9212	9217	9222	9227	9232	9238	1	1	2	2	3	3	4	4	5
84	9243	9248	9253	9258	9263	9269	9274	9279	9284	9289	1	1	2	2	3	3	4	4	5
85	9294	9299	9304	9309	9315	9320	9325	9330	9335	9340	1	1	2	2	3	3	4	4	5
86	9345	9350	9355	9360	9365	9370	9375	9380	9385	9390	1	1	2	2	3	3	4	4	5
87	9395	9400	9405	9410	9415	9420	9425	9430	9435	9440	0	1	1	2	2	3	3	4	4
88	9445	9450	9455	9460	9465	9469	9474	9479	9484	9489	0	1	1	2	2	3	3	4	4
89	9494	9499	9504	9509	9513	9518	9523	9528	9533	9538	0	1	1	2	2	3	3	4	4
90	9542	9547	9552	9557	9562	9566	9571	9576	9581	9586	0	1	1	2	2	3	3	4	4
91	9590	9595	9600	9605	9609	9614	9619	9624	9628	9633	0	1	1	2	2	3	3	4	4
92	9638	9643	9647	9652	9657	9661	9666	9671	9675	9680	0	1	1	2	2	3	3	4	4
93	9685	9689	9694	9699	9703	9708	9713	9717	9722	9727	0	1	1	2	2	3	3	4	4
94	9731	9736	9741	9745	9750	9754	9759	9763	9768	9773	0	1	1	2	2	3	3	4	4
95	9777	9782	9786	9791	9795	9800	9805	9809	9814	9818	0	1	1	2	2	3	3	4	4
96	9823	9827	9832	9836	9841	9845	9850	9854	9859	9863	0	1	1	2	2	3	3	4	4
97	9868	9872	9877	9881	9886	9890	9894	9899	9903	9908	0	1	1	2	2	3	3	4	4
98	9912	9917	9921	9926	9930	9934	9939	9943	9948	9952	0	1	1	2	2	3	3	4	4
99	9956	9961	9965	9969	9974	9978	9983	9987	9991	9996	0	1	1	2	2	3	3	3	4
N	0	1	2	3	4	5	6	7	8	9	1	2	3	4	5	6	7	8	9

TABLE A-5
Areas under the Standard Normal Probability Distribution between the Mean and
Successive Values of z

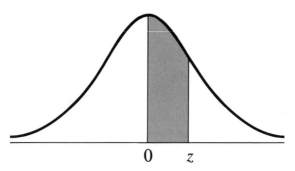

Example If $z = 1.00$, then the area between the mean and this value of z is
0.3413.

z	0.00	0.01	0.02	0.03	0.04	0.05	0.06	0.07	0.08	0.09
0.0	0.0000	0.0040	0.0080	0.0120	0.0160	0.0199	0.0239	0.0279	0.0319	0.0359
0.1	0.0398	0.0438	0.0478	0.0517	0.0557	0.0596	0.0636	0.0675	0.0714	0.0753
0.2	0.0793	0.0832	0.0871	0.0910	0.0948	0.0987	0.1026	0.1064	0.1103	0.1141
0.3	0.1179	0.1217	0.1255	0.1293	0.1331	0.1368	0.1406	0.1443	0.1480	0.1517
0.4	0.1554	0.1591	0.1628	0.1664	0.1700	0.1736	0.1772	0.1808	0.1844	0.1879
0.5	0.1915	0.1950	0.1985	0.2019	0.2054	0.2088	0.2123	0.2157	0.2190	0.2224
0.6	0.2257	0.2291	0.2324	0.2357	0.2389	0.2422	0.2454	0.2486	0.2518	0.2549
0.7	0.2580	0.2612	0.2642	0.2673	0.2704	0.2734	0.2764	0.2794	0.2823	0.2852
0.8	0.2881	0.2910	0.2939	0.2967	0.2995	0.3023	0.3051	0.3078	0.3106	0.3133
0.9	0.3159	0.3186	0.3212	0.3238	0.3264	0.3289	0.3315	0.3340	0.3365	0.3389
1.0	0.3413	0.3438	0.3461	0.3485	0.3508	0.3531	0.3554	0.3577	0.3599	0.3621
1.1	0.3643	0.3665	0.3686	0.3708	0.3729	0.3749	0.3770	0.3790	0.3810	0.3830
1.2	0.3849	0.3869	0.3888	0.3907	0.3925	0.3944	0.3962	0.3980	0.3997	0.4015
1.3	0.4032	0.4049	0.4066	0.4082	0.4099	0.4115	0.4131	0.4147	0.4162	0.4177
1.4	0.4192	0.4207	0.4222	0.4236	0.4251	0.4265	0.4279	0.4292	0.4306	0.4319
1.5	0.4332	0.4345	0.4357	0.4370	0.4382	0.4394	0.4406	0.4418	0.4429	0.4441
1.6	0.4452	0.4463	0.4474	0.4484	0.4495	0.4505	0.4515	0.4525	0.4535	0.4545
1.7	0.4554	0.4564	0.4573	0.4582	0.4591	0.4599	0.4608	0.4616	0.4625	0.4633
1.8	0.4641	0.4649	0.4656	0.4664	0.4671	0.4678	0.4686	0.4693	0.4699	0.4706
1.9	0.4713	0.4719	0.4726	0.4732	0.4738	0.4744	0.4750	0.4756	0.4761	0.4767
2.0	0.4772	0.4778	0.4783	0.4788	0.4793	0.4798	0.4803	0.4808	0.4812	0.4817
2.1	0.4821	0.4826	0.4830	0.4834	0.4838	0.4842	0.4846	0.4850	0.4854	0.4857
2.2	0.4861	0.4864	0.4868	0.4871	0.4875	0.4878	0.4881	0.4884	0.4887	0.4890
2.3	0.4893	0.4896	0.4898	0.4901	0.4904	0.4906	0.4909	0.4911	0.4913	0.4916
2.4	0.4918	0.4920	0.4922	0.4925	0.4927	0.4929	0.4931	0.4932	0.4934	0.4936
2.5	0.4938	0.4940	0.4941	0.4943	0.4945	0.4946	0.4948	0.4949	0.4951	0.4952
2.6	0.4953	0.4955	0.4956	0.4957	0.4959	0.4960	0.4961	0.4962	0.4963	0.4964
2.7	0.4965	0.4966	0.4967	0.4968	0.4969	0.4970	0.4971	0.4972	0.4973	0.4974
2.8	0.4974	0.4975	0.4976	0.4977	0.4977	0.4978	0.4979	0.4979	0.4980	0.4981
2.9	0.4981	0.4982	0.4982	0.4983	0.4984	0.4984	0.4985	0.4985	0.4986	0.4986
3.0	0.49865	0.4987	0.4987	0.4988	0.4988	0.4989	0.4989	0.4989	0.4990	0.4990
4.0	0.49997									

TABLE A-6
Student's *t*-Distribution

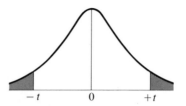

Example For 15 degrees of freedom, the *t*-value which corresponds to an area of 0.05 in both tails combined is 2131.

Degrees of Freedom	Area in Both Tails Combined			
	0.10	0.05	0.02	0.01
1	6.314	12.706	31.821	63.657
2	2.920	4.303	6.965	9.925
3	2.353	3.182	4.541	5.841
4	2.132	2.776	3.747	4.604
5	2.015	2.571	3.365	4.032
6	1.943	2.447	3.143	3.707
7	1.895	2.365	2.998	3.499
8	1.860	2.306	2.896	3.355
9	1.833	2.262	2.821	3.250
10	1.812	2.228	2.764	3.169
11	1.796	2.201	2.718	3.106
12	1.782	2.179	2.681	3.055
13	1.771	2.160	2.650	3.012
14	1.761	2.145	2.624	2.977
15	1.753	2.131	2.602	2.947
16	1.746	2.120	2.583	2.921
17	1.740	2.110	2.567	2.898
18	1.734	2.101	2.552	2.878
19	1.729	2.093	2.539	2.861
20	1.725	2.086	2.528	2.845
21	1.721	2.080	2.518	2.831
22	1.717	2.074	2.508	2.819
23	1.714	2.069	2.500	2.807
24	1.711	2.064	2.492	2.797
25	1.708	2.060	2.485	2.787
26	1.706	2.056	2.479	2.779
27	1.703	2.052	2.473	2.771
28	1.701	2.048	2.467	2.763
29	1.699	2.045	2.462	2.756
30	1.697	2.042	2.457	2.750
40	1.684	2.021	2.423	2.704
60	1.671	2.000	2.390	2.660
120	1.658	1.980	2.358	2.617
Normal Distribution	1.645	1.960	2.326	2.576

Source: Table A-6 is taken from Table III of Fisher and Yates: *Statistical Tables for Biological, Agricultural and Medical Research*, published by Oliver and Boyd Ltd., Edinburgh, and by permission of the authors and publishers.

TABLE A-7
Chi Square (χ^2) Distribution

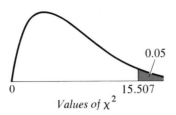

$$Values\ of\ \chi^2$$

0
15.507

0.05

Example In a chi square distribution with $\nu = 8$ degrees of freedom, the area to the right of a chi square value of 15.507 is 0.05.

Degrees of Freedom ν	Area in Right Tail				
	0.20	0.10	0.05	0.02	0.01
1	1.642	2.706	3.841	5.412	6.635
2	3.219	4.605	5.991	7.824	9.210
3	4.642	6.251	7.815	9.837	11.345
4	5.989	7.779	9.488	11.668	13.277
5	7.289	9.236	11.070	13.388	15.086
6	8.558	10.645	12.592	15.033	16.812
7	9.803	12.017	14.067	16.622	18.475
8	11.030	13.362	15.507	18.168	20.090
9	12.242	14.684	16.919	19.679	21.666
10	13.442	15.987	18.307	21.161	23.209
11	14.631	17.275	19.675	22.618	24.725
12	15.812	18.549	21.026	24.054	26.217
13	16.985	19.812	22.362	25.472	27.688
14	18.151	21.064	23.685	26.873	29.141
15	19.311	22.307	24.996	28.259	30.578
16	20.465	23.542	26.296	29.633	32.000
17	21.615	24.769	27.587	30.995	33.409
18	22.760	25.989	28.869	32.346	34.805
19	23.900	27.204	30.144	33.687	36.191
20	25.038	28.412	31.410	35.020	37.566
21	26.171	29.615	32.671	36.343	38.932
22	27.301	30.813	33.924	37.659	40.289
23	28.429	32.007	35.172	38.968	41.638
24	29.553	33.196	36.415	40.270	42.980
25	30.675	34.382	37.652	41.566	44.314
26	31.795	35.563	38.885	42.856	45.642
27	32.912	36.741	40.113	44.140	46.963
28	34.027	37.916	41.337	45.419	48.278
29	35.139	39.087	42.557	46.693	49.588
30	36.250	40.256	43.773	47.962	50.892

Source: Table A-7 is taken from Table IV of Fisher and Yates: *Statistical Tables for Biological, Agricultural and Medical Research*, published by Oliver and Boyd Ltd., Edinburgh, and by permission of the authors and publishers.

TABLE A-8
F Distribution

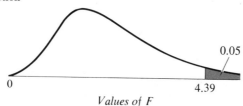

Values of F

Example In an F distribution with $v_1 = 5$ and $v_2 = 6$ degrees of freedom, the area to the right of an F value of 4.39 is 0.05. The value on the F scale to the right of which lies 0.05 of the area is in lightface type. The value on the F scale to the right of which lies 0.01 of the area is in boldface type. $v_1 =$ number of degrees of freedom for numerator; $v_2 =$ number of degrees of freedom for denominator.

v_2 \ v_1	1	2	3	4	5	6	7	8	9	10	20	30	40	50	100	200	∞	v_2
1	161	200	216	225	230	234	237	239	241	242	248	250	251	252	253	254	254	1
	4,052	**4,999**	**5,403**	**5,625**	**5,764**	**5,859**	**5,928**	**5,981**	**6,022**	**6,056**	**6,208**	**6,261**	**6,286**	**6,302**	**6,334**	**6,352**	**6,366**	
2	18.51	19.00	19.16	19.25	19.30	19.33	19.36	19.37	19.38	19.39	19.44	19.46	19.47	19.47	19.49	19.49	19.50	2
	98.49	**99.00**	**99.17**	**99.25**	**99.30**	**99.33**	**99.36**	**99.37**	**99.39**	**99.40**	**99.45**	**99.47**	**99.48**	**99.48**	**99.49**	**99.49**	**99.50**	
3	10.13	9.55	9.28	9.12	9.01	8.94	8.88	8.84	8.81	8.78	8.66	8.62	8.60	8.58	8.56	8.54	8.53	3
	34.12	**30.82**	**29.46**	**28.71**	**28.24**	**27.91**	**27.67**	**27.49**	**27.34**	**27.23**	**26.69**	**26.50**	**26.41**	**26.35**	**26.23**	**26.18**	**26.12**	
4	7.71	6.94	6.59	6.39	6.26	6.16	6.09	6.04	6.00	5.96	5.80	5.74	5.71	5.70	5.66	5.65	5.63	4
	21.20	**18.00**	**16.69**	**15.98**	**15.52**	**15.21**	**14.98**	**14.80**	**14.66**	**14.54**	**14.02**	**13.83**	**13.74**	**13.69**	**13.57**	**13.52**	**13.46**	
5	6.61	5.79	5.41	5.19	5.05	4.95	4.88	4.82	4.78	4.74	4.56	4.50	4.46	4.44	4.40	4.38	4.36	5
	16.26	**13.27**	**12.06**	**11.39**	**10.97**	**10.67**	**10.45**	**10.29**	**10.15**	**10.05**	**9.55**	**9.38**	**9.29**	**9.24**	**9.13**	**9.07**	**9.02**	
6	5.99	5.14	4.76	4.53	4.39	4.28	4.21	4.15	4.10	4.06	3.87	3.81	3.77	3.75	3.71	3.69	3.67	6
	13.74	**10.92**	**9.78**	**9.15**	**8.75**	**8.47**	**8.26**	**8.10**	**7.98**	**7.87**	**7.39**	**7.23**	**7.14**	**7.09**	**6.99**	**6.94**	**6.88**	
7	5.59	4.74	4.35	4.12	3.97	3.87	3.79	3.73	3.68	3.63	3.44	3.38	3.34	3.32	3.28	3.25	3.23	7
	12.25	**9.55**	**8.45**	**7.85**	**7.46**	**7.19**	**7.00**	**6.84**	**6.71**	**6.62**	**6.15**	**5.98**	**5.90**	**5.85**	**5.75**	**5.70**	**5.65**	
8	5.32	4.46	4.07	3.84	3.69	3.58	3.50	3.44	3.39	3.34	3.15	3.08	3.05	3.03	2.98	2.96	2.93	8
	11.26	**8.65**	**7.59**	**7.01**	**6.63**	**6.37**	**6.19**	**6.03**	**5.91**	**5.82**	**5.36**	**5.20**	**5.11**	**5.06**	**4.96**	**4.91**	**4.86**	
9	5.12	4.26	3.86	3.63	3.48	3.37	3.29	3.23	3.18	3.13	2.93	2.86	2.82	2.80	2.76	2.73	2.71	9
	10.56	**8.02**	**6.99**	**6.42**	**6.06**	**5.80**	**5.62**	**5.47**	**5.35**	**5.26**	**4.80**	**4.64**	**4.56**	**4.51**	**4.41**	**4.36**	**4.31**	
10	4.96	4.10	3.71	3.48	3.33	3.22	3.14	3.07	3.02	2.97	2.77	2.70	2.67	2.64	2.59	2.56	2.54	10
	10.04	**7.56**	**6.55**	**5.99**	**5.64**	**5.39**	**5.21**	**5.06**	**4.95**	**4.85**	**4.41**	**4.25**	**4.17**	**4.12**	**4.01**	**3.96**	**3.91**	
20	4.35	3.49	3.10	2.87	2.71	2.60	2.52	2.45	2.40	2.35	2.12	2.04	1.99	1.96	1.90	1.87	1.84	20
	8.10	**5.85**	**4.94**	**4.43**	**4.10**	**3.87**	**3.71**	**3.56**	**3.45**	**3.37**	**2.94**	**2.77**	**2.69**	**2.63**	**2.53**	**2.47**	**2.42**	
30	4.17	3.32	2.92	2.69	2.53	2.42	2.34	2.27	2.21	2.16	1.93	1.84	1.79	1.76	1.69	1.66	1.62	30
	7.56	**5.39**	**4.51**	**4.02**	**3.70**	**3.47**	**3.30**	**3.17**	**3.06**	**2.98**	**2.55**	**2.38**	**2.29**	**2.24**	**2.13**	**2.07**	**2.01**	
40	4.08	3.23	2.84	2.61	2.45	2.34	2.25	2.18	2.12	2.07	1.84	1.74	1.69	1.66	1.59	1.55	1.51	40
	7.31	**5.18**	**4.31**	**3.83**	**3.51**	**3.29**	**3.12**	**2.99**	**2.88**	**2.80**	**2.37**	**2.20**	**2.11**	**2.05**	**1.94**	**1.88**	**1.81**	
50	4.03	3.18	2.79	2.56	2.40	2.29	2.20	2.13	2.07	2.02	1.78	1.69	1.63	1.60	1.52	1.48	1.44	50
	7.17	**5.06**	**4.20**	**3.72**	**3.41**	**3.18**	**3.02**	**2.88**	**2.78**	**2.70**	**2.26**	**2.10**	**2.00**	**1.94**	**1.82**	**1.76**	**1.68**	
100	3.94	3.09	2.70	2.46	2.30	2.19	2.10	2.03	1.97	1.92	1.68	1.57	1.51	1.48	1.39	1.34	1.28	100
	6.90	**4.82**	**3.98**	**3.51**	**3.20**	**2.99**	**2.82**	**2.69**	**2.59**	**2.51**	**2.06**	**1.89**	**1.79**	**1.73**	**1.59**	**1.51**	**1.43**	
200	3.89	3.04	2.65	2.41	2.26	2.14	2.05	1.98	1.92	1.87	1.62	1.52	1.45	1.42	1.32	1.26	1.19	200
	6.76	**4.71**	**3.88**	**3.41**	**3.11**	**2.90**	**2.73**	**2.60**	**2.50**	**2.41**	**1.97**	**1.79**	**1.69**	**1.62**	**1.48**	**1.39**	**1.28**	
∞	3.84	2.99	2.60	2.37	2.21	2.09	2.01	1.94	1.88	1.83	1.57	1.46	1.40	1.35	1.24	1.17	1.00	∞
	6.64	**4.60**	**3.78**	**3.32**	**3.02**	**2.80**	**2.64**	**2.51**	**2.41**	**2.32**	**1.87**	**1.69**	**1.59**	**1.52**	**1.36**	**1.25**	**1.00**	

Source: This table is abridged by permission from *Statistical Methods*, 5th ed., by George W. Snedecor, © 1956 by The Iowa State University Press.

TABLE A-9
Squares, Square Roots, and Reciprocals

N	$\sqrt{N}$	N^2	$\sqrt{10N}$	$1000/N$	N	$\sqrt{N}$	N^2	$\sqrt{10N}$	$1000/N$
					50	7.07107	2500	22.36068	20.00000
1	1.00000	1	3.16228	1000.00000	51	7.14143	2601	22.58318	19.60784
2	1.41421	4	4.47214	500.00000	52	7.21110	2704	22.80351	19.23077
3	1.73205	9	5.47723	333.33333	53	7.28011	2809	23.02173	18.86792
4	2.00000	16	6.32456	250.00000	54	7.34847	2916	23.23790	18.51852
5	2.23607	25	7.07107	200.00000	55	7.41620	3025	23.45208	18.18182
6	2.44949	36	7.74597	166.66667	56	7.48331	3136	23.66432	17.85714
7	2.64575	49	8.36660	142.85714	57	7.54983	3249	23.87467	17.54386
8	2.82843	64	8.94427	125.00000	58	7.61577	3364	24.08319	17.24138
9	3.00000	81	9.48683	111.11111	59	7.68115	3481	24.28992	16.94915
10	3.16228	100	10.00000	100.00000	60	7.74597	3600	24.49490	16.66667
11	3.31662	121	10.48809	90.90909	61	7.81025	3721	24.69818	16.39344
12	3.46410	144	10.95445	83.33333	62	7.87401	3844	24.89980	16.12903
13	3.60555	169	11.40175	76.92308	63	7.93725	3969	25.09980	15.87302
14	3.74166	196	11.83216	71.42857	64	8.00000	4096	25.29822	15.62500
15	3.87298	225	12.24745	66.66667	65	8.06226	4225	25.49510	15.38462
16	4.00000	256	12.64911	62.50000	66	8.12404	4356	25.69047	15.15152
17	4.12311	289	13.03840	58.82353	67	8.18535	4489	25.88436	14.92537
18	4.24264	324	13.41641	55.55556	68	8.24621	4624	26.07681	14.70588
19	4.35890	361	13.78405	52.63158	69	8.30662	4761	26.26785	14.49275
20	4.47214	400	14.14214	50.00000	70	8.36660	4900	26.45751	14.28571
21	4.58258	441	14.49138	47.61905	71	8.42615	5041	26.64583	14.08451
22	4.69042	484	14.83240	45.45455	72	8.48528	5184	26.83282	13.88889
23	4.79583	529	15.16575	43.47826	73	8.54400	5329	27.01851	13.69863
24	4.89898	576	15.49193	41.66667	74	8.60233	5476	27.20294	13.51351
25	5.00000	625	15.81139	40.00000	75	8.66025	5625	27.38613	13.33333
26	5.09902	676	16.12452	38.46154	76	8.71780	5776	27.56810	13.15789
27	5.19615	729	16.43168	37.03704	77	8.77496	5929	27.74887	12.98701
28	5.29150	784	16.73320	35.71429	78	8.83176	6084	27.92848	12.82051
29	5.38516	841	17.02939	34.48276	79	8.88819	6241	28.10694	12.65823
30	5.47723	900	17.32051	33.33333	80	8.94427	6400	28.28427	12.50000
31	5.56776	961	17.60682	32.25806	81	9.00000	6561	28.46050	12.34568
32	5.65685	1024	17.88854	31.25000	82	9.05539	6724	28.63564	12.19512
33	5.74456	1089	18.16590	30.30303	83	9.11043	6889	28.80972	12.04819
34	5.83095	1156	18.43909	29.41176	84	9.16515	7056	28.98275	11.90476
35	5.91608	1225	18.70829	28.57143	85	9.21954	7225	29.15476	11.76471
36	6.00000	1296	18.97367	27.77778	86	9.27362	7396	29.32576	11.62791
37	6.08276	1369	19.23538	27.02703	87	9.32738	7569	29.49576	11.49425
38	6.16441	1444	19.49359	26.31579	88	9.38083	7744	29.66479	11.36364
39	6.24500	1521	19.74842	25.64103	89	9.43398	7921	29.83287	11.23596
40	6.32456	1600	20.00000	25.00000	90	9.48683	8100	30.00000	11.11111
41	6.40312	1681	20.24846	24.39024	91	9.53939	8281	30.16621	10.98901
42	6.48074	1764	20.49390	23.80952	92	9.59166	8464	30.33150	10.86957
43	6.55744	1849	20.73644	23.25581	93	9.64365	8649	30.49590	10.75269
44	6.63325	1936	20.97618	22.72727	94	9.69536	8836	30.65942	10.63830
45	6.70820	2025	21.21320	22.22222	95	9.74679	9025	30.82207	10.52632
46	6.78233	2116	21.44761	21.73913	96	9.79796	9216	30.98387	10.41667
47	6.85565	2209	21.67948	21.27660	97	9.84886	9409	31.14482	10.30928
48	6.92820	2304	21.90890	20.83333	98	9.89949	9604	31.30495	10.20408
49	7.00000	2401	22.13594	20.40816	99	9.94987	9801	31.46427	10.10101
50	7.07107	2500	22.36068	20.00000	100	10.00000	10000	31.62278	10.00000

Source: From *Statistics for Modern Business Decisions* by Lawrence L. Lapin, © 1973 by Harcourt Brace Jovanovich, Inc. and reproduced with their permission.

TABLE A-9 (continued)

N	$\sqrt{N}$	N^2	$\sqrt{10N}$	$1000/N$	N	$\sqrt{N}$	N^2	$\sqrt{10N}$	$1000/N$
100	10.00000	10000	31.62278	10.00000	150	12.24745	22500	38.72983	6.66667
101	10.04988	10201	31.78050	9.90099	151	12.28821	22801	38.85872	6.62252
102	10.09950	10404	31.93744	9.80392	152	12.32883	23104	38.98718	6.57895
103	10.14889	10609	32.09361	9.70874	153	12.36932	23409	39.11521	6.53595
104	10.19804	10816	32.24903	9.61538	154	12.40967	23716	39.24283	6.49351
105	10.24695	11025	32.40370	9.52381	155	12.44990	24025	39.37004	6.45161
106	10.29563	11236	32.55764	9.43396	156	12.49000	24336	39.49684	6.41026
107	10.34408	11449	32.71085	9.34579	157	12.52996	24649	39.62323	6.36943
108	10.39230	11664	32.86335	9.25926	158	12.56981	24964	39.74921	6.32911
109	10.44031	11881	33.01515	9.17431	159	12.60952	25281	39.87480	6.28931
110	10.48809	12100	33.16625	9.09091	160	12.64911	25600	40.00000	6.25000
111	10.53565	12321	33.31666	9.00901	161	12.68858	25921	40.12481	6.21118
112	10.58301	12544	33.46640	8.92857	162	12.72792	26244	40.24922	6.17284
113	10.63015	12769	33.61547	8.84956	163	12.76715	26569	40.37326	6.13497
114	10.67708	12996	33.76389	8.77193	164	12.80625	26896	40.49691	6.09756
115	10.72381	13225	33.91165	8.69565	165	12.84523	27225	40.62019	6.06061
116	10.77033	13456	34.05877	8.62069	166	12.88410	27556	40.74310	6.02410
117	10.81665	13689	34.20526	8.54701	167	12.92285	27889	40.86563	5.98802
118	10.86278	13924	34.35113	8.47458	168	12.96148	28224	40.98780	5.95238
119	10.90871	14161	34.49638	8.40336	169	13.00000	28561	41.10961	5.91716
120	10.95445	14400	34.64102	8.33333	170	13.03840	28900	41.23106	5.88235
121	11.00000	14641	34.78505	8.26446	171	13.07670	29241	41.35215	5.84795
122	11.04536	14884	34.92850	8.19672	172	13.11488	29584	41.47288	5.81395
123	11.09054	15129	35.07136	8.13008	173	13.15295	29929	41.59327	5.78035
124	11.13553	15376	35.21363	8.06452	174	13.19091	30276	41.71331	5.74713
125	11.18034	15625	35.35534	8.00000	175	13.22876	30625	41.83300	5.71429
126	11.22497	15876	35.49648	7.93651	176	13.26650	30976	41.95235	5.68182
127	11.26943	16129	35.63706	7.87402	177	13.30413	31329	42.07137	5.64972
128	11.31371	16384	35.77709	7.81250	178	13.34166	31684	42.19005	5.61798
129	11.35782	16641	35.91657	7.75194	179	13.37909	32041	42.30839	5.58659
130	11.40175	16900	36.05551	7.69231	180	13.41641	32400	42.42641	5.55556
131	11.44552	17161	36.19392	7.63359	181	13.45362	32761	42.54409	5.52486
132	11.48913	17424	36.33180	7.57576	182	13.49074	33124	42.66146	5.49451
133	11.53256	17689	36.46917	7.51880	183	13.52775	33489	42.77850	5.46448
134	11.57584	17956	36.60601	7.46269	184	13.56466	33856	42.89522	5.43478
135	11.61895	18225	36.74235	7.40741	185	13.60147	34225	43.01163	5.40541
136	11.66190	18496	36.87818	7.35294	186	13.63818	34596	43.12772	5.37634
137	11.70470	18769	37.01351	7.29927	187	13.67479	34969	43.24350	5.34759
138	11.74734	19044	37.14835	7.24638	188	13.71131	35344	43.35897	5.31915
139	11.78983	19321	37.28270	7.19424	189	13.74773	35721	43.47413	5.29101
140	11.83216	19600	37.41657	7.14286	190	13.78405	36100	43.58899	5.26316
141	11.87434	19881	37.54997	7.09220	191	13.82027	36481	43.70355	5.23560
142	11.91638	20164	37.68289	7.04225	192	13.85641	36864	43.81780	5.20833
143	11.95826	20449	37.81534	6.99301	193	13.89244	37249	43.93177	5.18135
144	12.00000	20736	37.94733	6.94444	194	13.92839	37636	44.04543	5.15464
145	12.04159	21025	38.07887	6.89655	195	13.96424	38025	44.15880	5.12821
146	12.08305	21316	38.20995	6.84932	196	14.00000	38416	44.27189	5.10204
147	12.12436	21609	38.34058	6.80272	197	14.03567	38809	44.38468	5.07614
148	12.16553	21904	38.47077	6.75676	198	14.07125	39204	44.49719	5.05051
149	12.20656	22201	38.60052	6.71141	199	14.10674	39601	44.60942	5.02513
150	12.24745	22500	38.72983	6.66667	200	14.14214	40000	44.72136	5.00000

TABLE A-9 (continued)

N	$\sqrt{N}$	N^2	$\sqrt{10N}$	$1000/N$	N	$\sqrt{N}$	N^2	$\sqrt{10N}$	$1000/N$
200	14.14214	40000	44.72136	5.00000	250	15.81139	62500	50.00000	4.00000
201	14.17745	40401	44.83302	4.97512	251	15.84298	63001	50.09990	3.98406
202	14.21267	40804	44.94441	4.95050	252	15.87451	63504	50.19960	3.96825
203	14.24781	41209	45.05552	4.92611	253	15.90597	64009	50.29911	3.95257
204	14.28286	41616	45.16636	4.90196	254	15.93738	64516	50.39841	3.93701
205	14.31782	42025	45.27693	4.87805	255	15.96872	65025	50.49752	3.92157
206	14.35270	42436	45.38722	4.85437	256	16.00000	65536	50.59644	3.90625
207	14.38749	42849	45.49725	4.83092	257	16.03122	66049	50.69517	3.89105
208	14.42221	43264	45.60702	4.80769	258	16.06238	66564	50.79370	3.87597
209	14.45683	43681	45.71652	4.78469	259	16.09348	67081	50.89204	3.86100
210	14.49138	44100	45.82576	4.76190	260	16.12452	67600	50.99020	3.84615
211	14.52584	44521	45.93474	4.73934	261	16.15549	68121	51.08816	3.83142
212	14.56022	44944	46.04346	4.71698	262	16.18641	68644	51.18594	3.81679
213	14.59452	45369	46.15192	4.69484	263	16.21727	69169	51.28353	3.80228
214	14.62874	45796	46.26013	4.67290	264	16.24808	69696	51.38093	3.78788
215	14.66288	46225	46.36809	4.65116	265	16.27882	70225	51.47815	3.77358
216	14.69694	46656	46.47580	4.62963	266	16.30951	70756	51.57519	3.75940
217	14.73092	47089	46.58326	4.60829	267	16.34013	71289	51.67204	3.74532
218	14.76482	47524	46.69047	4.58716	268	16.37071	71824	51.76872	3.73134
219	14.79865	47961	46.79744	4.56621	269	16.40122	72361	51.86521	3.71747
220	14.83240	48400	46.90416	4.54545	270	16.43168	72900	51.96152	3.70370
221	14.86607	48841	47.01064	4.52489	271	16.46208	73441	52.05766	3.69004
222	14.89966	49284	47.11688	4.50450	272	16.49242	73984	52.15362	3.67647
223	14.93318	49729	47.22288	4.48430	273	16.52271	74529	52.24940	3.66300
224	14.96663	50176	47.32864	4.46429	274	16.55295	75076	52.34501	3.64964
225	15.00000	50625	47.43416	4.44444	275	16.58312	75625	52.44044	3.63636
226	15.03330	51076	47.53946	4.42478	276	16.61325	76176	52.53570	3.62319
227	15.06652	51529	47.64452	4.40529	277	16.64332	76729	52.63079	3.61011
228	15.09967	51984	47.74935	4.38596	278	16.67333	77284	52.72571	3.59712
229	15.13275	52441	47.85394	4.36681	279	16.70329	77841	52.82045	3.58423
230	15.16575	52900	47.95832	4.34783	280	16.73320	78400	52.91503	3.57143
231	15.19868	53361	48.06246	4.32900	281	16.76305	78961	53.00943	3.55872
232	15.23155	53824	48.16638	4.31034	282	16.79286	79524	53.10367	3.54610
233	15.26434	54289	48.27007	4.29185	283	16.82260	80089	53.19774	3.53357
234	15.29706	54756	48.37355	4.27350	284	16.85230	80656	53.29165	3.52113
235	15.32971	55225	48.47680	4.25532	285	16.88194	81225	53.38539	3.50877
236	15.36229	55696	48.57983	4.23729	286	16.91153	81796	53.47897	3.49650
237	15.39480	56169	48.68265	4.21941	287	16.94107	82369	53.57238	3.48432
238	15.42725	56644	48.78524	4.20168	288	16.97056	82944	53.66563	3.47222
239	15.45962	57121	48.88763	4.18410	289	17.00000	83521	53.75872	3.46021
240	15.49193	57600	48.98979	4.16667	290	17.02939	84100	53.85165	3.44828
241	15.52417	58081	49.09175	4.14938	291	17.05872	84681	53.94442	3.43643
242	15.55635	58564	49.19350	4.13223	292	17.08801	85264	54.03702	3.42466
243	15.58846	59049	49.29503	4.11523	293	17.11724	85849	54.12947	3.41297
244	15.62050	59536	49.39636	4.09836	294	17.14643	86436	54.22177	3.40136
245	15.65248	60025	49.49747	4.08163	295	17.17556	87025	54.31390	3.38983
246	15.68439	60516	49.59839	4.06504	296	17.20465	87616	54.40588	3.37838
247	15.71623	61009	49.69909	4.04858	297	17.23369	88209	54.49771	3.36700
248	15.74802	61504	49.79960	4.03226	298	17.26268	88804	54.58938	3.35570
249	15.77973	62001	49.89990	4.01606	299	17.29162	89401	54.68089	3.34448
250	15.81139	62500	50.00000	4.00000	300	17.32051	90000	54.77226	3.33333

TABLE A-9 (continued)

N	√N	N²	√10N	1000/N	N	√N	N²	√10N	1000/N
300	17.32051	90000	54.77226	3.33333	350	18.70829	122500	59.16080	2.85714
301	17.34935	90601	54.86347	3.32226	351	18.73499	123201	59.24525	2.84900
302	17.37815	91204	54.95453	3.31126	352	18.76166	123904	59.32959	2.84091
303	17.40690	91809	55.04544	3.30033	353	18.78829	124609	59.41380	2.83286
304	17.43560	92416	55.13620	3.28947	354	18.81489	125316	59.49790	2.82486
305	17.46425	93025	55.22681	3.27869	355	18.84144	126025	59.58188	2.81690
306	17.49286	93636	55.31727	3.26797	356	18.86796	126736	59.66574	2.80899
307	17.52142	94249	55.40758	3.25733	357	18.89444	127449	59.74948	2.80112
308	17.54993	94864	55.49775	3.24675	358	18.92089	128164	59.83310	2.79330
309	17.57840	95481	55.58777	3.23625	359	18.94730	128881	59.91661	2.78552
310	17.60682	96100	55.67764	3.22581	360	18.97367	129600	60.00000	2.77778
311	17.63519	96721	55.76737	3.21543	361	19.00000	130321	60.08328	2.77008
312	17.66352	97344	55.85696	3.20513	362	19.02630	131044	60.16644	2.76243
313	17.69181	97969	55.94640	3.19489	363	19.05256	131769	60.24948	2.75482
314	17.72005	98596	56.03570	3.18471	364	19.07878	132496	60.33241	2.74725
315	17.74824	99225	56.12486	3.17460	365	19.10497	133225	60.41523	2.73973
316	17.77639	99856	56.21388	3.16456	366	19.13113	133956	60.49793	2.73224
317	17.80449	100489	56.30275	3.15457	367	19.15724	134689	60.58052	2.72480
318	17.83255	101124	56.39149	3.14465	368	19.18333	135424	60.66300	2.71739
319	17.86057	101761	56.48008	3.13480	369	19.20937	136161	60.74537	2.71003
320	17.88854	102400	56.56854	3.12500	370	19.23538	136900	60.82763	2.70270
321	17.91647	103041	56.65686	3.11526	371	19.26136	137641	60.90977	2.69542
322	17.94436	103684	56.74504	3.10559	372	19.28730	138384	60.99180	2.68817
323	17.97220	104329	56.83309	3.09598	373	19.31321	139129	61.07373	2.68097
324	18.00000	104976	56.92100	3.08642	374	19.33908	139876	61.15554	2.67380
325	18.02776	105625	57.00877	3.07692	375	19.36492	140625	61.23724	2.66667
326	18.05547	106276	57.09641	3.06748	376	19.39072	141376	61.31884	2.65957
327	18.08314	106929	57.18391	3.05810	377	19.41649	142129	61.40033	2.65252
328	18.11077	107584	57.27128	3.04878	378	19.44222	142884	61.48170	2.64550
329	18.13836	108241	57.35852	3.03951	379	19.46792	143641	61.56298	2.63852
330	18.16590	108900	57.44563	3.03030	380	19.49359	144400	61.64414	2.63158
331	18.19341	109561	57.53260	3.02115	381	19.51922	145161	61.72520	2.62467
332	18.22087	110224	57.61944	3.01205	382	19.54482	145924	61.80615	2.61780
333	18.24829	110889	57.70615	3.00300	383	19.57039	146689	61.88699	2.61097
334	18.27567	111556	57.79273	2.99401	384	19.59592	147456	61.96773	2.60417
335	18.30301	112225	57.87918	2.98507	385	19.62142	148225	62.04837	2.59740
336	18.33030	112896	57.96551	2.97619	386	19.64688	148996	62.12890	2.59067
337	18.35756	113569	58.05170	2.96736	387	19.67232	149769	62.20932	2.58398
338	18.38478	114244	58.13777	2.95858	388	19.69772	150544	62.28965	2.57732
339	18.41195	114921	58.22371	2.94985	389	19.72308	151321	62.36986	2.57069
340	18.43909	115600	58.30952	2.94118	390	19.74842	152100	62.44998	2.56410
341	18.46619	116281	58.39521	2.93255	391	19.77372	152881	62.52999	2.55754
342	18.49324	116964	58.48077	2.92398	392	19.79899	153664	62.60990	2.55102
343	18.52026	117649	58.56620	2.91545	393	19.82423	154449	62.68971	2.54453
344	18.54724	118336	58.65151	2.90698	394	19.84943	155236	62.76942	2.53807
345	18.57418	119025	58.73670	2.89855	395	19.87461	156025	62.84903	2.53165
346	18.60108	119716	58.82176	2.89017	396	19.89975	156816	62.92853	2.52525
347	18.62794	120409	58.90671	2.88184	397	19.92486	157609	63.00794	2.51889
348	18.65476	121104	58.99152	2.87356	398	19.94994	158404	63.08724	2.51256
349	18.68154	121801	59.07622	2.86533	399	19.97498	159201	63.16645	2.50627
350	18.70829	122500	59.16080	2.85714	400	20.00000	160000	63.24555	2.50000

TABLE A-9 (continued)

N	$\sqrt{N}$	N^2	$\sqrt{10N}$	$1000/N$	N	$\sqrt{N}$	N^2	$\sqrt{10N}$	$1000/N$
400	20.00000	160000	63.24555	2.50000	450	21.21320	202500	67.08204	2.22222
401	20.02498	160801	63.32456	2.49377	451	21.23676	203401	67.15653	2.21729
402	20.04994	161604	63.40347	2.48756	452	21.26029	204304	67.23095	2.21239
403	20.07486	162409	63.48228	2.48139	453	21.28380	205209	67.30527	2.20751
404	20.09975	163216	63.56099	2.47525	454	21.30728	206116	67.37952	2.20264
405	20.12461	164025	63.63961	2.46914	455	21.33073	207025	67.45369	2.19780
406	20.14944	164836	63.71813	2.46305	456	21.35416	207936	67.52777	2.19298
407	20.17424	165649	63.79655	2.45700	457	21.37756	208849	67.60178	2.18818
408	20.19901	166464	63.87488	2.45098	458	21.40093	209764	67.67570	2.18341
409	20.22375	167281	63.95311	2.44499	459	21.42429	210681	67.74954	2.17865
410	20.24846	168100	64.03124	2.43902	460	21.44761	211600	67.82330	2.17391
411	20.27313	168921	64.10928	2.43309	461	21.47091	212521	67.89698	2.16920
412	20.29778	169744	64.18723	2.42718	462	21.49419	213444	67.97058	2.16450
413	20.32240	170569	64.26508	2.42131	463	21.51743	214369	68.04410	2.15983
414	20.34699	171396	64.34283	2.41546	464	21.54066	215296	68.11755	2.15517
415	20.37155	172225	64.42049	2.40964	465	21.56386	216225	68.19091	2.15054
416	20.39608	173056	64.49806	2.40385	466	21.58703	217156	68.26419	2.14592
417	20.42058	173889	64.57554	2.39808	467	21.61018	218089	68.33740	2.14133
418	20.44505	174724	64.65292	2.39234	468	21.63331	219024	68.41053	2.13675
419	20.46949	175561	64.73021	2.38663	469	21.65641	219961	68.48357	2.13220
420	20.49390	176400	64.80741	2.38095	470	21.67948	220900	68.55655	2.12766
421	20.51828	177241	64.88451	2.37530	471	21.70253	221841	68.62944	2.12314
422	20.54264	178084	64.96153	2.36967	472	21.72556	222784	68.70226	2.11864
423	20.56696	178929	65.03845	2.36407	473	21.74856	223729	68.77500	2.11416
424	20.59126	179776	65.11528	2.35849	474	21.77154	224676	68.84766	2.10970
425	20.61553	180625	65.19202	2.35294	475	21.79449	225625	68.92024	2.10526
426	20.63967	181476	65.26868	2.34742	476	21.81742	226576	68.99275	2.10084
427	20.66398	182329	65.34524	2.34192	477	21.84033	227529	69.06519	2.09644
428	20.68816	183184	65.42171	2.33645	478	21.86321	228484	69.13754	2.09205
429	20.71232	184041	65.49809	2.33100	479	21.88607	229441	69.20983	2.08768
430	20.73644	184900	65.57439	2.32558	480	21.90890	230400	69.28203	2.08333
431	20.76054	185761	65.65059	2.32019	481	21.93171	231361	69.35416	2.07900
432	20.78461	186624	65.72671	2.31481	482	21.95450	232324	69.42622	2.07469
433	20.80865	187489	65.80274	2.30947	483	21.97726	233289	69.49820	2.07039
434	20.83267	188356	65.87868	2.30415	484	22.00000	234256	69.57011	2.06612
435	20.85665	189225	65.95453	2.29885	485	22.02272	235225	69.64194	2.06186
436	20.88096	190096	66.03030	2.29358	486	22.04541	236196	69.71370	2.05761
437	20.90454	190969	66.10598	2.28833	487	22.06808	237169	69.78539	2.05339
438	20.92845	191844	66.18157	2.28311	488	22.09072	238144	69.85700	2.04918
439	20.95233	192721	66.25708	2.27790	489	22.11334	239121	69.92853	2.04499
440	20.97618	193600	66.33250	2.27273	490	22.13594	240100	70.00000	2.04082
441	21.00000	194481	66.40783	2.26757	491	22.15852	241081	70.07139	2.03666
442	21.02380	195364	66.48308	2.26244	492	22.18107	242064	70.14271	2.03252
443	21.04757	196249	66.55825	2.25734	493	22.20360	243049	70.21396	2.02840
444	21.07131	197136	66.63332	2.25225	494	22.22611	244036	70.28513	2.02429
445	21.09502	198025	66.70832	2.24719	495	22.24860	245025	70.35624	2.02020
446	21.11871	198916	66.78323	2.24215	496	22.27106	246016	70.42727	2.01613
447	21.14237	199809	66.85806	2.23714	497	22.29350	247009	70.49823	2.01207
448	21.16601	200704	66.93280	2.23214	498	22.31591	248004	70.56912	2.00803
449	21.18962	201601	67.00746	2.22717	499	22.33831	249001	70.63993	2.00401
450	21.21320	202500	67.08204	2.22222	500	22.36068	250000	70.71068	2.00000

TABLE A-9 (continued)

N	$\sqrt{N}$	N^2	$\sqrt{10N}$	$1000/N$	N	$\sqrt{N}$	N^2	$\sqrt{10N}$	$1000/N$
500	22.36068	250000	70.71068	2.00000	550	23.45208	302500	74.16198	1.81818
501	22.38303	251001	70.78135	1.99601	551	23.47339	303601	74.22937	1.81488
502	22.40536	252004	70.85196	1.99203	552	23.49468	304704	74.29670	1.81159
503	22.42766	253009	70.92249	1.98807	553	23.51595	305809	74.36397	1.80832
504	22.44994	254016	70.99296	1.98413	554	23.53720	306916	74.43118	1.80505
505	22.47221	255025	71.06335	1.98020	555	23.55844	308025	74.49832	1.80180
506	22.49444	256036	71.13368	1.97628	556	23.57965	309136	74.56541	1.79856
507	22.51666	257049	71.20393	1.97239	557	23.60085	310249	74.63243	1.79533
508	22.53886	258064	71.27412	1.96850	558	23.62202	311364	74.69940	1.79211
509	22.56103	259081	71.34424	1.96464	559	23.64318	312481	74.76630	1.78891
510	22.58318	260100	71.41428	1.96078	560	23.66432	313600	74.83315	1.78571
511	22.60531	261121	71.48426	1.95695	561	23.68544	314721	74.89993	1.78253
512	22.62742	262144	71.55418	1.95313	562	23.70654	315844	74.96666	1.77936
513	22.64950	263169	71.62402	1.94932	563	23.72762	316969	75.03333	1.77620
514	22.67157	264196	71.69379	1.94553	564	23.74868	318096	75.09993	1.77305
515	22.69361	265225	71.76350	1.94175	565	23.76973	319225	75.16648	1.76991
516	22.71563	266256	71.83314	1.93798	566	23.79075	320356	75.23297	1.76678
517	22.73763	267289	71.90271	1.93424	567	23.81176	321489	75.29940	1.76367
518	22.75961	268324	71.97222	1.93050	568	23.83275	322624	75.36577	1.76056
519	22.78157	269361	72.04165	1.92678	569	23.85372	323761	75.43209	1.75747
520	22.80351	270400	72.11103	1.92308	570	23.87467	324900	75.49834	1.75439
521	22.82542	271441	72.18033	1.91939	571	23.89561	326041	75.56454	1.75131
522	22.84732	272484	72.24957	1.91571	572	23.91652	327184	75.63068	1.74825
523	22.86919	273529	72.31874	1.91205	573	23.93742	328329	75.69676	1.74520
524	22.89105	274576	72.38784	1.90840	574	23.95830	329476	75.76279	1.74216
525	22.91288	275625	72.45688	1.90476	575	23.97916	330625	75.82875	1.73913
526	22.93469	276676	72.52586	1.90114	576	24.00000	331776	75.89466	1.73611
527	22.95648	277729	72.59477	1.89753	577	24.02082	332929	75.96052	1.73310
528	22.97825	278784	72.66361	1.89394	578	24.04163	334084	76.02631	1.73010
529	23.00000	279841	72.73239	1.89036	579	24.06242	335241	76.09205	1.72712
530	23.02173	280900	72.80110	1.88679	580	24.08319	336400	76.15773	1.72414
531	23.04344	281961	72.86975	1.88324	581	24.10394	337561	76.22336	1.72117
532	23.06513	283024	72.93833	1.87970	582	24.12468	338724	76.28892	1.71821
533	23.08679	284089	73.00685	1.87617	583	24.14539	339889	76.35444	1.71527
534	23.10844	285156	73.07530	1.87266	584	24.16609	341056	76.41989	1.71233
535	23.13007	286225	73.14369	1.86916	585	24.18677	342225	76.48529	1.70940
536	23.15167	287296	73.21202	1.86567	586	24.20744	343396	76.55064	1.70648
537	23.17326	288369	73.28028	1.86220	587	24.22808	344569	76.61593	1.70358
538	23.19483	289444	73.34848	1.85874	588	24.24871	345744	76.68116	1.70068
539	23.21637	290521	73.41662	1.85529	589	24.26932	346921	76.74634	1.69779
540	23.23790	291600	73.48469	1.85185	590	24.28992	348100	76.81146	1.69492
541	23.25941	292681	73.55270	1.84843	591	24.31049	349281	76.87652	1.69205
542	23.28089	293764	73.62065	1.84502	592	24.33105	350464	76.94154	1.68919
543	23.30236	294849	73.68853	1.84162	593	24.35159	351649	77.00649	1.68634
544	23.32381	295936	73.75636	1.83824	594	24.37212	352836	77.07140	1.68350
545	23.34524	297025	73.82412	1.83486	595	24.39262	354025	77.13624	1.68067
546	23.36664	298116	73.89181	1.83150	596	24.41311	355216	77.20104	1.67785
547	23.38803	299209	73.95945	1.82815	597	24.43358	356409	77.26578	1.67504
548	23.40940	300304	74.02702	1.82482	598	24.45404	357604	77.33046	1.67224
549	23.43075	301401	74.09453	1.82149	599	24.47448	358801	77.39509	1.66945
550	23.45208	302500	74.16198	1.81818	600	24.49490	360000	77.45967	1.66667

TABLE A-9 (continued)

N	$\sqrt{N}$	N^2	$\sqrt{10N}$	$1000/N$	N	$\sqrt{N}$	N^2	$\sqrt{10N}$	$1000/N$
600	24.49490	360000	77.45967	1.66667	650	25.49510	422500	80.62258	1.53846
601	24.51530	361201	77.52419	1.66389	651	25.51470	423801	80.68457	1.53610
602	24.53569	362404	77.58866	1.66113	652	25.53429	425104	80.74652	1.53374
603	24.55606	363609	77.65307	1.65837	653	25.55386	426409	80.80842	1.53139
604	24.57641	364816	77.71744	1.65563	654	25.57342	427716	80.87027	1.52905
605	24.59675	366025	77.78175	1.65289	655	25.59297	429025	80.93207	1.52672
606	24.61707	367236	77.84600	1.65017	656	25.61250	430336	80.99383	1.52439
607	24.63737	368449	77.91020	1.64745	657	25.63201	431649	81.05554	1.52207
608	24.65766	369664	77.97435	1.64474	658	25.65151	432964	81.11720	1.51976
609	24.67793	370881	78.03845	1.64204	659	25.67100	434281	81.17881	1.51745
610	24.69818	372100	78.10250	1.63934	660	25.69047	435600	81.24038	1.51515
611	24.71841	373321	78.16649	1.63666	661	25.70992	436921	81.30191	1.51286
612	24.73863	374544	78.23043	1.63399	662	25.72936	438244	81.36338	1.51057
613	24.75884	375769	78.29432	1.63132	663	25.74879	439569	81.42481	1.50830
614	24.77902	376996	78.35815	1.62866	664	25.76820	440896	81.48620	1.50602
615	24.79919	378225	78.42194	1.62602	665	25.78759	442225	81.54753	1.50376
616	24.81935	379456	78.48567	1.62338	666	25.80698	443556	81.60882	1.50150
617	24.83948	380689	78.54935	1.62075	667	25.82634	444889	81.67007	1.49925
618	24.85961	381924	78.61298	1.61812	668	25.84570	446224	81.73127	1.49701
619	24.87971	383161	78.67655	1.61551	669	25.86503	447561	81.79242	1.49477
620	24.89980	384400	78.74008	1.61290	670	25.88436	448900	81.85353	1.49254
621	24.91987	385641	78.80355	1.61031	671	25.90367	450241	81.91459	1.49031
622	24.93993	386884	78.86698	1.60772	672	25.92296	451584	81.97561	1.48810
623	24.95997	388129	78.93035	1.60514	673	25.94224	452929	82.03658	1.48588
624	24.97999	389376	78.99367	1.60256	674	25.96151	454276	82.09750	1.48368
625	25.00000	390625	79.05694	1.60000	675	25.98076	455625	82.15838	1.48148
626	25.01999	391876	79.12016	1.59744	676	26.00000	456976	82.21922	1.47929
627	25.03997	393129	79.18333	1.59490	677	26.01922	458329	82.28001	1.47710
628	25.05993	394384	79.24645	1.59236	678	26.03843	459684	82.34076	1.47493
629	25.07987	395641	79.30952	1.58983	679	26.05763	461041	82.40146	1.47275
630	25.09980	396900	79.37254	1.58730	680	26.07681	462400	82.46211	1.47059
631	25.11971	398161	79.43551	1.58479	681	26.09598	463761	82.52272	1.46843
632	25.13961	399424	79.49843	1.58228	682	26.11513	465124	82.58329	1.46628
633	25.15949	400689	79.56130	1.57978	683	26.13427	466489	82.64381	1.46413
634	25.17936	401956	79.62412	1.57729	684	26.15339	467856	82.70429	1.46199
635	25.19921	403225	79.68689	1.57480	685	26.17250	469225	82.76473	1.45985
636	25.21904	404496	79.74961	1.57233	686	26.19160	470596	82.82512	1.45773
637	25.23886	405769	79.81228	1.56986	687	26.21068	471969	82.88546	1.45560
638	25.25866	407044	79.87490	1.56740	688	26.22975	473344	82.94577	1.45349
639	25.27845	408321	79.93748	1.56495	689	26.24881	474721	83.00602	1.45138
640	25.29822	409600	80.00000	1.56250	690	26.26785	476100	83.06624	1.44928
641	25.31798	410881	80.06248	1.56006	691	26.28688	477481	83.12641	1.44718
642	25.33772	412164	80.12490	1.55763	692	26.30589	478864	83.18654	1.44509
643	25.35744	413449	80.18728	1.55521	693	26.32489	480249	83.24662	1.44300
644	25.37716	414736	80.24961	1.55280	694	26.34388	481636	83.30666	1.44092
645	25.39685	416025	80.31189	1.55039	695	26.36285	483025	83.36666	1.43885
646	25.41653	417316	80.37413	1.54799	696	26.38181	484416	83.42661	1.43678
647	25.43619	418609	80.43631	1.54560	697	26.40076	485809	83.48653	1.43472
648	25.45584	419904	80.49845	1.54321	698	26.41969	487204	83.54639	1.43266
649	25.47548	421201	80.56054	1.54083	699	26.43861	488601	83.60622	1.43062
650	25.49510	422500	80.62258	1.53846	700	26.45751	490000	83.66600	1.42857

TABLE A-9 (continued)

N	$\sqrt{N}$	N^2	$\sqrt{10N}$	$1000/N$	N	$\sqrt{N}$	N^2	$\sqrt{10N}$	$1000/N$
700	26.45751	490000	83.66600	1.42857	750	27.38613	562500	86.60254	1.33333
701	26.47640	491401	83.72574	1.42653	751	27.40438	564001	86.66026	1.33156
702	26.49528	492804	83.78544	1.42450	752	27.42262	565504	86.71793	1.32979
703	26.51415	494209	83.84510	1.42248	753	27.44085	567009	86.77557	1.32802
704	26.53300	495616	83.90471	1.42045	754	27.45906	568516	86.83317	1.32626
705	26.55184	497025	83.96428	1.41844	755	27.47726	570025	86.89074	1.32450
706	26.57066	498436	84.02381	1.41643	756	27.49545	571536	86.94826	1.32275
707	26.58947	499849	84.08329	1.41443	757	27.51363	573049	87.00575	1.32100
708	26.60827	501264	84.14274	1.41243	758	27.53180	574564	87.06320	1.31926
709	26.62705	502681	84.20214	1.41044	759	27.54995	576081	87.12061	1.31752
710	26.64583	504100	84.26150	1.40845	760	27.56810	577600	87.17798	1.31579
711	26.66458	505521	84.32082	1.40647	761	27.58623	579121	87.23531	1.31406
712	26.68333	506944	84.38009	1.40449	762	27.60435	580644	87.29261	1.31234
713	26.70206	508369	84.43933	1.40252	763	27.62245	582169	87.34987	1.31062
714	26.72078	509796	84.49852	1.40056	764	27.64055	583696	87.40709	1.30890
715	26.73948	511225	84.55767	1.39860	765	27.65863	585225	87.46428	1.30719
716	26.75818	512656	84.61678	1.39665	766	27.67671	586756	87.52143	1.30548
717	26.77686	514089	84.67585	1.39470	767	27.69476	588289	87.57854	1.30378
718	26.79552	515524	84.73488	1.39276	768	27.71281	589824	87.63561	1.30208
719	26.81418	516961	84.79387	1.39082	769	27.73085	591361	87.69265	1.30039
720	26.83282	518400	84.85281	1.38889	770	27.74887	592900	87.74964	1.29870
721	26.85144	519841	84.91172	1.38696	771	27.76689	594441	87.80661	1.29702
722	26.87006	521284	84.97058	1.38504	772	27.78489	595984	87.86353	1.29534
723	26.88866	522729	85.02941	1.38313	773	27.80288	597529	87.92042	1.29366
724	26.90725	524176	85.08819	1.38122	774	27.82086	599076	87.97727	1.29199
725	26.92582	525625	85.14693	1.37931	775	27.83882	600625	88.03408	1.29032
726	26.94439	527076	85.20563	1.37741	776	27.85678	602176	88.09086	1.28866
727	26.96294	528529	85.26429	1.37552	777	27.87472	603729	88.14760	1.28700
728	26.98148	529984	85.32292	1.37363	778	27.89265	605284	88.20431	1.28535
729	27.00000	531441	85.38150	1.37174	779	27.91057	606841	88.26098	1.28370
730	27.01851	532900	85.44004	1.36986	780	27.92848	608400	88.31761	1.28205
731	27.03701	534361	85.49854	1.36799	781	27.94638	609961	88.37420	1.28041
732	27.05550	535824	85.55700	1.36612	782	27.96426	611524	88.43076	1.27877
733	27.07397	537289	85.61542	1.36426	783	27.98214	613089	88.48729	1.27714
734	27.09243	538756	85.67380	1.36240	784	28.00000	614656	88.54377	1.27551
735	27.11088	540225	85.73214	1.36054	785	28.01785	616225	88.60023	1.27389
736	27.12932	541696	85.79044	1.35870	786	28.03569	617796	88.65664	1.27226
737	27.14774	543169	85.84870	1.35685	787	28.05352	619369	88.71302	1.27065
738	27.16616	544644	85.90693	1.35501	788	28.07134	620944	88.76936	1.26904
739	27.18455	546121	85.96511	1.35318	789	28.08914	622521	88.82567	1.26743
740	27.20294	547600	86.02325	1.35135	790	28.10694	624100	88.88194	1.26582
741	27.22132	549081	86.08136	1.34953	791	28.12472	625681	88.93818	1.26422
742	27.23968	550564	86.13942	1.34771	792	28.14249	627264	88.99438	1.26263
743	27.25803	552049	86.19745	1.34590	793	28.16026	628849	89.05055	1.26103
744	27.27636	553536	86.25543	1.34409	794	28.17801	630436	89.10668	1.25945
745	27.29469	555025	86.31338	1.34228	795	28.19574	632025	89.16277	1.25786
746	27.31300	556516	86.37129	1.34048	796	28.21347	633616	89.21883	1.25628
747	27.33130	558009	86.42916	1.33869	797	28.23119	635209	89.27486	1.25471
748	27.34959	559504	86.48699	1.33690	798	28.24889	636804	89.33085	1.25313
749	27.36786	561001	86.54479	1.33511	799	28.26659	638401	89.38680	1.25156
750	27.38613	562500	86.60254	1.33333	800	28.28427	640000	89.44272	1.25000

TABLE A-9 (continued)

N	$\sqrt{N}$	N^2	$\sqrt{10N}$	1000/N	N	$\sqrt{N}$	N^2	$\sqrt{10N}$	1000/N
800	28.28427	640000	89.44272	1.25000	850	29.15476	722500	92.19544	1.17647
801	28.30194	641601	89.49860	1.24844	851	29.17190	724201	92.24966	1.17509
802	28.31960	643204	89.55445	1.24688	852	29.18904	725904	92.30385	1.17371
803	28.33725	644809	89.61027	1.24533	853	29.20616	727609	92.35800	1.17233
804	28.35489	646416	89.66605	1.24378	854	29.22328	729316	92.41212	1.17096
805	28.37252	648025	89.72179	1.24224	855	29.24038	731025	92.46621	1.16959
806	28.39014	649636	89.77750	1.24069	856	29.25748	732736	92.52027	1.16822
807	28.40775	651249	89.83318	1.23916	857	29.27456	734449	92.57429	1.16686
808	28.42534	652864	89.88882	1.23762	858	29.29164	736164	92.62829	1.16550
809	28.44293	654481	89.94443	1.23609	859	29.30870	737881	92.68225	1.16414
810	28.46050	656100	90.00000	1.23457	860	29.32576	739600	92.73618	1.16279
811	28.47806	657721	90.05554	1.23305	861	29.34280	741321	92.79009	1.16144
812	28.49561	659344	90.11104	1.23153	862	29.35984	743044	92.84396	1.16009
813	28.51315	660969	90.16651	1.23001	863	29.37686	744769	92.89779	1.15875
814	28.53069	662596	90.22195	1.22850	864	29.39388	746496	92.95160	1.15741
815	28.54820	664225	90.27735	1.22699	865	29.41088	748225	93.00538	1.15607
816	28.56571	665856	90.33272	1.22549	866	29.42788	749956	93.05912	1.15473
817	28.58321	667489	90.38805	1.22399	867	29.44486	751689	93.11283	1.15340
818	28.60070	669124	90.44335	1.22249	868	29.46184	753424	93.16652	1.15207
819	28.61818	670761	90.49862	1.22100	869	29.47881	755161	93.22017	1.15075
820	28.63564	672400	90.55385	1.21951	870	29.49576	756900	93.27379	1.14943
821	28.65310	674041	90.60905	1.21803	871	29.51271	758641	93.32738	1.14811
822	28.67054	675684	90.66422	1.21655	872	29.52965	760384	93.38094	1.14679
823	28.68798	677329	90.71935	1.21507	873	29.54657	762129	93.43447	1.14548
824	28.70540	678976	90.77445	1.21359	874	29.56349	763876	93.48797	1.14416
825	28.72281	680625	90.82951	1.21212	875	29.58040	765625	93.54143	1.14286
826	28.74022	682276	90.88454	1.21065	876	29.59730	767376	93.59487	1.14155
827	28.75761	683929	90.93954	1.20919	877	29.61419	769129	93.64828	1.14025
828	28.77499	685584	90.99451	1.20773	878	29.63106	770884	93.70165	1.13895
829	28.79236	687241	91.04944	1.20627	879	29.64793	772641	93.75500	1.13766
830	28.80972	688900	91.10434	1.20482	880	29.66479	774400	93.80832	1.13636
831	28.82707	690561	91.15920	1.20337	881	29.68164	776161	93.86160	1.13507
832	28.84441	692224	91.21403	1.20192	882	29.69848	777924	93.91486	1.13379
833	28.86174	693889	91.26883	1.20048	883	29.71532	779689	93.96808	1.13250
834	28.87906	695556	91.32360	1.19904	884	29.73214	781456	94.02127	1.13122
835	28.89637	697225	91.37833	1.19760	885	29.74895	783225	94.07444	1.12994
836	28.91366	698896	91.43304	1.19617	886	29.76575	784996	94.12757	1.12867
837	28.93095	700569	91.48770	1.19474	887	29.78255	786769	94.18068	1.12740
838	28.94823	702244	91.54234	1.19332	888	29.79933	788544	94.23375	1.12613
839	28.96550	703921	91.59694	1.19190	889	29.81610	790321	94.28680	1.12486
840	28.98275	705600	91.65151	1.19048	890	29.83287	792100	94.33981	1.12360
841	29.00000	707281	91.70605	1.18906	891	29.84962	793881	94.39280	1.12233
842	29.01724	708964	91.76056	1.18765	892	29.86637	795664	94.44575	1.12108
843	29.03446	710649	91.81503	1.18624	893	29.88311	797449	94.49868	1.11982
844	29.05168	712336	91.86947	1.18483	894	29.89983	799236	94.55157	1.11857
845	29.06888	714025	91.92388	1.18343	895	29.91655	801025	94.60444	1.11732
846	29.08608	715716	91.97826	1.18203	896	29.93326	802816	94.65728	1.11607
847	29.10326	717409	92.03260	1.18064	897	29.94996	804609	94.71008	1.11483
848	29.12044	719104	92.08692	1.17925	898	29.96665	806404	94.76286	1.11359
849	29.13760	720801	92.14120	1.17786	899	29.98333	808201	94.81561	1.11235
850	29.15476	722500	92.19544	1.17647	900	30.00000	810000	94.86833	1.11111

TABLE A-9 (continued)

N	$\sqrt{N}$	N^2	$\sqrt{10N}$	$1000/N$	N	$\sqrt{N}$	N^2	$\sqrt{10N}$	$1000/N$
900	30.00000	810000	94.86833	1.11111	950	30.82207	902500	97.46794	1.05263
901	30.01666	811801	94.92102	1.10988	951	30.83829	904401	97.51923	1.05152
902	30.03331	813604	94.97368	1.10865	952	30.85450	906304	97.57049	1.05042
903	30.04996	815409	95.02631	1.10742	953	30.87070	908209	97.62172	1.04932
904	30.06659	817216	95.07891	1.10619	954	30.88689	910116	97.67292	1.04822
905	30.08322	819025	95.13149	1.10497	955	30.90307	912025	97.72410	1.04712
906	30.09983	820836	95.18403	1.10375	956	30.91925	913936	97.77525	1.04603
907	30.11644	822649	95.23655	1.10254	957	30.93542	915849	97.82638	1.04493
908	30.13304	824464	95.28903	1.10132	958	30.95158	917764	97.87747	1.04384
909	30.14963	826281	95.34149	1.10011	959	30.96773	919681	97.92855	1.04275
910	30.16621	828100	95.39392	1.09890	960	30.98387	921600	97.97959	1.04167
911	30.18278	829921	95.44632	1.09769	961	31.00000	923521	98.03061	1.04058
912	30.19934	831744	95.49869	1.09649	962	31.01612	925444	98.08160	1.03950
913	30.21589	833569	95.55103	1.09529	963	31.03224	927369	98.13256	1.03842
914	30.23243	835396	95.60335	1.09409	964	31.04835	929296	98.18350	1.03734
915	30.24897	837225	95.65563	1.09290	965	31.06445	931225	98.23441	1.03627
916	30.26549	839056	95.70789	1.09170	966	31.08054	933156	98.28530	1.03520
917	30.28201	840889	95.76012	1.09051	967	31.09662	935089	98.33616	1.03413
918	30.29851	842724	95.81232	1.08932	968	31.11270	937024	98.38699	1.03306
919	30.31501	844561	95.86449	1.08814	969	31.12876	938961	98.43780	1.03199
920	30.33150	846400	95.91663	1.08696	970	31.14482	940900	98.48858	1.03093
921	30.34798	848241	95.96874	1.08578	971	31.16087	942841	98.53933	1.02987
922	30.36445	850084	96.02083	1.08460	972	31.17691	944784	98.59006	1.02881
923	30.38092	851929	96.07289	1.08342	973	31.19295	946729	98.64076	1.02775
924	30.39737	853776	96.12492	1.08225	974	31.20897	948676	98.69144	1.02669
925	30.41381	855625	96.17692	1.08108	975	31.22499	950625	98.74209	1.02564
926	30.43025	857476	96.22889	1.07991	976	31.24100	952576	98.79271	1.02459
927	30.44667	859329	96.28084	1.07875	977	31.25700	954529	98.84331	1.02354
928	30.46309	861184	96.33276	1.07759	978	31.27299	956484	98.89388	1.02249
929	30.47950	863041	96.38465	1.07643	979	31.28898	958441	98.94443	1.02145
930	30.49590	864900	96.43651	1.07527	980	31.30495	960400	98.99495	1.02041
931	30.51229	866761	96.48834	1.07411	981	31.32092	962361	99.04544	1.01937
932	30.52868	868624	96.54015	1.07296	982	31.33688	964324	99.09591	1.01833
933	30.54505	870489	96.59193	1.07181	983	31.35283	966289	99.14636	1.01729
934	30.56141	872356	96.64368	1.07066	984	31.36877	968256	99.19677	1.01626
935	30.57777	874225	96.69540	1.06952	985	31.38471	970225	99.24717	1.01523
936	30.59412	876096	96.74709	1.06838	986	31.40064	972196	99.29753	1.01420
937	30.61046	877969	96.79876	1.06724	987	31.41656	974169	99.34787	1.01317
938	30.62679	879844	96.85040	1.06610	988	31.43247	976144	99.39819	1.01215
939	30.64311	881721	96.90201	1.06496	989	31.44837	978121	99.44848	1.01112
940	30.65942	883600	96.95360	1.06383	990	31.46427	980100	99.49874	1.01010
941	30.67572	885481	97.00515	1.06270	991	31.48015	982081	99.54898	1.00908
942	30.69202	887364	97.05668	1.06157	992	31.49603	984064	99.59920	1.00806
943	30.70831	889249	97.10819	1.06045	993	31.51190	986049	99.64939	1.00705
944	30.72458	891136	97.15966	1.05932	994	31.52777	988036	99.69955	1.00604
945	30.74085	893025	97.21111	1.05820	995	31.54362	990025	99.74969	1.00503
946	30.75711	894916	97.26253	1.05708	996	31.55947	992016	99.79980	1.00402
947	30.77337	896809	97.31393	1.05597	997	31.57531	994009	99.84989	1.00301
948	30.78961	898704	97.36529	1.05485	998	31.59114	996004	99.89995	1.00200
949	30.80584	900601	97.41663	1.05374	999	31.60696	998001	99.94999	1.00100
950	30.82207	902500	97.46794	1.05263	1000	31.62278	1000000	100.00000	1.00000

TABLE A-10
Critical Values of T in the Wilcoxon Matched-Pairs Signed-Ranks Test
Critical Values of T at Various Levels of Probability

The symbol T denotes the smaller sum of ranks associated with differences that are all of the same sign. For any given N (number of ranked differences), the obtained T is significant at a given level if it is equal to or *less than* the value shown in the table.

	Level of Significance for One-tailed Test					Level of Significance for One-tailed Test			
	0.05	0.025	0.01	0.005		0.05	0.025	0.01	0.005
	Level of Significance for Two-tailed Test					Level of Significance for Two-tailed Test			
N	0.10	0.05	0.02	0.01	N	0.10	0.05	0.02	0.01
5	0	—	—	—	28	130	116	101	91
6	2	0	—	—	29	140	126	110	100
7	3	2	0	—	30	151	137	120	109
8	5	3	1	0	31	163	147	130	118
9	8	5	3	1	32	175	159	140	128
10	10	8	5	3	33	187	170	151	138
11	13	10	7	5	34	200	182	162	148
12	17	13	9	7	35	213	195	173	159
13	21	17	12	9	36	227	208	185	171
14	25	21	15	12	37	241	221	198	182
15	30	25	19	15	38	256	235	211	194
16	35	29	23	19	39	271	249	224	207
17	41	34	27	23	40	286	264	238	220
18	47	40	32	27	41	302	279	252	233
19	53	46	37	32	42	319	294	266	247
20	60	52	43	37	43	336	310	281	261
21	67	58	49	42	44	353	327	296	276
22	75	65	55	48	45	371	343	312	291
23	83	73	62	54	46	389	361	328	307
24	91	81	69	61	47	407	378	345	322
25	100	89	76	68	48	426	396	362	339
26	110	98	84	75	49	446	415	379	355
27	119	107	92	83	50	466	434	397	373

(Slight discrepancies will be found between the critical values appearing in the table above and in Table 2 of the 1964 revision of F. Wilcoxon and R. A. Wilcox, *Some Rapid Approximate Statistical Procedures*, New York, Lederle Laboratories, 1964. The disparity reflects the latter's policy of selecting the critical value nearest a given significance level, occasionally overstepping that level. For example, for $N = 8$,

the probability of a T of $3 = 0.0390$ (two-tail)

and

the probability of a T of $4 = 0.0546$ (two-tail).

Wilcoxon and Wilcox select a T of 4 as the critical value at the 0.05 level of significance (two-tail), whereas Table J reflects a more conservative policy by setting a T of 3 as the critical value at this level.)

SOURCE: Hughes, A. J. and Grawoig, D. E., *Statistics: A Foundation for Analysis*, Addison-Wesley Publishing Co., Reading, Mass., 1971.

TABLE A-11
Table of Exponential Functions

x	e^x	e^{-x}	x	e^x	e^{-x}
0.00	1.000	1.000	3.00	20.086	0.050
0.10	1.105	0.905	3.10	22.198	0.045
0.20	1.221	0.819	3.20	24.533	0.041
0.30	1.350	0.741	3.30	27.113	0.037
0.40	1.492	0.670	3.40	29.964	0.033
0.50	1.649	0.607	3.50	33.115	0.030
0.60	1.822	0.549	3.60	36.598	0.027
0.70	2.014	0.497	3.70	40.447	0.025
0.80	2.226	0.449	3.80	44.701	0.022
0.90	2.460	0.407	3.90	49.402	0.020
1.00	2.718	0.368	4.00	54.598	0.018
1.10	3.004	0.333	4.10	60.340	0.017
1.20	3.320	0.301	4.20	66.686	0.015
1.30	3.669	0.273	4.30	73.700	0.014
1.40	4.055	0.247	4.40	81.451	0.012
1.50	4.482	0.223	4.50	90.017	0.011
1.60	4.953	0.202	4.60	99.484	0.010
1.70	5.474	0.183	4.70	109.95	0.009
1.80	6.050	0.165	4.80	121.51	0.008
1.90	6.686	0.150	4.90	134.29	0.007
2.00	7.389	0.135	5.00	148.41	0.007
2.10	8.166	0.122	5.10	164.02	0.006
2.20	9.025	0.111	5.20	181.27	0.006
2.30	9.974	0.100	5.30	200.34	0.005
2.40	11.023	0.091	5.40	221.41	0.005
2.50	12.182	0.082	5.50	244.69	0.004
2.60	13.464	0.074	5.60	270.43	0.004
2.70	14.880	0.067	5.70	298.87	0.003
2.80	16.445	0.061	5.80	330.30	0.003
2.90	18.174	0.055	5.90	365.04	0.003
3.00	20.086	0.050	6.00	403.43	0.002

Symbols, Subscripts, and Summations

In statistics, *symbols* such as X, Y, and Z are used to represent different sets of data. Hence, if we have data for five families, we might let

$$X = \text{family income}$$
$$Y = \text{family clothing expenditures}$$
$$Z = \text{family savings}$$

Subscripts are used to represent individual observations within these sets of data. Thus, we write X_i to represent the income of the ith family, where i takes on the values 1, 2, 3, 4, and 5. In this notation X_1, X_2, X_3, X_4, and X_5 stand for the incomes of the first family, the second family, etc. The data are arranged in some order, such as by size of income, the order in which the data were gathered, or any other way suitable to the purposes or convenience of the investigator. The subscript i is a variable used to index the individual data observations. To continue with the example, X_i, Y_i, and Z_i represent the income, clothing expenditures, and savings of the ith family. For example, X_2 represents the income of the second family, Y_2 clothing expenditures of the second family (same family), and Z_5 the savings of the fifth family.

Now, let us suppose that we have data for two different samples, say the net worths of 100 corporations and the test scores of 20 students. To refer to individual observations in these samples, we can let X_i denote the net worth of the ith corporation, where i assumes values from 1 to 100. (This latter idea is indicated by the notation $i = 1, 2, 3, \ldots , 100$.) We can also let Y_j denote the test score of the jth student, where $j = 1, 2, 3, \ldots , 20$. Thus, the different subscript letters make it clear that different samples are involved. Letters such as X, Y, and Z are generally used to represent the different variables or types of measurements involved, whereas subscripts such as i, j, k, and l are used to designate individual observations.

We now turn to the method of expressing summations of sets of data. Suppose we want to add a set of four observations, denoted X_1, X_2, X_3, and X_4. A convenient way of designating this addition is

$$\sum_{i=4}^{4} X_i = X_1 + X_2 + X_3 + X_4$$

where the symbol Σ (Greek capital "sigma") means "the sum of." Hence, the symbol

$$\sum_{i=1}^{4} X_i$$

is read "the sum of the X_is, i going from 1 to 4." For example, if $X_1 = 3, X_2 = 1, X_3 = 10$, and $X_4 = 5$,

$$\sum_{i=1}^{4} X_i = 3 + 1 + 10 + 5 = 19$$

In general, if there are n observations, we write

$$\sum_{i=1}^{n} X_i = X_1 + X_2 + \cdots + X_n$$

EXAMPLE 1

Let $X_1 = -2, X_2 = 3, X_3 = 5$. Find

a.
$$\sum_{i=1}^{3} X_i$$

b.
$$\sum_{j=1}^{3} X_j^2$$

c.
$$\sum_{j=1}^{3} (2X_j + 3)$$

SOLUTION

a.
$$\sum_{i=1}^{3} X_i = X_1 + X_2 + X_3$$
$$= -2 + 3 + 5 = 6$$

b.
$$\sum_{j=1}^{3} X_j^2 = X_1^2 + X_2^2 + X_3^2$$
$$= (-2)^2 + (3)^2 + (5)^2 = 38$$

c.
$$\sum_{j=1}^{3} (2X_j + 3) = (2X_1 + 3) + (2X_2 + 3) + (2X_3 + 3)$$
$$= (-4 + 3) + (6 + 3) + (10 + 3)$$
$$= -1 + 9 + 13 = 21$$

EXAMPLE 2

Prove

a.
$$\sum_{i=1}^{n} aX_i = a \sum_{i=1}^{n} X_i$$

b.
$$\sum_{i=1}^{n} a = na$$

c.
$$\sum_{i=1}^{n} (X_i + Y_i) = \sum_{i=1}^{n} X_i + \sum_{i=1}^{n} Y_i$$

where a is a constant.

SOLUTION

a.
$$\sum_{i=1}^{n} aX_i = aX_1 + aX_2 + \cdots + aX_n$$
$$= a(X_1 + X_2 + \cdots + X_n)$$
$$= a \sum_{i=1}^{n} X_i$$

b.
$$\sum_{i=1}^{n} a = a \sum_{i=1}^{n} 1$$
$$= \underbrace{a(1 + 1 + \cdots + 1)}_{n \text{ terms}}$$
$$= na$$

c. $\displaystyle\sum_{i=1}^{n} (X_i + Y_i) = X_1 + Y_1 + X_2 + Y_2 + \cdots + X_n + Y_n$

$$= (X_1 + X_2 + \cdots + X_n) +$$
$$(Y_1 + Y_2 + \cdots + Y_n)$$

$$= \sum_{i=1}^{n} X_i + \sum_{i=1}^{n} Y_i$$

These three summation properties are listed as Rules 1, 2, and 3 at the end of Appendix B.

Double summations are used to indicate summations of more than one variable, where different subscript indexes are involved. For example, the symbol

$$\sum_{j=1}^{3} \sum_{i=1}^{2} X_i Y_j$$

means "the sum of the products of X_i and Y_j where $i = 1, 2$ and $j = 1, 2, 3$." Thus, we can write

$$\sum_{j=1}^{3} \sum_{i=1}^{2} X_i Y_j = X_1 Y_1 + X_2 Y_1 + X_1 Y_2 + X_2 Y_2 + X_1 Y_3 + X_2 Y_3$$

SIMPLIFIED SUMMATION NOTATIONS

In this text, simplified summation notations are often used in which subscripts are eliminated. Thus, for example, ΣX, ΣX^2, and ΣY^2 are used instead of

$$\sum_{i=1}^{n} X_i \qquad \sum_{i=1}^{n} X_i^2 \qquad \text{and} \qquad \sum_{i=1}^{n} Y_i^2$$

Also in this text, subscripts have ordinarily been dropped in the case of probability distributions. For example, consider the following discrete probability distribution:

x_i	$f(x_i)$
$x_1 = 0$	0.2
$x_2 = 1$	0.3
$x_3 = 2$	0.5
	1.0

The statement that the sum of the probabilities is equal to 1 is given by

$$\sum_{i=1}^{3} f(x_i) = 1$$

However, we have used the following customary simplified notation:

x	$f(x)$
0	0.2
1	0.3
2	<u>0.5</u>
	1.0

The corresponding summation statement is

$$\sum_{x} f(x) = 1$$

where $\sum_{x}$ means "sum over all values of x." The notation is also further simplified by writing

$$\sum f(x) = 1$$

SUMMATION PROPERTIES

RULE 1

$$\sum_{i=1}^{n} aX_i = a \sum_{i=1}^{n} X_i$$

RULE 2

$$\sum_{i=1}^{n} a = \underbrace{a + a + \cdots + a}_{n \text{ terms}} = na$$

RULE 3

$$\sum_{i=1}^{n} (X_i + Y_i) = \sum_{i=1}^{n} X_i + \sum_{i=1}^{n} Y_i$$

Properties of Expected Values and Variances

In keeping with notational conventions used in this text, a and b represent constants, whereas X represents random variables.

RULE 1

$E(a) = a$
The expected value of a constant is equal to that constant.

RULE 2

$E(bX) = bE(X)$
The expected value of a constant times a random variable is equal to the constant times the expected value of the random variable.

RULE 3

$E(a + bX) = a + bE(X)$

Rule 3 combines Rules 1 and 2. A brief proof is given for Rule 3 to illustrate a general method of proofs for expected values.

Let X denote a discrete random variable that takes on values x_1, $x_2, \ldots, x_i, \ldots, x_n$ with probabilities $f(x_1)$, $f(x_2)$, $\ldots$, $f(x_i)$, $\ldots$, $f(x_n)$. Then, using the definition of an expected value given in Equation 3.11 in Chapter 3, we have

$$E(a + bX) = \sum_{i=1}^{n} (a + bx_i)f(x_i) = \sum_{i=1}^{n} af(x_i) + \sum_{i=1}^{n} bx_i f(x_i)$$

$$= a \sum_{i=1}^{n} f(x_i) + b \sum_{i=1}^{n} x_i f(x_i)$$

$$= a(1) + bE(X) = a + bE(X)$$

RULE 4

$E(X_1 + X_2 + \cdots + X_n) = E(X_1) + E(X_2) + \cdots + E(X_n)$
where $X_1, X_2, \ldots, X_n$ are random variables.

The expected value of a sum equals the sum of the expected values. The X_is are not restricted in any way. That is, they may either be independent or dependent.

Expressing this rule in somewhat different symbols, we have

$$E\left[\sum_{i=1}^{n} (X_i) \right] = \sum_{i=1}^{n} [E(X_i)]$$

Treating the expected value and summation symbols as operators (that is, as defining specific operations on the X_is), we have the result that the summation sign and expected value symbol are interchangeable operators.

RULE 5

$\sigma^2(a) = 0$
The variance of a constant is equal to 0.

RULE 6

$\sigma^2(bX) = b^2\sigma^2(X)$
The variance of a constant times a random variable is equal to the constant squared times the variance of the random variable.

RULE 7

$\sigma^2(a + bX) = b^2\sigma^2(X)$
Rule 7 combines Rules 5 and 6. As in the case of Rule 3 for the expected

value, a simple application of the definition of a variance yields the desired result. The proof is left to the reader as an exercise.

RULE 8

$\sigma^2(X_1 + X_2 + \cdots + X_n) = \sigma^2(X_1) + \sigma^2(X_2) + \cdots + \sigma^2(X_n)$, where X_1, $X_2, \ldots, X_n$ are independent random variables, that is, every pair of X_is is independent.

Thus, if the X_is are independent, the variance of a sum is equal to the sum of the variances.

Expressing this rule in summation terminology, we obtain

$$\sigma^2 \left[\sum_{i=1}^{n} (X_i) \right] = \sum_{i=1}^{n} [\sigma^2(X_i)]$$

Hence, the variance and summation symbols are interchangeable operators if the X_is are independent.

RULE 9

$\sigma^2(a_1X_1 + a_2X_2) = a_1^2\sigma^2(X_1) + a_2^2\sigma^2(X_2)$ if X_1 and X_2 are independent.

Rule 9 is derived by applying Rules 7 and 8.

Special cases of Rule 9 are given as Rules 10 and 11.
In Rule 10, $a_1 = +1$, and $a_2 = +1$.
In Rule 11, $a_1 = +1$, but $a_2 = -1$.

RULE 10

$\sigma^2(X_1 + X_2) = \sigma^2(X_1) + \sigma^2(X_2)$ if X_1 and X_2 are independent.

RULE 11

$\sigma^2(X_1 - X_2) = \sigma^2(X_1) + \sigma^2(X_2)$ if X_1 and X_2 are independent.

RULE 12

$$\sigma^2(\bar{X}) = \frac{\sigma^2}{n}$$

where X is a random variable, μ and σ are its mean and standard deviation, respectively, and $\bar{X}$ is the arithmetic mean in a sample of n independent observations of X. If $X_1, X_2, \ldots, X_n$ denote the n observations, then

$$\bar{X} = \frac{1}{n} \sum_{i=1}^{n} X_i$$

This rule may be proven in a few steps.

$$(1) \qquad \sigma^2(\bar{X}) = \sigma^2 \left(\frac{1}{n} \sum_{i=1}^{n} X_i \right) = \frac{1}{n^2} \sigma^2 \left(\sum_{i=1}^{n} X_i \right) \qquad \text{(by Rule 6)}$$

$$(2) \qquad \qquad = \frac{1}{n^2} \sum_{i=1}^{n} [\sigma^2(X_i)] \qquad \text{(by Rule 8)}$$

But since every X_i has the same probability distribution as X, then $\sigma^2(X_i) = \sigma^2(X)$ for each i. Hence,

$$(3) \qquad \qquad \sum_{i=1}^{n} [\sigma^2(X_i)] = n\sigma^2$$

Substituting (3) into (2) gives

$$(4) \qquad \qquad \sigma^2(\bar{X}) = \left(\frac{1}{n^2} \right) (n\sigma^2) = \frac{\sigma^2}{n}$$

which completes the proof.

Let us express Rule 12 in the language and symbolism of sampling theory. If a simple random sample of size n is drawn from an infinite population (or a finite population with replacement), with standard deviation σ, the variance of the sample mean is given by

$$(5) \qquad \qquad \sigma_{\bar{X}}^2 = \frac{\sigma^2}{n}$$

RULE 13

$E(\bar{X}) = \mu$
where the same conditions prevail as in Rule 12, that is, X is a random variable, μ and σ are its mean and standard deviation, respectively, etc.

This rule is easily proven as follows:

$$(1) \qquad \qquad E(\bar{X}) = E \left(\frac{1}{n} \sum_{i=1}^{n} X_i \right)$$

$$\qquad \qquad = \frac{1}{n} \left[E \left(\sum_{i=1}^{n} X_i \right) \right] \qquad \text{(by Rule 2)}$$

$$(2) \qquad \qquad = \frac{1}{n} \left[\sum_{i=1}^{n} E(X_i) \right] \qquad \text{(by Rule 4)}$$

But since every X_i has the same probability distribution as X, then $E(X_i) = E(X)$ for each i. Hence,

$$(3) \qquad \qquad \sum_{i=1}^{n} E(X_i) = nE(X) = n\mu$$

Substituting (3) into (2) gives

$$(4) \qquad \qquad E(\bar{X}) = \left(\frac{1}{n} \right) (n\mu) = \mu$$

As in Rule 12, let us express this result in terms of sampling theory. If a simple random sample of size n is drawn from an infinite population (or a finite population with replacement) with mean μ, the expected value (arithmetic mean) of the sample mean is given by

$$(5) \qquad E(\bar{X}) = \mu_x = \mu$$

The expected value of the sample variance defined with divisor $n - 1$ is equal to the population variance. This result is expressed in symbols in Rule 14.

RULE 14

$$E(s^2) = E \left[\frac{\Sigma\,(X_i - \bar{X})^2}{n - 1} \right] = \sigma^2$$

where the same conditions prevail as in Rules 12 and 13.

Shortcut Formulas

SHORTCUT CALCULATION OF THE MEAN (STEP-DEVIATION METHOD)

A shortcut calculation known as the step-deviation method is useful when class intervals in a frequency distribution are of equal size. The method results in simpler arithmetic than the direct definitional formula, particularly if the class intervals and frequencies involve a large number of digits.

The step-deviation method of computing the mean involves three basic steps:

1. Selection of an assumed (or arbitrary) mean.

2. Calculation of an average deviation from this assumed mean.

3. Addition of this average deviation as a correction factor to the assumed mean to obtain the true mean. This correction factor is positive if the assumed mean lies below the true mean, negative if above.

To accomplish step 2, a midpoint of a class (preferably near the center of the distribution) is taken as the assumed mean. Then deviations of the midpoints of the other classes are taken from the assumed mean in class interval units. These deviations are denoted d. After the d values are averaged, the result must be multiplied by the size of the class interval to return to the units of the original data.

The formula for the step-deviation method is given in Equation D.1.

Step-Deviation Method for the Arithmetic Mean (Grouped Data)

$$(\textbf{D.1}) \qquad\qquad \bar{X} = \bar{X}_a + \left(\frac{\Sigma\,fd}{n}\right)(i)$$

where $\bar{X}$ = the arithmetic mean
$\quad\bar{X}_a$ = the assumed arithmetic mean
$\quad f$ = frequencies
$\quad d$ = deviations of midpoints from the assumed mean in class interval units
$\quad n$ = the number of observations
$\quad i$ = the size of a class interval

The step-deviation method is illustrated in Table D-1 for the same frequency distribution of earnings of 100 semiskilled workers given in

TABLE D-1
Calculation of the Arithmetic Mean for Grouped Data by the Step-Deviation Method: Weekly Earnings Data

Weekly Earnings	Number of Employees f	d	fd
$160.00 and under 170.00	4	−3	−12
170.00 and under 180.00	14	−2	−28
180.00 and under 190.00	18	−1	−18
190.00 and under 200.00	28	0	0
200.00 and under 210.00	20	1	20
210.00 and under 220.00	12	2	24
220.00 and under 230.00	4	3	12
	100		−2

$\bar{X}_a = \$195$

$\bar{X} = \bar{X}_a + \left(\dfrac{\Sigma\,fd}{n}\right)(i) = \$195 + \left(\dfrac{-2}{100}\right)(\$10) = \$194.80$

Table 1-5 on page 22, where the arithmetic mean was computed by the direct definitional method. In Table D-1, the assumed arithmetic mean is $195. As may be noted in the d column, these values indicate the number of class intervals below or above the one in which the assumed mean is taken.

SHORTCUT CALCULATION OF THE STANDARD DEVIATION (STEP-DEVIATION METHOD)

Just as in the case of the arithmetic mean, the step-deviation method of calculating the standard deviation is useful when class intervals in a frequency distribution are of the same size. The saving in computational effort accomplished by the use of the step-deviation method is illustrated for the case of the weekly earnings data given in Table D-1.

As in the case of the calculation for the arithmetic mean, the procedure involves taking deviations of midpoints of classes from an assumed mean and stating them in class interval units. Only one additional column of values, fd^2, is required to compute the standard deviation by the

TABLE D-2
Calculation of the Standard Deviation for Grouped Data by the Step-Deviation Method: Weekly Earnings Data

Weekly Earnings	Number of Employees f	d	fd	fd^2
$160.00 and under $170.00	4	−3	−12	36
170.00 and under 180.00	14	−2	−28	56
180.00 and under 190.00	18	−1	−18	18
190.00 and under 200.00	28	0	0	0
200.00 and under 210.00	20	1	20	20
210.00 and under 220.00	12	2	24	48
220.00 and under 230.00	4	3	12	36
	100		−2	214

$$\bar{X}_a = \$195$$

$$s = (i)\sqrt{\frac{\Sigma fd^2 - \frac{(\Sigma fd)^2}{n}}{n-1}} = (\$10)\sqrt{\frac{214 - \frac{(-2)^2}{100}}{99}} = \$10(1.47)$$

$$s = \$14.70.$$

step-deviation method as compared to the corresponding arithmetic mean computation given in Table D-1. The formula for the step-deviation method is given in Equation D.2. All of the symbols have the same meaning as in Equation D.1 for the arithmetic mean and the computation in Table D-1.

Step-Deviation Method for the Sample Standard Deviation (Grouped Data)

(D.2)
$$s = (i)\sqrt{\frac{\Sigma fd^2 - \dfrac{(\Sigma fd)^2}{n}}{n-1}}$$

The use of Equation D.2 is illustrated in Table D-2.
The same assumed mean, $\bar{X}_a = \$195$, was used in Table D-2 as in the calculation of the arithmetic mean given in Table D-1.

Solutions to Odd-Numbered Exercises

CHAPTER 1

Section 1.6

1. a. A possible table is the following:

Weekly Commission	Number of Representatives
$100.00 and under $125.00	1
125.00 and under 150.00	1
150.00 and under 175.00	2
175.00 and under 200.00	5
200.00 and under 225.00	4
225.00 and under 250.00	3
250.00 and under 275.00	3
275.00 and under 300.00	2
Total	21

Note: The units for the variance are not shown in the solutions in this and subsequent chapters. These units are squares of those for the mean and standard deviation.

b.

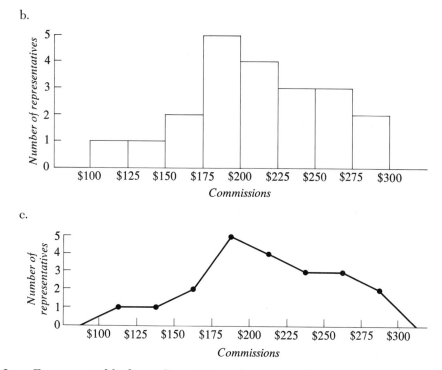

c.

3. a. Frequency table for credit positions of corporate clients:

Credit Outstanding (in Millions of Dollars)	Number of Companies	Midpoint (in Millions of Dollars)
0 and under 5	2	2.5
5 and under 10	4	7.5
10 and under 15	6	12.5
15 and under 20	5	17.5
20 and under 25	3	22.5
25 and under 30	4	27.5
30 and under 35	1	32.5

b.

Credit Outstanding	Number of Companies
Less than 5	2
Less than 10	6
Less than 15	12
Less than 20	17
Less than 25	20
Less than 30	24
Less than 35	25

5. a. The class intervals are not mutually exclusive, because the class limits are unclear and overlapping.
   b. There is a gap in values between classes; thus, the classes are not exhaustive.

Section 1.9

1. a. $\overline{X} = \$212.38$
   b. $\overline{X} = \$4437.50/21 = \$211.31$
   c. The answer to part (a) is precisely the mean. To calculate the arithmetic mean in frequency tables, the data are grouped into classes and then the midpoints are assumed to be the means of the values of items in each class. This procedure results in rounding-type error.

3. a. For strategy one, average cost of United Aerodynamics was $\$5275/125 = \$42.20$; average cost of Mitton Industries was $\$4675/125 = \$37.40$. For strategy two, average cost of United Aerodynamics was $\$5039/125 = \$40.31$; average cost of Mitton Industries was $\$4959/141 = \$35.17$.
   b. Strategy two achieved the lower average cost for both stocks.
   c. For strategy two, since the number of shares bought each time varied inversely with the stock price, higher weights were placed on the lower stock prices in the calculation of the average cost per share. This resulted in a lower overall average for each stock than in strategy one, where the average cost per share was the unweighted mean of the five share prices. That is, if equal weights are used in computing a weighted arithmetic mean, the resulting figure is the same as the unweighted mean.

5. a. $\dfrac{(3.6 + 3.8 + 2.5 + 6.5)}{4} = 4.1\%$
   b. $\overline{X}_w = (4118 + 8161 + 5158 + 54{,}805)/1{,}378{,}637 = 72{,}242/1{,}378{,}637 = 5.2\%$
   c. In the weighted average of part (b), the greatest weight is applied to the highest percentage unemployed, 6.5% (for County D). This pulls the weighted mean above the unweighted mean of part (a).

Section 1.12

1. Mean $= \frac{37}{12} = 3.08$ projects; mode $= 2$ projects; median $= 3$ projects

3. Agree, since in all three cases it is quite logical to assume that there would be extreme values at the upper end of the scale and that such values would tend to make the arithmetic mean larger than the median.

5. a.

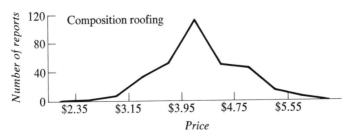

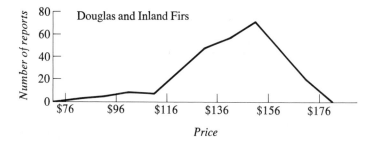

b. Composition roofing: Median class is $3.95 and under $4.35; modal class is $3.95 and under $4.35. Douglas and Inland Firs: Median class is $136 − $145; modal class is $146 − $155.

c.

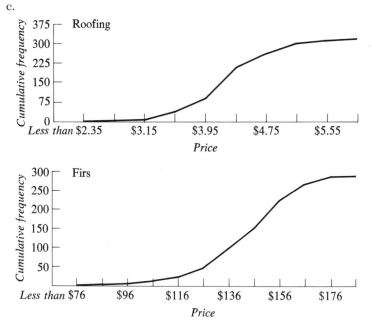

d. The distribution of composition roofing prices is nearly symmetrical, and the distribution of fir prices is skewed to the left.

Section 1.17

1. a. Both stocks have mean and median prices of $25. Thus, there is no difference simply on the basis of measures of central tendency.

   b. s (Highfly) $= \sqrt{164/4} = \$6.40$

   s (Stabil) $= \sqrt{20/4} = \$2.24$

   The standard deviation is a measure of price fluctuation. Hence, it can be considered a measure of the risk associated with the stock.

3. a. Using the short-cut formulas given in Appendix D, we have

   $\bar{X} = 615 + \frac{45}{150}(30) = 624$

$$s = 30 \sqrt{\frac{499 - 45^2/150}{149}} = 54.15$$

b. Yes, the variability in weights and grade averages can be compared in terms of the coefficient of variation, which is a relative measure of dispersion. The coefficient of variation for the test scores would be

$$CV = s/\bar{X} = 54.15/624 = 8.68\%$$

5. a. Tiny Tot: arithmetic mean $= \bar{X} = \$1.10$; standard deviation $= s = \sqrt{1.22/(10 - 1)} = \$0.37$. Gigantic Game: arithmetic mean $= \bar{X} = \$7.80$; standard deviation $= s = \sqrt{6.28/(10 - 1)} = \$0.84$. Gigantic Game Corporation earnings per share showed greater absolute variation.
   b. Tiny Tot: coefficient of variation $= s/\bar{X} = \$0.37/\$1.10 = 33.6\%$. Gigantic Game: coefficient of variation $= s/\bar{X} = \$0.84/\$7.80 = 10.77\%$. Tiny Tot earnings per share showed relatively greater variation than Gigantic Game, according to the coefficient of variation.

CHAPTER 2

Section 2.1

1. {11110, 11101, 11011, 10111, 01111}

3. {(F. B. Richgood, General Thrills), (F. B. Richgood, Pacific Pie), (F. B. Richgood, Multiplex), (General Thrills, Pacific Pie), (General Thrills, Multiplex), (Pacific Pie, Multiplex)}

5.

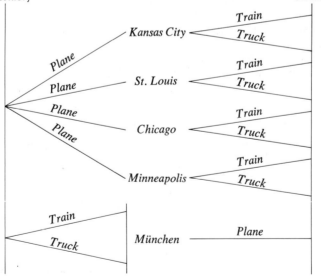

7. The events in (a), (b), and (e) are mutually exclusive; those in (c) and (d) are not mutually exclusive.

Section 2.2

1. a. No. $P(A$ and $B)$ must be 0 for A and B to be mutually exclusive events.
   b. Yes. $P(A$ and $B) = P(A)P(B) = (0.5)(0.4) = (0.2)$

3. a. $P(\text{electric stove or electric dryer}) = P(\text{electric stove}) + P(\text{electric dryer}) - P(\text{electric stove and electric dryer}) = 0.40 + 0.30 - 0.25 = 0.45$
   b. $P(\text{electric stove}|\text{electric dryer}) = P(\text{electric stove and electric dryer})/P(\text{electric dryer}) = 0.25/0.30 = 0.83$, but $P(\text{electric stove}) = 0.40$. Since $0.40 \neq 0.83$, the events are not independent.

5. a. $1 - \frac{1}{8} - \frac{7}{12} - \frac{1}{4} = \frac{1}{24}$
   b. $P(A) = \frac{1}{8} + \frac{7}{12} = \frac{17}{24}$; $P(B) = \frac{1}{8} + \frac{1}{4} = \frac{3}{8}$; $P(A$ and $B) = \frac{1}{8} \neq P(A)P(B)$. Hence, A and B are *not* independent.

7. a. $\frac{18}{38} = \frac{9}{19}$ d. $18{:}20 = 9{:}10$
   b. $\frac{1}{38}$ e. $1{:}37$
   c. $\frac{18}{38} = \frac{9}{19}$ f. Win: $\frac{18}{38} = \frac{9}{19}$
                                                         Lose: $\frac{20}{38} = \frac{10}{19}$

9. Let S_1 = sells to first client, S_2 = sells to second client, S_3 = sells to third client, S_4 = sells to fourth client. Then $P(S_1) = P(S_2) = P(S_3) = P(S_4) = 0.4$
   a. $P(S_1)P(S_2)P(S_3)P(S_4) = (0.4)(0.4)(0.4)(0.4) = 0.0256$
   b. $P(\bar{S}_1)P(\bar{S}_2)P(\bar{S}_3)P(\bar{S}_4) = (0.6)(0.6)(0.6)(0.6) = 0.1296$
   c. $P(S_1\bar{S}_2\bar{S}_3\bar{S}_4$ or $\bar{S}_1S_2\bar{S}_3\bar{S}_4$ or $\bar{S}_1\bar{S}_2S_3\bar{S}_4$ or $\bar{S}_1\bar{S}_2\bar{S}_3S_4) = (0.4)(0.6)(0.6)(0.6) + (0.6)(0.4)(0.6)(0.6) + (0.6)(0.6)(0.4)(0.6) + (0.6)(0.6)(0.6)(0.4) = 4(0.0864) = 0.3456$

11. Probability of all accepting the position is $(0.65)^8 = 0.0319$. Thus, the probability of not overhiring is 0.9681.

13. There are 700 workers in favor of the proposal; 460 of those are blue-collar workers. The total number of blue-collar workers is 650; thus, 190 blue-collar workers are against the proposal. There are 110 white-collar workers opposed to the proposal.
    a. $\frac{190}{1000} = \frac{19}{100}$ b. $\frac{700}{1000} = \frac{7}{10}$ c. $\frac{460}{650} = \frac{46}{65}$
    d. Dependent. Let A_1 = for proposal, A_2 = against proposal, B_1 = white-collar, B_2 = blue-collar. Then

    $$P(A_1 \text{ and } B_1) \neq P(A_1)P(B_1)$$
    $$240/1000 \neq (700/1000)(350/1000)$$

    Dependence of opinion about the proposal on job type is shown by the slightly higher proportion of blue-collar workers (0.71) favoring the proposal than of white-collar workers (0.69).

15. a. $\frac{190}{600} = \frac{19}{60}$ b. $\frac{200}{600} = \frac{1}{3}$ c. $\frac{55}{180} = \frac{11}{36}$
    d. No. The numbers would have been as follows.

Opinion	East	Region Midwest	West	Total
Opposed	45	60	75	180
Not opposed	105	140	175	420
Total	150	200	250	600

Section 2.3

1. $\dfrac{(0.7)(0.4)}{(0.7)(0.4) + (0.3)(0.8)} = 0.54$

3. $\dfrac{(0.85)(0.9)}{(0.85)(0.9) + (0.15)(0.2)} = 0.962$

5. $\dfrac{(0.4)(0.1)}{(0.4)(0.1) + (0.5)(0.4) + (0.1)(0.8)} = 0.125$

Section 2.4

1. a. Any combination of four of the twelve colors.

   b. $\dbinom{12}{4} = 495$

   c. $\dbinom{11}{3} = 165$

   d. $\frac{165}{495} = \frac{1}{3}$, or $\frac{4}{12} = \frac{1}{3}$

   e. $\dfrac{\dbinom{10}{4}}{\dbinom{12}{4}} = \dfrac{210}{495} = 0.424$

3. $\dbinom{7}{3} = 35$

5. $\dbinom{8}{3}\dbinom{6}{3} = (56)(20) = 1120$

   a. $\dbinom{8}{6} = 28$

   b. $\dbinom{6}{6} = 1$

7. a. $(7)(12) = 84$
   b. $(7)(12)(11) = 924$

9. $\dfrac{10!}{5!5!} = 252$

11. $\dbinom{13}{2}\dbinom{6}{1}\dbinom{4}{1}\dbinom{8}{2} = 52{,}416$

13. a. $\dbinom{5}{3} = 10$

    b. $(5)(4)(3) = 60$

    c. $\dfrac{(3)(2)(1)}{(5)(4)(3)} = \dfrac{6}{60} = \dfrac{1}{10}$, or $\dfrac{1}{\dbinom{5}{3}} = \dfrac{1}{10}$

CHAPTER 3

Section 3.1

Note: Although many solutions may be obtained from the Appendix tables, the theoretical solutions are usually shown here.

1. a. Continuous   c. Discrete
   b. Continuous   d. Discrete
   Although (a) and (b) are conceptually continuous random variables, in actual use they are discrete, because of the limited accuracy of the measuring instruments.

3. a. Yes
   b. Yes
   c. No; $f(x)$ cannot be a negative number
   d. No; $\sum_x f(x)$ must total to 1

5. Answers will depend on the subjective probabilities assigned. Note that these probabilities must sum to 1.

7. $f(1) = \frac{1}{25}; f(2) = \frac{3}{25}; f(3) = \frac{5}{25} = \frac{1}{5}; f(4) = \frac{7}{25}; f(5) = \frac{9}{25}$

Section 3.3

3. $f(x) = \frac{1}{51}; x = 150, 151, \ldots, 200.$
   a. $\frac{30}{51} = \frac{10}{17}$
   b. $\frac{5}{51}$
   c. $\frac{10}{51}$
   d. $\frac{1}{51}$

Section 3.4

1. a. Binomial distribution; the parameters are p and $n = 10$.
   b. Binomial distribution; the parameters are p and $n = 30$.

3. a. $\binom{8}{0} (0.70)^8 (0.30)^0 = 0.0576$

   b. $\binom{8}{8} (0.70)^0 (0.30)^8 = 0.0001$

   c. $\binom{8}{3} (0.70)^5 (0.30)^3 = 0.2541$

   d. $\sum_{x=3}^{8} \binom{8}{x} (0.70)^{8-x} (0.30)^x = 0.4482$

5. $\binom{5}{1} (0.80)^4 (0.20)^1 = 0.4096$

7. The probability that all 20 would be perfect is 0.1216. Since this does not seem excessively improbable, we would not agree with Jones.

9. a. $\displaystyle\sum_{x=9}^{15} \binom{15}{x} (0.50)^{15-x}(0.50)^x = 0.3036$

   b. $\displaystyle\sum_{x=12}^{15} \binom{15}{x} (0.50)^{15-x}(0.50)^x = 0.0176$

   c. $\displaystyle\sum_{x=0}^{7} \binom{15}{x} (0.50)^{15-x}(0.50)^x = 0.5000$

11. $\displaystyle\sum_{x=2}^{9} \binom{9}{x} (0.75)^{9-x}(0.25)^x = 0.6997$

13. $\displaystyle\binom{4}{0} (0.85)^4(0.15)^0 = 0.5220$

15. The distribution is symmetrical when p = 0.5, skewed to the right when p = 0.2, and skewed to the left when p = 0.8.

Section 3.5

1. a. $\dfrac{8!}{4!\ 1!\ 1!\ 1!\ 1!} (0.75)^4(0.10)^1(0.03)^1(0.07)^1(0.05)^1 = 0.0056$

   b. $\dbinom{8}{4} (0.75)^4(0.25)^4 = 0.0865$

   c. Multinomial in (a), binomial in (b)

3. $\dfrac{6!}{2!\ 2!\ 2!} (0.50)^2(0.30)^2(0.20)^2 = 0.081$

Section 3.6

1. $\dfrac{\dbinom{10}{4}\dbinom{5}{0}}{\dbinom{15}{4}} = \dfrac{210}{1365} = 0.1538$

3. $\dfrac{\dbinom{5}{2}\dbinom{5}{2}}{\dbinom{10}{4}} = 0.4762$

5. a. $\dfrac{\dbinom{4}{1}\dbinom{4}{1}\dbinom{4}{1}\dbinom{4}{1}\dbinom{4}{1}\dbinom{4}{1}}{\dbinom{24}{6}} = \dfrac{4096}{134{,}596} = 0.0304$

   b. $\dfrac{\dbinom{4}{4}\dbinom{20}{2}}{\dbinom{24}{6}} = \dfrac{190}{134{,}596} = 0.0014$

c. $\dfrac{\binom{4}{0}\binom{20}{6}}{\binom{24}{6}} = \dfrac{38{,}760}{134{,}596} = 0.2880$

7. a. $\dfrac{\binom{9}{5}\binom{9}{0}}{\binom{18}{5}} = \dfrac{126}{8568} = 0.0147$

b. $\dfrac{\binom{4}{0}\binom{14}{5}}{\binom{18}{5}} = \dfrac{2002}{8568} = 0.2337$

c. $\dfrac{\binom{9}{2}\binom{5}{2}\binom{4}{1}}{\binom{18}{5}} = \dfrac{1440}{8568} = 0.1681$

d. $\dfrac{\binom{9}{0}\binom{5}{3}\binom{4}{2}}{\binom{18}{5}} = \dfrac{60}{8568} = 0.0070$

Section 3.7

1. $\dfrac{4^0 e^{-4}}{0!} = 0.018$

3. a. $1 - \sum\limits_{x=0}^{7} \dfrac{6^x e^{-6}}{x!} = 0.256$

b. $\dfrac{6^0 e^{-6}}{0!} = 0.002$

c. $(0.256)^5 = 0.001$

5. $1 - \sum\limits_{x=0}^{10} \dfrac{6^x e^{-6}}{x!} = 0.043$

7. a. $\dfrac{20^{15} e^{-20}}{15!} = 0.052$

b. $1 - \sum\limits_{x=0}^{15} \dfrac{20^x e^{-20}}{x!} = 0.843$

9. a. $1 - f(0) = 1 - \left(\dfrac{1.9^0 e^{-1.9}}{0!}\right) = 1 - 0.15 = 0.85$

b. $\binom{5}{1}(0.15)^4 (0.85)^1 = 0.0022$

11. a. The probability that exactly one will not work is

$$\binom{20}{1}(0.95)^{19}(0.05)^1 = 0.3773$$

The probability that at least two will not work is

$$\sum_{x=2}^{20} \binom{20}{x}(0.95)^{20-x}(0.05)^x = 0.2642$$

b. The corresponding probabilities are

$$\frac{1^1 e^{-1}}{1!} = 0.368$$

and

$$\sum_{x=2}^{20} \frac{1^x e^{-1}}{x!} = 0.264$$

Section 3.11

1. $E(X) = (+\$4.26)(0.04) + (-\$.15)(0.96) = \$0.0264$

3. $f(x) = \frac{1}{11}$; $x = 10, 11, \ldots, 20$; $E(X) = \Sigma\, xf(x) = 15$ items.

5. $E(\text{gain}) = (\$36)(0.998) + (-\$14,964)(0.002) = \$6$.
   No. On any single policy, the company will either earn the premium of \$36 or pay out \$14,964 (\$15,000 − \$36). The \$6 represents the average return per policy if an infinite number of policies were issued.

7. $X_i = \begin{cases} +\$125 - \$55 = \$70 \text{ if numbers 1 through 12 occur on } i\text{th roll} \\ -\$60 \text{ if numbers other than 1 through 12 occur on } i\text{th roll} \end{cases}$

   $E(X_1) = (\$70)(\frac{12}{38}) + (-\$60)(\frac{26}{38}) = \$18.95$

   $X_1 + X_2 + \cdots + X_{500} = \text{Total profit (loss)}$

   $E(X_1 + X_2 + \cdots + X_{500}) = E(X_1) + E(X_2) + \cdots + E(X_{500}) = (500)E(X_i) = 500(-\$18.95) = -\$9,475$

9. $\mu = E(X) = \Sigma\, xf(x) = 0.0500$; $\sigma^2(X) = \Sigma\,(x - \mu)^2 f(x) = 0.000592$; $\sigma(X) = 0.0243$; expected percentage of users affected: 5%; standard deviation: 2.43%

11. $E(A) = (-\$5000)(0.1) + (0)(0.2) + (\$3000)(0.4) + (\$6000)(0.3)$
    $= \$2500$
    $E(B) = (0)(0.2) + (\$1000)(0.3) + (\$3000)(0.3) + (\$5000)(0.2)$
    $= \$2200$
    $\sigma^2(A) = [(-\$5000)^2(0.1) + (\$3000)^2(0.4) + (\$6000)^2(0.3)] - (\$2500)^2$
    $= 10,650,000$
    $\sigma^2(B) = [(\$1000)^2(0.3) + (\$3000)^2(0.3) + (\$5000)^2(0.2)] - (\$2200)^2$
    $= 3,160,000$
    $\sigma(A) = \$3263.43$
    $\sigma(B) = \$1777.64$

13. $E(\text{total time}) = E(I + II + III + IV) = E(I) + E(II) + E(III) + E(IV)$
$= 5 + 14 + 8 + 3 = 30$ weeks
$\sigma^2(\text{total time}) = \sigma^2(I + II + III + IV) = \sigma^2(I) + \sigma^2(II) + \sigma^2(III) + \sigma^2(IV)$
$= 4 + 25 + 9 + 1 = 39$
$\sigma(\text{total time}) = \sqrt{39} = 6.24$ weeks

Section 3.12

1.

y \\ x	0	1	2	3
0	0.03	0.09	0.12	0.06
1	0.03	0.09	0.12	0.06
2	0.02	0.06	0.08	0.04
3	0.02	0.06	0.08	0.04

3. A_1 = under 2000 cases, A_2 = at least 2000 cases, B_1 = ordinary bottle, B_2 = ordinary cans, B_3 = flip-top cans, B_4 = screw-top bottles.

a. $P(A_2 \text{ and } B_3) = \dfrac{10}{200} = \dfrac{1}{20}$

$P(B_3) = \dfrac{40}{200} = \dfrac{1}{5}$

$P(A_2|B_3) = \dfrac{10}{40} = \dfrac{1}{4}$

b. $P(A_1) = \dfrac{140}{200} = 0.7$

$P(A_2) = \dfrac{60}{200} = 0.3$

$P(B_1) = P(B_2) = \dfrac{60}{200} = 0.3$

$P(B_3) = P(B_4) = \dfrac{40}{200} = 0.2$

c. $P(A_1|B_1) = \dfrac{45}{60} = 0.75, \qquad P(A_2|B_1) = \dfrac{15}{60} = 0.25$

d. Yes; $f(\text{sales})g(\text{packaging}) \neq f(\text{sales, packaging})$

e.

Sales	Ordinary Bottles	Ordinary Cans	Flip-Top Cans	Screw-Top Bottles	Total
Under 2000 cases	42	42	28	28	140
At least 2000 cases	18	18	12	12	60
	60	60	40	40	200

5. Yes; $P(\text{for}|\text{skilled}) = \frac{275}{600} \neq P(\text{for}) = \frac{500}{1400}$. A higher proportion of skilled employees than of unskilled employees is in favor of the labor proposal.

CHAPTER 4

Section 4.2

1. See the discussion in Section 4.2.

3. See the discussion in Section 4.2.

Section 4.4

1. See the discussion in Section 4.4.

3. Conceptually, the population of interest is all male consumers in the New York City area. We are interested in the proportion of this population who possess the characteristic "would buy a certain type of man's suit." As a practical matter, the population might be defined as those male consumers who shop in the stores and departments of stores that sell this type of man's suit. This is a dynamic population, since men enter and leave this population primarily because of shifting shopping patterns and changing tastes.

5. a. Conceptually, all areas other than low-income urban areas would constitute a control group. The specific control group(s) used would depend on the types of comparison desired. For example, urban areas other than low-income ones might be used.
   b. Perhaps the most direct control group would be a part of the plant in which safety lights are not installed and in which similar types of work and working conditions prevail.

7. Random errors are attributable to chance fluctuations of sampling and tend to decrease as sample size increases. On the other hand, systematic errors are primarily attributable to faulty experimental design and measurement. Since these errors are not associated with the number of elements sampled, they do not decrease as sample size increases. Systematic error can and should be eliminated.

Section 4.5

1. a. The universe sampled was all those parents with children in public schools in Lower Fenwick. However, although a simple random sample of pupils was drawn, this did not constitute a simple random sample of the parent population. Parents with more than one child in the public schools had a higher probability of inclusion in the sample than did parents with only one child in these schools.
   b. See part (a).
   c. We would not approve the universe studied. If the school board wished to ascertain voter opinion, it should have sampled the population of voters in Lower Fenwick, many of whom may not have had children in the

public schools. A relevant sampling frame would have been an up-to-date list of registered voters in Lower Fenwick.

3. a. False. With judgment sampling, no objective measure of random error is possible because of the nonrandom method of sampling.
b. False. Sampling can reduce the time spent and the likelihood of error.
c. True.

CHAPTER 5

Section 5.2

1. The number of defectives, X, can have the values 0, 1, 2, . . . , 10.

$$\mu(X) = 1 \qquad \mu_{\bar{p}} = 0.1$$
$$\sigma^2(X) = 0.9 \qquad \sigma^2_{\bar{p}} = 0.009$$

3.

$\mu_{n\bar{p}}$	$\sigma_{n\bar{p}}$
4 persons	$\sqrt{(10)(0.4)(0.6)} = 1.55$ persons
8 persons	$\sqrt{(20)(0.4)(0.6)} = 2.19$ persons
16 persons	$\sqrt{(40)(0.4)(0.6)} = 3.10$ persons

5. a. $\mu_{\bar{p}} = p = 0.05$

$$\sigma_{\bar{p}} = \sqrt{pq/n} = \sqrt{(0.05)(0.95)/20} = 0.0487$$

b. $\mu_{\bar{p}} = p = 0.05$

$$\sigma_{\bar{p}} = \sqrt{pq/n} = \sqrt{(0.05)(0.95)/40} = 0.0345$$

Section 5.4

1. See pages 98 to 99 in Chapter 3.

3. a. 0.5   d. 0.2302
b. 0.0228   e. 0.6554
c. 0.8664   f. 0.7881

5. a. True, because

$$P\left(z < \frac{\$16{,}000 - \$18{,}000}{\$1000}\right) = P\left(z > \frac{\$20{,}000 - \$18{,}000}{\$1000}\right)$$

b. False, because $P\left(\dfrac{\$18{,}000 - \$18{,}000}{\$1000} \leqslant z \leqslant \dfrac{\$19{,}500 - \$18{,}000}{\$1000}\right)$

$$= P\left(\frac{\$16{,}500 - \$18{,}000}{\$1000} \leqslant z \leqslant \frac{\$18{,}000 - \$18{,}000}{\$1000}\right)$$

c. True, because

$$P\left(z < \frac{\$18{,}000 - \$18{,}000}{\$1000}\right) = P\left(z > \frac{\$18{,}000 - \$18{,}000}{\$1000}\right)$$

d. False, because $P\left(z < \frac{\$20{,}000 - \$18{,}000}{\$1000}\right) = P(z < 2) = 97.72\%$

7. a. $P\left(z > \frac{0 - 0}{1}\right) = 0.5$

b. $P(\epsilon_t \leqslant -1.50) + P(\epsilon_t \geqslant 1.50) = P\left(z \leqslant \frac{-1.50 - 0}{1}\right)$

$$+ P\left(z \geqslant \frac{1.50 - 0}{1}\right) = 0.1336$$

c. $P(P_t - P_{t-1} \leqslant 1.75) = P(\epsilon_t \leqslant 1.75) = P\left(z \leqslant \frac{1.75 - 0}{1}\right) = 0.9599$

9. a. 0.5

b. $P\left(z > \frac{13 - 12}{0.5}\right) = 0.0228$

c. $P\left(\frac{11.5 - 12}{0.5} \leqslant z \leqslant \frac{13 - 12}{0.5}\right) = 0.8185$

11. a. $P\left(\frac{1.5 - 2}{\sqrt{(0.1)(0.9)(20)}} \leqslant z \leqslant \frac{2.5 - 2}{\sqrt{(0.1)(0.9)(20)}}\right) = P(-0.37 \leqslant z \leqslant 0.37)$

$$= 0.2886$$

b. $P(z > 0.37) = 0.3557$

c. $P(z < 0.37) = 0.3557$

13. $P\left(z > \frac{0.07 - 0.05}{\sqrt{(0.05)(0.95)/400}}\right) = 0.0336$

15. $P(z > 2.33) = 0.01$. Hence,

$$\frac{x - 500}{\sqrt{(0.2)(0.8)(2500)}} = 2.33$$

$$x = 546.6$$

Therefore, the store must stock 547 boxes.

Section 5.5

1. a. $P\left(z > \frac{303 - 300}{30/\sqrt{900}}\right) = 0.00135$

b. Yes; it is a reasonable assertion.

3. $P\left(z \geqslant \frac{48 - 44}{16/\sqrt{64}}\right) = P(z \geqslant 2) = 0.0228$

5. a. $\mu_{\bar{x}} = \$502,680$; $\sigma_{\bar{x}} = \dfrac{\$78,000}{\sqrt{100}} \sqrt{\dfrac{257 - 100}{257 - 1}} \approx \$6108$

   b. $\mu_{\bar{x}} = \$626,540$; $\sigma_{\bar{x}} = \dfrac{\$78,000}{\sqrt{100}} \sqrt{\dfrac{8216 - 100}{8216 - 1}} \approx \$7753$

   c. The standard deviation in (a) is smaller because the population is smaller than in (b).

7. $P\left(z \geq \dfrac{30 - 28}{\sqrt{81/100}}\right) = 0.0132$

   No, because of the Central Limit Theorem.

9. $\$1200 \pm 3(\$200)$, or $\$600$ to $\$1800$

   $\$1200 \pm 3\left(\dfrac{\$200}{\sqrt{25}}\right)$, or $\$1080$ to $\$1320$

   $\$1200 \pm 3\left(\dfrac{\$200}{\sqrt{100}}\right)$, or $\$1140$ to $\$1260$

11. a. $\mu_{\bar{x}} = \$38,900$; $\sigma_{\bar{x}} = \dfrac{\$4210}{\sqrt{25}} \sqrt{\dfrac{100 - 25}{100 - 1}} = \$732.87$

    b. $\mu_{\bar{x}} = \$38,900$; $\sigma_{\bar{x}} = \dfrac{\$4210}{\sqrt{50}} \sqrt{\dfrac{100 - 50}{100 - 1}} = \$423.12$

    c. $\mu_{\bar{x}} = \$38,900$; $\sigma_{\bar{x}} = 0$, since only one sample is possible.

CHAPTER 6

Section 6.2

1. a. Interval estimate
   b. Estimators
   c. Point estimate
   d. Interval estimate

3. a. False. Unbiasedness is only one criterion of goodness of estimation. In one use of the term, a *best estimator* is the unbiased estimator with the smallest variance.
   b. True.
   c. False. If two competing estimators are both *unbiased,* then the one with the smaller variance (for a given sample size) is said to be relatively more efficient.

Section 6.3

1. Since $\bar{x}$ can take on different values depending on which sample is drawn from a population with mean μ, $\bar{x}$ is a random variable. As such, it has a

probability distribution, and the standard deviation of $\bar{x}$ is a measure of the dispersion of that probability distribution. This standard deviation is a measure of the spread of $\bar{x}$ values around μ, the value to be estimated. It can thus be interpreted as a measure of errors due to sampling. From this viewpoint, $\sigma_{\bar{x}}$ is a measure of the error involved in using $\bar{x}$ as an estimator of μ.

3. Only (c) and (f) are valid.

5. $1000 \pm 1.96(200/\sqrt{100})$, or 960.8 hours and 1039.2 hours

7. $0.6 \pm 2\sqrt{(0.6)(0.4)/1000}$, or from 0.569 to 0.631 in favor

9. a. $(\$5000 - \$5500) \pm (1.65) \sqrt{\dfrac{(\$1000)^2}{40} + \dfrac{(\$1000)^2}{40}} = -\$500$

   $\pm 1.65(\$223.61)$, or from $-\$868.96$ to $-\$131.04$

   b. $(0.20 - 0.15) \pm (1.65) \sqrt{\dfrac{(0.15)(0.85)}{40} + \dfrac{(0.20)(0.80)}{40}} = 0.50 \pm 1.65(0.085)$,

   or from -0.09 to 0.19

11. No, because, for example, 95% confidence intervals for pollster A and pollster B would overlap. These intervals, which are $0.402 \leqslant p \leqslant 0.598$ for A and $0.452 \leqslant p \leqslant 0.648$ for B, overlap in the interval from 0.452 to 0.598.

13. a. $\$23.50 \pm 2.58(\$6.00/\sqrt{100})$, or from \$21.95 to \$25.05. Note that the finite population correction is equal to $\sqrt{(10,000 - 100)/(10,000 - 1)} = 0.995 \approx$ 1. Since the finite population correction is approximately equal to 1, it has not been included in the calculation of the standard error of the mean.
    b. Note that the finite population correction is equal to $\sqrt{(10,000 - 100)/(10,000 - 1)} = 0.995 \approx 1$. Since the finite population correction is approximately equal to 1, it has not been included in the calculation of the standard error of the mean.

Section 6.4

1. a. True, where $s_{\bar{x}} = s/\sqrt{n}$ and $s = \sqrt{\Sigma (x - \bar{x})^2/(n - 1)}$. The t value is determined for $\nu = n - 1$ degrees of freedom.
   b. True. If the statistics are unbiased, then the more efficient statistic would have a smaller standard error.
   c. False. The 95% confidence means that 95% of all possible samples would yield intervals including the true population mean strength of this type of wood beam. A confidence interval does not refer to a range of values within any particular sample.

3. a. $9 \pm 2.306(4/\sqrt{9})$, or from 5.93 weeks to 12.07 weeks ($t = 2.306$ for $\nu = 8$ and a 0.95 confidence coefficient).
   b. As stated in the solution to 1(c) above, the 95% confidence refers to the percentage of all possible samples that would result in intervals including the true population mean. Since the company has already drawn a sample, we can only say that the interval estimate is either correct (one of the 95%) or incorrect (one of the 5%).

5. a. $51 \pm 3.355(16.10/\sqrt{9}) = 51 \pm 18.02$, or from 32.98 minutes to 69.02 minutes.

b. It is assumed in the derivation of the t distribution that the underlying population is normally distributed, and this assumption would be violated by a highly skewed population distribution. Hence, the range set up in part (a) would not have exactly a 99% confidence coefficient associated with it.

Section 6.5

1. $2.05\sigma_{\bar{p}} = 0.04$; $\sqrt{(0.50)(0.50)/n} = 0.0195$; $n = 657$. This would represent a "conservative" estimate for sample size, since $p = 0.50$ and $q = 0.50$ were used. If there were prior knowledge concerning the proportion of citizens favoring federal controls, this information might be used to estimate the values of p and q.

3. a. $2.33\sigma_{\bar{p}} = 0.04$

   $(2.33)\sqrt{(0.20)(0.80)/n} = 0.04$

   $n = 543$

   b. $0.15 \pm (2.58)\sqrt{(0.15)(0.85)/400} = 0.15 \pm 2.58(0.018)$, or from 0.104 to 0.196

5. $(1.96)\sqrt{(0.25)(0.75)/n} = 0.03$

   $n = 800$ companies

7. By increasing the sample size and by decreasing the level of confidence.

9. $(2.33)\sqrt{(0.05)(0.95)/n} = 0.005$

   $n = 10,326$

CHAPTER 7

Section 7.1

1. A parameter is a measure computed from a population, whereas a statistic is a measure computed from a sample drawn from the population. Hence, a statistic takes on various values depending on which sample is drawn.

3. a. In a one-tailed test, rejection of the null hypothesis takes place in only one tail of the sampling distribution. In a two-tailed test, rejection of the null hypothesis takes place in both tails of the sampling distribution.
   b. A Type I error is the error of rejecting H_0 when in fact H_0 is true. A Type II error is the error of accepting H_0 when in fact H_0 is not true.

Section 7.2

1. The power of a test is the probability of rejecting H_0 when H_1 is the true state of nature. Thus, power $= 1 - \beta$, where β is the probability of a Type II error.

3. a. Disagree. The operating characteristic curve is graphed with the probability of accepting the null hypothesis on the vertical axis and the possible values of the parameter on the horizontal axis.

   b. Disagree. The quantity β represents the probability of incorrectly accepting the null hypothesis.

5. a. H_0: $\mu \geq 40$
   H_1: $\mu < 40$

   $\alpha = 0.50$

   Decision Rule
   1. Reject H_0 if $\bar{x} < 37.52$ hours
   2. Accept H_0 otherwise
   Critical value $= 40 - 1.65(15/\sqrt{100}) = 37.52$ hours

   Since $\bar{x} = 38.5$ hours > 37.52 hours, accept H_0 and accept the shipment.
   b. Power curve:

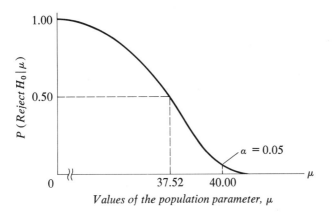

Values of the population parameter, μ

7. No. Since the assistant accepted the null hypothesis, the possible error is Type II. Therefore, if β was large, the evidence would not be "strong" in favor of the null hypothesis H_0: $\mu \geq \$10,000$.

9. a. H_0: $\mu_2 \leq 30,000$ miles
   H_1: $\mu_2 > 30,000$ miles

   $\alpha = 0.02$

   Decision Rule
   1. Reject H_0 if $\bar{x}_2 > 30,820$
   2. Accept H_0 if $\bar{x}_2 \leq 30,820$
   Critical value $= 30,000 + 2.05(4000/\sqrt{100}) = 30,000 + 820 = 30,820$

   b. 1. $P(\bar{x}_2 \leq 30,820 | \mu_2 = 30,200) = P\left(z \leq \dfrac{30,820 - 30,200}{4000/\sqrt{100}}\right) = P(z \leq 1.55)$

      $= 0.9394$
   2. Incorrectly accepting the null hypothesis is a Type II error.

11. H_0: $\mu = 6$
   H_1: $\mu \neq 6$

   $\alpha = 0.05$

   Critical values $= 6 \pm 1.96(2/\sqrt{100})\sqrt{1 - (100/500)} = 6 \pm 0.351$

   Decision Rule
   1. Reject H_0 if $\bar{x} < 5.649$ or $\bar{x} > 6.351$
   2. Accept H_0 if $5.649 \leqslant \bar{x} \leqslant 6.351$

   Since $\bar{x} = 5.5 < 5.649$, we reject H_0 and conclude that there has been a change.

13. a. $P(\text{accept } H_0 | \mu = \$61)$

   $$= P\left(\frac{\$58.72 - \$61}{\$1} < z < \frac{\$61.28 - \$61}{\$1}\right) = 0.5990$$

   b. The power of the test when $\mu = \$61$ is the probability of rejecting the null hypothesis, H_0: $\mu = \$60$, for $\mu = \$61$. The power for $\mu = \$61$ is $1 - \beta = 1 - 0.5990 = 0.4010$.

   c. $\$1.28 = z(\$20/\sqrt{400})$
   $z = 1.28$
   $\alpha = 0.20$ level of significance

15. Probability of Type I error $= P(\bar{p} \leqslant 0.15 | p = 0.20)$

   $$= P\left(z \leqslant \frac{0.15 - 0.20}{\sqrt{(0.2)(0.8)/200}}\right)$$

   $$= P(z \leqslant -1.77) = 0.0384$$

   The Type I error here is an incorrect decision to stop marketing the product.

17. H_0: $p \geqslant 0.05$
   H_1: $p < 0.05$

   $\alpha = 0.05$

   Critical value $= 0.05 - 1.65\sqrt{(0.05)(0.95)/1000} = 0.039$

   Decision Rule
   1. Reject H_0 if $\bar{p} < 0.039$
   2. Accept H_0 otherwise

   Decision: Reject H_0, since $0.036 < 0.039$, and thus decide that the system is effective.

19. $P\left(p \geqslant \dfrac{144}{400}\right) = P\left(z \geqslant \dfrac{0.36 - 0.30}{\sqrt{(0.30)(0.70)/400}}\right) = 0.0044$

   Yes, the bank concluded correctly.
   H_0: $p \leqslant 0.3$
   H_1: $p > 0.3$

   $\alpha = 0.05$

   Critical value $= 0.3 + (1.65)\sqrt{(0.3)(0.7)/400} = 0.338$

Decision Rule
    1. Reject H_0 if $\bar{p} > 0.338$
    2. Accept H_0 otherwise
Decision: Reject H_0, and thus open new bank

$f(\bar{p})$:

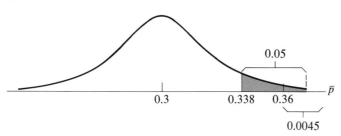

Note: The tail of the curve has been magnified to show the desired items more clearly.

21. H_0: $p \leqslant 0.35$
    H_1: $p > 0.35$

    $\alpha = 0.10$

    Critical value $= 0.35 + (1.28)\sqrt{(0.35)(0.65)/800} = 0.372$

    Decision Rule
        1. Reject H_0 if $\bar{p} > 0.372$
        2. Accept H_0 otherwise
    Decision: Reject H_0, since $\bar{p} = \frac{320}{800} = 0.4 > 0.372$; hence, conclude that the current sample percentage is significantly larger than 35%.

23. a. H_0: $p \leqslant 0.08$
      H_1: $p > 0.08$

     $\alpha = 0.05$

     Critical value $= 0.08 + (1.65)\sqrt{(0.08)(0.92)/200} = 0.1117$

     Decision Rule
        1. Reject H_0 if $\bar{p} > 0.1117$
        2. Accept H_0 otherwise
     Decision: Reject H_0, since $\bar{p} = 0.14$
   b. Yes, the manufacturer might actually be correct. This would represent a Type I error, or the incorrect rejection of the null hypothesis.

25. H_0: $p \geqslant 0.667$
    H_1: $p < 0.667$

    $\alpha = 0.05$

    Critical value $= 0.667 - (1.65)\left(\sqrt{\dfrac{(0.667)(0.333)}{90}}\right)\left(\sqrt{1 - \dfrac{90}{435}}\right) = 0.667 -$
    $0.073 = 0.594$

Decision Rule
  1. Reject H_0 if $\bar{p} < 0.594$
  2. Accept H_0 otherwise
Decision: Accept H_0, since $\bar{p} = \frac{57}{90} = 0.633 > 0.594$.   You cannot conclude that the veto will not be overridden.

Section 7.3

1. a. 5   c. 2
   b. 1   d. 4

3. a. $H_0: \mu_1 = \mu_2$
      $H_1: \mu_1 \neq \mu_2$

   $\alpha = 0.02$

   Critical value $= 0 \pm (2.33)\sqrt{(140)^2/100 + (120)^2/150} = 0 \pm 39.82$

   Decision Rule
     1. Reject H_0 if $\bar{x}_1 - \bar{x}_2 < -39.82$ or $\bar{x}_1 - \bar{x}_2 > 39.82$
     2. Accept H_0 if $-39.82 \leq \bar{x}_1 - \bar{x}_2 \leq 39.82$
   Decision: Reject H_0, since $1500 - 1450 = 50 > 39.82$

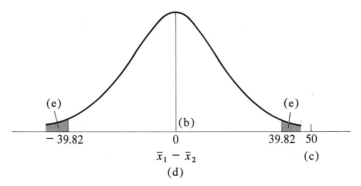

(e)   (e)
          (b)
   − 39.82   0   39.82   50
                 $\bar{x}_1 - \bar{x}_2$ (c)
                    (d)

5. a. $H_0: p_T = p_{PE}$ T = traders
      $H_1: p_T \neq p_{PE}$ PE = professors of economics

   Assume that $\alpha = 0.05$

   $$\bar{p} = \frac{135 + 154}{225 + 200} = \frac{289}{425} = 0.68$$

   Critical value $= 0 \pm (1.96)\sqrt{(0.68)(0.32)(\frac{1}{200} + \frac{1}{225})} = 0 \pm 1.96(0.045) = 0 \pm 0.088$

   Decision Rule
     1. Reject H_0 if $|\bar{p}_T - \bar{p}_{PE}| > 0.088$
     2. Accept H_0 if $|\bar{p}_T - \bar{p}_{PE}| \leq 0.088$

   $$\bar{p}_T = \frac{135}{225} = 0.60, \qquad \bar{p}_{PE} = \frac{154}{200} = 0.77$$

   Decision: Reject H_0, since $|\bar{p}_T - \bar{p}_{PE}| = |0.6 - 0.77| = 0.17 > 0.088$

b. The error of concluding that there is a difference in the proportions of foreign exchange traders and professors of economics who believe that flexible exchange rates will enhance international monetary stability, when in fact the proportions are the same.

7. H_0: $p_{NEW} \geq p_{OLD}$ (the new spray is not more effective)
H_1: $p_{NEW} < p_{OLD}$ (the new spray is more effective)

Assume that $\alpha = 0.01$

$$\bar{p} = \frac{86 + 74}{200 + 200} = \frac{160}{400} = 0.4$$

Critical value $= 0 - (2.33)\sqrt{(0.4)(0.6)(\frac{1}{200} + \frac{1}{200})} = 0 - 2.33(0.049) = -0.114$

Decision Rule

    1. Reject H_0 if $p_{NEW} - p_{OLD} < -0.114$
    2. Accept H_0 if $p_{NEW} - p_{OLD} \geq -0.114$

Decision: Accept H_0, since $p_{NEW} - p_{OLD} = 0.37 - 0.43 = -0.06 > -0.114$. The sample proportion p_{NEW} is not significantly lower than p_{OLD}; therefore, we cannot conclude that the new spray is more effective than the old spray.

Section 7.4

1. The test statistic should be a t statistic with 19 degrees of freedom.

3. a. $s^2 = \dfrac{14(10)^2 + 10(12)^2}{15 + 11 - 2} = 118.33$

$s_{\bar{x}_1 - \bar{x}_2} = \sqrt{118.33}(\sqrt{\frac{1}{15} + \frac{1}{11}}) = 10.878(0.397) = 4.32$

$t = \dfrac{60 - 55}{4.32} = 1.16$

For 24 degrees of freedom and $\alpha = 0.10$, the critical value of t is equal to 1.711. Since $1.16 < 1.711$, we conclude that there was not a statistically significant difference between the sample averages.

b. H_0: $\mu_1 - \mu_2 = 0$
H_1: $\mu_1 - \mu_2 \neq 0$

Decision Rule

    1. Reject H_0 if $t < -1.711$ or $t > 1.711$
    2. Accept H_0 if $-1.711 \leq t \leq 1.711$

5. $\bar{x} = 9$; $s = \sqrt{\dfrac{\Sigma (x - \bar{x})^2}{n - 1}} = \sqrt{\dfrac{34}{11}} = 1.76$; $n = 12$

H_0: $\mu = 10$
H_0: $\mu \neq 10$

For $12 - 1 = 11$ degrees of freedom, the critical value of t is 2.718.

Decision Rule

    1. Reject H_0 if $t < -2.718$ or $t > 2.718$
    2. Accept H_0 if $-2.718 \leq t \leq 2.718$

$s_{\bar{x}} = 1.76/\sqrt{12} = 0.508$; $t = (9 - 10)/0.508 = -1.97$. Since $-2.718 \leqslant -1.97 \leqslant 2.718$, we accept H_0 and conclude that the sample average is not significantly different from 10 (million dollars).

Section 7.6

1. Let μ_w and μ_{wo} be the mean population scores with payment and without payment, respectively.

H_0: $\mu_w = \mu_{wo}$
H_1: $\mu_w > \mu_{wo}$

$s_{\bar{d}} = 0.237/\sqrt{8} = 0.084$ points

Critical value of t is 1.895 for $\nu = 7$ in a one-tailed test.

Decision Rule
    1. Reject H_0 if $t > 1.895$
    2. Accept H_0 otherwise
Decision: Since $t = 0.06/0.084 < 1.895$, accept H_0. Conclude that payment does not affect grade average.

3. a. Let μ_A and μ_B equal the mean population rating for concern for production after and before the session, respectively.

H_0: $\mu_A = \mu_B$
H_1: $\mu_A > \mu_B$

$s_{\bar{d}} = 1.76/\sqrt{10} = 0.56$

The critical value of t is 2.262 for $\nu = 9$ in a one-tailed test.

Decision Rule
    1. Reject H_0 if $t > 2.262$
    2. Accept H_0 otherwise
Decision: Since $t = 1/0.56 = 1.79 < 2.262$, accept H_0. Conclude that the session does not have a positive effect on managerial ratings for production concern.

  b. Let μ_A and μ_B be the mean population rating for concern for people (sensitivity) after and before the session, respectively.

H_0: $\mu_A = \mu_B$
H_1: $\mu_A > \mu_B$

$s_{\bar{d}} = 1.58/\sqrt{10} = 0.50$

The critical value of t is 2.262 for $\nu = 9$ in a one-tailed test.

Decision Rule
    1. Reject H_0 if $t > 2.262$
    2. Accept H_0 otherwise
Decision: Since $t = 1.5/0.5 = 3 > 2.262$, reject H_0. Conclude that the session does have a positive effect on managerial ratings for sensitivity.

CHAPTER 8

Section 8.1

1. H_0: The distribution is binomial with $p = \frac{1}{2}$
   H_1: The distribution is not binomial with $p = \frac{1}{2}$

   χ^2 $= 10.934$

   ν $= 6$

   $\chi^2_{0.05} = 12.592$

   Hence, accept H_0 and conclude that the binomial distribution with $p = \frac{1}{2}$ is a good fit.

3. H_0: The probability distribution is uniform
   H_1: The probability distribution is not uniform

Brand Preferred	Observed Frequency f_0	Expected Frequency f_t
A	27	20
B	16	20
C	22	20
D	18	20
E	17	20
Total	100	100

   χ^2 $= 4.100$

   ν $= 4$

   $\chi^2_{0.05} = 9.488$

   $\chi^2_{0.01} = 13.277$

   Hence, accept H_0 at both $\alpha = 0.05$ and $\alpha = 0.01$. The data are consistent with the hypothesis of no real difference in taste preference among the five brands.

5. H_0: The bank's employees with MBA degrees are equally distributed among the four schools
   H_1: The bank's employees with MBA degrees are not equally distributed among the four schools

School	f_0	f_t
A	4	6
B	9	6
C	8	6
D	3	6
Total	24	24

$\chi^2 = 4.333$

$\nu = 3$

$\chi^2_{0.05} = 7.815$

Hence, accept H_0 at $\alpha = 0.05$. We cannot reject the null hypothesis.

Section 8.2

1. H_0: The proportions hired were the same for the three categories of applicants

   H_1: The proportions hired were not the same for the three categories of applicants

f_0	f_t
76	60
23	29
26	36
164	180
93	87
118	108
500	500

$\chi^2 = 11.048$

$\nu = 2$

$\chi^2_{0.05} = 5.991$

Hence, reject H_0 at $\alpha = 0.05$ and conclude that the proportions hired were not the same for the three categories of applicants.

3. H_0: Grade performance for the classes of 1973, 1974, and 1975 was independent of entrance exam performance

   H_1: Grade performance for the classes of 1973, 1974, and 1975 was not independent of entrance exam performance

f_0	f_t
8	7.4
35	39.9
22	17.7
10	13.6
83	73.6
27	32.7
7	4.0
17	21.5
11	9.6
220	220.0

$\chi^2 = 8.240$

$\nu = 4$

$\chi^2_{0.01} = 13.277$

Hence, accept H_0 at the 0.01 significance level. The data are consistent with the hypothesis that grade performance was independent of entrance exam performance.

5. H_0: Choice of newspaper and social class are independent
H_1: Choice of newspaper and social class are not independent

f_0	f_t
80	73.3
70	76.7
90	73.3
60	76.7
50	73.3
100	76.7
450	450.0

$\chi^2 = 23.122$

$\nu = 2$

$\chi^2_{0.05} = 5.991$

Hence, reject H_0 and conclude that choice of newspaper and social class are not independent.

7. H_0: Type of defect and supplier are independent
H_1: Type of defect and supplier are not independent

f_0	f_t
70	48
10	12
10	18
30	30
0	12
10	32
10	8
20	12
20	20
20	8
200	200

$\chi^2 = 64.929$

$\nu = 4$

$\chi^2_{0.01} = 13.277$

Hence, reject H_0 and conclude that type of defect and supplier are not independent.

Section 8.3

1. H_0: The seven samples were drawn randomly from the same population
H_1: The seven samples were not drawn randomly from the same population

Source of Variation	Sums of Squares	Degrees of Freedom	Mean Square
Between columns	200	6	33.33
Between rows	315	63	5
Total	515	69	

$$F(6, 63) = \frac{33.33}{5} = 6.67$$

$F_{0.05}(6, 63) \approx 2.26$ (by linear interpolation in Table A-8 of Appendix A). Since $6.67 > 2.26$, reject H_0.
$F_{0.01}(6, 63) \approx 3.13$ (by linear interpolation in Table A-8 of Appendix A). Since $6.67 > 3.13$, reject H_0.

3. H_0: The mean lives of the four brands are equal
H_1: The mean lives of the four brands are not equal

Source of Variation	Sums of Squares	Degrees of Freedom	Mean Square
Between columns	24	3	8
Between rows	42	12	3.5
Total	66	15	

$$F(3, 12) = \frac{8}{3.5} = 2.29$$

$F_{0.05}(3, 12) = 3.59$ (by linear interpolation in Table A-8 of Appendix A). Since $2.29 < 3.59$, accept H_0.

CHAPTER 9

Section 9.4

1. a = Y intercept = 10

   b = slope = $\frac{50}{4}$ = 12.5

   $Y_C = a + bX = 10 + 12.5X$

3. a. Y = amount of life insurance (thousands of dollars)

   X = income (thousands of dollars)

   $$b = \frac{7144 - (12)(16)(34)}{3370 - (12)(16)^2} = \frac{616}{298} = 2.067$$

   a = $34 - (2.067)(16) = 0.928$

   $Y_C = 0.928 + 2.067X$

   Y_C: 29.866, 40.201, 48.469, 25.732, 19.531, 31.933, 46.402, 52.603, 31.933, 21.598, 25.732, 34.000

   b. The regression coefficient $b = 2.067$ means that for two families whose income differs by $1000, the estimated difference in the amount of life insurance held by the heads of these families is $2067.

   c. $Y_C = 0.928 + (2.067)(20) = 42.268$ (thousands of dollars)

5. 1. The first purpose of regression analysis is to provide estimates of values of the dependent variable from values of the independent variable. The regression line and the equation of this line, the regression equation, are used for this purpose.

   2. The second objective of regression analysis is to obtain measures of the error involved in using the regression line for estimation. The standard error of estimate and related measures are useful for this purpose.

3. The third objective (correlation analysis) is to obtain a measure of the degree of association between the two variables. The coefficient of determination measures the strength of the relationship between the variables.

Section 9.5

1. a. Y = earnings (in millions of dollars) in 1976

X = research and development expenditures (in millions of dollars) in 1975

$$b = \frac{10{,}186.5 - (15)(6.2)(73)}{895 - (15)(6.2)^2} = 10.671$$

$a = 73 - (10.671)(6.2) = 6.840$

$Y_C = 6.840 + 10.671X$

b. $Y_C = 6.840 + 10.671(10) = 113.55$

c. $s_{Y.X} = \sqrt{\dfrac{122{,}825 - (6.84)(1095) - (10.671)(10{,}186.5)}{13}}$

$= \sqrt{510.389} = 22.592$ (millions of dollars)

The standard error of estimate measures the scatter of the observed values of Y around the corresponding computed Y_C values on the regression line.

d. $s_{Y_C} = s_{Y.X} \sqrt{\dfrac{1}{15} + \dfrac{(10 - 6.2)^2}{895 - (93)^2/15}} = (22.592)\sqrt{0.1121} = 7.564$

The estimated interval is $113.55 \pm (2.160)(7.564) = 113.55 \pm 16.338$, or from \$97.212 million to \$129.888 million

3. a. Y = first-year sales (in millions of dollars)

X = advertising expenditures (in millions of dollars)

$$b = \frac{1271.2 - (14)(1.3)(55)}{29.22 - (14)(1.3)^2} = \frac{270.2}{5.56} = 48.597$$

$a = 55 - (48.597)(1.3) = -8.176$

$Y_C = -8.176 + 48.597X$

b. $s_{Y.X} = \sqrt{\dfrac{56476 - (-8.176)(770) - (48.597)(1271.2)}{12}}$

$= \sqrt{82.918} = 9.106$ (millions of dollars)

c. $s_{Y_C} = s_{Y.X} \sqrt{\dfrac{1}{14} + \dfrac{(1 - 1.3)^2}{29.22 - (18.2)^2/14}} = (9.106)\sqrt{0.0876}$

$= 2.695$ (millions of dollars)

The estimated interval is $40.421 \pm (2.179)(2.695) = 40.421 \pm 5.872$, or from 34.549 to 46.293

d. $s_{IND} = s_{Y.X} \sqrt{1 + \dfrac{1}{14} + \dfrac{(1 - 1.3)^2}{29.22 - (18.2)^2/14}} = (9.106)\sqrt{1.0876}$

$= 9.496$ (millions of dollars)

The estimated interval is $40.421 \pm (2.179)(9.496) = 40.421 \pm 20.692$, or from \$19.729 million to \$61.113 million

e. Factors affecting the width of intervals are
   1. The larger the sample size, n, the smaller are the estimated standard error of the conditional mean and the standard error of forecast and the narrower are the widths of the intervals.
   2. The greater the deviation of X from $\overline{X}$, the greater the standard error and the wider are the intervals for the given X value.
   3. The larger the estimated standard error of estimate $s_{Y.X}$, the larger are the confidence and prediction intervals.
   4. The more variability there is in the sample of X values, the smaller will be the standard errors and the narrower will be the width of confidence and prediction intervals.

5. a. $b = \dfrac{14,900 - (122)(12)(10)}{18,000 - (122)(12)^2}$

   $= \dfrac{260}{432} = 0.602$

   $a = 10 - (0.602)(12) = 2.776$

   $Y_c = +2.776 + 0.602X$

b. $s_{Y.X} = \sqrt{\dfrac{12,475 - (2.776)(1220) - (0.602)(14,900)}{120}}$

   $= \sqrt{0.987} = 0.993$

c. For a family with an income of \$8000,

   $$2.776 + (0.602)(8) = 7.542$$

   Percentage of family annual income $= 7.542/80 = 9.43\%$

   For a family with an income of \$25,000

   $$2.776 + (0.602)(25) = 17.826$$

   Percentage of family annual income $= 17.826/250 = 7.13\%$

   The estimated percentage of family income spent on medical care decreases as family income increases across the sample.

d. For a family with an income of \$8000,

   $$s_{Y_c} = s_{Y.X} \sqrt{\dfrac{1}{122} + \dfrac{(8 - 12)^2}{18,000 - (1464)^2/122}} = (0.993)\sqrt{0.0452}$$

   $= 0.211$

   The estimated interval is $7.592 \pm (1.98)(0.211) = 7.592 \pm 0.418$, or from

7.174 to 8.010 (in hundreds of dollars). For a family with an income of $25,000,

$$s_{Y_c} = s_{Y.X}\sqrt{\frac{1}{122} + \frac{(25 - 12)^2}{18,000 - (1464)^2/122}} = (0.993)\sqrt{0.3994}$$

$$= 0.628$$

The estimated interval is $17.826 \pm (1.98)(0.628) = 17.826 \pm 1.243$, or from 16.583 to 19.069 (in hundreds of dollars)

e. The variance of Y_c is composed of two independent parts, the variance in the average height of the regression line $(\overline{Y})$ and the variance in the slope of the line (b).

Section 9.8

1. a. $s_{Y.X} = \sqrt{\dfrac{\Sigma (Y - Y_c)^2}{n - 2}} = \sqrt{\dfrac{2000}{40}} = 7.071$

The standard error of estimate, $s_{Y.X}$, measures the scatter of the observed values of Y (company profits) around the corresponding computed Y_c values on the regression line, $Y_c = 2.0 + 0.15X$

b. $r^2 = 1 - \dfrac{\Sigma (Y - Y_c)^2}{\Sigma (Y - \overline{Y})^2} = 1 - \dfrac{2000}{17,000} = 0.882$

$r_c^2 = 1 - \dfrac{\Sigma (Y - Y_c)^2/(n - 2)}{\Sigma (Y - \overline{Y})^2/(n - 1)} = 1 - \dfrac{2000/40}{17,000/41} = 0.879$

The coefficient of determination is a measure of the degree of association between Y and X. The calculated figures may be interpreted as the percentage of variation in the dependent variable, Y, that has been accounted for or "explained" by the relationship between Y and X expressed in the regression line.

c. An appropriate test is an hypothesis-testing procedure with the following null and alternative hypotheses:

$H_0: B = 0$
$H_1: B \neq 0$

The standard error of the regression coefficient is

$$s_b = \frac{s_{Y.X}}{\sqrt{\Sigma (X - \overline{X})^2}} = \frac{7.071}{\sqrt{500,000}} = 0.01$$

The t statistic is

$$t = \frac{b - B}{s_b} = \frac{0.15 - 0}{0.01} = 15 \text{ with 40 degrees of freedom}$$

Since $15 > 2.704$, we reject at the 1% level of significance the null hypothesis that the population coefficient is equal to 0. We conclude that the estimated relationship between company profits and company asset size is significant.

3. a. $b = \dfrac{0.3854 - (20)(0.125)(0.13)}{0.4024 - (20)(0.13)^2} = \dfrac{0.0604}{0.0644} = 0.9379$

   $a = 0.125 - (0.9379)(0.13) = 0.0031$

   $R_i = 0.0031 + (0.9379)R_m$

   Since $a = (1 - \beta)R_F$, we have $0.0031 = (1 - 0.9379)R_F$

   $R_F = \dfrac{0.0031}{0.0621} = 0.05$

   b. $s_{Y.X} = \sqrt{\dfrac{0.3812 - (0.0031)(2.5) - (0.9379)(0.3854)}{18}} = 0.0258$

   $s_Y = \sqrt{\dfrac{0.0687}{19}} = 0.0601$

   $r_C^2 = 1 - \dfrac{(0.0258)^2}{(0.0601)^2} = 1 - 0.1843 = 0.8157$

   $r^2 = \dfrac{(0.0031)(2.5) + (0.9379)(0.3854) - (20)(0.125)^2}{0.3812 - (20)(0.125)^2} = 0.8256$

   c. $s_b = \dfrac{0.0258}{\sqrt{0.0644}} = 0.1017$

   $t = \dfrac{0.9379 - 1}{0.1017} = -0.611$ with 18 degrees of freedom

   Since $-0.611 > -2.878$, we conclude that the estimated regression coefficient is not significantly different from 1.

   d. $s_{Y_c} = (0.0258)\sqrt{\dfrac{1}{20} + \dfrac{(0.093 - 0.13)^2}{0.4024 - \dfrac{(2.6)^2}{20}}} = (0.0258)(0.2669) = 0.00689$

   The estimated interval is $0.0903 \pm (2.101)(0.00689)$, or from 0.0758 to 0.1048

5. a. Y = company sales (millions of dollars)

   X = industry sales (millions of dollars)

   $b = \dfrac{1{,}627{,}595 - (10)(1181.5)(119)}{16{,}203{,}625 - (10)(1181.5)^2} = \dfrac{221{,}610}{2{,}244{,}203} = 0.099$

   $a = 119.0 - (0.099)(1181.5) = 2.032$

   $Y_C = 2.032 + 0.099X$

   b. $r^2 = \dfrac{(2.032)(1190) + (0.099)(1{,}627{,}595) - (10)(119)^2}{163{,}976 - (10)(119)^2}$

   $= \dfrac{21{,}939.98}{22{,}366} = 0.981$

$$r_C^2 = 1 - \frac{s^2_{Y.X}}{s^2_Y} = 1 - \frac{53.246}{2485.111} = 0.979$$

$$s^2_Y = \frac{\Sigma (Y - \bar{Y})^2}{n - 1} = \frac{22,366}{9} = 2485.111$$

$$s_{Y.X} = \sqrt{\frac{163,976 - (2.032)(1190) - (0.099)(1,627,595)}{8}}$$

$$= \sqrt{\frac{426.02}{8}} = 7.297$$

c. $s_{Y_C} = s_{Y.X} \sqrt{\frac{1}{10} + \frac{(1070 - 1181.5)^2}{16,203,625 - (11,815)^2/10}}$

$$= (7.297)(0.325) = 2.372$$

The estimated interval is $Y_C \pm t s_{Y_C} = 107.962 \pm (2.306)(2.372) = 107.962 \pm 5.470$, or from \$102.492 million to \$113.432 million.

d. $s_{IND} = (7.297)\sqrt{1 + 0.1 + \frac{12432.25}{2,244,203}} = (7.297)(1.0515) = 7.673$

The estimated interval is $Y_C \pm t s_{IND} = 107.962 \pm (2.306)(7.673) = 107.962 \pm 17.694$, or from \$90.268 million to \$125.656 million.

7. a. Any interpretation of the intercept (-12) would be meaningless, because $X = -12$ is outside the range of the sample of observations (X ranges from 25 to 55 years). However, if the relationship between X and Y is linear, then the regression equation is a valid one within the observed range of the X variable.

b. For two sales representatives whose ages differed by 1 year, the older representative earned an estimated \$1000 more.

$$s_b = 3.5/\sqrt{625} = 0.14$$

$$t = \frac{1 - 0}{0.14} = 7.143, \text{ with 60 degrees of freedom}$$

Since $7.143 > 2.66$, we conclude that the estimated coefficient is significantly different from 0 at the 1% level of significance.

c. $r_C^2 = 1 - \frac{(3.5)^2}{(7.5)^2} = 0.782$

d. $s_{IND} = (3.5)\sqrt{1 + \frac{1}{62} + \frac{(7)^2}{625}}$

$$= (3.5)(1.0462) = 3.662$$

An estimated 95% prediction interval for the annual sales commission for a 45-year-old sales representative would be $33 \pm (2)(3.662)$, or from 25.676 to 40.324 (in thousands of dollars). Since the commission of the sales representative under consideration is considerably below the lower limit of this interval, we may conclude that this is a poor performer.

Section 9.9

1. a. For b_1, $t_1 = (-3.875 - 0)/0.655 = -5.916$, with 120 degrees of freedom. Conclude that b_1 is significantly different from 0. For b_2, $t_2 = (7.955 - 0)/4.118 = 1.932$, with 120 degrees of freedom. Conclude that b_2 is not significantly different from 0.

   b. Family income (X_1) has a statistically significant effect on the percentage of cash rebate spent within a year (Y), but after this income effect has been accounted for, family size (X_2) does not have a statistically significant influence.

   c. An increment of one unit in X_1 results in an increase of b_1 units in Y_C, regardless of the values of the other independent variables. Second, the inclusion of other independent variables "nets out" the effects of these variables; the net regression coefficient b_1 then will represent the effects of X_1 and any other variables that may be correlated with X_1 but not explicitly included in the analysis.

   d. $R^2_{Y.12} = 1 - (22.042)^2/(40.035)^2 = 0.697$
   Approximately 69.7% of the variance of Y is explained by the regression equation relating Y to X_1 and X_2.

   e. $Y_C = 94.210 + (12)(-3.875) + (6)(7.955) = 95.44\%$

   f. $F = \dfrac{137{,}237.891/2}{58{,}302.667/120} = 141.233$, with $(2, 120)$ degrees of freedom. The regression is significant at the 1% level of significance.

3. a. $S_{Y.12} = \sqrt{25/(21 - 3)} = \sqrt{1.389} = 1.179$

   There are 18 degrees of freedom.

   b. $R^2_{Y.12} = 1 - (25/100) = 0.75$

   Seventy-five percent of the variation in the unemployment rate (Y) has been "explained" by the regression equation.

   c. $s_{b_1} = 1.179/\sqrt{(60{,}000)(0.36)} = 0.008$

   $s_{b_2} = 1.179/\sqrt{(16)(0.36)} = 0.4913$

   For b_1, $t_1 = -0.04/0.008 = -5$, with 18 degrees of freedom. Conclude that b_1 differs significantly from 0 at the 1% level of significance. For b_2, $t_2 = -3/0.4913 = -6.106$, with 18 degrees of freedom. Conclude that b_2 differs significantly from 0 at the 1% level of significance.

   d. $Y_C = 18 - (0.04)(200) - (3)(2.00) = 18 - 14 = 4$
   The estimate is 4% unemployed.

   e. In this case, $F = \dfrac{75/2}{25/18} = \dfrac{37.5}{1.389} = 26.998$, with $(2, 18)$ degrees of freedom. The regression is significant at the 1% level of significance.

5. a. $r^2_C = 1 - (5^2/12^2) = 0.826$

   b. $t_1 = 0.8/0.15 = 5.333$, with 14 degrees of freedom. Conclude that b_1 differs significantly from 0 at the 1% level of significance.

   c. $R^2_{Y.12} = 1 - (4.8)^2/12^2 = 0.840$

The inclusion of X_2 has explained very little of the variance in Y beyond that already explained by X_1 alone.

d. The value $b_1 = 0.9$ indicates that if a house has an assessed value \$1000 greater than another house of the same age, then the estimated selling price of the first house exceeds that of the second by \$900.

e. For b_1, $t_1 = 0.9/0.25 = 3.60$, with 13 degrees of freedom. Conclude that b_1 is significantly different from 0 at the 1% level of significance. For b_2, $t_2 = -0.15/0.1 = -1.5$, with 13 degrees of freedom. Conclude that b_2 is not significantly different from 0.

Conclude that assessed value has a statistically significant effect on selling price but that after this assessed value effect has been accounted for, the age of the house in years does not have a statistically significant effect. At least two situations are possible: (1) There is simply no relationship between selling price and age of home. (2) The two independent variables, X_1 and X_2, may be highly correlated. This would reduce the reliability of the estimated coefficients.

CHAPTER 10

Section 10.4

1. See Section 10.2.

3. a $= \dfrac{\Sigma Y}{n} = \dfrac{4305.1}{11} = 391.37$

$b = \dfrac{\Sigma xY}{\Sigma x^2} = \dfrac{3730.2}{110} = 33.91$

$Y_t = 391.37 + 33.91x$ (see graph on p. 769)

$x = 0$ in 1969

x is in one-year intervals; Y is in billions of current dollars

The trend line for the constant dollar series seems to fit the data more closely, since that series has relatively more constant annual amounts of change than does the current dollar series. Without more data, however, it is impossible to tell whether the second series merely exhibits a stronger cyclical component.

5. a. $a = \dfrac{\Sigma Y}{n} = \dfrac{224.1}{5} = 44.82$

$b = \dfrac{\Sigma xY}{\Sigma x^2} = \dfrac{49.7}{10} = 4.97$

$Y_t = 44.82 + 4.97x$ ($x = 0$ in 1935)

Trend values (billions of dollars): 34.88, 39.85, 44.82, 49.79, 54.76

b. No, because the time period of five years is too short to get a good descrip-

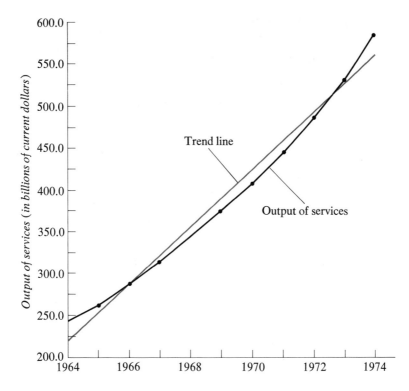

tion of trend. Since 1933 through 1937 were depression years, the trend values of this problem doubtless would be below those for a longer period that included both periods of economic recovery and periods of contraction.

c. $\left(\dfrac{Y - Y_t}{Y_t}\right)(100) = \left(\dfrac{44 - 44.82}{44.82}\right)(100) = -1.83\%$

The relative cyclical residual for 1935 is -1.83%, indicating that the actual national income figure is 1.83% below the trend figure because of cyclical and irregular factors.

7. a. The labor force of this county is increasing by decreasing amounts and at a decreasing percentage rate.

b. $Y_t = 74.62 + 6.83(8) - 0.31(8)^2 = 109.42$ (in thousands). The deviation of 126,430 from the computed trend figure of 109,420 does not necessarily imply a poor fit. It is conceivable that the fit is adequate and that the deviation represents an unusually strong cyclical influence.

9. a. Straight line:

$$a = \frac{\Sigma Y}{n} = \frac{1398.8}{13} = 107.60$$

$$b = \frac{\Sigma xY}{\Sigma x^2} = \frac{2049.3}{182} = 11.26$$

$Y_t = 107.60 + 11.26x$

$x = 0$ in 1968

x is in one-year intervals; Y is in billions of dollars

b. Second-degree parabola:
From part (a),

$\Sigma Y = 1398.8$

$\Sigma xY = 2049.3$

$\Sigma x^2 = 182$

$n = 13$

$1398.8 = 13a + 182c \qquad a = 98.682$

$20{,}858.5 = 182a + 4550c \qquad c = 0.637$

$$b = \frac{2049.3}{182} = 11.260$$

$Y_t = 98.682 + 11.260x + 0.637x^2$

$x = 0$ in 1968

x is in one-year intervals; Y is in billions of dollars

c. The second-degree trend line appears to be the more appropriate.

11.

Year	Y_t
1975	3808.9
1976	3657.6
1977	3512.4

$$a = \frac{\Sigma \log Y}{n} = \frac{55.8238}{15} = 3.7216$$

$$b = \frac{\Sigma x \log Y}{\Sigma x^2} = \frac{-4.9272}{280} = -0.0176$$

$\log Y_t = 3.7216 - 0.0176x$

$x = 0$ in 1967

x is in one-year intervals; Y is in thousands

Section 10.5

1. a. $\dfrac{Y}{SI} = \dfrac{\$132{,}000}{1.08} = \$122{,}222.22$

   $\dfrac{Y/SI}{Y_t} = \dfrac{\$122{,}222.22}{\$125{,}000} = 97.8\%$ of trend

   b. The trend value for December 1976 indicates the level of sales for that month had cyclical, seasonal, and random factors not been present.

3. a. See the discussion in Section 10.5.
   b. 1. Seasonal variations would have to be removed from the true production values by dividing the latter figures by the seasonal indices. These values would then indicate whether there had been a decline not attributable to seasonal variations.
      2. The true production values would again have to be divided by the seasonal indices to remove seasonal variations. Then these deseasonalized (Y/SI) values would have to be divided by a series of trend figures. To obtain the trend values we would need more data on milk production for months of earlier years. We could compute a trend line from these data to yield trend values. The deseasonalized figures would then be divided by the trend values. The resulting series of figures reflects primarily the effect of cyclical fluctuations, since trend and seasonal influences have been removed.

5. a. Total of modified means = 400.82

   Adjustment factor = $\dfrac{400}{400.82} = 0.998$

   The seasonal indexes are 89.82 (I), 96.01 (II), 111.20 (III), and 102.99 (IV).
   b. Yes, constant seasonal indexes seem appropriate, because the percentage of moving average figures are relatively stable for each quarter.
   c. See column (7) of the table.
   d. There is some evidence of cycles. Cyclical peaks appear in 1971, quarter II, and 1974, quarter I, with corresponding cyclical troughs in 1972, quarter III, and 1975, quarter I.

7. Total of modified means = 399.73

   Adjustment factor = $\dfrac{400}{399.73} = 1.0007$

   The seasonal indexes are 103.86 (I), 96.24 (II), 89.25 (III), and 110.66 (IV).

9. a. 3   e. 1
   b. 2   f. 1
   c. 4   g. 3
   d. 2

Section 10.6

1.

Day	Next Forecast $w = 0.1$	Next Forecast $w = 0.7$
1	$10 = F_2$	$10 = F_2$
2	$10 = F_3$	$10 = F_3$
3	$10 = F_4$	$10 = F_4$
4	$11 = F_5$	$17 = F_5$
5	$10.9 = F_6$	$12.1 = F_6$
6	$10.81 = F_7$	$10.63 = F_7$
7	$10.729 = F_8$	$10.189 = F_8$
8	$10.6561 = F_9$	$10.0567 = F_9$
9	$10.59049 = F_{10}$	$10.01701 = F_{10}$
10	$10.531441 = F_{11}$	$10.005103 = F_{11}$

3. Forecasts derived from low values of w tend to vary less from period to period than corresponding forecasts made with the use of a high w value. Hence in Exercise 1, where there was a one-time jump in demand to 20 in Day 4, the $w = 0.1$ forecasts changed very little from the basic level of 10. On the other hand, the forecasts using $w = 0.7$ increased sharply to a fifth-day forecast of 17 and then dropped quickly to values close to the basic demand of 10 units. In Exercise 2, the actual demand data shifted to a higher basic level of 20 units in Day 4. Forecasts using $w = 0.1$ reacted very sluggishly, rising only to a level of 15 units for the forecast for Day 11. Forecasts using $w = 0.7$ rose rapidly to levels close to the new basic level of 20 units and then tracked the demand series very closely.

5.

Year	Next Forecast $w = 0.4$
1960	$\$2.50 \quad = F_2$
1961	$2.4840 = F_3$
1962	$2.4684 = F_4$
1963	$2.3570 = F_5$
1964	$2.3522 = F_6$
1965	$2.5393 = F_7$
1966	$2.9656 = F_8$
1967	$3.4914 = F_9$
1968	$3.5708 = F_{10}$
1969	$3.8025 = F_{11}$
1970	$3.7375 = F_{12}$
1971	$3.7345 = F_{13}$
1972	$3.6007 = F_{14}$
1973	$4.0644 = F_{15}$
1974	$5.7826 = F_{16}$

The forecast of $5.7826 for 1975 is 93.9% of the actual figure of $6.16.

CHAPTER 11

Section 11.2

1. a. Weighted aggregate index, with 1970 weights:

$$\frac{\Sigma \, P_{76}Q_{70}}{\Sigma \, P_{70}Q_{70}} \cdot 100 = \frac{(\$25)(20) + (\$2)(100) + (\$6)(50)}{(\$20)(20) + (\$1)(100) + (\$5)(50)} \cdot 100$$

$$= \frac{\$1000}{\$750} \cdot 100 = 133.3$$

b. We cannot conclude that the Jones Metal Company paid more per unit in 1976 than its competitor. We know only that in 1976 it costs Jones 133.3% of what it did in 1970, whereas in 1976 it costs the competitor 120% of the cost to purchase the 1970 quantities. However, the absolute level of prices as measured by price per unit may very well be lower for Jones than for its competitor. The prices paid by Jones have risen more, on the average, than its competitor's, but its prices per unit in 1970 may have been at a much lower level than its competitor's.

Section 11.3

1. a. For 1973 on a 1976 base,

$$\frac{\Sigma \, \left(\dfrac{P_{73}}{P_{76}}\right)P_{76}Q_{76}}{\Sigma \, P_{76}Q_{76}} \cdot 100 = \frac{\Sigma \, P_{73}Q_{76}}{\Sigma \, P_{76}Q_{76}} \cdot 100 = \frac{\$2012}{\$2440} \cdot 100 = 82.5$$

b. If consumers had purchased the same quantities of these three models of fans in 1973 as were purchased in 1976, then it would have cost the consumers only 82.5% of what it did in the later year, or 17.5% less than the 1973 cost.

3. For 1976 on a 1975 base,

$$\frac{\Sigma \, \left(\dfrac{P_{76}}{P_{75}}\right)P_{75}Q_{75}}{\Sigma \, P_{75}Q_{75}} \cdot 100 = \frac{\$10,\!400}{\$8200} \cdot 100 = 126.8$$

Section 11.6

1. a. For 1975 on a 1974 base,

$$\frac{\Sigma \, \left(\dfrac{Q_{75}}{Q_{74}} \cdot 100\right)Q_{74}P_{74}}{\Sigma \, Q_{74}P_{74}} = \frac{\$1874.25}{\$15.77} = 118.8$$

For 1976 on a 1974 base,

$$\frac{\Sigma \, \left(\dfrac{Q_{76}}{Q_{74}} \cdot 100\right)Q_{74}P_{74}}{\Sigma \, Q_{74}P_{74}} = \frac{\$1902.50}{\$15.77} = 120.6$$

b. Same numerical results as in part (a).

3. Deflated sales for 1971: $20,000,000/1.101 = $18,165,304
   Deflated sales for 1973: $32,000,000/1.221 = $26,208,026
   Deflated sales for 1975: $43,000,000/1.704 = $25,234,742
   "Real gross sales" increased by 38.9% from 1971 to 1975 and decreased by 3.7% from 1973 to 1975.

5. Deflated average weekly earnings for 1972: $154.69/1.253 = $123.46; for 1973: $166.06/1.331 = $124.76; for 1974: $176.40/1.477 = 119.43; for 1975: $189.51/1.612 = 117.56.
   Change from 1972 to 1973: $+1.1\%$; from 1973 to 1974: -4.3%; from 1974 to 1975: -1.6%.

CHAPTER 12

Section 12.2

1. H_0: $p = 0.50$
   H_1: $p > 0.50$

   $\alpha = 0.01$

   $\sigma_{\bar{p}} = \sqrt{(0.50)(0.50)/138} = 0.043$

   $\bar{p} = \frac{96}{138} = 0.70$

   $z = (0.70 - 0.50)/0.043 = 4.65$

   Since $4.65 > 2.33$, we reject the null hypothesis of no difference in effectiveness of the two incentive programs. We conclude that the new program is better.

3. H_0: $p \leq 0.50$
   H_1: $p > 0.50$

   $\alpha = 0.05$

   $\sigma_{\bar{p}} = \sqrt{(0.50)(0.50)/32} = 0.088$

   $\bar{p} = \frac{20}{32} = 0.625$

   $z = (0.625 - 0.50)/0.088 = 1.42$

   Since $1.42 < 1.65$, we accept the null hypothesis that $p \leq 0.50$. Hence, we conclude that the proportion of sales representatives who rate the new program as more effective than the old does not exceed 50%.

Section 12.3

1.

Student	Signed Rank Rank (+)	Signed Rank Rank (−)
A	1.5	
B	6.5	
C		5
D	11	
E		1.5
F	3.5	
G		6.5
H		9
I	10	
J	8	
K		3.5
L		
	40.5	25.5 = T

$n = 11$; $T = 25.5$; $T_{0.05} = 13$. Since $T = 25.5 > T_{0.05} = 13$, we cannot reject the null hypothesis of identical population distributions. Hence, we cannot conclude that the scores on the second test are significantly higher than the scores on the first test.

3.

City	Signed Rank Rank (+)	Signed Rank Rank (−)
Atlanta		5
Boston		2
Chicago		9
Detroit	7	
Los Angeles		3
Miami		1
New Orleans		4
New York		10
Philadelphia		8
San Francisco		6
	T = 7	48

a. *Wilcoxon Test:* $n = 10$; $T = 7$; $T_{0.05} = 10$. Since $T = 7 < T_{0.05} = 10$, we reject the null hypothesis of identical population distributions. Hence, we conclude that the product improvement led to a decrease in the numbers of product returns.

b.

$d = x_2 - x_1$	$(d - \bar{d})^2$
-68	246.49
-52	1004.89
-166	6773.29
$+119$	41,087.29
-57	712.89
-38	2088.49
-59	610.09
-246	26,341.29
-162	6130.89
-108	590.49
	85,586.10

$$\bar{d} = -837/10 = -83.7$$

$$s_d = \sqrt{\frac{\Sigma (d - \bar{d})^2}{n - 1}} = \sqrt{\frac{85,586.10}{10 - 1}} = 97.52$$

$$s_{\bar{d}} = \frac{s_d}{\sqrt{n}} = \frac{97.52}{\sqrt{10}} = 30.84$$

$$t = \frac{\bar{d} - 0}{s_{\bar{d}}} = \frac{-83.7}{30.84} = -2.71$$

$$\nu = n - 1 = 10 - 1 = 9$$

$t_{0.05} = -1.833$ (Table A-6 under heading 0.10)

Since $t = -2.71 < t_{0.05} = -1.833$, we reject the null hypothesis of no difference between the average numbers of returns before and after the product improvement. We conclude that the average number of returns was lower after the product improvement.

Section 12.4

In the solutions in Section 12.4, the data are ranked from highest to lowest, with rank 1 assigned to the highest value.

1. $R_1 = 198$, $R_2 = 208$, $n_1 = 14$, $n_2 = 14$, $U = 103$, $\mu_U = 98$.

$$\sigma_U = \sqrt{(14)(14)(29)/12} = 21.8$$

$$z = (103 - 98)/21.8 = 0.23$$

Since $0.23 < 2.58$, we accept the hypothesis that the samples were drawn from populations having the same average profit.

3. $R_1 = 168$, $R_2 = 238$, $n_1 = 13$, $n_2 = 15$, $U = 118$, $\mu_U = 97.5$, $\alpha = 0.05$.

$\sigma_U = \sqrt{(13)(15)(29)/12} = 21.7$

$z = (118 - 97.5)/21.7 = 0.94$

Since $0.94 < 1.96$, we accept the hypothesis of no difference between the true average times for the two different methods.

Section 12.5

1. $n_1 = 22$, $n_2 = 28$, $r = 26$.

$\mu_r = \dfrac{2(22)(28)}{22 + 28} + 1 = 25.64$

$\sigma_r = \sqrt{\dfrac{2(22)(28)[2(22)(28) - 22 - 28]}{(22 + 28)^2(22 + 28 - 1)}} = 3.44$

$z = (26 - 25.64)/3.45 = 0.10$

Since $0.10 < 1.96$ and $0.10 < 2.58$, we accept the hypothesis of randomness of runs of even and odd digits at the 0.05 and 0.01 levels of significance.

3. $n_1 = 12$, $n_2 = 12$, $r = 6$.

$\mu_r = \dfrac{2(12)(12)}{12 + 12} + 1 = 13$

$\sigma_r = \sqrt{\dfrac{2(12)(12)[2(12)(12) - 12 - 12]}{(12 + 12)^2(12 + 12 - 1)}} = 2.40$

$z = (6 - 13)/2.40 = -7/2.40 = -2.92$

Since $-2.92 < -2.58$, we reject the hypothesis of randomness of runs above and below the median. Since runs below the median tend to occur in the earlier part of the series, and runs above the median in the later part, there is evidence of an increasing trend in the monthly number of on-the-job accidents.

Section 12.6

1.

Number Produced:	65	68	72	74	81	82	85	88	92	97	101
	104	105	108	110	111	114	116	117	119		
	120	121	125								

Rank:	1	2	3	4	5	6	7	8	9	10	11
	12	13	14	15	16	17	18	19	20		
	21	22	23								

$\nu = 3 - 1 = 2$; $\chi^2_{0.05} = 5.991$.

$$K = \frac{12}{23(23 + 1)} \left(\frac{92^2}{8} + \frac{93^2}{7} + \frac{91^2}{8} \right) - 3(23 + 1) = 0.362$$

Since $K = 0.363 < 5.991$, accept the null hypothesis of identically distributed populations.

3. a.

Number of Units Sold: 30 36 38 43 47 52 55 59 61 64 66 67
69 73 75 77 80 82

Rank: 1   2   3   4   5   6   7   8   9 10 11 12
13 14 15 16 17 18

$\nu = 3 - 1 = 2;$ $\chi^2_{0.01} = 9.210$

$$K = \frac{12}{18(18 + 1)} \left(\frac{31^2}{6} + \frac{68^2}{6} + \frac{72^2}{6} \right) - 3(18 + 1) = 5.977$$

Since $5.977 < 9.210$, we cannot reject the null hypothesis of no difference in effectiveness among the three advertisement designs at the 1% level of significance.

b.

	Analysis of Variance Table		
(1) *Source of* *Variation*	(2) *Sum of* *Squares*	(3) *Degrees of* *Freedom*	(4) *Mean* *Square*
Between columns	1425	2	712.50
Between rows	2831	15	188.73
Total	4256	17	

$$F(2, 15) \quad = \frac{712.50}{188.73} = 3.78$$

$F_{0.01}(2, 15) = 6.705$ (by linear interpolation in Table A-8)

Since $3.78 < 6.705$, we cannot reject the null hypothesis of no difference in average numbers of sales resulting from the three advertisements.

Section 12.7

1. $r_r = 1 - \dfrac{6(208)}{10(10^2 - 1)} = -0.26$

3. $r_r = 1 - \dfrac{6(15)}{10(10^2 - 1)} = 0.91$

CHAPTER 13

Section 13.4

1. The sum of the prior probabilities is not equal to 1.

3. Gain from cost reduction = 1% of $2,000,000 = $20,000

   Expected gain (campaign) = (0.8)(0.12)(0.10)($2,000,000) = $19,200

   The better act is the cost reduction program.

5. Money figures are in units of $100,000:
   Expected profit (16) = (0.4)(1.6) + (0.3)(1.6) + (0.2)(1.6) + (0.1)(1.6) = $1.6

   Expected profit (17) = (0.4)(0) + (0.3)(2.6) + (0.2)(2.6) + (0.1)(2.6) = $1.56

   Expected profit (18) = (0.4)(0) + (0.3)(0) + (0.2)(3.6) + (0.1)(3.6) = $1.08

   Expected profit (19) = (0.4)(0) + (0.3)(0) + (0.2)(0) + (0.1)(4.6) = $.46

   The best bid is $1,600,000.

7. The cost per year is $4850. The expected gain from the information is (0.05)($75,000) = $3750. Since cost exceeds the expected gain, do not include newsletter.

9.

	Irrigate This Year	*Irrigate Next Year*
No drought	$ 0	$ 0
Mild drought	900	450
Severe drought	2700	1350

Hence, the expected net gains are

Irrigate this year: (0.35)($0) + (0.40)($900) + (0.25)($2700) − $600 = $435.00

Irrigate next year: (0.35)($0) + (0.40)($450) + (0.25)($1350) − $150 = $367.50

Therefore, it is preferable to irrigate this year.

Section 13.6

1. See Section 13.5.

3. No, because the opportunity loss table gives for each state of nature only the difference between payoffs for different acts. We would need to know the actual payoff for at least one act for each state of nature to compute payoffs.

5. Money figures are in units of $1000:
   a. Expected profit under certainty = $(0.5)(18) + (0.3)(12) + (0.2)(8) = 14.2$
   b. Expected profits:

   $A_1 = (0.5)(18) + (0.3)(8) + (0.2)(2) = 11.8$

   $A_2 = (0.5)(15) + (0.3)(12) + (0.2)(5) = 12.1$

   $A_3 = (0.5)(16) + (0.3)(12) + (0.2)(3) = 12.2$

   $A_4 = (0.5)(11) + (0.3)(10) + (0.2)(8) = 10.1$

   The optimal act is A_3. Hence, the expected profit under uncertainty is 12.2
   c. EVPI $= 14.2 - 12.2 = 2$

7.
Opportunity Loss Table

Campaign Received	Accept Proposal	Reject Proposal
Well	0	$175,000
Moderately well	0	55,000
Poorly	0	0

9. a. Expected gain (hire) $= (0.3)(\$10,000) + (0.4)(\$3000) + (0.3)(-\$13,000) = \$300$
   Since the expected gain is positive, you should hire the new sales representative.
   b. Expected gain under certainty is $(0.3)(10,000) + (0.4)(3000) + (0.3)(0) = \$4200$

   EVPI is thus $4200 - 300 = \$3900$

11. EOL(accept) $= (0.4)(0) + (0.2)(0) + (0.4)(0) = \$0$

   EOL(reject) $= (0.4)(175,000) + (0.2)(55,000) + (0.4)(0) = \$81,000$
   The optimal act is to accept the campaign proposal.

Section 13.8

1. a. Disagree. Although the expected monetary return of purchasing life insurance may be negative, the expected utility will tend to be positive because of the shape of most individuals' utility functions.
   b. The premium cost for automobile insurance exceeds the company's expected losses. If a company has abundant assets, it will be rational for the firm to make decisions based on expected monetary value for alternatives

involving only a small percentage of total assets. On the other hand, a single loss of a few thousand dollars may be relatively catastrophic for an individual, leading to a risk avoider's utility function for the individual.

3. a. Expected utility (buy the stock) = $(0.4)(45) + (0.6)(-15) = 9$

   Expected utility (interest in bank) = 10
   b. You should put the money in the bank.

5. a. Expected monetary loss (dispute) = $(0.95)(245) + (0.05)(20) = \$233.75$
   Since $\$233.75 > \$225.00$, do not dispute.
   b. Expected utility [don't dispute] = -425

   Expected utility [dispute]   = $(0.95)(-440) + (0.05)(-4) = -418.2$

   Hence, the best act is to dispute the bill.

7. a. The utility function appears to be that of a risk avoider until a figure between $50,000 and $75,000. From then on the curve is that of a risk preferrer.
   b. $E[U(\text{buy})] = (0.5)(-40) + (0.2)(6) + (0.1)(10) + (0.2)(40) = -9.8$
   Therefore, the investor should not buy the franchise.
   c. $E[U(\frac{1}{4})] = (0.5)(-2) + (0.2)(1.2) + (0.1)(2.5) + (0.2)(6.2) = 0.73$

   $E[U(\frac{1}{2})] = (0.5)(-4.2) + (0.2)(2.5) + (0.1)(6.0) + (0.2)(12.5) = 1.5$

   $E[U(\frac{3}{4})] = (0.5)(-10) + (0.2)(5.8) + (0.1)(7.5) + (0.2)(26) = 2.11$

   Therefore, the best act is to buy a three-fourths interest.

CHAPTER 14

Section 14.2 (After the subsection "Decision Making after the Observation of Sample Evidence")

1. $P(\theta_1|X) = 0.6; P(\theta_2|X) = 0.4$

3. $P(S_1|X) = \frac{1}{3}; P(S_2|X) = \frac{2}{3}$

5. $P(\theta_1) + P(\theta_2) = 1.$ $P(\theta_1)P(X|\theta_1) = P(\theta_2)P(X|\theta_2).$ Therefore, $P(\theta_1) = 0.25$ and $P(\theta_2) = 0.75.$

7. $P_1(\text{High}|\text{High}) = 0.71; P_1(\text{Average}|\text{High}) = 0.24; P_1(\text{Low}|\text{High}) = 0.05$

9. Money figures are in thousands of dollars.
   a. $\text{EOL}(A_1) = (0.3)(0) + (0.7)(20) = 14$

   $\text{EOL}(A_2) = (0.3)(80) + (0.7)(0) = 24$

   $\text{EVPI} = 14$

   Therefore, the maximum amount the firm should pay is $14,000.

b. $P(S_1|M_1) = 0.66$; $P(S_2|M_1) = 0.34$. The better action is A_1.

$\text{EOL}(A_1) = (0.66)(0) + (0.34)(20) = 6.80$

$\text{EOL}(A_2) = (0.66)(80) + (0.34)(0) = 52.80$

Therefore, the revised EVPI is $6,800.

c. $P(S_1|M_1) = 0.39$; $P(S_2|M_1) = 0.61$. The better action is A_1.

$\text{O.L.}(A_1) = (0.39)(0) + (0.61)(20) = 12.20$

$\text{O.L.}(A_2) = (0.39)(80) + (0.61)(0) = 31.20$

Therefore, the revised EVPI is $12,200.

Section 14.2 (After the subsection "Prior and Posterior Means")

1. a. $P_1(0.05) = 0.75$; $P_1(0.15) = 0.25$
   b. $P_1(0.05) = 0.47$; $P_1(0.15) = 0.53$
   c. $P_1(0.05) = 0.21$; $P_1(0.15) = 0.79$

3. a. $P_1(0.01|4) = 0.008$; $P_1(0.05|4) = 0.890$; $P_1(0.10|4) = 0.100$;
   $P_1(0.15|4) = 0.002$
   b. $P_1(0.01|12) = 0.000$; $P_1(0.05|12) = 0.006$; $P_1(0.10|12) = 0.837$;
   $P_1(0.15|12) = 0.157$

5. $P_1(\text{Success}) = 0.54$; $P_1(\text{Failure}) = 0.46$

   $\text{EOL}(\text{do not publish}) = (0.54)(8) = 4.32$ (million dollars)

   $\text{EOL}(\text{publish}) = (0.46)(4) = 1.84$ (million dollars)

   Therefore, the better decision is to publish.

7. a. $P_1(0.10) = 0.15$; $P_1(0.20) = 0.45$; $P_1(0.30) = 0.40$
   b. $P_1(0.10) = 0.01$; $P_1(0.20) = 0.89$; $P_1(0.30) = 0.10$

CHAPTER 15

Section 15.2

1. See the discussion in Section 15.2.

3. Let $X_1 =$ Forecast that the price will advance

   $X_2 =$ Forecast that the price will not advance

   $P(\theta_1|X_1) = 0.857$; $P(\theta_2|X_1) = 0.143$; $P(\theta_1|X_2) = 0.273$; $P(\theta_2|X_2) = 0.727$

5. a. 1. $\text{EOL}(\text{build}) = (0.5)(5) = 2.5$

      $\text{EOL}(\text{do not build}) = (0.5)(6) = 3.0$

      Therefore, the better act is to build.

2. $P(\theta_1|X_1) = 0.524$; $P(\theta_2|X_1) = 0.476$

   EOL(build) $= (0.476)(5) = 2.38$

   EOL(do not build) $= (0.524)(6) = 3.144$

   Therefore, the better act is still to build.
3. $P(\theta_1|X_2) = 0.474$; $P(\theta_2|X_2) = 0.526$

   EOL(build) $= (0.526)(5) = 2.63$

   EOL(do not build) $= (0.474)(6) = 2.844$

   Therefore, the better act is still to build.
4. $P(X_1) = 0.525$

   $P(X_2) = 0.475$
5. EVSI $= 2.5 - [(0.525)(2.38) + (0.475)(2.63)] = 0$

b. No, since neither X_1 nor X_2 changes the optimal act. Alternatively, the answer is "no" because EVSI = 0.

7.

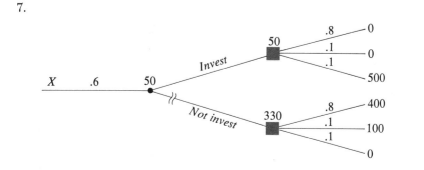

9. EVSI = 14 − 8.64 = 5.36 utility units

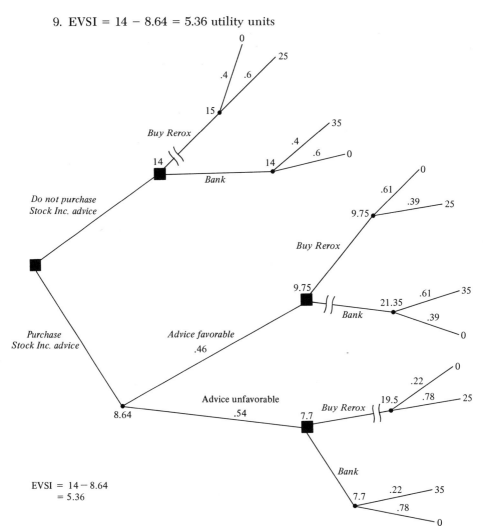

EVSI = 14 − 8.64
     = 5.36

11. EVSI = 7.50 − 2.24 = 5.26 (in tens of thousands of dollars)

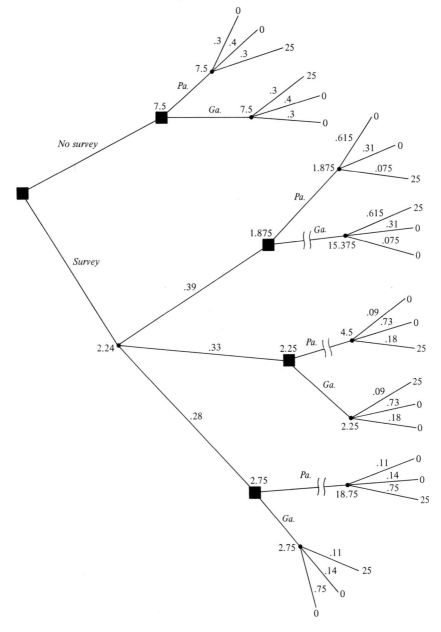

13. a. EVSI = 1.80 − 1.56 = 0.24 (in tens of thousands of dollars)
    b. EVSI = 1.80 − 1.36 = 0.44 (in tens of thousands of dollars)

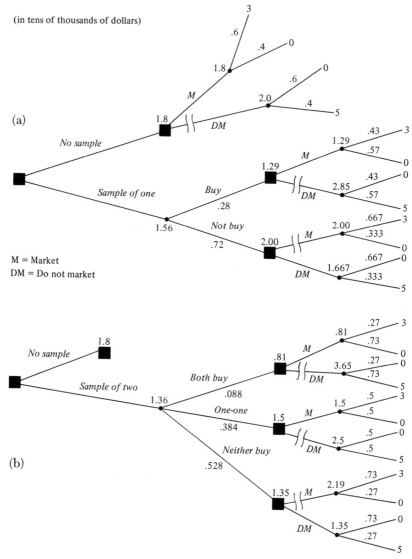

(in tens of thousands of dollars)

(a)

No sample

M = Market
DM = Do not market

(b)

No sample

Sample of two

Section 15.4

1. See Sections 15.3 and 15.4.

3. The missing strategies are $s_1(a_1, a_2, a_2)$ and $s_5(a_2, a_2, a_2)$. The optimal strategy is s_3.

5. The missing strategies are $s_1(a_1, a_1, a_2)$, $s_4(a_1, a_2, a_1)$, $s_5(a_2, a_1, a_2)$, and $s_7(a_2, a_2, a_2)$.

$R(s_1; 300) = (0.8)(0) \quad + (0.1)(0) \quad + (0.1)(25) = 2.5$

$R(s_1; 350) = (0.3)(0) \quad + (0.6)(0) \quad + (0.1)(0) \quad = 0$

$R(s_1; 400) = (0.1)(25) + (0.2)(25) + (0.7)(0) \quad = 7.5$

$R(s_3; 300) = (0.8)(0) \quad + (0.1)(25) + (0.1)(25) = 5.0$

$R(s_3; 350) = (0.3)(0) \quad + (0.6)(0) \quad + (0.1)(0) \quad = 0$

$R(s_3; 400) = (0.1)(25) + (0.2)(0) \quad + (0.7)(0) \quad = 2.5$

$R(s_4; 300) = (0.8)(0) \quad + (0.1)(25) + (0.1)(0) \quad = 2.5$

$R(s_4; 350) = (0.3)(0) \quad + (0.6)(0) \quad + (0.1)(0) \quad = 0$

$R(s_4; 400) = (0.1)(25) + (0.2)(0) \quad + (0.7)(25) = 20$

$\text{EOL}(s_1) \quad = (0.3)(2.5) + (0.4)(0) \quad + (0.3)(7.5) = 3.0$

$\text{EOL}(s_3) \quad = (0.3)(5.0) + (0.4)(0) \quad + (0.3)(2.5) = 2.25$

$\text{EOL}(s_4) \quad = (0.3)(2.5) + (0.4)(0) \quad + (0.3)(20) \quad = 6.75$

(Figures are in tens of thousands of dollars.) The optimal strategy is s_3.

Section 15.6

1. Let X = sample outcome, $P_0(\theta_1)$ = prior probability of success

$$\frac{P_0(\theta_1)(0.2051)}{P(X)}(8) = \frac{(1 - P_0(\theta_1))(0.0881)}{P(X)}(4)$$

$1.6408\, P_0(\theta_1) = 0.3524 - 0.3524\, P_0(\theta_1)$

$1.9932\, P_0(\theta_1) = 0.3524$

$P_0(\theta_1) \qquad = 0.177$

Therefore, if the prior probability of success drops below 17.7%, the better action changes to "do not publish."

3. $(p)\ \text{EOL}(s_1) \quad = (1 - p)\ \text{EOL}(s_2)$

$p(\$4500) \qquad = (1 - p)(\$6400)$

$p = \dfrac{\$6400}{\$10,900} = 0.587$

CHAPTER 16

Section 16.2

3. Let X_1 = Prices will advance according to SB

   X_1' = Prices will advance according to SPFK & B

   $P(\theta_1|X_1X_1') = 0.996$; $P(\theta_2|X_1X_1') = 0.004$

5. The final portion of the tree diagram is shown below.

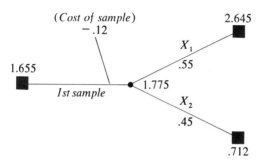

7. As the tree diagram will show, it is not advisable to purchase the second study. EOL without any study is $(0.35)(8) + (0.65)(0) = 2.8$. Since $2.40 < 2.8$, it is better to purchase the first study than not to purchase any study at all.

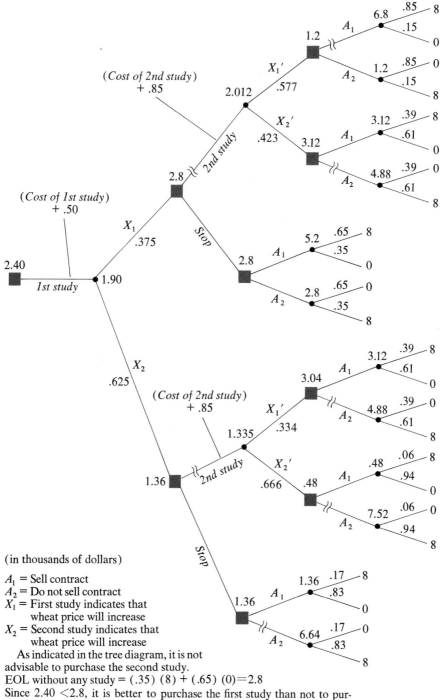

(in thousands of dollars)

A_1 = Sell contract
A_2 = Do not sell contract
X_1 = First study indicates that
　　wheat price will increase
X_2 = Second study indicates that
　　wheat price will increase
　As indicated in the tree diagram, it is not
advisable to purchase the second study.
EOL without any study = $(.35)(8) + (.65)(0) = 2.8$
Since $2.40 < 2.8$, it is better to purchase the first study than not to purchase any study at all.

9. EOL for all three branches = $(0.6)(50) + (0.3)(77.34) + (0.1)(80) = 61.202$

EOL + cost of first investigation = $61.202 + 10 = 71.202$

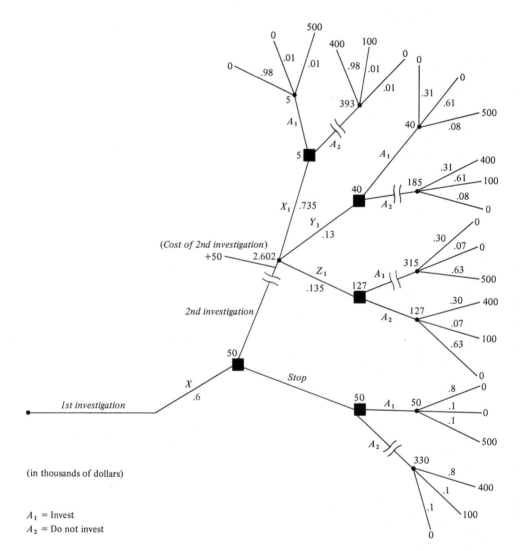

(in thousands of dollars)

A_1 = Invest
A_2 = Do not invest

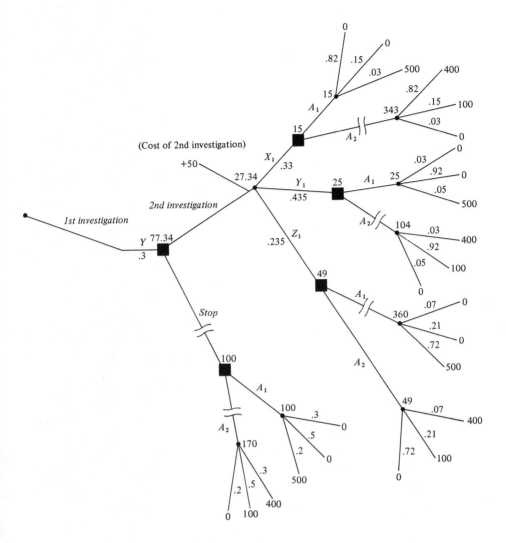

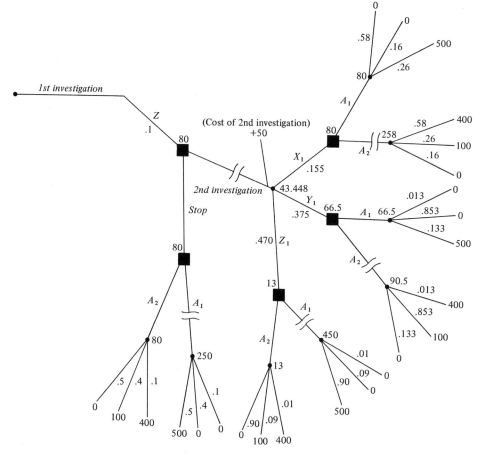

EOL for all three branches
$= (.6)(50) + (.3)(77.34) + (.1)(80)$
$= 61.202$
$+$ cost of first investigation (10)
$= 71.202$

CHAPTER 17

Section 17.2

3.

State of Nature	Sample Result			Total
	Type L	*Type M*	*Type H*	
θ_1	.15	.12	.06	.33
θ_2	.07	.23	.08	.38
θ_3	.04	.08	.17	.29
	.26	.43	.31	1.00

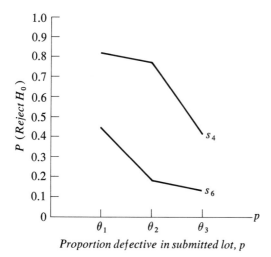

Proportion defective in submitted lot, p

Neither strategy satisfies the requirement that the probability of a Type I error be no larger than 0.10.

Index

A

Acceptance region, 260–261
Acceptance sampling:
    illustrations, 597–605, 635–645
    procedure, 263
Addition rule, 60–63
Alternative hypothesis, 258–259
Analysis of variance, 312, 337-355
    by ranks, 535–539
    decomposition of total variation,
        340–345
    degrees of freedom, 345–346
    hypothesis tested, 340
    notation, 339–340
    short-cut formulas, 344–345
    table, 346
    two-factor, 351–353
Arithmetic mean, 20–21
    characteristics of, 29–30
    grouped data, 21–22
    symbolism, 21
    uses of, 29–30
    weighted, 22–24
Averages, problems of interpretation,
    45–46

B

Backward induction, 560–561,
    657–660
Bayesian decision theory, 309,
    544–683
    comparison with classical statistics,
        665–676
    optimal sample size, 645–647
    posterior analysis, 589–607, 609
    preposterior analysis, 609–633
        extensive form analysis, 624–625,
            636–641
        normal form analysis, 625–630,
            641–644
    prior analysis, 544–587
    sensitivity analysis, 633–635
    sequential decision analysis,
        649–664
Bayesian estimation
    comparison with classical estimation,
        676–681
Bayesian statistics
    comparison with classical statistics,
        665–683
Bayes' strategy, 629

Bayes' theorem, 78–83, 544, 656
Bayes, Thomas, 78
Bernoulli, James, 106n
Bernoulli process, 106–117, 219, 230, 677–679
Bernoulli trials, 106–117
Binomial distribution, 106–117
 table of cumulative probabilities, 690–694
Bracket medians, 571–574
Bureau of Labor Statistics. *See* U.S. Bureau of Labor Statistics

C

Central Limit Theorem, 215–216, 219
Central tendency, 19
Chance fluctuations, 2
Chi-square:
 decision procedure, 319–322
 degrees of freedom, 318–319, 331–332
 goodness of fit tests, 313–328
 rule for size of frequencies, 323
 tests of independence, 328–337
Chi-square distribution, 219, 317–318
 equation of, 317
 graph of, 317
 table of areas, 705
Class limits, 11–12
Cluster sampling, 222–223
Cochran, William G., 334n
Coefficient of
 correlation, 394–396
 determination, 388–394
 multiple correlation, 424
 multiple determination, 419–420
 partial correlation, 425
 variation, 43–44
Combinations, 87–90
 definition of, 88
 of n things, x at a time, 88
Confidence intervals, 226, 231–249
 arithmetic mean:
  large sample, 235-236
  small sample, 244–248
 conditional mean, 377–380
 difference between two means:
  large samples, 238–240
 difference between two proportions:
  large samples, 240–241
 interpretation and use, 234–235

proportion:
  large sample, 236–238
  rationale, 231–234
Consistency, 227–229
Consumer Price Index (CPI), 493, 498, 505–506
Control groups, 172–175
Correlation analysis, 333, 356, 388–396
Council of Economic Advisors, 485
Counting principles, 83–91
 and techniques, 83–91
Cumulative distribution function, 100–104, 570–571
 graph of, 101

D

Decision making:
 criteria of choice, 548–556
 structure of the problem, 545–548
 using prior and sample information, 588–607
 using prior information, 544–587
Decision rules, 266–271
de Fermat, Pierre, 50
de Finetti, B., 52n
Degrees of freedom, 244
de Méré, Chevalier. *See* Gombould, Antoine, 49
Design of an investigation, 163–167
 direct experimentation studies, 165–166
 ideal research design, 166–167
 observational studies, 164–165
Dispersion, 19
 average deviation measures, 38–43
 distance measures, 37–38
Distribution–free tests. *See* nonparametric statistics

E

Efficiency, 227, 229–230
Errors:
 of prediction, 44–45
 random (sampling), 176–177
 systematic (bias), 177–179
 Type I, 258
 Type II, 258
Estimation, statistical, 224–256
 criteria of goodness of, 226–231
 types of estimates, 225–226
Estimator, 225

Events, 56–63
  certain, 57
  complementary, 57–58
  dependent, 70
  elementary, 57–58
  impossible, 51
  independent, 71–75
  mutually exclusive, 51, 57–58
Expected opportunity loss, 552–554
Expected profit:
  of perfect information, 556–559
  under uncertainty, 550–552
Expected utility, 574–583
Expected value, 141–143
  of a random variable, 141–143
  of a sum of random variables
  of sample information, 616–619,
    645, 647
  of perfect information, 556–559,
    647, 652–653
  prior, 556–559, 592, 594
  posterior, 592–594
Experimental designs
  completely randomized, 351–352
  randomized block, 351–352
Extensive form analysis, 624–625,
    636–641
  comparison with normal form,
    630–631

F

F distribution
  equation of, 348
  graph of, 350
  table of areas, 706
Federal Reserve Board (FRB) Index
    of Industrial Production, 509–510
Finite population multiplier, 216–219
Frequency curve, 14–15
Frequency distribution:
  construction of, 9–11
  cumulative, 15–16
  descriptive measures, 19–20
  graphic presentation of, 13–15
Frequency polygon, 14–15
F test, 348–350

G

Galton, Sir Francis, 366–367
Gauss, Karl F., 198

Gombauld, Antoine. *See* de Méré,
    Chevalier, 49
Gompertz curve, 449
Good, I. J., 52n
Gossett, W. S. [pseud. student], 246

H

Histogram, 14, 196–197
Homoscedasticity, 349, 360
Hypergeometric distribution, 121–127
Hypothesis testing, 257–311
  procedure of, 259–262
  rationale of, 258–259

I

Independence of random variables,
    154–160
Index numbers, 491–517
  aggregative price indices, 493–500
    fixed weights, 498–499
    unweighted, 493–495
    weighted, 495–499
  average of relatives price indices,
    500–505
    unweighted mean, 500–501
    weighted mean, 501–503
  deflation of value series, 510–512
  problems of construction, 505–507
    base period, 506–507
    items to be included, 505–506
    quality changes, 517
  quantity indices, 507–510
  shifting the base, 514
  splicing, 516–517
  uses of, 491–493, 517
Internal Revenue Service, 179
Interval estimation, comparison
    between classical and Bayesian
    statistics, 679–681

K

Koopman, B. O., 52n
Kruskal, W. H., 525n
Kruskal–Wallis test. *See* Analysis of
    variance by ranks.

L

Laspeyres price index, 497, 502–503
Laspeyres quantity index, 508

Least squares, method of, 367–369
Linear loss function, 677–678
Literary Digest poll, 178
Logistic curve, 449

M

Mann–Whitney U test, 528–532, 535
Maximum criterion, 548–550
Median, 26–29
    characteristics of, 29–30
    uses of, 29–30
Miller, D. W., 550n
Mode, 30–32
Morgenstern, O., 577n, 584
Multimodal (bimodal) distributions, 32–33
Multinomial distribution, 118–121
Multiple regression analysis, 410–442
    analysis of variance in, 428–431
    beta coefficients, 425
    case study, 435–442
    coefficients of correlation, 420–421
    coefficient of determination, 419–420
    equation in, 412–414, 417
    inferences about net regression coefficients, 422–424
    standard error of estimate, 418–419
    use of computers in, 426–431
Multiplication principle, 84
Multiplication rule, 67–70

N

National Bureau of Economic Research, 484
Nonparametric statistics, 518–543
Normal distribution, 198–211
    approach of binomial to, 198
    areas under, 201–205
    as approximation to binomial, 206–211
    equation of, 199
    graph of, 198, 200
    properties of, 199–201
    standard normal deviate, 201
    table of areas, 703
Normal equations, 369–371
Normal form analysis, 625–630, 641–644
    comparison with extensive form, 630–631
Null hypothesis, 258–259, 262–263

O

Odds ratios, 59–60
Ogive, 16
One-sample tests:
    for means, 263–275
    for proportions, 275–280
    large samples, 263–286
One-tailed test, 261–262
    for a mean, 263–274
    for a proportion, 279–280
    for differences between proportions, 294–297
Operating characteristic curve, 271n
Optimal sample size, 645–647

P

Paasche price index, 497–498, 503
Parameter, 270
Pascal, Blaise, 50
Payoff, 545
Payoff table, 546
Permutations, 85–87
    definition of, 86
    of n things, x at a time, 86
Point estimation, 225–226, 677–679
Poisson distribution, 127–140
    as an approximation to binomial, 135–139
    considered in its own right, 127–135
    nature of, 128–131
    table of cumulative probabilities, 696
Poisson, Siméon Denis, 127n
Population (universe), 2
    and sample, 2, 168–170
    infinite, 110, 170
    sampling frame, 169
    target, 171–172
Posterior analysis, 589–607, 609
    effect of sample size, 600–602
Posterior mean, 602–605
Power curve, 271–275
    for strategies, 673–674
    graph of, 272, 273, 274
    to control Type I and Type II errors, uses of, 274–275
Power of a test, 271
Precision, 214
Preposterior analysis, 609–633
    expected net gain of sample information, 617–619

expected value of sample
   information, 616–619, 645, 647
Prior analysis, 544–587
Prior mean, 602–605
Probability:
   classical, 50–51
   conditional, 65–67
   joint, 61, 64–65
   marginal, 65
   meaning of, 49–60
   posterior, 79
   relative frequency, 50, 51–52
   rules, 60–63
   subjective, 50, 52–53
   theory of, 2
Probability distribution(s):
   assessment of, 564–574
   binomial, 106–117
   characteristics of, 99–100
   conditional, 153–154
   hypergeometric, 121–127
   joint, 150–160
   marginal, 152
   multinomial, 118–121
   multivariate, 151
   Poisson, 127–140
   uniform, 104–106
Purchasing power index, 512

R

Ramsey, F. P., 52n
Random numbers, tables of, 185–186
Random process, 106
Random sample, 2
Random variable(s)
   and probability distributions,
      93–162
   continuous, 98–100, 195–199
   definition, 94
   discrete, 98–100
   types of, 98
Rank correlation, 539–543
   t test for, 541–542
Rank sum tests. *See* Mann–Whitney *U*
      test
Regression analysis, 333, 356–388
   assumptions of linear model, 360
   cause and effect relationships,
      402–403
   conditional probability distributions,
      359
   confidence intervals in, 374–384

curvilinear models, 405–406
dispersion around regression line,
   374–376
extrapolation, 403
inference about correlation
   coefficient, 397–399
inference about regression
   coefficient, 399–401
limitations of, 401–406
multiple, 348, 406
normal equations, 369–370, 406,
   412–414
prediction intervals in, 374–384
purposes of, 364–365
scatter diagrams, 361–364
simple two-variable linear model,
   358–361
Regression (slope) coefficient, 371
Rejection region, 260–261
Risk, 582–583
Robustness, 348n
Runs test, 532–535

S

Sample:
   and population, 2
Sample size:
   for estimating
      arithmetic mean, 252–255
      proportion, 249–252
Sample space, 53–58, 260
Sampling:
   cluster, 222–223
   from nonnormal populations,
      215–216
   from normal populations, 212–215
   nonrandom, 182
   purposes, 181–182
   random, 182–184
   simple random, 184–187
   stratified random, 222–223
   without replacement, 70, 121, 123,
      183
   with replacement, 70, 110, 123, 183
Sampling distribution:
   empirical, 212
   of a proportion, 190–195, 237
      graph of, 193, 194
   of number of occurrences, 189–190,
      237
      graph of, 193

of the difference between two means,
287–288
graph of, 288
of the difference between two
proportions, 291–293
graph of, 214
of the mean, 212–222
graph of, 214
theoretical, 212
Sampling errors, 214–215
Savage, L. J., 52n
Scatter diagrams, 361–364
Schell, Emil, D., 112n
Schlaifer, R. S., 572n, 580n
Seasonal variations. *See* Time series,
seasonal variations
Selected points, method of, 449n
Sensitivity analysis, 633–635
Sequential decision analysis, 649–664
Signed rank test. *See* Wilcoxon
matched pair test
Significant difference, 265
Significance level, 265
Sign test, 519–523
Skewness, 19–20
Snedecor, George W., 334n
Spetzler, C. S., 564n
Squared error loss function. *See*
quadratic loss function
Staël Von Holstein, C. S., 564n
Standard and Poor's common stock
price index, 484
Standard deviation:
calculation of, 40–41
definition, 39, 41
uses of, 42–43
Standard normal distribution, 202–203
Standard scores. *See* Standard units
Standard error:
of a proportion, 219, 237
for finite populations, 219
of estimate:
multiple regression, 418, 428
two-variable regression, 374–376
of forecast, 381
of the difference between two
means, 288
of the difference between two
proportions, 292
of the mean, 214, 232
for finite populations, 217–219
Standard units (standard scores),
142–143, 198–199

Starr, M. K., 550n
Statistical independence, 70–75
Statistical inference:
estimation, 2, 224–256
tests of hypotheses, 2
Statistics:
nature of, 1–3
role of, 3–4
Stochastic process, 106
Strategy, 625–626
Stratified random sampling, 222–223
Student. *See* Gossett, W. S., 246

T

t distribution, 219, 299–304
equation of, 245
graph of, 246
hypothesis testing with, 299–311
nature of, 245–246
table of areas, 704
t statistic, 244, 300
Time series, 443–490
classical model, 444–446
cyclical fluctuations, 445, 456–458
exponential smoothing, 486–490
forecasting methods, 482–489
business indicators, 483–485
cyclical forecasting, 483–485
irregular movements, 445–446
seasonal variations, 445, 468–477
purposes, 468–469
ratio-to-moving average method,
469–477
trend analysis:
fitting a second degree line,
458–461
fitting logarithmic lines, 461–463
fitting of trend lines, 450–455
projection of trend line, 455–456
purposes, 447–448
types of movements, 448–449
Transitivity, 584
t test for paired observations, 305–311
Two-sample tests, for large samples,
286–299
Two-tailed test, 261
for a proportion, 275–279
for difference between means,
287–290
for difference between proportions,
290–294

Type II error, 258–259
Type I error, 258–259

U

Unbiasedness, 227–228
Uncertainty, 2, 545
Unit of association, 358
Universe. *See* Population
U.S. Bureau of Labor Statistics (BLS), 493, 498, 506
U.S. Bureau of the Census, 164, 169
U.S. Department of Commerce, 484
Utility functions:
  characteristics of, 582–583
  construction of, 577–582
    five point procedure, 580–582
    reference consequences, 580–582
  types of, 582–583
Utility scales, 584–585
Utilty theory, assumptions, 583–584

V

Variables:
  discrete, 7
  continuous, 8
Variance:
  definition of, 39
  of a random variable, 144
  of a sum of independent random variables, 146–147
Von Neumann, J., 577n, 584

W

Wald, Abraham, 548, 548n
Wallis, W. A., 525n
Weaver, Warren, 86n
Wilcoxon matched-pairs test, 523–527

Y

Yates, F., 334n

F 1
G 2
H 3
I 4
J 5

Symbol	Definition
ν	Number of degrees of freedom (6.4)
$P(A)$	Probability of event A (2.1)
$P(A_1 \text{ or } A_2)$	Probability of the occurrence of at least one of events A_1 and A_2 (2.2)
$P(A_1 \text{ and } A_2)$	Joint probability of events A_1 and A_2 (2.2)
$P(B_1\|A_1)$	Conditional probability of event B_1 given A_1 (2.2)
P_n	Price in a nonbase period in an index number formula (11.2)
P_0	Price in a base period in an index number formula (11.2)
$P_0(p)$	Prior probability distribution of random variable p (14.2)
$P_1(p)$	Posterior probability distribution of random variable p (14.2)
$P(X = x\|Y = y)$	Conditional probability that random variable X is equal to the value x given that random variable Y is equal to y (3.12)
$P(X = x \text{ and } Y = y)$	Joint probability that X takes on the value x and Y takes on the value y (3.12)
p	Probability of a success on a given trial (binomial distribution); also used as population proportion of successes (3.4)
$\bar{p} = \dfrac{x}{n}$	Proportion of successes in a sample of size n (5.2)
$\bar{p}$	Weighted mean of two sample proportions (7.3)
Q_f	A fixed set of weights in a price index number formula (11.2)
Q_n	Quantity in a nonbase period in an index number formula (11.2)
Q_0	Quantity in a base period in an index number formula (11.2)
$q = 1 - p$	Probability of failure on a given trial (binomial distribution); also used as population proportion of failures (3.4)
R_1, R_2	Sum of ranks of the items in samples 1 and 2, respectively, in a rank sum test (12.4)
$R(s_1; \theta_1)$	Risk or expected opportunity loss associated with the use of strategy s_1, given that state of nature θ_1 occurs (15.3)
$R^2_{Y.12...(k-1)}$	Sample coefficient of multiple determination for a regression equation involving $k - 1$ independent variables $X_1, X_2, \ldots, X_{k-1}$ and dependent variable Y (9.9)
r	Number of rows in an arrangement of data to which analysis of variance is applied (8.3)
r	Sample correlation coefficient (9.6)
r^2	Sample coefficient of determination (9.6)
r^2_C	Corrected or adjusted sample coefficient of determination (9.6)
r_j	Number of observations in the jth column (8.3)
r_r	Rank correlation coefficient (12.7)
r_{12}	Correlation coefficient for variables X_1 and X_2 (9.9)
ρ	Population correlation coefficient (9.6)
ρ^2	Population coefficient of determination (9.6)
S	Sample space (2.1)
SI	Seasonal index (10.5)
SS_c	Between-column sum of squares (8.3)
SS_r	Between-row sum of squares (8.3)
SS_t	Total sum of squares (8.3)
$S^2_{Y.12...(k-1)}$	Sample variance around a regression equation involving $K - 1$ independent variables $X_1, X_2, \ldots, X_{k-1}$ and dependent variable Y (9.9)
s	Standard deviation of a sample (1.14)
s^2	Variance of a sample (1.14)
$s_1, s_2, \ldots$	Strategies (15.3)
s_b	Standard error of regression coefficient b (9.7)
s_d	Standard deviation of differences between pairs of observations made on the same individuals or objects (7.5)
$s_{\bar{d}}$	Standard error of $\bar{d}$, the mean difference of pairs of observations made on the same individuals or objects (7.5)
$s_{\bar{p}}$	Estimated standard error of a proportion (6.3)
s_{IND}	Standard error of forecast in two-variable regression analysis. Used to establish prediction intervals for individual Y values (9.5)
$s_{\bar{p}_1 - \bar{p}_2}$	Estimated or approximate standard error of the difference between two proportions (6.3)